Biomaterials and Implant Biocompatibility

Biomaterials and Implant Biocompatibility

Special Issue Editors

Anişoara Cîmpean
Florin Miculescu

MDPI • Basel • Beijing • Wuhan • Barcelona • Belgrade

Special Issue Editors
Anişoara Cîmpean
Universitatea din Bucuresti,Bucuresti
Romania

Florin Miculescu
Politehnica University of Bucharest
Romania

Editorial Office
MDPI
St. Alban-Anlage 66
4052 Basel, Switzerland

This is a reprint of articles from the Special Issue published online in the open access journal *Materials* (ISSN 1996-1944) from 2018 to 2019 (available at: https://www.mdpi.com/journal/materials/special_issues/biomaterials_implant_biocompatibility).

For citation purposes, cite each article independently as indicated on the article page online and as indicated below:

LastName, A.A.; LastName, B.B.; LastName, C.C. Article Title. *Journal Name* **Year**, *Article Number*, Page Range.

ISBN 978-3-03928-216-6 (Pbk)
ISBN 978-3-03928-217-3 (PDF)

Contents

About the Special Issue Editors

Anişoara Cîmpean is full Professor in the Department of Biochemistry and Molecular Biology at the University of Bucharest (UB). She set up the first cell culture laboratory (1999) from UB and has been intensively involved in didactic and scientific activity in cell cultures and their applications in bio-engineering, regenerative medicine, and tissue engineering. Her current research areas of interest relate to the investigation of the in vitro effects of newly developed biomaterials on different cell types, in order to prove if these biomaterials are suitable and efficient for targeted biomaterials application. Much of her work has been focused on analyzing the biocompatibility of various materials designed for bone implant applications and on defining the mechanisms by which osteoblast– and macrophage–material interactions might guide implant success. She has participated in postdoctoral research stages in Italy and Belgium. She published over 70 papers indexed in Web of Science Clarivate Analytics, seven books and book chapters, and two patents. She is a reviewer for over 20 WOS-quoted journals in the Materials Science and Life Sciences domains. She has received more than 30 awards for her contributions to science and is constantly involved in supervising BSc, MSc, and PhD students.

Florin Miculescu is Full Professor in the Metallic Materials Science and Physical Metallurgy Department at the Politehnica University of Bucharest (UPB). He leads a research group working on advanced characterization methods for materials and coordinates some research laboratories (Electron Microscopy and Microanalysis, Materials Characterization and Analysis Methods, Biomaterials and Nanomaterials Synthesis and Processing, and Transmission Electron Microscopy). He has participated in some postdoctoral stages in Europe and USA, and applied his expertise to various research projects related to materials science, engineering, and technology. His research activities in the fields of biomaterials, nanomaterials, and materials synthesis, processing, and characterization have also been presented in over 110 papers indexed in Web of Science Clarivate Analytics, 13 books and book chapters, and two patents. He is a reviewer for over 20 WOS quoted journals in the Materials Science domain. He is a former President of the Romanian Society for Biomaterials (2014-2017), Founding member of the Romanian Microscopy Society, and Member in Romanian Commission CNATDCU of Materials Engineering. He has received more than 50 awards for his contribution in science (for published papers, presentations, and patents). He constantly supervises a large team of BSc, MSc, and PhD students.

Preface to "Biomaterials and Implant Biocompatibility"

This book is the result of the publication of the MDPI journal *Materials*' Special Issue "Biomaterials and Implant Biocompatibility" (https://www.mdpi.com/journal/materials/special_issues/biomaterials_implant_biocompatibility). We are honored by the academics, researchers, and specialists' interest in this Special Issue, which was finalized with a total of 23 published articles. The release of this book is due to interdisciplinary and international cooperation between scientists from Austria, Belgium, Brazil, China, Dominican Republic, Germany, Hong Kong, India, Ireland, Japan, Korea, New Zeeland, Romania, Russia, Slovakia, Spain, United Kingdom, Uruguay, and USA. The topics covered in the afroementioned review or research articles comprise different interest areas in the field of biomaterial synthesis, processing, and complex characterization. The included articles cover polymers, ceramics, metals, and composites, from synthesis to compositional, morphological, structural, in vitro, and in vivo characterization.

Anişoara Cîmpean, Florin Miculescu
Special Issue Editors

Review

Mechanisms of Bioactive Glass on Caries Management: A Review

Lin Lu Dai [1], May Lei Mei [2], Chun Hung Chu [1] and Edward Chin Man Lo [1,*]

[1] Faculty of Dentistry, University of Hong Kong, Pok Fu Lam, Hong Kong; doreen07@hku.hk (L.L.D.); chchu@hku.hk (C.H.C.)

[2] Faculty of Dentistry, University of Otago, 9054 Dunedin, New Zealand; may.mei@otago.ac.nz

* Correspondence: edward-lo@hku.hk; Tel.: +85-228-590-292

Received: 29 October 2019; Accepted: 10 December 2019; Published: 12 December 2019

Abstract: This review investigates the mechanisms of bioactive glass on the management of dental caries. Four databases (PubMed, Web of Science, EMBASE (via Ovid), Medline (via Ovid)) were systematically searched using broad keywords and terms to identify the literature pertaining to the management of dental caries using "bioactive glass". Titles and abstracts were scrutinized to determine the need for full-text screening. Data were extracted from the included articles regarding the mechanisms of bioactive glass on dental caries management, including the aspect of remineralizing effect on enamel and dentine caries, and antimicrobial effect on cariogenic bacteria. After removal of duplicates, 1992 articles were identified for screening of the titles and abstracts. The full texts of 49 publications were scrutinized and 23 were finally included in this review. Four articles focused on the antimicrobial effect of bioactive glass. Twelve papers discussed the effect of bioactive glass on demineralized enamel, while 9 articles investigated the effect of bioactive glass on demineralized dentine. In conclusion, bioactive glass can remineralize caries and form apatite on the surface of enamel and dentine. In addition, bioactive glass has an antibacterial effect on cariogenic bacteria of which may help to prevent and arrest dental caries.

Keywords: bioactive glass; mechanism; caries; review

1. Introduction

Dental caries is a prevalent oral disease worldwide and can occur in both primary and permanent dentitions throughout an individual's life. It is a biofilm-mediated disease, resulting in mineral loss and destruction of dental hard tissues. Cariogenic bacteria will produce acids, causing prolonged periods of low pH environment in the oral cavity which leads to demineralization of dental hard tissues [1]. The process of dental caries starts with chemical dissolution of enamel and dentine caused by the acids produced by the bacteria that adhere onto the tooth surface. If sufficient time is allowed for progression, the carious lesions on the surface will progress to cavity formation in the affected tooth [2,3].

The current approaches of caries management aim to 1) stop or control the progression of caries, 2) preserve dental hard tissue as much as possible, and 3) avoid the re-restoration process [4]. Management of carious lesions with varying severity is outlined below. For initial lesions, nonsurgical approaches are commonly used. Fluoride-containing products are delivered in different forms onto teeth to promote remineralization and the mineral contents of the lesions are recovered by penetration of calcium and phosphate from a higher concentration into the lesions. Casein phosphopeptide–amorphous calcium phosphate (CPP-ACP) is a stabilized system of Ca–P that has superior remineralization potential on carious lesions. A modification is to add fluoride into the system (CPP-ACPF), which can improve the remineralization efficacy compared to that of the original system [5]. The approaches listed above are mainly due to their remineralization effect on the tooth surface. Some anticaries agents possess antibacterial properties, which can inhibit the growth of cariogenic bacteria. Chlorhexidine (CHX) is a type of antibacterial agent which can reduce the *Streptococcus mutans* level in the oral cavity [6]. Triclosan, which can affect the acid production of biofilm, is another anticaries agent. Previous studies have shown that amino acid arginine has an anticaries effect because of its effect on oral biofilms. Furthermore, xylitol is a natural substitute for sugar and has antibacterial potential on dental caries. Similarly to the other two previously mentioned agents, they possess the ability to control bacterial level, thus promoting the process of remineralization [3,5].

Placement of pit and fissure sealant is another minimally invasive therapy for managing initial lesions of the tooth surface. For moderate lesions, mechanical blocking or sealing off the lesion is an effective method to arrest caries after applying resin-based fissure sealant. Topical application of silver diamine fluoride (SDF) is an alternative way to arrest moderate carious lesions due to its antibacterial effect and remineralization effect [2,3]. Besides, the classic standard treatment of extensive lesions is removing all the demineralized tissues of the tooth and placing dental restorative material like composite resin to fill up the prepared cavity. In a recent development, stepwise or partial removal of caries is a new trend to preserve more dental tissues and it can reduce the incidence of pulpal exposure and favor the formation of tertiary dentine after restoration [2,3]. For the restorative method mentioned above, various materials are used. These include chemically bonded ceramic cements set by acid–base reaction, such as zinc phosphate, silicate, polycarboxylate, and glass ionomers. Composite resin is another type of cement that is set by a polymerization reaction. In addition, resin-modified glass ionomer cement is made by combining the two reactions [7].

Bioactive glass is a relatively new agent with an ability to heal bone defects caused by trauma or diseases and lead to bone regeneration. It has been applied in many healthcare fields. The first bioactive glass introduced in 1969 was a sodium, calcium, and phosphorus silicate glass. Currently, there are different types of bioactive glass, such as silicate-based glass and phosphate-based glass. Bioactive glass is an excellent material from the perspective of material properties. Because of its bioactivity and biocompatibility, the basic concept of applying bioactive glass in bone repair is to use a scaffold to act as a 3-dimensional template to guide bone regeneration [8]. It has been applied in wide-ranging fields, especially in the use of bone grafts, scaffold, disinfectant of the dental root canal and coating materials of dental implants [9]. The main advantage of bioactive glass in bone augmentation and repair is its high reactivity when in contact with bone surface and the most well-known capability of bioactive glass is the bonding ability to bone as well as stimulation of bone growth [10]. Firstly, when the material is in contact with an aqueous solution, the particles will change to mesoporous shape. Then, the particles will form an enrichment layer to produce an apatite-like layer on bone surface, similar to the component of bone or other hard tissues [11]. The formation of a hydroxyapatite (HA) layer involves the exchange of ions between the bioactive glass and the bone surface. The deposition of bone-like precipitates on bone surface plays a key role in the healing of bone defects [9]. The action on tooth is similar to that on bone. Bioactive glass can mineralize dentine tubules to relieve tooth sensitivity. The process is as follows: The glass material dissolves into an aqueous solution, followed by a pH rise. The pH rise promotes precipitation of hydroxyapatite (HA), the main component of mineralized enamel and dentine. Calcium and phosphate ions from bioactive glass and mineralizing agents in saliva may enhance the process of mineralization [8]. The most successful commercial product derived from a type of noncrystalline amorphous bioactive glass (Bioglass 45S5) with the name of NovaMin (GlaxoSmithKline, UK) is used in dentine repairing toothpaste, which can relieve the symptoms of dentine hypersensitivity. Bioglass 45S5 is silica-based and composed of 45 wt% SiO_2, 24.5 wt% CaO, 24.5 wt% Na_2O, and 6.0 wt% P_2O_5. It can appear in the form of particulates or granules [12–14].

Although studies have shown that bioactive glass has an ability to promote regeneration of bone and mineralization of dental hard tissues, it is not known whether bioactive glass is effective in preventing and arresting dental caries. Literature reviews conducted so far focus mainly on the mechanisms of bioactive glass on bone regeneration, tissue engineering, or dentine hypersensitivity, and very few have reviewed the mechanisms of action of bioactive glass on caries management.

The purpose of this study was to review the literature on the actions of bioactive glass on dental caries management regarding its effects on the caries process and cariogenic bacteria.

2. Materials and Methods

2.1. Searching Strategy

The literature search was conducted on four databases, namely PubMed, Web of Science, EMBASE (via Ovid), and Medline (via Ovid). Articles in these databases were searched using the keywords ("bioglass" OR "bioactive glass" OR "bioceramic") AND ("dentistry" OR "dental caries").

2.2. Study Inclusion and Exclusion

The lists of publications from the four databases were checked to remove duplications. Afterwards, the titles and abstracts of the identified articles were screened. This review aimed to summarize the mechanisms of bioactive glass on caries management. The inclusion and exclusion criteria were as follows.

2.2.1. Inclusion Criteria:

1.   Laboratory studies
2.   Studies related to the antimicrobial effect of bioactive glass
3.   Studies on the remineralization effect of bioactive glass on dental hard tissues (enamel and dentine)

2.2.2. Exclusion Criteria:

1.   Studies on root canal therapy and pulp regeneration
2.   Studies on periodontal disease
3.   Studies on orthodontic treatment
4.   Studies on tissue engineering
5.   Studies on bioactive composites or other bioactive materials

Two reviewers independently performed the screening to select potentially relevant articles. An independent reviewer was consulted on studies that were not able to be determined. The information extracted after reading the full text of the selected articles included basic publication details (authors and year), methods and materials used, measurement of outcomes, and main results.

3. Results

A total of 1992 potentially eligible articles published up to July 2019 (1051 articles in PubMed, 437 in Medline, 253 in Web of Science, and 251 in Embase) were identified (Figure 1). After checking for duplications, 748 records were removed. For the remaining 1244 articles, titles and abstracts were screened and they were classified into randomized clinical trial (RCT), case report, literature review, and laboratory study. Only laboratory studies were selected, and studies not related to the mechanisms of bioactive glass on caries management were excluded. Full-text readings were carried out on 49 articles and only 23 articles met the study eligibility criteria to be included in the final review. Among these 23 publications, there were 4 studies which examined the action of bioactive glass on cariogenic bacteria (Table 1), 12 studies focused on the remineralizing effect of bioactive glass on enamel (Table 2), while 9 studies investigated the effect of bioactive glass on dentine mineral contents (Table 3).

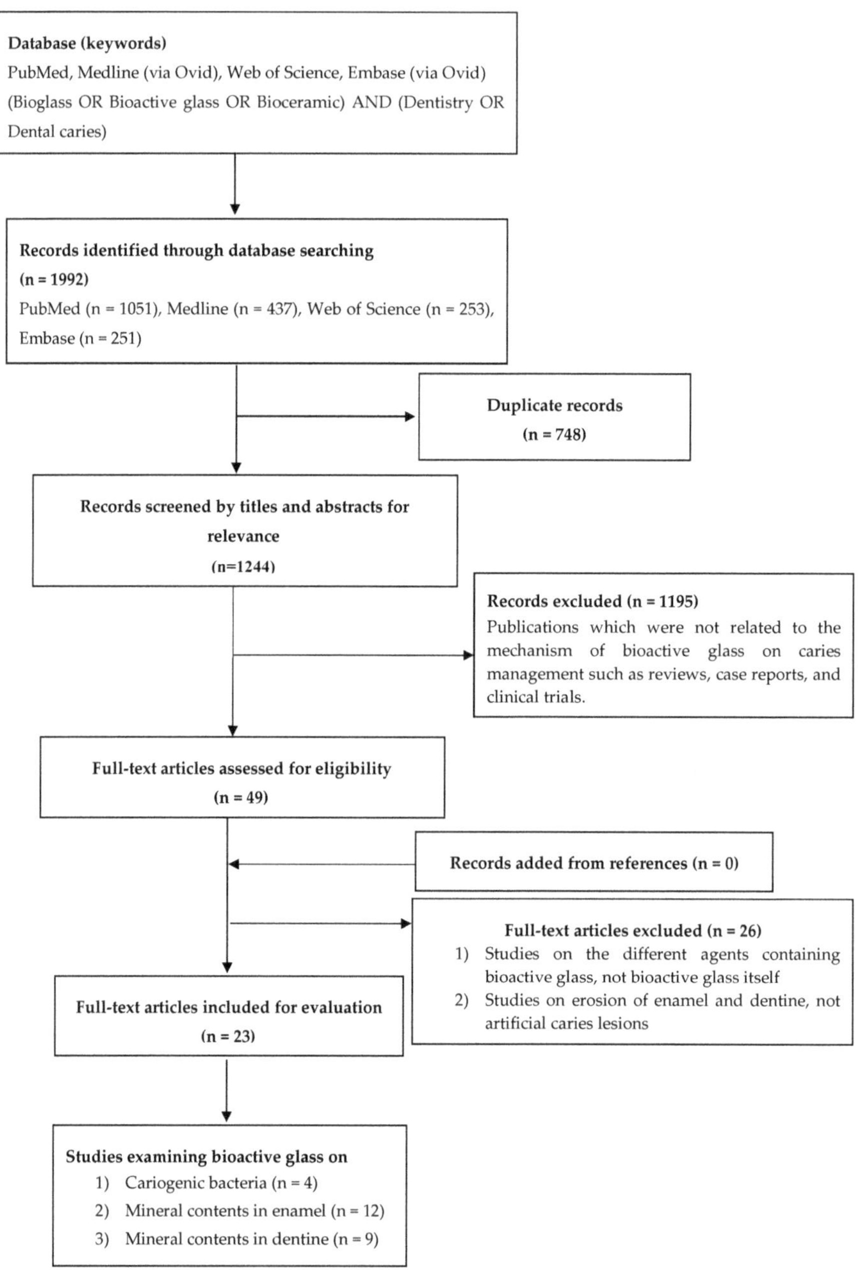

Figure 1. Flowchart of literature search of bioactive glass.

Table 1. Summary of studies on antimicrobial effect of bioactive glass.

Author (Year)	Methods	Main Findings
Xu et al. (2015) [15]	MIC and MBC were determined to test the antibacterial effect of a bioactive glass against *Streptococcus mutans*.	The MBC and MIC of bioactive glass was 37.5 and 18.75 mg/mL, respectively.
Martins et al. (2011) [16]	Three methods (agar diffusion, direct contact, and MIC) were used to determine the antibacterial effect of a bioactive glass-ceramic (Biosilicate) against a wide spectrum of bacteria. The assessed cariogenic species were *Streptococcus mutans*, *Lactobacillus casei*, *Actinomyces naeslundii*).	The MIC of Biosilicate ranged from ≤ 2.5 mg/mL to 20 mg/mL in different bacterial species. The best antibacterial effect of Biosilicate was against *S. mutans* (inhibition halo: 19.0 ± 2.0 mm) and *S. mutans* clinical isolate (MIC ≤ 2.5 mg/mL).
Jung et al. (2018) [17]	Light absorbance was used to evaluate the antibacterial effect of silver-doped bioglass MSN against *Lactobacillus casei*.	The increasing density of silver-doped bioglass MSN induced reduction of light absorbance. It illustrated that bacterial growth was inhibited.
Siqueira et al. (2019) [18]	Agar dilution method was used to determine the MIC values. The assessed cariogenic species were *Streptococcus mutans* and *Lactobacillus casei*.	Both the MIC of Bioglass and Biosilicate against *S. mutans* were 4mg/mL, which was the same as the MIC against *L. casei*. Bio-FP doped with different cations had different MIC against *S. mutans* and *L. casei*: Ag (8 and 4 mg/mL), Mg (2 and 4 mg/mL), Sr (2 and 4 mg/mL), Zn (2 and 4 mg/mL), Ga (2 and 4 mg/mL).

CFU, colony-forming units; MIC: minimal inhibitory concentration; MBC: minimal bactericidal concentration; TCS, triclosan; MSN, mesoporous silica nanoparticle.

Table 2. Summary of effect of bioactive glass on enamel mineral content.

Author (Year)	Methods	Main Findings
Palaniswamy et al. (2015) [19]	Demineralized enamel was treated with ACP-CPP and BAG, followed by microhardness test. BAG and ACP-CPP were applied on samples for 10 days in the first remineralization cycle and applied for another 5 days in the second remineralizing cycle.	Microhardness of dentine treated with ACP-CPP and BAG both increased but showed no significant difference between the 1st and 2nd remineralization cycles (BAG after 10 days: 346 ± 45; BAG after 15 days: 363 ± 65).
Rajan et al. (2015) [20]	Demineralized teeth were allocated into five groups as follows: fluoridated toothpaste, CPP-ACPF, ReminPro, SHY-NM and control group. Micro-CT was used to measure lesion depth.	Lesion depth after remineralization in SHY-NM group showed the least mean score of 987 μm compared to other groups.
Soares et al. (2017) [21]	Enamel samples with artificial lesions were treated with CPP-ACP, BAG, ReminPro, and self-assembling peptide. The recovery rate of microhardness was assessed.	Microhardness recovery rate of enamel treated with peptide was the highest (62.1%), followed by CPP-ACPF (48.4%) and BAG group (28.8%).

Table 2. *Cont.*

Author (Year)	Methods	Main Findings
Prabhakar et al. (2009) [22]	Teeth with artificial carious lesions were divided into 2 experimental groups (sodium fluoride films, bioactive glass films) and 2 control groups (control films placed interproximally and no treatment group).	Percentages of regain of lesion depth after remineralization in BAG were more in the experimental groups (NaF films: 67.7% ± 3.8%; and BAG films: 73.0% ± 3.0%) than those in the control groups (control film: 21.1% ± 3.3%; and no treatment: 30.7% ± 2.5%).
Chinelatti et al. (2017) [23]	Artificial caries lesions were formed on enamel fragments and either treated with Biosilicate or acidulated phosphate fluoride (APF), or had no treatment (control), followed by microhardness test.	Biosilicate group had higher microhardness on enamel surface (265 ± 10 KHN) than APF and control group. CLSM also displayed shallower lesions in Biosilicate group when compared to APF and control group.
Milly et al. (2013) [24]	Enamel samples with artificial WSLs were assigned to 4 groups: BAG slurry, PAA-BAG slurry, remin solution, and deionized water; the surface and cross-sectional microhardness of enamel was assessed.	BAG group illustrated the highest surface microhardness (138 ± 5 KHN), but there were no significant differences among the other groups.
Bakry et al. (2014) [25]	Demineralized enamel specimens were divided into 4 groups: (1) no intervention, (2) only bioglass, (3) only brushing abrasion challenge, and (4) bioglass + brushing abrasion. After demineralizing and application of bioglass, all specimens were stored in remineralizing medium for 24 h, followed by removing the thin layer of bonding agent on bioglass in Groups 2 and 4, and then Groups 3 and 4 were sent to brushing abrasion challenge.	Hydroxyapatite was detected using XRD on the surface of enamel in Group 2 and Group 4 and these two groups also exhibited 100% coverage of crystalline structures on enamel surface.
Zhang et al. (2018) [26]	Artificial enamel WSLs were assigned to BG slurry, BG+PAA, CS-BG, CS-BG+PAA, remin solution, and deionized water groups. Microhardness was assessed and the intensity of surface mineral content was measured by Raman intensity mapping.	Intensity increase in BG group was significantly greater when compared to those without BG. CS-BG+PAA group showed the highest microhardness (222 ± 38 KHN) of enamel surface. Other groups with BG also exhibited higher microhardness than the control group.
Narayana et al. (2014) [27]	Enamels with artificial carious lesions were treated with bioactive glass, fluoride toothpaste, CPP-ACP, or CPP-ACPF and the control had no treatment. EDS was used to test the weight change of different elements.	BAG group showed significant difference when compared with control group for elements Ca and P. The mean weight percentage of Ca was 40.0% (BAG) and 31.1% (control), while the percentage of P was 14.0% (BAG) and 13.2% (control).

Table 2. *Cont.*

Author (Year)	Methods	Main Findings
Mehta et al. (2014) [28]	Enamel specimens were randomly distributed into two groups: BAG and CPP-ACP dentifrice. Vickers microhardness test was used.	Mean microhardness values were 372 VHN in BAG group and 357 VHN in CPP-ACP group afterremineralization, but the difference was not significant.
EI-Wassefy et al. (2016) [29]	Demineralized enamels were treated with no treatment, fluoride varnish, cold plasma, bioglass paste, cold plasma + bioglass paste. Microhardness was assessed by Vickers hardness tester.	Microhardness of enamel surface become higher in PB groups (175 VHN and 221 VHN) when compared with bioglass groups (153 VHN and 201 VHN) at 30 and 50 µm depth, but with no significant difference between the two groups at 70–200 µm depth.
Zhang et al. (2019) [30]	Enamel slabs with artificial WSL were assigned into 4 groups: bioglass (chitosan pre-treated lesions), chitosan-bioglass slurry, remin solution (PC), and deionized water (NC). Subsurface microhardness was assessed.	Mean hardness of bioglass group and chitosan–bioglass group were 56.7 ± 8.7 and 65.1 ± 8.9 KHN, which were significantly higher than those of NC group (12.7 ± 1.3 KHN) and PC group (18.6 ± 5.8 KHN).

CPP-ACP, calcium phosphate–casein phosphopeptide; SHY-NM: name of a bioactive glass; HA, hydroxyapatite; PAA-BAG, bioactive glass containing polyacrylic acid; WSL, white spot lesions; EDS: energy dispersive X-ray spectroscopy; CLSM, confocal laser scanning microscopy analysis; APF, acidulated phosphate fluoride; XRD, X-ray diffraction.

Table 3. Summary of effect of bioactive glass on dentine mineral content.

Author (Year)	Methods	Main Findings
Sleibi et al. (2018) [31]	Teeth with root caries were divided into 4 groups and treated with different agents (CPP-ACP+fluoride, bioglass+fluoride, fluoride only, no treatment). Severity index of root caries was evaluated through visual–tactile examinations. X-ray microtomography was used to measure mineral change.	The bioglass and fluoride group had the maximum reduction (100%) in severity index of root caries and it also had the highest percentage (60%) increase in mineral deposition.
Rajan et al. (2015) [20]	Demineralized teeth were treated with fluoridated toothpaste, CPP-ACPF, ReminPro, SHY-NM (bioglass), and no treatment (negative control). Lesion depth was measured after application.	SHY-NM (bioglass) group showed the lowest mean lesion depth after remineralization procedure.
Sauro et al. (2011) [32]	Dentine segments were treated with bioactive glass (Sylc), NaH C_2O_4 H_2O, Cavitron Prophy Powder, EMS Perio, CPP-ACP, Colgate Sensitive Pro-Relief, NUPRO Solution Prophy Paste. Microhardness and EDX were evaluated.	The dentine surface hardness increased after treated with bioactive glass (Sylc). There was no significant change in Ca and P/O ratios.

Table 3. *Cont.*

Author (Year)	Methods	Main Findings
Saffarpour et al. (2017) [33]	Demineralized dentine discs were treated with 3 agents: bioactive glass (BG), BG modified with 5% strontium, BG modified with 10% strontium and followed by evaluation of morphology.	BG with 10% strontium showed highest rate of remineralization and completely occluded dentinal tubules.
Forsback et al. (2004) [34]	Dentine discs were treated with bioactive glass S53P4 and control glass (CG). Weight loss of dentine discs was measured by weighing before and after remineralization.	Weight loss was less when discs were pretreated with BAG (21.0 ± 7.4 µg/mm^2) than without BAG (49.1 ± 6.5 µg/mm^2).
Vollenweider et al. (2007) [35]	Demineralizing dentine bars were applied by nanometric bioactive glass (NBG) and PeriGlas (PG) suspension. SEM was used to observe the dentine surface.	Dentine specimens treated with NBG showed apatite depositions on the surface after 10 or 30 days.
Jung et al. (2018) [17]	Demineralized dentine discs were divided into four groups: bioglass, MSN, silver-doped bioglass MSN, and no treatment, followed by acid resistance test.	Silver-doped bioglass MSN group had dentinal tubules completely occluded to a depth of 2–3 µm and the highest proportion ($83.4\% \pm 7.5\%$) of occluded area after acid challenge.
Cardoso et al. (2018) [36]	Root dentine slices were allocated into four groups: MTA, ERRM, Bioglass 45S5, and NbG. Microhardness was assessed.	Bioglass 45S5 group showed an increase in microhardness.
Zhang et al. (2019) [37]	Dentine discs treated with EDTA were allocated to 4 groups: AS (artificial saliva), Asp, BAG, Asp-BAG, and followed by 6% citric acid challenge. The mineral matrix ratio was measured.	Compared to AS and Asp group, BAG group (17.8 ± 2.3) and Asp-BAG group (12.5 ± 2.3) had significantly higher mineral matrix area ratio.

EDX, energy dispersive X-ray spectroscopy; Ag-BGN, silver-doped bioactive glass; MSN, mesoporous silica nanoparticle; DW: deionized water; EDTA, ethylene diamine tetraacetic acid; Asp, DL-aspartic amino; MTA, mineral trioxide aggregate; ERRM, EndoSequence Root Repair Material; NbG, niobophosphate glass.

3.1. Effect of Bioactive Glass on Cariogenic Bacteria

Table 1 shows the main findings of the four studies that investigated the effect of bioactive glass on cariogenic bacteria. A study found that the minimal inhibitory concentration (MIC) and minimal bactericidal concentration (MBC) of bioactive glass powder (45S5; Datsing Bio-Tech Co. Ltd, Beijing, China) were 18.8 and 37.5 mg/mL, respectively. The study showed that when bioactive glass dissolved in water, alkaline ions were released to raise the pH of the solution and this could kill *Streptococcus mutans* [15]. Another type of bioactive glass-ceramic (Biosilicate) was shown to exhibit antimicrobial properties that could inhibit a wide spectrum of microorganism. It was found that Biosilicate possessed antibacterial action against multi-species cariogenic bacteria strains, such as *Streptococcus mutans, Actinomyces naeslundii, Lactobacillus casei,* through agar diffusion and direct contact [16]. Another investigation found that growth of *L. casei* incubated in a silver-doped bioactive glass (Ag-BGN@MSN) was inhibited. Silver melted into bioactive glass displays a synergistic effect on microorganisms as it can inhibit the growth of cariogenic bacteria [17]. A study investigated cation-doped (Ag, Mg, Sr, Zn, and Ga) bioactive ceramics and revealed a bacterial inhibitory effect on *Streptococcus mutans* and *Lactobacillus casei* with varied MIC [18].

3.2. Effect of Bioactive Glass on the Mineral Content of Enamel and Dentine

Table 2 shows the main findings of the 12 published papers that investigated the effect of bioactive glass on the mineral content of enamel. Demineralized enamel was treated with different types of bioactive glass. Surface microhardness of the enamel tissue decreased after the demineralization procedure and increased after application of bioactive glass. The value of microhardness was found to be higher in bioactive glass group when compared to the control group or application of other agents [19–21,23,29,38]. A study reported the recovery rate of microhardness on demineralized enamel surface after treatment with bioactive glass was 28.8% [21]. Further investigation showed that a combination of bioglass paste and cold plasma had a synergistic effect on increasing the surface microhardness of demineralized enamel [29]. Apart from assessing microhardness, the mean carious lesion depth in specimens treated with bioactive glass were significantly lower than those of specimens without bioactive glass treatment [20,26,38]. A study assessed the percentage of regain in lesion depth after remineralization and the experimental group with bioactive glass had the highest regain percentage (73.0 ± 3.0%) of lesion depth in enamel [22]. In addition, energy-dispersive X-ray spectroscopy (EDX) analysis indicated that Ca/P ratio was higher in the region treated with bioactive glass than in other regions not covered by bioactive glass particles [26,29,30]. One study found that compared to just application of deionized water, enamel lesions after treated with two bioactive glass had significantly higher Ca/P ratios [30].

Table 3 shows the main findings of the nine studies on the effect of bioactive glass on the mineral content of dentine. Microhardness measurement was a commonly used method to evaluate the surface of the demineralized dentine. The remineralization process induced an increase in surface microhardness of carious lesions [32]. In addition, it was found that application of Bioglass 45S5 significantly increased root dentine microhardness [36]. Dentine lesion depth decreased after the application of bioactive glass in two in vitro studies [17,20]. Another study used visual–tactile examination to assess the severity of root caries and found that there was a significant reduction in the group combining bioglass and fluoride and that group also had the highest percentage (60%) increase in mineral deposition [31]. Dentine discs treated with bioactive glass had significantly higher mineral matrix area ratio when compared to that of discs in the artificial saliva and DL-aspartic amino groups [37]. Furthermore, weight loss of dentine slices treated with BAG S53P4, an amorphous glass with the composition of 53 wt% SiO_2, 23 wt% Na_2O, 20 wt% CaO, and 4 wt% P_2O_5, was less than that of slices without such treatment [34]. EDX was used in a study to analyze the elements in the occluding materials within dentine tubules. The results indicated that the ratio of Ca/P of hydroxyapatite was not significantly different between the bioactive glass and control groups [39].

Apart from the approaches mentioned above, qualitative parameters were also used to measure the mineralization effect of bioactive glass on enamel and dentine. Most of these studies [25–27,29,33,35,37] analyzed the morphology of enamel and dentine surface using scanning electron microscopy (SEM). The deposits newly formed on the surface of dental hard tissue were crystal-like hydroxyapatite (HAP) and rich in calcium and phosphate with the presence of silica. A layer of mineral formed by the particles of bioactive glass covered the lesion surface in the remineralized enamel group [24]. Different from that seen on enamel, a layer of particles of bioactive glass not only deposited on the dentine surface, but also partially or completely occluded dentine tubules during remineralization [33,39]. Figure 2 shows two SEM images of demineralized dentine with or without treatment with bioactive glass. After observing the morphology of the remineralized enamel and dentine, the content of the new deposition was assessed by X-ray diffraction (XRD) [25]. XRD results showed that the bioactive glass (Ag-BGN@MSN) particle used had an amorphous two-dimensional hexagonal structure [17]. Another XRD study found that both nanoparticles and conventional bioactive glass were in amorphous state [35]. Furthermore, XRD results of another two studies matched the standard diffraction peak of hydroxyapatite crystal on the enamel and dentine surfaces [31,37]. Strontium-modified bioactive glass displayed a higher intensity of XRD peaks than that of the original bioactive glass [33]. Another study used a qualitative method to assess the mineral concentration by using X-ray microtomography. The result showed that the highest percentage increase of mineral content in the lesion area that was treated with bioglass and fluoride [31]. Raman spectroscopy was another method used in the studies to confirm the content of remineralized enamel and dentine tissues. The phosphate peak of sound enamel and dentine appeared in a specific wavelength (around 960 cm^{-1}) of Raman spectra, while demineralized hard tissues showed no peaks. In two studies, the dental tissues treated with bioactive glass displayed the intensity of phosphate peak [26,32], while another study illustrated that there was a reduction of the intensity of phosphate peak in demineralized enamel compared to sound enamel [24]. Demineralized dentine showed phosphate peak after one-day treatment with nanoparticle bioactive glass, but no phosphate peak appeared after treatment with conventional bioglass, though all the dentine specimens immersed in the two types of bioactive glass had deposition of apatite on the surface after 10 or 30 days [35].

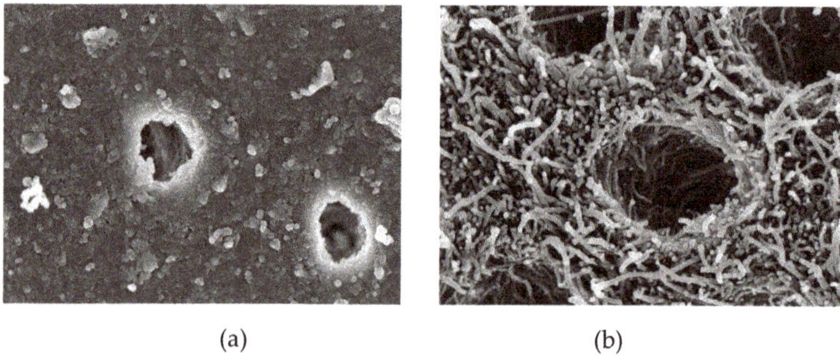

(a)   (b)

Figure 2. SEM images of the morphology of demineralized dentine: (**a**) 10,000× magnification image of demineralized dentine treated with bioactive glass; (**b**) 10,000× magnification image of demineralized dentine without treatment with bioactive glass.

3.3. Effect of Bioactive Glass on the Organic Content of Dentine

Only two studies mentioned changes in the organic content of dentine. One of them found that the dentine remineralized in nanometric bioactive glass suspension, compared to the dentine in PeioGlas® (Millipore, Bedford, MA, USA), Bioglass 45S5 with particles size ranging from 90 to 710 μm showed a significantly lower protein content due to the removal of organic contents [35]. As illustrated in another study, Raman spectra showed no peak for hydroxyapatite but only a high intensity of organic components in the demineralized dentine without treatment by the bioactive glass [32]. Lower organic content indicates better remineralization.

4. Discussion

After screening and analyzing the results of all selected laboratory studies, a number of possible mechanisms of how bioactive glass act on dental caries were found. The mode of action of bioactive glass for arresting caries is related to two aspects: 1) the antibacterial properties of bioactive glass on cariogenic bacteria, and 2) the remineralizing effect on the mineral content of dental hard tissues.

In the oral cavity, oral microbiota and dental biofilms are commonly present. Formation of dental plaque (dental biofilm) involves several stages. First, the acquired pellicle on tooth surface provides sites for bacterial colonizers. The oral microorganisms then grow and form a conditioning film of bacteria, proteins, and other bacterial products covering the tooth surface. *Streptococci, Lactobacilli,* and *Actinomycetes* are recognized as the main species of bacteria contributing to caries progression. *Streptococci* have high incidence and proportions and in the dental biofilms covering early caries lesions [2]. The key microorganism in initiating and developing dental caries is *Streptococcus mutans* (*S. mutans*). *Lactobacillus casei* (*L. casei*) is a type of cariogenic bacteria strains that commonly appear in deep or advanced caries lesions. More recently, another type of acid-producing and acid-tolerating species, called *Actinomycetes*, has been found to be associated with caries [40]. This systematic review found that very few studies investigated the antimicrobial effect of bioactive glass. This may be because the most obvious advantage of bioactive glass is its remineralization effect on bone and teeth rather than its bactericidal efficacy. Xu et al. assayed plaque biofilm of *S. mutans* and applied bioactive glass 45S5 at a concentration twice of the minimal bactericidal concentration to show that the bioactive glass had a great inhibitory effect on *S. mutans* biofilm [15]. This suggests that the concentration of antimicrobial agent needed for inhibiting biofilm may be many times higher than that for inhibiting planktonic bacteria. The possible action of bioactive glass acting on cariogenic bacteria is release of alkaline ions, followed by pH elevation that builds an environment in which bacteria cannot grow. This is similar to the mechanism of action of arginine, an amino acid, in which the arginine deiminase system has been identified as a novel technology to prevent initiation of the dental caries process by increasing pH around the biofilm on tooth surface [3]. Apart from the process of pH elevation, the presence of antibacterial ions can also control bacterial growth. Cation-doped bioactive ceramics, such as Ag, Mg, Sr, and Zn, have good inhibitory effect on *S. mutans* and *L. casei* [18]. Two literature reviews proposed that silver diamine fluoride (SDF), in which silver ion is the major antimicrobial agent, is an effective treatment to arrest established dental caries [3,40]. It has been shown by utilizing bacterial and biofilm models, that SDF can inhibit the growth both *Streptococcus mutans* and *Actinomyces naeslundii* [41]. Therefore, bioactive glass with silver may have additional inhibition effect against cariogenic bacteria.

The various compositions in bioactive glass have different roles in the remineralization process. The proportion of calcium and phosphate in dental tissues is identical to that in bone. Phosphate has a great contribution to hydroxyapatite formation and increases biocompatibility significantly. Formation of hydroxyapatite layer promotes remineralization in enamel and dentine. The physical occlusion on the lesion surface begins with the bioactive glass particles exposed to the aqueous environment, along with ion release and pH elevation [37]. When the biomaterial is exposed to an aqueous environment, sodium ions will exchange with H^+ (hydrogen ions). Meanwhile, Ca^{2+} (calcium ions) in the particles as well as PO_4^{3-} (phosphate ions) are released from the biomaterial. Thus, a localized pH rise will allow the precipitates of calcium and phosphate ions, together with the ions from saliva to form a calcium phosphate (Ca–P) layer on the lesion surface [20]. The silica network from bioactive glass can react with hydroxyl ions from aqueous solution and form soluble silanol compounds. It can be observed that the increase in Ca and P content would induce a decrease in Si content [29]. The newly formed layer displays good resistance to abrasion and transforms to a hydroxyapatite layer ultimately, which is structurally similar to those of original enamel and dentine [32].

Topical fluoride has already been proved to be effective in treating dental caries. The mechanism of fluoride is to inhibit demineralization and promote remineralization, which conducts a similar procedure with bioactive glass. The fluoride in oral fluid or solution can penetrate along with the acid at the subsurface and protect the minerals from dissolution, and thus prevents the demineralization process. After acidic challenge, fluoride will be adsorbed to the demineralized crystals and attract calcium ions, thus making the solution highly supersaturated with respect to fluorohydroxyapatite, which can promote the remineralization process [42]. A recent review found that SDF can inhibit the demineralization and promote remineralization of the mineral content of enamel and dentine and protect collagen matrix from degradation [40]. An in vitro study showed that the fluoride in bioactive glass could be switched to fluorapatite on the tooth surface, which leads to higher resistance to acid dissolution [20]. The precipitation of mineral deposits occurs mostly in the superficial layer, particularly when fluoride is present [31]. The deposition of a fluoride-contained mineral layer on dentine surface can occlude dentine tubules and reduce permeability [21].

A study stated that strontium can be a substitution of calcium in bioactive glass which may show a better bonding ability [9]. Strontium can supply ions for hydroxyapatite formation. Incorporation of strontium and fluoride can inhibit hydroxyapatite dissolution by the acids produced by cariogenic bacteria. Strontium can be a substitute for calcium for precipitate formation and it has synergistic caries inhibition effect with fluoride. The remineralization effect can last for different periods due to the addition of various proportions of strontium into the bioactive glass, which shows that strontium may be a beneficial factor in preventing caries through remineralizing dental hard tissues [33]. Besides, nanometric particles of bioactive glass have better remineralization potential compared to the conventional ones because of its larger surface area and higher Ca/P ratio [22]. The experiments conducted by Meret showed a greater effect of remineralization on dentine surface due to the nanosize of bioactive glass [35], while another study also demonstrated that Biosilicate microparticles were more effective in slowing down progression of caries lesions and promoting remineralization [38]. Smaller particles may completely block the porosity of enamel and dentine lesions. These microstructures are capable of penetrating from the tooth surface to the whole lesion and enhancing the remineralization of carious lesions [24].

An advantageous aspect of bioactive glass is its bioactivity and biocompatibility. Previous studies adopted the direct contact cell viability method to evaluate the biocompatibility of bioactive glass and showed a high cell survival rate [43,44]. As a very safe material and based on the merits stated above, a potential new application of bioactive glass is for dental caries prevention and remineralization of early caries lesions [45]. Further research should pay more attention to how the bioactive glass work in treating dental caries in the real oral environment.

5. Conclusions

Based on the findings of the present review, it is concluded that bioactive glass is able to inhibit the growth of cariogenic bacteria. Bioactive glass can promote remineralization by forming apatite on the surface of demineralized enamel and dentine. The main mechanisms of bioactive glass for caries management include an antibacterial effect on cariogenic bacteria, prohibition of mineral demineralization, and promotion of remineralization.

Author Contributions: L.L.D. conducted the literature search and drafted the manuscript. M.L.M. helped to design the manuscript structure and double confirmed the including articles. C.H.C. reviewed the article and gave some suggestions. E.C.M.L. revised and finalized the manuscript of this review.

Funding: This review received no external funding.

Conflicts of Interest: The authors declare no conflict of interests.

References

1. Kutsch, V.K. Dental caries: An updated medical model of risk assessment. *J. Prosthet. Dent.* **2014**, *111*, 280–285. [CrossRef] [PubMed]
2. Pitts, N.B.; Zero, D.T.; Marsh, P.D.; Ekstrand, K.; Weintraub, J.A.; Ramos-Gomez, F.; Tagami, J.; Twetman, S.; Tsakos, G.; Ismail, A. Dental caries. *Nat. Rev. Dis. Primers* **2017**, *3*, 17030. [CrossRef] [PubMed]
3. Wong, A.; Subar, P.E.; Young, D.A. Dental Caries: An Update on Dental Trends and Therapy. *Adv. Pediatr.* **2017**, *64*, 307–330. [CrossRef] [PubMed]
4. Schwendicke, F.; Frencken, J.E.; Bjorndal, L.; Maltz, M.; Manton, D.J.; Ricketts, D.; Van Landuyt, K.; Banerjee, A.; Campus, G.; Domejean, S.; et al. Managing Carious Lesions: Consensus Recommendations on Carious Tissue Removal. *Adv. Dent. Res.* **2016**, *28*, 58–67. [CrossRef] [PubMed]
5. Gonzalez-Cabezas, C.; Fernandez, C.E. Recent Advances in Remineralization Therapies for Caries Lesions. *Adv. Dent. Res.* **2018**, *29*, 55–59. [CrossRef] [PubMed]
6. Walsh, T.; Oliveira-Neto, J.M.; Moore, D. Chlorhexidine treatment for the prevention of dental caries in children and adolescents. *Cochrane Database Syst. Rev.* **2015**, CD008457. [CrossRef]
7. Jefferies, S.R. Bioactive and biomimetic restorative materials: A comprehensive review. Part I. *J. Esthet. Restor. Dent. Off. Publ. Am. Acad. Esthet. Dent.* **2014**, *26*, 14–26. [CrossRef]
8. Jones, J.R. Review of bioactive glass: From Hench to hybrids. *Acta Biomater.* **2013**, *9*, 4457–4486. [CrossRef]
9. Ali, S.; Farooq, I.; Iqbal, K. A review of the effect of various ions on the properties and the clinical applications of novel bioactive glasses in medicine and dentistry. *Saudi Dent. J.* **2014**, *26*, 1–5. [CrossRef]
10. Hench, L.L. The story of Bioglass. *J. Mater. Sci. Mater. Med.* **2006**, *17*, 967–978. [CrossRef]

11. Izquierdo-Barba, I.; Salinas, A.J.; Vallet-Regi, M. Bioactive Glasses: From Macro to Nano. *Int. J. Appl. Glass Sci.* **2013**, *4*, 149–161. [CrossRef]

12. Shivaprasad, B.M.; Padmavati, P.; Nehal, N.S. Chair Side Application of NovaMin for the Treatment of Dentinal Hypersensitivity—A Novel Technique. *J. Clin. Diagn. Res. JCDR* **2014**, *8*, Zc05-8.

13. Burwell, A.; Jennings, D.; Muscle, D.; Greenspan, D.C. NovaMin and dentin hypersensitivity—In vitro evidence of efficacy. *J. Clin. Dent.* **2010**, *21*, 66–71. [PubMed]

14. Fiume, E.; Barberi, J.; Verne, E.; Baino, F. Bioactive Glasses: From Parent 45S5 Composition to Scaffold-Assisted Tissue-Healing Therapies. *J. Funct. Biomater.* **2018**, *9*, 24. [CrossRef]

15. Xu, Y.T.; Wu, Q.; Chen, Y.M.; Smales, R.J.; Shi, S.Y.; Wang, M.T. Antimicrobial effects of a bioactive glass combined with fluoride or triclosan on Streptococcus mutans biofilm. *Arch. Oral Biol.* **2015**, *60*, 1059–1065. [CrossRef]

16. Martins, C.H.G.; Carvalho, T.C.; Souza, M.G.M.; Ravagnani, C.; Peitl, O.; Zanotto, E.D.; Panzeri, H.; Casemiro, L.A. Assessment of antimicrobial effect of Biosilicate against anaerobic, microaerophilic and facultative anaerobic microorganisms. *J. Mater. Sci. Mater. Med.* **2011**, *22*, 1439–1446. [CrossRef]

17. Jung, J.H.; Kim, D.H.; Yoo, K.H.; Yoon, S.Y.; Kim, Y.; Bae, M.K.; Chung, J.; Ko, C.C.; Kwon, Y.H.; Kim, Y.I. Dentin sealing and antibacterial effects of silver-doped bioactive glass/mesoporous silica nanocomposite: An in vitro study. *Clin. Oral Investig.* **2018**, *23*, 253–266. [CrossRef]

18. Siqueira, R.L.; Alves, P.F.S.; Moraes, T.D.; Casemiro, L.A.; da Silva, S.N.; Peitl, O.; Martins, C.H.G.; Zanotto, E.D. Cation-doped bioactive ceramics: In vitro bioactivity and effect against bacteria of the oral cavity. *Ceram. Int.* **2019**, *45*, 9231–9244. [CrossRef]

19. Palaniswamy, U.K.; Prashar, N.; Kaushik, M.; Lakkam, S.R.; Arya, S.; Pebbeti, S. A comparative evaluation of remineralizing ability of bioactive glass and amorphous calcium phosphate casein phosphopeptide on early enamel lesion. *Dent. Res. J.* **2016**, *13*, 297–302. [CrossRef]

20. Rajan, R.; Krishnan, R.; Bhaskaran, B.; Kumar, S.V. A Polarized Light Microscopic Study to Comparatively evaluate Four Remineralizing Agents on Enamel viz CPP-ACPF, ReminPro, SHY-NM and Colgate Strong Teeth. *Int. J. Clin. Pediatr. Dent.* **2015**, *8*, 42–47. [CrossRef]

21. Soares, R.; De Ataide, I.N.; Fernandes, M.; Lambor, R. Assessment of Enamel Remineralisation After Treatment with Four Different Remineralising Agents: A Scanning Electron Microscopy (SEM) Study. *J. Clin. Diagn. Res. JCDR* **2017**, *11*, Zc136–zc141. [CrossRef] [PubMed]

22. Ramashetty Prabhakar, A.; Arali, V. Comparison of the remineralizing effects of sodium fluoride and bioactive glass using bioerodible gel systems. *J. Dent. Res. Dent. Clin. Dent. Prospects* **2009**, *3*, 117–121. [PubMed]

23. Chinelatti, M.A.; Tirapelli, C.; Corona, S.A.M.; Jasinevicius, R.G.; Peitl, O.; Zanotto, E.D.; Pires-de-Souza, F.C.P. Effect of a Bioactive Glass Ceramic on the Control of Enamel and Dentin Erosion Lesions. *Braz. Dent. J.* **2017**, *28*, 489–497. [CrossRef] [PubMed]

24. Milly, H.; Festy, F.; Watson, T.F.; Thompson, I.; Banerjee, A. Enamel white spot lesions can remineralise using bio-active glass and polyacrylic acid-modified bio-active glass powders. *J. Dent.* **2014**, *42*, 158–166. [CrossRef] [PubMed]

25. Bakry, A.S.; Takahashi, H.; Otsuki, M.; Tagami, J. Evaluation of new treatment for incipient enamel demineralization using 45S5 bioglass. *Dent. Mater.* **2014**, *30*, 314–320. [CrossRef]

26. Zhang, J.; Boyes, V.; Festy, F.; Lynch, R.J.M.; Watson, T.F.; Banerjee, A. In-vitro subsurface remineralisation of artificial enamel white spot lesions pre-treated with chitosan. *Dent. Mater. Off. Publ. Acad. Dent. Mater.* **2018**, *34*, 1154–1167. [CrossRef]

27. Narayana, S.S.; Deepa, V.K.; Ahamed, S.; Sathish, E.S.; Meyappan, R.; Satheesh Kumar, K.S. Remineralization efficiency of bioactive glass on artificially induced carious lesion an in-vitro study. *J. Ind. Soc. Pedodon. Prevent. Dent.* **2014**, *32*, 19–25.

28. Mehta, A.B.; Kumari, V.; Jose, R.; Izadikhah, V. Remineralization potential of bioactive glass and casein phosphopeptide-amorphous calcium phosphate on initial carious lesion: An in-vitro pH-cycling study. *J. Conserv. Dent. JCD* **2014**, *17*, 3–7. [CrossRef]

29. El-Wassefy, N.A. Remineralizing effect of cold plasma and/or bioglass on demineralized enamel. *Dent. Mater. J.* **2017**, *36*, 157–167. [CrossRef]

30. Zhang, J.; Lynch, R.J.M.; Watson, T.F.; Banerjee, A. Chitosan-bioglass complexes promote subsurface remineralisation of incipient human carious enamel lesions. *J. Dent.* **2019**, *84*, 67–75. [CrossRef]

31. Sleibi, A.; Tappuni, A.R.; Davis, G.R.; Anderson, P.; Baysan, A. Comparison of efficacy of dental varnish containing fluoride either with CPP-ACP or bioglass on root caries: Ex vivo study. *J. Dent.* **2018**, *73*, 91–96. [CrossRef] [PubMed]

32. Sauro, S.; Thompson, I.; Watson, T.F. Effects of common dental materials used in preventive or operative dentistry on dentin permeability and remineralization. *Oper. Dent.* **2011**, *36*, 222–230. [CrossRef] [PubMed]

33. Saffarpour, M.; Mohammadi, M.; Tahriri, M.; Zakerzadeh, A. Efficacy of Modified Bioactive Glass for Dentin Remineralization and Obstruction of Dentinal Tubules. *J. Dent.* **2017**, *14*, 212–222.

34. Forsback, A.P.; Areva, S.; Salonen, J.I. Mineralization of dentin induced by treatment with bioactive glass S53P4 in vitro. *Acta Odontol. Scand.* **2004**, *62*, 14–20. [CrossRef]

35. Vollenweider, M.; Brunner, T.J.; Knecht, S.; Grass, R.N.; Zehnder, M.; Imfeld, T.; Stark, W.J. Remineralization of human dentin using ultrafine bioactive glass particles. *Acta Biomater.* **2007**, *3*, 936–943. [CrossRef]

36. Santos Cardoso, O.; Coelho Ferreira, M.; Moreno Carvalho, E.; Campos Ferreira, P.V.; Bauer, J.; Carvalho, C.N. Effect of Root Repair Materials and Bioactive Glasses on Microhardness of Dentin. *Iran. Endod. J.* **2018**, *13*, 337–341.

37. Zhang, Y.; Wang, Z.; Jiang, T.; Wang, Y. Biomimetic regulation of dentine remineralization by amino acid in vitro. *Dent. Mater. Off. Publ. Acad. Dent. Mater.* **2019**, *35*, 298–309. [CrossRef]

38. De Morais, R.C.; Silveira, R.E.; Chinelatti, M.; Geraldeli, S.; Pires-de-Souza, F.D.P. Bond strength of adhesive systems to sound and demineralized dentin treated with bioactive glass ceramic suspension. *Clin. Oral Investig.* **2018**, *22*, 1923–1931. [CrossRef]

39. Ma, Q.; Wang, T.D.; Meng, Q.F.; Xu, X.; Wu, H.Y.; Xu, D.J.; Chen, Y.M. Comparison of in vitro dentinal tubule occluding efficacy of two different methods using a nano-scaled bioactive glass-containing desensitising agent. *J. Dent.* **2017**, *60*, 63–69. [CrossRef]

40. Zhao, I.S.; Gao, S.S.; Hiraishi, N.; Burrow, M.F.; Duangthip, D.; Mei, M.L.; Lo, E.C.; Chu, C.H. Mechanisms of silver diamine fluoride on arresting caries: A literature review. *Int. Dent. J.* **2018**, *68*, 67–76. [CrossRef]

41. Peng, J.J.; Botelho, M.G.; Matinlinna, J.P. Silver compounds used in dentistry for caries management: A review. *J. Dent.* **2012**, *40*, 531–541. [CrossRef] [PubMed]

42. Buzalaf, M.A.R.; Pessan, J.P.; Honorio, H.M.; Ten Cate, J.M. Mechanisms of action of fluoride for caries control. *Monogr. Oral. Sci.* **2011**, *22*, 97–114. [PubMed]

43. De Caluwe, T.; Vercruysse, C.W.J.; Declercq, H.A.; Schaubroeck, D.; Verbeeck, R.M.H.; Martens, L.C. Bioactivity and biocompatibility of two fluoride containing bioactive glasses for dental applications. *Dent. Mater.* **2016**, *32*, 1414–1428. [CrossRef] [PubMed]

44. Daguano, J.; Milesi, M.T.B.; Rodas, A.C.D.; Weber, A.F.; Sarkis, J.E.S.; Hortellani, M.A.; Zanotto, E.D. In vitro biocompatibility of new bioactive lithia-silica glass-ceramics. *Mater. Sci. Eng. C Mater. Biol. Appl.* **2019**, *94*, 117–125. [CrossRef] [PubMed]

45. Polini, A.; Bai, H.; Tomsia, A.P. Dental applications of nanostructured bioactive glass and its composites. *Wiley Interdiscip. Rev. Nanomed. Nanobiotechnol.* **2013**, *5*, 399–410. [CrossRef] [PubMed]

Review

Coating Techniques for Functional Enhancement of Metal Implants for Bone Replacement: A Review

Amir Dehghanghadikolaei [1] **and Behzad Fotovvati** [2,*]

[1] School of Mechanical, Industrial and Manufacturing Engineering, Oregon State University, Corvallis, OR 97331, USA; dehghana@oregonstate.edu
[2] Department of Mechanical Engineering, The University of Memphis, Memphis, TN 38152, USA
* Correspondence: bftvvati@memphis.edu

Received: 14 May 2019; Accepted: 31 May 2019; Published: 3 June 2019

Abstract: To facilitate patient healing in injuries and bone fractures, metallic implants have been in use for a long time. As metallic biomaterials have offered desirable mechanical strength higher than the stiffness of human bone, they have maintained their place. However, in many case studies, it has been observed that these metallic biomaterials undergo a series of corrosion reactions in human body fluid. The products of these reactions are released metallic ions, which are toxic in high dosages. On the other hand, as these metallic implants have different material structures and compositions than that of human bone, the process of healing takes a longer time and bone/implant interface forms slower. To resolve this issue, researchers have proposed depositing coatings, such as hydroxyapatite (HA), polycaprolactone (PCL), metallic oxides (e.g., TiO_2, Al_2O_3), etc., on implant substrates in order to enhance bone/implant interaction while covering the substrate from corrosion. Due to many useful HA characteristics, the outcome of various studies has proved that after coating with HA, the implants enjoy enhanced corrosion resistance and less metallic ion release while the bone ingrowth has been increased. As a result, a significant reduction in patient healing time with less loss of mechanical strength of implants has been achieved. Some of the most reliable coating processes for biomaterials, to date, capable of depositing HA on implant substrate are known as sol-gel, high-velocity oxy-fuel-based deposition, plasma spraying, and electrochemical coatings. In this article, all these coating methods are categorized and investigated, and a comparative study of these techniques is presented.

Keywords: surface modification; biocompatible metals; coating techniques; hydroxyapatite

---

1. Introduction

Metallic biomaterials have been mostly used for body implants thanks to their various properties, such as mechanical strength, corrosion resistance, and biocompatibility. Although there are several metallic elements and alloys, a few of them (such as Titanium (Ti), Ti-based alloys, Platinum (Pt), and austenitic stainless steel (316L)) are implemented for orthopedic and biomedical applications [1–3]. However, due to the nature of metal-corrosive media interaction, degradation takes place after the implementation of these materials inside the human body. Some of the products of the corrosion reactions are harmful to the living organs adjacent to the implants. Nickel (Ni) ions released from corrosion of nickel–titanium (NiTi) alloy implants are one of the examples of these byproducts [4,5]. One of the most important aspects of the suitability of a material for bio-applications is to have higher corrosion resistance and, consequently, lower toxicity due to released metallic ions [6]. Any difference between the chemical composition of the bone structure and the metallic implants causes bone/implant bonding issues and subsequent problems for a patient [7,8]. To solve the corrosion and bone/implant bonding issues, many researchers have suggested surface treatment by bioactive hydroxyapatite (HA) ceramic coating. This coating consists of Ca/P components ($Ca_{10}(PO_4)_6OH_2$), enhances the bone/implant bonding properties, and increases the corrosion resistance of the substrate [9,10]. Corrosion measurements are mostly done in prepared simulated body fluids (SBF) and one of the most common types of SBF is Hank's solution which is mostly consisted of NaCl [11]. HA coatings have a close composition match with that of the bones as the components are the major inorganic portion of the bone composition. Thanks to this match, these coatings allow fast and selective bone ingrowth and enhanced osseointegration [12,13].

Corrosion resistance, biocompatibility, wear resistance, stiffness match with bone, and enhanced bone ingrowth are the most effective features of bone implant materials [14–16]. Corrosion resistance and biocompatibility are the most crucial indicators of suitability of metallic implants while exposed to harsh environments. Human body environment and physiological fluids are corrosive saline mediums which cause noticeable corrosion of implants if they are not protected by oxide layers or protective covers. In some of the metallic materials which have specific elements such as nickel (Ni), the byproducts of the corrosion reactions happening inside the human body can be severely toxic and lethal to their adjacent living tissues. Concurrently, the corrosion process degrades the implant materials and reduces their mechanical stability which eventually causes a premature failure (before complete healing of the patient). Even if none of the above effects harm the patient, a secondary surgery is needed in order to remove the implant out of the patient body after complete recovery [17,18].

Other than corrosion resistance, biocompatibility is the most important functional feature of implant materials. Biocompatibility is defined as the reaction of living organs to the implant material around them. If the tissues present positive feedback and live in contact with the implants, the material is biocompatible, and cells can grow and sustain on or close to the implant surface. However, if there are toxic ions released from the implant material, cell growth will be prevented. This characteristic relies on the substrate microstructure and chemical composition as well as the quality of the surface of the material such as surface roughness, which depends on the manufacturing processes such as machining [19–23].

All being said, the coating processes and especially the ones offering deposition of HA, metallic oxides, and polymers are reliable solutions. In the following discussion, the most applicable techniques are introduced in detail, i.e., sol-gel, high-velocity oxy-fuel plasma coating (HVOF), plasma spraying, and electrochemical coatings. Although the mentioned processes are widely used in the deposition of protective layers and surface treatment, there are other processes which are more advanced and are mostly used for specific applications. Instances of these processes are laser beam melting (LBM) electron beam melting (EBM), and ion beam melting (IBM) processes that utilize the emitted energy of the electron, ion, and laser beams to melt materials and deposit the melt on a substrate surface. In addition, these processes are known as high-energy coating techniques which are less used for common applications [24–29]. The coating quality in these methods is affected by the melt pool characteristics, which depend on the process parameters [30]. Moreover, different modeling and simulation techniques can be implemented to achieve a deeper understanding of the deposition processes. These methods can utilize continuum mechanics principles, numerical solutions, and use of software to release the highest possible accuracy in their predictions [31–34].

In summary, this study intends to introduce the most efficient means of surface protection and coating for biomedical materials, especially the ones used as bone implants, and evaluate their potential as a reliable way to deposit the desired materials on the surface of implants with the least possible side-effects. The following sections are talking about the coating techniques and the deposited layers with the most significant advantage they provide.

2. Bioactive Material Deposition Techniques

One of the most common surface modification processes is the deposition of a set of selected materials, called coating. However, since the range of these materials is wide, different methods have been introduced. These processes are selected based on the substrate material, applications of the coated material, and the coating layer thickness [35,36]. There are numerous coating methods offering different capabilities each, however, only a few of these techniques are sufficiently reliable to be applied for bio-application purposes [37,38]. These techniques simultaneously provide corrosion resistance and biocompatibility enhancement for the substrate. Among many materials for these means, using HA shows a high increase in biocompatibility and bone/implant interface formation. The following discussion presents reliable methods (i.e., sol-gel technique, High-Velocity Suspension Flame-Spraying (HVSFS), plasma spray coating, and electrochemical deposition) of HA deposition [39,40].

2.1. Sol-gel

Sol-gel has been at the center of attention in recent years thanks to its simplicity, flexibility, and low cost of the process. This process provides a reliable enhancement of coating adhesion on the substrate of metallic biomaterials [35,41]. Sol-gel consists of two distinct parts known as sol and gel. For the sol part, calcium-phosphorus-based (CaP) precursors are solved in ethanol and distilled water to produce phosphorus pentoxide (P_2O_5)/triethyl phosphate ($C_6H_{15}O_4P$) and increase hydrolysis of the sol part, respectively [42–44]. Ethanol plays a key role in solving Ca part of the precursors, as well [45]. In the next step, the two distinct parts are mixed carefully, and after undergoing an evaporation process, the liquid medium goes away. This process repeats for many times until the desired viscosity of the sol-gel medium is achieved. Besides the viscosity, the chemical concentration of the ingredients is crucial in achieving a high HA formation on substrates [46,47]. The sol-gel technique is a dipping process that undergoes three different steps: Dipping, withdrawing, and air drying. The first two steps are mostly done in a controlled constant speed to prevent entrapping air bubbles and a non-uniform layer thickness of the coating medium, respectively. To obtain such a quality of uniformity, many researchers utilize controlled speed motors or servo motors. However, there has not been a significant difference between parts coated manually and automatically since the process is a dipping method and covers the whole geometry of the substrate regardless of its complexity [48–50]. On the other hand, this process can perform composite and multi-layer coatings by changing sol-gel medium (i.e., composition and viscosity) and iteration of dipping, withdrawing, and drying steps, respectively. With this method, it has been reported that HA coating of 0.05–15 mm thickness has been achieved [35,51]. Based on drying and the method of applying a sol-gel coating (dipping and rotation) to the substrate, different structure of coating (rods and spheres) can be achieved [35]. Figure 1 represents a summary of the steps of the sol-gel coating process.

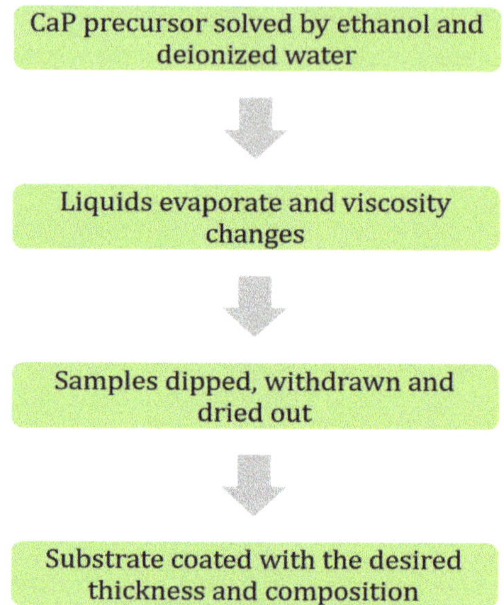

Figure 1. Flowchart of sol-gel coating process coating steps, in brief.

Another advantage of sol-gel coatings can be the ability to undergo annealing in a furnace to further stabilize the deposited layers with less deep thermal cracks, which cause discontinuity and less protection [52]. Compared to the thermal coating processes, there is no significant change in the composition of the deposited layer of coating. Kuntin et al. [53], reported that deposited HA layer on the substrate by plasma spraying decomposes to calcium oxide (CaO), tetra-calcium phosphate, and tricalcium phosphate. The high temperatures of these types of coating processes (above 5000 °C) may burn or delaminate the deposited layers, as well. However, annealing temperatures used for curing and stabilization of sol-gel deposited layers are in the range of 375 to 500 °C that is notably lower than those of thermal coatings [54]. Liu et al. [55], reported a significant adhesion enhancement of HA layers after annealing in an atmospheric protected furnace in the mentioned temperature range. In other scientific reports, it has been mentioned that post-processes such as heat treatment facilitates densification and apatite formation of the deposited HA materials on the substrates and increases adhesion between the substrate and the deposited materials. However, the heat treatment temperatures must be below the melting point of the weakest material in order not to force any collapse in the mechanical and biological properties [7,56,57]. Other considerations have been applied to improve the quality of sol-gel deposited HA layers. For instance, a new polymeric material known as poly ε-caprolactone (PCL) has been introduced [58]. Hanas et al. [59], reported the formation of a porous microstructure of the coating layer and increased osseointegration of coated substrates by addition of PCL to HA. Alemon et al. [60], claimed that a porous coating layer of 184 μm thickness was formed on a Ti6Al4V substrate that provided enhanced adhesion between coating and substrate with fewer micro-cracks on the surface of the coating layer. In a similar study by Catauro et al. [61], they reported an increased wear and corrosion resistance of the coated substrates resulting in less metallic ion release as corrosion byproducts. Figure 2 shows a composite sol-gel coating microstructure. In this figure, the sol-gel coated samples were post-processed in 600 and 1000 °C and then were exposed to SBF media for 21 days. The difference in morphology of the coating surface can be clearly seen for different heat treatment temperatures and medium compositions.

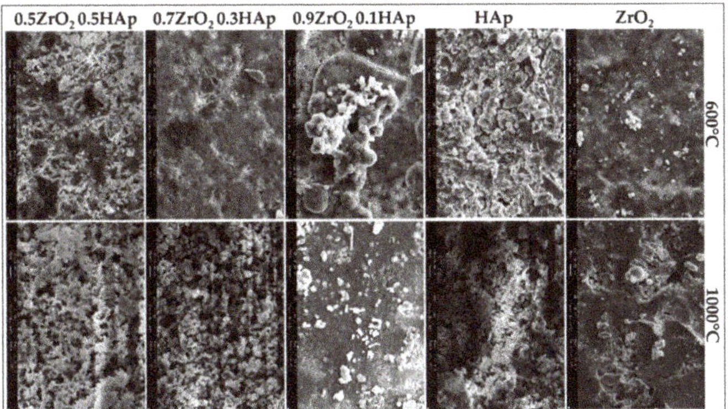

Figure 2. SEM micrographs of composite coated samples heated to 600 °C and 1000 °C after 21 days of exposure to simulated body fluids (SBF) with different compositions [62].

2.2. High-Velocity Suspension Flame-Spray (HVSFS) Coating

HVSFS is a modified type of high-velocity oxy-fuel (HVOF) coating process that utilizes suspensions with the desired composition to deposit a coating layer on substrates [63,64]. Figure 3 represents a schematic view of an experimental HVSFS setup. In this method, an inlet pushes the coating materials to the stream of hot flames and accelerates them toward the target substrate. The advantage of using this process is a significantly increased coating speed and a larger area of coverage [65,66]. The principles of this process are straightforward, and the equipment are not expensive compared to other advanced coating technologies such as EBM and IBM processes. Ghosh et al. [67], reported that the materials through the hot stream can be mixed and new composite materials can be produced prior to deposition on substrates. This feature increases the flexibility of HVSPS coating process. On the other hand, there is no need for the deposited HA particles to perform a heat treatment as the particles have been already passed through a flame stream [68]. It has been reported by Forg et al. [69], that with increasing the velocity of the flame stream, the porosity of the coating layer increases while the velocity and porosity can be controlled in real-time. Based on research by Bernstein et al. [70], finer powders can be utilized in specifically designed fluid media in order as carriers. In this case, a finer coating structure with less unintentional porosity forms on substrates. Taking advantage of HVSPS characteristics in determining coating thickness, the gap between thin-film fabrication (chemical and physical vapor methods) and thick layers created by thermal spraying can be covered [71,72].

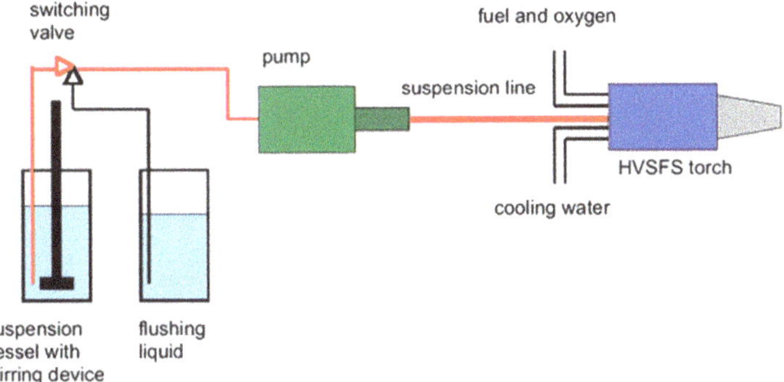

Figure 3. Schematic illustration of a high-velocity suspension flame-spray (HVSFS) experimental setup for nano-oxide ceramic coatings [73].

Although this process offers interesting advantages, the nature of HVSFS is a thermal coating method, and residual stresses will remain in the deposited structure. Norouzi et al. [74], suggested that with a decreased thickness of the deposited coating layer the chance of thermal cracking and coating failure decreases. They reported that the coating/substrate bonding is under the direct influence of the working parameters of HVSFS, i.e., flame stream velocity, oxygen/fuel ratio, and the distance between the nozzle and the target substrate. Gadow et al. [75], deposited bio-ceramic materials and reposted that with a change in the suspension composition, carrier liquids, and particle sizes, HA deposition takes place in a higher quality. They suggested that use of diethylene glycol (DEG) as an alternative to water-based media, significantly enhances coating adhesion to the substrate and offers a less porous microstructure. Chen et al. [76], reported that introducing metal oxide particles to the coating stream enhances the thermal stability of the substrate because the first layer of the deposited materials acts as a thermal barrier and protects the substrate against thermal exchange to the successive coating layers. In another study by Gadow et al. [73], they implemented metal oxide materials suspended in isopropanol to coat pure Ti substrate in order to form a protective layer. They reported an increase in the coating surface quality and a significantly finer microstructure than the one obtained by conventional thermal coating processes. However, they found slightly lower microhardness values of the top surface of the coating layer compared to those of the conventional thermal techniques. Table 1 presents their findings in brief.

Table 1. Summary of characteristics of different materials used for coating of pure Ti with HVSFS process [73].

Spray Material	Microhardness (HV)	Porosity (%)	Roughness (Ra, μm)	Phase Composition
Al_2O_3	620–880	5	0.58	Mainly γ
TiO_2	1000	0–0.05	0.65	Mainly anatase
Cr_2O_3 (in propane)	1400–1800	4–5	0.47	Hexagonal
Cr_2O_3 (in ethane)	1100–2000	7–10	0.48	Hexagonal
3YSZ	715–816	<1	1.75	Tetragonal

Figure 4 represents an HVSFS-treated Ti substrate. HVSFS process can be optimized based on different working parameters. As with any other experimental method, these parameters can affect the performance of the process. It has been claimed that an increase in the rate of oxygen flow, increases the rate of coating/substrate adhesion [75]. The same trend has been observed regarding the simultaneous increase of oxygen and fuels rate [77]. With an increase in these two parameters, the temperature of the flame stream rises melting the involved particles inside the flame stream. Due to the acceleration of these particles, they may penetrate to the top layers of the substrate. At the same time, after partial melting, they become softer. Combining these two characteristics, they form a strong bonding to the substrate. However, excessive increase of the fuel and oxygen rate may increase the flame temperature to a point that decomposes the deposited materials and alter their functionality. The other affecting parameter is the distance of the nozzle head to the substrate. It has been reported that a shorter distance results in an increased bonding of the deposited materials to the substrate. However, with an inappropriate decrease in the distance, the formed turbulence may cause less deposition of materials due to splashing [78,79].

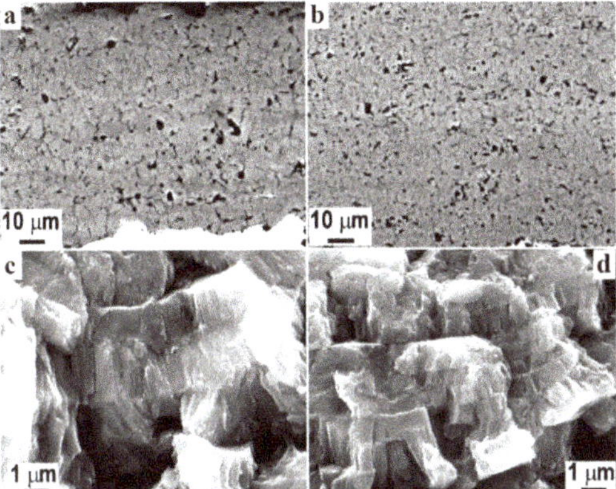

Figure 4. Polished cross-section of atmospheric plasma spray (APC) (**a**) and high-velocity oxy-fuel plasma coating (HVOF) (**b**) TiO$_2$ coatings and respective fracture sections (**c,d**) [80].

2.3. Plasma Spray Coating

One of the most investigated methods of material deposition is a plasma spray coating. Thanks to its ability to deposit metallic alloys, oxides, and ceramics, this process offers a wide range of applications in biomaterial and body implants coatings [81,82]. This process is similar to HVOF and HVSFS coating techniques. Figure 5 represents a schematic overview of a working plasma spray setup and the position of the targeted substrate to perform an effective HA coating experiment. A flow of solid particles is merged into a hot and accelerated plasma stream. Due to the high temperature of this process, there is a possibility of forming composite coating materials. The accelerated particles get hot and soft through the plasma stream and cover the surface upon impact to the substrate. The mechanism of coating is either by penetrating to the substrate or by deforming the particles due to the impact of energy. In the latter case, the flattened particles accumulate after a series of spraying and cover the surface of the substrate [83]. In addition, this process is capable of holding different atmospheric protections, i.e., air, inert gas, or vacuum. Narayanan et al. [84], reported that although the plasma jet reaches the high temperatures of 10,000 K or higher, a drastic temperature drop occurs after the plasma jet exits the nozzle tip. This process enjoys a high range of applications from biomedical implant coatings to fabricating thermal barrier coats. It has been reported that the plasma-treated surfaces present a high adhesion of coating/substrate [85]. Oehr et al. [86] coated a Ti plate with plasma spraying and deposited HA particles on the substrate. They observed an enhanced coating adhesion with improved osseointegration properties. Fatigue performance of titanium alloys is important regarding various applications [87], therefore, many studies were carried out to enhance the fatigue properties of these components. Yoshinari et al. [88] investigated the effect of HA coating thickness on the fatigue of Ti6Al4V alloy. They reported coating thicknesses of 25–100 µm do not show any negative effect, while an increase in thickness to 150 µm reduces the fatigue properties of the mechanical part. According to their results, increasing the thickness of the coating increases the difference between the material behavior of the substrate and the coating layer. Moreover, they discussed the excessive thermal shock, which is applied to the substrate during a thicker coating formation, causes micro-crack formation in both the coating layer and the substrate, reducing the fatigue life of the whole component.

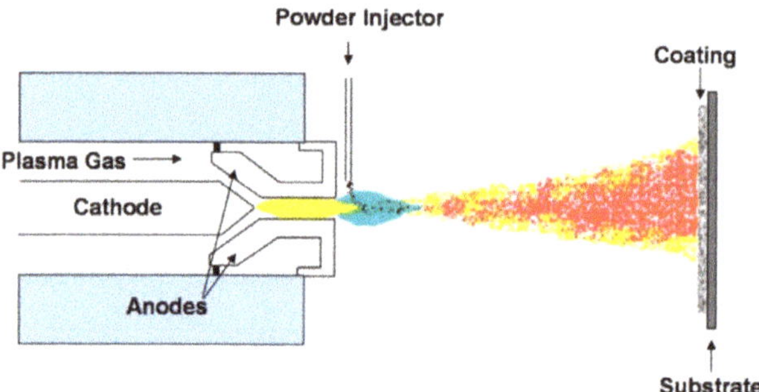

Figure 5. Schematic plasma spray coating technique and the targeted substrate [89].

Similar to HVOF, a modification can be made to plasma spray coating to change the feedstock state. Gross et al. [90] introduced a suspension intake to the plasma jet and reached to thin-film coatings of 5–50 µm thickness. The process of utilizing suspension in the plasma spray coating is known as suspension plasma spray (SPS). The thickness of SPS-treated layer is significantly lower than the ones made by conventional plasma spray coating [91]. Different working parameters can affect the deposition quality. These parameters are particle size and feed rate of the feedstock, and atmosphere and acceleration of the plasma jet. Based on different combinations of the mentioned parameters, different deposition thicknesses ranging from sub-micron to 300 µm can be achieved [92,93]. For instance, Reitman et al. [94], discussed a decrease in coating/substrate adhesion in higher plasma jet temperatures due to higher amorphous HA content of the coating layer. Besides optimizing the working parameters, post-processing is a solution to improve the coating quality. By annealing the coated samples at 700 °C for 1 h, Lynn et al. [95] increased the coating purity and observed a crystalline structure of HA. Basu et al. [96] investigated the effect of elevated annealing temperature (up to 1100 °C) of HA-coated Ti substrate and claimed that annealing at higher temperatures, results in the formation of more Ca and Ti oxides in the coating. Zheng et al. [97] utilized the addition of Ti particles to HA particles in plasma spray coating and reported a significant enhancement of coating/substrate adhesion with increasing Ti content of the feedstock. Figure 6 represents a composite HA-Zirconia coating layer on a stainless-steel substrate. As can be seen, the heat treatment process seals a number of micro-pores, but there is no significant change in the appearance of micro-cracks. Singh et al. [98] modified the composition of the feedstock and proposed a new composition of 10 wt% ($80Al_2O_3$-$20TiO_2$) on Ti6Al4V substrate and reported an enhanced bonding strength of 30 MP. Table 2 summarizes the properties of different plasma spray coatings.

Table 2. Comparison between characteristics of different modes of plasma spray coating.

Process	Speed	Quality	Cost	Thickness
Air plasma	High	Low	Low	High
Inert gas plasma	High	Moderate	Moderate	High
Vacuum plasma	Low	High	High	High
Suspension plasma	Moderate	Moderate	Moderate	Low

Figure 6. Morphology of HA-Zirconia coating stainless steel 316L substrate at different magnifications (a) as-sprayed (b) heat treated [99].

2.4. Electrochemical Coating Techniques

In the electrochemical coating technique, substrates undergo a series of electrochemical processes prior to being completely coated. Figure 7 represents a schematic overview of an electrophoretic coating setup. This process utilizes the potential difference between cathodic and anodic poles of an electrical circuit to form micro-arcs or to exchange ions between anion and cation sides [100]. The coating process takes place in two steps called electrophoretic and electrolytic deposition. The former is responsible for depositing large suspension particles existing inside the electrolyte, while the latter is in charge of depositing fine materials and structure. These two steps may be considered as two distinct processes or as separate steps of a single coating process [101,102]. In electrochemical processes, desired coating materials, e.g., CaP precursors, are dissolved in the working electrolyte. The highest number of applications of electrodeposition coatings done on Ti and Ti alloys substrates [103]. One of the characteristics of the electrochemical coatings is the uniformity in thickness of the deposited layer throughout the substrate [104]. Zhao et al. [105], investigated the applications of Pt and graphite anodic electrodes and reported an increased rate of deposition and enhanced coating quality. Although this process is considered a low-temperature method, the coated substrates have to undergo a series of densification and sintering in the furnace. The reason is that the surface of the coated substrate is not a compact structure due to the large suspension particle deposition during the electrophoretic step [106]. Due to electrochemical facts, the sharp edges, such as micro-cracks, are more prone to ion exchange, and as a result, they undergo a higher rate of material deposition. This characteristic guarantees the highest value of homogeneity and integrity in the coating layers [107].

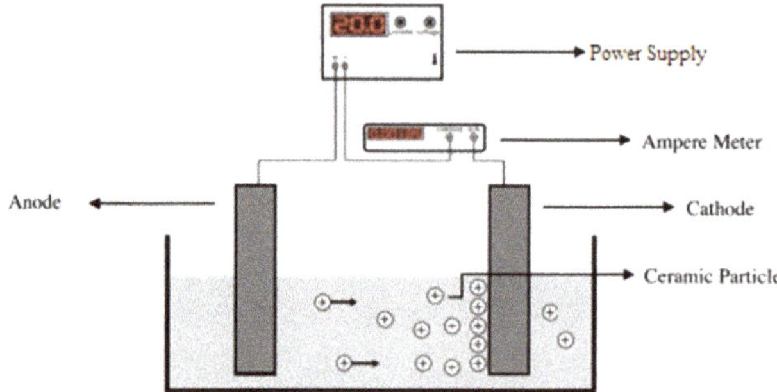

Figure 7. Schematic view of simple electrophoretic deposition process [108].

Peng et al. [109] investigated electrochemical coating of CoCrMo substrates and proved that although the thickness of the coating layer was decreased after sintering and annealing processes, the adhesion of HA deposited layer was significantly enhanced. The same results were reported by Zhang et al. [110] after annealing HA coated substrates at 500 °C for 1 h. However, they reported that due to the nature of electrochemical reactions in water-based electrolytes, a notable volume of hydrogen and air bubbles form on the surface of the cathode (the substrate) and prevent the complete HA deposition process. This discontinuity in the deposition step results in less uniformity and integrity of the coating layer. Moreover, after the sintering and annealing processes, these defects may get more intense due to shrinkage [111–113]. To solve this issue, it was recommended to implement H_2O_2 as a replacement for water. Because peroxide cancels the deteriorating effects of released H_2 gas [114]. In addition, utilizing pulsed power supplies can result in enhanced coating quality and thickness while the uniformity is improved [115]. They suggested a higher off-time of the pulse to let the HA structures nucleate and grow easily. Xavier et al. [116] reported an increased apatite crystal formation after immersing the samples in SBF. There are different types of electrochemical coating processes known as micro-arc oxidation (MAO) and anodization. The voltage range, at which MAO occurs is comparably higher than that of anodization [117]. However, anodization is mostly considered as a pretreatment to electrochemical coating processes, whereas MAO is known as a separate coating technique [118–120]. He et al. [121], reported a successful deposition of a porous Al_2O_3 structure on Ti substrate utilizing anodization pretreatment. They observed an increase in CaP structure nucleation inside the pores of the coating layer. Figure 8 depicts the effect of different parameters on the morphology of the electrodeposited HA coating layer. In constant voltage mode, with an increase in coating voltage, the needle structure increases in size. While the process is run under the constant current mode, in the current density of 5 mA/cm^2, the deposited material structure is similar to snowflakes while with increasing the current density to 10 mA/cm^2, the materials deposition takes place in needle structures again.

Figure 8. SEM micrographs of HA-coated samples by electrodeposition at (**a**) 1 V, (**b**) 2 V, (**c**) 3 V, (**d**,**e**) 5 mA/cm^2 and (**f**) 10 mA/cm^2 [122].

According to many research reports, the addition of materials such as ZrO_2, carbon nano-tubes (CNT), and titanium dioxide (TiO_2) enhances HA coating performance [123,124]. In addition, it was observed that addition of other coating layers on the initial one (increasing the thickness by deposition of multilayered coating) increases the corrosion resistance and enhances ion release behavior of the coated substrate. This technique is called multi-walled coatings [125]. Henriques et al. [126] observed that a post heat treatment significantly improves the coating/substrate adhesion and the nucleation of CaP microstructure. Park et al. [127] reported an enhancement in corrosion resistance and stability of composite CNT/HA coating resulted in significantly increased biocompatibility. Yuan et al. [128] observed higher thickness homogeneity and enhanced coating/substrate adhesion in HA layers deposited on stainless steel. More information can be found in the literature [129,130]. Table 3 represents a brief comparison between different characteristics of the different methods of deposition.

Table 3. A qualitative comparison of properties of different deposition processes.

Method	Nature of Coating	Thickness	Porosity	Adhesion	Flexibility	Speed	Cost	References
Sol-gel	Physical	Low-medium	Medium-high	Medium-high	Very high	Very low	Low	[35]
HVOF	Thermal	High	High	High	Moderate	high	Moderate	[131]
Plasma spray	Thermal	High	High	High	High	High	Moderate-high	[132]
Electrochemical	Chemical	Low-medium	Low-medium	High	Very high	Moderate	Low	[133]

3. Properties of Coating Layers

Coating processes and the materials deposited on any substrate are assessed based on their performance. Some of these important criteria are the deposition-substrate bonding strength, the coating thickness, the corrosion resistance of the coating layers, and the stability of coatings in different working conditions. Bonding between deposited layers and substrates, called adhesion strength, is evaluated by applying a stress to the coating layer and measuring the highest strength value, at which no breakage or delamination occurs. The coating layers adhesion have to be equal or higher than the human bone stiffness [134]. To enhance the coating layer adhesion, several studies are carried out to investigate the chemical composition of the deposited materials. As an instance, Zhang et al. [135] reported up to 35% enhancement in adhesion of sol-gel deposited HA layers on Ti substrates with an increase in fluorine content of the coating medium. In addition, they claimed that with an increase in the heat treatment temperature, the adhesion increases, as well. Sopcak et al. [136] utilized the addition of polyethylene glycol (PEG) and polyvinylpyrrolidone (PVP) polymers into HA containing coating medium and coated Ti samples with electrostatic spraying. In a comparison between substrates coated using different process parameters, they observed a significant increase of the coating adhesion strength with an increase in PEG and PVP compounds (up to 400% of the sole HA-coated samples). Figure 9 represents the surface conditions of different coated samples by HA, PEG, and PVP. In an investigation by Rocha et al. [71], Ti6Al4V substrate was coated by thermal spraying, and HA-TiO$_2$ ceramic was deposited on the surface. They observed the adhesion strength of 30 ± 2 MPa, which was higher than what was reported in the literature. Fujihara et al. [137] compared thermal spraying-coated samples with ion/laser beam-treated samples and reported that after long-term in vivo tests, 90% of the thick coatings of the first group failed, while ion/laser beam-deposited HA provided higher durability despite their considerably lower thickness.

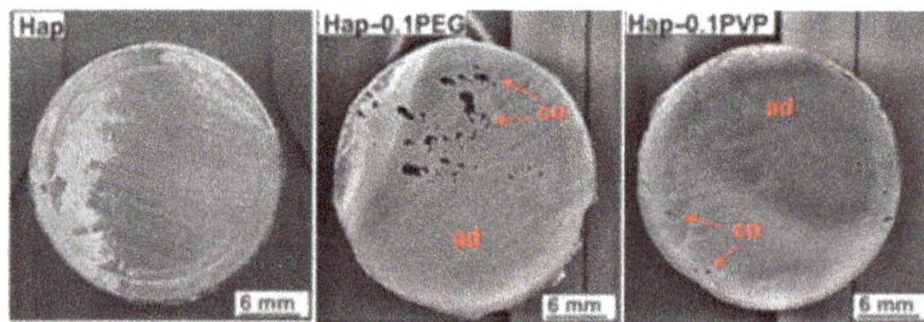

Figure 9. Surface of the coated samples and the detached coatings after adhesion strength tests [136].

Another factor affecting the protective properties of coating layers is the layer thickness. It has been reported that hot-isostatic press can form thick coatings of up to 200 μm that is close to the range of thermal spray coating thicknesses. High-temperature processes such as ion/laser beam material deposition provide the least thickness of coatings, while they have a reliable adhesion and protection. In addition, low-temperature processes such as electrochemical processes and sol-gel have variable thicknesses and offer multi-layered coating layers. Although the last two types of the material deposition processes are very flexible in coating composition and thickness, they cannot offer a very thick coating layer, and they are categorized as moderate thicknesses coating techniques [138–141]. Based on the previous discussion, it has been revealed that the protective purposes of ion/laser beam coating layers have maintained their functionality during long-term applications offering the best performance. However, the disadvantages of these two processes are the high costs of the process and equipment and their inability to coat complex geometries [84]. Based on the coating layer adhesion and thickness, it is important to have a higher physical and chemical stability of the deposited materials. Stability of a coating means the interaction of the deposited materials under different chemical and mechanical conditions while the coated implants are in use. Degradation of the coating material in physiological fluids can be an example of these conditions [142,143]. Figure 10 depicts the degradation of body implants in different conditions after a specific number of days of application. The samples are coated by plasma electrolytic oxidation (PEO), then sealed with polycaprolactone (PCL), and finally dipped in polydopamine (PDAM). The final sample was then washed to remove loosely bonded PDAM from the coating surface. In many applications, the implants are needed to degrade in the human body after a specific time of utilization, e.g., after a fractured bone is healed, but it is important to maintain the minimum function time [144,145]. Mg implants are very vulnerable to the body fluids, and they corrode quickly after exposure. However, these materials are good candidates for implants due to their biocompatibility properties. Many researchers have tried to coat Mg substrates in an engineered way to provide enough time for the implant to serve inside the human body and to dissolve gradually and disappear after the healing period. However, if Mg samples are coated with durable coatings, such as the ones provided by ion beam deposition, the implants remain inside the patient body for a long time, and another surgery is needed to remove them [146–148]. This example highlights the importance of stability of the coating materials.

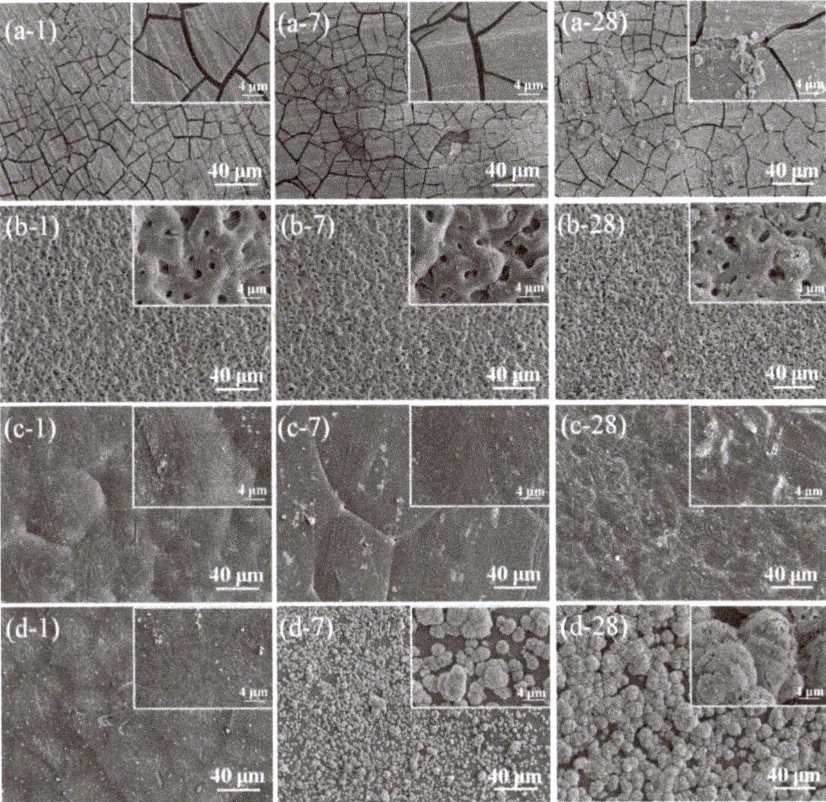

Figure 10. Degraded Mg-1.2Zn-0.5Ca alloy coupons aged at different durations after in vitro immersion in the SBF solution for different times of 1, 7, and 28 days. (**a**) Degradation of Mg-1.2Zn-0.5Ca (**b**) degradation of PEO coated Mg-1.2Zn-0.5Ca samples, (**c**) degradation of PEO-coated Mg-1.2Zn-0.5Ca samples sealed with PCL, and (**d**) degradation of PEO/PCL-coated Mg-1.2Zn-0.5Ca samples dipped in PDAM. In the non-coated samples, a severe cracking scheme is visible which is a result of dehydration of $Mg(OH)_2$ as the main byproduct of corrosion. PEO-treated samples represent no severe sign of corrosion after 28 days. The PEO/PCL-treated samples provide higher corrosion resistance with a slight degradation of PCL layers on the PEO-deposited layer. Finally, the PEO/PCL/PDAM samples provide high corrosion resistance with no signs of pitting or cracking on the surface while sites of HA formation can be seen obviously on the outer layer which is resulted by addition of PDAM [149].

One of the most important applications of the coating layers is to increase the corrosion resistance of a substrate, which is exposed to a harsh environment. This means that not all of the bio-applications need a coating layer, which degrades gradually and disappears after a period of time [150]. The corrosion resistance of the coating layers is desired for two reasons: (1) The protection integrity of the implants and (2) to prevent ion release into the body fluids. Metallic ions are the products of corrosion of metallic implants and can be harmful to human health, especially when they are higher than a specific dosage. A significant instance of these materials can be the released Ni ions from NiTi body implants [4]. Based on the literature, although CaP-based coatings enhance the bone ingrowth and form a quick and strong interface between bones and implants, they suffer from low corrosion resistance, compared to metallic oxide coatings, such as TiO_2, Al_2O_3, etc. As a result, solid implants are exposed to corrosive media and lose their mechanical properties that can be considered as a malfunction in healing a fractured bone [151]. However, various techniques, such as surface finish, surface blast, heat treatment, change in composition, etc., have been proposed to modify the surface quality and the deposited materials structure, thus to enhance the corrosion resistance of CaP-based coatings [152,153]. Zhao et al. [154] have utilized PCL to seal porous structure of HA coating. They obtained HA/PCL composite scaffolds with higher corrosion resistance. Based on the surface finish and porosity of the deposited layer, many considerations can be considered to improve the corrosion properties of a coating layer. Heat treatment is another method of coating modification that improves the crystallization and enhances corrosion resistance. Xia et al. [155] investigated an HA-coated Ti6Al4V substrate by the thermal spraying technique with a successive hydrothermal treatment (HT) resulted in an increased crystallinity. They reported an enhancement in coating degradation after the HT process. Moskalewicz et al. [156] implemented TiO_2 nanoparticles to enhance the electrochemical corrosion of Ca/P-based coatings. They introduced a specific ratio of TiO_2 nanoparticles to HA solutions prior to implementing the coating process. Corrosion testing of the coated substrates revealed an enhancement in the corrosion resistance and bioactivity of the deposited layer. In another study, Ionita et al. [157] achieved a TiO_2/HA coating layer with enhanced corrosion properties and bioactivity, while the deposited materials presented active antibacterial properties. Santos-Coquillat et al. [158] modified Ti substrate by utilizing MAO. They observed an increased corrosion resistance after depositing Ca/P structures on the substrates. In addition, they implemented cell culture experiments and reported a good cell adhesion with no or lower signs of toxicity after MAO surface treatment. Figure 11 represents the MAO-treated surface. The difference in coating structure results from different exposure time of specimens. In the two groups, the microstructure is porous with different pore sizes reflecting the voltage of current density applied to the sample surface. After exposure of the coated parts to the cell platform, it can be seen that in both cases, the cells are growing well, which reveals the suitability of the coating composition and microstructure.

Overall, various parameters are effective in obtaining a reliable coating layer, which is suitable for bio-purposes. Although at first glance, some of these parameters look more important than the others, they are all in a direct interrelationship and ignoring one may affect the overall performance of the coated surface. In most of the bio-applications, corrosion resistance, stability, composition, and bioactivity of the coating layers are simultaneously in action to offer the best protection and healing properties.

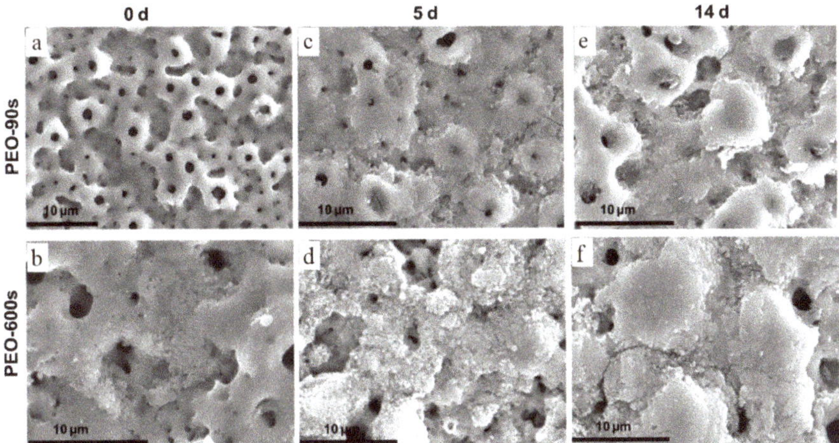

Figure 11. SEM micrographs of MAO-treated Ti CP before and after immersion in α-MEM without cells: (**a**) MAO-90s, (**b**) MAO-600s, (**c**) MAO-90s 5-day immersion, (**d**) MAO-600s 5-day immersion, (**e**) MAO-90s 14-day immersion, (**f**) MAO-600s 14-day immersion [158].

4. Summary

Many metallic and non-metallic materials are used as body implants to facilitate patient healing. However, to date, the ratio of implementation of non-metallic to metallic implants is negligible. As a result, a deep investigation is needed to select the best metallic compound or alloy to derive the best performance for implants. Unfortunately, metallic implants suffer from different deficiencies, such as low/high corrosion resistance, releasing toxic ions, and low biocompatibility. To resolve this issue, many surface modification methods are proposed that coating techniques are among the most important ones. Each of these techniques demonstrates different capabilities and one has to select a coating method based on their needs and applications. Among these techniques the most common ones are sol-gel, high-velocity suspension spray, plasma spraying, and electrochemical coating processes. Table 4 represents a summary of the materials deposited on different substrates using sol-gel, HVOFS, plasma spraying, and electrochemical processes.

I.   Sol-gel is the cheapest method that has high flexibility in the composition of the deposited materials and can form deposition on complex substrates.
II. Electrochemical processes are in the same run while they are applicable only to conductive materials. This limits their applications on surfaces that are not electrically conductive.
III. High-velocity flame and plasma spraying processes in different forms can produce considerably thick depositions. However, these processes implement a high-temperature thermal process to melt the feedstock or change them into a semi-solid form. A high thermal gradient may affect the properties of a metallic or ceramic substrate while makes these processes unable to coat polymeric and plastic substrates.

Table 4. Summary of materials deposited on different substrates via various coating techniques.

Process	Deposited Materials	Substrates	Reference
Sol-gel	HA, TiO_2, PCL	Ti alloys, ceramic, Stainless steel	[35,37,59,60]
HVOFS	Metal oxides such as TiO_2, Al_2O_3, Cr_2O_3, ceramics	Ti alloys, Stainless steel, ceramics	[77–80]
Plasma spray	HA, HA-Zr_2O_3, Al_2O_3-TiO_2	Ti alloys, Zr alloys, ceramics, metallic alloys	[85–90]
electrochemical	ZrO_2, TiO_2, MgO, CNT, HA, HA-TiO_2, SWCNT, MWCNT	Metallic alloys, ceramics, Ti alloys	[119–125]

Therefore, a method should be selected based on the desired needs and functionalities. Many scientific reports have been revealed that multi-layered coatings are achieved to employ the advantages of different deposition processes while diminishing their drawbacks. Although current deposition techniques are reliable means of surface protection, there is still a need for finding better solutions by introducing new techniques and materials for coating. Perspective studies can focus on developing combined processes with fewer side effects, higher control on deposition rate, less cost and higher ease of use.

Funding: This research received no external funding.

Conflicts of Interest: The authors declare no conflict of interest.

References

1. Niinomi, M.; Nakai, M.; Hieda, J. Development of new metallic alloys for biomedical applications. *Acta Biomater.* **2012**, *8*, 3888–3903. [CrossRef]
2. Breme, H.; Biehl, V.; Reger, N.; Gawalt, E. A Metallic Biomaterials: Introduction. In *Handbook of Biomaterial Properties*; Springer: Berlin, Germany, 2016; pp. 151–158.
3. Chen, Q.; Thouas, G.A. Metallic implant biomaterials. *Mater. Sci. Eng. R Rep.* **2015**, *87*, 1–57. [CrossRef]
4. Ibrahim, H.; Jahadakbar, A.; Dehghan, A.; Moghaddam, N.S.; Amerinatanzi, A.; Elahinia, M. In Vitro Corrosion Assessment of Additively Manufactured Porous NiTi Structures for Bone Fixation Applications. *Metals* **2018**, *8*, 164. [CrossRef]
5. Yamamoto, A.; Honma, R.; Sumita, M. Cytotoxicity evaluation of 43 metal salts using murine fibroblasts and osteoblastic cells. *J. Biomed. Mater. Res.* **1998**, *39*, 331–340. [CrossRef]
6. Espallargas, N.; Torres, C.; Muñoz, A. A metal ion release study of CoCrMo exposed to corrosion and tribocorrosion conditions in simulated body fluids. *Wear* **2015**, *332*, 669–678. [CrossRef]
7. Singh, A.; Singh, G.; Chawla, V. Influence of post coating heat treatment on microstructural, mechanical and electrochemical corrosion behaviour of vacuum plasma sprayed reinforced hydroxyapatite coatings. *J. Mech. Behav. Biomed. Mater.* **2018**, *85*, 20–36. [CrossRef] [PubMed]
8. Dehghanghadikolaei, A. Additive Manufacturing as A New Technique of Fabrication. *J. 3D Print. Appl.* **2018**, *1*, 3–4.
9. Luo, L.; Jiang, Z.Y.; Wei, D.B.; He, X.F. Surface modification of titanium and its alloys for biomedical application. *Adv. Mater. Res.* **2014**, *887*, 1115–1120. [CrossRef]
10. Dorozhkin, S.V. Calcium orthophosphate deposits: Preparation, properties and biomedical applications. *Mater. Sci. Eng. C* **2015**, *55*, 272–326. [CrossRef]
11. Shi, P.; Ng, W.F.; Wong, M.H.; Cheng, F.T. Improvement of corrosion resistance of pure magnesium in Hanks' solution by microarc oxidation with sol–gel TiO_2 sealing. *J. Alloy. Compd.* **2009**, *469*, 286–292. [CrossRef]
12. Ayu, H.M.; Izman, S.; Daud, R.; Krishnamurithy, G.; Shah, A.; Tomadi, S.H.; Salwani, M.S. Surface modification on CoCrMo alloy to improve the adhesion strength of hydroxyapatite coating. *Procedia Eng.* **2017**, *184*, 399–408. [CrossRef]
13. Chen, Q.; Cao, L.; Wang, J.; Jiang, L.; Zhao, H.; Yishake, M.; Ma, Y.; Zhou, H.; Lin, H.; Hong, D.; et al. Bioinspired Modification of Poly (L-lactic acid)/Nano-Sized β-Tricalcium Phosphate Composites with Gelatin/Hydroxyapatite Coating for Enhanced Osteointegration and Osteogenesis. *J. Biomed. Nanotechnol.* **2018**, *14*, 884–899. [CrossRef] [PubMed]

14. Al-Tamimi, A.A.; Fernandes, P.R.A.; Peach, C.; Cooper, G.; Diver, C.; Bartolo, P.J. Metallic bone fixation implants: A novel design approach for reducing the stress shielding phenomenon. *Virtual Phys. Prototyp.* **2017**, *12*, 141–151. [CrossRef]

15. Johansson, P. *On Hydroxyapatite Modified Peek Implants for Bone Applications*; Malmö University, Faculty of Odontology, Department of Prosthodontics: Malmo, Sweden, 2017.

16. Fitz, W.; Bojarski, R.A.; Lang, P. Implants for Altering Wear Patterns of Articular Surfaces. U.S. Patent No. 9,700,420, 2017.

17. Gretzer, C.; Petersson, I. Bone Tissue Implant Comprising Strontium Ions. U.S. Patent No. 9,889,227, 2018.

18. Bistolfi, A.; Cimino, A.; Lee, G.C.; Ferracini, R.; Maina, G.; Berchialla, P.; Massazza, G.; Massè, A. Does metal porosity affect metal ion release in blood and urine following total hip arthroplasty? A short term study. *Hip Int.* **2018**, *28*, 522–530. [CrossRef] [PubMed]

19. Dehghanghadikolaei, A.; Mohammadian, B.; Namdari, N.; Fotovvati, B. Abrasive machining techniques for biomedical device applications. *J. Mater. Sci.* **2018**, *5*, 1–11.

20. Jansen, J.A.; Dhert, W.J.A.; Van der Waerden, J.P.; Von Recum, A.F. Semi-quantitative and qualitative histologic analysis method for the evaluation of implant biocompatibility. *J. Investig. Surg.* **1994**, *7*, 123–134. [CrossRef]

21. Hatamleh, M.M.; Wu, X.; Alnazzawi, A.; Watson, J.; Watts, D. Surface characteristics and biocompatibility of cranioplasty titanium implants following different surface treatments. *Dent. Mater.* **2018**, *34*, 676–683. [CrossRef] [PubMed]

22. Dehghanghadikolaei, A.; Fotovvati, B.; Mohammadian, B.; Namdari, N. Abrasive Flow Finishing of Stainless Steel 304 Biomedical Devices. *Res. Dev. Mater. Sci.* **2018**, *8*, 1–8. [CrossRef]

23. Ghoreishi, R.; Roohi, A.H.; Ghadikolaei, A.D. Analysis of the influence of cutting parameters on surface roughness and cutting forces in high speed face milling of Al/SiC MMC. *Mater. Res. Express* **2018**, *5*, 086521. [CrossRef]

24. Fotovvati, B.; Asadi, E. Size Effects on Geometrical Accuracy for Additive Manufacturing of Ti-6Al-4V ELI Parts. *Int. J. Adv. Manuf. Technol.* **2019**. submitted for publication.

25. Mehrpouya, M.; Dehghanghadikolaei, A.; Fotovvati, B.; Vosooghnia, A.; Emamian, S.S.; Gisario, A. The potential of additive manufacturing in the smart factory industrial 4.0: A review. *J. Intell. Manuf.* **2019**, *31*. submitted for publication.

26. Dehghanghadikolaei, A.; Namdari, N.; Mohammadian, B.; Fotovvati, B. Additive Manufacturing Methods: A Brief Overview. *J. Sci. Eng. Res.* **2018**, *5*, 123–131.

27. Ataee, A.; Li, Y.; Fraser, D.; Song, G.; Wen, C. Anisotropic Ti-6Al-4V gyroid scaffolds manufactured by electron beam melting (EBM) for bone implant applications. *Mater. Des.* **2018**, *137*, 345–354. [CrossRef]

28. Binder, M.; Illgner, M.; Anstaett, C.; Kindermann, P.; Kirchbichler, L.; Seidel, C. Automated Manufacturing of Sensor-Monitored Parts: Enhancement of the laser beam melting process by a completely automated sensor integration. *Laser Tech. J.* **2018**, *15*, 36–39. [CrossRef]

29. Momeni, S.; Guschlbauer, R.; Osmanlic, F.; Körner, C. Selective electron beam melting of a copper-chrome powder mixture. *Mater. Lett.* **2018**, *223*, 250–252. [CrossRef]

30. Fotovvati, B.; Wayne, S.F.; Lewis, G.; Asadi, E. A review on melt-pool characteristics in laser welding of metals. *Adv. Mater. Sci. Eng.* **2018**, *2018*, 4920718. [CrossRef]

31. Namdari, N.; Mohammadian, B.; Dehghanghadikolaei, A.; Alidad, S.; Abbasi, M. A Numerical Study on Two-Dimensional Fins with Non-Constant Heat Flux. *Int. J. Sci. Eng. Sci.* **2018**, *2*, 12–16.

32. Namdari, N.; Dehghan, A. Natural frequencies and mode shapes for vibrations of rectangular and circular membranes: A numerical study. *Int. Res. J. Adv. Eng. Sci.* **2018**, *3*, 30–34.

33. Dehghanghadikolaei, A.; Namdari, N.; Mohammadian, B.; Ghoreishi, S.R. Deriving one dimensional shallow water equations from mass and momentum balance laws. *Int. Res. J. Eng. Technol.* **2018**, *5*, 408–419.

34. Namdari, N.; Abdi, M.; Chaghomi, H.; Rahmani, F. Numerical Solution for Transient Heat Transfer in Longitudinal Fins. *Int. Res. J. Adv. Eng. Sci.* **2018**, *3*, 131–136.

35. Dehghanghadikolaei, A.; Ansary, J.; Ghoreishi, R. Sol-gel process applications: A mini-review. *Proc. Nat. Res. Soc.* **2018**, *2*, 02008. [CrossRef]

36. Velten, D.; Biehl, V.; Aubertin, F.; Valeske, B.; Possart, W.; Breme, J. Preparation of TiO_2 layers on cp-Ti and Ti6Al4V by thermal and anodic oxidation and by sol-gel coating techniques and their characterization. *J. Biomed. Mater. Res.* **2002**, *59*, 18–28. [CrossRef] [PubMed]

37. Fotovvati, B.; Namdari, N.; Dehghanghadikolaei, A. On Coating Techniques for Surface Protection: A Review. *J. Manuf. Mater. Process.* **2019**, *3*, 28. [CrossRef]

38. Fotovvati, B.; Namdari, N.; Dehghanghadikolaei, A. Laser-Assisted Coating Techniques and Surface Modifications: A Short Review. *Part. Sci. Technol.* **2019**, *30*. submitted for publication.

39. Catauro, M.; Papale, F.; Sapio, L.; Naviglio, S. Biological influence of Ca/P ratio on calcium phosphate coatings by sol-gel processing. *Mater. Sci. Eng. C* **2016**, *65*, 188–193. [CrossRef]

40. Barrere, F.; Layrolle, P.; Van Blitterswijk, C.A.; De Groot, K. Biomimetic calcium phosphate coatings on Ti6Al4V: A crystal growth study of octacalcium phosphate and inhibition by Mg^{2+} and HCO_3^-. *Bone* **1999**, *25*, 107–111. [CrossRef]

41. Zhang, J.; Guan, R.; Zhang, X. Synthesis and characterization of sol–gel hydroxyapatite coatings deposited on porous NiTi alloys. *J. Alloy. Compd.* **2011**, *509*, 4643–4648. [CrossRef]

42. Costa, O.D.; Dixon, S.J.; Rizkalla, A.S. One-and three-dimensional growth of hydroxyapatite nanowires during sol–gel–hydrothermal synthesis. *ACS Appl. Mater. Interfaces* **2012**, *4*, 1490–1499. [CrossRef]

43. Cardoso, A.D.; Jansen, J.; Leeuwenburgh, S.G. Synthesis and application of nanostructured calcium phosphate ceramics for bone regeneration. *J. Biomed. Mater. Res. Part. B Appl. Biomater.* **2012**, *100*, 2316–2326. [CrossRef]

44. Kessler, V.G.; Spijksma, G.I.; Seisenbaeva, G.A.; Håkansson, S.; Blank, D.H.; Bouwmeester, H.J. New insight in the role of modifying ligands in the sol-gel processing of metal alkoxide precursors: A possibility to approach new classes of materials. *J. Sol. Gel Sci. Technol.* **2006**, *40*, 163–179. [CrossRef]

45. Agrawal, K.; Singh, G.; Prakash, S.; Puri, D. Synthesis of HA by various sol-gel techniques and their comparison: A review. In *National Conference on Advancements and Futuristic Trends in Mechanical and Materials Engineering*; Indian Institute of Technology Roorkee: Uttarakhand, India, 2011.

46. Dorozhkin, S.V. Calcium orthophosphate coatings on magnesium and its biodegradable alloys. *Acta Biomater.* **2014**, *10*, 2919–2934. [CrossRef]

47. Olding, T.; Sayer, M.; Barrow, D. Ceramic sol–gel composite coatings for electrical insulation. *Thin Solid Film.* **2001**, *398*, 581–586. [CrossRef]

48. Guo, L.; Li, H. Fabrication and characterization of thin nano-hydroxyapatite coatings on titanium. *Surf. Coat. Technol.* **2004**, *185*, 268–274. [CrossRef]

49. Motealleh, A.; Eqtesadi, S.; Perera, F.H.; Pajares, A.; Guiberteau, F.; Miranda, P. Understanding the role of dip-coating process parameters in the mechanical performance of polymer-coated bioglass robocast scaffolds. *J. Mech. Behav. Biomed. Mater.* **2016**, *64*, 253–261. [CrossRef] [PubMed]

50. Yuan, J.; Zhao, K.; Cai, T.; Gao, Z.; Yang, L.; He, D. One-step dip-coating of uniform γ-Al2O3 layers on cordierite honeycombs and its environmental applications. *Ceram. Int.* **2016**, *42*, 14384–14390. [CrossRef]

51. Mohseni, E.; Zalnezhad, E.; Bushroa, A.R. Comparative investigation on the adhesion of hydroxyapatite coating on Ti–6Al–4V implant: A review paper. *Int. J. Adhes. Adhes.* **2014**, *48*, 238–257. [CrossRef]

52. Ivanova, T.; Harizanova, A.; Koutzarova, T.; Vertruyen, B.; Stefanov, B. Structural and morphological characterization of sol-gel ZnO: Ga films: Effect of annealing temperatures. *Thin Solid Film.* **2018**, *646*, 132–142. [CrossRef]

53. Kuntin, D.; Gosling, N.; Wood, D.; Genever, P. Wnt signalling in mesenchymal stem cells is heightened in response to plasma sprayed hydroxyapatite coatings. *Osteoarthr. Cart.* **2018**, *26*, 146. [CrossRef]

54. Tang, H.; Tao, W.; Wang, C.; Yu, H. Fabrication of hydroxyapatite coatings on AZ31 Mg alloy by micro-arc oxidation coupled with sol–gel treatment. *RSC Adv.* **2018**, *8*, 12368–12375. [CrossRef]

55. Liu, D.-M.; Yang, Q.; Troczynski, T. Sol–gel hydroxyapatite coatings on stainless steel substrates. *Biomaterials* **2002**, *23*, 691–698. [CrossRef]

56. Besinis, A.; Hadi, S.D.; Le, H.R.; Tredwin, C.; Handy, R.D. Antibacterial activity and biofilm inhibition by surface modified titanium alloy medical implants following application of silver, titanium dioxide and hydroxyapatite nanocoatings. *Nanotoxicology* **2017**, *11*, 327–338. [CrossRef] [PubMed]

57. Dinda, G.; Shin, J.; Mazumder, J. Pulsed laser deposition of hydroxyapatite thin films on Ti–6Al–4V: Effect of heat treatment on structure and properties. *Acta Biomater.* **2009**, *5*, 1821–1830. [CrossRef]

58. Koenig, M.; Huang, S. Evaluation of crosslinked poly (caprolactone) as a biodegradable, hydrophobic coating. *Polym. Degrad. Stab.* **1994**, *45*, 139–144. [CrossRef]

59. Hanas, T.; Kumar, T.S.; Perumal, G.; Doble, M.; Ramakrishna, S. Electrospun PCL/HA coated friction stir processed AZ31/HA composites for degradable implant applications. *J. Mater. Process. Technol.* **2018**, *252*, 398–406.

60. Alemón, B.; Flores, M.; Ramírez, W.; Huegel, J.C.; Broitman, E. Tribocorrosion behavior and ions release of CoCrMo alloy coated with a TiAlVCN/CNx multilayer in simulated body fluid plus bovine serum albumin. *Tribol. Int.* **2015**, *81*, 159–168. [CrossRef]

61. Catauro, M.; Bollino, F.; Papale, F.; Giovanardi, R.; Veronesi, P. Corrosion behavior and mechanical properties of bioactive sol-gel coatings on titanium implants. *Mater. Sci. Eng. C* **2014**, *43*, 375–382. [CrossRef] [PubMed]

62. Bollino, F.; Armenia, E.; Tranquillo, E. Zirconia/hydroxyapatite composites synthesized via Sol-Gel: Influence of hydroxyapatite content and heating on their biological properties. *Materials* **2017**, *10*, 757. [CrossRef]

63. Li, H.; Khor, K.; Cheang, P. Titanium dioxide reinforced hydroxyapatite coatings deposited by high velocity oxy-fuel (HVOF) spray. *Biomaterials* **2002**, *23*, 85–91. [CrossRef]

64. Melero, H.C.; Sakai, R.T.; Vignatti, C.A.; Benedetti, A.V.; Fernández, J.; Guilemany, J.M.; Suegama, P.H. Corrosion Resistance Evaluation of HVOF Produced Hydroxyapatite and TiO_2-hydroxyapatite Coatings in Hanks' Solution. *Mater. Res.* **2018**, *21*. [CrossRef]

65. Bolelli, G.; Rauch, J.; Cannillo, V.; Killinger, A.; Lusvarghi, L.; Gadow, R. Microstructural and tribological investigation of high-velocity suspension flame sprayed (HVSFS) Al_2O_3 coatings. *J. Spray Technol.* **2009**, *18*, 35. [CrossRef]

66. Zhang, S.L.; Li, C.X.; Li, C.J.; Yang, G.J.; Liu, M. Application of high velocity oxygen fuel flame (HVOF) spraying to fabrication of $La_{0.8}Sr_{0.2}Ga_{0.8}Mg_{0.2}O_3$ electrolyte for solid oxide fuel cells. *J. Power Sources* **2016**, *301*, 62–71. [CrossRef]

67. Ghosh, G.; Sidpara, A.; Bandyopadhyay, P. High efficiency chemical assisted nanofinishing of HVOF sprayed WC-Co coating. *Surf. Coat. Technol.* **2018**, *334*, 204–214. [CrossRef]

68. Fantozzi, D.; Matikainen, V.; Uusitalo, M.; Koivuluoto, H.; Vuoristo, P. Effect of Carbide Dissolution on Chlorine Induced High Temperature Corrosion of HVOF and HVAF Sprayed Cr_3C_2-NiCrMoNb Coatings. *J. Therm. Spray Technol.* **2018**, *27*, 220–231. [CrossRef]

69. Förg, A.; Konrath, G.; Popa, S.; Kailer, A.; Killinger, A.; Gadow, R. Tribological properties of high velocity suspension flame sprayed (HVSFS) ceramic coatings. *Surf. Coat. Technol.* **2018**, *349*. [CrossRef]

70. Bernstein, A.; Suedkamp, N.; Mayr, H.O.; Gadow, R.; Burtscher, S.; Arhire, I.; Killinger, A.; Krieg, P. Thin Degradable Coatings for Optimization of Osseointegration Associated with Simultaneous Infection Prophylaxis. In *Nanostructures for Antimicrobial Therapy*; Elsevier: Amsterdam, The Netherlands, 2017; pp. 117–137.

71. Rocha, R.C.; Galdino, A.G.D.S.; Silva, S.N.D.; Machado, M.L.P. Surface, microstructural, and adhesion strength investigations of a bioactive hydroxyapatite-titanium oxide ceramic coating applied to Ti-6Al-4V alloys by plasma thermal spraying. *Mater. Res.* **2018**, *21*. [CrossRef]

72. Korostynska, O.; Gigilashvili, G.; Mason, A.; Tofail, S.A. Hydroxyapatite Thick Films as Pressure Sensors. In *Electrically Active Materials for Medical Devices*; World Scientific Publishing: Singapore, 2016; pp. 417–434.

73. Gadow, R.; Killinger, A.; Rauch, J. Introduction to high-velocity suspension flame spraying (HVSFS). *J. Spray Technol.* **2008**, *17*, 655–661. [CrossRef]

74. Nourouzi, S.; Azizpour, M.J.; Salimijazi, H. Parametric study of residual stresses in HVOF thermally sprayed WC–12Co coatings. *Mater. Manuf. Process.* **2014**, *29*, 1117–1125. [CrossRef]

75. Gadow, R.; Killinger, A.; Stiegler, N. Hydroxyapatite coatings for biomedical applications deposited by different thermal spray techniques. *Surf. Coat. Technol.* **2010**, *205*, 1157–1164. [CrossRef]

76. Chen, X.; Zhang, B.; Gong, Y.; Zhou, P.; Li, H. Mechanical properties of nanodiamond-reinforced hydroxyapatite composite coatings deposited by suspension plasma spraying. *Appl. Surf. Sci.* **2018**, *439*, 60–65. [CrossRef]

77. Killinger, A.; Kuhn, M.; Gadow, R. High-velocity suspension flame spraying (HVSFS), a new approach for spraying nanoparticles with hypersonic speed. *Surf. Coat. Technol.* **2006**, *201*, 1922–1929. [CrossRef]

78. Shahien, M.; Suzuki, M. Low power consumption suspension plasma spray system for ceramic coating deposition. *Surf. Coat. Technol.* **2017**, *318*, 11–17. [CrossRef]

79. Morks, M. Fabrication and characterization of plasma-sprayed HA/SiO_2 coatings for biomedical application. *J. Mech. Behav. Biomed. Mater.* **2008**, *1*, 105–111. [CrossRef] [PubMed]

80. Bolelli, G.; Cannillo, V.; Gadow, R.; Killinger, A.; Lusvarghi, L.; Rauch, J. Properties of high velocity suspension flame sprayed (HVSFS) TiO_2 coatings. *Surf. Coat. Technol.* **2009**, *203*, 1722–1732. [CrossRef]

81. Pillai, S.R.; Frasnelli, M.; Sglavo, V.M. HA/β-TCP plasma sprayed coatings on Ti substrate for biomedical applications. *Ceram. Int.* **2018**, *44*, 1328–1333. [CrossRef]

82. Jemat, A.; Ghazali, M.J.; Razali, M.; Otsuka, Y.; Rajabi, A. Effects of TiO_2 on microstructural, mechanical properties and in-vitro bioactivity of plasma sprayed yttria stabilised zirconia coatings for dental application. *Ceram. Int.* **2018**, *44*, 4271–4281. [CrossRef]

83. Heimann, R.B. *Plasma-Spray Coating: Principles and Applications*; John Wiley & Sons: Hoboken, NJ, USA, 2008.

84. Narayanan, R.; Seshadri, S.K.; Kwon, T.Y.; Kim, K.H. Calcium phosphate-based coatings on titanium and its alloys. *J. Biomed. Mater. Res. Part. B* **2008**, *85*, 279–299. [CrossRef]

85. Ahmed, K.; Shoeib, M. Development of Air Plasma Thermal Spray Coating for Thermal Barrier Coating and Oxidation Resistance Applications on Ni-Base Super Alloys. *Sch. J. Appl. Sci. Res.* **2018**, *1*, 20–35.

86. Oehr, C. Plasma surface modification of polymers for biomedical use. *Nucl. Instrum. Methods Phys. Res. Sect. B Beam Interact. Mater.* **2003**, *208*, 40–47. [CrossRef]

87. Fotovvati, B.; Namdari, N.; Dehghanghadikolaei, A. Fatigue performance of selective laser melted Ti6Al4V components: State of the art. *Mater. Res. Express* **2018**, *6*, 012002. [CrossRef]

88. Yoshinari, M.; Ozeki, K.; Sumii, T. Properties of hydroxyapatite-coated Ti-6Al-4V alloy produced by the ion-plating method. *Bull. Tokyo Dent. Coll.* **1991**, *32*, 147–156.

89. Levingstone, T.J.; Ardhaoui, M.; Benyounis, K.; Looney, L.; Stokes, J.T. Plasma sprayed hydroxyapatite coatings: Understanding process relationships using design of experiment analysis. *Surf. Coat. Technol.* **2015**, *283*, 29–36. [CrossRef]

90. Gross, A.K.; Saber-Samandari, S. Revealing mechanical properties of a suspension plasma sprayed coating with nanoindentation. *Surf. Coat. Technol.* **2009**, *203*, 2995–2999. [CrossRef]

91. Kozerski, S.; Pawlowski, L.; Jaworski, R.; Roudet, F.; Petit, F. Two zones microstructure of suspension plasma sprayed hydroxyapatite coatings. *Surf. Coat. Technol.* **2010**, *204*, 1380–1387. [CrossRef]

92. Ke, D.; Robertson, S.F.; Dernell, W.S.; Bandyopadhyay, A.; Bose, S. Effects of MgO and SiO_2 on plasma-sprayed hydroxyapatite coating: An in vivo study in rat distal femoral defects. *Acs Appl. Mater. Interfaces* **2017**, *9*, 25731–25737. [CrossRef]

93. Liu, X.; He, D.; Zhou, Z.; Wang, Z.; Wang, G. The Influence of Process Parameters on the Structure, Phase Composition, and Texture of Micro-Plasma Sprayed Hydroxyapatite Coatings. *Coatings* **2018**, *8*, 106. [CrossRef]

94. Reitman, R.; Buch, R.; Temple, T.; Eberle, R.W.; Kerzhner, E. Silver Coatings Protect against Infection in Salvage Total Joint Arthroplasty. In *Orthopaedic Proceedings*; The British Editorial Society of Bone & Joint Surgery: London, UK, 2017.

95. Lynn, A.; DuQuesnay, D. Hydroxyapatite-coated Ti–6Al–4V: Part 1: The effect of coating thickness on mechanical fatigue behaviour. *Biomaterials* **2002**, *23*, 1937–1946. [CrossRef]

96. Basu, B.; Ghosh, S. Case Study: Hydroxyapatite–Titanium Bulk Composites for Bone Tissue Engineering Applications. In *Biomaterials for Musculoskeletal Regeneration*; Springer: Berlin, Germany, 2017; pp. 15–44.

97. Zheng, X.; Huang, M.; Ding, C. Bond strength of plasma-sprayed hydroxyapatite/Ti composite coatings. *Biomaterials* **2000**, *21*, 841–849. [CrossRef]

98. Singh, G.; Singh, S.; Prakash, S. Surface characterization of plasma sprayed pure and reinforced hydroxyapatite coating on Ti6Al4V alloy. *Surf. Coat. Technol.* **2011**, *205*, 4814–4820. [CrossRef]

99. Singh, A.; Singh, G.; Chawla, V. Characterization and mechanical behaviour of reinforced hydroxyapatite coatings deposited by vacuum plasma spray on SS-316L alloy. *J. Mech. Behav. Biomed. Mater.* **2018**, *79*, 273–282. [CrossRef]

100. Shalom, H.; Feldman, Y.; Rosentsveig, R.; Pinkas, I.; Kaplan-Ashiri, I.; Moshkovich, A.; Perfilyev, V.; Rapoport, L.; Tenne, R. Electrophoretic Deposition of Hydroxyapatite Film Containing Re-Doped MoS_2 Nanoparticles. *Int. J. Mol. Sci.* **2018**, *19*, 657. [CrossRef]

101. He, D.H.; Wang, P.; Liu, P.; Liu, X.K.; Ma, F.C.; Zhao, J. HA coating fabricated by electrochemical deposition on modified Ti6Al4V alloy. *Surf. Coat. Technol.* **2016**, *301*, 6–12. [CrossRef]

102. Yan, L.; Xiang, Y.; Yu, J.; Wang, Y.; Cui, W. Fabrication of antibacterial and antiwear hydroxyapatite coatings via in situ chitosan-mediated pulse electrochemical deposition. *Acs Appl. Mater. Interfaces* **2017**, *9*, 5023–5030. [CrossRef] [PubMed]

103. Bakin, B.; Delice, T.K.; Tiric, U.; Birlik, I.; Azem, F.A. Bioactivity and corrosion properties of magnesium-substituted CaP coatings produced via electrochemical deposition. *Surf. Coat. Technol.* **2016**, *301*, 29–35. [CrossRef]

104. Chakraborty, R.; Seesala, V.S.; Sengupta, S.; Dhara, S.; Saha, P.; Das, K.; Das, S. Comparison of Osteoconduction, cytocompatibility and corrosion protection performance of hydroxyapatite-calcium hydrogen phosphate composite coating synthesized in-situ through pulsed electro-deposition with varying amount of phase and crystallinity. *Surf. Interfaces* **2018**, *10*, 1–10. [CrossRef]

105. Zhao, X.; Hu, T.; Li, H.; Chen, M.; Cao, S.; Zhang, L.; Hou, X. Electrochemically assisted co-deposition of calcium phosphate/collagen coatings on carbon/carbon composites. *Appl. Surf. Sci.* **2011**, *257*, 3612–3619. [CrossRef]

106. Rad, A.T.; Solati-Hashjin, M.; Osman, N.A.A.; Faghihi, S. Improved bio-physical performance of hydroxyapatite coatings obtained by electrophoretic deposition at dynamic voltage. *Ceram. Int.* **2014**, *40*, 12681–12691.

107. Sankar, M.; Suwas, S.; Balasubramanian, S.; Manivasagam, G. Comparison of electrochemical behavior of hydroxyapatite coated onto WE43 Mg alloy by electrophoretic and pulsed laser deposition. *Surf. Coat. Technol.* **2017**, *309*, 840–848. [CrossRef]

108. Sa'adati, H.; Raissi, B.; Riahifar, R.; Yaghmaee, M.S. How preparation of suspensions affects the electrophoretic deposition phenomenon. *J. Eur. Ceram. Soc.* **2016**, *36*, 299–305. [CrossRef]

109. Peng, P.; Kumar, S.; Voelcker, N.H.; Szili, E.; Smart, R.S.C.; Griesser, H.J. Thin calcium phosphate coatings on titanium by electrochemical deposition in modified simulated body fluid. *J. Biomed. Mater. Res. Part. A* **2006**, *76*, 347–355. [CrossRef]

110. Zhang, Y.Y.; Jie, T.A.O.; Pang, Y.C.; Wei, W.A.N.G.; Tao, W.A.N.G. Electrochemical deposition of hydroxyapatite coatings on titanium. *Trans. Nonferrous Met. Soc. China* **2006**, *16*, 633–637. [CrossRef]

111. Blanda, G.; Brucato, V.; Pavia, F.C.; Greco, S.; Piazza, S.; Sunseri, C.; Inguanta, R. Galvanic deposition and characterization of brushite/hydroxyapatite coatings on 316L stainless steel. *Mater. Sci. Eng. C* **2016**, *64*, 93–101. [CrossRef] [PubMed]

112. Wu, P.P.; Zhang, Z.Z.; Xu, F.J.; Deng, K.K.; Nie, K.B.; Gao, R. Effect of duty cycle on preparation and corrosion behavior of electrodeposited calcium phosphate coatings on AZ91. *Appl. Surf. Sci.* **2017**, *426*, 418–426. [CrossRef]

113. Wei, D.; Du, Q.; Guo, S.; Jia, D.; Wang, Y.; Li, B.; Zhou, Y. Structures, bonding strength and in vitro bioactivity and cytotoxicity of electrochemically deposited bioactive nano-brushite coating/TiO$_2$ nanotubes composited films on titanium. *Surf. Coat. Technol.* **2018**, *340*, 93–102. [CrossRef]

114. Blackwood, D.; Seah, K. Galvanostatic pulse deposition of hydroxyapatite for adhesion to titanium for biomedical purposes. *Mater. Sci. Eng. C* **2010**, *30*, 561–565. [CrossRef]

115. Furko, M.; May, Z.; Havasi, V.; Kónya, Z.; Grünewald, A.; Detsch, R.; Boccaccini, A.R.; Balázsi, C. Pulse electrodeposition and characterization of non-continuous, multi-element-doped hydroxyapatite bioceramic coatings. *J. Solid State Electrochem.* **2018**, *22*, 555–566. [CrossRef]

116. Xavier, A.S.S.; Vijayalakshmi, U. Electrochemically grown functionalized-Multi-walled carbon nanotubes/hydroxyapatite hybrids on surgical grade 316L SS with enhanced corrosion resistance and bioactivity. *Colloids Surf. B Biointerfaces* **2018**, *171*, 186–196. [CrossRef] [PubMed]

117. Nie, X.; Leyland, A.; Matthews, A. Deposition of layered bioceramic hydroxyapatite/TiO$_2$ coatings on titanium alloys using a hybrid technique of micro-arc oxidation and electrophoresis. *Surf. Coat. Technol.* **2000**, *125*, 407–414. [CrossRef]

118. Kar, A.; Raja, K.; Misra, M. Electrodeposition of hydroxyapatite onto nanotubular TiO$_2$ for implant applications. *Surf. Coat. Technol.* **2006**, *201*, 3723–3731. [CrossRef]

119. Dehghanghadikolaei, A. Enhance its Corrosion Behavior of Additively Manufactured NiTi by Micro-Arc Oxidation Coating. Ph.D. Thesis, University of Toledo, Toledo, OH, USA, 2018.

120. Dehghanghadikolaei, A.; Ibrahim, H.; Amerinatanzi, A.; Hashemi, M.; Moghaddam, N.S.; Elahinia, M. Improving corrosion resistance of additively manufactured nickel–titanium biomedical devices by micro-arc oxidation process. *J. Mater. Sci.* **2019**, *54*, 7333–7355. [CrossRef]

121. He, L.-P.; Wu, Z.J.; Chen, Z.Z. In-situ growth of nanometric network calcium phosphate/porous Al$_2$O$_3$ biocomposite coating on Al Ti substrate. *Chin. J. Nonferrous Met.* **2004**, *14*, 460–464.

122. Chakraborty, R.; Saha, P. A comparative study on surface morphology and electrochemical behaviour of hydroxyapatite-calcium hydrogen phosphate composite coating synthesized in-situ through electro chemical process under various deposition conditions. *Surf. Interfaces* **2018**, *12*, 160–167. [CrossRef]

123. Catauro, M.; Bollino, F.; Giovanardi, R.; Veronesi, P. Modification of Ti6Al4V implant surfaces by biocompatible TiO$_2$/PCL hybrid layers prepared via sol-gel dip coating: Structural characterization, mechanical and corrosion behavior. *Mater. Sci. Eng. C* **2017**, *74*, 501–507. [CrossRef] [PubMed]

124. Khoshsima, S.; Yilmaz, B.; Tezcaner, A.; Evis, Z. Structural, mechanical and biological properties of hydroxyapatite-zirconia-lanthanum oxide composites. *Ceram. Int.* **2016**, *42*, 15773–15779. [CrossRef]

125. Rojaee, R.; Fathi, M.; Raeissi, K.; Sharifnabi, A. Biodegradation assessment of nanostructured fluoridated hydroxyapatite coatings on biomedical grade magnesium alloy. *Ceram. Int.* **2014**, *40*, 15149–15158. [CrossRef]

126. Henriques, P.C.; Borges, I.; Pinto, A.M.; Magalhães, F.D.; Gonçalves, I.C. Fabrication and antimicrobial performance of surfaces integrating graphene-based materials. *Carbon* **2018**, *132*, 709–732. [CrossRef]

127. Park, J.E.; Jang, Y.S.; Park, I.S.; Jeon, J.G.; Bae, T.S.; Lee, M.H. The effect of multi-walled carbon nanotubes/hydroxyapatite nanocomposites on biocompatibility. *Adv. Compos. Mater.* **2018**, *27*, 53–65. [CrossRef]

128. Yuan, Q.D.; Golden, T. Electrochemical study of hydroxyapatite coatings on stainless steel substrates. *Thin Solid Films* **2009**, *518*, 55–60. [CrossRef]

129. Yang, B.; Uchida, M.; Kim, H.M.; Zhang, X.; Kokubo, T. Preparation of bioactive titanium metal via anodic oxidation treatment. *Biomaterials* **2004**, *25*, 1003–1010. [CrossRef]

130. Lee, K.; Ko, Y.M.; Choe, H.C. Electrochemical Deposition of Hydroxyapatite Substituted with Magnesium and Strontium on Ti–6Al–4V Alloy. *J. Nanosci. Nanotechnol.* **2018**, *18*, 1449–1452. [CrossRef]

131. Sidhu, T.; Agrawal, R.; Prakash, S. Hot corrosion of some superalloys and role of high-velocity oxy-fuel spray coatings—A review. *Surf. Coat. Technol.* **2005**, *198*, 441–446. [CrossRef]

132. Sun, L.; Berndt, C.C.; Gross, K.A.; Kucuk, A. Material fundamentals and clinical performance of plasma-sprayed hydroxyapatite coatings: A review. *J. Biomed. Mater. Res.* **2001**, *58*, 570–592. [CrossRef] [PubMed]

133. Asri, R.I.M.; Harun, W.S.W.; Hassan, M.A.; Ghani, S.A.C.; Buyong, Z. A review of hydroxyapatite-based coating techniques: Sol–gel and electrochemical depositions on biocompatible metals. *J. Mech. Behav. Biomed. Mater.* **2016**, *57*, 95–108. [CrossRef] [PubMed]

134. Lee, I.S.; Whang, C.N.; Kim, H.E.; Park, J.C.; Song, J.H.; Kim, S.R. Various Ca/P ratios of thin calcium phosphate films. *Mater. Sci. Eng. C* **2002**, *22*, 15–20. [CrossRef]

135. Zhang, S.; Xianting, Z.; Yongsheng, W.; Kui, C.; Wenjian, W. Adhesion strength of sol–gel derived fluoridated hydroxyapatite coatings. *Surf. Coat. Technol.* **2006**, *200*, 6350–6354. [CrossRef]

136. Sopcak, T.; Medvecky, L.; Zagyva, T.; Dzupon, M.; Balko, J.; Balázsi, K.; Balázsi, C. Characterization and adhesion strength of porous electrosprayed polymer–hydroxyapatite composite coatings. *Resolut. Discov.* **2018**, 1–7. [CrossRef]

137. Fujihara, T.; Tsukamoto, M.; Abe, N.; Miyake, S.; Ohji, T.; Akedo, J. Hydroxyapatite film formed by particle beam irradiation. *Vacuum* **2004**, *73*, 629–633. [CrossRef]

138. Zhu, X.; Son, D.W.; Ong, J.L.; Kim, K. Characterization of hydrothermally treated anodic oxides containing Ca and P on titanium. *J. Mater. Sci. Mater. Med.* **2003**, *14*, 629–634. [CrossRef]

139. Das, A.; Chikkala, A.K.; Bharti, G.P.; Behera, R.R.; Mamilla, R.S.; Khare, A.; Dobbidi, P. Effect of thickness on optical and microwave dielectric properties of Hydroxyapatite films deposited by RF magnetron sputtering. *J. Alloy. Compd.* **2018**, *739*, 729–736. [CrossRef]

140. Lin, J.; Tian, Q.; Aslani, A.; Liu, H. Characterization of Hydroxyapatite Coated Mg for Biomedical Applications. *MRS Adv.* **2018**, 1–5. [CrossRef]

141. Levingstone, T.J.; Barron, N.; Ardhaoui, M.; Benyounis, K.; Looney, L.; Stokes, J. Application of response surface methodology in the design of functionally graded plasma sprayed hydroxyapatite coatings. *Surf. Coat. Technol.* **2017**, *313*, 307–318. [CrossRef]

142. Amerinatanzi, A.; Mehrabi, R.; Ibrahim, H.; Dehghan, A.; Shayesteh Moghaddam, N.; Elahinia, M. Predicting the biodegradation of magnesium alloy implants: Modeling, parameter identification, and validation. *Bioengineering* **2018**, *5*, 105. [CrossRef] [PubMed]

143. Ghoreishi, R.; Roohi, A.H.; Ghadikolaei, A.D. Evaluation of tool wear in high-speed face milling of Al/SiC metal matrix composites. *J. Braz. Soc. Mech. Sci. Eng.* **2019**, *41*, 146. [CrossRef]

144. Hornberger, H.; Virtanen, S.; Boccaccini, A. Biomedical coatings on magnesium alloys—A review. *Acta Biomater.* **2012**, *8*, 2442–2455. [CrossRef] [PubMed]

Materials **2019**, *12*, 1795

145. Massaro, C.; Baker, M.A.; Cosentino, F.; Ramires, P.A.; Klose, S.; Milella, E. Surface and biological evaluation of hydroxyapatite-based coatings on titanium deposited by different techniques. *J. Biomed. Mater. Res.* **2001**, *58*, 651–657. [CrossRef]
146. Ibrahim, H.; Esfahani, S.N.; Poorganji, B.; Dean, D.; Elahinia, M. Resorbable bone fixation alloys, forming, and post-fabrication treatments. *Mater. Sci. Eng. C* **2017**, *70*, 870–888. [CrossRef] [PubMed]
147. Ibrahim, H.; Klarner, A.D.; Poorganji, B.; Dean, D.; Luo, A.A.; Elahinia, M. Microstructural, mechanical and corrosion characteristics of heat-treated Mg-1.2Zn-0.5Ca (wt %) alloy for use as resorbable bone fixation material. *J. Mech. Behav. Biomed. Mater.* **2017**, *69*, 203–212. [CrossRef] [PubMed]
148. Ibrahim, H.; Moghaddam, N.; Elahinia, M. Me-chanical and In Vitro Corrosion Properties of a Heat-Treated Mg-Zn-Ca-Mn Alloy as a Potential Bioresorbable Material. *Sci. Pages Met. Mater. Eng.* **2017**, *1*, 1–7.
149. Tian, P.; Xu, D.; Liu, X. Mussel-inspired functionalization of PEO/PCL composite coating on a biodegradable AZ31 magnesium alloy. *Colloids Surf. B Biointerfaces* **2016**, *141*, 327–337. [CrossRef]
150. Weber, J.; Atanasoska, L.; Eidenschink, T. Corrosion Resistant Coatings for Biodegradable Metallic Implants. US. Patent Application No 11/387,032, 27 September 2007.
151. Cabrini, M.; Cigada, A.; Rondell, G.; Vicentini, B. Effect of different surface finishing and of hydroxyapatite coatings on passive and corrosion current of Ti6Al4V alloy in simulated physiological solution. *Biomaterials* **1997**, *18*, 783–787. [CrossRef]
152. Rivero, D.P.; Fox, J.; Skipor, A.K.; Urban, R.M.; Galante, J.O. Calcium phosphate-coated porous titanium implants for enhanced skeletal fixation. *J. Biomed. Mater. Res.* **1988**, *22*, 191–201. [CrossRef]
153. Ghadikolaei, A.D.; Vahdati, M. Experimental study on the effect of finishing parameters on surface roughness in magneto-rheological abrasive flow finishing process. *Proc. Inst. Mech. Eng. Part B J. Eng. Manuf.* **2015**, *229*, 1517–1524. [CrossRef]
154. Zhao, J.; Guo, L.Y.; Yang, X.B.; Weng, J. Preparation of bioactive porous HA/PCL composite scaffolds. *Appl. Surf. Sci.* **2008**, *255*, 2942–2946. [CrossRef]
155. Xia, L.; Xie, Y.; Fang, B.; Wang, X.; Lin, K. In situ modulation of crystallinity and nano-structures to enhance the stability and osseointegration of hydroxyapatite coatings on Ti-6Al-4V implants. *Chem. Eng. J.* **2018**, *347*, 711–720. [CrossRef]
156. Moskalewicz, T.; Łukaszczyk, A.; Kruk, A.; Kot, M.; Jugowiec, D.; Dubiel, B.; Radziszewska, A. Porous HA and nanocomposite nc-TiO$_2$/HA coatings to improve the electrochemical corrosion resistance of the Co-28Cr-5Mo alloy. *Mater. Chem. Phys.* **2017**, *199*, 144–158. [CrossRef]
157. Ionita, D.; Necula, L.; Prodana, M.; Totea, G.; Demetrescu, I. Corrosion of an Active Antibacterial Nanostructured Coating on Titanium. *Rev. De Chim.* **2018**, *69*, 1115–1121.
158. Santos-Coquillat, A.; Martínez-Campos, E.; Mohedano, M.; Martínez-Corriá, R.; Ramos, V.; Arrabal, R.; Matykina, E. In vitro and in vivo evaluation of PEO-modified titanium for bone implant applications. *Surf. Coat. Technol.* **2018**, *347*, 358–368. [CrossRef]

Review

Contact Lens Materials: A Materials Science Perspective

Christopher Stephen Andrew Musgrave [1] and Fengzhou Fang [1,2,*]

1   Centre of MicroNano Manufacturing Technology (MNMT-Dublin), University College Dublin, D14 YH57 Dublin, Ireland; christopher.musgrave@ucd.ie
2   State Key Laboratory of Precision Measuring Technology and Instruments, Centre of MicroNano Manufacturing Technology (MNMT), Tianjin University, Tianjin 300072, China
*   Correspondence: fengzhou.fang@ucd.ie; Tel.: +353-1-716-1810

Received: 20 December 2018; Accepted: 7 January 2019; Published: 14 January 2019

Abstract: More is demanded from ophthalmic treatments using contact lenses, which are currently used by over 125 million people around the world. Improving the material of contact lenses (CLs) is a now rapidly evolving discipline. These materials are developing alongside the advances made in related biomaterials for applications such as drug delivery. Contact lens materials are typically based on polymer- or silicone-hydrogel, with additional manufacturing technologies employed to produce the final lens. These processes are simply not enough to meet the increasing demands from CLs and the ever-increasing number of contact lens (CL) users. This review provides an advanced perspective on contact lens materials, with an emphasis on materials science employed in developing new CLs. The future trends for CL materials are to graft, incapsulate, or modify the classic CL material structure to provide new or improved functionality. In this paper, we discuss some of the fundamental material properties, present an outlook from related emerging biomaterials, and provide viewpoints of precision manufacturing in CL development.

Keywords: contact lens; materials; biomedical implant

---

1. Introduction

The market for contact lenses (CLs) is ever-growing, with over 125 million consumers as of 2004 [1], and an estimated global market size worth $7.1 billion in 2015 [2]. A quick search of patent literature shows that over 100 patents have been filed since 2000, representing the growing popularity of CLs. The applications of CLs range from corrective vision and therapeutics to cosmetic appearance [3,4]. Within these applications comes the demands from the end user of the lenses, including length of wear, comfort, durability, practically of handling, stability of vision, etc. This also means that within the applications of contact lenses comes the demands from manufacturers, such as material costs, ease of production, and reliability of the CLs, etc. Finally, the demands from manufacturers determine the parameters of the material which scientists must focus their research on for developing CL materials. This premise has guided materials scientists, from the creation of glass scleral lenses in the 1930s to rigid, non-gas-permeable polymethyl methacrylate (PMMA) in the 1940s. The 1960s and 70s ushered in hydrogel (polymer and silicone) lenses, with silicone hydrogel proving to be the most dominant kind of CL material today.

Commonly, the labels of "hard" or "soft" are used as blanket definitions of CLs [3,4]. Hard CLs are rigid (durable), gas-permeable lenses, whereas soft contact lenses are made of flexible, high-water-content material. Hard lenses are often interchangeably referred to as rigid gas-permeable lenses (RGPs); however, this is not strictly true. The first PMMA lenses could be classified as hard, whereas modern RGP lenses are, in fact, more flexible, due to the incorporation of low-modulus components—hence, they are more rigid, rather than hard. Another defining characteristic is that

PMMA hard lenses have no oxygen permeability, whereas RGP lenses are permeable. On the other hand, a soft contact lens (SCL) is a highly flexible, oxygen-permeable material with often high water-content. This flexibility means that SCLs fit the shape of a user's eye much faster than a rigid lens. SCLs can be disposed daily, weekly, or monthly. These blanket definitions of CLs can hint at its material properties, though not for certain. There is often an overlap in materials used between hard and soft lenses, such as silicone hydrogels and RGPs. Although both use silicone materials, the difference lies in factors such as the gel network and water content. Derivatives therein can further diversify the range of possible CLs and their properties. The requirements for CLs are quite extensive, and there are a huge number of existing CLs on the market to reflect this; daily-disposable lenses, weekly/monthly lenses, special fitting lenses, and even lenses that can be worn overnight. Users' demands can be described by several general parameters, such as comfort, wear time, handling (cleaning, ease of use), cost, and vision specifications (Figure 1) [3,4].

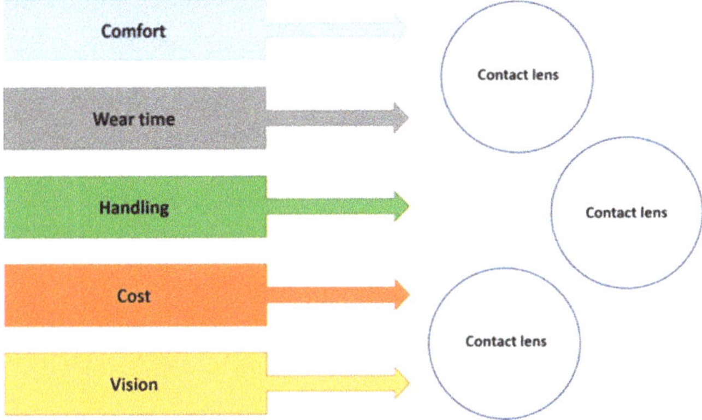

Figure 1. General user demands from CLs. Each parameter can be sub-divided into many categories, which is why the CL market is so vastly populated.

PMMA is a durable, optically transparent polymer with limited hydrophilic character. However, PMMA has negligible oxygen permeability, which may lead to several eye health issues, such as hypoxia [5,6]. Researchers quickly discovered polymer hydrogels, which were typically based on hydroxy ethyl methacrylate (HEMA) [7,8]. These polymer hydrogels were composed of hydrophilic monomers, meaning they contained electrochemical polarity allowing interaction with water. This provided much greater biocompatibility than PMMA. These hydrogels were also an oxygen-permeable and flexible class of material that could hold a large percentage of water within the polymer network. These factors improved the comfort, oxygen permeability, and wear time of CLs, which resulted in many CL derivatives based on HEMA [9,10]. Progress did not stop, as the oxygen permeability of HEMA hydrogels was not sufficient for extended CL wear (>24 h) [3,4]. The next evolution of CLs was silicone-based rigid lenses and hydrogels. These materials have very high gas-permeability, meaning extended-wear lenses could be fabricated. However, silicone materials are inherently hydrophobic, meaning they are not comfortable due to factors such as poor wetting and abrasiveness. Polymer scientists overcame this by using copolymerization of a silicone monomer (e.g., siloxymethacrylates and fluoromethacrylates) with hydrophilic comonomers to add the desired hydrophilic character [11–14]. This was particularly successful for silicone hydrogel lenses, which is the most dominant CL material on the market today (64% in the US) [15]. Newer CL materials appeared in the 1990s, such as polyvinyl alcohol (PVA), which is a low-cost and very hydrophilic polymer [16,17]. In fact, the PVA hydrogel was the subject of investigation in the early 1990s as a potential CL material [18]. In addition, new surface coatings, such as polyethylene glycol (PEG) have appeared in recent years to improve the hydrophilicity of silicone-based CLs [19]. These are just two examples amongst a vast pool of relevant literature.

Despite this significant progress, more is demanded from CLs today. CLs provide a route for improving the quality of life. This could simply be for cosmetic reasons or for corrective vision in place of traditional spectacles, which remain the two most common uses of CLs today [20]. Cosmetic CLs can include anything from pigmented to prosthetic CLs. However, CLs are increasingly considered a platform for more proactive ophthalmic treatments. As the number of people developing myopia, glaucoma, and other eye conditions is increasing globally, more effective treatments are required [21]. One review paper estimated that over 1400 million people worldwide are currently suffering from myopia [22]. Myopia control using CLs has been a subject of debate [23]; however, there is mounting research that specialty lenses can control the progression of myopia [24]. CLs can offer one such treatment route for glaucoma; the drug-loaded lens can simply be placed onto the eye, and the drug is released onto the eye [25–27]. This is currently a hot topic, and has been the subject of several excellent review papers [27–29]. Therefore, it is clear that new and improved CL materials are required to deliver more effective treatments for these growing issues.

Many sources detail the general synthesis of hydrogels or discuss their end application, such as bioavailability, eye-fitting, CLs in practice, manufacturing, and much more [3,4,30,31]. There have been some articles that have talked specifically about materials for contact lenses, which provided important insights at the time [32,33]. Other sources have discussed general bioavailable materials [34,35] and the properties of CL brands [33], with a lot of modern research being concerned about drug delivery using CLs [27,36–38]. This review brings together research specifically about the latest development of CL materials, and touches on details from a materials science perspective. We aim to bring together the innovations in CL materials and explain the impact they have. We discuss general polymerization mechanisms and monomers used to produce CL materials, which are all considerations for designing new CLs. We then discuss new evolutions to the main classes of CL materials, such as RGP lenses, HEMA- and silicone-hydrogels, and their future perspectives. Finally, we highlight some particularly impactful materials, emerging materials' technologies, and future manufacturing viewpoints.

2. Contact Lens Materials

2.1. Overview

To manufacture CLs, there must be a suitable polymeric material. This opens an incredible number of possibilities, not only from the range of polymers but to the formula of components within a given recipe. In addition, there can be considerations for the different types of polymerization mechanisms to form the same polymer, such as radical vs. catalytic polymerizations and derivatives. Within this, the polymerization conditions (temperature, initiator type, vessel used, etc.) can be altered to produce the same polymer but with different properties. Finally, the material must be suitable for the manufacturing stages, which include the synthesis, inspection, and packaging processes. The manufacturing stages have their own intricacies, which will be discussed later; thus, it is easy to quickly get lost in the search for the most suitable CL material. The extent of variation in materials is why there is such a wide range of CLs available today, and this has been extensively researched. Figure 2 contains some of the most important factors from a materials science perspective when designing CLs. From this, an assessment of current CL materials' pros and cons is given in Table 1.

Table 1. General pros and cons of current CL material classes.

CL Material	Pros	Cons
PMMA	Inexpensive, well-understood polymer	No oxygen permeability, inflexible on the eye
RGP	High oxygen permeability, durable	Expensive regents, requires hydrophilic comonomer, can be abrasive
HEMA hydrogel	Inexpensive, biocompatible, abundant copolymer possibilities	Low oxygen permeability, protein deposition issues
Silicone hydrogel	High oxygen permeability, durable, comfortable	Expensive regents, requires hydrophilic comonomer, can be abrasive
PVA	Inexpensive, straight-forward manufacturing, biocompatible	Low oxygen permeability, fixed water content

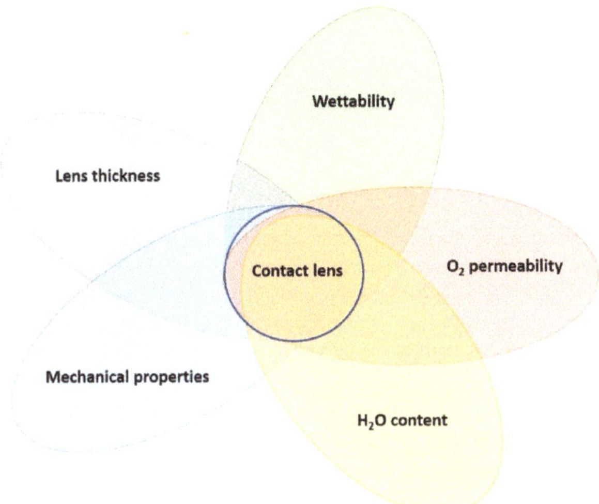

Figure 2. A CL lens is dependent on many parameters from a material science perspective. Stronger emphasis on specific characteristics are required, depending on the specific demand placed on the CL. The final CL material accounts for wear time and comfort. These characteristics are often dependent on the materials, but also includes manufacturing processes, such as plasma treatment.

For clarity, a polymer is a large macromolecule composed of hundreds or thousands of a repeating molecule, called a monomer. Each monomer has a covalent bond between each connecting unit. Some common monomers and polymers used in the production of CLs can be seen in Figure 3. The process by which the polymer is formed is called polymerization. Another reactive molecule, the initiator, is used to begin the polymerization. Polymerization initiators are typically selected based on the interaction with the reactive functional groups within the monomers. Radicals are formed on the breakdown of the initiator, which then induce the polymerization by stripping a radical from the monomer functional group. Now the monomer has a free radical within itself, which reacts with a neighboring monomer functional group by stripping a radical from this group. The original monomer would have then formed a new bond with the new monomer unit, which now has a new radical to continue polymerizing. This propagation step is continuously repeated, thus forming the polymer over time, which extinguishes when there are no more monomers to consume. The polymerization terminates when the radical is quenched in another manner, such as by a radical scavenger. The initiator can be chosen for their practicality, such as ultra-violet (UV) or thermal initiators, which is a huge consideration for the manufacturing method in producing CLs.

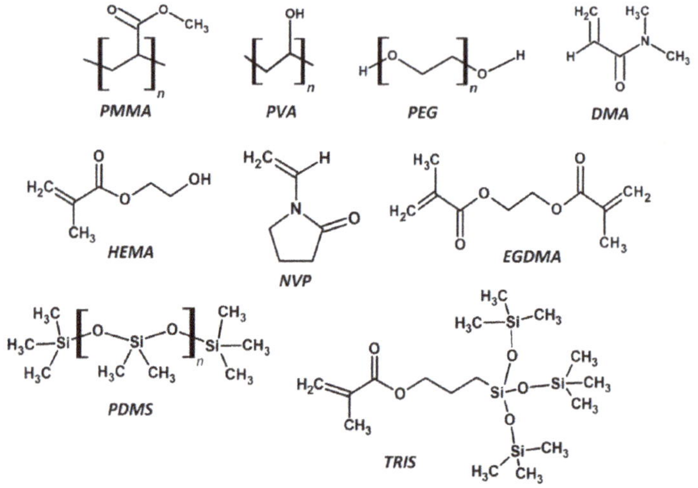

Figure 3. The chemical structures of common monomers and polymers used to produce CLs. This includes some macromonomers and cross-linking agents. PMMA—poly methyl methacrylate, PVA—poly vinyl alcohol, PEG—poly ethylene glycol, DMA—dimethyl methacrylate, HEMA—hydroxy ethyl methacrylate, NVP—N-vinyl pyrrolidone, EGDMA—ethylene glycol dimethacrylate, PDMS—poly dimethyl siloxane, TRIS—3-[tris(trimethylsiloxy)silyl]propyl methacrylate.

2.2. Polymerization Mechanisms

Free-radical polymerization (FRP) is a type of chain-growth polymerization mechanism. Typically, free-radical polymerizations are often not easily controlled, producing high dispersity (large variations in molecular weights) of the resulting polymers. This can lead to a distribution of the polymer properties. High strain and tensile modulus polymers are often determined by the length of the polymer chains [39]. One advantage of FRP is that it is easy to form gelling networks with, as the polymerization has many initiation sites within the vessel. This allows for the simultaneous growth of many chains which can physically entangle or cross-link to form the gel network. Cross-linking is dependent on one of the monomers containing two functional groups, which enables chemical bonding to two different polymer chains. Cross-linking polymer chains often improves gelling and increases the modulus of the material. To date, most CLs are produced using free-radical polymerization [11,13,40–44]. It is facile and doesn't require expensive reagents, such as catalysts. Furthermore, unwanted/unreacted chemicals can be removed from the material on post-fabrication cleaning processes. Catalysts are often composed of heavy metals, which is something to avoid for human health.

Today, most CLs are produced from a polymerization of two or more monomers [42,45]. Copolymers incorporate the properties of the individual polymers; consequently, copolymerization is often the first method used in overcoming issues with a single polymer. This principle has governed the development of CLs for many years. For example, silicone polymers are very hydrophobic, despite their high oxygen permeability. Therefore, they are not ideal as a homopolymer CL material. However, copolymers of silicones with a hydrophilic (highly polar) monomer can solve this problem. Copolymerization may also be used to enhance physical properties through the cross-linking of polymer chains by adding molecular weight to the chain. In some cases, adding a soft polymer can reduce the modulus of particularly tough materials (often silicone-based materials). Hydrophobic comonomers add oxygen permeability to materials that require improvement to this property, and silicone-based monomers are often used for this purpose. Table 2 contains the properties of common CL materials and related copolymers, including oxygen permeability, water content, moduli, and wear time. Wear time is from the perspective of the maximum wear time of a CL before eye health issues arise. More exhaustive resources comparing these properties are given elsewhere [3,4]. The effect of copolymerization is particularly noticeable for rigid materials derived from PMMA with a large reduction in the modulus, whereas the hydrogel classes of lenses have a much wider stable range of moduli, but large variations in water content and oxygen permeability.

Table 2. Generalized properties of some common CL materials.

Material	Oxygen Permeability (Dk/t)	Water Content (wt %)	Modulus (MPa)	Wear Time (days) *
PMMA	0	0	1000	<1
PMMA-silicone	15	0		
Silicone-HEMA (rigid)	10–100	0	10	
HEMA hydrogel	10–50	30–80	0.2–2	1–7
HEMA-NVP				
HEMA-MMA				
Silicone (PDMS) hydrogel	60–200	20–55	0.2–2	~7–28
TRIS-DMA				
PDMS-HEMA				
PVA	10–30	60–70		<1

* Maximum wear time without extensive complications to the eye before lens disposal.

Full-density polymers utilize molecular weight and intermolecular forces to provide the strength of the material; this is partly why PMMA has a high modulus. Other factors, such as the lens thickness of the final lens, is important too, whereas the properties of very low-density polymers (10% of full density, 90% air), such as porous materials are sensitive to small changes in cross-linking [46,47]. Another study demonstrated a reduction the Youngs modulus of up to 40% in a 90% porous material due to inefficient cross-linking [48]. This could be relevant to very high-water-content hydrogels on hydration from the final material properties and manufacturing considerations [49]. In fact, Maldonado-Codina and Efron highlighted the need for improving processes between polymerization batches and manufacturing methods. Other factors that are important include homogeneity of the wall vertices, uniform pore size, etc.; all of which are not guaranteed to be consistent with FRP. Moreover, if we consider SCLs to be porous cellular solids, these factors must be considered in future designs [50]. Although FRP has been the workhorse for fabricating excellent CL materials, they may not be producing entirely efficient networks for functional materials [51]. This could be another reason why there are no drug-delivery CLs on the market today, in addition to the many reasons stated by Dixon et al. [52]. Alternative polymerization mechanisms could be of interest to improve the physical properties of CL materials. Other kinds of methods include catalysts or controlled radical polymerization. One controlled method for radical polymerization is chain-transfer polymerization. This has been a popular area of growth in polymer science, including examining the potential in other bio-applications [53–55]. The chain-transfer agent accurately mediates the growth of the polymer chain, so that the molecular weight can be pre-designed [39,56,57]. This results in a low-dispersity polymer, meaning the resulting properties and structure are more reliable than FRP (Figure 4). There are many examples of hydrogels produced using chain-transfer agents [58–60]. One specific example of reverse addition fragmentation chain transfer (RAFT) polymerization encompasses silicone-based polymers [61]. Other researchers also used RAFT to modify polyacrylic acid pH-responsive hydrogels for drug delivery [60]. This opens the potential of RAFT-synthesized silicone- and conventional hydrogels as CLs. Recently, Zhang et al. synthesized a promising RAFT-polymerized soft contact lens based on polyallyl methacrylate and PEG components [62]. The lenses had low contact angles (<80°), high Dk values (>100 barrers), and elastic moduli ranging from 0.5–1.5 MPa, which are in the range of the CL parameters given in Table 1.

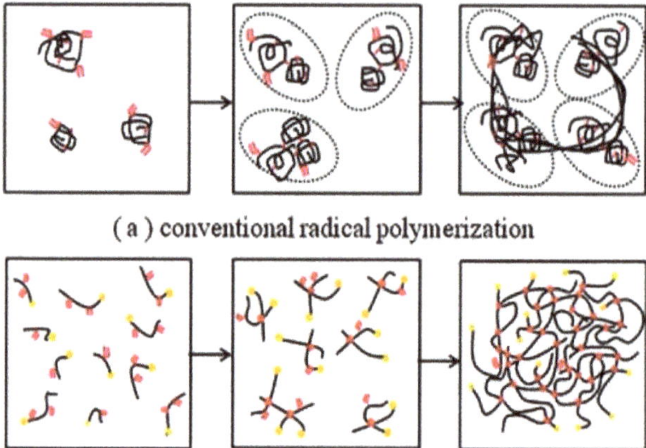

Figure 4. Schematic for the macromolecular structure differences between FRP and RAFT-polymerized materials. The RAFT polymer can be used to create a more ordered structure compared to FRP, which could be important to the macromolecular structure and final application of the material. Republished with the permission of The Royal Society of Chemistry, from Pushing the mechanical strength of PolyHIPEs up to the theoretical limit through living radical polymerization, Y. Luo, A-N. Wang, X. Gao, 8, 2012; permission conveyed through Copyright Clearance Center, Inc. [51].

2.3. PMMA and Rigid Contact Lenses

2.3.1. PMMA

Today, PMMA CLs occupy a market share of about 1% [15]; however, they are a useful place to begin to appreciate CLs materials—the properties of polymers that are suitable for ocular wear. One of the biggest issues with PMMA is that it has little to no oxygen permeability. This is due to the lack of mobility of polymer chains preventing the flow of oxygen or internal water to mediate the flow of O_2 (Figure 5). This occurs in PMMA due to intermolecular forces, such as dipole–dipole bonding and physical entanglement that is prevalent between polymer chains. The dipoles are created by the negatively charged (electrochemical negative) oxygen compared with the adjacent positively charged (electrochemical positive) carbon and hydrogen atoms. Therefore, neighboring polymer chains can attract each other to provide thermodynamic stability to the polymer. These intermolecular forces also mean that PMMA has low free volume (space between polymer chains), meaning the chains do not rotate or move easily. In addition, PMMA does not contain large pendant chains that prevent the interaction of neighboring chains. All these factors together prevent the flow of oxygen through the polymer. However, functionalization of the PMMA surface can improve the hydrophilicity [63], which would be useful to these CLs.

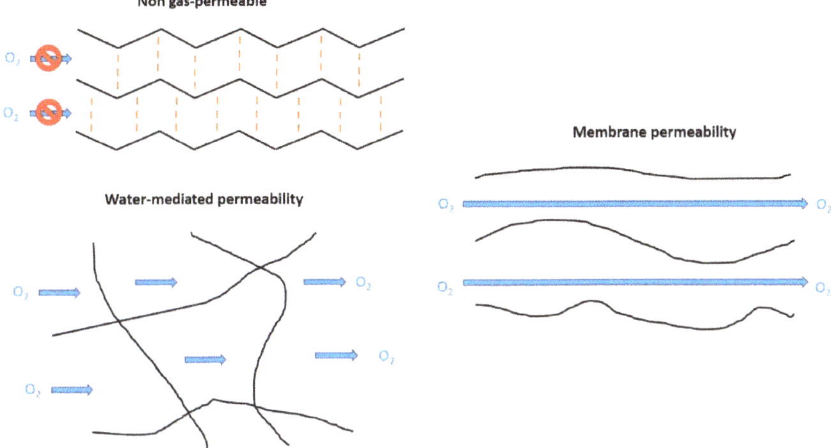

Figure 5. Schematic for gas-permeability mechanisms of CL materials from the perspective of polymer chains. These 2D models represent how oxygen passes (or not) through the molecular structure of the CL material. These schematics do not show factors such as extensive cross-linking or the macromolecular structure that would be present in a 3D structure.

In the recent literature, PMMA has typically been utilized as a reference material for investigating various effects of CLs on the eye [64–67]. These works are some of many with a particular focus on the effect of the PMMA lens on eye conditions, such as astigmatism [64], strabismus [68], and blepharoptosis [65]. Alió et al. fitted patients who had received post-corneal refractive surgery with a variety of lens types, with RGP lenses showing the best results [64]. An interesting work by Li et al. showed fabrication of a possible new PMMA hybrid lens with zinc oxide quantum dots to reduce UV exposure [69]. Very low loadings of ZnO quantum dots (0.017 wt %) reduced the transmittance of UV light by 50%, yet retained suitable optical transparency expected of a contact lens (Figure 6). Perhaps the most novel advancement for PMMA lenses was the possibility of developing nanophotonic lenses [70,71]. Acid- and hydroxyl-functionalized fullerenes (C_{60}) were attached to PMMA to try to harness the photo- and electro-activity of the fullerene. However, these nanophotonic lenses have now been a focus of SCL materials instead [72].

However, all may not be bad for PMMA, as van der Worp discussed the relevance of PMMA as scleral lenses [73]. In addition, PEG grafted onto PMMA is seen as a prosthetic eye replacement [74]. Overall, these are only a few works amongst an enormous amount of literature on contact lenses. The fact remains that PMMA is a very well-studied polymer; as such, the interest in developing new materials derived from PMMA is not exciting at present. Simply, the lack of many publications involving PMMA in the modern literature is strong evidence that PMMA is not relevant in today's climate for CLs. The only hope is that PMMA can continue to be used for fundamental understanding, along with the factors that affect the eye gained when using PMMA as a reference CL material. For this reason, PMMA will remain a marginalized contact lens material for both research and commercial purposes.

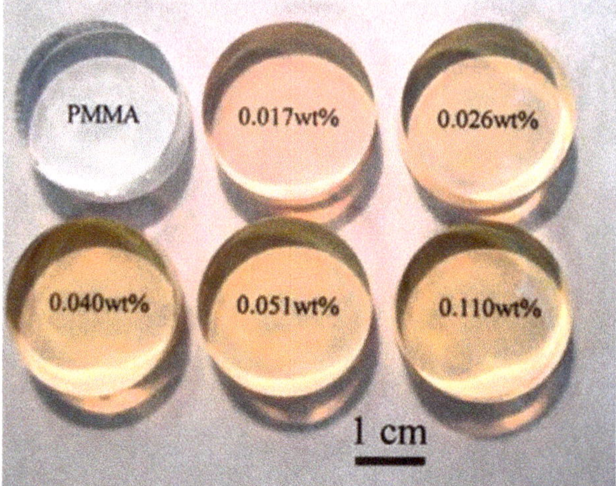

Figure 6. PMMA buttons loaded with ZnO quantum dots. The fabricated buttons would be suitable for lathe-cutting manufacturing to form a final CL. Reprinted (and adapted) with permission from John-Wiley and Sons [69].

2.3.2. Other Rigid CLs

Modern rigid gas-permeable (RGP) lenses have moved away from PMMA, or significantly reduced the molar fraction of PMMA in many copolymers. Efron wrote an obituary for RGP lenses stating the reasons why these lenses are obsolete [75]. Nevertheless, RGP lenses (including PMMA) still account for about 14% of all lenses fit in the US as of 2017 [15]. Modern research concerning RGP lenses is in regard to their application, treatment, and effect on eye conditions, rather than innovations on the material [76–78]. Eggnik et al. reported that a large-diameter RGP lens of unknown composition was used to improve post-surgery treatment for laser in situ keratomileusis (LASIK) patients [79]. The success of RGP lenses lies in their rigidity, causing a reshaping of the cornea, which was useful for some post-surgery treatment [80]. Soft lenses are sometimes unsuitable for treatment as they are very malleable, and therefore will shape to the user's eye. These other works often use commercially manufactured lenses to carry out their research; therefore, it is more difficult to know the exact composition and methods by which these materials were synthesized. However, it is known that the typical composition of these lenses are based on fluoro silicone or siloxane (such as PDMS and TRIS) acrylate moieties, alongside hydrophilic monomers such as HEMA, NVP, and methacrylic acid (MAA) [4,76]. Baucsh & Lomb were assigned a patent for polysiloxane, copolymerized with urea moieties prepolymers to from RGP lenses [81]. The urea (or carbamides ($CO(NH_2)_2$) moieties added hydrophilic properties. These vinyl end-capped prepolymers were then copolymerized with well-known monomers such as NVP, MAA, MMA, and TRIS to form the RGP lens.

Although Efron's prediction may not have come to pass, what is true is that there is a lack of material innovation for RGP lenses. This simply could be due to the fact that RGP lenses are unfashionable from a research perspective, particularly with the booming field of hydrogels. In fact, Efron first alludes to this by commenting on the marketing for SCLs over RGP lenses, and that the lack of finance directed at improving these materials has caused their decline. Smart or wearable biosensors show promise for RGP lens materials [82], but even this field is being dominated by SCL materials [83]. Perhaps the route to reinvigorating RGP lenses is through the growth of augmented reality (AR) technology. Patents for the design of an AR contact lens have been assigned to companies such as Samsung and RaayonNova in 2016 and 2017, respectively [84,85]. However, it is likely that the first lenses will borrow heavily from existing RGP lens materials. This is in contrast to other smart CLs, which are often based on SCL materials [86].

2.4. HEMA-Derived Hydrogels

HEMA and related hydrogels are a high-water content, oxygen-permeable polymeric material. These hydrogels can have water content between 20–80% depending on the comonomers, with a hydrogel composed of only HEMA containing about 38% water. HEMA's highly polar properties mean these CLs have generally suitable wetting properties, meaning they are comfortable. The oxygen permeability of these gels is suitable for longer wear, but not to the same extent of silicone-based lenses. HEMA-derived hydrogels form an important part of the market, occupying about 22% [15]. HEMA is commonly copolymerized with monomers such as EGDMA, MAA, and NVP. NVP and MAA increase the water content of hydrogels due the strong hydrophilic character arising from amine, carboxylic acid, hydroxyl groups, etc. Therefore, these comonomers also influence the wettability of the surface [87,88]. The mechanical properties can be improved by using a cross-linking molecule, such as ethylene glycol dimethacrylate (EGDMA). EGDMA has two functional groups allowing the formation of covalent bonds between two individual polymer chains. This increases the mass of the polymer dramatically, and improves its ability to form a gel network. However, the crosslinks also reduce the polymer-chain motion, which can be a factor in swelling and oxygen transport [88]. The water content and cross-linking affect the modulus and oxygen permeability of the hydrogels (Figure 7). As such, a balance must be reached between these parameters when designing a CL for a particular application.

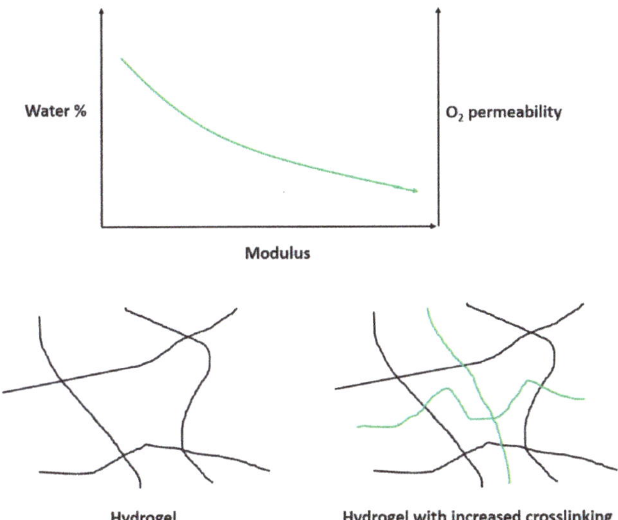

Figure 7. Schematic on the effect of cross-linking on the modulus, water-content percentage, and oxygen permeability. The increased cross-linking prevents (green) the polymer chains from swelling, compared with lower-modulus gels.

To improve CL materials, it is vital to better understand the properties of these hydrogels. The properties of HEMA-based hydrogels were shown to vary as a function of use by patients [89]. Tranoudis et al. measured the properties of lenses, such as total diameter, oxygen permeability, and the back optic zone radius, at temperatures ranging from 20–35° and before/after 6 h of wear. All the materials were altered with statistical significance; however, according to the summary tables, HEMA-VP 70% (meaning 70% water content) was moderately stable. In another study, the authors showed that free-to-bound water content was an important factor for the rate of dehydration and oxygen transport in SCLs [90]. The bound water to the polymer may be a factor in stability of hydrogels, such as HEMA-VP 70%, compared with other gels. Tranoudis et al. published another paper to more accurately characterize the tensile properties of the same organic hydrogels [91]. They concluded that hydrogels with high water content did not necessarily correlate with poor mechanical properties. This is largely true, and is an uncomplicated way of summarizing the data. However, comparing different polymer systems can be difficult due to the different intermolecular forces that exist to provide stability (or lack of) based on different chemical compositions. There are properties of the polymer structure which could reveal more factors that affect the mechanical properties of hydrogels. Explanations by Ashby may reveal the reasons for inconsistent properties between batches of the same polymeric material [50]. Porous cellular solids with varying wall thicknesses and lengths between vertices are some of the factors that influence the mechanical properties of porous materials. Therefore, it is important to study the polymer's physical structure and polymerization mechanisms. Use of ^{1}H Nuclear Magnetic Resonance (NMR) relaxation times by Woźniak-Braszak et al. was a novel characterization method used to probe the dynamics of free water in contact lenses [92], where they showed an increase in free water in four-week-old lenses compared with new lenses (83% to 71%, respectively). These studies together provide a more detailed understanding of hydrogels, which in turn can be used for the better design of CLs.

Currently, modification of the HEMA-hydrogel structure is paving the way for new lenses, innovations, and applications. Protein deposition on the eye remains a concern, particularly for SCL materials based on HEMA [93,94]. Common comonomers, such as MAA and NVP, increase the protein deposition. Both comonomers introduce hydrophilicity and electrochemical charge to the hydrogel, which attracts the proteins from the tear film [93]. Lord et al. also showed that the protein uptake could affect the water content of a HEMA-MAA hydrogel. This could be related to observations by Tranoudis et al. where a HEMA-MAA hydrogel lost 15% water content across a 6 h period [89]. An interesting report by Borazjani et al. regarding the bacteria adhesion of *Pseudomonas aeruginosa*, a risk factor in keratitis, to a HEMA-hydrogel was no less than that of a silicone hydrogel [95]. Borazjani et al. also reported that adhesion of the bacteria was more related to the strain of bacteria than the lens material. Another study by Szczotka-Flynn et al. demonstrated that biofilms of bacteria (particularly *Pseudomonas aeruginosa*) were resistant to the antimicrobial action of CL solutions [96]. Dutta et al. looked into the literature of this area by considering the material type, length of wear, and bacterial strains on the cumulation of the bacteria on contact lenses [97]. New strategies to prevent this have been in development since the early 2000s, including further modification of the lens material. A more detailed and extensive article on antimicrobial strategies was published by Xiao et al. [98].

Grafting additional substances to HEMA has proven to be an effective method of modification for improved lens functionality [99–101]. Some specific examples include grafts to reduce the protein deposition or increase the anti-microbial action of contact lenses. A successful product based on a HEMA hydrogel lens incorporated with silver nanoparticles (AgNPs) has been on the market in the UK (MicroBlock®). Other researchers looked at the impact of the monomer composition of the HEMA hydrogel contact lens on AgNP uptake and performance [102]. They noticed that HEMA-MAA-EGDMA gels had the strongest affinity to the nanoparticles, soaking in AgNPs from 10 to 20 ppm. Other interesting antimicrobial graft materials include melimine and polymyxin B. Melimine is a synthesized antimicrobial peptide consisting of 29 amino acid units, and was covalently bonded to HEMA-MAA-EGDMA using an acidic (pH = 5) buffer to induce the reaction to the pendant acid groups on the lens [103]. This post-hydrogel modification may not be entirely practical for manufacturing, but is interesting due to the excellent properties of the resulting hydrogels. Polymyxin B is another anti-microbial macromolecule, and was grafted using an azobisisobutyronitrile (AIBN) free-radical initiator during hydrogel synthesis [104]. The action of polymyxin B induces greater water permeability of bacterial cells, eventually leading to bursting through the increased water uptake. Sato et al. modified HEMA by copolymerization with moieties such as 2-methacryloxy ethyl phosphate (MOEP) to facilitate drug delivery [105]. The MOEP adds additional anionic character, which was used to bind a model drug to the hydrogel. The drug release profile was a superior HEMA-MAA hydrogel due the stronger ionic group based on the chemical changes with pH.

Other modification materials include the incorporation or grafting of surfactants. A surfactant consists of a hydrophobic and hydrophilic component, and is primarily used to promote the reduction in surface tension between two immiscible liquids. Bengani et al. employed the novel use of polymerizable surfactants attached to HEMA hydrogels to enhance wettability and lubricating properties [106]. They achieved a 10° reduction in water-contact angle with about 2.4 wt % surfactant, which was covalently bonded to the hydrogel by UV polymerization. The low-surfactant loading meant that the HEMA-hydrogels remained below 45% water content, which indicates the power of such a technique. In this case, the hydrophilic component interacts with the aqueous tear film and the hydrophobic part remains in the hydrogel (Figure 8). There is much modern literature that still reports high rates of contact lens discontinuation due to discomfort and dryness [107–109]. This emphasizes the importance of new techniques, such as surfactant loading/grafting employed by Bengani et al. Other uses of surfactants include use as a drug delivery system for cyclosporine A in modified HEMA-hydrogels [110]. In this case, the surfactant leached out of the hydrogel with the drug encapsulated within surfactant aggregates (Figure 9). This is now becoming a more researched topic, using a larger number of surfactants to deliver drugs [111–114]. These works have used a diverse range of surfactants, from cationic to non-ionic, with molecular weights ranging from about 400–12,500 g mol^{-1}. This wide variation in surfactant properties is interesting, suggesting that the hydrogel structure is highly adaptable to such modifications.

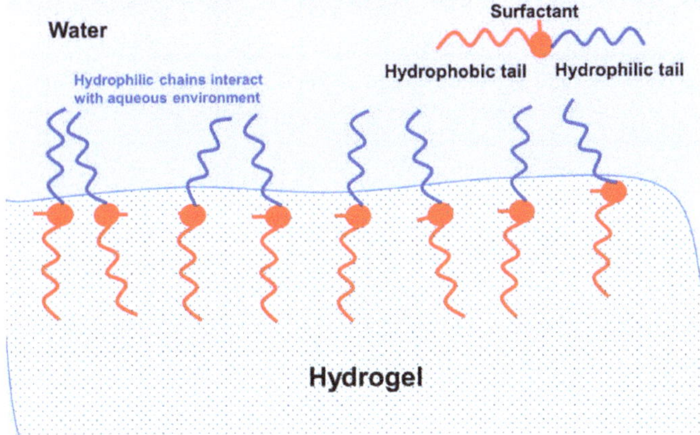

Figure 8. Schematic for the interaction between the aqueous tear film and a surfactant bonded with a hydrogel lens. Reprinted from Journal of Colloid and Interface Science, 445, Bengani, L.C.; Scheiffele, G.W.; Chauhan, A.; Incorporation of polymerizable surfactants in hydroxyethyl methacrylate lenses for improving wettability and lubricity, 60, Copyright 2014, with permission from Elsevier [106].

In summary, HEMA-hydrogel CLs are in an exciting place. The ability to design new materials is due to the flexible hydrogel structure composed from many functional groups. Essentially, the core of the hydrogel has remained relatively unchanged, due to the principal properties that matter to CLs, such as oxygen permeability, water content, wettability, etc. Modern HEMA hydrogel lenses have evolved through modification of the hydrogel structure through techniques such as encapsulation and grafting. Some of the materials used for modification were surfactants, nanoparticles, and anti-microbial agents, all of which are diverse in themselves. As such, the works presented here are only the beginning of an emerging field. There were also efforts to better understand the mechanical properties of the HEMA-hydrogels, such as dehydration rates, bound and free water content, and factors influencing protein absorption into the hydrogels. From a better understanding of hydrogel properties, new manufacturing processes can evolve to further improve the material specifications. This is encouraging progress for designing better materials, which will only keep evolving as more and more chemical modification techniques and materials are employed.

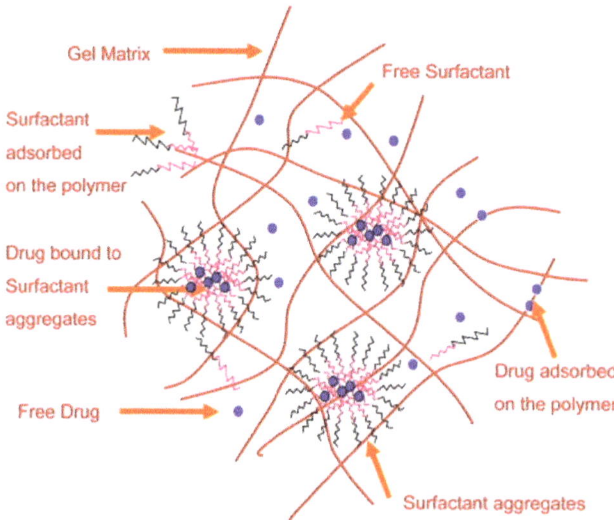

Figure 9. Drug-loaded hydrogel loaded with surfactants. The aggregation and leaching of the surfactant acts as a drug delivery vehicle. Reprinted from Biomaterials, 30, Kapoor, Y.; Thomas, J.C.; Tan, G.; John, V.T.; Chauhan, A. Surfactant-laden soft contact lenses for extended delivery of ophthalmic drugs, 867, Copyright 2008, with permission from Elsevier [110].

2.5. Silicone-Derived Hydrogels

Silicone-based hydrogels are the most common type of CL material today, occupying about 64% of the US market [15]. This includes silicone, silioxanes, fluorosiloxanes, and derivative materials. Their popularity is linked to the fact that these CLs have the highest oxygen permeability of all contact lens materials (>100 Dk, typically). Silicone CLs are often durable, originating from the high Si–O bonding energy, often with a higher modulus than conventional polymer hydrogels. This typically applies to RGP lenses, as the modern silicone hydrogel has a similar modulus to HEMA-derived hydrogels. The high modulus is linked to causing irritation to the eye, such as the conjunctiva of the inner eyelid [4]. In fact, the discomfort and dryness of silicone-based lenses are two of the main reasons for users' discontinuation [107,109,115,116]. Specifically, the wettability of the lens and the incompatibility with the cornea environment in vivo requires detailed study to improve these parameters [117]. The patent literature contains much about improving the wettability of lenses, with companies such as CooperVision [118] and Johnson and Johnson [119] having filed for patents to develop such lenses. This includes using many combinations of comonomers with silicone monomers. However, similarly to the preceding sections, modern silicone lenses have not deviated from the base materials since they were first developed. Nowadays, new chemical techniques look to evolve the silicone CL material beyond its predecessors.

Silicone hydrogels have commonly been subject to post-fabrication processes. Some, such as plasma treatment, are used to improve lens wettability by manufacturers [120–122]. This technique has been proven very effective; however, there are still a number of issues relating to silicone CLs [123]. Santos et al. suggested that protein absorption of commercial silicone hydrogel lenses was independent of the plasma treatment [124]. They summarized that the inherent hydrophobic character of the lenses helped in reducing the deposition. With this in mind, new chemical modification techniques, with or without plasma treatment, are becoming of interest. The use of hydrophilic and hydrophobic lipids by Bhamla et al. investigated the wetting and dewetting character of a Balafilicon A lens [125]. They showed that a hydrophobic lipid (meibum) could reduce the wetting area on the lens surface by nearly 70% in 50 s. Lin et al. modified the lens surface to adhere polyelectrolyte monolayers (PEMs) of chitosan and hyaluronic acid [126]. Chitosan is a naturally derived polymer (from chitin) with high bioavailability originating from the hydroxyl and amine groups within the structure, lending itself to lens modification. They modified the plasma-treated silicone surface to create negatively charged groups, which allowed the formation of the positively charged chitosan layer. Then, the negatively charged hyaluronic acid formed a layer on this surface to form alternative layers. The contact angle was reduced from 90° in the mother lens to up to 50° in the PEM-modified lenses. Similarly, a modified chitosan was also used successfully by Tian et al. to form a chitosan layer on the surface of the lens (Figure 10) [127]. Hydroxypropyl trimethyl ammonium chloride chitosan is a quaternary ammonium salt and is positively charged, which facilitates the formation of self-assembled layers. These pendant chemical groups could lead to further modification of the lens. Thissen et al. modified silicone lenses with a poly ethylene oxide (PEO) graft to the surface for anti-biofouling properties [128]. First, they used plasma polymerization to adhere allyl amine to the surface of the lens, and then used a reducing solution of sodium cyanoborohydrate and PEO to complete the PEO graft. A similar process using acrylic acid was performed by Dutta et al. to attach a melamine-derived peptide (Mel4) to the surface of a range of silicone lenses (lotrafilcon A, lotrafilcon B, somofilcon A, senofilcon A, and comfilcon A) [129]. They also used an acid buffer solution to covalently bond the Mel4 peptide to the surface of the lenses [130]. The peptide improved the anti-microbial properties of the lenses without causing other irritations to the tested subjects (rabbits).

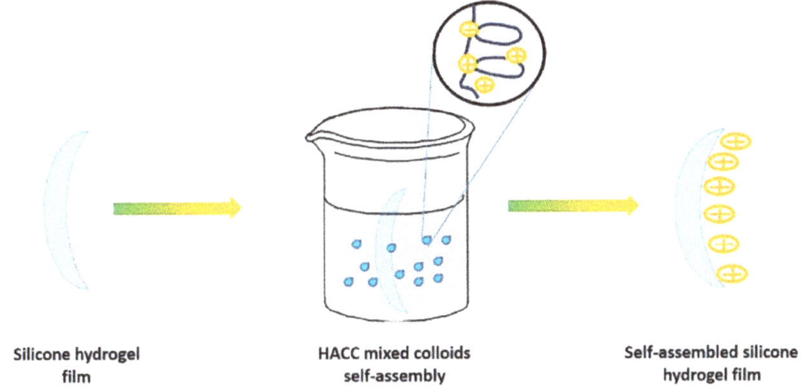

Figure 10. Schematic for the adhesion of a self-assembled layer onto the surface of silicone hydrogel. Reprinted (and adapted) from Colloids and Surfaces A, 558, Tian, L.; Wang, X.; Qi, J.; Yao, Q.; Oderinde, O.; Yao, C.; Song, W.; Shu, W.; Chen, P.; Wang, Y. Improvement of the surface wettability of silicone hydrogel films by self-assembled hydroxypropyltrimethyl ammonium chloride chitosan mixed colloids, 422, Copyright 2018, with permission from Elsevier [127].

Other materials were of interest as new grafting or encapsulation components into silicone-based hydrogels. Tuby et al. used Zn-CuO particles as an anti-microbial film on CLs composed of silicone hydrogels (Bausch&Lomb) [131]. They used a base bath (pH = 8) loaded with the nanoparticles to soak into the lenses to induce uptake of the nanoparticles by the hydrogel. They also commented that the nanoparticles did not leech easily from the structure, which would be an important safety concern. Jung et al. used a polymer nanoparticle (propoxylated glyceryl triacrylate) to deliver the drug timolol, which is used for glaucoma treatment [132]. The lenses also incorporated the nanoparticles through a soaking method, which was a phosphate buffer solution (pH ≈ 7.4). Although the nanoparticles did not affect optical transparency, they noted an undesirable decline in the water content, oxygen permeability, and modulus. Willis et al. grafted hydrophilic phosphorylcholine (PC) to the surface of silicone lenses using several Michael-type addition steps [133]. The coating reduced the water contact angle by nearly 50°, which was an impressive result. One concern they had was the lowering modulus by 0.5 MPa in the PC-treated lenses compared with unmodified lenses. Wang et al. used an interpenetrating network of silicone and PC to synthesize the hydrogel to achieve similar wettability results [134]. A 35–80 nm layer of PC was also grafted onto silicone using UV light (305 and 365 nm) to induce the polymerization graft [135]. The surface graft did not affect the oxygen permeability and modulus, but reduced the lysozyme and fibrinogen deposition and water contact angle. Wang et al. grafted polyethylene glycol methacrylate (PEGMA) to silicone using UV light again, and again the hydrogels retained excellent properties, such as >140 Dk and >1 MPa elastic modulus [136]. Synthesis of silicone hydrogels also incorporating PEGMA by Lin et al. successfully reduced the lysozyme and human serum albumin (HSA) deposition by 82% and 77%, respectively [45]. However, the increasing volumes of PEGMA affected other properties of the lens, such as increasing brittleness above 20% PEGMA content. The PEGMA polymer brush is a known anti-biofouling material [137,138]. Its low cost and easy incorporation is likely worth pursing for other CLs. These studies highlight the impact of surface modification techniques on improving CL properties. However, manufacturing aspects need to be considered to realize these modifications in a new generation of CLs.

In summary, silicone hydrogel lenses are targets for highly biocompatible coatings that can be grafted/encapsulated to the surface of, or within, the lenses by chemistry techniques. In these experiments, the performance of the lens was improved, particularly towards anti-biofouling applications. Hydrophilic anti-biofouling materials are being applied more often to silicone hydrogel CLs [139]. This emphasizes the innovations which silicone-based hydrogel lens materials are undergoing. For silicone hydrogel, there is a two-fold benefit of these anti-biofouling materials: (1) These hydrophilic grafts/modifications are the route to anti-biofouling behavior, and; (2) they are hydrophilic, increasing the wetting character of the lens. This effect is also relevant to HEMA-based lenses; however, it is more impactful to silicone lenses, given their inherent wettability disadvantages and market size. Continued improvements to these materials will result in continued market dominance for silicone-based hydrogel CLs.

2.6. Other Contact Lens Materials

2.6.1. Polyvinyl Alcohol Hydrogels

PVA is a synthetic polymer that contains many hydroxy (–OH) groups, one in each repeating monomer unit. This is the source of PVA's excellent hydrophilic and biocompatible properties [140]. As such, it is clear why this material is an interesting contact lens material. PVA hydrogels were of attention to several researchers in the early 1990s [17,18]. Even without modification, PVA CL hydrogels were also shown to have lower protein absorption rates than HEMA and MMA/VP hydrogels [18]. However, it wasn't until the late 1990s that this material broke into the market in the form of the Nelfilcon A lens [16]. These lenses have a Dk of about 26 barrers and high wettability, which is acceptable for daily wear. Produced by CIBA vision, the PVA hydrogel was synthesized using water as the solvent and produced inside a transparent mold to allow UV initiation of the polymer solution. Water was an environmentally friendly choice of solvent which also did not hinder the polymerization stages. Buhler et al. formed the PVA hydrogel by adding a new functional group to facilitate cross-linking between chains [16]. In acidic conditions and a suitably reactive dialdehyde group, PVA can be functionalized with a stable cyclic-acetal group. This can then crosslink with other PVA chains to form a crosslinked hydrogel. More recently, PVA has been used as a tool for producing more comfortable lenses [141,142] or to facilitate the loading of a colored pigment into the lens [143]. One of the comfort mechanisms include elution of an unbound PVA polymer from the contact lens into the tear film. Non-crosslinked PVA with an approximately 47,000 molecular weight was soaked into a cross-linked PVA lens. The high molecular weight and bioavailability meant that PVA leeched out of the lens to lubricate the eye as a sort of artificial tear component [142,144]. Researchers identified that the higher molecular weight species took longer to elute from the lens than lower molecular weight species. Although these examples are strictly not using PVA hydrogels, they highlight the importance of this material for other ophthalmic applications. This even extends to areas such as corneal replacement [145].

There has been some investigation into the modification of PVA hydrogel CLs. Xu et al. modified the PVA structure with β-cyclodextrins (β-CDs) to improve drug-loading of puerarin and acetazolamide [146]. They functionalized both the β-CDs and PVA with a methacrylate moiety with *N,N*-dimethyl-4-pyridinamine and glycidyl methacrylate to achieve the necessary functionality. Then, both methacrylated species were copolymerized to form the hydrogel. This method also resulted in incorporation of 30 wt % of β-cyclodextrin. Sun et al. also used a functionalized β-CD polymer incorporated into a PVA film for the delivery of voriconazole [147]. A β-CD solution containing dissolved voriconazole was mixed with a PVA solution and then used to form electrospun nanofibers. PVA cross-linked with cellulose offers another a potential route to improved contact lens materials [148,149]. Cellulose is a naturally occurring polymer with enormous potential for bio-applications [150]. Mihranyan chemically bound the cellulose microcrystal "whiskers" onto the surface of PVA using a 2,2,6,6-tetramethyl-1-piperidine oxoammonium salt (TEMPO)-mediated surface oxidation [149]. The TEMPO technique is a special technique used for regioselective oxidation of cellulose to preserve crystallinity [151]. The technique only functionalizes the surface of the material. Although the physical properties of the PVA hydrogels improved, the opaque material would not be ideal as a CL material. Tummala et al. incorporated nanocellulose, a transparent version of cellulose, as nanofibrils or nanocrystals into a PVA hydrogel, improving the physical properties [148]. In addition, the lens retained excellent optical properties, with >90% transparency to visible light (Figure 11). They used a mixed solvent system of water:dimethyl sulfoxide (DMSO) at ratios between 60–80% DMSO content to facilitate the solvation of the nanocellulose components. These nanocellulose PVA hydrogels were investigated for their light-scattering properties by Tummala et al. in another work [152]. The hydrogel composition and angle of light affected the scattering of light in the range of 3% to 40%, with the nanocellulose-functionalized PVA hydrogel reducing the scattering. The nanocellulose hydrogels were then used to encapsulate acrylic-acid-functionalized chitosan nanoparticles as a vehicle for a lysozyme-triggered release of a model drug. Lysozyme (0.2 mM)-mediated hydrolysis of the chitosan-functionalized nanoparticle would trigger the release of the drug.

The PVA hydrogel is a relatively new class of hydrogel lens compared with silicone- and HEMA-hydrogels. The low cost, high bioavailability, and wettability properties mean PVA hydrogels will remain important to the CL industry. This means there is a lot of design space for investigation into modifying the structure to improve lens functionality. There are an increasing number of modification techniques and materials that have yet to be explored with PVA hydrogels. The secondary alcohol groups within the polymer chain are very easily modified for grafting, as shown by cellulose or β-cyclodextrin incorporation. As with HEMA- and silicone-hydrogels, the grafting or encapsulation of anti-microbial agents is just one of many interesting avenues of research that is possible for PVA hydrogels.

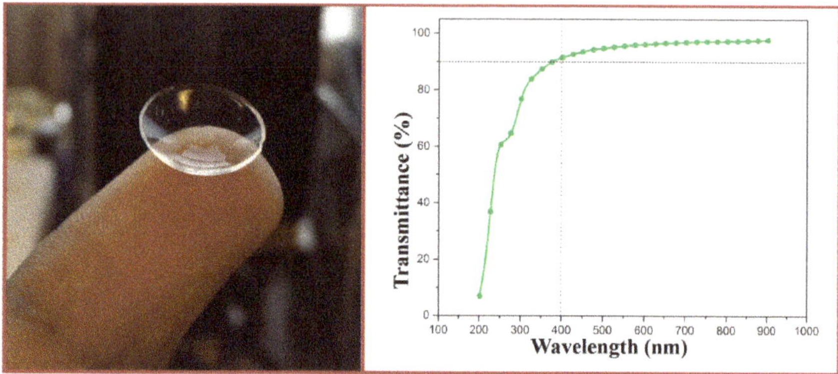

Figure 11. Optical image (**left**) and the transmittance data (**right**) for PVA hydrogels incorporating nanocellulose. Reprinted (adapted) with permission from Tummala, G.K.; Rojas, R.; Mihranyan, A.; Poly(vinyl alcohol) hydrogels reinforced with nanocellulose for ophthalmic applications: general characteristics and optical properties, Journal of Physical Chemistry B, 2016, 120, 13094. Copyright 2016 American Chemical Society [148].

2.6.2. Hyaluronic-Acid-Modified Hydrogels

Hyaluronic acid (HA) is a hydroscopic biopolymer that naturally occurs within the human body. HA has already been mentioned in previous sections. Here, we explain why this material is significantly impacting CL research. This is an important material for a wide range of tissue-engineering purposes [153]. From the chemical structure, we can see why it is useful for incorporating into CL materials (Figure 12). The amino acid and hydroxyl groups present in each repeating unit provide the necessary hydrophilic character, leading to high biocompatibility. As such, this material has been of interest in developing ophthalmic treatments, such as lubricating solution [154] or contact lens modification [155]. The incorporation of HA was shown not to affect the surface morphology of the CL even after 12 h of wear, showing the stability of these modifications [156]. HA is typically a graft/encapsulating material to other established CL hydrogels. The commercial success of HA is apparent with the Bausch&Lomb Biotrue™ solution and Open30 (by Safilens) lenses incorporating HA as a lubricating agent. Other ophthalmic treatments and modifications include HA use as a treatment for dry eyes, [157,158] modification of the lens material [126,159,160], and drug-delivery mechanism [161–163].

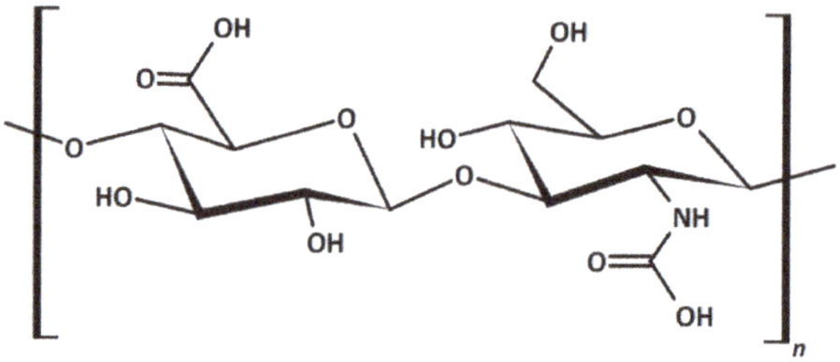

Figure 12. Chemical structure of hyaluronic acid. The repeating unit within the polymer is hydrophilic with many highly polar groups, such as the amine, acid, and hydroxy groups.

HA was investigated as a reusable wetting agent, and considered as a vehicle for the delivery of timolol within a silicone hydrogel lens [164]. HA is hydroscopic, meaning it was used to add water content to a material without a significant reduction in the other hydrogel components of DMA and TRIS [160]. All of the silicone hydrogels Paterson et al. produced had a water contact angle of less than 50°, emphasizing the wettability qualities of HA. The wetting agent was a copolymer of PEG and HA, which was first copolymerized before being soaked into the HA-modified silicone hydrogels. Van Beek et al. crosslinked/physically entrapped HA into HEMA hydrogels by using 1-ethyl-3-(3-dimethylaminopropyl)-carbondiimide (EDC)/diaminobutane-4 dendimer to provide the necessary functionality for a reaction [159]. Furthermore, HA in very low amounts (5 g/L solution) reduced the water contact angle of the hydrogels by about 15–25°. These HA-modified hydrogels also showed less lysozyme absorption, which can often be a criticism of HEMA-derived hydrogels. However, large molecular weight (M_w) HA (169 kDa) had reduced optical transparency between 500–800 nm versus the lower M_w HA (35 kDa). On the other hand, the 169 kDa HA was responsible for the lowest amount of protein deposition. Korogiannaki et al. covalently bonded HA to the surface of HEMA by using a nucleophilic Michael-addition thiol-ene "click" chemical reaction [165]. These lenses retained an optical transparency level of >92%, and decreased the water contact angle and rate of lens dehydration. A more detailed mechanism of the covalent bonding of HA to the HEMA hydrogel structure was presented by Deng et al. [166] "Click" chemistry was used with adipic acid dihydrazde (ADH) to anchor HA to HEMA. HEMA was first oxidized to provide the functionality for bonding the ADH-HA anchor. The lens properties virtually remained unchanged, such as the modulus and optical properties (Figure 13). They also measured a reduction in the rate of dehydration of the lens compared with the unmodified lenses. Weeks et al. used UV light to covalently bond a methacrylate modified HA to HEMA and silicone hydrogel lenses [167]. In all cases, the HA-modified hydrogels had a greatly reduced water-contact angle and reduction of lysozyme absorption. They also pointed out that HA with little or no methacrylation on the surface would not prevent the interaction of the protein with the surface of the hydrogel, whereas if HA was entrapped inside the hydrogel, the protein could not interact with the surface at as many points, preventing its deposition.

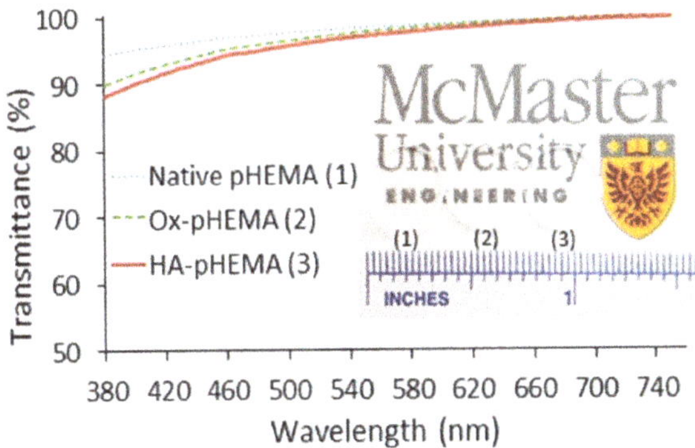

Figure 13. Optical properties of HA-HEMA, oxidized-HEMA (OX-HEMA), and unmodified HEMA lenses. The transparent hydrogels can be seen just above the ruler markings. Reprinted (adapted) with permission from Deng, X.; Korogiannaki, M.; Rastegari, B.; Zhang, J.; Chen, M.; Fu, Q.; Sheardown, H.; Filipe, C.D.M.; Hoare, T. "Click" chemistry-tethered hyaluronic acid-based contact lens coatings improve lens wettability and lower protein adsorption, ACS Applied Materials and Interfaces, 2016, 8, 22064. Copyright 2016 American Chemical Society [166].

HA is an increasingly popular material for the modification of both HEMA-hydrogel and silicone-hydrogel CLs. The high bioavailability and straightforward incorporation methods have already seen recent commercial success for HA. In these cases, HA was a lubricating agent rather than any physical/chemical modification of the lens material. Recent works regarding HA are exciting, demonstrating the ability that HA can alter, tune, or improve the properties of contact lenses. HA can be entrapped or chemical-bound to the hydrogel to provide increased wettability or resistance to protein deposition. One of the most limiting factors for the wider application of HA is the high cost.

Although dilution can reduce the cost, the effectiveness of HA will diminish. Once this issue has been overcome, HA will be commonly incorporated into new lenses.

3. Outlook for CL Materials

3.1. Drug Delivery

Today, CLs are increasingly seen as a tool for specific ophthalmic treatments that require delivery of a drug to the eye. The low bioavailability (about 5%) of common ophthalmic treatments, such as eye drops, is a concern to delivering effective treatments [168]. In turn, practitioners rely on new lenses to be developed to improve these treatments, which is reliant on material scientists developing new materials. This is evident by the publication of review papers in the last few years [27,28,169]. These reviews contain significant detail on the latest drug delivery mechanisms, materials, and relevant literature of modern drug delivery, including general drug delivery within the body. The majority of the referenced literature is after the year 2000, indicating the significant advances and interest in the field within the last 20 years. In this paper, we highlight the trends in the CL materials used for ocular drug delivery. Some of the methods investigated include drug-loading by nanoparticle- or polymeric encapsulation [170–173], β-cyclodextrin delivery [40,174–176], molecular imprinting [177–179], and solution soaking [37,180–182]. Molecular imprinting involves fabrication of the hydrogel alongside the drug. Typically, increasing crosslinking and addition of a porogenic solvent facilitates this by creating more cavities for residing the drug [179]. The nanoparticles used in these examples, which were all polymeric, encapsulated the drug. The biocompatible polymers responded to the ocular environment to release the drug. Ethyl cellulose and lactic-co-glycolic acid are two examples [171,172] These are all of scientific interest; however, more should be done to access their practical application. This can include clinical trials or practicality of manufacturing.

The properties of an incorporated material into CLs can be used as improved drug-binding mechanisms [183]. For example, incorporation of ionic monomers can create binding sites for a polar drug to bind to (Figure 14), meaning the lens can retain the drug until it is placed onto the eye. MAA, for example, is one comonomer commonly used for the formation of hydrogels, but which also has an anionic group to facilitate drug-binding. Another clever use of new materials was the use of pendent- or copolymerized cyclodextrins for introducing both hydrophilic and hydrophobic character to the CL [174,176]. The inside of the 7-unit ring of cyclodextrin is hydrophobic, ideal for drug complexing, whereas the outside of the molecule is hydrophilic to provide the lens with suitable biocompatibility (Figure 15). Dos Santos et al. loaded hydrocortisone and acetazoamide into a HEMA-hydrogel lens, achieving a release profile of several days [174]. An alternative method has involved, using vitamin E to essentially create a barrier to improve the release profile of loaded drugs [25,184,185]. The hydrophobic barrier interacts with the drug, which greatly increases the permeation time rather than simply eluting from the lens. Vitamin E contains a large hydrocarbon side chain and methyl groups to provide the necessary hydrophobic character. In another experiment by Kim et al., vitamin E was easily incorporated into commercial silicone-hydrogel lenses such as ACUVUE®, OASYS™, O$_2$OPTIX™, and NIGHT&DAY™ [184]. Some of the delivered drugs included cysteamine [186,187] and pirfenidone [188]. Typically, vitamin E was incorporated into the hydrogel through soaking, and the hydrophobic character means it is unlikely to leech out, compared with a hydrophilic component such as PVA. This is a straightforward process that could be adopted into more widespread usage to improve elution times of drugs from CLs. Although, strictly speaking, there was no fundamental change to the hydrogel networks, the work is worthy of inclusion given the profound impact on drug-release profiles. Some examples used straightforward incorporation methods, which could be a manufacturing consideration.

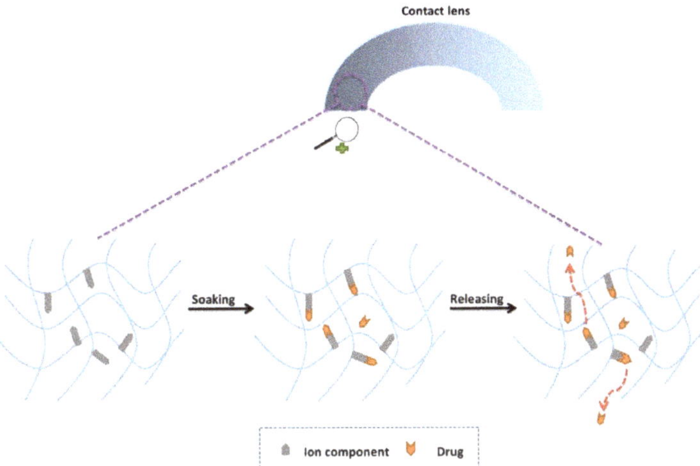

Figure 14. Drug-binding and releasing mechanism for a contact lens with ionic sites within the structure. Reprinted from Journal of Controlled Release, 281, Xu, J.; Xue, Y.; Hu, G.; Lin, T.; Gou, J.; Yin, T.; He. H.; Zhang, Y.; Tang, X. A comprehensive review on contact lens for ophthalmic drug delivery, 97, Copyright 2018, with permission from Elsevier [27].

Figure 15. Chemical structure of beta-cyclodextrin. The inner environment of cyclodextrin is hydrophobic, suitable for drug-binding, whereas the outside is hydrophilic, providing suitable bioavailability.

Modern CL materials have shown much-improved drug delivery capacity than current ophthalmic treatments, such as eye drops. Further modification of the base CL polymer structure is required to improve drug delivery. Molecular grafts, particle encapsulation, and soaking are the most common methods for modification. Despite evidence of the improving performance of these materials, more must be done in order to take these materials to the point of commerciality. The biggest barrier to the market is the cost of clinical trials, as well as the manufacturing requirements. Another note of interest is that most of these successful works have been based on HEMA-hydrogels [189]. HEMA-based hydrogels have an abundance of easily accessible chemical functional groups for modification, rather than silicone. In some cases, silicone-based hydrogels were more suitable to host drugs, such as the vitamin E barrier [27]. This represents an interesting position for CL materials, where the requirement of increased hydrophilicity for comfort and oxygen permeability has driven the increased usage of silicone-hydrogels over HEMA-hydrogel lenses. Conventionally, many drugs are chosen based on a strong hydrophobic character [190]. Development of suitable drug-delivery lenses will mostly depend on the intended time of treatment, and the length of wear that facilitates this. In turn, the length of wear will best define what type of CL material should be used and what modification should support the drug release.

3.2. Emerging CL Materials Technologies

In this section, we will discuss several emerging materials technologies that could be relevant to future CL materials. Biocompatible materials have become of great interest to the scientific community, particularly as new materials have been innovated [191–194]. These biocompatible materials could be an avenue for CL researchers to develop novel materials. The necessity for biocompatibility is an obvious reason why these materials could be of interest. Two such areas that have seen particular growth are double-network/interpenetrating hydrogels and pH-responsive polymers. Although some interest has been expressed in areas such as liquid crystal CLs [195,196], the number of research publications remains somewhat limited compared to these other technologies. Briefly, the liquid crystal contact lens is of interest to treating presbyopia, due to the liquid crystal's on/off character to alter the lens power in the region of +2.00 D. The liquid crystal lens has anisotropic refractive indexes, realigning the molecules when a voltage is applied. Commonly, the lens design often involves PMMA to support the liquid crystal phases [197]. Perhaps the biggest difficulties with these lenses is ensuring ocular health, comfort, and a power source.

Double-network/interpenetrating hydrogels combine two gels by interconnecting the gel networks together, resulting in a new gel composite (Figure 16). Essentially, the network of one gel is intertwined with the other, rather than the formation of a traditional copolymer gel. This is a subtle, but significant, difference—e.g., a double network gel could be formed by two polymers with different functional groups, meaning they could not be copolymerized. Figure 16 also shows that these hydrogels can be formed pre- and post-formation of one gel. Formation of these new materials has been shown to improve the bioavailability/compatibility of hydrogels [198,199]. Some of the materials investigated are commonly known CL- or modification-materials, such as HA, chitosan, thiol-functionalized PEG, poly-2-hydroxyethly acrylate (HEA), and polyacrylamide (PAA). With the large number of materials used in CL synthesis, this method could produce a unique double-network contact lens with improved functionality.

Double-network hydrogel formation is fundamentally a chemical technique to improve the bulk properties of the material. These hydrogels could be manufactured in the same manner as current CL hydrogels, yet enhance the CL properties. This area is of interest to the CL community; a patent was assigned to Stanford University in 2010 for interpenetrating polymer-network contact lenses [200]. One example by Yañez et al. used interpenetrating networks composed of HEMA and polyvinyl pyrrolidone (PVP) to develop new comfortable SCLs [201]. PVP was semi-interpenetrating and leeched from the lens as a comfort mechanism rather than a true double-network hydrogel. These gels have been tailored for applications involving tissue engineering, emphasizing the robustness of the gel [202]. The inner eyelid (conjunctiva) and cornea are two different environments sensitive to the properties of the CL. Castellino et al. combined PDMS-MAA-HEMA to form an interpenetrating hydrogel network [203]. Their purpose was to form a general bioavailable material, rather than a material for specific ophthalmic applications. Wang and Li synthesized a double-network hydrogel of TRIS and HEMA-phosphorylcholine (PC) as a potential ophthalmic material [134]. At higher PC (20%) content, the double-network gels had a low water-contact angle (50°) and a high modulus (7 MPa). These are very attractive qualities for CL hydrogels, which could be further developed. Other double networks of silicone have been investigated for other research fields, suggesting it could be more widely used for CLs [204–206]. These works incorporated organic components, which would be essential for ocular biocompatibility. This type of hydrogel could be the next evolution of contact lenses.

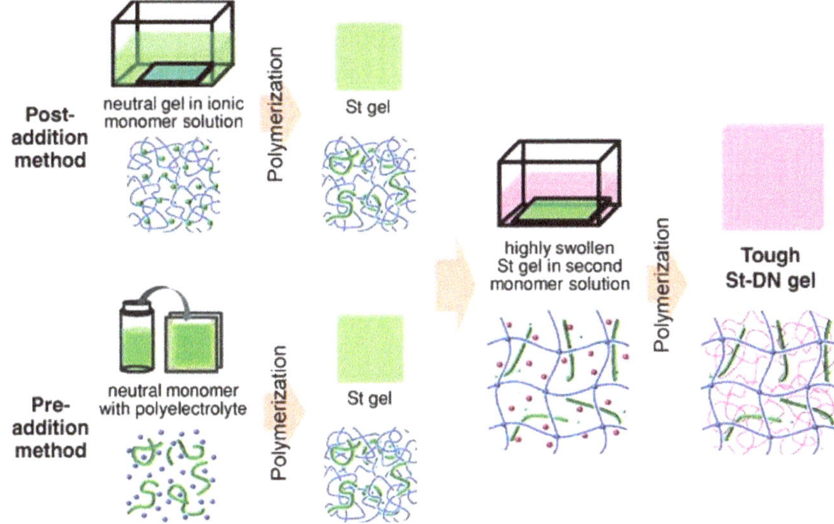

Figure 16. A double-network hydrogel combines the properties of two hydrogels. The gel networks are interpenetrating, meaning they form a new gel that retains the properties of each individual gel. Reprinted with permission from John-Wiley and Sons [199].

We have already discussed some of the more popular drug-delivery materials in a previous section. However, there are new concepts within biocompatible hydrogels as potential drug-delivery mechanisms [207,208]. Temperature- and pH-responsive polymers are materials that change their macromolecular conformation when exposed to a specific temperature or acidic/alkaline conditions (hence, pH-responsive) [209]. For pH-responsive polymers, there are four principal mechanisms by which the conformation is changed, and this is summarized in Figure 17. The acidic/alkaline conditions of the body can induce protonation, stripping the coating and conformation changes to the polymer and allowing release of the drug. As such, they are becoming exciting as drug-delivery mechanisms within the human body [210]. The concept was recently investigated for drug-delivery contact lenses [211]. Researchers here used a film of cellulose acetate to deliver betaxolol hydrochloride over a sustained period. They showed that 25% of the drug was delivered to rabbits in the first 3 h of use, and 66% had been delivered within 72 h [211]. The cellulose acetate film was attached to a CL composed of HEMA-NVP-TRIS. Cellulose is a biopolymer and therefore has high biocompatibility, meaning it is suitable for the ocular environment.

Temperature-responsive polymers, when exposed to sufficient heat, undergo conformation changes, which allows the release of the embedded/encapsulated drug. Once the conformation change is complete, the drug will be released in the highly aqueous environment [212,213]. These papers describe in detail the mechanisms of these polymers [212,213]. This could be relevant to contact lenses, as typically there is a significant difference between the storage/room temperature and the eye. A lens composed of such a polymer would undergo a conformation change on insertion to the eye. This temperature difference could be harnessed for drug-delivery CLs. Polyethylene glycol (PEG) has seen growth in this area as a block-copolymer graft to other moieties, such as poly(*N*-isopropylacrylamide) (p(NIPAAm)) and HEMA. Silicone rubbers have even been mentioned as potential drug-delivery vehicles by Fenton et al. [207]. These materials are known for their ocular compatibility, meaning there is design space to be explored for improved drug-delivery contact lenses. Jung et al. studied more common contact materials, such as HEMA, due to the ability to modify conformation with differing conditions, such as nanoparticle loading into the lens [214].

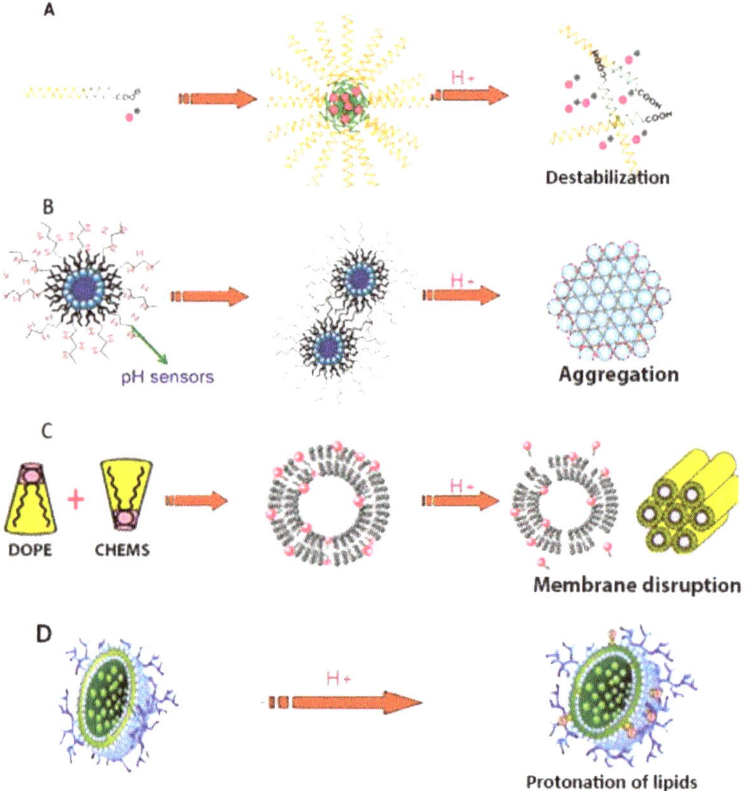

Figure 17. Mechanisms of action for pH-responsive drug-delivery polymers. Reprinted from Biomaterials, 85, Kanamala, M.; Wilson, W.R.; Yang, M.; Palmer, B.D.; Wu, Z. Mechanisms and biomaterials in pH-responsive tumour targeted drug delivery; a review, 152, Copyright 2016, with permission from Elsevier [210].

These emerging materials technologies offer interesting new routes for future CL development. We have provided some approaches, such as novel materials in the form of liquid crystals, or evolving the CL material through methods such as double-network hydrogels. In addition, temperature- and pH-responsive polymers are growing drug-delivery vehicles, but have not yet been widely explored for CLs. What is encouraging is the fact that many of these methods involve many of the same materials used for current CLs, meaning adapting to CLs is a clear possibility in the near future. However, despite these encouraging signs, there are significant challenges for the ocular environment, which remain prominent given the aging population worldwide [215]. This includes specific aspects, such as anti-fungal drug delivery, which are not yet ready for commercial specifications [216]. The biggest challenge for these materials is the efforts in other CL research, as previously discussed.

4. Manufacturing Perspectives

The final part of developing a CL is how the new material will be manufactured. New technologies have emerged to enhance CL properties, such as plasma processing, to introduce hydrophilicity to silicone lenses. CL manufacturing is an ever-evolving area, with manufacturers constantly improving their methods. It is probably fair to say there is much the public does not know about the specific CL manufacturing methodologies. This includes operating conditions, speed of lathes, tip-edge parameters, and exact chemical recipes, amongst a huge number of parameters. For researchers, this can be a difficult area to evaluate, given such lack of information. However, researchers can still analyze the products from manufacturers to help shape the direction of their research. There is literature concerning the differences in the mechanical properties of common HEMA-derived hydrogel lenses based on the manufacturing technique (Figure 18) [49]. The polymerization and manufacturing method were responsible for the differences in the hydrogel properties. Improving the polymerization steps during manufacturing could improve the properties of the lens materials. The manufacturing- or process-driven development is likely pursued for economical or environmental reasons, or both. For example, Nelfilcon A PVA lenses use water as the solvent, which is much easier to handle than organic solvents and does not require further processing [16]. Here, we will discuss some specialist manufacturing techniques seen in other applications of optics that could be used for CL manufacturing. We also discuss some aspects of manufacturing, such as molding, and the materials that have been investigated in recent literature.

Materials are an important consideration for the molding processes, i.e., both cast- and spin-molding. The interaction between the mold surface and polymer solution will affect the surface finish of the CL. Some examples of common molding materials are polypropylene [42] and quartz [16]. Polypropylene is useful due to the ease of handling and high melt temperatures. The polymer is hydrophobic, meaning manufacturers choose this with the intent that the contact lens will not adhere strongly to the mold surface on polymerization (particularly hydrophilic materials). Similarly, quartz (SiO_2) has non-sticking qualities that are useful for manufacturing lenses. This is a practical consideration to reduce defect occurrence in the final CLs. The recent patent literature has many examples of the new materials being considered as a lens mold material [217–220]. These patents are assigned to CLs manufacturers, showing the commercial implications of improving this aspect of manufacturing. Some of the mold materials include alicyclic polymers or polyoxymethylene, which are non-polar and polar, respectively. The mold material can influence the adhesion to the mold, which, in turn, could influence the resulting wettability properties of the hydrogels.

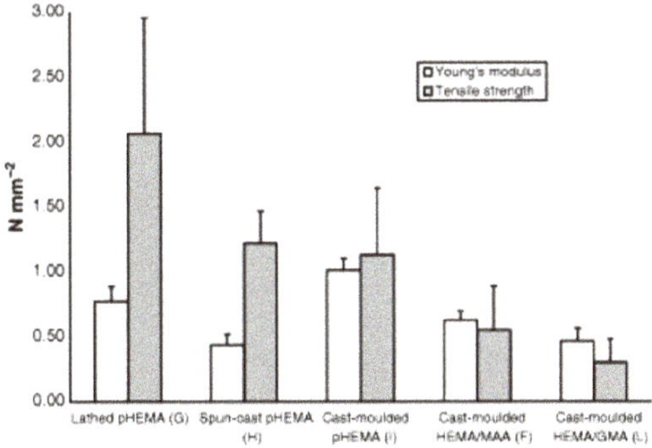

Figure 18. Tensile and Young's modulus data for HEMA-based hydrogel lenses produced by different manufacturing methods. Reprinted with permission from John-Wiley and Sons [49].

Ultra-precision- and nano-manufacturing is an area of research that could be applied to CLs to enhance the physical and optical properties of the material [221–223]. These techniques have been successfully implemented in a number of areas, including biomedical device manufacturing, optics, and more [224]. Ultra-precision lathes are now used in the production of specialty CLs to obtain nanometer control of the surface, which can control the wettability characteristics of polymers [225] and glass surfaces [226]. However, to date, this has not been extensively applied to contact lenses, or at least cannot be seen in the published literature. However, the control of wettability is possible by using machining techniques [227]. Freeform optics are a special type of optical lenses with irregular contouring on a non-symmetrical surface. Improvements in optical performance have seen their use in applications relating to imaging, illumination, and aerospace interests [228,229]. This is an increasingly important area of precision manufacturing, requiring consideration of the design, as well as machining and evaluation aspects [230]. Furthermore, servo-control machining could be feasible to fabricate freeform optical CLs for specialty applications, such as myopia control. Fast servo-manufacturing of freeform optics has even been shown to be feasible on polymers such as polystyrene [231]. This suggests that the use of a similar technique could be employed on a pre-hydrated CL material. The usage of such precision manufacturing techniques to intraocular lens manufacturing was also reported [232]. All of these techniques are subject to the effect which the resulting material will have on the ocular environment, such as the effect on the tear film, protein deposition, optical properties, etc. Finally, there is a cost component to manufacturing in a commercial space that would also need to be considered.

5. Conclusions

The purpose of this review was to bring together state-of-the-art research on contact lenses and their modification. Modern CL research aims to better deliver ophthalmic treatments or improve existing issues with contact lenses. Overall, modern CL materials are an evolution of the well-known lens materials based on HEMA and silicone hydrogels. These hydrogels grafted or incorporated various bioavailable components, such as PEG, HA, chitosan, β-cyclodextrin, cellulose, and other moieties. These biomaterials are very biocompatible due to their inherent chemistry. The biomaterials field is exciting and dynamic, where new and improved biocompatible, drug-delivering materials are constantly being developed. One of the biggest hurdles to overcome is the cost and practicality of incorporating such species to a new lens. We highlighted some emerging fields related to CLs due to the materials involved or which have only recently been applied to CLs. The application of new manufacturing methods can offer other solutions to specific challenges for CLs: new mold materials, surface contouring, etc. CL materials are now resilient and modifiable, and are thus suitable for these improving manufacturing processes. Advances in precision manufacturing, such as the production of freeform optics or modification of surface properties to improve wettability could be applied to future CLs. In summary, future CL materials will continue to push the boundaries of the biocompatibility and materials sciences to better tailor to the needs of a growing CL-using population.

Author Contributions: C.S.A.M. conducted the literature survey and prepared the manuscript. F.F. designed the manuscript structure and finalized the paper.

Funding: This work was supported by the Science Foundation Ireland (Nos. 15/RP/B3208 & 16RC3872) and the National Natural Science Foundation of China (Nos. 51320105009 & 61635008).

Acknowledgments: In this section you can acknowledge any support given which is not covered by the author contribution or funding sections. This may include administrative and technical support, or donations in kind (e.g., materials used for experiments).

References

1. Barr, J. 2004 Annual Report. Available online: https://www.clspectrum.com/issues/2005/january-2005/2004-annual-report (accessed on 13 October 2018).
2. Nichols, J. Contact Lenses 2015. In *Contact Lens Spectrum*; PentaVision LLC: Ambler, PA, USA, 2016; pp. 18–23.
3. Philips, A.J.; Speedwell, L.; Morris, J. (Eds.) *Contact Lenses*, 5th ed.; Butterworth-Heinemann: Oxford, UK, 2007.

4. Efron, N. (Ed.) *Contact Lens Practice*, 3rd ed.; Elsevier: Edinburgh, UK, 2018.

5. Bruce, A.S.; Brennan, N.A.; Lindsay, R.G. Diagnosis and mangement of ocular changes during contact lens wear, Part II. *Clin. Signs Ophthalmol.* **1995**, *17*, 2–11.

6. McMahon, T.T.; Zadnik, K. Twenty-five years of contact lenses: the impact on the cornea and ophthalmic practice. *Cornea* **2000**, *19*, 730–740. [CrossRef] [PubMed]

7. Wichterle, O.; Lím, D. Hydrophilic Gels for Biological Use. *Nature* **1960**, *185*, 117. [CrossRef]

8. Wichterle, O. Method of Manufacturing Soft and Flexible Contact Lenses. U.S. Patent 3498254, 17 February 1970.

9. Lim, D.; Kopecek, J.; Sverinova, H.B.N.; Vacik, J. Hydrophilic N,N-Diethyl Acrylamide Copolymers. U.S. Patent 4074039, 14 February 1978.

10. Yokohama, Y.; Masuhara, E.; Kadoma, Y.; Tarumi, N.; Tsuchiya, M. Soft Contact Lens. U.S. Patent 4650843, 17 March 1987.

11. Mitchell, D.D. Wettable Silicone Resin Optical Devices and Curable Compositions Therefor. U.S. Patent 4487905, 11 December 1984.

12. Chromecek, R.C.; Deichert, W.G.; Falcetta, J.J.; VanBuren, M.F. Polysiloxane/Acrylic Acid/Polycyclic Esters of Methacrylic Acid Polymer Contact Lens. U.S. Patent 4276402, 30 June 1981.

13. Gaylord, N.G. Oxygen-Permeable Contact Lens Composition, Methods and Article of Manufacture. U.S. Patent 3808178, 30 April 1974.

14. Keogh, P.; Kunzler, J.F.; Niu, G.C.C. Hydrophilic Contact Lens Made from Polysiloxanes Which Are Thermally Bonded to Polymerizable Groups and Which Contain Hydrophilic Sidechains. U.S. Patent 4260725, 7 April 1981.

15. Nichols, J. Contact Lenses 2017. In *Contact Lens Spectrum*; PentaVision LLC: Ambler, PA, USA, 2018; pp. 20–25.

16. Buhler, N.; Haerri, H.P.; Hofmann, M.; Irrgang, C.; Muhlebach, A.; Muller, B.; Stockinger, F. Nelfilcon A, a new material for contact lenses. *Chimia (Aarau)* **1999**, *53*, 269–274.

17. Goldenberg, M. Polyoxirane Crosslinked Polyvinyl alcohol Hydrogel Contact Lens. Patent EP0189375B1, 1 June 1994.

18. Kita, M.; Ogura, Y.; Honda, Y.; Hyon, S.-H.; Cha, W.-I.; Ikada, Y. Evaluation of polyvinyl alcohol hydrogel as a soft contact lens material. *Graefe's Arch. Clin. Exp. Ophthalmol.* **1990**, *228*, 533–537. [CrossRef]

19. Imafuku, S. Silicone Hydrogel Soft Contact Lens Having Wettable Surface. U.S. Patent 2014/0362339, 11 December 2014.

20. Bui, T.H.; Cavanagh, H.D.; Robertson, D.M. Patient compliance during contact lens wear: Perceptions, awareness, and behavior. *Eye Contact Lens* **2010**, *36*, 334–339. [CrossRef] [PubMed]

21. Kirchhof, S.; Goepferich, A.M.; Brandl, F.P. Hydrogels in ophthalmic applications. *Eur. J. Pharm. Biopharm.* **2015**, *95*, 227–238. [CrossRef]

22. Holden, B.A.; Fricke, T.R.; Wilson, D.A.; Jong, M.; Naidoo, K.S.; Sankaridurg, P.; Wong, T.Y.; Naduvilath, T.J.; Resnikoff, S. Global Prevalence of Myopia and High Myopia and Temporal Trends from 2000 through 2050. *Ophthalmology* **2016**, *123*, 1036–1042. [CrossRef]

23. Walline, J.J.; Jones, L.A.; Sinnott, L.; Manny, R.E.; Gaume, A.; Rah, M.J.; Chitkara, M.; Lyons, S. A randomized trial of the effect of soft contact lenses on myopia progression in children. *Investig. Ophthalmol. Vis. Sci.* **2008**, *49*, 4702–4706. [CrossRef] [PubMed]

24. Sankaridurg, P. Contact lenses to slow progression of myopia. *Clin. Exp. Optom.* **2017**, *100*, 432–437. [CrossRef]

25. Peng, C.C.; Kim, J.; Chauhan, A. Extended delivery of hydrophilic drugs from silicone-hydrogel contact lenses containing Vitamin E diffusion barriers. *Biomaterials* **2010**, *31*, 4032–4047. [CrossRef]

26. Ciolino, J.B.; Hoare, T.R.; Iwata, N.G.; Behlau, I.; Dohlman, C.H.; Langer, R.; Kohane, D.S. A drug-eluting contact lens. *Investig. Ophthalmol. Vis. Sci.* **2009**, *50*, 3346–3352. [CrossRef]

27. Xu, J.; Xue, Y.; Hu, G.; Lin, T.; Gou, J.; Yin, T.; He, H.; Zhang, Y.; Tang, X. A comprehensive review on contact lens for ophthalmic drug delivery. *J. Control. Release* **2018**, *281*, 97–118. [CrossRef] [PubMed]

28. Hsu, K.H.; Gause, S.; Chauhan, A. Review of ophthalmic drug delivery by contact lenses. *J. Drug Deliv. Sci. Technol.* **2014**, *24*, 123–135. [CrossRef]

29. Carvalho, I.M.; Marques, C.S.; Oliveira, R.S.; Coelho, P.B.; Costa, P.C.; Ferreira, D.C. Sustained drug release by contact lenses for glaucoma treatment - A review. *J. Control. Release* **2015**, *202*, 76–82. [CrossRef]

30. Efron, N. *Contact lens Complications*, 3rd ed.; Elsevier: Amsterdam, The Netherlands, 2012; ISBN 9780702042690.

31. Phillips, A.J.; Speedwell, L. *Contact Lenses*; Butterworth-Heinemann: Oxford, UK, 1997; ISBN 9780750618199.

32. Nicolson, P.C.; Vogt, J. Soft contact lens polymers: An evolution. *Biomaterials* **2001**, *22*, 3273–3283. [CrossRef]

33. Alderson, A. Soft Contact Lens Material Properties. *Bausch Lomb Acad. Vis. Care* **2008**, 1–10. Available online: http://www.academyofvisioncare.com/files/documents/Contact%20Lens%20Article%20080513.pdf (accessed on 3 October 2018).

34. Caló, E.; Khutoryanskiy, V.V. Biomedical applications of hydrogels: A review of patents and commercial products. *Eur. Polym. J.* **2015**, *65*, 252–267. [CrossRef]

35. Vidal-Rohr, M.; Wolffsohn, J.S.; Davies, L.N.; Cerviño, A. Effect of contact lens surface properties on comfort, tear stability and ocular physiology. *Contact Lens Anterior Eye* **2018**, *41*, 117–121. [CrossRef] [PubMed]

36. Ciolino, J.B.; Stefanescu, C.F.; Ross, A.E.; Salvador-Culla, B.; Cortez, P.; Ford, E.M.; Wymbs, K.A.; Sprague, S.L.; Mascoop, D.R.; Rudina, S.S.; et al. In vivo performance of a drug-eluting contact lens to treat glaucoma for a month. *Biomaterials* **2014**, *35*, 432–439. [CrossRef]

37. Lee, D.; Cho, S.; Park, H.S.; Kwon, I. Ocular drug delivery through pHEMA-Hydrogel contact lenses Co-loaded with lipophilic vitamins. *Sci. Rep.* **2016**, *6*, 1–8. [CrossRef]

38. Xinming, L.; Yingde, C.; Lloyd, A.W.; Mikhalovsky, S.V.; Sandeman, S.R.; Howel, C.A.; Liewen, L. Polymeric hydrogels for novel contact lens-based ophthalmic drug delivery systems: A review. *Contact Lens Anterior Eye* **2008**, *31*, 57–64. [CrossRef] [PubMed]

39. Perkins, W.G.; Capiati, N.J.; Porter, R.S. The effect of molecular weight on the physical and mechanical properties of ultra-drawn high density polyethylene. *Polym. Eng. Sci.* **1976**, *16*, 200–203. [CrossRef]

40. Rosa dos Santos, J.F.; Alvarez-Lorenzo, C.; Silva, M.; Balsa, L.; Couceiro, J.; Torres-Labandeira, J.J.; Concheiro, A. Soft contact lenses functionalized with pendant cyclodextrins for controlled drug delivery. *Biomaterials* **2009**, *30*, 1348–1355. [CrossRef] [PubMed]

41. Seinder, L.; Spinelli, H.J.; Ali, M.I.; Weintraub, L. Silicone-Containing Contact Lens Polymers, Oxygen Permeable Contact Lenses and Methods for Making There Lenses and Treating Patients with Visual Impairment. U.S. Patent 5331067, 19 July 1994.

42. Hahn, D.; Johansson, G.A.; Ruscio, D.V.; Blank, C.E. Method for Manufacturing Hydrophilic Contact Lenses. U.S. Patent 4983332, 1 January 1991.

43. Tanaka, K.; Takahashi, K.; Kanada, M.; Kanome, S.; Nakajima, T. Copolymer for Soft Contact Lens, Its Preparation and Soft Contact Lens Made Thereof. U.S. Patent 4139513, 13 February 1979.

44. Keeley, E.M. Method of Manufacturing Soft Contact Lens Buttons. U.S. Patent 4931228, 5 June 1990.

45. Lin, C.H.; Yeh, Y.H.; Lin, W.C.; Yang, M.C. Novel silicone hydrogel based on PDMS and PEGMA for contact lens application. *Colloids Surf. B Biointerfaces* **2014**, *123*, 986–994. [CrossRef]

46. Zhao, J.; Mayumi, K.; Creton, C.; Narita, T. Rheological properties of tough hydrogels based on an associating polymer with permanent and transient crosslinks: Effects of crosslinking density. *J. Rheol. (N. Y.)* **2017**, *61*, 1371–1383. [CrossRef]

47. Narita, T.; Mayumi, K.; Ducouret, G.; Hébraud, P. Viscoelastic properties of poly(vinyl alcohol) hydrogels having permanent and transient cross-links studied by microrheology, classical rheometry, and dynamic light scattering. *Macromolecules* **2013**, *46*, 4174–4183. [CrossRef]

48. Musgrave, C.S.A.; Nazarov, W.; Bazin, N. The effect of para-divinyl benzene on styrenic emulsion-templated porous polymers: A chemical Trojan horse. *J. Mater. Sci.* **2017**, *52*, 3179–3187. [CrossRef]

49. Maldonado-Codina, C.; Efron, N. Impact of manufacturing technology and material composition on the mechanical properties of hydrogel contact lenses. *Ophthalmic Physiol. Opt.* **2004**, *24*, 551–561. [CrossRef]

50. Ashby, M.F. The properties of foams and lattices. *Philos. Trans. R. Soc. A Math. Phys. Eng. Sci.* **2006**, *364*, 15–30. [CrossRef]

51. Luo, Y.; Wang, A.-N.; Gao, X. Pushing the mechanical strength of PolyHIPEs up to the theoretical limit through living radical polymerization. *Soft Matter* **2012**, *8*, 1824–1830. [CrossRef]

52. Dixon, P.; Shafor, C.; Gause, S.; Hsu, K.-H.; Powell, K.C.; Chauhan, A. Therapeutic contact lenses: A patent review. *Expert Opin. Ther. Pat.* **2015**, *25*, 1117–1129. [CrossRef] [PubMed]

53. Fairbanks, B.D.; Gunatillake, P.A.; Meagher, L. Biomedical applications of polymers derived by reversible addition—Fragmentation chain-transfer (RAFT). *Adv. Drug Deliv. Rev.* **2015**, *91*, 141–152. [CrossRef] [PubMed]

54. Pai, T.S.C.; Barner-Kowollik, C.; Davis, T.P.; Stenzel, M.H. Synthesis of amphiphilic block copolymers based on poly(dimethylsiloxane) via fragmentation chain transfer (RAFT) polymerization. *Polymer (Guildf.)* **2004**, *45*, 4383–4389. [CrossRef]

55. Abdollahi, E.; Khalafi-Nezhad, A.; Mohammadi, A.; Abdouss, M.; Salami-Kalajahi, M. Synthesis of new molecularly imprinted polymer via reversible addition fragmentation transfer polymerization as a drug delivery system. *Polymer (U. K.)* **2018**, *143*, 245–257. [CrossRef]

56. Junkers, T.; Lovestead, T.M.; Barner-Kowollik, C. The RAFT process as a kinetic tool: Accessing fundamental parameters of free radical polymerization. In *Handbook of RAFT Polymerization*; Barner-Kowollik, C., Ed.; Wiley-VCH: Weinheim, Germany, 2008; pp. 105–144.

57. Krstina, J.; Moad, G.; Rizzardo, E.; Winzor, C.L.; Berge, C.T.; Fryd, M. Narrow Polydispersity Block Copolymers by Free-Radical Polymerization in the Presence of Macromonomers. *Macromolecules* **1995**, *28*, 5381–5385. [CrossRef]

58. Bivigou-Koumba, A.M.; Görnitz, E.; Laschewsky, A.; Müller-Buschbaum, P.; Papadakis, C.M. Thermoresponsive amphiphilic symmetrical triblock copolymers with a hydrophilic middle block made of poly(N-isopropylacrylamide): Synthesis, self-organization, and hydrogel formation. *Colloid Polym. Sci.* **2010**, *288*, 499–517. [CrossRef]

59. Hemp, S.T.; Smith, A.E.; Bunyard, W.C.; Rubinstein, M.H.; Long, T.E. RAFT polymerization of temperature- and salt-responsive block copolymers as reversible hydrogels. *Polymer (Guildf.)* **2014**, *55*, 2325–2331. [CrossRef]

60. Liu, J.; Cui, L.; Kong, N.; Barrow, C.J.; Yang, W. RAFT controlled synthesis of graphene/polymer hydrogel with enhanced mechanical property for pH-controlled drug release. *Eur. Polym. J.* **2014**, *50*, 9–17. [CrossRef]

61. Guan, C.M.; Luo, Z.H.; Qiu, J.J.; Tang, P.P. Novel fluorosilicone triblock copolymers prepared by two-step RAFT polymerization: Synthesis, characterization, and surface properties. *Eur. Polym. J.* **2010**, *46*, 1582–1593. [CrossRef]

62. Zhang, C.; Liu, Z.; Wang, H.; Feng, X.; He, C. Novel Anti-Biofouling Soft Contact Lens: L -Cysteine Conjugated Amphiphilic Conetworks via RAFT and Thiol—Ene Click Chemistry. *Macromol. Biosci.* **2017**, *17*, 1600444. [CrossRef]

63. Ozgen, O.; Hasirci, N. Modification of Poly(methyl methacrylate) Surfaces with Oxygen, Nitrogen and Argon Plasma. *J. Biomater. Tissue Eng.* **2014**, *4*, 479–487. [CrossRef]

64. Alió, J.L.; Belda, J.I.; Artola, A.; García-Lledó, M.; Osman, A. Contact lens fitting to correct irregular astigmatism after corneal refractive surgery. *J. Cataract Refract. Surg.* **2002**, *28*, 1750–1757. [CrossRef]

65. Thean, J.H.J.; Mcnab, A.A. Blepharoptosis in RGP and PMMA hard contact lens wearers. *Clin. Exp. Optom.* **2004**, *87*, 11–14. [CrossRef]

66. Subbaraman, L.N.; Glasier, M.-A.; Senchyna, M.; Sheardown, H.; Jones, L. Kinetics of In Vitro Lysozyme Deposition on Silicone Hydrogel, PMMA, and FDA Groups I, II, and IV Contact Lens Materials. *Curr. Eye Res.* **2006**, *31*, 787–796. [CrossRef]

67. Wang, J. Topographical Thickness of the Epithelium and Total Cornea after Hydrogel and PMMA Contact Lens Wear with Eye Closure. *Investig. Ophthalmol. Vis. Sci.* **2003**, *44*, 1070–1074. [CrossRef]

68. Asgharzadeh Shishavan, A.; Nordin, L.; Tjossem, P.; Abramoff, M.D.; Toor, F. PMMA-based ophthalmic contact lens for vision correction of strabismus. *Proc. SPIE* **2016**, *9918*, 99180C.

69. Li, S.; Toprak, M.S.; Jo, Y.S.; Dobson, J.; Kim, D.K.; Muhammed, M. Bulk Synthesis of Transparent and Homogeneous Polymeric Hybrid Materials with ZnO Quantum Dots and PMMA. *Adv. Mater.* **2007**, *19*, 4347–4352. [CrossRef]

70. Koruga, D.; Stamenković, D.; Djuricic, I.; Mileusnic, I.; Šakota, J.; Bojović, B.; Golubović, Z. Nanophotonic Rigid Contact Lenses: Engineering and Characterization. *Adv. Mater. Res.* **2013**, *633*, 239–252. [CrossRef]

71. Tomić, M.; Munćan, J.; Stamenković, D.; Jokanović, M.; Matija, L. Biocompatibility and cytotoxicity study of nanophotonic rigid gas permeable contact lens material. *J. Phys. Conf. Ser.* **2013**, *429*. [CrossRef]

72. Mitrović, A.D.; Miljković, V.M.; Popović, D.P.; Koruga, D.L. Mechanical properties of nanophotonic soft contact lenses based on poly (2-hydrohzethil methacrylate) and fullerenes. *Struct. Integr. Life* **2016**, *16*, 3–6.

73. Van der Worp, E.; Bornman, D.; Ferreira, D.L.; Faria-Ribeiro, M.; Garcia-Porta, N.; González-Meijome, J.M. Modern scleral contact lenses: A review. *Contact Lens Anterior Eye* **2014**, *37*, 240–250. [CrossRef] [PubMed]

74. Ko, J.; Cho, K.; Han, S.W.; Sung, H.K.; Baek, S.W.; Koh, W.-G.; Yoon, J.S. Hydrophilic surface modification of poly(methyl methacrylate)-based ocular prostheses using poly(ethylene glycol) grafting. *Colloids Surf. B Biointerfaces* **2017**, *158*, 287–294. [CrossRef] [PubMed]

75. Efron, N. Obituary—Rigid contact lenses. *Contact Lens Anterior Eye* **2010**, *33*, 245–252. [CrossRef] [PubMed]

76. Haluk, E.; Nazan, E. Corneal endothelial polymegethism and pleomorphism induced by daily-wear rigid gas-permeably contact lenses. *CLAO J.* **2002**, *28*, 40–43.

77. Shokrollahzadeh, F.; Hashemi, H.; Jafarzadehpur, E.; Mirzajani, A.; Khabazkhoob, M.; Yekta, A.; Asgari, S. Corneal aberration changes after rigid gas permeable contact lens wear in keratokonic patients. *J. Curr. Ophthalmol.* **2016**, *28*, 194–198. [CrossRef] [PubMed]

78. Yuksel Elgin, C.; Iskeleli, G.; Aydin, O. Effects of the rigid gas permeable contact lense use on tear and ocular surface among keratoconus patients. *Contact Lens Anterior Eye* **2018**, *41*, 273–276. [CrossRef]

79. Eggink, F.A.G.J.; Beekhuis, W.H.; Nuijts, R.M.M.A. Rigid gas-permeable contact lens fitting in LASIK patients for the correction of multifocal corneas. *Graefe's Arch. Clin. Exp. Ophthalmol.* **2001**, *239*, 361–366. [CrossRef]

80. Vincent, S.J.; Alonso-Caneiro, D.; Collins, M.J. Corneal changes following short-term miniscleral contact lens wear. *Contact Lens Anterior Eye* **2014**, *37*, 461–468. [CrossRef] [PubMed]

81. Lai, Y.-C.; Bonafini, J.A., Jr. Rigid Gas Permeable Lens Material. U.S. Patent 7344731B2, 18 March 2008.

82. Ruan, J.L.; Chen, C.; Shen, J.H.; Zhao, X.L.; Qian, S.H.; Zhu, Z.G. A gelated colloidal crystal attached lens for noninvasive continuous monitoring of tear glucose. *Polymers* **2017**, *9*, 125. [CrossRef]

83. Tseng, R.C.; Chen, C.C.; Hsu, S.M.; Chuang, H.S. Contact-lens biosensors. *Sensors* **2018**, *18*, 2651. [CrossRef]

84. Shtukater, A. Smart Contact Lens with Orientation Sensor. U.S. Patent 2017/0371184, 28 December 2017.

85. Kim, T.; Hwang, S.; Kim, S.; Ahn, H.; Chung, D. Smart Contact Lenses for Augmented Reality and Methods of Manufacturing and Operating the Same. U.S. Patent 2016/0091737A1, 31 March 2016.

86. Kim, J.; Kim, M.; Lee, M.S.; Kim, K.; Ji, S.; Kim, Y.T.; Park, J.; Na, K.; Bae, K.H.; Kim, H.K.; et al. Wearable smart sensor systems integrated on soft contact lenses for wireless ocular diagnostics. *Nat. Commun.* **2017**, *8*, 14997. [CrossRef]

87. Maldonado-Codina, C.; Efron, N. Dynamic wettability of pHEMA-based hydrogel contact lenses. *Ophthalmic Physiol. Opt.* **2006**, *26*, 408–418. [CrossRef] [PubMed]

88. Seo, E.; Kumar, S.; Lee, J.; Jang, J.; Park, J.H.; Chang, M.C.; Kwon, I.; Lee, J.S.; Huh, Y. il Modified hydrogels based on poly(2-hydroxyethyl methacrylate) (pHEMA) with higher surface wettability and mechanical properties. *Macromol. Res.* **2017**, *25*, 704–711. [CrossRef]

89. Tranoudis, I.; Efron, N. Parameter stability of soft contact lenses made from different materials. *Contact Lens Anterior Eye* **2004**, *27*, 115–131. [CrossRef]

90. Tranoudis, I.; Efron, N. Water properties of soft contact lens materials. *Contact Lens Anterior Eye* **2004**, *27*, 193–208. [CrossRef] [PubMed]

91. Tranoudis, I.; Efron, N. Tensile properties of soft contact lens materials. *Contact Lens Anterior Eye* **2004**, *27*, 177–191. [CrossRef] [PubMed]

92. Woźniak-Braszak, A.; Kaźmierczak, M.; Baranowski, M.; Hołderna-Natkaniec, K.; Jurga, K. The aging process of hydrogel contact lenses studied by 1H NMR and DSC methods. *Eur. Polym. J.* **2016**, *76*, 135–146. [CrossRef]

93. Lord, M.S.; Stenzel, M.H.; Simmons, A.; Milthorpe, B.K. The effect of charged groups on protein interactions with poly(HEMA) hydrogels. *Biomaterials* **2006**, *27*, 567–575. [CrossRef]

94. Luensmann, D.; Jones, L. Protein deposition on contact lenses: The past, the present, and the future. *Contact Lens Anterior Eye* **2012**, *35*, 53–64. [CrossRef]

95. Borazjani, R.N.; Levy, B.; Ahearn, D.G. Relative primary adhesion of Pseudomonas aeruginosa, Serratia marcescens and Staphylococcus aureus to HEMA-type contact lenses and an extended wear silicone hydrogel contact lens of high oxygen permeability. *Contact Lens Anterior Eye* **2004**, *27*, 3–8. [CrossRef]

96. Szczotka-flynn, L.B.; Imamura, Y.; Chandra, J.; Yu, C.; Mukherjee, P.K.; Pearlman, E.; Ghannoum, M.A. Increased resistance of contact lens related bacterial biofilms to antimicrobial activity of soft contact lens care solutions. *Cornea* **2009**, *28*, 918–926. [CrossRef] [PubMed]

97. Dutta, D.; Cole, N.; Willcox, M. Factors influencing bacterial adhesion to contact lenses. *Mol. Vis.* **2012**, *18*, 14–21. [PubMed]

98. Xiao, A.; Dhand, C.; Leung, C.M.; Beuerman, R.W.; Ramakrishna, S.; Lakshminarayanan, R. Strategies to design antimicrobial contact lenses and contact lens cases. *J. Mater. Chem. B* **2018**, *6*, 2171–2186. [CrossRef]

99. Jung, Y.P.; Kim, J.; Lee, D.S.; Kim, Y.H. Preparation and Properties of Modified PHEMA Hydrogel with Sulfonated PEG Graft. *J. Appl. Polym. Sci.* **2007**, *104*, 2484–2489. [CrossRef]

100. Dutta, D.; Vijay, A.K.; Kumar, N.; Willcox, M.D.P. Melimine-Coated Antimicrobial Contact Lenses Reduce Microbial Keratitis in an Animal Model. *Investig. Ophthalmol. Vis. Sci.* **2016**, *57*, 5616–5624. [CrossRef]

101. Dutta, D.; Ozkan, J.; Willcox, M.D.P. Biocompatibility of Antimicrobial Melimine lenses: Rabbit and Human studies. *Optom. Vis. Sci.* **2014**, *91*, 570–581. [CrossRef]

102. Shaynai Rad, M.; Khameneh, M.; Mohajeri, S.A.; Fazly Bazzaz, B.S. Antibacterial activity of silver nanoparticle-loaded soft contact lens materails: the effect of monomer composition. *Curr. Eye Res.* **2016**, *41*, 1286–1293. [CrossRef]

103. Willcox, M.D.P.; Hume, E.B.H.; Aliwarga, Y.; Kumar, N.; Cole, N. A novel cationic-peptide coating for the prevention of microbial colonization on contact lenses. *J. Appl. Microbiol.* **2008**, *105*, 1817–1825. [CrossRef]

104. Malakooti, N.; Alexander, C.; Alvarez-lorenzo, C. Imprinted Contact Lenses for Sustained Release of Polymyxin B and Related Antimicrobial Peptides. *J. Pharm. Sci.* **2015**, *104*, 3386–3394. [CrossRef]

105. Sato, T.; Uchida, R.; Tanigawa, H.; Uno, K.; Murakami, A. Application of polymer gels containing side-chain phosphate groups to drug-delivery contact lenses. *J. Appl. Polym. Sci.* **2005**, *98*, 731–735. [CrossRef]

106. Bengani, L.C.; Scheiffele, G.W.; Chauhan, A. Incorporation of polymerizable surfactants in hydroxyethyl methacrylate lenses for improving wettability and lubricity. *J. Colloid Interface Sci.* **2015**, *445*, 60–68. [CrossRef] [PubMed]

107. Chalmers, R. Overview of factors that affect comfort with modern soft contact lenses. *Contact Lens Anterior Eye* **2014**, *37*, 65–76. [CrossRef] [PubMed]

108. Sulley, A.; Young, G.; Hunt, C. Factors in the success of new contact lens wearers. *Contact Lens Anterior Eye* **2017**, *40*, 15–24. [CrossRef] [PubMed]

109. Dumbleton, K.; Woods, C.A.; Jones, L.; Fonn, D. The impact of contemporary contact lenses on contact lens discontinuation. *Eye Contact Lens Sci. Clin. Pract.* **2013**, *39*, 93–99. [CrossRef] [PubMed]

110. Kapoor, Y.; Thomas, J.C.; Tan, G.; John, V.T.; Chauhan, A. Surfactant-laden soft contact lenses for extended delivery of ophthalmic drugs. *Biomaterials* **2009**, *30*, 867–878. [CrossRef] [PubMed]

111. Maulvi, F.A.; Desai, A.R.; Choksi, H.H.; Patil, R.J.; Ranch, K.M.; Vyas, B.A.; Shah, D.O. Effect of surfactant chain length on drug release kinetics from microemulsion-laden contact lenses. *Int. J. Pharm.* **2017**, *524*, 193–204. [CrossRef] [PubMed]

112. Bengani, L.C.; Chauhan, A. Extended delivery of an anionic drug by contact lens loaded with a cationic surfactant. *Biomaterials* **2013**, *34*, 2814–2821. [CrossRef]

113. Kapoor, Y.; Bengani, L.C.; Tan, G.; John, V.; Chauhan, A. Aggregation and transport of Brij surfactants in hydroxyethyl methacrylate hydrogels. *J. Colloid Interface Sci.* **2013**, *407*, 390–396. [CrossRef]

114. Sahoo, R.K.; Biswas, N.; Guha, A.; Sahoo, N.; Kuotsu, K. Nonionic surfactant vesicles in ocular delivery: Innovative approaches and perspectives. *Biomed Res. Int.* **2014**, *2014*. [CrossRef]

115. Wolffsohn, J.; Hall, L.; Mroczkowska, S.; Hunt, O.A.; Bilkhu, P.; Drew, T.; Sheppard, A. The influence of end of day silicone hydrogel daily disposable contact lens fit on ocular comfort, physiology and lens wettability. *Contact Lens Anterior Eye* **2015**, *38*, 339–344. [CrossRef]

116. Stapleton, F.; Stretton, S.; Papas, E.; Skotnitsky, C.; Sweeney, D.F. Silicone hydrogel contact lenses and the ocular surface. *Ocul. Surf.* **2006**, *4*, 24–43. [CrossRef]

117. Keir, N.; Jones, L. Wettability and Silicone hydrogel lenses: A review. *Eye Contact Lens Sci. Clin. Pract.* **2013**, *39*, 100–108. [CrossRef]

118. Chen, C.; Hong, Y.; Manesis, N. Wettable Silicone Hydrogel Contact Lenses and Related Compositions and Methods. U.S. Patent 7572841, 11 August 2009.

119. Steffen, R.; Mccabe, K.; Turner, D.; Alli, A.; Young, K.; Schnider, C.; Hill, G.A. Soft Contact Lenses Displaying Superior on-Eye Comfort. U.S. Patent 7461937B2, 9 December 2008.

120. Valint, P.L.; Ammon, D.M.; McGee, J.A.; Grobe, G.L.; Ozark, R.M. Surface Treatment for Silicone Hydrogel Contact Lenses Comprising Hydrophilic Polymer Chains Attached to an Intermediate Carbon Coating. U.S. Patent 6902812, 7 June 2005.

121. Valint, P.L.; Grobe, G.L.; Ammon, D.M.; Moorehead, M.J. Plasma Surface Treatment of Silicone Hydrogel Contact Lenses. U.S. Patent 6193369B1, 27 Februry 2001.

122. Iwata, J.; Hoki, T.; Ikawa, S.; Back, A. Silicone Hydrogel Contact Lens. U.S. Patent 2006/0063852A1, 23 March 2006.

123. Lin, M.C.; Yeh, T.N. Mechanical complications induced by silicone hydrogel contact lenses. *Eye Contact Lens* **2013**, *39*, 115–124. [CrossRef] [PubMed]

124. Santos, L.; Rodrigues, D.; Lira, M.; Oliveira, M.E.C.D.R.; Oliveira, R.; Vilar, E.Y.P.; Azeredo, J. The influence of surface treatment on hydrophobicity, protein adsorption and microbial colonisation of silicone hydrogel contact lenses. *Contact Lens Anterior Eye* **2007**, *30*, 183–188. [CrossRef] [PubMed]

125. Bhamla, M.S.; Nash, W.L.; Elliott, S.; Fuller, G.G. Influence of lipid coatings on surface wettability characteristics of silicone hydrogels. *Langmuir* **2015**, *31*, 3820–3828. [CrossRef] [PubMed]

126. Lin, C.-H.; Cho, H.-L.; Yeh, Y.-H.; Yang, M.-C. Improvement of the surface wettability of silicone hydrogel contact lenses via layer-by-layer self-assembly technique. *Colloids Surf. B Biointerfaces* **2015**, *136*, 735–743. [CrossRef]

127. Tian, L.; Wang, X.; Qi, J.; Yao, Q.; Oderinde, O.; Yao, C.; Song, W.; Shu, W.; Chen, P.; Wang, Y. Improvement of the surface wettability of silicone hydrogel films by self- assembled hydroxypropyltrimethyl ammonium chloride chitosan mixed colloids. *Colloids Surf. A* **2018**, *558*, 422–428. [CrossRef]

128. Thissen, H.; Gengenbach, T.; du Toit, R.; Sweeney, D.F.; Kingshott, P.; Griesser, H.J.; Meagher, L. Clinical observations of biofouling on PEO coated silicone hydrogel contact lenses. *Biomaterials* **2010**, *31*, 5510–5519. [CrossRef] [PubMed]

129. Dutta, D.; Kamphuis, B.; Ozcelik, B.; Thissen, H.; Pinarbasi, R.; Kumar, N.; Willcox, M.D.P. Development of silicone hydrogel antimicrobial contact lenses with Mel4 peptide coating. *Optom. Vis. Sci.* **2018**, *95*, 937–946. [CrossRef]

130. Dutta, D.; Cole, N.; Kumar, N.; Willcox, M.D.P. Broad spectrum antimicrobial activity of melimine covalently bound to contact lenses. *Investig. Ophthalmol. Vis. Sci.* **2013**, *54*, 175–182. [CrossRef]

131. Tuby, R.; Gutfreund, S.; Perelshtein, I.; Mircus, G.; Ehrenberg, M.; Mimouni, M.; Gedanken, A.; Bahar, I. Fabrication of a Stable and Efficient Antibacterial Nanocoating of Zn-CuO on Contact Lenses. *Chem. Nano Mater.* **2016**, *2*, 547–551. [CrossRef]

132. Jung, H.J.; Abou-Jaoude, M.; Carbia, B.E.; Plummer, C.; Chauhan, A. Glaucoma therapy by extended release of timolol from nanoparticle loaded silicone-hydrogel contact lenses. *J. Control. Release* **2013**, *165*, 82–89. [CrossRef] [PubMed]

133. Willis, S.L.; Court, J.L.; Redman, R.P.; Wang, J.H.; Leppard, S.W.; O'Byrne, V.J.; Small, S.A.; Lewis, A.L.; Jones, S.A.; Stratford, P.W. A novel phosphorylcholine-coated contact lens for extended wear use. *Biomaterials* **2001**, *22*, 3261–3272. [CrossRef]

134. Wang, J.; Li, X. Interpenetrating polymer network hydrogels based on silicone and poly(2-methacryloyloxyethyl phosphorylcholine). *Polym. Adv. Technol.* **2011**, *22*, 2091–2095. [CrossRef]

135. Wang, J.J.; Liu, F. Photoinduced graft polymerization of 2-methacryloyloxyethyl phosphorylcholine on silicone hydrogels for reducing protein adsorption. *J. Mater. Sci. Mater. Med.* **2011**, *22*, 2651–2657. [CrossRef] [PubMed]

136. Wang, J.J.; Liu, F. Imparting Antifouling Properties of Silicone Hydrogels by Grafting Poly (ethylene glycol) Methyl Ether Acrylate Initiated by UV Light. *J. Appl. Polym. Sci.* **2012**, *125*, 548–554. [CrossRef]

137. Kingshott, P.; Thissen, H.; Griesser, H.J. Effects of cloud-point grafting, chain length, and density of PEG layers on competitive adsorption of ocular proteins. *Biomaterials* **2002**, *23*, 2043–2056. [CrossRef]

138. Kingshott, P.; Wei, J.; Bagge-Ravn, D.; Gadegaard, N.; Gram, L. Covalent attachment of poly(ethylene glycol) to surfaces, critical for reducing bacterial adhesion. *Langmuir* **2003**, *19*, 6912–6921. [CrossRef]

139. Chen, S.; Li, L.; Zhao, C.; Zheng, J. Surface hydration: Principles and applications toward low-fouling/nonfouling biomaterials. *Polymer (Guildf.)* **2010**, *51*, 5283–5293. [CrossRef]

140. Baker, M.I.; Walsh, S.P.; Schwartz, Z.; Boyan, B.D. A review of polyvinyl alcohol and its uses in cartilage and orthopedic applications. *J. Biomed. Mater. Res. Part B Appl. Biomater.* **2012**, *100B*, 1451–1457. [CrossRef]

141. Winterton, L.C. Contact Lenses with Improved Wearing Comfort. U.S. Patent US8030369B2, 4 October 2011.

142. Winterton, L.C.; Lally, J.M.; Sentell, K.B.; Chapoy, L.L. The Elution of Poly (vinyl alcohol) From a Contact Lens: The Realization of a Time Release Moisturizing Agent/Artificial Tear. *J. Biomed. Mater. Res. Part B Appl. Biomater.* **2007**, *80B*, 424–432. [CrossRef] [PubMed]

143. Francis, C.A.; Turek, R.C.; Keeley, D.E. Contact Lens with PVA Cover Layer. U.S. Patent 6890075B2, 10 May 2005.

144. Peterson, R.C.; Wolffsohn, J.S.; Nick, J.; Winterton, L.; Lally, J. Clinical performance of daily disposable soft contact lenses using sustained release technology. *Contact Lens Anterior Eye* **2006**, *29*, 127–134. [CrossRef]

145. Hou, Y.; Chen, C.; Liu, K.; Tu, Y.; Zhang, L.; Li, Y. Preparation of PVA hydrogel with high-transparence and investigations of its transparent mechanism. *RSC Adv.* **2015**, *5*, 24023–24030. [CrossRef]

146. Xu, J.; Li, X.; Sun, F.; Cao, P. PVA Hydrogels Containing β-Cyclodextrin for Enhanced Loading and Sustained Release of Ocular Therapeutics. *J. Biomater. Sci. Polym. Ed.* **2010**, *21*, 1023–1038. [CrossRef] [PubMed]

147. Sun, X.; Yu, Z.; Cai, Z.; Yu, L.; Lv, Y. Voriconazole Composited Polyvinyl Alcohol/Hydroxypropyl-β-Cyclodextrin Nanofibers for Ophthalmic Delivery. *PLoS ONE* **2016**, *11*, e0167961. [CrossRef]

148. Tummala, G.K.; Rojas, R.; Mihranyan, A. Poly(vinyl alcohol) Hydrogels Reinforced with Nanocellulose for Ophthalmic Applications: General Characteristics and Optical Properties. *J. Phys. Chem. B* **2016**, *120*, 13094–13101. [CrossRef]

149. Mihranyan, A. Viscoelastic properties of cross-linked polyvinyl alcohol and surface-oxidized cellulose whisker hydrogels. *Cellulose* **2013**, *20*, 1369–1376. [CrossRef]

150. Klemm, D.; Heublein, B.; Fink, H.P.; Bohn, A. Cellulose: Fascinating biopolymer and sustainable raw material. *Angew. Chem. Int. Ed.* **2005**, *44*, 3358–3393. [CrossRef] [PubMed]

151. Chang, P.S.; Robyt, J.F. Oxidation of Primary Alcohol Groups of Naturally Occurring Polysaccharides with 2,2,6,6-Tetramethyl-1-Piperidine Oxoammonium Ion. *J. Carbohydr. Chem.* **1996**, *15*, 819–830. [CrossRef]

152. Ubholz, B.; Chröder, S.V.E.N.S.; Ihranyan, A.L.M. Light scattering in poly (vinyl alcohol) hydrogels reinforced with nanocellulose for ophthalmic use. *Opt. Mater. Express* **2017**, *7*, 2824–2837.

153. Collins, M.N.; Birkinshaw, C. Hyaluronic acid based scaffolds for tissue engineering—A review. *Carbohydr. Polym.* **2013**, *92*, 1262–1279. [CrossRef]

154. White, C.J.; Thomas, C.R.; Byrne, M.E. Bringing comfort to the masses: A novel evaluation of comfort agent solution properties. *Contact Lens Anterior Eye* **2014**, *37*, 81–91. [CrossRef]

155. Rah, M.J. A review of hyaluronan and its ophthalmic applications. *Optometry* **2011**, *82*, 38–43. [CrossRef] [PubMed]

156. Stach, S.; Ţălu, Ş.; Trabattoni, S.; Tavazzi, S.; Głuchaczka, A.; Siek, P.; Zając, J.; Giovanzana, S. Morphological Properties of Siloxane-Hydrogel Contact Lens Surfaces. *Curr. Eye Res.* **2017**, *42*, 498–505. [CrossRef] [PubMed]

157. Salzillo, R.; Schiraldi, C.; Corsuto, L.; D'Agostino, A.; Filosa, R.; De Rosa, M.; La Gatta, A. Optimization of hyaluronan-based eye drop formulations. *Carbohydr. Polym.* **2016**, *153*, 275–283. [CrossRef]

158. Johnson, M.E.; Murphy, P.J.; Boulton, M. Effectiveness of sodium hyaluronate eyedrops in the treatment of dry eye. *Graefe's Arch. Clin. Exp. Ophthalmol.* **2006**, *244*, 109–112. [CrossRef] [PubMed]

159. Van Beek, M.; Jones, L.; Sheardown, H. Hyaluronic acid containing hydrogels for the reduction of protein adsorption. *Biomaterials* **2008**, *29*, 780–789. [CrossRef]

160. Paterson, S.M.; Liu, L.; Brook, M.A.; Sheardown, H. Poly(ethylene glycol)-or silicone-modified hyaluronan for contact lens wetting agent applications. *J. Biomed. Mater. Res. Part A* **2015**, *103*, 2602–2610. [CrossRef] [PubMed]

161. Barbault-Foucher, S.; Gref, R.; Russo, P.; Guechot, J.; Bochot, A. Design of poly-ε-caprolactone nanospheres coated with bioadhesive hyaluronic acid for ocular delivery. *J. Control. Release* **2002**, *83*, 365–375. [CrossRef]

162. Moustafa, M.A.; Elnaggar, Y.S.R.; El-Refaie, W.M.; Abdallah, O.Y. Hyalugel-integrated liposomes as a novel ocular nanosized delivery system of fluconazole with promising prolonged effect. *Int. J. Pharm.* **2017**, *534*, 14–24. [CrossRef]

163. Li, C.; Chen, R.; Xu, M.; Qiao, J.; Yan, L.; Guo, X.D. Hyaluronic acid modified MPEG-*b*-PAE block copolymer aqueous micelles for efficient ophthalmic drug delivery of hydrophobic genistein. *Drug Deliv.* **2018**, *25*, 1258–1265. [CrossRef] [PubMed]

164. Korogiannaki, M.; Guidi, G.; Jones, L.; Sheardown, H. Timolol maleate release from hyaluronic acid-containing model silicone hydrogel contact lens materials. *J. Biomater. Appl.* **2015**, *30*, 361–376. [CrossRef] [PubMed]

165. Korogiannaki, M.; Zhang, J.; Sheardown, H. Surface modification of model hydrogel contact lenses with hyaluronic acid via thiol-ene "click" chemistry for enhancing surface characteristics. *J. Biomater. Appl.* **2017**, *32*, 446–462. [CrossRef] [PubMed]

166. Deng, X.; Korogiannaki, M.; Rastegari, B.; Zhang, J.; Chen, M.; Fu, Q.; Sheardown, H.; Filipe, C.D.M.; Hoare, T. "Click" Chemistry-Tethered Hyaluronic Acid-Based Contact Lens Coatings Improve Lens Wettability and Lower Protein Adsorption. *ACS Appl. Mater. Interfaces* **2016**, *8*, 22064–22073. [CrossRef] [PubMed]

167. Weeks, A.; Morrison, D.; Alauzun, J.G.; Brook, M.A.; Jones, L.; Sheardown, H. Photocrosslinkable hyaluronic acid as an internal wetting agent in model conventional and silicone hydrogel contact lenses. *J. Biomed. Mater. Res. Part A* **2012**, *100 A*, 1972–1982. [CrossRef]

168. Lang, J.C. Ocular drug delivery conventional ocular formulations. *Adv. Drug Deliv. Rev.* **1995**, *16*, 39–43. [CrossRef]

169. Choi, S.W.; Kim, J. Therapeutic Contact Lenses with Polymeric Vehicles for Ocular Drug Delivery: A Review. *Materials* **2018**, *11*, 1125. [CrossRef]

170. Cheng, R.; Meng, F.; Deng, C.; Klok, H.A.; Zhong, Z. Dual and multi-stimuli responsive polymeric nanoparticles for programmed site-specific drug delivery. *Biomaterials* **2013**, *34*, 3647–3657. [CrossRef]

171. Maulvi, F.A.; Lakdawala, D.H.; Shaikh, A.A.; Desai, A.R.; Choksi, H.H.; Vaidya, R.J.; Ranch, K.M.; Koli, A.R.; Vyas, B.A.; Shah, D.O. In vitro and in vivo evaluation of novel implantation technology in hydrogel contact lenses for controlled drug delivery. *J. Control. Release* **2016**, *226*, 47–56. [CrossRef]

172. El Shaer, A.; Mustafa, S.; Kasar, M.; Thapa, S.; Ghatora, B.; Alany, R.G. Nanoparticle-laden contact lens for controlled ocular delivery of prednisolone: Formulation optimization using statistical experimental design. *Pharmaceutics* **2016**, *8*, 14. [CrossRef]

173. Behl, G.; Iqbal, J.; O'Reilly, N.J.; McLoughlin, P.; Fitzhenry, L. Synthesis and Characterization of Poly(2-hydroxyethylmethacrylate) Contact Lenses Containing Chitosan Nanoparticles as an Ocular Delivery System for Dexamethasone Sodium Phosphate. *Pharm. Res.* **2016**, *33*, 1638–1648. [CrossRef] [PubMed]

174. Dos Santos, J.F.R.; Couceiro, R.; Concheiro, A.; Torres-Labandeira, J.J.; Alvarez-Lorenzo, C. Poly(hydroxyethyl methacrylate-co-methacrylated-β-cyclodextrin) hydrogels: Synthesis, cytocompatibility, mechanical properties and drug loading/release properties. *Acta Biomater.* **2008**, *4*, 745–755. [CrossRef]

175. Phan, C.-M.; Subbaraman, L.N.; Jones, L. In vitro drug release of natamycin from β-cyclodextrin and 2-hydroxypropyl-β-cyclodextrin-functionalized contact lens materials. *J. Biomater. Sci. Polym. Ed.* **2014**, *25*, 1907–1919. [CrossRef] [PubMed]

176. Xu, J.; Li, X.; Sun, F. Cyclodextrin-containing hydrogels for contact lenses as a platform for drug incorporation and release. *Acta Biomater.* **2010**, *6*, 486–493. [CrossRef] [PubMed]

177. Alvarez-Lorenzo, C.; Yañez, F.; Barreiro-Iglesias, R.; Concheiro, A. Imprinted soft contact lenses as norfloxacin delivery systems. *J. Control. Release* **2006**, *113*, 236–244. [CrossRef] [PubMed]

178. Malaekeh-Nikouei, B.; Vahabzadeh, S.A.; Mohajeri, S.A. Preparation of a Molecularly Imprinted Soft Contact Lens as a New Ocular Drug Delivery System for Dorzolamide. *Curr. Drug Deliv.* **2013**, *10*, 279–285. [CrossRef] [PubMed]

179. Hiratani, H.; Alvarez-Lorenzo, C. Timolol uptake and release by imprinted soft contact lenses made of N,N-diethylacrylamide and methacrylic acid. *J. Control. Release* **2002**, *83*, 223–230. [CrossRef]

180. Li, C.C.; Chauhan, A. Ocular transport model for ophthalmic delivery of timolol through p-HEMA contact lenses. *J. Drug Deliv. Sci. Technol.* **2007**, *17*, 69–79. [CrossRef]

181. Schrader, S.; Wedel, T.; Moll, R.; Geerling, G. Combination of serum eye drops with hydrogel bandage contact lenses in the treatment of persistent epithelial defects. *Graefe's Arch. Clin. Exp. Ophthalmol.* **2006**, *244*, 1345–1349. [CrossRef] [PubMed]

182. Lesher, G.A.; Gunderson, G.G. Continuous drug delivery through the use of disposable contact lenses. *Optom. Vis. Sci.* **1993**, *70*, 1012–1018. [CrossRef]

183. Hu, X.; Hao, L.; Wang, H.; Yang, X.; Zhang, G.; Wang, G.; Zhang, X. Hydrogel contact lens for extended delivery of ophthalmic drugs. *Int. J. Polym. Sci.* **2011**. [CrossRef]

184. Kim, J.; Peng, C.C.; Chauhan, A. Extended release of dexamethasone from silicone-hydrogel contact lenses containing vitamin E. *J. Control. Release* **2010**, *148*, 110–116. [CrossRef] [PubMed]

185. Peng, C.C.; Burke, M.T.; Chauhan, A. Transport of topical anesthetics in vitamin E loaded silicone hydrogel contact lenses. *Langmuir* **2012**, *28*, 1478–1487. [CrossRef] [PubMed]

186. Dixon, P.; Fentzke, R.C.; Bhattacharya, A.; Konar, A.; Hazra, S.; Chauhan, A. In vitro drug release and in vivo safety of vitamin E and cysteamine loaded contact lenses. *Int. J. Pharm.* **2018**, *544*, 380–391. [CrossRef] [PubMed]

187. Hsu, K.H.; Fentzke, R.C.; Chauhan, A. Feasibility of corneal drug delivery of cysteamine using vitamin E modified silicone hydrogel contact lenses. *Eur. J. Pharm. Biopharm.* **2013**, *85*, 531–540. [CrossRef] [PubMed]

188. Dixon, P.; Ghosh, T.; Mondal, K.; Konar, A.; Chauhan, A.; Hazra, S. Controlled delivery of pirfenidone through vitamin E-loaded contact lens ameliorates corneal inflammation. *Drug Deliv. Transl. Res.* **2018**, *8*, 1114–1126. [CrossRef] [PubMed]

189. Hui, A.; Bajgrowicz-Cieslak, M.; Phan, C.M.; Jones, L. In vitro release of two anti-muscarinic drugs from soft contact lenses. *Clin. Ophthalmol.* **2017**, *11*, 1657–1665. [CrossRef] [PubMed]

190. Zuegg, J.; Cooper, M.A. Drug-likeness and increased hydrophobicity of commercially available compound libraries for drug screening. *Curr. Top. Med. Chem.* **2012**, *12*, 1500–1513. [CrossRef]

191. O'Brien, F.J. Biomaterials & scaffolds for tissue engineering. *Mater. Today* **2011**, *14*, 88–95.

192. Tang, D.; Tare, R.S.; Yang, L.Y.; Williams, D.F.; Ou, K.L.; Oreffo, R.O.C. Biofabrication of bone tissue: Approaches, challenges and translation for bone regeneration. *Biomaterials* **2016**, *83*, 363–382. [CrossRef]

193. Madaghiele, M.; Sannino, A.; Ambrosio, L.; Demitri, C. Polymeric hydrogels for burn wound care: Advanced skin wound dressings and regenerative templates. *Burn. Trauma* **2014**, *2*, 153. [CrossRef] [PubMed]

194. Roseti, L.; Parisi, V.; Petretta, M.; Cavallo, C.; Desando, G.; Bartolotti, I.; Grigolo, B. Scaffolds for Bone Tissue Engineering: State of the art and new perspectives. *Mater. Sci. Eng. C* **2017**, *78*, 1246–1262. [CrossRef] [PubMed]

195. Bailey, J.; Morgan, P.; Gleeson, H.; Jones, J. Switchable Liquid Crystal Contact Lenses for the Correction of Presbyopia. *Crystals* **2018**, *8*, 29. [CrossRef]

196. Syed, I.M.; Kaur, S.; Milton, H.E.; Mistry, D.; Bailey, J.; Morgan, P.B.; Jones, J.C.; Gleeson, H.F. Novel switching mode in a vertically aligned liquid crystal contact lens. *Opt. Express* **2015**, *23*, 9911. [CrossRef] [PubMed]

197. Bailey, J.; Kaur, S.; Morgan, P.B.; Gleeson, H.F.; Clamp, J.H.; Jones, J.C. Design considerations for liquid crystal contact lenses. *J. Phys. D. Appl. Phys.* **2017**, *50*. [CrossRef]

198. Nonoyama, T.; Gong, J.P. Double-network hydrogel and its potential biomedical application: A review. *Proc. Inst. Mech. Eng. Part H J. Eng. Med.* **2015**, *229*, 853–863. [CrossRef] [PubMed]

199. Nakajima, T.; Sato, H.; Zhao, Y.; Kawahara, S.; Kurokawa, T.; Sugahara, K.; Gong, J.P. A Universal Molecular Stent Method to Toughen any Hydrogels Based on Double Network Concept. *Adv. Funct. Mater.* **2012**, *22*, 4426–4432. [CrossRef]

200. Myung, D.; Noolandl, J.; Ta, C.; Frank, C.W. Interpenetrating Polymer Network Hydrogel Contact Lenses. U.S. Patent 7857447B2, 28 December 2010.

201. Yañez, F.; Concheiro, A.; Alvarez-Lorenzo, C. Macromolecule release and smoothness of semi-interpenetrating PVP-pHEMA networks for comfortable soft contact lenses. *Eur. J. Pharm. Biopharm.* **2008**, *69*, 1094–1103. [CrossRef]

202. Macdougall, L.J.; Perez-Madrigal, M.; Shaw, J.; Inam, M.; Hoyland, J.A.; O'Reilly, R.K.; Richardson, S.M.; Dove, A.P. Self-healing, stretchable and robust interpenetrating network hydrogels. *Biomater. Sci.* **2018**, *6*, 2932–2937. [CrossRef]

203. Castellino, V.; Acosta, E.; Cheng, Y.L. Interpenetrating polymer networks templated on bicontinuous microemulsions containing silicone oil, methacrylic acid, and hydroxyethyl methacrylate. *Colloid Polym. Sci.* **2013**, *291*, 527–539. [CrossRef]

204. Riber, L.; Burmølle, M.; Alm, M.; Milani, S.M.; Thomsen, P.; Hansen, L.H.; Sørensen, S.J. Enhanced plasmid loss in bacterial populations exposed to the antimicrobial compound irgasan delivered from interpenetrating polymer network silicone hydrogels. *Plasmid* **2016**, *87–88*, 72–78. [CrossRef] [PubMed]

205. Tugui, C.; Cazacu, M.; Sacarescu, L.; Bele, A.; Stiubianu, G.; Ursu, C.; Racles, C. Full silicone interpenetrating bi-networks with different organic groups attached to the silicon atoms. *Polymer (Guildf)* **2015**, *77*, 312–322. [CrossRef]

206. Vuillequez, A.; Moreau, J.; Garda, M.R.; Youssef, B.; Saiter, J.M. Polyurethane methacrylate/silicone interpenetrating polymer networks synthesis, thermal and mechanical properties. *J. Polym. Res.* **2008**, *15*, 89–96. [CrossRef]

207. Fenton, O.S.; Olafson, K.N.; Pillai, P.S.; Mitchell, M.J.; Langer, R. Advances in Biomaterials for Drug Delivery. *Adv. Mater.* **2018**, *30*, 1–29. [CrossRef] [PubMed]

208. Aydin, D.; Alipour, M.; Kizilel, S. Design of Stimuli-responsive drug delivery hydrogels. In *Functional Hydrogels in Drug Delivery: Key Features and Future Perspectives*; Spizzirri, U.G., Cirillo, G., Eds.; Taylor & Francis Group: Boca Raton, FL, USA, 2017; pp. 1–23. ISBN 9781498747998.

209. Schmaljohann, D. Thermo- and pH-responsive polymers in drug delivery. *Adv. Drug Deliv. Rev.* **2006**, *58*, 1655–1670. [CrossRef] [PubMed]

210. Kanamala, M.; Wilson, W.R.; Yang, M.; Palmer, B.D.; Wu, Z. Mechanisms and biomaterials in pH-responsive tumour targeted drug delivery: A review. *Biomaterials* **2016**, *85*, 152–167. [CrossRef]

211. Zhu, Q.; Cheng, H.; Huo, Y.; Mao, S. Sustained ophthalmic delivery of highly soluble drug using pH-triggered inner layer-embedded contact lens. *Int. J. Pharm.* **2018**, *544*, 100–111. [CrossRef]

212. Vanparijs, N.; Nuhn, L.; De Geest, B.G. Transiently thermoresponsive polymers and their applications in biomedicine. *Chem. Soc. Rev.* **2017**, *46*, 1193–1239. [CrossRef] [PubMed]

213. Zhang, Q.; Weber, C.; Schubert, U.S.; Hoogenboom, R. Thermoresponsive polymers with lower critical solution temperature: From fundamental aspects and measuring techniques to recommended turbidimetry conditions. *Mater. Horizons* **2017**, *4*, 109–116. [CrossRef]

214. Jung, H.J.; Chauhan, A. Temperature sensitive contact lenses for triggered ophthalmic drug delivery. *Biomaterials* **2012**, *33*, 2289–2300. [CrossRef]

215. Barar, J.; Aghanejad, A.; Fathi, M.; Omidi, Y. Advanced drug delivery and targeting technologies for the ocular diseases. *BioImpacts* **2016**, *6*, 49–67. [CrossRef] [PubMed]

216. Phan, C.-M.; Subbaraman, L.N.; Jones, L. Contact lenses for antifungal ocular drug delivery: A review. *Expert Opin. Drug Deliv.* **2014**, *11*, 537–546. [CrossRef] [PubMed]

217. Yin, C.; Ansell, S.F. Molds for Producing Contact Lenses. U.S. Patent 8292256B2, 23 October 2012.

218. Samuel, N.T.; Huang, H.; Wu, D.; Haken, U.; Pruitt, J.D.; Domschke, A.M.; Matsuzawa, Y. Method for Making Silicone Hydrogel Contact Lenses. Patent WO2012/078457, 14 June 2012.

219. Norris, L.D.; Bialek, E.S.; Almond, S.; Morsley, D.R.; Siddiqui, A.K.M.S.; Rogers, R.C.; Bruce, I.; Sheader, B.S. Polar Thermoplastic Ophthalmic Lens Molds, Ophthalmic Lenses Molded Therin, and Related Methods. U.S. Patent 9156214B2, 13 October 2015.

220. Hopson, P.; Pegram, S.C.; Nitin, N.; Hanson, H.S.; Rubal, M.; Hanson, D.P. Apparatus for Trating an Ophthalmic Lens Mold Part. Patent WO2011/119945, 29 September 2011.

221. Fang, F.; Xu, F. Recent Advances in Micro/Nano-cutting: Effect of Tool Edge and Material Properties. *Nanomanuf. Metrol.* **2018**, *1*, 4–31. [CrossRef]

222. Fang, F.Z.; Zhang, X.D.; Gao, W.; Guo, Y.B.; Byrne, G.; Hansen, H.N. Nanomanufacturing—Perspective and applications. *CIRP Ann. Manuf. Technol.* **2017**, *66*, 683–705. [CrossRef]

223. Zhu, L.; Li, Z.; Fang, F.; Huang, S.; Zhang, X. Review on fast tool servo machining of optical freeform surfaces. *Int. J. Adv. Manuf. Technol.* **2018**, *95*, 2071–2092. [CrossRef]

224. Kang, C.; Fang, F. State of the art of bioimplants manufacturing: part II. *Adv. Manuf.* **2018**, *6*, 137–154. [CrossRef]

225. Lee, W.; Jin, M.K.; Yoo, W.C.; Lee, J.K. Nanostructuring of a polymeric substrate with well-defined nanometer-scale topography and tailored surface wettability. *Langmuir* **2004**, *20*, 7665–7669. [CrossRef]

226. Arora, H.S.; Xu, Q.; Xia, Z.; Ho, Y.H.; Dahotre, N.B.; Schroers, J.; Mukherjee, S. Wettability of nanotextured metallic glass surfaces. *Scr. Mater.* **2013**, *69*, 732–735. [CrossRef]

227. Encinas, N.; Pantoja, M.; Abenojar, J.; Martínez, M.A. Control of wettability of polymers by surface roughness modification. *J. Adhes. Sci. Technol.* **2010**, *24*, 1869–1883. [CrossRef]

228. Kong, L.B.; Cheung, C.F.; Jiang, J.B.; To, S.; Lee, W.B. Characterization of freeform optics in automotive lighting systems using an Optical-Geometrical Feature Based Method. *Optik (Stuttg)* **2011**, *122*, 358–363. [CrossRef]

229. Forbes, G.W. Characterizing the shape of freeform optics. *Opt. Express* **2012**, *20*, 2483. [CrossRef] [PubMed]

230. Fang, F.Z.; Zhang, X.D.; Weckenmann, A.; Zhang, G.X.; Evans, C. Manufacturing and measurement of freeform optics. *CIRP Ann. Manuf. Technol.* **2013**, *62*, 823–846. [CrossRef]

231. Bono, M.J.; Hibbard, R.L. Fabrication and metrology of micro-scale sinusoidal surfaces in polymer workpiece materials. In Proceedings of the American Society for Precision Engineering 2004 Annual Meeting, Orlando, FL, USA, 24–29 October 2004.

232. Yu, N.; Fang, F.; Wu, B.; Zeng, L.; Cheng, Y. State of the art of intraocular lens manufacturing. *Int. J. Adv. Manuf. Technol.* **2018**, *98*, 1103–1130. [CrossRef]

 *materials*

Article

Naturally-Derived Biphasic Calcium Phosphates through Increased Phosphorus-Based Reagent Amounts for Biomedical Applications

Aura-Cătălina Mocanu [1,2], George E. Stan [3,*], Andreea Maidaniuc [1,4], Marian Miculescu [1], Iulian Vasile Antoniac [1], Robert-Cătălin Ciocoiu [1], Ștefan Ioan Voicu [5], Valentina Mitran [6], Anișoara Cîmpean [6] and Florin Miculescu [1,*]

[1]   Department of Metallic Materials Science, Physical Metallurgy, University Politehnica of Bucharest, 313 Splaiul Independentei, J Building, District 6, 060042 Bucharest, Romania; mcn_aura@hotmail.com (A.-C.M.); andreea.maidaniuc@gmail.com (A.M.); m_miculescu@yahoo.com (M.M.); antoniac.iulian@gmail.com (I.V.A.); ciocoiurobert@gmail.com (R.-C.C.)

[2]   Department of Research, Development and Innovation, S.C. Nuclear NDT Research & Services S.R.L, 104 Berceni Str., Central Laboratory Building, District 4, 041919 Bucharest, Romania

[3]   National Institute of Materials Physics, Laboratory of Multifunctional Materials and Structures, 405A Atomistilor Str., 077125 Măgurele-Ilfov, Romania

[4]   Destructive and Nondestructive Testing Laboratory, S.C. Nuclear NDT Research & Services S.R.L, 104 Berceni Str., Central Laboratory Building, District 4, 041919 Bucharest, Romania

[5]   Department of Analytical Chemistry and Environmental Engineering, University Politehnica of Bucharest, 1-7 Gh. Polizu Str., Polizu campus, L 015 Building, District 1, 011061 Bucharest, Romania; svoicu@gmail.com

[6]   Department of Biochemistry and Molecular Biology, University of Bucharest, 91-95 Splaiul Independentei, 050095 Bucharest, Romania; valentinamitran@yahoo.com (V.M.); anisoara.cimpean@bio.unibuc.ro (A.C.)

*   Correspondence: george_stan@infim.ro (G.E.S.); f_miculescu@yahoo.com (F.M.); Tel.: +40-21-3169563 (F.M.)

Received: 21 December 2018; Accepted: 23 January 2019; Published: 25 January 2019

Abstract: Calcium carbonate from marble and seashells is an eco-friendly, sustainable, and largely available bioresource for producing natural bone-like calcium phosphates (CaPs). Based on three main objectives, this research targeted the: (i) adaptation of an indirect synthesis route by modulating the amount of phosphorus used in the chemical reaction, (ii) comprehensive structural, morphological, and surface characterization, and (iii) biocompatibility assessment of the synthesized powdered samples. The morphological characterization was performed on digitally processed scanning electron microscopy (SEM) images. The complementary 3D image augmentation of SEM results also allowed the quantification of roughness parameters. The results revealed that both morphology and roughness were modulated through the induced variation of the synthesis parameters. Structural investigation of the samples was performed by Fourier transform infrared spectroscopy and X-ray diffraction. Depending on the phosphorus amount from the chemical reaction, the structural studies revealed the formation of biphasic CaPs based on hydroxyapatite/brushite or brushite/monetite. The in vitro assessment of the powdered samples demonstrated their capacity to support MC3T3-E1 pre-osteoblast viability and proliferation at comparable levels to the negative cytotoxicity control and the reference material (commercial hydroxyapatite). Therefore, these samples hold great promise for biomedical applications.

Keywords: dolomitic marble; seashell; $CaCO_3$ derived-calcium phosphates; modulated synthesis set-up; SEM; image analysis; pre-osteoblasts

---

1. Introduction

Orthopedic surgery advancements outlined a new, challenging, and necessary era for bone-loss reconstruction. Apart from small skeletal fractures when bone can repair itself, extensive bone defects that are above the critical size and result from accidents, trauma impact, or bone diseases, require filling treatment techniques [1–4].

Currently, the standard technique for bone reconstruction involves bone tissue harvesting from different body parts of the patient, process known as autografting/autologous bone grafting [5,6]. This method has several downsides including morbidity, supplementary surgery and reduced bone graft quantities [7,8], and therefore, is incompatible for massive bone defects repair. As an alternative, various synthetic grafts, including calcium phosphate-based materials, stirred up interest for almost three decades due to their biocompatibility and osteoconductivity properties [5,9,10]. Now, along with the high demand for reconstruction materials and the technological advancements which allow the industrial production of biomimetic calcium phosphates (CaPs), research aims turned to the bio-functionalization of natural resources (seashells, bovine bone, marble) as an environmentally friendly, sustainable and cost-effective alternative [11–18].

It is stated that an ideal biomaterial destined for skeletal repair applications should mimic the biological, compositional, and mechanical properties of the host bone and also create the necessary niche for further functionalization [1,19,20]. This is the reason why recent studies were focused on the synthesis of hydroxyapatite (HA) and acidic CaPs cements consisting of brushite (DCPD) and monetite (DCPA)—an anhydrous form of brushite [21–23], naturally found in bone, teeth, and renal calculi [24,25]. All three-candidate materials are crystalline forms of calcium phosphates. Hydroxyapatite has been widely reported as the preferred bone grafting material, and brushite and monetite were involved mainly in bone cement preparation due to their resorbable, self-regenerating, and osteoconductive character [1,25]. Under physiological conditions, brushite is metastable and highly reactive and was shown to reprecipitate into hydroxyapatite [21]. In addition, it is less soluble and forms first throughout cement reactions, even though monetite is a more stable phase [5,26,27]. Recent in vivo results, obtained after monetite-based granules implantation, showed an improved degradation and bone regeneration than the hydroxyapatite-based ones [23].

The morphological aspects such as texture, roughness and topographic patterns stand as well as essential factors for the biological success of ceramic materials/ceramic materials-based structures [28–30]. Along with chemical composition, surface features dictate the cellular behavior in terms of adhesion, differentiation, migration and proliferation (both in vitro and in vivo), and the degree of bone formation [19,31–36].

It was observed that a micrometric texture consisting of alternate valleys and peaks is relevant for the cells cytoskeleton organization [31,34,37]. Currently, 3D digital topographic reconstruction based on morphological analysis can provide enhanced insight on surface texture and topographic patterns. Also, it allows for quantification of surface roughness parameters which are classified as amplitude, spatial, and hybrid parameters [34,38,39]. A rather moderate rough surface with microporosity or grooves promotes in vivo the biological mediators secretion, which leads to cell adhesion and migration and new bone matrix formation [37,40]. It was reported that a depth of 2–5 nm on the structure's surface is required for material–cell interaction, attachment, and development [37].

Most CaPs inherit an osteoconductive behavior due to their surface features. Biodegradable forms such as brushite and monetite are difficult to investigate in terms of surface topography influence on resorption mechanism, over a prolonged period of time due to ionic release in the in vitro culture medium [1,27].

Given the high frequency of orthopedic problems, we aimed for a resolution based on sustainable raw resources (marble and seashells) for a facile, cost-efficient, and direct synthesis (with less intermediary technological stages) of biomaterials with biological potential at least similar with that of commercial synthetic available materials. We followed this route to induce directly the synthesis of biphasic calcium phosphates compounds. Therefore, the main purpose of this research study resides on the development of novel naturally-derived biphasic CaPs with different phase composition, based on incipient results previously reported [11,12]. Further, the attention falls on the assessment of the adequate necessary reagent amount ranges, capable to allow for the synthesis of HA/DCPD and

DCPD/DCPA biphasic materials through the developed methodology. This involved an adapted synthesis set-up [12] based on the conversion of natural calcium carbonate ($CaCO_3$) precursors to biomimetic calcium phosphates (CaPs). An extensive characterization of the raw materials (i.e. marble and seashell) was presented in a previously published article [11]. This article further investigates the chemical reactions dynamics induced by the variation of phosphoric acid amount involved in the synthesis reaction. The modulation involved the gradually increase of phosphoric acid quantity starting with the stoichiometrically amount (considered as starting point of modulation -0%) up to 90% acid addition above the stoichiometry. Apart from a preliminary 0–30% reagent amount modulation [12], for which the new extended experimental sample set of investigations provided reproducible results, the key insights of this article rely on the further acid increment (40–90%). Thus, a complex characterization, focused on physico-chemically and digitally enhanced topographic features, was conducted along with an in vitro evaluation of the biological performance of all synthesized powdered samples in terms of pre-osteoblast viability and proliferation.

2. Materials and Methods

2.1. Ceramic Materials Synthesis

Dolomitic marble and *Mytilus galloprovincialis* seashells were thermally treated at 1300 °C for 6 h for $CaCO_3$ dissociation to calcium oxide (CaO). Previously reported investigations regarding the thermal transformations of both precursors [11] confirm that after the CO_2 loss, the obtained CaO phase is stable at 1100–1200 °C and therefore the thermal treatment conducted at 1300 °C ensures a complete decomposition. The resulting powder was further hydrated with distilled water, filtered, deposited in thin layer on watch glass and dried for 144 hours at room temperature (RT) resulting the calcium hydroxide ($Ca(OH)_2$) powder and no residual water [11]. Then, the $Ca(OH)_2$ compound was weighed on a calibrated four decimal analytical balance (Kern & Sohn GmbH, Balingen, Germany) and treated with various amounts of phosphoric acid (H_3PO_4, Sigma-Aldrich, St. Louis, MO, USA). According to stoichiometrically calculated amounts (S.C.A.), 10 g of $Ca(OH)_2$ were mixed with 200 mL distilled water and 5.5 mL of phosphoric acid, drop-wise added at a rate of 1 mL/min at room temperature (RT). The modulation of the products final chemical composition required a controlled addition of H_3PO_4 with respect to the S.C.A. ratio, incrementally increasing the acid volume in steps of 10% up to a maximum value of 90% (e.g. 0-M/S = S.C.A = 5.5 mL, 10-M/S = 5.5 mL + 10% × 5.5 mL = 6.05 mL, 20-M/S = 5.5 mL + 20% × 5.5 mL = 6.6 mL). Consequently, the Ca/P molar ratios of the solution were modified as indicated in Table 1, such to explore/enable the formation of other CaP-like phases than HA. The resulted slurries were stirred at 25 °C for 2 h, followed by aging for 72 h at RT, and drying at 100 °C for 2 h. The synthesized powders were deposited in Petri dishes and sealed in a desiccator. Further, the dried ceramic powders were ground in a planetary mill with agate balls, granulometric sorted with standardized sieves (<20–100 μm particle size) and transformed in cylindrical pressed samples (Φ 30 mm) by cold isostatic pressing with a force of 10 MPa. After 24 h at RT, pressed samples' with parallel planar surfaces were obtained by polishing with P2500 grade abrasive sandpaper.

For an easy tracking the sample codes included the amount of phosphoric acid added above the S.C.A, expressed in percents (0–90%) and the precursor's abbreviation—M for marble; S for seashell (e.g., 20-M represents the sample prepared from marble in which a chemical treatment with S.C.A. + 20% phosphoric acid was used).

For structural and cellular investigations, commercial hydroxyapatite (Merck KGaA, Darmstadt, Germany) was used as a reference material (Ref.).

Table 1. Denomination of samples and Ca/P molar ratios of the precursor solutions.

H$_3$PO$_4$ Increment	0%	10%	20%	30%	40%	50%	60%	70%	80%	90%
Sample Batch	0-M;	10-M;	20-M;	30-M;	40-M;	50-M;	60-M;	70-M;	80-M;	90-M;
Code	0-S	10-S	20-S	30-S	40-S	50-S	60-S	70-S	80-S	90-S
Ca/P Molar Ratio	1.67	~1.52	~1.39	~1.28	~1.19	~1.11	~1.04	~0.98	~0.93	~0.88

2.2. Characterization Techniques

2.2.1. XRD Analysis

The crystalline status and phase composition of the synthesized materials was investigated by X-ray diffraction (XRD) with a Bruker D8 Advance diffractometer (Bruker Corporation, Billerica, MA, USA) with Cu K$_\alpha$ (λ = 1.5418 Å) radiation, equipped with a Lynx Eye linear detector type, in Bragg–Brentano geometry. The samples were scanned in the 2θ angular range of 9–55° with a step size of 0.04° and 2 s acquisition time/step.

2.2.2. FT-IR Spectroscopy Measurements

The analysis of the bonding architecture and identification of functional groups present in the samples was performed by Fourier transform infrared (FTIR) spectroscopy with a Perkin Elmer Spectrum BX II spectrometer (PerkinElmer, Inc., Waltham, MA, USA), in attenuated total reflectance (ATR) mode using a Pike-Miracle head with diamond-ZnSe crystal. The spectra were recorded in the range 4000–500 cm^{-1}, at a resolution of 4 cm^{-1} and 32 scans/experiment.

2.2.3. Morphological and Compositional Evaluation and 3D Image Augmentation

The morphological evaluation of the ceramic green bodies was performed by scanning electron microscopy (SEM) with a Philips XL 30 ESEM TMP microscope (FEI/Phillips, Hillsboro, OR, USA). Micrographs were acquired at an acceleration voltage of 25 kV and 10 mm working distance. SEM investigations were performed in five randomly chosen areas. Topographic reconstruction of pressed samples along with the quantification of the surface roughness parameters were possible through SEM 3D top view image analysis via MountainsMap software (Digital Surf, Besançon, France). Six of the most relevant roughness parameters for morphological surface texture were graphically displayed.

The compositional evaluation was performed with a portable X-ray fluorescence spectrometer (SPECTRO xSORT, Kleve, Germany). Synthesized samples were analyzed without further preparation and results are reported as average of three measurements/sample.

2.2.4. Biocompatibility Experiments

Indirect contact studies were performed using mouse pre-osteoblasts MC3T3-E1 (ATCC®, CRL-2593™) grown in Dulbecco's Modified Eagle's Medium-DMEM (Sigma-Aldrich Co., St. Louis, MO, USA) supplemented with 10% fetal bovine serum (FBS) (Gibco (Life Technologies Corporation, Grand Island, NY, USA)) and 1% antibiotic-antimycotic mixture (Sigma-Aldrich Co., St. Louis, MO, USA). Marble- and seashell-derived powdered samples were sterilized for 1 h at 180 °C and afterwards subjected to extraction in culture medium at a concentration of 0.02 g/mL. The extracts obtained after incubation for 24 h at 37 °C were centrifuged, collected, and filtered using a filter with pore size 0.22 μm.

The pre-osteoblasts were seeded at a density of 1×10^4 cells cm^{-2} in a 96-well plate and incubated at 37 °C in a humidified atmosphere of 5% CO$_2$/95% air for 24 h. After that, the cell culture medium was discarded and the cell monolayer was incubated for further 1 day and 3 days in 100 μL samples' extracts. In parallel, the cells were incubated in DMEM containing 10% FBS without (cytotoxicity negative control) or with 5% dimethyl sulfoxide (DMSO) (Sigma-Aldrich Co., St. Louis, MO, USA) (cytotoxicity positive control).

A qualitative cell viability analysis consisting of cell staining with LIVE/DEAD Cell Viability/Cytotoxicity Assay Kit (Molecular Probes, Eugene, OR, USA) was performed after 1 and 3 days of cell incubation in the samples' extracts. The labeled cells were visualized using an inverted microscope equipped with epifluorescence (Olympus IX71, Olympus, Tokyo, Japan) and representative fields were captured with fluorescence imaging software Cell F. This assay was accompanied by a quantitative analysis of cell viability/proliferation, namely MTT [3-(4,5-dimethyl-2-thiazolyl)-2,5-diphenyl-2H-tetrazolium bromide) tetrazolium salt] assay, performed as previously described [41]. Briefly, cell monolayer was incubated with 1 mg mL^{-1} MTT solution for 3 h at 37 °C. The formazan produced by metabolically active viable cells was solubilized with DMSO and the absorbance of the dye was recorded at 550 nm using a microplate reader (Thermo Scientic Appliskan, Vantaa, Finland).

3. Results and Discussion

3.1. Structural and Chemical Characterization

3.1.1. XRD Analysis

The XRD patterns of all the naturally synthesized ceramic powders are comparatively presented in Figure 1. A single phase material (i.e., monophasic HA) was obtained only in the case of stoichiometrically seashell-derived sample (0-S). The broad diffraction maxima (appertaining to a hexagonal HA, ICDD: 00-009-0432) with respect to the reference commercial HA material, indicates the nanostructured nature of the 0-S-type material. In the case of 0-M-type sample, the nano-sized HA is accompanied by a secondary dicalcium phosphate dehydrate (DCPD) monoclinic phase (CaHPO$_4$·2H$_2$O, brushite, ICDD: 01-072-1240).

Further acid addition strongly influenced the composition of the samples. It facilitated the formation of biphasic CaPs mixture—HA/DCPD for (0–10)-M and (10–30)-S samples. If in the case of M-derived samples (Figure 1a), the HA presence lingers up to 10%, for the S-derived samples (Figure 1b), the HA content progressively decreases for acid amounts in the range of 10–30%, such as at 40% it becomes extinct. Onward, the emergence of a dicalcium phosphate anhydrous (DCPA or monetite, CaHPO$_4$, ICDD: 01-070-1425) triclinic phase was signaled. Consequently, for higher acid additions (more than 10% and 30% in the case of M- and S-derived samples, respectively), the phase composition shifted from biphasic HA/DCPD to biphasic DCPD/DCPA. At higher acid increments, the DCPD/DCPA ratio remained seemingly similar for the M-type materials, contrary to the S-type ones, for which DCPA was found to predominate with acid additions situated over 60%.

The coexistence of HA and DCPD is not unexpected, nor unprecedented. DCPD is more likely to precipitate in neutral or moderate acidic solutions at temperatures up to 40 °C as primary or secondary compound [42–44], and given the metastable thermodynamic character of the reaction, it is possible that in a saturated calcium and phosphate media the incipient HA crystals formation inhibited the precipitation and rapid development of DCPD crystals. It is also known that DCPD crystals can act as nuclei for HA evolution [45]. Nonetheless, the additional acid amount cannot only induce the abrupt (M-derived samples) or progressive (S-derived sample) reduction of HA content, but can promote DCPA formation based on its faster kinetic reaction [45,46]. In M-derived calcium carbonate, the natural Mg^{2+} dopant [11] stands as a possible inhibitory factor for HA precipitation even in stoichiometric conditions [42,47], and stabilized the biphasic mixture in acidic conditions. No unreacted Ca(OH)$_2$ was transferred to final ceramic products structure, which indicated its complete conversion to CaPs.

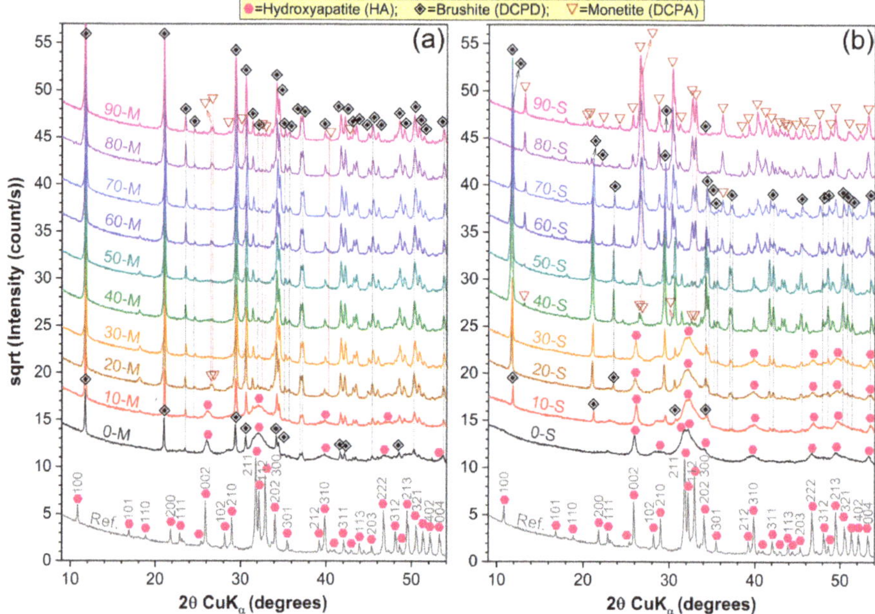

Figure 1. Comparative XRD patterns of the calcium phosphate powders synthesized using natural resources (i.e., (**a**) marble and (**b**) seashells) as calcium precursors and phosphoric acid amounts situated in the range 0–90%.

3.1.2. FTIR-ATR Measurements

Figure 2 presents the FTIR-ATR comparative spectra for the products obtained using both type of calcium natural precursors and phosphoric acid addition in the range of 0–90%.

The FTIR-ATR measurements confirmed the calcium phosphates formation and phase evolution previously disclosed by the XRD investigations. All the vibration bands characteristic to orthophosphate tetrahedral units into a hydroxyapatite-type structure were found to be prominent in the case of 0-M and 0-S samples: ν_4 bending (~560–559 and ~600 cm^{-1}) (Figure 2a,b), ν_1 symmetric stretching (~962 cm^{-1}) (Figure 2c,d), and ν_3 asymmetric stretching (~1018 and ~1087 cm^{-1}) (Figure 2c,d) [48]. To the difference of the pure, highly crystalline commercial HA, the spectra of the 0-M and 0-S samples elicited (i) broader peaks (testimony of their nanostructuring), (ii) the emergence of ν_2 bending (~874 cm^{-1}) and ν_3 asymmetric stretching (~1419 and ~1457 cm^{-1}) modes of carbonate groups (carbonatation of calcium phosphates synthesized under normal atmosphere conditions is to be expected), and (iii) a diminution of the vibration bands appertaining to structural hydroxyl units: libration (~629 cm^{-1}) and stretching (~3573 cm^{-1}) [48]. The higher than expected intensity of the (~874 cm^{-1} band (i.e. the intensity of ν_2 $(CO_3)^{2-}$ should be one fifth of the ν_3 $(CO_3)^{2-}$) ones [48]) recorded in the case of 0-M sample suggested the additive contribution of the vibration modes of acid phosphate [35,48], which emphasizes the concurrent formation of non-apatitic environments. This is in agreement with the XRD results (Figure 1a), which highlighted the simultaneous formation of DCPD along HA for the 0-M sample. In congruence with the XRD analyses, the FTIR-ATR results indicated that dramatic structural modifications occur at the same thresholds: 20-M, 40-S and 60-S. In the case of 20-M and 40-M samples, a series of newly emerged bands, characteristic to DCPD (brushite) [45,49–53] were emphasized: bending of (H–O)–P=O bonds in $(HPO_4)^{2-}$ (~538–535 and ~576–575 cm^{-1}), H_2O librations (~663 cm^{-1}), P–O–H out-of-plane bending (~788–787 cm^{-1}), stretching of P–O(H) bonds in $(HPO_4)^2$ molecules (~874 cm^{-1}), ν_1 symmetric stretching of phosphate (~984 and

~1004 cm^{-1}), ν_3 asymmetric stretching of $(PO_4)^{3-}$ (~1055–1053, ~1119–1118, and ~1133–1130 cm^{-1}), $(OH)^-$ in-plane bending (~1209 cm^{-1}), H_2O bending (~1649 cm^{-1}), $(P)O–H$ stretching (~2888 cm^{-1}), two O–H doublet bands (~3162–3158 cm^{-1} and ~3270–3269 & ~3478–3475 and ~3537–3533 cm^{-1}). The emergence of these two latter doublet bands represents a testimony for the presence of two types of water molecules in the structure of brushite [50,51]: the higher and the lower wavenumbers doublet bands appertaining to the bound and free water molecules, respectively. The issue is still disputed by spectroscopy specialists, the reverse assignment being sometimes endorsed [49,51]. The vibration bands typical for a DCPA (monetite)-type structure were clearly emphasized for the S-derived products starting with 60-S sample [49–53]: bending of O–P–O(H) bonds (~561 and ~582 cm^{-1}), stretching of P–O(H) bonds in $(HPO_4)^{2-}$ molecules (~888 and ~862 cm^{-1}), ν_1 symmetric stretching of phosphate (~987 cm^{-1}), ν_3 asymmetric stretching of phosphate (~1057 and ~1125 cm^{-1}), and in-plane bending of P–O(H) bonds (~1350 and ~1410 cm^{-1}). The formation and development of DCPA is also marked by the extinction of the $(OH)^-$ vibrations linked with the presence of water molecules present in the DCPD structure.

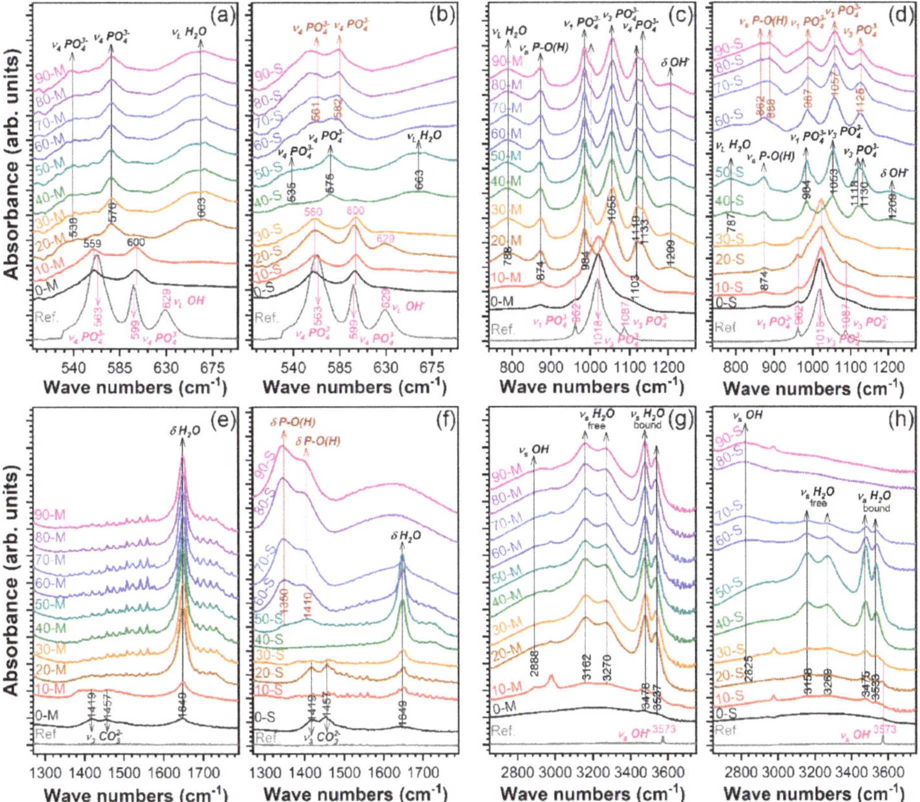

Figure 2. Comparative FTIR-ATR (attenuated total reflectance) spectra of the calcium phosphate powders synthesized using natural resources (i.e., (**a,c,e,g**) marble and (**b,d,f,h**) seashells) as calcium precursors and phosphoric acid amounts situated in the range 0–90%, collected in the four relevant wave numbers regions: (**a,b**) 700–500 cm^{-1}; (**c,d**) 1300–700 cm^{-1}; (**e,f**) 1800–1300 cm^{-1}; and (**g,h**) 3750–2650 cm^{-1}. To facilitate a better visual evaluation, the FTIR-ATR spectra were normalized to the intensity of the most prominent band region situated at ~1100–900 cm^{-1}.

3.1.3. XRF Evaluation. Ca/P Molar Ratio

The compositional evaluation of natural precursors ($CaCO_3$) and both CaO and $Ca(OH)_2$ powders is presented in Table 2 below. The samples consists initially of Ca, O and C. Results further revealed (i) a complete thermal dissociation of $CaCO_3$ sustained by the absence of C content identified for CaO powder derived from both precursors [11] and (ii) minor concentrations of Mg^{2+} preserved along the thermal decomposition and hydration processes in case of marble precursor and marble derived CaO and $Ca(OH)_2$ powders. Therefore, its influence during the synthesis process is still up for discussion, as described above. Regarding the calcium content from the calcium hydroxide derived from marble and the one derived from seashells, results revealed minor differences. This confirms our previously reported results [11].

Table 2. XRF characterization of raw precursors and CaO, $Ca(OH)_2$ powders.

Chemical Element (wt. %)		Ca	O	Mg	C
Marble	Raw precursor	33.30	42.86	0.83	22.71
	CaO	77.59	21.43	0.58	–
	$Ca(OH)_2$	50.72	48.76	0.22	–
Seashell	Raw precursor	39.03	41.03	–	19.64
	CaO	75.06	24.54	–	–
	$Ca(OH)_2$	51.21	48.39	–	–

The XRF compositional evaluation performed on synthesized samples revealed that all investigated samples contain chemical elements characteristic to CaP phases: Ca, P, and O as major components, and the absence of other elemental traces. Ca/P molar ratios calculated on the basis of XRF results are graphically presented in Figure 3. The Ca/P values varied inversely proportional with the acid share used in the synthesis process, and their progressive decrease down to values situated in the vicinity of ~1, conferred further evidences of DCPD and DCPA phase formation. Compared to the reference sample, only 0-S sample elicited a Ca/P molar ratio close to the 1.67 theoretical value, specific to the stoichiometric hydroxyapatite [11,13,54].

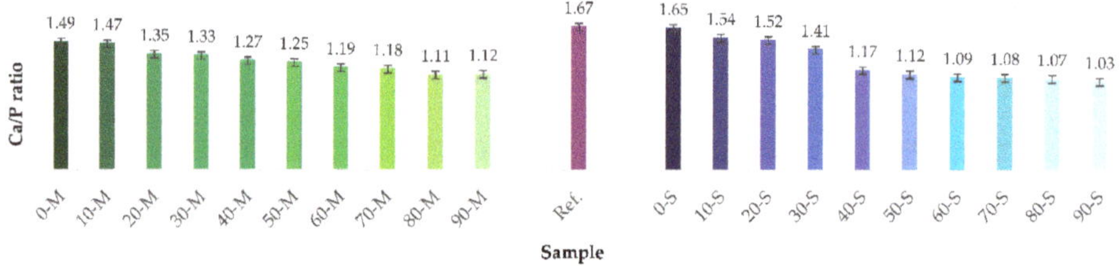

Figure 3. Ca/P molar ratio of the powders synthesized from both natural precursors (M—marble, S—seashell) with acid addition in the 0–90% range.

3.2. Morphology and Roughness Evaluation

Digitally processed SEM images and their correspondent roughness profile are presented in Figure 4. For each 3D image, the following quantitative parameters were extracted from roughness profiles (Figure 5): R_{sk}—profile skewness, R_a—profile arithmetic mean deviation, R_{ku}—profile kurtosis, R_p—maximum profile peak height, R_v—maximum profile valley depth, R_z—maximum height of profile, R_t—total height of profile. Their mathematical expressions are defined in ISO 4287:1997 [39].

Morphologically, the samples consisted of grains with different shapes and size distributions. An evolutionary tendency from well-defined polyhedral- to round-shaped grains, accompanied by a random and irregular grain size distribution with the increase of the acid amount, was observed. Initially, the grains conglomerate in larger, uniform, compact and centered isles (0–10%-M, 0–30%-S).

Starting with 20%-M and 40%-S, the isles tend to disperse towards the corners of each morphological map until the 90% acid amount. Also, a shape transition from round grains/conglomerates to thin needle-like ones was evidenced up to 90% acid amount.

Higher acid increments induced the formation of either mixed or distinctive isles formed from the two grain types. The morphology of the synthesized samples differed from the reference sample, which revealed uniform grain shapes and size distributions.

The 3D reconstruction allowed for a colorimetric distinction of two areas: the navy blue and dark orange colors highlight the deepest valleys and highest mountain peaks. Their presence is strongly related to either the positive (mountains) or negative (valleys) values of R_{sk} parameter. The presence of valleys indicates the micro-porous character of the granular material, highly important for cell adhesion [34,39]. Both R_{sk} and R_{ku} parameters revealed a surface texture characterized by symmetry, sharpness and curvature of heights profile distributions [39]. According to the R_{sk} values (Figure 4), both M- and S-derived samples presented similar microporous patterns (>-1 µm), which is consistent with the profile valley depth (R_v) values. Further, the R_{ku} graphic displayed as well cases of accentuated peak sharpness ($R_{ku} > 3$ µm) for 30-M and 50-S samples, which are associated with the starting points of the major phase composition shifts revealed by the XRD (Figure 1) and FTIR-ATR (Figure 2) investigations. At higher acid concentrations, the morphological assessments disclosed surfaces with predominantly uniform heights distribution and curvature, which correlates with the topological texture observed from 3D reconstruction images.

Another important roughness parameter—R_a—is related to the profile's roughness amplitude and stochastic surface roughness. Slightly smoother surfaces were obtained for the S-derived samples (R_a values in the range ~1.3–2.5 µm), independent of the acid share used in the synthesis process. These results are endorsed by the total and maximum height profile—R_t, R_p, and R_z—parameters: for a maximum peak height there is also a maximum valley depth, but graphically smoother.

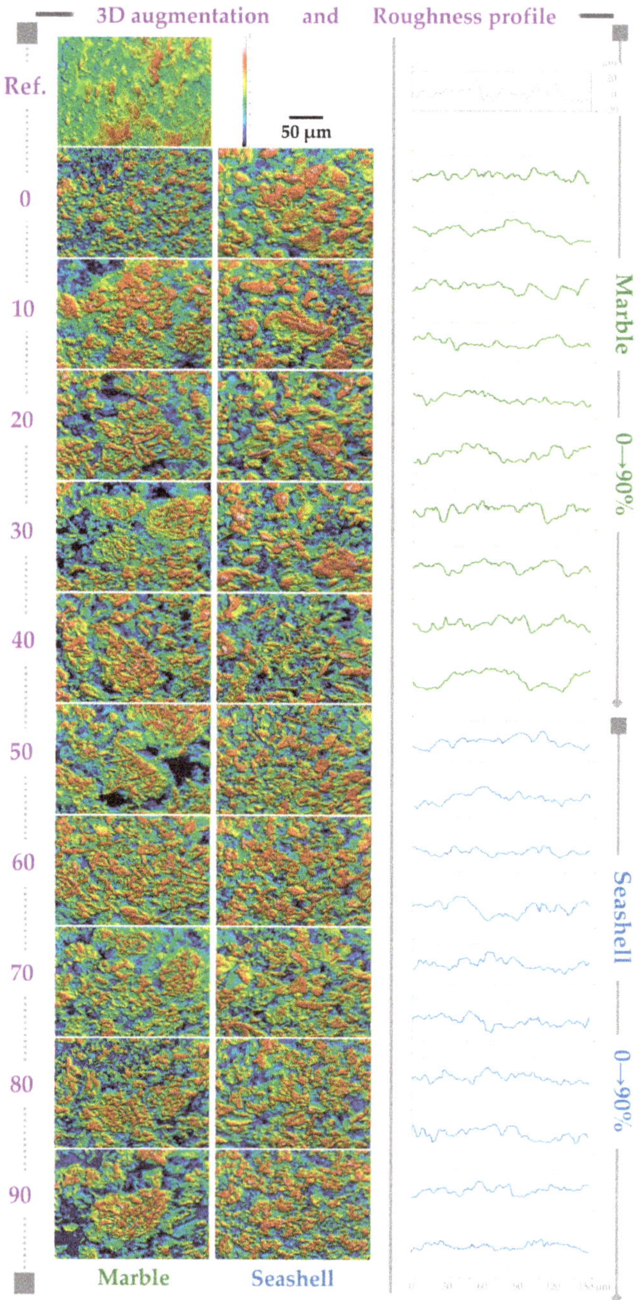

Figure 4. 3D surface topology reconstruction on the basis of SEM micrographs processing via Mountains Map software and roughness profiles assessment. Images were recorded at 500× magnification. Top center image: Scale bar (50 µm) for all micrographs. For all roughness profiles, the scale bar is provided on the right side (up and down) of the figure.

Overall, the morphological observations suggested that independent of the natural precursor and acid share, the powder products were characterized by a randomly distributed microporous surface texture with conglomerate isles. The morphological features are recognized as a key factor for both the in vitro and in vivo performance of biomaterials [28]. On a cellular level, the topographic surface parameters dictate cell adhesion and proliferation, along with their further proper functioning in what concerns the cytoskeletal organization and cell differentiation. These are directly related to the intracellular signaling mechanisms between cell receptors and material's surface [34,37]. Therefore, a micro-scale texture coupled with a moderate surface roughness are the preferred characteristics for an appropriate cell anchorage and development. Also, one can expect that the conglomerate isles can act as center points for first cell-biomaterial contact and for further cell attachment, growth and proliferation. From this point of view, a suitable cellular behavior is anticipated for all synthesized bioceramic samples.

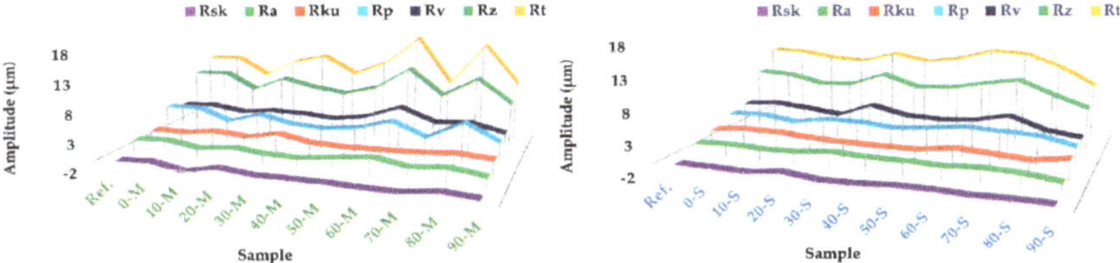

Figure 5. Roughness parameters quantification for marble (M) and seashell (S) derived samples: R_{sk}—profile skewness, R_a—profile arithmetic mean deviation, R_{ku}—profile kurtosis, R_p—maximum profile peak height, R_v—maximum profile valley depth, R_z—maximum height of profile, R_t—total height of profile.

3.3. In Vitro Pre-Osteoblast Behavior

The pre-osteoblast behavior in the extracts of the developed powdered samples was evaluated in terms of cell viability and proliferation by combining the results of the LIVE/DEAD Cell Viability/Cytotoxicity Qualitative assay and MTT assay. As shown in Figure 6, the extraction media of both M- and S-derived powdered samples sustained the cell viability regardless of acid increment. Thus, a high percentage of viable cells (green fluorescence) and a reduced number of dead cells (red fluorescence) were noticed at both incubation time points. Moreover, an increasing number of viable cells could be observed along the culture period suggesting the potential of these samples to support the pre-osteoblast proliferation. Likewise, typical healthy cell morphology was displayed except for the cytotoxicity positive control that exhibited mainly near-round cell shapes. The most probably, these cells are in progress of detachment from the underlying substrate as result of the toxicity elicited by DMSO. That also explains the progressive decrease in cell density along the culture period. Overall, the tested extracts showed cell morphologies and densities similar to the ones recorded for the reference sample extract and cytotoxicity negative control. Therefore, all analyzed powdered samples proved to be biocompatible. Interestingly, although no red-stained dead cells were noticed in case of (40–70)-S samples, lower cell densities were noticed.

Quantification by MTT assay of the cellular survival and proliferation capacity of MC3T3-E1 pre-osteoblasts demonstrated that the optical densities (OD) values, expressing the number of metabolically active viable cells, after 1 day of incubation in the extraction media of the analyzed M- and S-derived powdered samples were almost similar to the reference sample's extract and cytotoxicity negative control (Figure 7). On the contrary, in the case of the cytotoxicity positive control, the number of viable cells was significantly reduced. The prolonged incubation of MC3T3-E1 cells (i.e. 3 days) led to higher OD values than those expressed at 1 day-time point with the exception of the 40-S extract that exhibited almost a similar number of metabolically active viable cells and the cytotoxicity positive control showing a decrease in OD values.

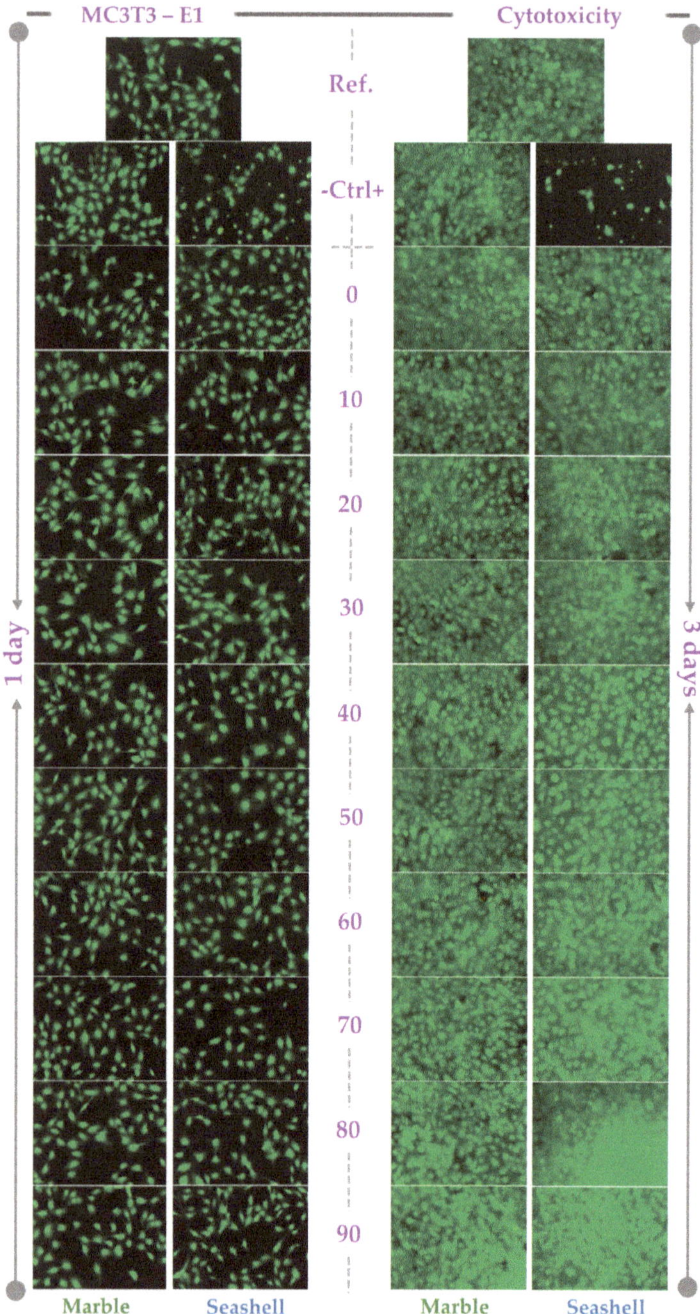

Figure 6. Fluorescence micrographs of the MC3T3-E1 pre-osteoblasts grown in the extracts of marble and seashell-derived powdered samples for 1 day and 3 days. Cell staining with the LIVE/DEAD Cell Viability/Cytotoxicity Assay Kit (green fluorescence: live cells; red fluorescence: dead cells). Scale bar: 100 μm.

However, the fluorescence images acquired after performing the LIVE/DEAD Cell Viability/Cytotoxicity assay indicated that, except for the cytotoxicity positive control, all analyzed extracts exhibited higher cell densities at 3 days-time point than after 1 day of incubation. In comparison with the cytotoxicity negative control, reference sample and the extraction media of all other powdered samples, these findings conducted us to the conclusion that the lower OD values recorded at 3 days of cell incubation for (40–70)-S powdered samples could be rather due to the inhibition of the cell metabolic activity than to a decrease in cell viability and samples proliferation potential. Collectively, the results of the LIVE/DEAD Cell Viability/Cytotoxicity and MTT assays suggest that the extraction media of the M- and S-derived powdered samples exhibit good cytocompatibility and support the viability and proliferation of pre-osteoblast cells to a similar extent with the reference material extraction media and the cytotoxicity negative control.

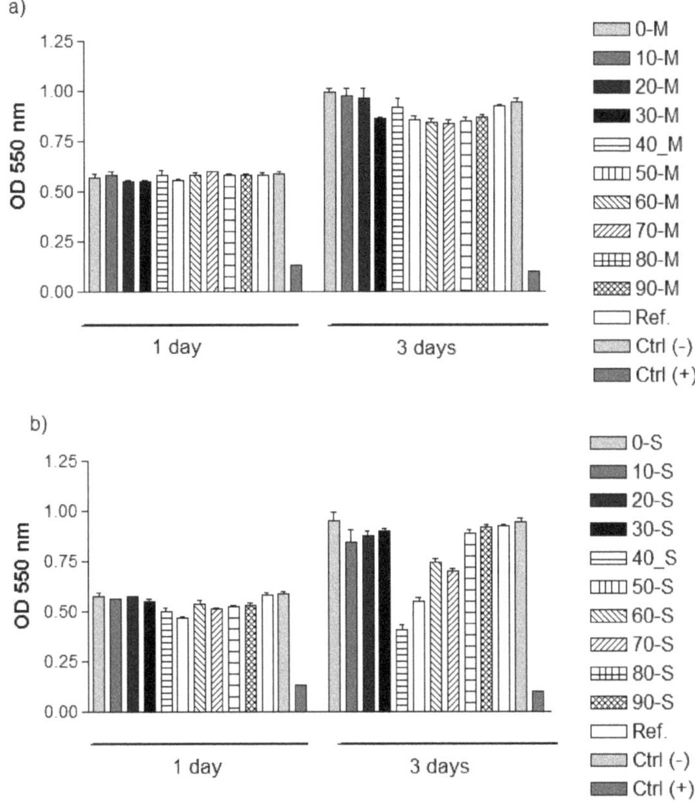

Figure 7. Cell viability/ proliferation of MC3T3-E1 pre-osteoblasts grown in the extraction media of marble- (**a**) and seashell- (**b**) derived powdered samples, as assessed by MTT assay (n = 3, mean $\pm$ SD).

4. Conclusions

This research study focused on the complex structural, morphological and biological characterization of a series of bioceramic samples derived from natural sources as dolomitic marble and seashells with respect to commercial highly crystalline hydroxyapatite.

The adapted indirect synthesis route targeted the maximum additional phosphoric acid amount at which biomimetic biphasic CaPs can be obtained. The results demonstrated that biphasic hydroxyapatite/brushite mix is stable for 0–10% and 10–30% additional to stoichiometric acid amounts in the case of marble- and seashell-derived samples, respectively. Above these concentrations, the biphasic ceramic shifted to brushite/monetite mix, which demonstrated that a minimum of 20% and 40% additional acid amount is necessary for its precipitation.

Since the synthesized materials are intended for reconstructive orthopedics, as bone fillers or cements, biological investigations coupled with digital topographic reconstruction were considered indispensable. The investigated roughness parameters revealed microporous surface textures independent of additional acid amounts or natural precursor. Further, indirect contact in vitro assessment of the MC3T3-E1 pre-osteoblast behavior proved that the extraction media of the powdered samples supported cell viability and proliferation at comparable levels to the ones recorded for the cytotoxicity negative control and commercial hydroxyapatite. The presented experimental approach could represent a first step towards the development of inexpensive yet promising biomedical solutions based on bioceramic biphasic powder systems with applications in bone regeneration.

Author Contributions: Conceptualization, F.M., A.-C.M.; Methodology, F.M., A.-C.M., G.E.S., I.V.A., M.M.; V.M.; Software, F.M., S.I.V., I.V.A.; Validation, F.M., A.C., G.E.S., S.I.V; Formal Analysis, A.-C.M., F.M.; Investigation, F.M., G.E.S., I.V.A., M.M., R.-C.C., S.I.V; V.M.; Resources, A.-C.M., G.E.S., A.M., M.M., I.V.A., A.C., F.M., R.-C.C.; Data Curation, F.M., A.C.; Writing—Original Draft Preparation, A.-C.M., A.M.; V.M.; Writing—Review and Editing, A.-C.M., A.M, F.M., M.M; A.C.; Visualization, A.-C.M, A.M, F.M.; Supervision, F.M., G.E.S., A.C.; Project Administration, G.E.S.; Funding Acquisition, G.E.S.

Funding: G.E.S. acknowledge the support of Romanian Ministry of Research and Innovation, CCCDI—UEFISCDI, in the framework of project PN-III-P1-1.2-PCCDI-2017-0062 (contract no. 58)/ component project no. 2.

Acknowledgments: The authors are thankful to Digital Surf, Besançon, France, for their technical support and for providing the software Mountains Map used in processing of SEM images for evaluating the roughness parameters.

Conflicts of Interest: The authors declare no conflict of interest.

References

1. Idowu, B.; Cama, G.; Deb, S.; Di Silvio, L. In vitro osteoinductive potential of porous monetite for bone tissue engineering. *J. Tissue Eng.* **2014**, *5*, 2041731414536572. [CrossRef]

2. Duta, L.; Mihailescu, N.; Popescu, A.; Luculescu, C.; Mihailescu, I.; Çetin, G.; Gunduz, O.; Oktar, F.; Popa, A.; Kuncser, A. Comparative physical, chemical and biological assessment of simple and titanium-doped ovine dentine-derived hydroxyapatite coatings fabricated by pulsed laser deposition. *Appl. Surf. Sci.* **2017**, *413*, 129–139. [CrossRef]

3. Tite, T.; Popa, A.-C.; Balescu, L.; Bogdan, I.; Pasuk, I.; Ferreira, J.; Stan, G. Cationic substitutions in hydroxyapatite: Current status of the derived biofunctional effects and their in vitro interrogation methods. *Materials* **2018**, *11*, 2081. [CrossRef] [PubMed]

4. Antoniac, I.; Negrusoiu, M.; Mardare, M.; Socoliuc, C.; Zazgyva, A.; Niculescu, M. Adverse local tissue reaction after 2 revision hip replacements for ceramic liner fracture: A case report. *Medicine* **2017**, *96*, 6687. [CrossRef] [PubMed]

5. Montazerolghaem, M.; Ott, M.K.; Engqvist, H.; Melhus, H.; Rasmusson, A. Resorption of monetite calcium phosphate cement by mouse bone marrow derived osteoclasts. *Mater. Sci. Eng. C* **2015**, *52*, 212–218. [CrossRef] [PubMed]

6. Cama, G.; Nkhwa, S.; Gharibi, B.; Lagazzo, A.; Cabella, R.; Carbone, C.; Dubruel, P.; Haugen, H.; Di Silvio, L.; Deb, S. The role of new zinc incorporated monetite cements on osteogenic differentiation of human mesenchymal stem cells. *Mater. Sci. Eng. C* **2017**, *78*, 485–494. [CrossRef] [PubMed]

7. O'Halloran, M. Cellular responses to chondroitin-6-sulphate releasing brushite bone cements. *J. Res. Pract. Dent.* **2014**, *2014*, 1–19. [CrossRef]

8. Richard, R.C.; Sader, M.S.; Dai, J.; Thiré, R.M.; Soares, G.D. Beta-type calcium phosphates with and without magnesium: From hydrolysis of brushite powder to robocasting of periodic scaffolds. *J. Biomed. Mater. Res. Part A* **2014**, *102*, 3685–3692. [CrossRef]

9. Trbakovic, A.; Hedenqvist, P.; Mellgren, T.; Ley, C.; Hilborn, J.; Ossipov, D.; Ekman, S.; Johansson, C.B.; Jensen-Waern, M.; Thor, A. A new synthetic granular calcium phosphate compound induces new bone in a sinus lift rabbit model. *J. Dent.* **2018**, *70*, 31–39. [CrossRef]

10. Fernandes, H.R.; Gaddam, A.; Rebelo, A.; Brazete, D.; Stan, G.E.; Ferreira, J.M. Bioactive glasses and glass-ceramics for healthcare applications in bone regeneration and tissue engineering. *Materials* **2018**, *11*, 2530. [CrossRef]

11. Miculescu, F.; Mocanu, A.-C.; Dascălu, C.A.; Maidaniuc, A.; Batalu, D.; Berbecaru, A.; Voicu, S.I.; Miculescu, M.; Thakur, V.K.; Ciocan, L.T. Facile synthesis and characterization of hydroxyapatite particles for high value nanocomposites and biomaterials. *Vacuum* **2017**, *146*, 614–622. [CrossRef]

12. Miculescu, F.; Mocanu, A.C.; Stan, G.E.; Miculescu, M.; Maidaniuc, A.; Cîmpean, A.; Mitran, V.; Voicu, S.I.; Machedon-Pisu, T.; Ciocan, L.T. Influence of the modulated two-step synthesis of biogenic hydroxyapatite on biomimetic products' surface. *Appl. Surf. Sci.* **2017**, *438*, 147–157. [CrossRef]

13. Miculescu, F.; Stan, G.; Ciocan, L.; Miculescu, M.; Berbecaru, A.; Antoniac, I. Cortical bone as resource for producing biomimetic materials for clinical use. *Dig. J. Nanomater. Biostruct.* **2012**, *7*, 1667–1677.

14. Onoda, H.; Yamazaki, S. Homogenous hydrothermal synthesis of calcium phosphate with calcium carbonate and corbicula shells. *J. Asian Ceram. Soc.* **2016**, *4*, 403–406. [CrossRef]

15. Maidaniuc, A.; Miculescu, F.; Voicu, S.I.; Andronescu, C.; Miculescu, M.; Matei, E.; Mocanu, A.C.; Pencea, I.; Csaki, I.; Machedon-Pisu, T. Induced wettability and surface-volume correlation of composition for bovine bone derived hydroxyapatite particles. *Appl. Surf. Sci.* **2018**, *438*, 158–166. [CrossRef]

16. Maidaniuc, A.; Miculescu, M.; Voicu, S.; Ciocan, L.; Niculescu, M.; Corobea, M.; Rada, M.; Miculescu, F. Effect of micron sized silver particles concentration on the adhesion induced by sintering and antibacterial properties of hydroxyapatite microcomposites. *J. Adhes. Sci. Technol.* **2016**, *30*, 1829–1841. [CrossRef]

17. Miculescu, F.; Ciocan, L.; Miculescu, M.; Ernuteanu, A. Effect of heating process on micro structure level of cortical bone prepared for compositional analysis. *Dig. J. Nanomater. Biostruct.* **2011**, *6*, 225–233.

18. Pandele, A.; Comanici, F.; Carp, C.; Miculescu, F.; Voicu, S.; Thakur, V.; Serban, B. Synthesis and characterization of cellulose acetate-hydroxyapatite micro and nano composites membranes for water purification and biomedical applications. *Vacuum* **2017**, *146*, 599–605. [CrossRef]

19. Miculescu, F.; Bojin, D.; Ciocan, L.; Antoniac, I.; Miculescu, M.; Miculescu, N. Experimental researches on biomaterial-tissue interface interactions. *J. Optoelectron. Adv. Mater.* **2007**, *9*, 3303–3306.

20. Vranceanu, M.; Antoniac, I.; Miculescu, F.; Saban, R. The influence of the ceramic phase on the porosity of some biocomposites with collagen matrix used as bone substitutes. *J. Optoelectron. Adv. Mater.* **2012**, *14*, 671–677.

21. Rentsch, B.; Bernhardt, A.; Henß, A.; Ray, S.; Rentsch, C.; Schamel, M.; Gbureck, U.; Gelinsky, M.; Rammelt, S.; Lode, A. Trivalent chromium incorporated in a crystalline calcium phosphate matrix accelerates materials degradation and bone formation in vivo. *Acta Biomater.* **2018**, *69*, 332–341. [CrossRef] [PubMed]

22. Kanter, B.; Geffers, M.; Ignatius, A.; Gbureck, U. Control of in vivo mineral bone cement degradation. *Acta Biomater.* **2014**, *10*, 3279–3287. [CrossRef] [PubMed]

23. Kruppke, B.; Farack, J.; Wagner, A.-S.; Beckmann, S.; Heinemann, C.; Glenske, K.; Rößler, S.; Wiesmann, H.-P.; Wenisch, S.; Hanke, T. Gelatine modified monetite as a bone substitute material: An in vitro assessment of bone biocompatibility. *Acta Biomater.* **2016**, *32*, 275–285. [CrossRef] [PubMed]

24. Higuita, L.P.; Vargas, A.F.; Gil, M.J.; Giraldo, L.F. Synthesis and characterization of nanocomposite based on hydroxyapatite and monetite. *Mater. Lett.* **2016**, *175*, 169–172. [CrossRef]

25. Parvinzadeh Gashti, M.; Stir, M.; Bourquin, M.; Hulliger, J. Mineralization of calcium phosphate crystals in starch template inducing a brushite kidney stone biomimetic composite. *Cryst. Growth Des.* **2013**, *13*, 2166–2173. [CrossRef]

26. Liu, B.; Guo, Y.-Y.; Xiao, G.-Y.; Lu, Y.-P. Preparation of micro/nano-fibrous brushite coating on titanium via chemical conversion for biomedical applications. *Appl. Surf. Sci.* **2017**, *399*, 367–374. [CrossRef]

27. Schamel, M.; Barralet, J.E.; Groll, J.; Gbureck, U. In vitro ion adsorption and cytocompatibility of dicalcium phosphate ceramics. *Biomater. Res.* **2017**, *21*, 10. [CrossRef] [PubMed]

28. Ross, A.M.; Jiang, Z.; Bastmeyer, M.; Lahann, J. Physical aspects of cell culture substrates: Topography, roughness, and elasticity. *Small* **2012**, *8*, 336–355. [CrossRef]

29. Pandele, A.M.; Andronescu, C.; Ghebaur, A.; Garea, S.A.; Iovu, H. New biocompatible mesoporous silica/polysaccharide hybrid materials as possible drug delivery systems. *Materials* **2019**, *12*, 15. [CrossRef]

30. Iulian, A.; Cosmin, S.; Aurora, A. *Adhesion Aspects in Biomaterials and Medical Devices*; Taylor & Francis: Boca Raton, FL, USA, 2016.

31. Ventre, M.; Natale, C.F.; Rianna, C.; Netti, P.A. Topographic cell instructive patterns to control cell adhesion, polarization and migration. *J. R. Soc. Interface* **2014**, *11*, 20140687. [CrossRef]

32. Barbosa, T.P.; Naves, M.M.; Menezes, H.H.M.; Pinto, P.H.C.; de Mello, J.D.B.; Costa, H.L. Topography and surface energy of dental implants: A methodological approach. *J. Braz. Soc. Mech. Sci. Eng.* **2017**, *39*, 1895–1907. [CrossRef]

33. Wang, K.; Zhou, C.; Hong, Y.; Zhang, X. A review of protein adsorption on bioceramics. *Interface Focus* **2012**, *2*, 259–277. [CrossRef] [PubMed]

34. Salerno, M.; Reverberi, A.P.; Baino, F. Nanoscale topographical characterization of orbital implant materials. *Materials* **2018**, *11*, 660. [CrossRef] [PubMed]

35. Popa, A.; Stan, G.; Husanu, M.; Mercioniu, I.; Santos, L.; Fernandes, H.; Ferreira, J. Bioglass implant-coating interactions in synthetic physiological fluids with varying degrees of biomimicry. *Int. J. Nanomed.* **2017**, *12*, 683–707. [CrossRef] [PubMed]

36. Miculescu, F.; Jepu, I.; Lungu, C.; Miculescu, M.; Bane, M. Researches regarding the microanalysis results optimisation on multilayer nanostructures investigations. *Dig. J. Nanomater. Biostruct.* **2011**, *6*, 769–778.

37. Feller, L.; Jadwat, Y.; Khammissa, R.A.; Meyerov, R.; Schechter, I.; Lemmer, J. Cellular responses evoked by different surface characteristics of intraosseous titanium implants. *Biomed Res. Int.* **2015**, *2015*, 171945. [CrossRef]

38. Wennerberg, A.; Albrektsson, T. Suggested guidelines for the topographic evaluation of implant surfaces. *Int. J. Oral Maxillofac. Implant.* **2000**, *15*, 331–344.

39. ISO:4287. *Geometrical Product Specifications (GPS). Surface Texture: Profile Method. Terms, Definitions and Surface Texture Parameters*; International Organization for Standardization: Geneva, Switzerland, 1997.

40. Yoon, H.I.; Yeo, I.S.; Yang, J.H. Effect of a macroscopic groove on bone response and implant stability. *Clin. Oral Implant. Res.* **2010**, *21*, 1379–1385. [CrossRef]

41. Ion, R.; Drob, S.I.; Ijaz, M.F.; Vasilescu, C.; Osiceanu, P.; Gordin, D.M.; Cimpean, A.; Gloriant, T. Surface characterization, corrosion resistance and in vitro biocompatibility of a new Ti-Hf-Mo-Sn alloy. *Materials* **2016**, *9*, 818. [CrossRef]

42. Yanovska, A.; Kuznetsov, V.; Stanislavov, A.; Danilchenko, S.; Sukhodub, L. A study of brushite crystallization from calcium-phosphate solution in the presence of magnesium under the action of a low magnetic field. *Mater. Sci. Eng. C* **2012**, *32*, 1883–1887. [CrossRef]

43. Rosa, S.; Madsen, H.E.L. Influence of some foreign metal ions on crystal growth kinetics of brushite ($CaHPO_4 \cdot 2H_2O$). *J. Cryst. Growth* **2010**, *312*, 2983–2988. [CrossRef]

44. Dorozhkin, S.V. Calcium orthophosphate-based bioceramics. *Materials* **2013**, *6*, 3840–3942. [CrossRef] [PubMed]

45. Cama, G.; Barberis, F.; Capurro, M.; Di Silvio, L.; Deb, S. Tailoring brushite for in situ setting bone cements. *Mater. Chem. Phys.* **2011**, *130*, 1139–1145. [CrossRef]

46. Tamimi, F.; Sheikh, Z.; Barralet, J. Dicalcium phosphate cements: Brushite and monetite. *Acta Biomater.* **2012**, *8*, 474–487. [CrossRef] [PubMed]

47. Ginebra, M.-P.; Canal, C.; Espanol, M.; Pastorino, D.; Montufar, E.B. Calcium phosphate cements as drug delivery materials. *Adv. Drug Deliv. Rev.* **2012**, *64*, 1090–1110. [CrossRef] [PubMed]

48. Markovic, M.; Fowler, B.O.; Tung, M.S. Preparation and comprehensive characterization of a calcium hydroxyapatite reference material. *J. Res. Natl. Inst. Stand. Technol.* **2004**, *109*, 553. [CrossRef] [PubMed]

49. Xu, J.; Butler, I.S.; Gilson, D.F. Ft-raman and high-pressure infrared spectroscopic studies of dicalcium phosphate dihydrate ($CaHPO_4 \cdot 2H_2O$) and anhydrous dicalcium phosphate ($CaHPO_4$). *Spectrochim. Acta Part A Mol. Biomol. Spectrosc.* **1999**, *55*, 2801–2809. [CrossRef]

50. Boulle, A.; Lang-Dupont, M. Infrared study of the dehydration and rehydration of $CaHPO_4$. *Compt. Rend* **1955**, *241*, 1927. [CrossRef]

51. Petrov, I.; Šoptrajanov, B.; Fuson, N.; Lawson, J. Infra-red investigation of dicalcium phosphates. *Spectrochim. Acta Part A: Mol. Spectrosc.* **1967**, *23*, 2637–2646. [CrossRef]

52. Tortet, L.; Gavarri, J.; Nihoul, G.; Dianoux, A. Study of protonic mobility in $CaHPO_4 \cdot 2H_2O$ (brushite) and cahpo4 (monetite) by infrared spectroscopy and neutron scattering. *J. Solid State Chem.* **1997**, *132*, 6–16. [CrossRef]

53. Rajendran, K.; Dale Keefe, C. Growth and characterization of calcium hydrogen phosphate dihydrate crystals from single diffusion gel technique. *Cryst. Res. Technol.* **2010**, *45*, 939–945. [CrossRef]
54. Stuart, B.W.; Murray, J.W.; Grant, D.M. Two step porosification of biomimetic thin-film hydroxyapatite/alpha-tri calcium phosphate coatings by pulsed electron beam irradiation. *Sci. Rep.* **2018**, *8*, 14530. [CrossRef] [PubMed]

Article

3D Superparamagnetic Scaffolds for Bone Mineralization under Static Magnetic Field Stimulation

Irina Alexandra Paun [1,2,*], Bogdan Stefanita Calin [1,2], Cosmin Catalin Mustaciosu [3], Mona Mihailescu [2], Antoniu Moldovan [4], Ovidiu Crisan [5], Aurel Leca [5] and Catalin Romeo Luculescu [1]

[1]   Center for Advanced Laser Technologies (CETAL), National Institute for Laser, Plasma and Radiation Physics, RO-077125 Magurele-Ilfov, Romania
[2]   Physics Department, Faculty of Applied Sciences, University Politehnica of Bucharest, RO-060042 Bucharest, Romania
[3]   Horia Hulubei National Institute for Physics and Nuclear Engineering IFIN-HH, RO-077125 Magurele-Ilfov, Romania
[4]   National Institute for Laser, Plasma and Radiation Physics, RO-077125 Magurele-Ilfov, Romania
[5]   National Institute of Materials Physics, RO-077125 Magurele-Ilfov, Romania
*   Correspondence: irina.paun@inflpr.ro

Received: 7 July 2019; Accepted: 26 August 2019; Published: 3 September 2019

Abstract: We reported on three-dimensional (3D) superparamagnetic scaffolds that enhanced the mineralization of magnetic nanoparticle-free osteoblast cells. The scaffolds were fabricated with submicronic resolution by laser direct writing via two photons polymerization of Ormocore/magnetic nanoparticles (MNPs) composites and possessed complex and reproducible architectures. MNPs with a diameter of 4.9 ± 1.5 nm and saturation magnetization of 30 emu/g were added to Ormocore, in concentrations of 0, 2 and 4 mg/mL. The homogenous distribution and the concentration of the MNPs from the unpolymerized Ormocore/MNPs composite were preserved after the photopolymerization process. The MNPs in the scaffolds retained their superparamagnetic behavior. The specific magnetizations of the scaffolds with 2 and 4 mg/mL MNPs concentrations were of 14 emu/g and 17 emu/g, respectively. The MNPs reduced the shrinkage of the structures from 80.2 ± 5.3% for scaffolds without MNPs to 20.7 ± 4.7% for scaffolds with 4 mg/mL MNPs. Osteoblast cells seeded on scaffolds exposed to static magnetic field of 1.3 T deformed the regular architecture of the scaffolds and evoked faster mineralization in comparison to unstimulated samples. Scaffolds deformation and extracellular matrix mineralization under static magnetic field (SMF) exposure increased with increasing MNPs concentration. The results are discussed in the frame of gradient magnetic fields of ~3 × 10^{-4} T/m generated by MNPs over the cells bodies.

Keywords: superparamagnetic scaffold; composite; laser direct writing; static magnetic field; extracellular matrix mineralization

---

1. Introduction

Bone is the second most commonly transplanted tissue, preceded only by blood transfusion [1]. The cost of osteoporotic fractures is estimated to reach 77 billion euros by 2050 [2]. Within this context, it is imperative to achieve the functional and structural restoration of damaged bone tissue [3–5]. A major difficulty in bone tissue engineering arises from the fact that the bone regeneration process requires a long time for achieving a completely functional tissue [6]. Generally, cells are seeded ex vivo into a three-dimensional (3D) biocompatible and sometimes biodegradable structure called scaffold, where they attach and grow. After the implantation into the injured site, the scaffolds should allow proper host cell colonization for regeneration purposes [7–10].

Magnetic scaffolds emerged as promising solution for this purpose. Activation of the magnetic scaffolds using external static magnetic fields (SMF) prevents the decrease of bone mineral density [11] and promotes the bone regeneration in bone fractures [12]. The significant alterations in cell behaviors stimulated by the externally applied magnetic fields has been demonstrated in numerous studies [8,11]. For example, it has been shown that an externally applied SMF using a magnet accelerates the osteogenic differentiation of osteoblasts-like cells in vitro and triggers peri-implant bone formation in vivo.

The magnetism can also be used through scaffolding materials with magnetic properties. For example, biomaterials that incorporate magnetic nanoparticles (MNPs) are being developed [6,13–17]. The superparamagnetic behavior of the MNPs increased the adhesion and differentiation of osteoblastic cells in vitro and the bone formation in vivo [13–18]. Structures with such intrinsic magnetic properties represent a promising biomatrix for bone tissue engineering [13–17,19,20]. It was also shown that the changes in the magnetic properties of MNPs in the presence of a magnetic field had no influence on cellular toxicity [6]. Additionally, magnetic scaffolds with incorporated MNPs increased the mechanical strength of the scaffolds and promoted the osteogenic differentiation of the seeded cells [5,13–15]. Moreover, the use of MNPs results in superior physiochemical properties of the material and closer replication of the hierarchical nanostructure of bone tissue [16,17]. Furthermore, the iron metabolism facilitates the proliferation of bone or non-bone cell lines [18–20] and has a positive influence on the bone density [21,22].

Previous attempts to fabricate magnetically-active scaffolds employed ceramics, gelatin or polymers that were impregnated with MNPs by freeze drying, deep coating and direct nucleation of fiber deposition [23]. Porous polycaprolactone scaffolds loaded with MNPs stimulated in external SMF promoted the osteoblastic differentiation of primary mouse calvarial osteoblasts [24]. A time-dependent magnetic field applied on 3D cylindrical poly(ε-caprolactone)/iron-doped hydroxyapatite nanocomposite scaffold fabricated by fiber deposition had osteogenic effects on seeded human mesenchymal stem cells [25]. The newest approaches assign the benefic role of the MNPs on the cellular behavior to the existence of high magnetic field gradients that traverse the cell bodies [26,27]. In scaffolds with incorporated MNPs, the nanoparticles concentrate the externally applied magnetic field and produce high gradients magnetic fields across the cells bodies [26,27]. It has been shown that in SMFs with gradients above 10^4 T/m the magnetic force magnitudes are comparable with the gravitational forces and affect the cell machinery [26,27]. Such magnetic field gradients promote the cell migration to the areas with the strongest magnetic field gradient. In particular, enhanced bone regeneration in osteoblast-like cells seeded on scaffolds with incorporated MNPs has been be explained through the integrins- and bone morphogenetic proteins-mediated signaling pathways, which improve the osteoblasts' functions and is beneficial for bone formation [24,28]. Despite these advantages, the fabrication of scaffolds containing MNPs for orthopedic applications has been restricted to few studies and the mechanism of action of SMFs on the bone regeneration process remains unknown [29,30]. Furthermore, the composite magnetic scaffolds reported previously provide with no control over the amount of loaded MNPs [31,32].

Currently, the major challenge is to fabricate magnetic scaffolds with reproducible architectures that contain precise MNPs concentrations and have a homogenous distribution of the nanoparticles over the scaffolds' structure [23,31,32]. In this study, we report a new method for fabricating innovative magnetic scaffolds with incorporated MNPs having unique advantages compared to the scaffolds reported by previous works. Specifically, the scaffolds developed in our study possess fully controllable 3D architectures, the MNPs are distributed in the scaffolds in precise concentrations, they have a homogenous distribution in the whole scaffolds' structure and preserve their superparamagnetic behavior. The combination of materials (photopolymer and MNPs) and the fact that the photopolymer/MNPs composite is processed by laser direct writing via two photons polymerization represent the original aspects of the work.

The scaffolds were fabricated from photopolymer/MNPs composites by laser direct writing via two photons polymerization (LDW via TPP) and tested in respect with osteogenic potential [33–37]. The photopolymer Ormocore was employed as 3D structurable material because of its biocompatibility and suitability for bone tissue engineering [38,39]. LDW via TPP technique is a sort of 3D printing that creates objects from 3D model data. To date, it has been used for processing magnetic nanocomposites mostly in combination with other techniques, such as electrodeposition and selective electroless magnetite plating [40–42]. While it was possible to create structures that demonstrate a proof-of-principle, the results were generally unreliable for practical applications. The scaffolds were seeded with nanoparticle-free osteoblast-like cells and exposed to static magnetic field of 1.3 T. The scaffolds' ability to control the cells behavior in terms of cells attachment and early extracellular matrix mineralization was assessed. The results were discussed in the frame of high gradient magnetic fields generated by the MNPs over the cells bodies.

2. Materials and Methods

2.1. Materials

The photopolymer (Ormocore) and the developer (Ormodev) were purchased from Micro resist technology GmbH (Berlin, Germany). The superparamagnetic nanoparticles with 4.9 ± 1.5 nm diameters and maghemite structure (gamma–Fe_2O_3) were produced by laser pyrolysis in identical experimental conditions as those reported in [33]. The laser pyrolysis technique relies on the laser-driven heating of an iron precursor in vapor phase in presence of oxygen [33,34]. The experimental parameters used for producing the MNPs used in this study are reported in [33]: laser power (CO_2 laser) 55 W, beam diameter 1.5 mm, $Fe(CO)_5$ flux 19 sccm, carrier gas flux 100 C_2H_4 + 70 Air sccm, productivity about 3.3 g/h. The saturation magnetization was 30 emu/g at room temperature, as determined by [34].

2.2. Scaffolds Design and Fabrication

Ormocore/MNPs composites were prepared by adding MNPs in Ormocore, in 0, 2 and 4 mg/mL concentrations. The homogeneous dispersion of MNPs in Ormocore viscous liquid formulation is essential for obtaining 3D scaffolds by proposed method. The unpolymerized Ormocore/MNPs composite was homogenized by 1000 W powerful ultrasonicator at 20 kHz (Hielscher Ultrasonics GmbH, Model UIP1000hdT) for about 30 s. Dispersions with MNPs concentrations in Ormocore up to 32 mg/mL showed good stability for several months. The stability of the unpolymerized i.e. liquid Ormocore/MNPs composite is important for the laser direct writing process, since any inhomogeneity of the irradiated material causes irregularities in the morphology of the scaffolds or it can even impede the photopolymerization process. In general, the stability of a dispersion is evaluated on a case-by-case basis, because it depends on how long we need the system to remain stable. In our experimental conditions, the unpolymerized, i.e., liquid Ormocore/MNPs composite only needs to be stable for several minutes, because this is how long the laser direct writing of the scaffolds lasts. Since the evaluation of the long-term stability of a dispersion is a rather complicated process and since we do not need such long time scales for the stability of our dispersions, in our experimental conditions we resumed to monitor the stability of the unpolymerized Ormocore/MNPs composite by visual inspection. For this, drops of unpolymerized composite were placed on glass slides and visualized under the optical microscope of the Nanoscribe system that was able to image any clusters formed by MNPs aggregation.

The scaffolds design was calculated using Python 3.6.6. All information related to structure geometry was delivered as a list of carthesian points, appropriately configured for the 3D lithography installation (Nanoscribe Photonic Professional). The design of the microstructures is presented in Figure 1. As a basis, we started from the optimized geometry reported in [35] that provided suitable porosity and mechanical resilience for the attachment and growth of osteoblast cells. In our recent study [35], we reported that when consecutive layers of ellipsoidal units were not separated on the vertical axis, the cells were not able to penetrate inside the structure of the scaffold and covered only the outer areas. For populating the whole volume of the scaffolds with interconnecting cells, the spacing between neighboring layers had to be increased with respect to the Z-axis and this was achieved by separating the consecutive layers of ellipsoidal units using cylindrical pillars.

The scaffolds were fabricated by laser direct writing via two-photon polymerization (LDW via TPP) [36]. The typical processing methodology consists in drop-casting several μL of photopolymerizable material on a glass substrate, followed by laser irradiation and sample development. We used 170 μm thick BK7 glass slides as substrates. The glass slides were cleaned using isopropanol. The Ormocore/MNPs composites were irradiated with 120 fs pulses, with a central wavelength $\lambda = 780$ nm, and a frequency of 80 MHz. Both the laser focus and the sample were mobile (sample on X-Y axes, laser beam on Z-axis). For high resolution sample positioning, the laser processing system uses a set of three synchronized piezoelectric stages. After the laser writing, the obtained Ormocore/MNPs composite scaffolds require no additional pre- or post-processing steps other than immersion in Ormodev solution for 3 min, to wash away the non-polymerized material.

2.3. Scaffolds Characterization

Scanning Electron Microscopy (SEM): the morphology of the magnetic scaffolds was investigated by Scanning Electron Microscopy (SEM, FEI InspectS model, Thermo Fisher Scientific, Waltham, MA, USA), using a 5 kV voltage. Prior to examination, the scaffolds were coated with a 10 nm layer of gold. Scaffolds shrinkage was calculated as [(bottom area − top area)/bottom area] × 100, where the top and bottom areas were determined from SEM images.

Enhanced Dark-Field Microscopy (EDFM): the location and distribution of the MNPs inside the scaffolds were investigated using CytoViva system (CytoViva Inc., Auburn, AL, USA), without any prior special preparation and in a nondestructive manner. CytoViva comprises a dark-field set illuminator that focuses at diagonal inclinations over the sample and is suitable to investigate translucent materials, based on the scattered light by the nanometric details of the sample. The technique has the capability of high signal-to-noise optical performance based on patent-pending deconvolution and particle location routines providing three dimensional optical image of the sample. The Z stacks images were collected at 100 nm between slices using a 60× oil immersion objective on Q-imaging Exi Blue Charged Coupled Device (CCD) (6.45 × 6.45 μm pixel pitch) at different exposure times, depending on the sample scattering. Two series of stacks (using a piezo-driven Z-axis stage) were acquired for each sample: one with fluorescein (FITC) excited filter with emission at 530 nm, to reconstruct the polymeric structures which are fluorescent at this wavelength, and one in white light, used to locate the nanoparticles in the polymerized Ormocore/MNPs composite.

To process the stacks of images, dedicated plugins were developed by the producer (CytoViva Inc., Auburn, AL, USA), under ImageJ software. The processing procedure started with the synchronization step for all stacks acquired for a given zone of the sample, in order to delimitate the region of interest (which is about 500 × 500 pixels). After this, the processing was different for the stacks acquired in fluorescence (which included the generation of point spread function, iterations for deconvolution until a threshold value was reached, all these being done using parameters like magnification, wavelength, refractive index of immersed oil, x, y, z voxel spacing, mean delta between consecutive iterations) and for the stacks acquired in white light (achieved by using the routine "Just locate nanoparticles", establishing the scattered intensity threshold and the number of pixels to represent one nanoparticle). For the investigations, we fabricated the samples in the same conditions as those used for fabricating the scaffolds (MNPs concentration, laser parameters for LDW via TPP process), but with only one layer of ellipses to avoid unnecessary scattering from multiple layers. This did not affect the material behavior or the nanoparticles distribution. The following settings were employed: magnification 60×, pixel dimension 107.5 nm in x–y transversal plane and 100 nm in z direction, oil refractive index 1.516.

We maintained the same parameters for all samples. The point spread functions were generated for each stack and the deconvolution routine was run until the mean delta was above 0.001 (for stacks acquired in fluorescence). After that, we generated 3D images only with the ellipsoidal units for each region of interest. The routine "Just locate nanoparticles" was run for all slices, in a stack acquired in white light. It returned 3D images where the MNPs were represented in red and a table with their number and location. Finally, the two 3D images (ellipsoidal units and MNPs respectively) were superposed in ImageJ.

Magnetic Force Microscopy (MFM): the MFM analysis was carried out using a commercial AFM (XE100, Park Systems, Suwon, Korea) with magnetic coated tips (PPP-MFMR, Nanosensors, Thermo Fisher Scientific, Waltham, MA, USA). The MFM images were recorded during a second pass, at a height of 100 nm from the topography scan, using the MFM phase signal. The lift height was selected to be 100 nm because of the specific topography of the samples.

Energy-Dispersive X-ray Spectroscopy (EDS) was performed at 5 kV acceleration voltage inside Scanning Electron Microscopy (FEI InspectS model, Thermo Fisher Scientific, Waltham, MA, USA) using a Si(Li) detector (EDAX Inc., Thermo Fisher Scientific, Waltham, MA, USA). In order to avoid errors in EDS measurements of porous samples/scaffolds, the rectangular structures of $200 \times 200 \times 20$ μm^2 were fabricated by LDW via TPP of Ormocore/MNPs composites with 0, 2 and 4 mg/mL MNPs concentrations, in identical experimental conditions as the scaffolds. The EDS results are obtained from the average of three different measurements over 40×50 μm^2 areas of polymerized composites, using standardless ZAF analysis. The trace analysis for iron provided errors under 0.5 percent.

Magnetization Measurements have been done using a vibrating sample magnetometer (VSM) module of a Physical Property Measurement System (PPMS) from Quantum Design, Inc., Bucharest, Romania. Initial magnetization versus applied magnetic field as well as major hysteresis loops have been recorded for the scaffolds with MNPs at 300 K in applied magnetic field of up to 5 T. The measurements have been taken with the applied field perpendicular to the scaffold basal plane.

2.4. Biological Assessments

Cells seeding: MG-63 osteoblast-like cells were purchased from ECACC (European Collection of Cell Cultures, Salisbury, UK). The cells were cultured in a 25 cm^2 flask, incubated in an atmosphere of 5% CO_2 at 37 °C for 24 h and cultured in Minimal Essential Medium, Biochrom containing 10% fetal bovine serum (FBS, Biochrom), 2 mM L-glutamine and 1% non-essential amino acids (complete medium). 100 IU/mL of penicillin/streptomycin was added to the solution. After confluency, the cells were detached with trypsin and seeded on the scaffolds. A cell density of 5000 cells/sample from the 16th cell passage was used. The cells in normal medium were seeded on top of the scaffolds with the aid of a sterile syringe. All chemicals were purchased from Sigma-Aldrich, unnless otherwise specified. Prior cell seeding, the scaffolds were sterilized for 3 h under a UV lamp.

Static Magnetic Field Stimulation (SMF) of the Cell-Seeded Scaffolds: nickel-plated NdFeB rectangular magnets ($40 \times 40 \times 20$ mm^3) with residual magnetism of 1.3 T were purchased from Supermagnete (Gottmadingen, Germany). For SMF stimulation, each cell-seeded scaffold was placed in close vicinity of a magnet. The magnetic stimulation ranged from 3 to 20 days. According to Zablotskii et al., 2016, these timescales of SMF exposure most likely lead to changes at the level of cell shape and size [26]. Control experiments were carried out on scaffolds without SMF exposure. The heating effects in superparamegnetic nanoparticles occur only in the presence of an alternating external magnetic field. Otherwise, like in our experimental conditions where only static magnetic fields are employed, the MNPs act as fillers that reinforce the scaffolds structure and become magnetized only in the presence of the magnetic field, without any thermal effects.

Cells Morphological Investigations: the cell-seeded scaffolds were washed with PBS and fixed for 1 h at 37 °C with 2.5% glutaraldehyde prepared in PBS. The samples were then washed with PBS and dehydrated using a two-steps protocol. In the first step, the samples were dehydrated/washed in ethanol (EtOH) solutions as follows: 2×15 min in EtOH 70%, 2×15 min in EtOH 90% and 2×15 min in EtOH 100%. In the second step, the samples were washed for 3 min in EtOH:HMDS solutions, prepared in 50%:50%; 25%:75% and 0%:100% ratios. Prior to SEM analysis, the samples were left to dry and sputtered with 10 nm of gold. Scanning electron micrographs were recorded with FEI InspectS model. The cells morphology was investigated after 3 days of cultivation.

Early Mineralization Assay by Alizarin Red S Staining: the cell-seeded scaffolds were analyzed via Alizarin Red S osteogenic differentiation assay that provides qualitative information about the calcium deposits formed in the samples [37]. The cell-seeded scaffolds were washed twice with double-distilled water. Next, 1 mL of 40 mM Alizarin Red S (Sigma Aldrich) (pH 4.1) were added per well. The samples were incubated at room temperature for 20 min and then washed three times with double-distilled water, while shaking. Images of the samples were recorded under a Nikon Eclipse Ti-U microscope equipped with a fluorescence module. The quantification of mineralization was achieved by extracting the calcified mineral at low pH, followed by neutralization with ammonium hydroxide and absorbance measurement at 405 nm. The measurements were performed after 20 days of incubation.

MTS Assay: 5000 cells/ sample were cultured in complete Minimum Essential Medium (MEM) for 3 days in standard conditions of temperature and humidity. The culture medium was then replaced with 16.67% MTS (Cell Titer 96®Aqueous One Solution Cell Proliferation Assay, Promega) and 83.33% MEM (5% FBS). The supernatant was collected after 3 h of incubation. 100 µL from each sample were distributed in a 96-well plate and the absorbance was measured at 490 nm using a Mitras LB 940 (Berthold Technologies, Bad Wildbad, Germany) spectrophotometer. The viability was calculated as percent from control (cells seeded on glass slides).

Statistical Analysis: for MTS, Alizarin Red Staining (ARS) fluorescence intensity and mineralization assays, the statistical analysis was carried out on five different measurements, with student's t test, where $p < 0.05$ indicates a significant result.

3. Results

3.1. Scaffolds Fabrication and Characterization

The scaffolds are composed of elliptical elements of 10 µm in high, disposed in a rectangular matrix. Consecutive levels of ellipses were separated by cylindrical pillars with a diameter of 5 µm and a height of 20 µm. The pillars were placed at the overlap of neighboring elliptical elements on the Y axis (Figure 1).

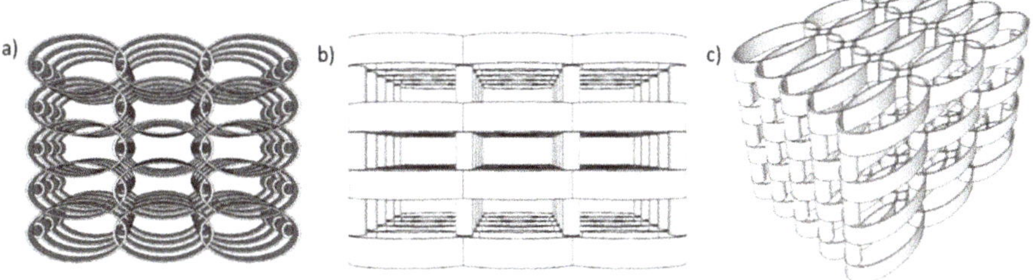

Figure 1. Line plot example of the optimized scaffold design with five layers: (**a**) top view; (**b**) lateral view; (**c**) inclined view.

Figure 2 displays scanning electron micrographs of scaffolds containing different MNPs concentrations. Dispersions with concentrations up to 32 mg/mL and good stability for several months were prepared. However, increasing the concentration of MNPs above 4 mg/mL impeded the photopolymerization process. Most probably, the high density of the MNPs in the photopolymer overheated the material, as proved by extensive bubbles formation followed by local microexplosions in the irradiated volume observed during the laser direct writing process. To provide evidence for this experimental observation, we studied the images recorded with the CCD camera that followed in real time what happened when we irradiated with the focused laser beam the Ormocore/MNPs composite having MNPs concentration above 4 mg/mL (please see the Supplementary information-Figure S1, at the end of the manuscript). There was no trace of polymerized material visible on the glass slide, only the laser spot appears as a small bright zone (Figure S1a). Few seconds later during the laser direct writing process, bubbles were formed (Figure S1b), followed by local micro-explosions of the irradiated material (Figure S1c). Most probably, this happened because the high density of MNPs in the polymer increased significantly the laser absorption at the irradiation spot overheating the material, followed by bubble formation and local micro-explosions. At the end of the laser writing process, no traces of polymerized material were found on the glass slide.

The scaffolds without MNPs underwent a strong shrinkage of top surface and the whole structure reorganized in the shape of a tent (Figure 2a). With increasing MNPs concentrations, the scaffolds' structural integrity was much improved, indicating that the MNPs play a significant role in structure reinforcement (Figure 2b,c). The shrinkage of scaffolds with different MNPs concentrations are listed in Table 1.

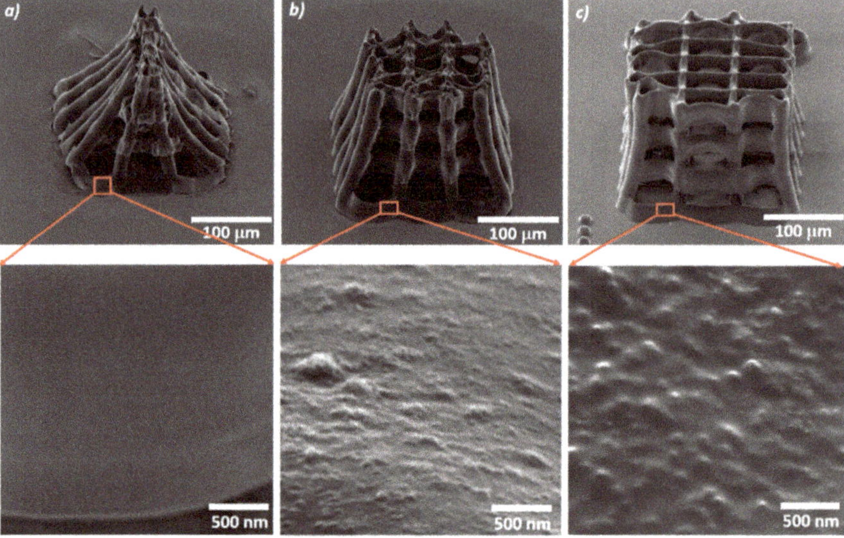

Figure 2. Scanning electron micrographs of scaffolds with magnetic nanoparticles (MNPs) concentrations of (**a**) 0 mg/mL; (**b**) 2 mg/mL; (**c**) 4 mg/mL. Upper panel: scaffolds overview (samples tilted at 45 grd). Lower panel: insets showing close views of the scaffolds' surfaces.

Table 1. Scaffolds shrinkage, number of MNPs from an ellipsoidal unit of a scaffold as determined from enhanced dark field microscopy images using the dedicated plug-ins for nanoparticles counting and scaffolds porosity determined as in [36].

MNPs Concentration in the Unpolimerized Composite (mg/mL)	Scaffolds Shrinkage (%)	Number of MNPs from an Ellipsoidal Unit of a Scaffold	Scaffolds Porosity (%)
0	80.2 ± 5.3	0	46.2 ± 4
2	35.6 ± 4.2	178 ± 5	87.6 ± 2
4	20.7 ± 4.7	332 ± 8	94.6 ± 1

The scaffolds porosities are listed in Table 1 and were determined using Solid Works, similar to as we described in detail in [36]. Except for the scaffolds without MNPs that had porosities below 50% and where the cells were not able to penetrate inside the structure, all the other scaffolds had porosities above 85% that allowed the cell migration from pore to pore. As the structures were highly complex (Figure 1), the pore diameters varied within a broad range, i.e., from 5 to 70 µm.

The location and the spatial distribution of the MNPs inside the scaffolds were monitored by enhanced dark field microscopy (Figure 3). The polymer is represented in yellow and the MNPs as red dots (false colors). To avoid scattering from multiple layers, for this analysis we fabricated particular structures with a single layer of ellipsoidal units. Figure 3b,c show that the MNPs were embedded in the scaffolds and have a homogeneous distribution in the whole volume. The number of nanoparticles from an ellipsoidal unit with 4 mg/mL MNPs concentration was twice the number of nanoparticles from a similar ellipsoid containing 2 mg/mL MNPs (Table 1). This result proves that the polymer/MNPs composites preserved their stoichiometry after the laser direct writing process.

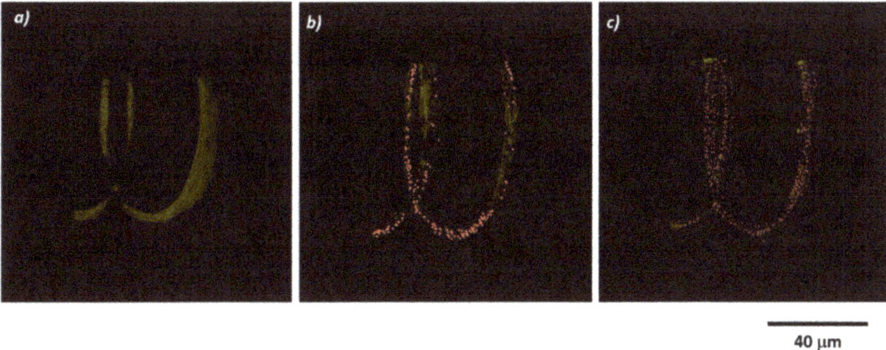

Figure 3. Images obtained by enhanced dark field microscopy using the Cytoviva three-dimensional (3D) module for scaffolds with MNPs concentration of: (**a**) 0 mg/mL; (**b**) 2mg/mL; (**c**) 4 mg/mL.

The EDS analysis provided evidence that iron was present in the photopolymerized composites (Figure 4a). The iron was uniformly distributed over the entire investigated areas (Figure 4b), in concordance with the enhanced dark field microscopy findings from Figure 3. Furthermore, the iron concentrations in the polymerized Ormocore/MNPs composites were similar with the MNPs concentrations from the unpolymerized composites (Table 2).

Table 2. Elemental composition of the polymerized Ormocore/MNPs composites as determined by EDS.

MNPs Concentration in Unpolymerized Composite (mg/mL)	Element	Atomic %	Error %
0	C K	73.6	4.7
	O K	25.4	7.4
	Fe L	0.01	-
2	C K	75.1	4.7
	O K	23.7	7.4
	Fe L	1.2	6.6
4	C K	69.0	5.0
	O K	28.3	7.1
	Fe L	2.7	6.2

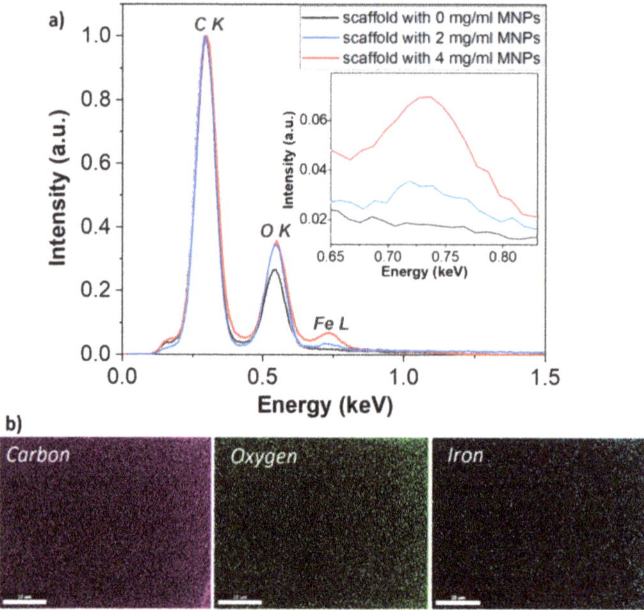

Figure 4. (**a**) Energy-Dispersive X-ray Spectroscopy (EDS) spectra of polymerized Ormocore/MNPs composite with MNPs concentrations of 0, 2 and 4 mg/mL; (**b**) EDS mapping of oxygen, carbon and iron from the polymerized Ormocore/MNPs composite with 4 mg/mL MNPs concentration.

To prove the magnetic nature of the MNPs in the polymerized composites, magnetic force microscopy (MFM) was carried out on scaffolds with different MNPs concentrations (Figure 5). To avoid the magnetic needle to be stacked inside the free spaces of the complex structure of the scaffolds, the analysis was performed on flat areas of the structures. As expected, the topography image of the scaffold without MNPs did not show the presence of any particles (Figure 5a). The surface was locally smooth and continuous and the MFM image revealed only contrast originating from the large topography features. For the scaffolds with 2 and 4 mg/mL MNPs concentrations, the topography images showed the presence of nanoparticles, with a relatively uniform distribution (Figure 5b,c). The particles stand out from the surface between 10 nm and 50 nm. The corresponding MFM images showed some contrast, which can be attributed both to topography effects and to magnetic interaction (Figure 5e,f). A more clear contrast can be noticed in the upper-left part of the MFM image of the magnetic scaffold containing 4 mg/mL MNPs, (Figure 5f), where the particles were spread on the scaffolds surface, with only traces of the embedding polymer.

The magnetic characteristics of the scaffolds with 2 and 4 mg/mL MNPs concentrations have been investigated using the vibration sample magnetometer (VSM) module of the Physical Property Measurement Systems (PPMS). Full major hysteresis loops were recorded on a single scaffold with MNPs (2 mg/mL and 4 mg/mL, respectively) at 300 K under applied magnetic fields up to 5 T. As there was quite low coercivity (less than 15 Oe) observed for each of the measured scaffolds, we show only the descending branch of the loop. Raw magnetization data have been corrected for the significant diamagnetic signal coming from the scaffold without MNPs. The corrected descending branches of the magnetization versus applied field are shown in Figure 6. A significant magnetic moment of the order of 10^{-4} emu was obtained for both scaffolds. The allure is typical for Fe-rich soft magnetic nanoparticles with fast approach to saturation (both samples saturate at applied fields as low as 9000 Oe), virtually zero remanence and high saturation magnetization. Taking into account the level of doping with MNPs and the single scaffold volume, a total specific magnetization of about 17 emu/g and 14 emu/g for the 4 mg/mL and 2 mg/mL doped scaffolds, respectively, has been determined per single scaffold, in good agreement with the estimated concentrations of nanoparticles per scaffold, listed in Table 1. This estimation is affected by differences between designed and real dimensions of the scaffolds as they are affected by resolution and shrinkage, but nevertheless, it shows that at least half of magnetic moments are preserved during laser photopolymerization.

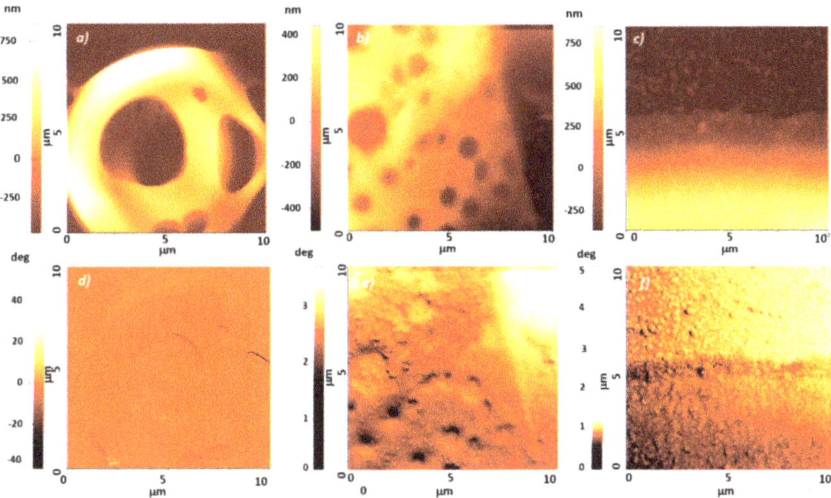

Figure 5. Topographical (upper panel) and magnetic force microscopy (lower panel) images of scaffolds with MNPs concentrations of: (**a**,**d**) 0 mg/mL; (**b**,**e**) 2 mg/mL; (**c**,**f**) 4 mg/mL.

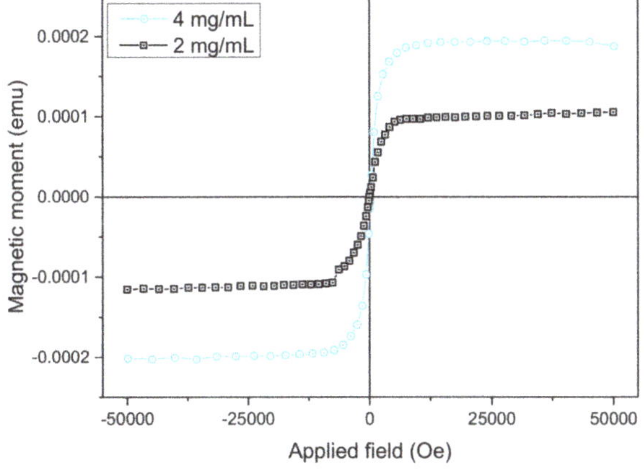

Figure 6. Magnetization versus applied magnetic field (descending branch of the hysteresis loop) for the scaffolds with 2 mg/mL and 4 mg/mL MNPs concentrations, respectively; 1 emu = 10^{-3} Am^2; 1 Oe = 79.5775 A/m.

3.2. Biological Assessments

3.2.1. Cells Morphology and Attachment

For separating the influence of the scaffolds architecture form that of the magnetic field on the cell behavior, Figure 7a illustrates SEM micrographs of cell-seeded scaffolds without MNPs. Figure 7b,c show SEM micrographs of cells growing on scaffolds with different MNPs concentrations. All SEM images were recorded after 3 days of cultivation, both in the absence and in the presence of SMF. At longer time scales, because of cell growth and division, the morphological insight becomes irrelevant because of the multiple layers of cells overlapping over the entire structure. Several experimental observations are worthy to be mentioned.

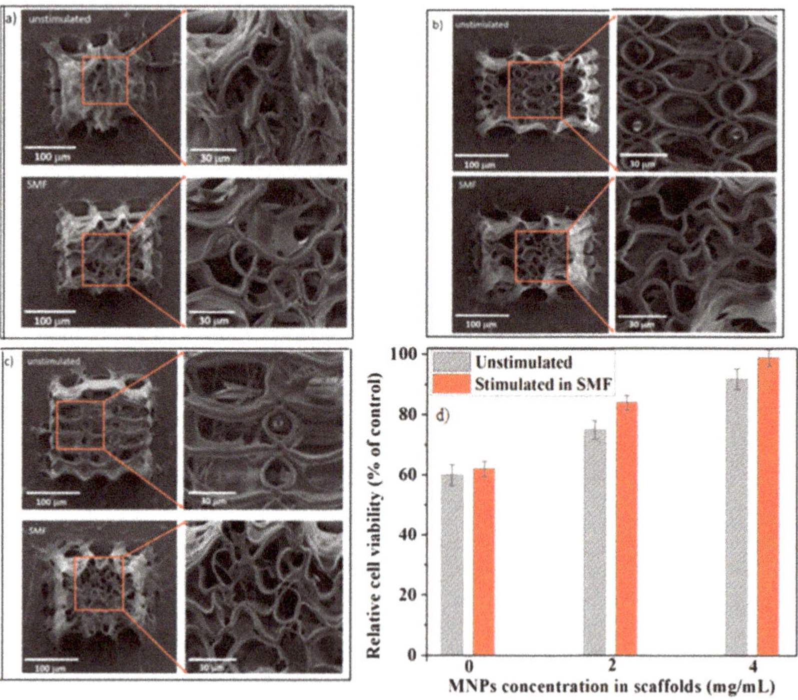

Figure 7. Scanning Electron Microscopy (SEM) micrographs illustrating the cells attached on scaffolds with MNPs concentrations of: (**a**) 0 mg/mL; (**b**) 2 mg/mL; (**c**) 4 mg/mL, after 3 days of cultivation, in the absence (upper panel) and in the presence (lower panel) of static magnetic field (SMF). Left panels: cells growing on the scaffolds. Right panels: insets; (**d**) relative cell viability as a function of MNPs concentration in the scaffolds; except for 0 mg/mL MNPs concentration, the results were statistically significant ($p < 0.05$).

First, on all scaffolds, either exposed or unexposed to SMF, the cells were stretched and had a mature osteoblast phenotype similar with the one from the bone surface. On the scaffolds without MNPs (Figure 7a) the cells mostly grew on the lateral walls of the scaffold, most probably because the tightened scaffolds architecture hampered the cells penetration inside the structure. In contrast, on the scaffolds containing MNPs the cells were able to invade the whole volume of the scaffolds (Figure 7b,c).

Second, the number of cells attached on the scaffolds increased with increasing MNPs concentration. Given that the MNPs are superparamagnetic and therefore activated only in the presence of external SMFs, the increase of the cell attachment with increasing MNPs concentration in scaffolds not exposed to SMF can be ascribed to the nanostructuring of the scaffolds' surface (insets from Figure 2). This could explain the low number of cells on the scaffolds without MNPs (Figure 7a upper panel), given that these scaffolds have smooth surfaces at nanoscale (inset from Figure 2a). In contrast, the scaffolds with 4 mg/mL MNPs concentration showed numerous cells penetrating inside the scaffolds structure, where they formed an interconnected network (Figure 7c upper panel). Most probably, this happened because their nanostructured surfaces provided more attachment points for the cells (inset from Figure 2c).

A third observation is that, excepting the scaffolds without MNPs, the SMF exposure of the cell-seeded scaffolds caused a dramatic change of the cellular behavior, with more cells attached as compared with the corresponding scaffolds in the unstimulated regime. In addition, the number of attached cells increased with increasing MNPs concentration (Figure 7b,c lower panels).

Another interesting finding is that, following SMF exposure of the cell-seeded scaffolds containing MNPs, the scaffolds architecture changed dramatically. The initial regular architecture of the scaffolds, comprising of ellipsoidal units with precise positioning spaced by vertical microtubes, shrank and changed into a highly disordered structure (Figure 7b,c upper panel versus Figure 7b,c lower panels). The scaffold structural disorder increased with increasing MNPs content (Figure 7b lower panel versus Figure 7b lower panel). The scaffolds without MNPs showed a different behavior: the seeded cells "opened up" the initial "tent-like" architecture of the scaffolds (Figure 7a upper panel versus Figure 7a lower panel).

The qualitative analysis of the SEM micrographs was confirmed quantitatively by MTS viability assay (Figure 7d). Except the scaffolds without MNPs, where the viability was low for both SMF-stimulated and unstimulated samples, for the scaffolds containing MNPs the relative cells viability was above 75% and increased with increasing MNPs concentration, followed by an additional increase up to 98% following SMF exposure.

For easier visualization of the cell adhesion and morphology on the scaffolds, the insets from Figure 7 were magnified and presented separately in Figure S3 from the Supplementary information file, where, for better viewing, the cells are indicated by red arrows.

3.2.2. Extracellular Matrix Mineralization by Alizarin Red Staining (ARS)

ARS was monitored by fluorescence microscopy for detecting the presence of calcium in the cellular deposits, which is generally used as indicative of early matrix mineralization [37]. The ARS fluorescence intensity increased with increasing MNPs concentration in the scaffolds, indicating more mineralized deposits (Figure 8a–c). The presence of SFM further increased the fluorescence signal (Figure 8d–f), proving the positive role of magnetic stimulation for early extracellular matrix mineralization. The ARS fluorescence intensity was measured using ImageJ software and supports the above findings (Figure 8g). Except for the scaffolds without MNPs, the results were statistically significant, indicating a significant increase of the mineral deposits with increasing MNPs concentration in the scaffolds (Figure 8h). Further increase of the mineral deposits was observed in the samples exposed to SMF i.e., an increase of the mineral deposits up to 50% was found in the cell-seeded scaffolds with MNPs concentration of 4 mg/mL.

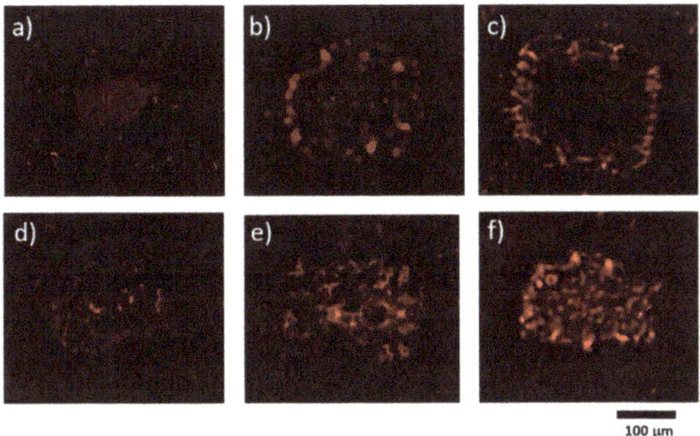

Figure 8. *Cont.*

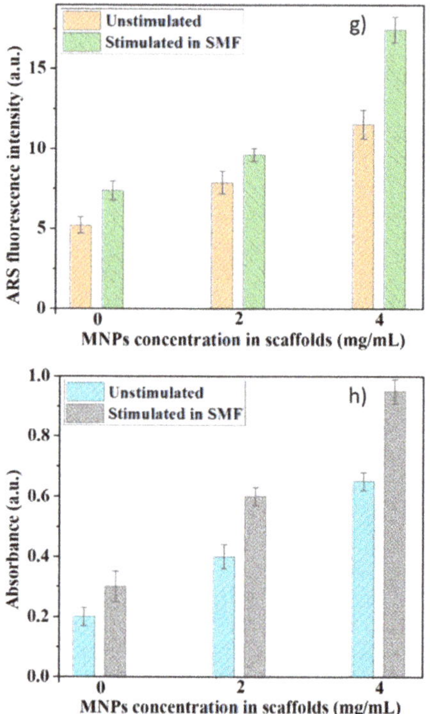

Figure 8. Alizarin Red Staining (ARS) staining of the cell-seeded scaffolds with MNPs concentration of: (**a,d**) 0 mg/mL; (**b,e**) 2mg/mL; (**c,f**) 4 mg/mL, after 20 days of incubation. Upper panel), unstimulated scaffolds. Lower panel, scaffolds exposed to SMF; (**g**) ARS fluorescence intensity as determined by ImageJ; (**h**) absorbance measurements for ARS marking of the mineral deposits in cells growing on scaffolds with different MNPs concentrations (except for the scaffolds without MNPs, the results were statistically significant ($p < 0.05$)).

4. Discussion

The attempts to control the cellular behavior in magnetic scaffolds face the major challenge of fabricating 3D structures with controlled architectures and homogenous distribution of the MNPs in the whole volume of the scaffold [23]. In the present study, we report the fabrication of 3D magnetic scaffolds with submicronic spatial resolution, high reproducibility and uniform distribution of the MNPs in the whole scaffolds structure, that promote the cell attachment and early mineralization under static magnetic field (SMF) stimulation. The scaffolds were fabricated by laser direct writing via two photons polymerization (LDW via TPP) of Ormocore/MNPs composites. The MNPs with diameters of 4.9 ± 1.5 nm were added to the photopolymer in concentrations of 0, 2 and 4 mg/mL.

In this paper, we brought several major improvements as compared to the fabrication methods used thus far. This is the first time that LDW via TPP is used for building magnetic scaffolds, which brings significant advantages over the methods previously employed. One is that LDW via TPP technique has undoubtable superiority as compared to other techniques used thus far, in terms of high spatial resolution of about 90 nm [43] and full reproducibility of the structures, which are both essential for systematic in vitro studies. Moreover, the MNPs were directly incorporated into the scaffolds during the photopolymerization process, without any additional processing steps. Most importantly, the homogenous distribution and the superparamagnetic behavior of the MNPs from the unpolymerized composite were preserved after the photopolymerization process, with MNPs uniformly dispersed within the entire structure of the scaffolds. The MNPs also improved the mechanical resilience of the scaffolds by significant reduction of the scaffolds' shrinkage. Of course, one must also keep in mind the limitations of the technique, such as long production time for large volume fabrication for scaffolds to be used clinically. Additionally, the expected mechanical stability of the scaffolds at large volume should be investigated. Moreover, to assess the origin of the SMF effects on the cellular behavior, controlled and quantitative biological investigations are required.

The lack of cellular toxicity of the MNPs in the presence of a magnetic field has been already proven [6]; therefore, we could consider that there are practically no limitations concerning the number of MNPs from the biological point of view. In our experimental conditions, the MNPs concentration and thus the number of MNPs per ellipsoidal unit of the scaffold was selected based on a tradeoff: on one side, we had to obtain a magnetic response form the scaffolds during exposure to SMF and thus a high enough number of MNPs was required; on the other side, the number of MNPs had to be low enough for allowing the photopolymerization process, since, as we state in the Results section, concentrations of MNPs higher that 4 mg/mL impeded the photopolymerization.

In the absence of MNPs, the scaffolds collapsed and shrunk in the shape of a tent (Figure 2a). As the concentration of MNPs in the composite increased, the shrinkage of the scaffolds became less significant (Figure 2b,c). The basis of the scaffolds was not collapsing, as the substrate adherence was sufficient to hold the scaffolds in place.

The shrinkage is a serious problem when fabricating micro/nanofeatured structures over a large area. This is caused mainly by the material densification as compared to the material before polymerization and results in volume reduction [44]. The shrinkage depends strongly on the type of architecture, since the geometrical deformations appear when the structure has not sufficient rigidity to withstand the developing and drying process. Another important factor is the hardness of the bulk material to be polymerized. Over the last years, there were several attempts to synthetize photopolymerizable materials with ultra-low shrinkage and negligible geometrical distortions during the development, by introducing in the photopolymer non-linear chromophores, quantum dots or organic dyes, for photonics and metamaterial production [44,45]. Within this context, our experimental results indicate that the MNPs added to the Ormocore reinforced the scaffolds' structure.

In general, the influence of the matrix stiffness on the cell behavior cannot be excluded. The facts that the dimensions of the scaffolds are very small i.e., of the order of hundreds if m^3 and that their architecture is very complex make their mechanical characterization very difficult by standard methods. Instead, what we can certainly state in the particular case of our experimental conditions is that the photopolymer used for building the scaffolds (Ormocore) has high mechanical and chemical stability, as reported by the producer (Micro resist technology GmbH, Berlin, Germany). Additionally, given that the MNPs concentration in the scaffolds was very low, their influence on scaffold stiffness when the scaffolds were exposed to SMF is less to be expected. Moreover, to demonstrate that the SMF exposure does not change the stiffness of the scaffolds, we recorded SEM images of scaffolds immediately after the fabrication process and after 20 days of exposure to SMF of 1.3 T. (Figure S2 in the Supplementary information file shows an example for a scaffold with MNPs concentration of 4 mg/mL, but the same observation stands for the 0 and 2 mg/mL concentrations that were used in our study).

Further investigations by enhanced dark field microscopy, magnetic force microscopy and Energy-Dispersive X-ray Spectroscopy showed that the MNPs were uniformly dispersed in the entire structure of the scaffolds (Figures 3 and 4b), preserved the stoichiometry of the composite (Tables 1 and 2) and retained their superparamagnetic behavior (Figure 6).

We also investigated the functionality of the scaffolds by assessing the effect of an externally applied static magnetic field (SMF) of 1.3 T on the cells behavior, in terms of cells attachment and extracellular matrix mineralization.

The magnetic field of 1.3 T was provided by the magnets used in our experimental conditions (as described in the Experimental section). We considered that this value of the magnetic field strength is appropriate for the experiments based on the fact that previous studies with significant relevance have already proven the ability of SMF of the order of 1.2 T to control the cells behavior [26,27]. In addition, one must underline that the main point of interest in not the strength of the SMF, but rather the magnetic field gradient is the main factor accounting for the cellular behavior in experimental conditions similar as ours. For example, a SMF of approximately 1 T with a large gradient (up to 1 GT/m) generated by micromagnet arrays was capable of assisting the cells migration [27], having a significant impact on the biological functionality of the cells [26]. Similarly with previous studies on magnetic scaffolds exposed to SMFs, in the superparamagnetic scaffolds reported in our study the MNPs acted as field concentrators of the SMF and produced high gradients magnetic fields within the cells bodies [26,27] that further promoted the cells differentiation process [24,28].

The study was carried out comparatively with scaffolds unexposed to SMF. In order to discriminate the influence of the scaffolds architecture form that of the magnetic field, scaffolds without MNPs were also investigated. It is worth mentioning that, although the concentrations up to 4 mg/mL used in this study were higher than those tested in previous works [5], the scaffolds provided a biocompatible 3D environment for the seeded cells as shown by SEM investigations (Figure 7).

On all the scaffolds from our study, regardless of the presence or the absence of SMF, the cells were stretched and had a mature osteoblast phenotype similar with the one from the bone surface [46]. The number of attached cells increased with increasing MNPs concentration (Figure 7). Given the superparamagnetic behavior of these MNPs that excludes the presence of magnetic forces in the absence of an external magnetic field, this trend can be attributed to the nanostructuring of scaffolds surfaces (insets from Figure 2) that increased the surface area and provided mode contact points for focal adhesions.

For the scaffolds without MNPs, the effect of SMF on the cell attachment was not significant. In contrast, on the scaffolds with 2 and 4 mg/mL MNPs concentrations, the applied SMF increased significantly the number of the attached cells (Figure 7b,c upper panels versus Figure 7b,c lower panels). For the scaffolds with 4 mg/mL MNPs concentration, the cells were even able to penetrate down to the inner parts of the structure where they formed an interlaced fibrous network (Figure 7c lower panel).

An interesting finding was that all cell-seeded scaffolds were highly deformed when exposed to SMF. The scaffolds' architecture changed from regular ellipsoidal units with precise positioning, characteristic for the unstimulated samples, to a highly disordered architecture (Figure 7a–c upper panels versus Figure 7a–c lower panels). The scaffolds' deformation increased with increasing MNPs concentration. The reason for the "opening-up" of the scaffolds without MNPs under SMF exposure (Figure 7a lower panel) is yet unknown. A possible explanation could be the absence of high gradient fields in these samples, which eliminates the role of magnetic forces in modulating the cells behavior [27].

The existing studies and the comparative analysis of the relationship between magnetic scaffolds and cell behavior remain unresolved because of the diversity in scaffolds architectures, theoretical models and investigation methods. In general, the influence of cells on scaffolds are described in terms of the contractile forces they generate, which further induce deformations of the scaffolds' structure [47–49]. Several studies succeeded to guide the establishment of cell networks via cellular response to high gradients magnetic fields [26,27]. Positive influence of external static magnetic field on magnetic nanoparticle-incorporated scaffolds on osteoblast differentiation and bone formation has been reported [24]. The MNPs acted as concentrators of the externally applied magnetic field and generated high gradient magnetic fields over the cells bodies, thus modulating their behavior [26,27].

In our experimental conditions, the MNPs added to the scaffolds enhanced the magnetic response, as investigated by VSM magnetometry. The scaffolds with 2 mg/mL and 4 mg/mL MNPs concentrations have both shown a detectable magnetization signal (10^{-4} emu) (Figure 6). The calculated specific magnetization yielded good results that are in agreement with the nanoparticles counting per scaffold, specifically 14 and 17 emu/g for scaffolds with 2 and 4 mg/mL MNPs concentrations, respectively. One must keep in mind that the magnetic properties of the composite scaffolds are determined by the size and magnetization of the MNPs, by their homogenization in the photopolymer and by the porous structure of the scaffolds [24].

In order to compute the field gradient generated by the MNPs along a distance (r), we employed the following formula [26]:

$$\frac{dB}{dr} = \frac{2\mu_0 M_S R^2}{r^4} \tag{1}$$

where M_S is the saturation magnetization, R is the MNP radius and $\mu_0 = 4 \times 10^{-7}$ H/m is the vacuum permeability. For our case, $Ms \approx 30$ emu/g, $R_{MNP} = 2.45$ nm. For 2 mg/mL and 4 mg/mL concentration, given that $\rho_{\gamma\text{-Fe2O3}} = 4.86$ g/cm^3, the average distance between two nanoparticles is of 50.8 and 40.3 nm, respectively. According with Equation (1), the field gradient between two adjacent MNPs in the first approximation limit is of the order of 3×10^4 T/m and is presented in Figure 9.

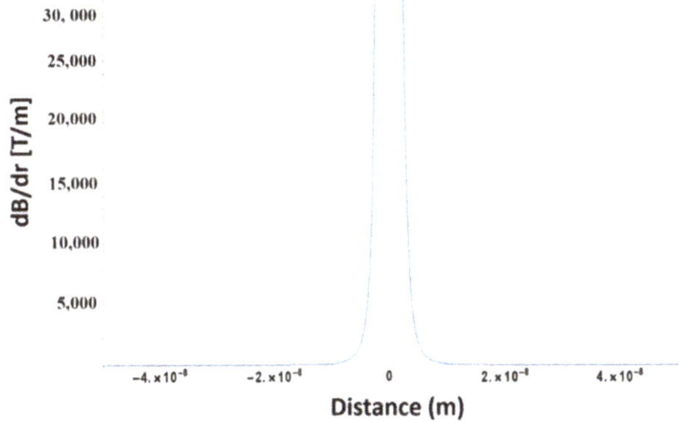

Figure 9. Magnetic field gradient between two adjacent MNPs.

Under high magnetic gradients, the cells are subjected to magnetic compressive or tensile stresses that cause membrane deformation, reorganization of the cytoskeleton and increase the tension of the actin filaments [26]. It is known that in moderate magnetic fields with gradient larger than 10^4 T/m the magnetic force magnitudes are comparable with those of gravity and are sufficient to affect the cell machinery [26,27]. Within this framework, the field gradients reached in our experimental conditions explain the preferential cell attachment and the dramatic changes of the scaffolds architecture for the cell-seeded scaffolds exposed to SMF. The fact that the cells attachment and the scaffolds' deformation increased with increasing MNPs concentration were likely caused by the higher magnetic field gradients exerting stronger magnetic stresses on the cells. The disordered scaffold structure induced by the SMF exposure and the anisotropy of the structural changes observed in Figure 2a–c lower panels are likely determined by the distribution of the magnetic gradient across the cell volume.

To validate the proposed concept, we monitored the extracellular matrix mineralization for cells seeded on the scaffolds. For this, we employed alizarin that emits a red signal under fluorescent green light and has been widely used for detailed identification of early mineralization events, with a good signal/noise ratio [37]. Alizarin detected under fluorescence at the absorbance at 405 nm increased with increasing MNPs concentration (Figure 8g,h), which confirms the presence of more mineralized deposits in these samples. The fluorescence intensity and the 405 nm absorbance were further increased by SMF exposure, indicating that the applied magnetic field fastened the extracellular matrix mineralization.

Our experimental results provide evidence that the static magnetic field and the magnetic scaffolds acted in synergy and generated favorable conditions for bone cells attachment and early mineralization. These findings complete the diverse scenarios reported by previous studies. SMF stimulation with 0.4 T of osteoblasts seeded on magnetic scaffolds resulted in an increase in the Alkaline Phosphatase activity and induced changes in cell morphology [49]. Human osteosarcoma cells seeded on poly(l-lactic acid) scaffolds exposed to SMF had a more differentiated phenotype depending on cell type and field strength [8]. Polycaprolactone/magnetic nanoparticles scaffolds used in combination with SMF stimulated the osteoblasts to reach a mature stage earlier and to deposit mineral phase more rapidly [24].

Given that the architectures of the scaffolds, the MNPs concentrations and their distributions inside the scaffolds differ between those studies and were much less controllable than in our experimental conditions, a straight comparison between previously published results and those reported by the present study are not straightforward. Nevertheless, our results provide evidence that the stronger deformation of the cell-seeded scaffolds and the faster cell mineralization with increasing MNPs concentration in the scaffolds exposed to SMF are due to local effects of magnetic forces.

Preliminary in vivo studies are currently carried out (Figure S4 from the Supplementary information). Wistar rats with scaffolds having 4 mg/mL MNPs concentration and implanted at femoral level were maintained in the static magnetic field by placing two powerful magnets under the cages in which they were accommodated. Images of computed tomography (CT) recorded at different time points after the implantation procedure did not highlight inflammatory processes. At 15 days post-implantation, for both SMF and non-stimulated groups, the CT evaluation showed that the scaffolds were in the right position, without signs of hematoma, edema or infection. Tissue necrosis has not been detected. The bone tissue was visible around the scaffolds, providing evidence that the scaffolds had a strong osteointegration. Importantly, the groups stimulated in SFM have shown a faster bone regeneration than the unstimulated specimens. These preliminary results provide evidence about the histocompatibility of these new magnetic scaffolds that have been implanted for the first time in vivo and provide great potential for the development of a long-term in vivo study. The results and the conclusions of the in vivo study will be the subject of a future report.

5. Conclusions

We designed and built homogeneous 3D superparamagnetic scaffolds and we proved their potential for biological applications. The scaffolds were fabricated by laser direct writing via two photons polymerization (LDW via TPP) of Ormocore/magnetic nanoparticles (MNPs) composites. The proposed concept provided unique advantages that were not achievable with other methods and materials. The LDW via TPP technique allowed us to fabricate scaffolds with complex and controlled 3D architectures, with MNPs directly incorporated into the scaffolds during the photopolymerization process. The LDW via TPP process was carried out on a homogenous dispersion of MNPs of pre-established concentrations in Ormocore of 0, 2 and 4 mg/mL. The homogenous distribution and the superparamagnetic behavior of the MNPs from the unpolymerized composite were preserved after the photopolymerization process. A uniform dispersion of the MNPs within the entire structure of the scaffolds was obtained. The MNPs also improved the mechanical resilience of the scaffolds by significant reduction of the scaffolds' shrinkage. An enhanced magnetic response (10^{-4} emu) has been obtained for all scaffolds containing MNPs, as seen in VSM magnetization measurements. Moreover, the specific magnetizations were found to be in agreement with the nanoparticles counting per scaffold. The intrinsically magnetic cues represented by the MNPs incorporated in the scaffolds acted in synergy with the externally magnetic cues represented by a SMF of 1.3 T and promoted the attachment and the early stage mineralization of nanoparticle-free osteoblast-like cells. The stronger scaffolds' deformation and the faster extracellular matrix mineralization occurred in the scaffolds having the highest MNPs concentration. The results were explained in the frame of high gradient magnetic fields of the order of 3×10^{-4} T/m locally generated by the MNPs over the cells bodies. The proposed method is suitable for other applications that require remote manipulation of magnetic field with submicronic resolution, with great potential for magnetically-driven tissue regeneration.

Supplementary Materials: The following are available online at http://www.mdpi.com/1996-1944/12/17/2834/s1, Figure S1: Optical images of the laser irradiated material at the laser focus, at different time points during the laser direct writing process: (a) at the beginning of the process, the laser spot is visible on the sample; (b) bubble formation at the irradiation spot, after few seconds of interaction of the focused laser beam with the material; (c) local micro-explosion at the laser-material interaction spot. Figure S2: SEM micrographs of a scaffold with MNPs concentration of 4 mg/mL: (a) immediately after the fabrication process and (b) after 20 days of exposure to

SMF of 1.3 T. Figure S3: SEM micrographs illustrating cells attached on scaffolds with MNPs concentrations of: (a) 0 mg/mL; (b) 2 mg/mL; (c) 4 mg/mL, after 3 days of cultivation, in the absence (upper panel) and in the presence (lower panel) of SMF. The cells are indicated by red arrows. Figure S4: Preliminary results in vivo: CT scans of Wistar rats with scaffolds implanted at femoral level, at different time points after the implantation procedure.

Author Contributions: Conceptualization: I.A.P.; Methodology: I.A.P. and C.R.L.; Fabrication: I.A.P. and B.S.C.; Software: B.S.C.; Validation: C.C.M., M.M., O.C. and A.L.; Investigation: C.R.L, M.M., A.M., O.C. and A.L.; Writing: I.A.P.

Funding: This work was supported by a grant of the Romanian National Authority for Scientific Research and Innovation, CNCS/CCCDI-UEFISCDI, project number PN-III-P2-2.1-PED-2016-1787, within PNCD III. A part of this work was performed in CETAL facility, supported by the National Program National Program PN 16 47-LAPLAS IV.3D imaging on Cytoviva^R equipment were possible due to European Regional Development Fund through Competitiveness Operational Program 2014-2020, Priority axis 1, Project No. P_36_611, MySMIS code 107066, Innovative Technologies for Materials Quality Assurance in Health, Energy and Environmental - Center for Innovative Manufacturing Solutions of Smart Biomaterials and Biomedical Surfaces — INOVABIOMED.

Acknowledgments: Catalin Romeo Luculescu acknowledges the MNPs received from Florian Dumitrache.

Conflicts of Interest: The authors declare no conflict of interest. The funders had no role in the design of the study; in the collection, analyses, or interpretation of data; in the writing of the manuscript, or in the decision to publish the results.

References

1. Liu, Y.; Lim, J.; Teoh, S.-H. Review: Development of clinically relevant scaffolds for vascularised bone tissue engineering. *Biotechnol. Adv.* **2013**, *31*, 688–705. [CrossRef] [PubMed]
2. Borgström, F.; Sobocki, P.; Ström, O.; Jönsson, B. The societal burden of osteoporosis in Sweden. *Bone* **2007**, *40*, 1602–1609. [CrossRef] [PubMed]
3. Oryan, A.; Bigham-Sadegh, A.; Abbasi-Teshnizi, F. Effects of osteogenic medium on healing of the experimental critical bone defect in a rabbit model. *Bone* **2014**, *63*, 53–60. [CrossRef] [PubMed]
4. Hesse, E.; Kluge, G.; Atfi, A.; Correa, D.; Haasper, C.; Berding, G.; Shin, H.; Viering, J.; Länger, F.; Vogt, P.M.; et al. Repair of a segmental long bone defect in human by implantation of a novel multiple disc graft. *Bone* **2010**, *46*, 1457–1463. [CrossRef] [PubMed]
5. Panseri, S.; Cunha, C.; D'Alessandro, T.; Sandri, M.; Giavaresi, G.; Marcacci, M.; Hung, C.T.; Tampieri, A. Intrinsically superparamagnetic Fe-hydroxyapatite nanoparticles positively influence osteoblast-like cell behaviour. *J. Nanobiotechnol.* **2012**, *10*, 32. [CrossRef] [PubMed]
6. Bock, N.; Riminucci, A.; Dionigi, C.; Russo, A.; Tampieri, A.; Landi, E.; Goranov, V.A.; Marcacci, M.; Dediu, V. A novel route in bone tissue engineering: Magnetic biomimetic scaffolds. *Acta Biomater.* **2010**, *6*, 786–796. [CrossRef] [PubMed]
7. Whitaker, M.J.; Quirk, R.A.; Howdle, S.M.; Shakesheff, K.M. Growth factor release from tissue engineering scaffolds. *J. Pharm. Pharmacol.* **2001**, *53*, 1427–1437. [CrossRef] [PubMed]
8. Feng, S.-W.; Lo, Y.-J.; Chang, W.-J.; Lin, C.-T.; Lee, S.-Y.; Abiko, Y.; Huang, H.-M. Static magnetic field exposure promotes differentiation of osteoblastic cells grown on the surface of a poly-l-lactide substrate. *Med. Biol. Eng. Comput.* **2010**, *48*, 793–798. [CrossRef]
9. Milkiewicz, M.; Ispanovic, E.; Doyle, J.L.; Haas, T.L. Regulators of angiogenesis and strategies for their therapeutic manipulation. *Int. J. Biochem. Cell Biol.* **2006**, *38*, 333–357. [CrossRef]
10. Causa, F.; Netti, P.A.; Ambrosio, L. A multi-functional scaffold for tissue regeneration: The need to engineer a tissue analogue. *Biomaterials* **2007**, *28*, 5093–5099. [CrossRef]
11. Ba, X.; Hadjiargyrou, M.; DiMasi, E.; Meng, Y.; Simon, M.; Tan, Z.; Rafailovich, M.H. The role of moderate static magnetic fields on biomineralization of osteoblasts on sulfonated polystyrene films. *Biomaterials* **2011**, *32*, 7831–7838. [CrossRef] [PubMed]
12. Bruce, J.N.; Criscuolo, G.R.; Merrill, M.J.; Moquin, R.R.; Blacklock, J.B.; Oldfield, E.H. Vascular permeability induced by protein product of malignant brain tumors: Inhibition by dexamethasone. *J. Neurosurg.* **1987**, *67*, 880–884. [CrossRef] [PubMed]
13. Jain, T.K.; Reddy, M.K.; Morales, M.A.; Leslie-Pelecky, D.L.; Labhasetwar, V. Biodistribution, Clearance, and Biocompatibility of Iron Oxide Magnetic Nanoparticles in Rats. *Mol. Pharm.* **2008**, *5*, 316–327. [CrossRef] [PubMed]

14. Prijic, S.; Scancar, J.; Romih, R.; Cemazar, M.; Bregar, V.B.; Znidarsic, A.; Sersa, G. Increased Cellular Uptake of Biocompatible Superparamagnetic Iron Oxide Nanoparticles into Malignant Cells by an External Magnetic Field. *J. Membr. Biol.* **2010**, *236*, 167–179. [CrossRef] [PubMed]

15. Sun, C.; Du, K.; Fang, C.; Bhattarai, N.; Veiseh, O.; Kievit, F.; Stephen, Z.; Lee, D.; Ellenbogen, R.G.; Ratner, B.; et al. PEG-Mediated Synthesis of Highly Dispersive Multifunctional Superparamagnetic Nanoparticles: Their Physicochemical Properties and Function In Vivo. *ACS Nano* **2010**, *4*, 2402–2410. [CrossRef] [PubMed]

16. Dvir, T.; Timko, B.P.; Kohane, D.S.; Langer, R. Nanotechnological strategies for engineering complex tissues. *Nat. Nanotechnol.* **2011**, *6*, 13–22. [CrossRef]

17. Zhang, L.; Webster, T.J. Nanotechnology and nanomaterials: Promises for improved tissue regeneration. *Nano Today* **2009**, *4*, 66–80. [CrossRef]

18. Le, N.T.V.; Richardson, D.R. Iron chelators with high antiproliferative activity up-regulate the expression of a growth inhibitory and metastasis suppressor gene: A link between iron metabolism and proliferation. *Blood* **2004**, *104*, 2967–2975. [CrossRef]

19. Parelman, M.; Stoecker, B.; Baker, A.; Medeiros, D. Iron Restriction Negatively Affects Bone in Female Rats and Mineralization of hFOB Osteoblast Cells. *Exp. Biol. Med. (Maywood)* **2006**, *231*, 378–386. [CrossRef]

20. Pareta, R.A.; Taylor, E.; Webster, T.J. Increased osteoblast density in the presence of novel calcium phosphate coated magnetic nanoparticles. *Nanotechnology* **2008**, *19*, 265101. [CrossRef]

21. Abraham, R.; Walton, J.; Russell, L.; Wolman, R.; Wardley-Smith, B.; Green, J.R.; Mitchell, A.; Reeve, J. Dietary determinants of post-menopausal bone loss at the lumbar spine: A possible beneficial effect of iron. *Osteoporos. Int.* **2006**, *17*, 1165–1173. [CrossRef] [PubMed]

22. Harris, M.M.; Houtkooper, L.B.; Stanford, V.A.; Parkhill, C.; Weber, J.L.; Flint-Wagner, H.; Weiss, L.; Going, S.B.; Lohman, T.G. Dietary Iron Is Associated with Bone Mineral Density in Healthy Postmenopausal Women. *J. Nutr.* **2003**, *133*, 3598–3602. [CrossRef] [PubMed]

23. Zhang, J.; Ding, C.; Ren, L.; Zhou, Y.; Shang, P. The effects of static magnetic fields on bone. *Prog. Biophys. Mol. Biol.* **2014**, *114*, 146–152. [CrossRef] [PubMed]

24. Yun, H.-M.; Ahn, S.-J.; Park, K.-R.; Kim, M.-J.; Kim, J.-J.; Jin, G.-Z.; Kim, H.-W.; Kim, E.-C. Magnetic nanocomposite scaffolds combined with static magnetic field in the stimulation of osteoblastic differentiation and bone formation. *Biomaterials* **2016**, *85*, 88–98. [CrossRef] [PubMed]

25. De Santis, R.D.; Russo, A.; Gloria, A.; D'Amora, U.; Russo, T.; Panseri, S.; Sandri, M.; Tampieri, A.; Marcacci, M.; Dediu, V.A.; et al. Towards the design of 3D fiber-deposited poly(ε-caprolactone)/iron-doped hydroxyapatite nanocomposite magnetic scaffolds for bone regeneration. *J. Biomed. Nanotechnol.* **2015**, *11*, 1236–1246. [CrossRef]

26. Zablotskii, V.; Polyakova, T.; Lunov, O.; Dejneka, A. How a High-Gradient Magnetic Field Could Affect Cell Life. *Sci. Rep.* **2016**, *6*, 37407. [CrossRef]

27. Zablotskii, V.; Dejneka, A.; Kubinová, Š.; Le-Roy, D.; Dumas-Bouchiat, F.; Givord, D.; Dempsey, N.M.; Syková, E. Life on Magnets: Stem Cell Networking on Micro-Magnet Arrays. *PLoS ONE* **2013**, *8*, e70416. [CrossRef]

28. Zhao, Y.; Fan, T.; Chen, J.; Su, J.; Zhi, X.; Pan, P.; Zou, L.; Zhang, Q. Magnetic bioinspired micro/nanostructured composite scaffold for bone regeneration. *Colloids Surf. B Biointerfaces* **2019**, *174*, 70–79. [CrossRef]

29. Buyukhatipoglu, K.; Chang, R.; Sun, W.; Clyne, A.M. Bioprinted Nanoparticles for Tissue Engineering Applications. *Tissue Eng. Part C Methods* **2009**, *16*, 631–642. [CrossRef]

30. Kim, K.; Fisher, J.P. Nanoparticle technology in bone tissue engineering. *J. Drug Target.* **2007**, *15*, 241–252. [CrossRef]

31. D'Amora, U.; Russo, T.; Gloria, A.; Rivieccio, V.; D'Antò, V.; Negri, G.; Ambrosio, L.; De Santis, R. 3D additive-manufactured nanocomposite magnetic scaffolds: Effect of the application mode of a time-dependent magnetic field on hMSCs behavior. *Bioact. Mater.* **2017**, *2*, 138–145. [CrossRef] [PubMed]

32. De Santis, R.; D'Amora, U.; Russo, T.; Ronca, A.; Gloria, A.; Ambrosio, L. 3D fibre deposition and stereolithography techniques for the design of multifunctional nanocomposite magnetic scaffolds. *J. Mater. Sci. Mater. Med.* **2015**, *26*, 250. [CrossRef] [PubMed]

33. Alexandrescu, R.; Bello, V.; Bouzas, V.; Costo, R.; Dumitrache, F.; García, M.A.; Giorgi, R.; Morales, M.P.; Morjan, I.; Serna, C.J.; et al. Iron Oxide Materials Produced by Laser Pyrolysis. *AIP Conf. Proc.* **2010**, *1275*, 22–25.

34. García, M.A.; Bouzas, V.; Costo, R.; Veintemillas, S.; Morales, P.; García-Hernández, M.; Alexandrescu, R.; Morjan, I.; Gasco, P. Magnetic Properties of Fe Oxide Nanoparticles Produced by Laser Pyrolysis for Biomedical Applications. *AIP Conf. Proc.* **2010**, *1275*, 26–29.

35. Paun, I.A.; Popescu, R.C.; Calin, B.S.; Mustaciosu, C.C.; Dinescu, M.; Luculescu, C.R. 3D Biomimetic Magnetic Structures for Static Magnetic Field Stimulation of Osteogenesis. *Int. J. Mol. Sci.* **2018**, *19*, 495. [CrossRef] [PubMed]

36. Paun, I.A.; Popescu, R.C.; Mustaciosu, C.C.; Zamfirescu, M.; Calin, B.S.; Mihailescu, M.; Dinescu, M.; Popescu, A.; Chioibasu, D.; Soproniy, M.; et al. Laser-direct writing by two-photon polymerization of 3D honeycomb-like structures for bone regeneration. *Biofabrication* **2018**, *10*, 025009. [CrossRef] [PubMed]

37. Bensimon-Brito, A.; Cardeira, J.; Dionísio, G.; Huysseune, A.; Cancela, M.L.; Witten, P.E. Revisiting in vivo staining with alizarin red S—A valuable approach to analyse zebrafish skeletal mineralization during development and regeneration. *BMC Dev. Biol.* **2016**, *16*, 2. [CrossRef] [PubMed]

38. Malinauskas, M.; Baltriukiene, D.; Kraniauskas, A.; Danilevicius, P.; Jarasiene, R.; Sirmenis, R.; Zukauskas, A.; Balciunas, E.; Purlys, V.; Gadonas, R.; et al. In vitro and in vivo biocompatibility study on laser 3D microstructurable polymers. *Appl. Phys. A* **2012**, *108*, 751–759. [CrossRef]

39. Marino, A.; Filippeschi, C.; Genchi, G.G.; Mattoli, V.; Mazzolai, B.; Ciofani, G. The Osteoprint: A bioinspired two-photon polymerized 3-D structure for the enhancement of bone-like cell differentiation. *Acta Biomater.* **2014**, *10*, 4304–4313. [CrossRef]

40. Williams, G.; Hunt, M.; Boehm, B.; May, A.; Taverne, M.; Ho, D.; Giblin, S.; Read, D.; Rarity, J.; Allenspach, R.; et al. Two-photon lithography for 3D magnetic nanostructure fabrication. *Nano Res.* **2018**, *11*, 845–854. [CrossRef]

41. Zandrini, T.; Taniguchi, S.; Maruo, S. Magnetically Driven Micromachines Created by Two-Photon Microfabrication and Selective Electroless Magnetite Plating for Lab-on-a-Chip Applications. *Micromachines* **2017**, *8*, 35. [CrossRef]

42. Xia, H.; Wang, J.; Tian, Y.; Chen, Q.-D.; Du, X.-B.; Zhang, Y.-L.; He, Y.; Sun, H.-B. Ferrofluids for Fabrication of Remotely Controllable Micro-Nanomachines by Two-Photon Polymerization. *Adv. Mater.* **2010**, *22*, 3204–3207. [CrossRef] [PubMed]

43. Farsari, M.; Vamvakaki, M.; Chichkov, B.N. Multiphoton polymerization of hybrid materials. *J. Opt.* **2010**, *12*, 124001. [CrossRef]

44. Ovsianikov, A.; Shizhou, X.; Farsari, M.; Vamvakaki, M.; Fotakis, C.; Chichkov, B.N. Shrinkage of microstructures produced by two-photon polymerization of Zr-based hybrid photosensitive materials. *Opt. Express* **2009**, *17*, 2143–2148. [CrossRef] [PubMed]

45. Li, J.; Jia, B.; Zhou, G.; Gu, M. Fabrication of three-dimensional woodpile photonic crystals in a PbSe quantum dot composite material. *Opt. Express* **2006**, *14*, 10740–10745. [CrossRef]

46. Aubin, J.E.; Liu, F.; Malaval, L.; Gupta, A.K. Osteoblast and chondroblast differentiation. *Bone* **1995**, *17*, S77–S83. [CrossRef]

47. Dado, D.; Levenberg, S. Cell–scaffold mechanical interplay within engineered tissue. *Semin. Cell Dev. Biol.* **2009**, *20*, 656–664. [CrossRef]

48. Corin, K.A.; Gibson, L.J. Cell contraction forces in scaffolds with varying pore size and cell density. *Biomaterials* **2010**, *31*, 4835–4845. [CrossRef]

49. Chiu, K.-H.; Ou, K.-L.; Lee, S.-Y.; Lin, C.-T.; Chang, W.-J.; Chen, C.-C.; Huang, H.-M. Static Magnetic Fields Promote Osteoblast-Like Cells Differentiation Via Increasing the Membrane Rigidity. *Ann. Biomed. Eng.* **2007**, *35*, 1932–1939. [CrossRef]

 *materials*

Article

New Biocompatible Mesoporous Silica/Polysaccharide Hybrid Materials as Possible Drug Delivery Systems

Andreea Madalina Pandele [1], Corina Andronescu [1,2], Adi Ghebaur [1], Sorina Alexandra Garea [1] and Horia Iovu [1,3,*]

[1] Advanced Polymer Materials Group, Faculty of Applied Chemistry and Material Science, University Polytehnica of Bucharest, str. Gheorghe Polizu 1-7, 0011601 Bucharest, Romania; pandele.m.a@gmail.com (A.M.P.); Corinaandronescu@yahoo.com (C.A.); ghebauradi@yahoo.com (A.G.); garea_alexandra@yahoo.co.uk (S.A.G.)

[2] Analytical Chemistry–Center for Electrochemical Science, Ruhr–Universitat Bochum, Universitätstraße 150, D-44780 Bochum, Germany

[3] Academy of Romanian Scientist, 0011601 Bucharest, Romania

* Correspondence: iovu@tsocm.pub.ro; Tel.: +4-021-402-3922

Received: 28 November 2018; Accepted: 19 December 2018; Published: 20 December 2018

Abstract: A high number of studies support the use of mesoporous silica nanoparticles (MSN) as carriers for drug delivery systems due to its high biocompatibility both in vitro and in vivo, its large surface area, controlled pore size and, more than this, its good excretion capacity from the body. In this work we attempt to establish the optimal encapsulation parameters of benzalkonium chloride (BZC) into MSN and further study its drug release. The influence of different parameters towards the drug loading in MSN such as pH, contact time and temperature were considered. The adsorption mechanism of the drug has been determined by using the equilibrium data. The modification process was proved using several methods such as Fourier transform-infrared (FT-IR), ultraviolet-visible (UV-VIS), X-ray photoelectron spectroscopy (XPS) and thermogravimetric analysis (TGA). Since MSN shows a lower drug release amount due to the agglomeration tendency, in order to increase MSN dispersion and drug release amount from MSN, two common biocompatible and biodegradable polymers were used as polymer matrix in which the MSN-BZC can be dispersed. The drug release profile of the MSN-BZC and of the synthesized hybrid materials were studied both in simulated gastric fluid (SGF) and simulated intestinal fluid (SIF). Polymer-MSN-BZC hybrid materials exhibit a higher drug release percent than the pure MSN-BZC when a higher dispersion is achieved. The dispersion of MSN into the hybrid materials was pointed out in scanning electron microscope (SEM) images. The release mechanism was determined using four mathematic models including first-order, Higuchi, Korsmeyer–Peppas and Weibull.

Keywords: MSN; biopolymer; drug delivery system; in vitro kinetic studies

---

1. Introduction

In recent years, drug delivery systems have been rapidly developed and become an important field in medical applications [1,2]. Among them, oral administration is the most widely used system exhibiting many advantages including easily self-administration, painless and low cost [3].

Vallet-Regi proposed for the first time mesoporous silica nanoparticles (MSN) as controlled delivery systems [4]. Since then it was found that this type of material may be one of the greatest carrier materials for hydrophobic or hydrophilic drugs [5] due to their controlled pore size, large surface area, pore volume and, moreover, it has been demonstrated that it has very high biocompatibility both in vitro and in vivo [6]. Unlike other carriers, MSN exhibit a higher resistance to temperature and pH variation, mechanical stress and hydrolysis-induced degradations thus making a stable and rigid framework. Another important advantage of MSN for the medical field is their degradability in aqueous solution which can avoid further problems related to the removal of the material after use, being easily excreted from the body. Since for the medical field it is very important to control the pore size, geometry and shape of the carrier, in the case of MSN the size and the surface chemistry of the pore could be easily controlled and changed depending on the drug which should be encapsulated to obtain the proper loading and release of the drug. Moreover, MSN proved to exhibit a higher versatility compared with other systems like polymer nanoparticles and liposomes.

However, despite the potential benefit of this carrier for drug delivery applications, some important challenges were reported. The agglomeration tendency of the particles exhibits a strong influence on the drug-loading capacity by decreasing the amount of drug loaded due to the steric hindrance. The agglomeration tendency is specific for inorganic nanoparticles and it is a very important factor that must be taken into account [7]. Although MSN were accepted as having a low toxicity and a good biocompatibility at nano-scale, some biocompatibility studies showed that MSN particles with diameters ranging from 150 nm have significant toxicity at high concentrations in vitro, and cause severe systemic toxicity in vivo after intraperitoneal and intravenous injections [8]. Synthesis of hybrid materials based on biopolymers and MSN may solve some of the aforementioned problems since polymer/silica composites may encapsulate large amounts of guest molecules and subsequently release them at later stages in an optimal way [9]. Biopolymers such as chitosan (CS) and alginate (Al) exhibit many advantages for developing an ideal drug delivery system [10,11]. They have high biocompatibility, biodegradability, bioadhesivity, antibacterial activity, etc. So the introduction of MSN within de polymer matrices will exceed their biocompatility for the human body.

Benzalkonium chloride (BZC), a quaternary ammonium salt with the general formula $[C_6H_5CH_2N(CH_3)_2R^+]Cl^-$ where R range from C_8H_{17} to $C_{19}H_{39}$ is a bacteriostatic agent used as a preservative and disinfectant in the pharmaceutical industry. It is often utilized as an antiseptic and medical equipment disinfectant similar to other cationic surfactants [12,13]. It has different physical, chemical and microbiological properties.

In the present work, the optimal parameters for BZC adsorption into mesoporous silica nanoparticles and its drug release are discussed. For that, the influence of the contact time, pH of the solution and temperature were considered. The adsorption mechanism of the drug has been determinated by using the equilibrium data. The MSN and BZC were further incorporated into two common biopolymers and the drug release profile and the release mechanism have been also pointed out. The biopolymers were chosen in order to decrease MSN agglomeration and increase the amount of drug release. The dispersion of MSN into the two biopolymers was observed into scanning elecron microscope (SEM) images for the synthesized hybrid materials.

2. Materials and Methods

2.1. Materials

BZC, CS with medium molecular weight, Al, glutaraldehyde (GA), calcium chloride used as gelling agents and mesoporous silica (MCM-41) with a pore size of about 2.1–2.7 nm, 0.98 cm^3/g pore volume and a specific surface area ~1000 m^2/g were purchased from Sigma Aldrich.

Sodium hydroxide, potassium phosphate monobasic, hydrochloric acid, potassium chloride were received from Sigma Aldrich.

2.2. Immobilization of Benzalkonium Chloride (BZC) to Mesoporous Silica Nanoparticles (MSN)

BZC was used as a model drug in order to evaluate the possible drug delivery capacity of different hybrid materials (Al-MSN; CS-MSN).

0.015 g BZC were dissolved in 2.5 mL pH 5 solutions and then 0.1 g of MSN were added under magnetic stirring. The stirring was maintained for 2 h at RT. The obtained suspension was centrifuged and dried at 35 °C for 24 h in a vacuum oven.

2.3. Synthesis of Chitosan (CS)-BZC and CS-MSN-BZC Hybrid Materials

50 mg of CS powder was dissolved in 10 wt.% acetic acid solution for 24 h at RT to form a homogenous viscous solution. The CS-BZC (20 mL CS and 0.015 g BZC) and CS-MSN-BZC (20 mL CS solution, 0.015 g BZC and 0.1 g MSN) suspension were obtained by mixing the two or three compounds and mechanically stirring at room temperature (RT) for 2 h. Then 0.0225 mL aqueous solution of GA (25 wt.%) was added as crosslinking agent and the stirring was continued for another one hour. CS-BZC and CS-MSN-BZC hybrid materials were cast onto transparent Petri dish and left undisturbed for 72 h at RT for solvent evaporation and thus, allowing to form thin films.

2.4. Synthesis of Alginate (Al)-BZC and Al-MSN-BZC Hybrid Materials

A solution of Al was prepared by dissolving 50 mg of Al in 50 mL of water for 3 h. The Al-BZC and Al-MSN-BZC suspensions were obtained following the same procedure described above (see Section 2.3). After solvent evaporation, the films were peeled off from the mold and impregnated in 1% $CaCl_2$ aqueous solution for 1 h. The samples were washed several times with water to remove $CaCl_2$ excess and dried for 24 h at RT.

2.5. Characterization

Fourier transform-infrared (FT-IR) measurements were performed on a Bruker VERTEX 70 spectrometer. The FT-IR spectra were recorded in $400 \div 4000$ cm^{-1} range with 4 cm^{-1} resolution. The samples were analyzed from KBr pellets.

The X-ray photoelectron spectroscopy (XPS) spectra were registered on a Thermo Scientific K-Alpha equipment, fully integrated, with an aluminum anode monochromatic source. Charging effects were compensated by a flood gun. Pass energy of 200 eV and 20 eV were used for surgery and high resolution spectra aquisition respectively.

Thermogravimetric analysis (TGA) was done on a Q500 TA Instruments equipment. 2 mg of sample was heated from RT to 700 °C using a heating rate of 10 °C/min under constant nitrogen flow rate.

UV adsorption measurements of BZC were performedat $\lambda = 262$ nm on a UV 3600 Shimadzu equipment provided with aquartz cell having a light path of 10 mm.

The morphological characterization of the CS/Al-MSN-BZC composite films was evaluated from the micrograph recorded using a Philips Xl 30 ESEM TMP scanning electron microscope (SEM).

2.6. Adsorption Experiments

For adsorption experiments, the influences of contact time, temperature, concentration of the drug and buffer pH were investigated. To establish the influence of contact time on the BZC adsorption on MSN, 0.015 g BZC and 0.1 g MSN were mixed in various buffer solutions with pH 5 for 10, 30, 60, 120, 240, 360 min. The influence of the temperature reaction was studied using the same quantities of drug and MSN. The reactions were kept for 1 h, at room temperature, 40 °C, 60 °C and 80 °C. The effect of pH was studied by maintaining the temperature reaction for 1 h at 80 °C in solutions with different pH values: 3, 4, 5, 6, 7, 8, 9, 10, 11. Another important parameter that was studied was the initial drug concentration: 3 g/L, 6 g/L, 11 g/L, 18 g/L, 24 g/L, 36 g/L, 47 g/L, 100 g/L, 200 g/L.

The unabsorbed drug concentration was determined from ultraviolet (UV) spectra at 262 nm, after the centrifugation of MSN-BZC suspension. The amounts of drug adsorbed at time t (Q_t, mg/g) and at equilibrium (Q_e, mg/g) were calculated using the following equations:

$$Q_t = \frac{(C_0 - C_t)V}{W} \tag{1}$$

$$Q_e = \frac{(C_0 - C_e)V}{W} \tag{2}$$

where C_0, C_t, C_e (mg/L) are the initial, the time and the equilibrium concentrations of BZC solution; V (L) is the volume of BZC solution, W (g) is the mass of MSN employed

2.7. In Vitro Drug Release Studies

The drug release studies were accomplished into a fully automated dissolution bath USP Apparatus 1 (708-DS Agilent) connected to an autocontrolled multi-channel peristaltic pump (810 Agilent) and at a UV-VIS spectrophotometer (Cary 60) with 1 mm flow cell and UV-Dissolution software. In a dialysis membrane bag was introduced certain amount of BZC-MSN, CS/Al-BZC and CS/AL-MSN-BZC hybrid materials and 4 mL buffer solution of pH 7.4 (simulated intestinal fluid, SIF) and pH 1.2 (simulated gastric fluid, SGF) respectively prepared as described by A. Ghebaur and coworkers [14]. These dialysis membranes were caught by the Apparatus 1 rods and immersed in 200 mL buffer solution.

The dissolution bath temperature was kept constant at 37 °C and the spindle rotation speed was set 100 rpm. At various time intervals the dissolution media were automatically extracted and the BZC concentration was calculated from the UV adsorption at 262 nm.

2.8. In Vitro Kinetic Evaluation

In vitro kinetic evaluation of BZC from different types of materials was analyzed by 4 various kinetic models: first order, Higuchi, Kormeyer–Peppas, Weibull.

First order model describes the release of drug from pharmaceutical forms that encapsulate water-soluble drugs in porous matrices [15] according to this model

$$\log C = \log C_0 - \frac{Kt}{2.303} \tag{3}$$

where C_0 is the initial concentration of drug, K is the first order rate constant and t is the time. The obtained data are plotted as log cumulative percent of drug remaining vs. time. A straight line with the slope $-K/2.303$ will be obtained.

Higuchi model descries the release of water soluble and low soluble drugs from semisolid and/or solid matrices [16,17].

The equations that describes the Higuchi model is:

$$Q = K_H \times t^{1/2} \tag{4}$$

where Q is the amount of drug release at time t, K_H is the Higuchi dissolution constant. The data obtained are plotted as cumulative percentage drug release versus square root of time.

Korsmeyer–Peppas model describes the release of a drug from a polymeric system [18–20].

The equation for this model is:

$$\frac{M_t}{M_\infty} = kt^n \tag{5}$$

where M_t/M_∞ is fraction of drug released at time t, k (min^{-n}) is the release rate constant and n is the release exponent. The data obtained are plotted as log cumulative percentage drug release versus log time. The n value is used to determine the release mechanism (see Table 1).

Table 1. Interpretation of drug transport mechanism for film-like materials.

n	Transport mechanism
0.5	Fickian diffusion
0.5 < n < 1	Anomalous Transport
1	Case II transport
1 < n	Super case II transport

The *Weibull model* is use to compare the release profile of different types of drug delivery matrixes [21]. The equation that describes this model is:

$$\frac{M_t}{M_\infty} = 1 - \exp\left(-at^b\right) \tag{6}$$

where M_t is accumulated fraction of drug in solution at time t, M_∞ is total amount of drug being released, a is the scale parameter that defines the time scale process, b is the parameter that describes the shape of the dissolution curve progression. For $b = 1$ the curve shape is exponential, $b > 1$ the curve shape is sigmoidal and if $b < 1$ the curve shape is parabolic.

3. Results and Discussion

3.1. Characterization of the Modified MSN with BZC

FT-IR analysis

The FT-IR spectra were recorded to confirm the interaction between MSN and BZC (see Figure 1). The MSN spectrum exhibits the bands at 453 cm^{-1}, 861 cm^{-1}, 1087 cm^{-1} and 1643 cm^{-1} assigned to the characteristic vibrations of the silica substrate [22]. The band at 1643 cm^{-1} can be attributed to the adsorbed water molecules while the bands at 453 cm^{-1}, 861 cm^{-1}, 1087 cm^{-1} were assigned to Si-O-Si bending and stretching vibrations. From the FT-IR spectrum of modified MSN with BZC some additional peak can be observed which confirm the presence of BZC. Thus the bands from 2966 cm^{-1}, 2928 cm^{-1} and 2858 cm^{-1} are attributed to symmetric and asymmetric stretching vibration of the C–H bond of the BZC tail. The peak from 701 cm^{-1} corresponds to the C-H bending vibration from aromatic ring while the 1458 cm^{-1} and 1488 cm^{-1} peaks are assigned to the C-H bending from methyl (-CH$_3$) and methylene (-CH$_2$) groups [12].

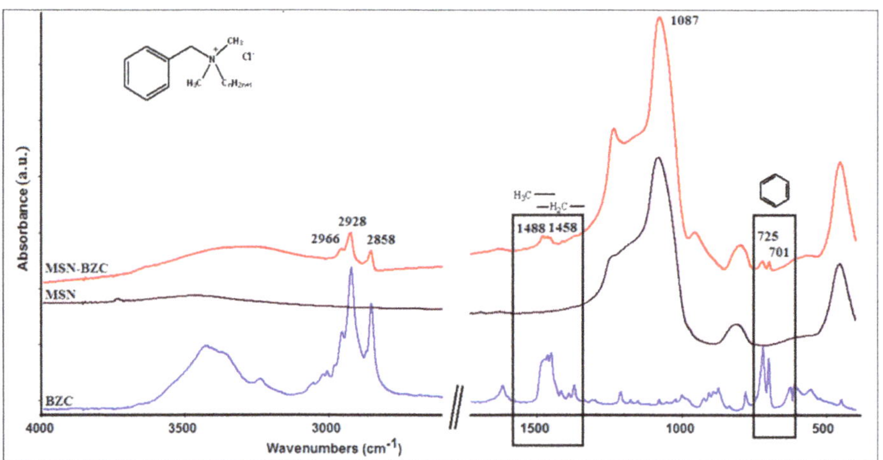

Figure 1. The Fourier transform-infrared (FT-IR) spectra for mesoporous silica nanoparticles (MSN), benzalkonium chloride (BZC) and MSN modified with BZC.

XPS analysis

The XPS analysis of MSN modified with BZC was done to show the chemical composition of the surface after the modification process (Figure 2). All the samples show the Si 2p and O 1s peaks which are assigned to silica framework. In the XPS survey spectra of the modified MSN, the presence of C 1s (BE = 282 eV) and N 1s (BE = 402 eV) can be clearly observed indicating that the modification process occurred.

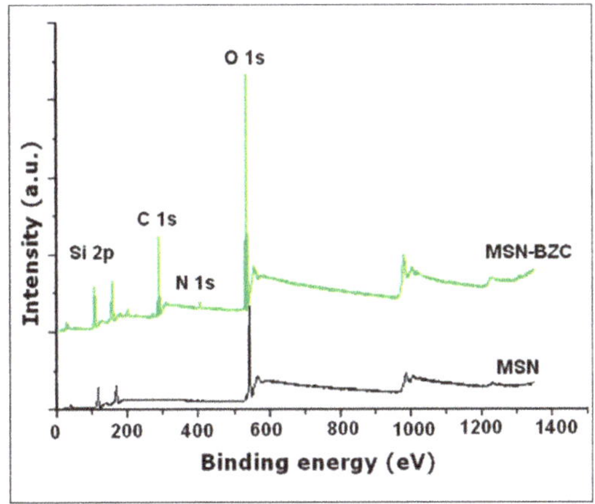

Figure 2. The X-ray photoelectron spectroscopy (XPS) survey spectra for MSN and modified MSN with BZC.

TGA data

The adsorption of BZC onto MSN was also confirmed from TGA curves (Figure 3). The MSN modified with BZC shows two thermal decomposition steps. The first step, around 200 °C, is attributed to the thermal degradation of the loaded drug and the second step which occurs at a higher temperature, around 300 °C, is assigned to the thermal degradation of the inorganic fraction [23].

The MSN modified with BZC exhibits also an increase of weight loss compared to unmodified MSN which is due to the thermal degradation of the organic compound adsorbed onto the MSN surface or/and within MSN pores.

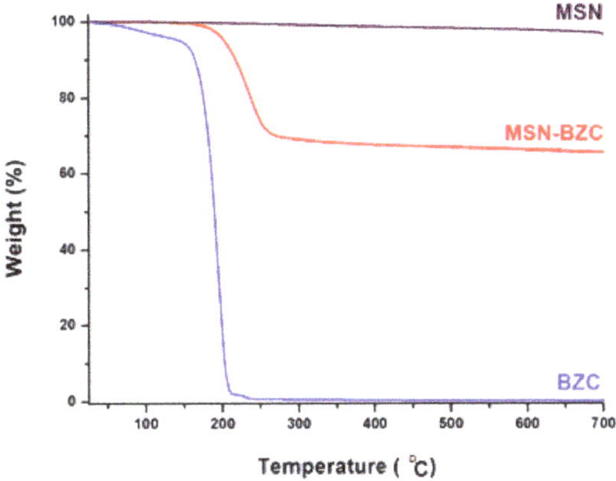

Figure 3. Thermogravimetric analysis (TGA) curves for MSN, BZC and MSN modified with BZC.

3.2. The Influence of Contact Time, pH, Temperature and Concentration of BZC

Contact time influence

The effect of the contact time for the adsorption of BZC onto MSN is presented in Figure 4. MSN has a good adsorption capacity, within 60 min the equilibrium being achieved. There was no significant change between the samples from 1 h to 6 h.

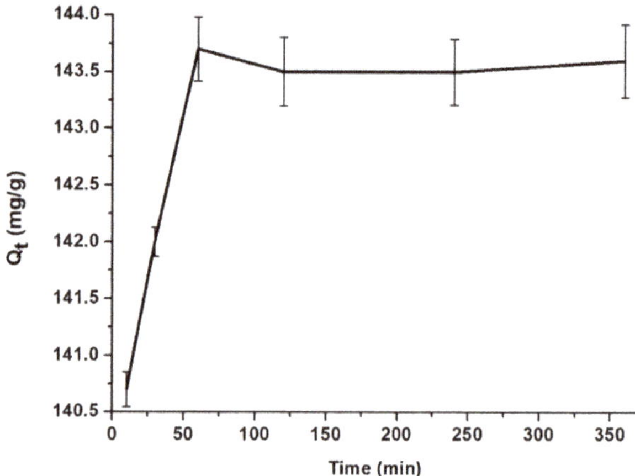

Figure 4. The influence of the contact time of BZC with MSN.

In order to determine the adsorption process type and to predict the adsorption rate, the kinetic parameters were determined. The adsorption of BZC onto MSN has been calculated using the pseudo-second order equation:

$$\frac{dq_t}{d_t} = k_2(q_e - q_t)^2 \tag{7}$$

where k_2 is the rate constant of second-order adsorption in (g mg^{-1} min^{-1}).

Equation (7) can be integrated using boundary conditions $t = 0$ to $t = t$ and $q = 0$ to $q = q$ and gives:

$$\frac{1}{(q_e - q)} = \frac{1}{q_e} + k_2 t \tag{8}$$

Equation (8) can be linear, written as:

$$\frac{t}{(q_t)} = \frac{1}{k_2 \times q_e^2} + \frac{1}{q_e} t \tag{9}$$

The straight-line plots of (t/q) versus t have been drawn to obtain rate parameters, k_2 and q_e [24].

The high correlation coefficient ($R^2 = 1$) suggests that the adsorption process of BZC onto MSN follows the pseudo-second-order kinetic model (see Table 2). Also, the calculated q_e has almost the same value as the q_e determined experimental [25].

Table 2. Kinetic parameters for BZC adsorption onto MSN.

Kinetic Model	
Pseudo-Second Order	
q_e, calc (mg/g)	142.85
k_2 (g/mg min)	0.037692
R2	1

Table 2. *Cont.*

Kinetic Model	
Intr-Particle Diffusion	
k_p (mg/g min)	0.6245
R2	0.9914

The temperature influence

The temperature effect on drug adsorption onto MSN was studied at RT, 40 °C, 60 °C and 80 °C. As can be observed in Figure 5, the highest amount of drug, 144.9 mg/g, was adsorbed at RT. At higher temperature values BZC starts too degraded, according to TG analysis (Figure 3).

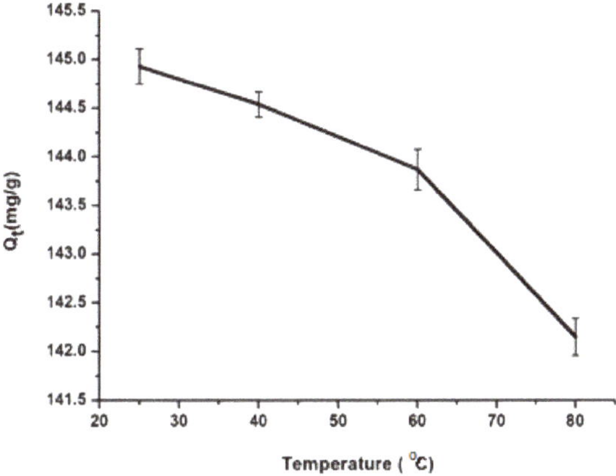

Figure 5. Adsorption of BZC at different temperature onto MSN.

Also, the thermodynamic parameters were calculated. The free energy (ΔG^0), enthalpy (ΔH^0) and entropy (ΔS^0) were calculated using the following equations:

$$\Delta G^0 = -RT \ln (Kc) \tag{10}$$

$$\Delta G^0 = \Delta H^0 - T\Delta S^0 \tag{11}$$

$$Kc = Qe/Ce \tag{12}$$

where Kc (L/g) is the adsorbed capacity to retain the active substance, R (8.314 J/mol K) is the universal gas constant and T (K) is the temperature. The values of ΔH^0 and ΔS^0 are calculated from the slope and intercept of the plot of ln (Kc) versus 1/T (Table 3). The negative values of the free energy indicate a spontaneous and physical process. This is confirmed by ΔH^0 values, which are smaller than 25 kJ/mol [26]. The positive values of ΔS^0 indicate a higher disorder of MSN as the adsorbed drug onto their surface increase. Since ΔG^0 is negative and ΔS^0 positive, the adsorption process is spontaneous with high affinity for BZC [27].

Table 3. Thermodynamic parameters for BZC adsorption onto MSN.

Temperature (K)	298	313	333	353
ΔG^0 (kJ/mol)	−2.3	−2.77	−3.4	−4.02
ΔH^0 (kJ/mol)		7.04		
ΔS^0 (kJ/molK)		0.031		

The pH influence

The pH value of the solution is one of the most important parameters for the adsorption of pharmaceutics onto mesoporous silica nanoparticles surface or pores. Solutions with the pH values in the range between 3 to 11 were employed, these values being adjusted with HCl 0.1 N or NaOH 0.1 N solutions.

The pH influence on the adsorption process of BZC onto the MSN surface or pores is shown in Figure 6. Thus, a BZC solution with an initial concentration of 6 g/L was used. Up to pH 5, the amount of the adsorbed BZC increases. After, a slowly decreases is observed to pH 6 and furthermore an abrupt decreases at higher pH values was noticed [28]. This phenomenon takes place because, at this concentration, the pH solution is 5–6 and the drug solubility is maximum [29]. Also, the molecule activity in solvent at this concentration and pH value is maximum.

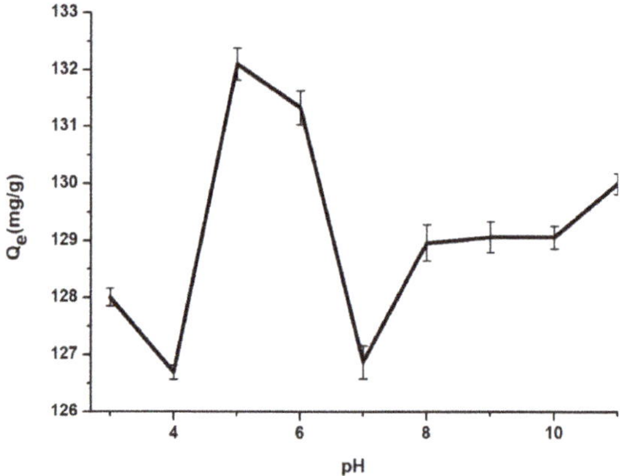

Figure 6. Adsorption of BZC at different pH values onto MSN.

The influence of initial drug concentration

Figure 7 illustrates the effect of initial BZC concentration against adsorption onto MSN. An increase of the equilibrium values from 18.56 to 352 mg/g for the adsorbed amount of the drug onto MSN was noticed when the initial drug concentration was increased from 3000 to 50,000 mg/L. For values of C_0 higher than 50,000 mg/L, the values of q_e are almost unchanged.

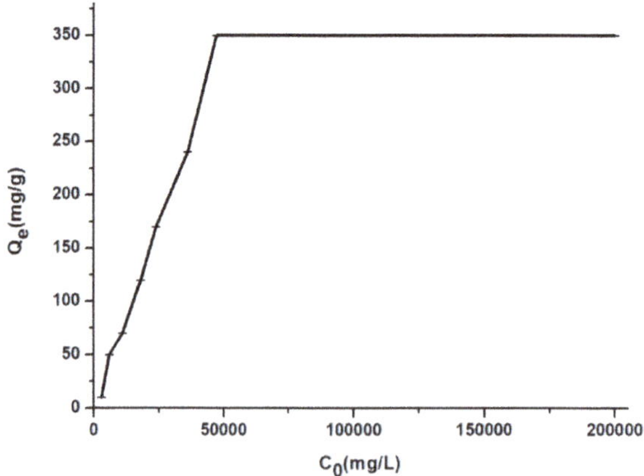

Figure 7. The influence of drug concentration onto MSN.

3.3. In Vitro Release Studies

In vitro release of BZC was investigated both in SIF (pH 7.4) and SGF (pH 1.2). Figures 8 and 9 show the amount of BZC released from MSN, from the two biopolymers and from the corresponding hybrid systems. At pH 1.2 (Figure 8), the systems CS-BZC and CS-MSN-BZC release a higher amount of drug than Al-BZC and Al-MSN-BZC systems due to the high swelling degree of CS in SIF [30]. At pH 7.4, the amount of the drug released within 24 h from the CS was higher than for Al also due to the good swelling behavior of the biopolymer in SIF which favors the diffusion of the drug molecules. Conversely, the lower release amount of BZC from Al matrix was attributed to a higher crosslinking density of the polymer which slows down considerably the diffusion of the drug and inhibits the crossing of water molecules through the polymer chains. These results are in good agreement with Xiujuan Huang and coworkers who report that the delivery of the drug might be determinate by varying the amount of sodium alginate and $CaCl_2$ concentration [30]. Moreover, Figures 8 and 9 show that the amount of BZC released from MSN-BZC systems is lower than the amount of drug released from the CS-MSN-BZC and Al-MSN-BZC hybrid materials. These results were attributed to a good dispersion of MSN in polymer matrices which reduced nanoparticles agglomeration and enabled a better drug diffusion.

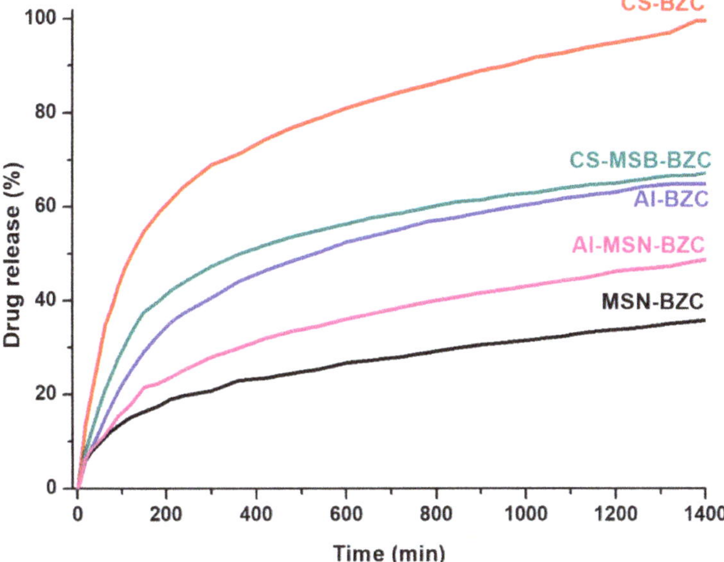

Figure 8. The release profile of BZC from different hybrid materials in simulated gastric fluid (SGF).

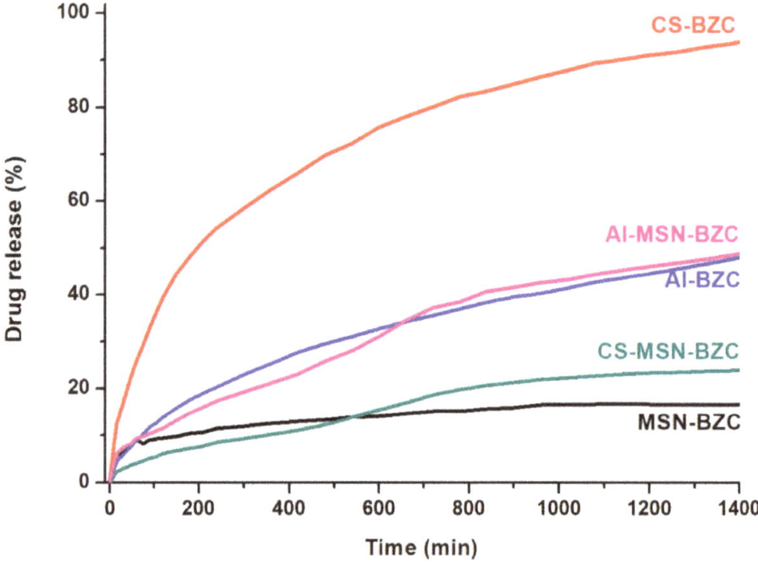

Figure 9. The release profile of BZC from different hybrid materials in simulated intestinal fluid (SIF).

3.4. Scanning Electron Microscopy (SEM) Analysis

In order to prevent the already reported agglomeration and sedimentation of MSN into aqueous solution [31], MSN was dispersed into the polymer matrix to achieve a more stable colloidal system. The dispersion of MSN into the polymer matrix was investigated by SEM. Figure 10 displays the surface morphology of both CS-MSN-BZC and Al-MSN-BZC films. According to the micrographs, the MSN dispersion in CS and Al is different. While a good dispersion is observed for Al-MSN-BZC, in the case of CS-MSN-BZC agglomerates are noticed. This result supports our initial claim, that a better MSN-BZC dispersion will facilitate the drug release and explains the higher drug release percent registered for Al-MSN-BZC in comparison with the CS-MSN-BZC hybrid material.

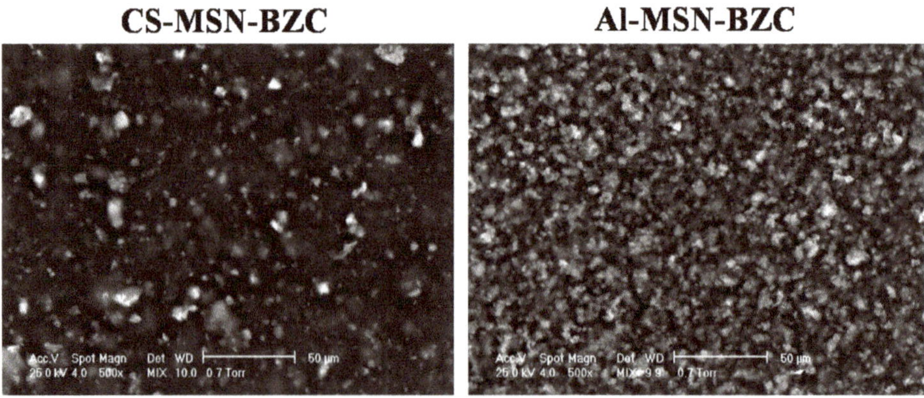

Figure 10. Scanning electron microscope (SEM) images recorded on the CS-MSN-BZC and AL-MSN BZC films.

3.5. In Vitro Kinetic Evaluation

The release mechanism of drug from different type of materials depends by the physic-chemical properties of host materials and the pore size from the materials or by the size of microparticles or nanoparticles [32]. Four types of mathematic models were used in this paper to determine the release mechanism of BZC from MSN-BZC, AL-BZC, CS-BZC, AL-MSN-BZC and CS-MSN-BZC in two different release media (pH 1.2 and pH 7.4). The calculated parameters from these models are presented in Table 4. The mathematic model which shows a correlation coefficient (R) near 1 is the model that is suitable to characterize the release mechanism.

Table 4. Kinetic release parameters of BZC release process from MSN-BZC, AL-BZC, CS-BZC, AL-MSN-BZC, CS-MSN-BZC in different simulated body fluids.

Sample		First Order			Higuchi		Korsmayer-Peppas			Weibull		
		C_0, mg/g	K, h^{-1}	R^2	K_H, $h^{-1/2}$	R^2	K (min^{-n})	n	R^2	a	b	R^2
MSN-BZC	pH 1.2	94.8	0.00069	0.89652	0.138	0.9339	0.4428	0.3848	0.9923	0.001	0.0024	0.9754
	pH 7.4	93.22	1.15×10^{-6}	0.9007	0.093	0.8327	0.4203	0.2916	0.9337	0.0005	0.0011	0.8017
AL-BZC	pH 1.2	98.81	0.00184	0.9796	0.2925	0.9456	0.4913	0.5	0.9594	0.0023	0.0067	0.9875
	pH 7.4	90.78	0.00046	0.9681	0.1951	0.9976	0.9806	0.5379	0.9964	0.0017	0.0054	0.9466
CS-BZC	pH 1.2	83.69	0.00276	0.885	0.526	0.8266	0.1418	0.3782	0.9419	0.0034	0.0079	0.9974
	pH 7.4	96.74	0.00184	0.991	0.433	0.9442	0.2283	0.4401	0.9801	0.0031	0.0085	0.9845
AL-MSN-BZC	pH 1.2	96.83	0.001152	0.9407	0.2026	0.9795	0.5025	0.45	0.9892	0.0016	0.0046	0.9736
	pH 7.4	96.6	0.00046	0.88	0.1684	0.8716	1.7196	0.5924	0.9559	0.0015	0.0055	0.7707
CS-MSN-BZC	pH 1.2	74.1	0.00069	0.8743	0.6255	0.8451	0.2614	0.4146	0.9304	0.0043	0.0106	0.9949
	pH 7.4	97.19	1.15×10^{-9}	0.977	0.9418	0.9555	3.8884	0.6314	0.9911	0.0098	0.0376	0.8927

The release of BZC from polymer obeys the Weibull model in pH 1.2 with a R2 values of 0.9974 for CS-BZC and 0.9875 for Al-BZC meaning that the properties of the release medium have a high impact on drug release. Al tends to be more stable at pH 1.2 compared with CS. At pH 7.4, the release of BZC obeys the first order or Higuchi model having a coefficient R2 value of 0.991 for CS-BZC and of 0.9681 for Al-BZC, meaning that the drug was transported from the matrix by diffusion for both CS and Al.

The introduction of MSN in the polymer matrix does not induce a significant modification in the release mechanism. At pH 1.2, the Weibull or Higuchi model describes the drug release and also the diffusion is the transport mechanism due to the different stability of the two polymers in SGF. Conversely, at pH 7.4 the mathematic model that describes the release mechanism is the first order model meaning that MSN induce its porosity in the hybrid material. The results obtained fit well the synthesized materials because the model is similar to a drug delivery system that contains watersoluble drugs which are encapsulated into porous matrices [33]. This time the drug respects the case II mechanism transport from the hybrid material (swelling followed by erosion).

4. Conclusions

The successful modification of MSN with different amounts of BZC was proved by FT-IR spectra in which distinctive bands assigned to BZC structure into the MSN-BZC spectra were identified, as well from the TGA data where a significant mass loss is obtained for MSN-BZC compared with pure MSN. Moreover, XPS was used to confirm the presence of BZC on the MSN surface.

The pH value of the environment, the contact time or the temperature used during the adsorption experiments proved to be important factors in the encapsulation process of BZC into the MSN. The highest encapsulation degree was recorded in solution having pH value of 5, at room temperature when the MSN was immersed for 60 min into the BZC containing solution. The adsorption process type as well as the adsorption rate, the kinetic parameters and thermodynamic parameters were determined.

Using a casting method, we obtained CS/AL-BZC and CS/AL-MSN-BZC composite films, which were further tested as potential drug delivery systems. The release profile of BZC from different systems, was studied in both SGF and SIF. The biopolymers are intended to increase the amount of the drug release by improving the dispersion of the MSN and allow a better diffusion of the drug. This is evident from the drug release curves of the MSN-BZC hybrid materials with and without the polymer. SEM proved the dispersion of the MSN within the polymer matrices whereby we observed a good distribution of the inorganic filler within the Al-MSN-BZC and the formation of some agglomeration in the case of CS-MSN-BZC. The dissolution media and the presence of MSN in the polymer matrix significantly influence the release mechanism. In SIF the release mechanism of BZC obeys the first order model because MSN increases the porosity of the hybrid materials.

Author Contributions: A.M.P. and C.A. were involved in laboratory experiments regarding the modification process and FT-IR, XPS and TGA characterization of the modified MSN. A.M.P. also performed the experimental protocol in order to study the influence of different parameters for the adsorption experiments. A.G. was involved in the calculation of the mathematic models including first-order, Higuchi, Korsmeyer–Peppas and Weibull of the release mechanism. S.A.G. is involved on synthesis of the hybrid materials and study of the in vitro drug release. H.I. is the mentor of this original idea and supervises the whole research work. All authors were involved in the results discussion, data interpretation and finalizing the manuscript.

Funding: The work has been funded by the Sectoral Operational Programme Human Resources Development 2007-2013 of the Ministry of European Funds through the Financial Agreement POSDRU/159/1.5/S/132397.

Conflicts of Interest: The authors declare no conflict of interest.

References

1. Digge, M.S.; Moon, R.S.; Gattani, S.G. Application of Carbon Nanotubes in Drug Delivery: A Review. *Int. J. PharmTech Res.* **2012**, *4*, 839–847.

2. Alhamdi, J.; Jacobs, E.; Gronowicz, G.; Benkirane-Jessel, N.; Hurley, M.; Kuhn, L. Cell Type Influences Local Delivery of Biomolecules from a Bioinspired Apatite Drug Delivery System. *Materials* **2018**, *11*, 1703. [CrossRef] [PubMed]

3. Bernkop-Schnürch, A. Nanocarrier systems for oral drug delivery: Do we really need them? *Eur. J. Pharm. Sci.* **2013**, *49*, 272–277. [CrossRef] [PubMed]

4. Vallet-Regí, M.; Rámila, A.; del Real, P.R.; Pérez-Pariente, J. A New Property of MCM-41: Drug Delivery System. *Chem. Mater.* **2001**, *13*, 308–311. [CrossRef]

5. Zhao, Y.Z.; Sun, C.Z.; Lu, C.T.; Dai, D.D.; Lv, H.F.; Wu, Y.; Wan, C.W.; Chen, L.J.; Lin, M.; Li, X.K. Characterization and anti-tumor activity of chemical conjugation of doxorubicin in polymeric micelles (DOX-P) in vitro. *Cancer Lett.* **2011**, *311*, 187–194. [CrossRef] [PubMed]

6. Marcelo, G.; Ariana-Machado, J.; Enea, M.; Carmo, H.; Rodriguez-Gonzales, B.; Capelo, J.L.; Lodeiro, C.; Oliviera, E. Toxicological Evaluation of Luminescent Silica Nanoparticles as New Drug Nanocarriers in Different Cancer Cell Lines. *Materials* **2018**, *11*, 1310. [CrossRef] [PubMed]

7. Miculescu, F.; Mocanu, A.C.; Dascalu, C.A.; Maidaniuc, A.; Batalu, D.; Berbecaru, A.; Voicu, S.I.; Miculescu, M.; Thakur, V.K.; Ciocan, L.T. Facile synthesis and characterization of hydroxyapatite particles for high value nanocomposites and biomaterials. *Vacuum* **2017**, *146*, 614–622. [CrossRef]

8. Chen, J.; Xia, N.; Zhou, T.; Tan, S.; Jiang, F. Mesoporous carbon spheres: Synthesis, characterization and supercapacitance. *Int. J. Electrochem. Sci.* **2009**, *4*, 1063–1073.

9. Vallet-Regi, M.; Balas, F.; Arcos, D. Mesoporous materials for drug delivery. *Chem. Mater.* **2007**, *46*, 7548–7558. [CrossRef]

10. Swatantra, K.K.S.; Awani, R.K.; Satyawan, S. Chitosan: A Platform for Targeted Drug Delivery. *Int. J. PharmTech Res.* **2010**, *2*, 2271–2282.

11. Tonnesen, H.H.; Karlsen, J. Alginate in drug delivery systems. *Drug Dev. Ind. Pharm.* **2002**, *28*, 621–630. [CrossRef] [PubMed]

12. Farías, T.; Charles de Ménorval, L. Benzalkonium chloride and sulfamethoxazole adsorption onto natural clinoptilolite: Effect of time, ionic strength, pH and temperature. *J. Colloid Interface Sci.* **2011**, *363*, 465–475. [CrossRef] [PubMed]

13. Kostić, D.A.; Mitić, S.S.; Nasković, D.C.; Zarubica, A.R.; Mitic, M.N. Determination of Benzalkonium Chloride in Nasal Drops by High-Performance Liquid Chromatography. *E-J. Chem.* **2012**, *9*, 1599–1604. [CrossRef]

14. Ghebaur, A.; Garea, S.A.; Iovu, H. New polymer–halloysite hybrid materials—Potential controlled drug release system. *Int. J. Pharm.* **2012**, *436*, 568–573. [CrossRef] [PubMed]

15. Mhlanga, N.; Ray, S.S. Kinetic model for the release of the anticancer drug doxorubicin from biodegradable polylactide/metal oxide- based hybrid. *Int. J. Biol. Macromol.* **2015**, *72*, 1301–1307. [CrossRef] [PubMed]

16. Petropoulous, J.H.; Papadokostaki, K.G.; Sanopoulou, M. Higuchi's equations and beyond: Overview of the formulation and application of generalized model of drug release from polymeric matrices. *Int. J. Pharm.* **2012**, *437*, 178–191. [CrossRef]

17. Bruschi, M. (Ed.) Mathematical models of drug release. In *Strategies to Modify the Drug Release from Pharmaceutical Systems*; Woodhead Publishing: Cambridge, UK, 2015; pp. 63–86.

18. Jose, S.; Fangueiro, J.F.; Smitha, J.; Cinu, T.A.; Chacko, A.J.; Premaletha, K.; Souto, E.B. Predictive modeling of insulin release profile from cross-linked chitosan microspheres. *Eur. J. Med. Chem.* **2013**, *60*, 249–253. [CrossRef]

19. Lungan, M.A.; Popa, M.; Racovita, S.; Hitruc, G.; Doroftei, F.; Desbieres, J.; Vasiliu, S. Surface characterization and drug release from porous microparticles based on methacrylic monomers and xanthan. *Carbohydr. Polym.* **2015**, *125*, 323–333. [CrossRef]

20. Korsmeyer, R.W.; Gurny, R.; Doelker, E.; Buri, P.; Peppas, N.A. Mechanisms of solute release from porous hydrophilic polymers. *Int. J. Pharm.* **1983**, *15*, 25–35. [CrossRef]

21. Tang, J.; Slowing, I.I.; Huang, Y.; Trewyn, B.G.; Hu, J.; Liu, H.; Lin, V.S.Y. poly(lactic acid)-coated mesoporous silica nanosphere for controlled release of venlafaxine. *J. Colloid Interface Sci.* **2011**, *360*, 488–496. [CrossRef]

22. Li, Z.; Su, K.; Cheng, B.; Deng, Y. Organically modified MCM-type material preparation and its usage in controlled amoxicillin delivery. *J. Colloid Interface Sci.* **2010**, *342*, 607–613. [CrossRef]

23. Havasi, F.; Ghorbani-Choghamarani, A.; Nikpour, F. Pd-Grafted functionalized mesoporous MCM-41: A novel, green and heterogeneous nanocatalyst for the selective synthesis of phenols and anilines from aryl halides in water. *New J. Chem.* **2015**, *39*, 6504–6512. [CrossRef]

24. Najafi, M.; Yousefi, Y.; Rafati, A.A. Synthesis, characterization and adsorption studies of several heavy metal ions on amino-functionalized silica nano hollow sphere and silica gel. *Sep. Purif. Technol.* **2012**, *85*, 193–205. [CrossRef]

25. Bui, T.X.; Choi, H. Adsorptive removal of selected pharmaceuticals by mesoporous silica SBA-15. *J. Hazard. Mater.* **2009**, *168*, 602–608. [CrossRef]

26. Huang, C.H.; Chang, K.P.; Ou, H.D.; Chiang, Y.C.; Wang, C.F. Adsorption of cationic dyes onto mesoporous silica. *Microporous Mesoporous Mater.* **2011**, *141*, 102–109. [CrossRef]

27. Kuo, C.Y. Comparison with as-grown and microwave modified carbon nanotubes to removal aqueous bisphenol A. *Desalination* **2009**, *249*, 976–982. [CrossRef]

28. Mohammadi, N.; Khani, H.; Gupta, V.K.; Amereh, E.; Agarwal, S. Adsorption process of methyl orange dye onto mesoporous carbon material–kinetic and thermodynamic studies. *J. Colloid Interface Sci.* **2011**, *362*, 457–462. [CrossRef]

29. Zanini, G.P.; Ovesen, R.G.; Hansen, H.C.B.; Strobel, B.W. Adsorption of the disinfectant benzalkonium chloride on montmorillonite. Syntergetic effect in mixture of molecules with different chain lengths. *J. Environ. Manag.* **2013**, *128*, 100–105. [CrossRef]

30. Huang, X.; Xiao, Y.; Lang, M. Micelles/sodium-alginate composite gel beads: A new matrix for oral drugdelivery of indomethacin. *Carbohydr. Polym.* **2012**, *87*, 790–798. [CrossRef]

31. Miculescu, F.; Bojin, D.; Ciocan, L.T.; Antoniac, I.A.; Miculescu, M.; Niculescu, N. Experimental Research on Biomaterial-Tissue Interface Interactions. *J. Optoelectron. Adv. Mater.* **2007**, *9*, 3303–3306.

32. Martin, A.; Morales, V.; Ortiz-Bustos, J.; Perez-Garnes, J.; Bautista, L.F.; Garcia-Munoz, R.A.; Sanz, R. Modelling the adsorption and controlled release of drug from the pure and amino surface-functionalized mesoporous silica host. *Microporous Mesoporous Mater.* **2018**, *262*, 23–34. [CrossRef]

33. Dash, S.; Murthy, P.N.; Nath, L.; Chowdhury, P. Kinetic modeling on drug release from controlled drug delivery systems. *Acta Pol. Pharm.* **2010**, *67*, 217–223. [PubMed]

 *materials*

Article

Preparing Sodium Alginate/Polyethyleneimine Spheres for Potential Application of Killing Tumor Cells by Reducing the Concentration of Copper Ions in the Lesions of Colon Cancer

Ru Xu [1], Chen Su [1], Longlong Cui [1], Kun Zhang [1,*] and Jingan Li [2,*]

[1]   School of Life Science, Zhengzhou University, Zhengzhou 450000, China; 18341349663@163.com (R.X.); suchenqwer@163.com (C.S.); whitebear24@163.com (L.C.)
[2]   School of Material Science and Engineering & Henan Key Laboratory of Advanced Magnesium Alloy & Key Laboratory of materials processing and mold technology (Ministry of Education), Zhengzhou University, Zhengzhou 450001, China
*   Correspondence: zhangkun@zzu.edu.cn (K.Z.); lijingan@zzu.edu.cn (J.L.)

Received: 8 April 2019; Accepted: 10 May 2019; Published: 13 May 2019

Abstract: Inhibition of residual malignant tumors in patients with colon cancer after operation is one of the difficulties in rehabilitation treatment. At present, using biocompatible materials to remove the copper ion which is the growth dependence of malignant tumors in the lesion site is considered to be the frontier means to solve this problem. In this work, we developed a sodium alginate (SA)/polyethyleneimine (PEI) hydrogel sphere via cross-linking method (SA/SP/SA; SP = SA/PEI) as an oral biomaterial for adsorbing and removing copper ions from colon cancer lesions. The evaluated results showed that the SA/PEI/SA (SPS) hydrogel sphere obtained the largest swelling rate at pH 8.3 which was the acid-base value of colon microenvironment and absorbed more copper ions compared with the SA control. The cell experiment presented that the SPS hydrogel sphere owned better compatibility on normal fibroblasts and promoted higher death of colon cancer cells compared with SA/PEI (SP) and SA control. Our data suggested that the SA/PEI hydrogel sphere had the potentiality as an oral biomaterial for inhibiting colon cancer cells.

Keywords: colon cancer cells; copper ions; hydrogel sphere; sodium alginate; polyethyleneimine

---

1. Introduction

Colon cancer is a kind of malignant gastrointestinal cancer with a high incidence among people aged 40–50 years; it is the third most common cancer of that age group [1,2]. The main treatment method for colon cancer is surgical resection, supplemented by chemotherapy and drug treatment [3,4]. However, about half of patients may suffer from postoperative metastasis and recurrence that endanger their life [5]. In addition, some patients show no significant improvement on preoperative and postoperative radiotherapy [6]. Thus, there is an urgent need for a targeted and effective method to inhibit and kill colon cancer cells as an adjuvant or alternative therapy for surgery and chemotherapy. Studies have shown that the accumulation of copper ions in cancerous sites is an important factor for the survival and proliferation of cancer cells [7,8]; hence, removal of copper ions accumulated in the colon may be effective in killing cancer cells.

The technical bottleneck, then, is how to remove the copper ions from the focal site without damaging healthy cells. Hydrogels with good swelling properties are undoubtedly preferred, and preparing the hydrogels as oral biomaterials (e.g., hydrogel spheres) is a good adjuvant therapy for colon cancer patients after operation. However, the oral hydrogel spheres will go through different organs of the digestive system, such as stomach (pH 1.2), rectum (pH 6.8), blood (pH 7.4) and colon (pH 8.3), the pH microenvironment of which will affect the swelling properties of the materials [9,10]. Thus, choosing a hydrogel material with maximum swelling rate in colon microenvironment is the key to solve this problem. Sodium alginate (SA) hydrogel is a kind of pH-sensitive hydrogel, but its swelling rate is stable above pH 7.4 [11]. We also found that SA hydrogel has good plasticity and biocompatibility in our previous work [12]. However, the porous structures of tens of micron scales on the surface of SA hydrogels may not be conducive to the implementation of this function: The porous connectivity structure of hydrogels may cause the exudation of inhaled copper ions and reduce the efficiency of carrying copper [13]; on the other hand, pores with a diameter of tens of micron-scales may allow most cells to migrate and aggregate, and the migrated cells will cause long-term hydrogel retention in the body, delaying the excretion of adsorbed copper ions outside [14]. In the previous research, polyethyleneimine (PEI) molecules were modified on the materials surface via layer-by-layer self-assembly and were proved to inhibit series of pathologically related cells of esophageal cancer [15,16]. Thus, modifying the SA hydrogel with PEI and SA molecules may improve its surface compactness and swelling degree, further endowing the hydrogel function of inhibiting cancer cells.

In this contribution, the SA/PEI/SA (SPS) hydrogel sphere was prepared via the cross-linking method aimed at absorbing and removing the copper ion and further killing the colon cancer cell. The SPS hydrogel sphere's surface morphology, inner structure, swelling ratio, and ability on removing copper ion and promoting colon cancer cell apoptosis were investigated systematically. We hope this SPS hydrogel sphere may have potential application as an oral biomaterial for postoperative rehabilitation of patients with colon cancer.

2. Materials and Methods

2.1. Preparation of Sodium Alginate/Polyvinylimide Hydrogel Spheres via Crosslinking

Sodium alginate (SA, 3.2×10^4 Da–2.5×10^5 Da, Sigma) was dissolved in deionized water (dH$_2$O) to prepare a solution of 20 mg/mL. Thereafter, the SA solution was added to 100 mg/mL calcium chloride (CaCl$_2$) supersaturated solution (dissolved in the dH$_2$O) drop by drop using an injector with 1 mL capacity, and followed with 30 min magnetic stirring. After filtering residual solution and cleaning, SA hydrogel spheres that would be placed in the core position were obtained. Next, the SA hydrogel spheres were poured into polyethyleneimine (PEI, 2.5×10^4 Da, Sigma) solution (20 mg/mL, dissolved in the dH$_2$O) and stirred for 30 min. After filtering residual solution and clear operation once more, the obtained samples were SA spheres covered with a layer of PEI molecule and were labeled as SP. Then the SP spheres were put into 20 mg/mL SA solution again and stirred for another 30 min. By the final filtering of residual solution and cleaning step, the acquired samples were SP spheres overlapped with another layer of SA and were named as SPS. The preparation process of SPS hydrogel spheres is exhibited in Figure 1.

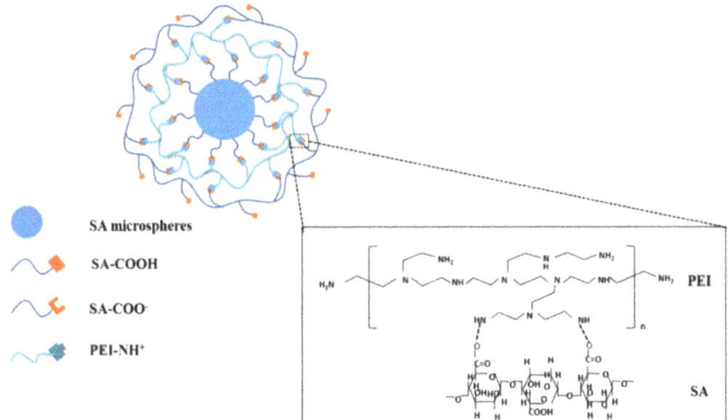

Figure 1. Preparation of SA/PEI/SA (SPS) spheres. SA: sodium alginate; PEI: polyethyleneimine.

2.2. Characterization of the SPS Hydrogel Sphere

The infrared absorption spectra of the SPS, SP and SA samples were gained using a Fourier transform infrared (FTIR, NICOLET 5700, Waltham, MA, USA) spectrometer in the scanning range of 4000–400 cm^{-1} after being freeze-dried, crushed, ground and tablet compressed with KBr [17]. The SPS, SP and SA hydrogel spheres were photographed with Huawei Mate 10 Pro camera equipment to directly observe their entire object and sizes, and their diameters were also detected from at least 100 spheres for each sample [18]. Surface and cross-section morphologies as well as surface energy dispersive spectrum (EDS) of the SA, SP and SPS hydrogel spheres were observed by scanning electron microscopy (SEM, FEI Quanta200, The Netherlands) after freezing at −80 °C, being fully dried with a freeze-dryer, and gold spraying [19].

The swelling ratios (SR) of the SPS hydrogel sphere within 18 h were determined and calculated in the following formula: SR = m1 − m0/m0, wherein m0 indicated the dry weight of the samples and m1 indicated the wet weight of the samples [12]. In addition, the SR values and degradation ratio of the SPS under condition of pH 1.2, pH 6.8, pH 7.4 and pH 8.3 were detected for the consideration of the pH difference in vivo microenvironment that the oral colon cancer drugs may go through: The pH value of stomach is 1.2, the pH value of rectum is 6.8, the pH value of blood is 7.4 and the pH value of colon is 8.3 [20,21]. The absorption ratios of the SA, SP and SPS samples after exposure to different cumulative pH (pH 1.2 for 4 h, pH 7.4 for 8 h, and pH 8.3 for 60 h) were also investigated.

To investigate the ability of SPS, SP and SA hydrogels spheres on absorbing copper ion we performed the experiment as follow [22]: The concentrations of copper sulphate solution (dissolved in dH$_2$O) were designed as 22.0 mg/L, with pH 8.3. The SA, SP and SPS spheres (200 mg per group) were immersed in the copper sulfate solution separately, and the solution with spheres were magnetic stirred by 150 rpm at 37 °C. Two mL solutions were taken out from each group after 1 h, 2 h, 3 h, 4 h, 5 h, 6 h, 18 h and 24 h to determine their absorbance by ultraviolet spectrophotometer, and the concentration of copper ion in solution was determined according to the standard curve.

2.3. Biocompatibility Testing of the Developed Hydrogels Spheres

A first set of experiments was performed to evaluate in vitro biocompatibility of the SPS, SP and SA hydrogel spheres by using L929 cell line. These fibroblast-like cells were seeded at an initial cell density of 4000 cells/well and incubated at 37 °C in culture medium (90% Dulbecco's Modified Eagle medium high glucose, 10% fetal bovine serum, 100 units/mL penicillin and 0.1 mg/mL streptomycin solution) until they reached 60% confluence. At this time point, the cells were treated with SPS, SP and SA hydrogel spheres (4 spheres per well) for 1 day, 2 days and 3 days. For comparative purposes, the cells cultured with no hydrogels spheres were also investigated and considered as control (CON) group. After each incubation period the cells were stained with the cell-permeable acridine orange (AO) in combination with the plasma membrane-impermeable DNA-binding dye propidium iodide (PI). AO and PI excite green and red fluorescence, respectively, when they are intercalated into the cells' DNA, which represents living or dead cells, because only AO was able to cross the plasma membrane of living cells, while PI can only cross the plasma membrane of dead or dying cells [23]. L929 survival rates in each group were counted and calculated from at least 15 images [15].

2.4. Investigation of the Hydrogel Spheres' Capacity to Kill Colon Cancer Cells

Human colon cancer cells (HT-29, Shanghai Zhong Qiao Xin Zhou Biotechnology Co., Ltd., Shanghai, China) were seeded in the 24-well cell culture plate with a density of 4000 cells/well, and incubated in a humidified incubator with 95% air and 5% CO$_2$ at 37 °C. When the cells reach 60% of the confluence, the SPS hydrogel spheres prepared under aseptic conditions were placed in the culture plate in a density of 4 per well, and co-cultured with the colon cancer cells for 1 day, 2 days and 3 days, respectively [24]. The SP, SA and non-hydrogel sphere co-cultured cells (the group was labeled as CON) were also incubated as comparison. The function of SPS on suppressing colon cancer cells was also evaluated by staining with AO/PI. The inhibition rate of each group on colon cancer cells was also counted from the fluorescence images [15].

The study was conducted in accordance with the Declaration of Helsinki, and the protocol was approved by the Ethics Committee of Zhengzhou University.

3. Results and discussion

3.1. Physical and Chemical Properties of the SPS Hydrogel Sphere

The chemical structures of SPS sample and the SA and PEI controls were analyzed by FTIR spectroscopy, and are presented in Figure 2: The absorption peaks of 1672 cm^{-1} in SPS spheres came from the bending vibration peaks of N-H groups in PEI, and there were obvious double absorption peaks at 3700–3100 cm^{-1}, which were obvious amino characteristic peaks; the hydroxyl absorption peaks in the SA curve disappeared in the SPS spheres, which may be due to the destruction of hydroxyl groups by Ca^{2+} ion bonding; in SPS spheres, 2648–3030 cm^{-1} was the stretching and bending vibration peaks of CH$_2$ group in PEI molecule and 1143–1359 cm^{-1} was the stretching and retracting absorption peaks of carboxyl group in SA molecule; SA and PEI participated in the reaction by forming amide bond, so the peak value of N-H (3500–3400cm^{-1}) single bond of PEI molecule in SPS decreased; at the same time, due to the spatial change of N-H bonds in the reaction, the absorption peaks of N-H bonds shifted; in SPS, the characteristic peaks of amide bond and the absorption peaks of amide I bond C=O (1690–1630 cm^{-1}) were enhanced. Thus, it could be summarized that SPS spheres were bound together by carboxyl and amino groups to form amide bonds.

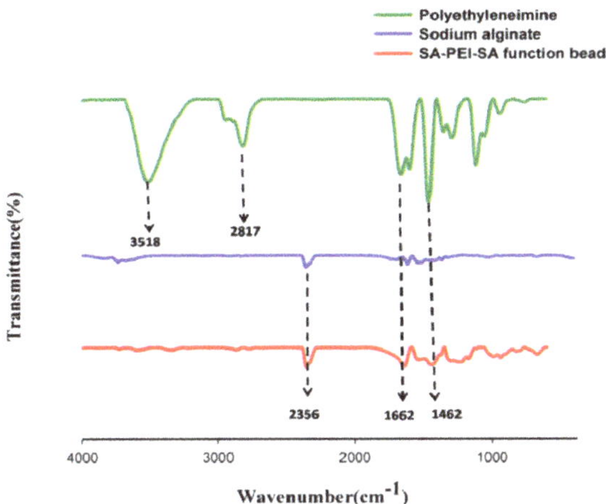

Figure 2. Fourier transform infrared (FTIR) spectra of PEI, SA and SPS samples.

Figure 3 depicts a design of the abbreviation of our institution Zhengzhou University, "ZZU", which was composed of three kinds of hydrogel spheres, SA, SP and SPS: From left to right, the first Z is made of SA balls, the second Z is made of SP balls and the U is made of SPS balls. It could be seen from the figure that the sizes of SA and SP spheres were relatively the same, and they were transparent. SPS spheres were larger than the first two kinds of spheres and were light yellow. Because the color of sodium alginate solution was light yellow, the spheres encapsulated with sodium alginate showed the same color. Statistical results of the sphere diameters in Table 1 further verified the visual images, displaying their trend of SPS > SP > SA.

Figure 3. Photographs of the abbreviation of Zhengzhou University, "ZZU", which was composed of the SA (the first letter "Z"), SA/PEI (SP) (the second letter "Z") and SPS (the third letter "U") hydrogel spheres.

Table 1. Diameter statistics of the SA, SP and SPS hydrogel spheres (n = 100, mean ± SD).

Samples	SA	SP	SPS
Diameters	2.4 ± 0.0 mm	2.5 ± 0.0 mm	2.7 ± 0.1 mm

After freeze-drying of SA, SP and SPS hydrogel and breaking by liquid nitrogen, the micro-morphologies and internal structure of its surface and cross section was observed. The cross-sectional graph of SA, SP and SPS exhibits obvious three-dimensional network cross-linking structure and porous morphology (Figure 4a), and this is a typical hydrogel inner structure which contributed to better swelling property and further absorbing more copper ion from the lesion location [25]. Wherein, SA and SP show larger pore diameters and thicker pore walls compared with SPS. The surface morphology of the SA, SP and SPS hydrogel sphere is compact in the micro sizes and there are obvious wrinkles (Figure 4b), which is completely different from the morphology of the existing porous, coherent and open hydrogels [26]. The latter, as scaffolds for tissue engineering, often promotes cell migration, proliferation and differentiation [27], while the former, as a targeted anti-tumor biomaterial, should inhibit cell growth to avoid prolonged retention of SPS spheres and their absorbed copper ion in vivo. Figure 4c displays the content distribution of C, O and N elements in the surface analysis of each sample: The surface of SA spheres contains a lot of C and O elements, but a very small amount of N elements also appears on the surface, possibly due to a false positive result; in addition to a large number of C and O elements, N elements on the surface of SP spheres increase significantly compared with SA spheres, indicating that the PEI molecule has been successfully encapsulated on the surface of SA spheres; the surface composition of SPS spheres is similar to that of SA spheres, indicating that the sodium alginate solution successfully wraps the PEI on the surface of SP spheres.

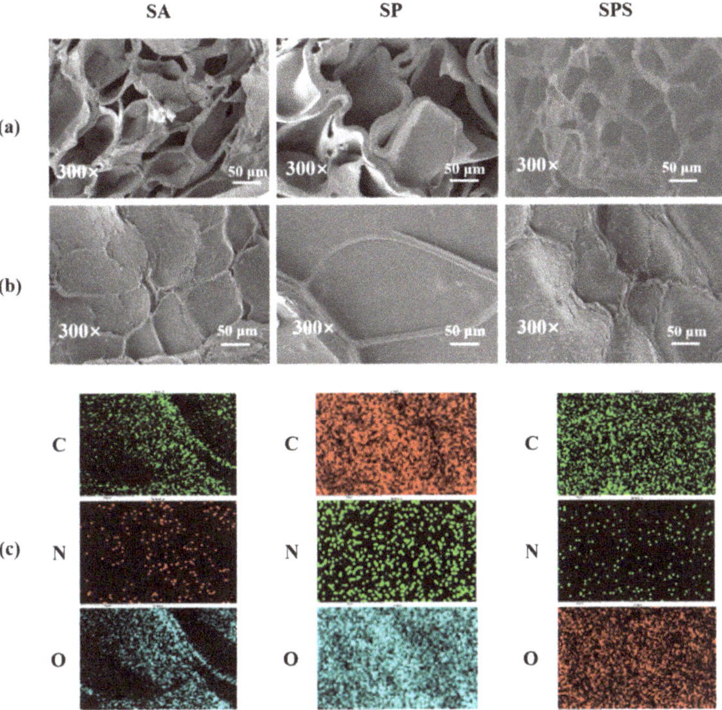

Figure 4. The (**a**) cross-section, (**b**) surface microstructure and (**c**) surface energy dispersive spectrum (EDS) of the SA, SP and SPS hydrogel sphere observed by scanning electron microscopy (SEM).

Oral colon cancer medicines pass through the stomach and rectum before reaching the site of the lesion, where the pH environment is quite different from that of the colon [16,28,29]. In this research, we measured the SR values of the SPS hydrogel sphere under different pH values according to the corresponding microenvironments over time (Figure 5a). The results show that the SR values of SPS hydrogel sphere increase with time in colon (pH = 8.3), rectum (pH = 6.8) and blood (pH = 7.4). From the 4th hour, the SR values of SPS hydrogel sphere in colon and blood microenvironment are significantly higher than that of other groups, and reach 586.3% at the 6th hour (pH = 8.3). These results indicate that SPS spheres have good swelling properties in intestinal environment (including blood microenvironment of wound after operation), and the increase of sphere volume is beneficial to its better adsorption of copper ions in vivo. In simulated gastric solution (pH = 1.2), the SR value of SPS spheres is much lower than that of other microenvironments, and there is no significant change over time. This is mainly due to the acidic environment that makes the carboxylic group of sodium alginate protonated, and the gel mesh is tightly adsorbed by the interaction of intermolecular electrostatic forces and hydrogen bonds. The degradation ratio result (Figure 5b) showed that pH value was a crucial factor which affects the SPS degradation behaviors. Under pH 1.2 condition, SPS maintains a stable degradation ratio of about 60%; while under other pH values, the SPS shows very low degradation ratio within the first 8 h; after 16 h at pH 6.8, 7.4 and 8.3, the spheres show a swelling trend, which is correlated with pH; the higher the pH value, the more obvious the swelling is; however, after 16 h at pH 1.2, the volume of the SPS spheres remain stable, and there is no obvious phenomenon of sharp increase or decrease; it can be inferred that the spheres are wrinkled under the influence of pH in the gastric environment and gradually show a swelling trend in the intestinal fluid environment. The absorption ratios of the SA, SP and SPS samples after exposure to different cumulative pH (Figure 5c) show that SA, SP and SPS spheres present different degrees of water loss in pH 1.2 environment; during the period of pH 7.4, the spheres begin to slowly absorb water and recover their original state; in the pH 8.3 environment, the spheres begin to swell, and the SP and SPS spheres possess higher swelling ratios compared with SA, wherein the swelling effect of SP spheres is the most obvious.

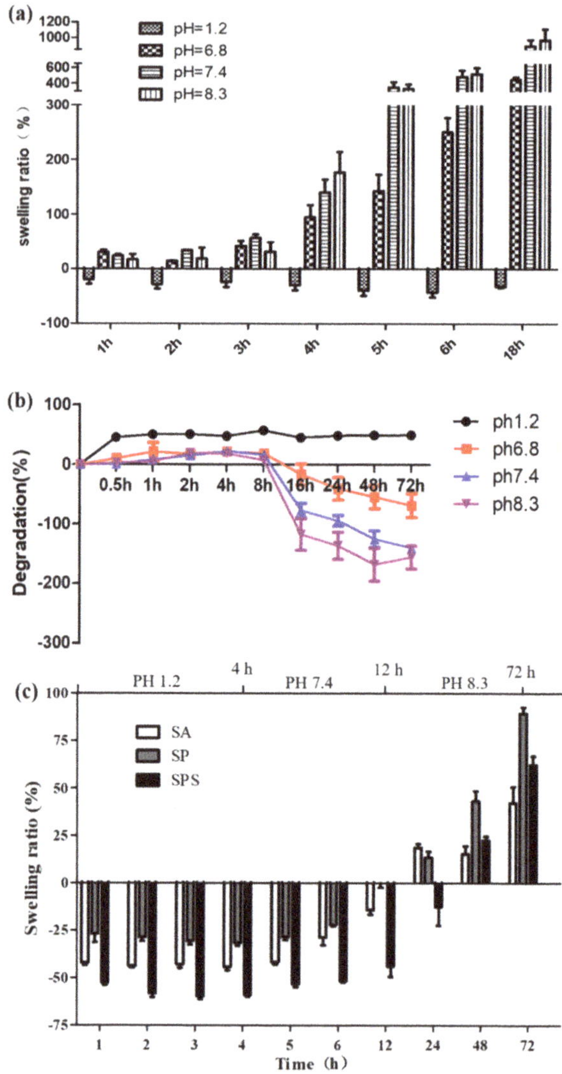

Figure 5. (a) The swelling ratio of the SPS hydrogel sphere after immersed in the phosphate buffer solution for 1 h, 2 h, 3 h, 4 h, 5 h, 6 h and 18 h in the condition of pH 1.2, pH 6.8, pH 7.4 and pH 8.3, separately; (b) the degradation ratio of the SPS hydrogel sphere for 0.5 h, 1 h, 2 h, 4 h, 8 h, 16 h, 24 h, 48 h and 72 h in the condition of pH 1.2, pH 6.8, pH 7.4 and pH 8.3, separately; (c) the swelling ratio of the SA, SP and SPS hydrogel spheres after immersion in the PBS solution in the condition of pH 1.2 for 4 h, and then pH 7.4 for 8 h, and finally pH 8.3 for 60 h. (mean ± SD, n = 5).

Figure 6 displays the percent of the absorbed copper ions of SPS, SP and SA hydrogel sphere within 24 h. It is obvious that SPS and SP samples absorb more copper ions compared with the SA sample which may be attributed to the good complexation ability of PEI with copper ions and the compact microstructures of the spheres. In addition, the SPS sample presents rapider absorption compared to SP within the first 6 h, and this will make more contribution to killing and inhibiting colon cancer cells, because the hydrogel sphere may be deferred for only several hours in the colon and then be excreted in vitro through feces.

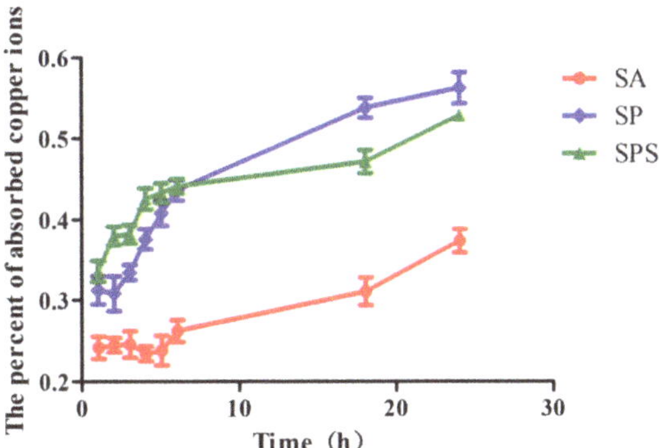

Figure 6. The ability of the SPS, SP and SA hydrogel spheres to absorb copper ions at pH 8.3 (mean ± SD, n = 5).

3.2. Hydrogel Spheres' Cytotoxicity against Normal Cells

To investigate the biocompatibility of the SPS hydrogel spheres, the SPS, SP and SA samples were co-cultured with the L929 cell line (Fibroblasts). The live/dead staining images (Figure 7) and the following cell counting results (Figure 8) indicate that the cells treated with SPS hydrogel spheres showed higher cell survival rates than those treated with SA and SP spheres. This finding suggests that the SPS spheres are the most biocompatible and well-tolerated by normal cells.

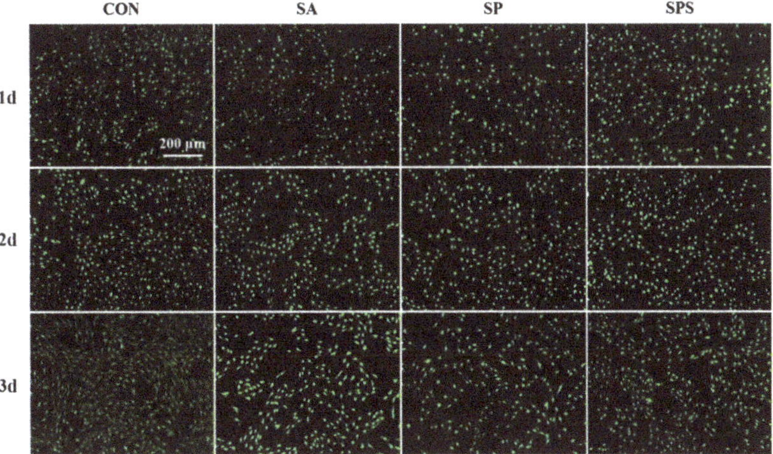

Figure 7. Acridine orange (AO) and propidium iodide (PI) staining images of L929 cell in the SPS, SP, SA and control (CON) groups. (The green dots stained by AO indicate living cells, and the red dots stained by PI indicate dead or dying cells).

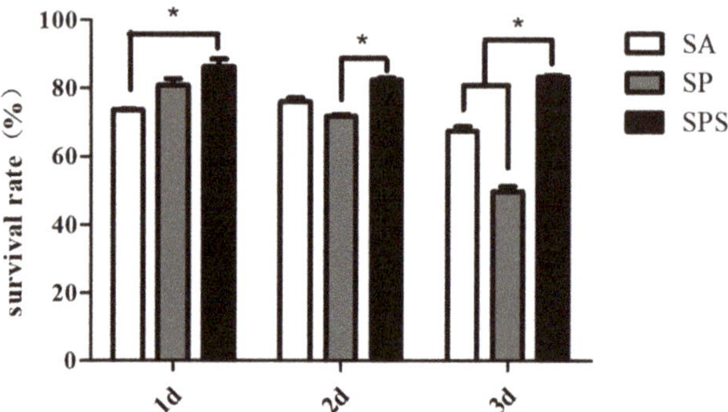

Figure 8. Inhibition rate of the SPS, SP and SA hydrogel spheres on the L929 cells (* $p < 0.05$, mean ± SD, n = 3).

3.3. The Ability of SPS Hydrogel Spheres to Kill Colon Cancer Cells

Pathological metastasis and proliferation of colon cancer cells after tissue resection is an important cause of recurrence of the disease [30]. Therefore, effective growth inhibition or killing of the colon cancer cells represent the main objective of researchers involved in the development of related drugs and biomaterials. Figure 9 presents the AO/PI staining images of colon cancer cells that co-cultured with the SPS, SP and SA hydrogel spheres, and the cancer cells cultured alone were also stained as control (the CON group). Wherein, the green fluorescence represented the living cells and the red fluorescence represent dead or dying cells. The SPS group shows obviously more red-labeled cells compared to other groups: At the 2nd day more than half of the colon cancer cells in the SPS group show red color, and at the 3rd day almost all the colon cancer cells present red color; only a few cells maintain green; this phenomenon indicates that the SPS hydrogel sphere had a strong function on inhibiting or killing the colon cancer cells.

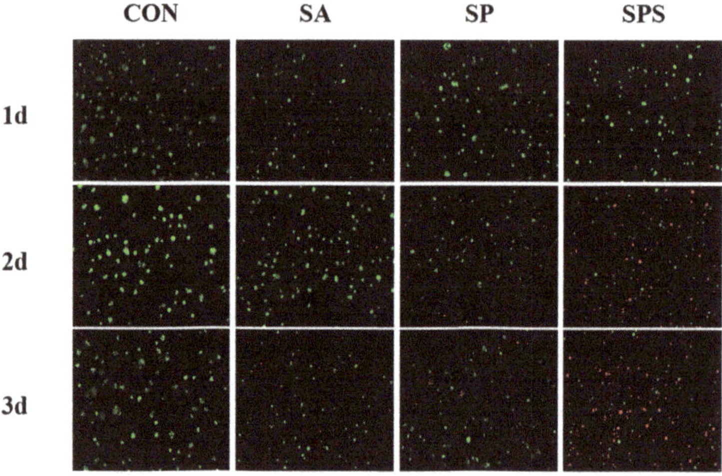

Figure 9. AO and PI staining images of colon cancer cells in the SPS, SP, SA and CON groups. (The green dots stained by AO indicate living cells, and the red dots stained by PI indicate dead or dying cells).

The statistical results are consistent with the trend of AO/PI images (Figure 10). While the SA and SP group also exhibit certain properties on inhibiting the colon cancer cells, their inhibition rates are significantly lower compared with the SPS group's value, and the trend is: SA < SP < SPS. The reason may be the reduced copper ions and the existing free carboxyl group [7,19]: SA has a certain amount of free carboxyl group, while SP reduces the copper ions; judging from the inhibition rate, reducing copper ions plays a more important role for inhibiting colon cancer cells; SPS has both the reduced copper ions and the free carboxyl group, thus showing the best function on killing colon cancer cells.

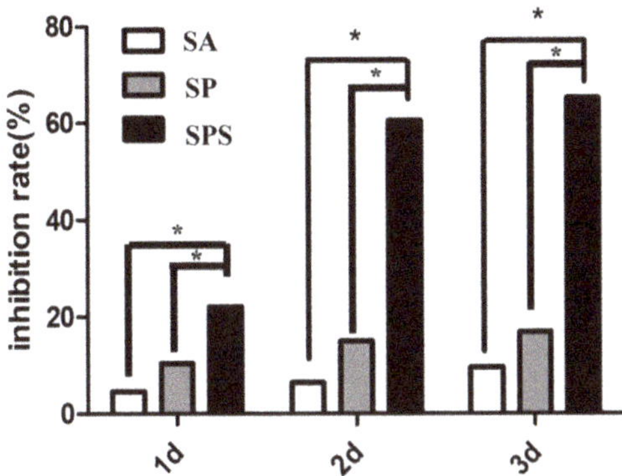

Figure 10. Inhibition rate of the SPS, SP and SA hydrogel spheres on the colon cancer cells (* $p < 0.05$).

4. Conclusions

In this contribution, we developed an SPS hydrogel sphere as an adjuvant or alternative therapeutic material for post-operative radiotherapy of colon cancer. The aim of designing the SPS hydrogel was to reduce the aggregated copper ion from colon cancer lesions, which was considered as the dependence of the colon cancer cells pathological metastasis and hyperplasia. The SPS possessed an internal structure of porous networks, but a dense surface structure, which might endow it with a better swelling property, contributing to absorbing more copper ion from the focus. However, the SPS also had dense surface morphology at micron scale, and this would inhibit the cell growth into the sphere and avoid the retention of loaded copper ions in vivo. As an oral target biomaterial, the SPS had a lower swelling ratio in the gastric juice pH environment but a significantly higher swelling ratio in the intestinal pH environment. As we expected, compared with other hydrogel spheres, the SPS absorbed and removed more copper ions and further inhibited and killed colon cancer cells more effectively.

Author Contributions: Conceptualization, K.Z. and J.L.; methodology, K.Z.; software, R.X.; validation, K.Z., R.X., C.S. and L.C.; formal analysis, J.L.; investigation, J.L.; resources, K.Z.; data curation, C.S.; writing—original draft preparation, R.X.; writing—review and editing, J.L.; visualization, C.S.; supervision, L.C.; project administration, J.L.; funding acquisition, K.Z.

Funding: This research was funded by The Fostering Talents of National Natural Science Foundation of China and Henan Province, grant number U1504310; Key Scientific Research Projects of Higher Education Institutions in Henan Province, grant number 16A430030; Key Scientific and Technological Research Projects in Henan Province, grant number 182102310076; Top Doctor Program of Zhengzhou University, grant number 32210475; and The Joint Fund for Fostering Talents of NCIR-MMT & HNKL-MMT, grant number MMT2017-01.

Conflicts of Interest: The authors declare no conflict of interest.

References

1. Grothey, A.; Sobrero, A.F.; Shields, A.F.; Yoshino, T.; Paul, J.; Taieb, J.; Souglakos, J.; Shi, Q.; Kerr, R.; Labianca, R.; et al. Duration of Adjuvant Chemotherapy for Stage III Colon Cancer. *N. Engl. J. Med.* **2018**, *378*, 1177–1188. [CrossRef] [PubMed]
2. Ulintz, P.J.; Greenson, J.K.; Wu, R.; Fearon, E.R.; Hardiman, K.M. Lymph Node Metastases in Colon Cancer Are Polyclonal. *Clin. Cancer Res.* **2018**, *24*, 2214–2224. [CrossRef]
3. West, N.; Morris, E.; Finan, P.; Ingeholm, P.; Kennedy, R.; Sugihara, K.; Hohenberger, W.; Quirke, P. Study to identify the optimum surgical technique in colon cancer. *Lancet* **2014**, *383* (Suppl. 1), s107. [CrossRef]
4. Macdonald, J.S. Adjuvant therapy of colon cancer. *CA Cancer J. Clin.* **1999**, *49*, 202–219. [PubMed]
5. Lin, C.; Zhang, Y.; Chen, Y.; Bai, Y.; Zhang, Y. Long noncoding RNA LINC01234 promotes serine hydroxymethyltransferase 2 expression and proliferation by competitively binding miR-642a-5p in colon cancer. *Cell Death Dis.* **2019**, *10*, 137. [CrossRef] [PubMed]
6. Markowitz, S.D.; Dawson, D.M.; Willis, J.; Willson, J.K.V. Focus on colon cancer. *Cancer Cell* **2002**, *1*, 233–236. [CrossRef]
7. Li, J.; Si, X.; Li, X.; Wang, N.; An, Q.; Ji, S. Preparation of acid-resistant PEI/SA composite membranes for the pervaporation dehydration of ethanol at low pH. *Sep. Purif. Technol.* **2018**, *192*, 205–212. [CrossRef]
8. Ramdhan, T.; Ching, S.H.; Prakash, S.; Bhandari, B. Time dependent gelling properties of cuboid alginate gels made by external gelation method: Effects of alginate-CaCl$_2$ solution ratios and pH. *Food Hydrocoll.* **2019**, *90*, 232–240. [CrossRef]
9. Gan, J.; Robinson, R.C.; Wang, J.; Krishnakumar, N.; Manning, C.J.; Lor, Y.; Breck, M.; Barile, D.; German, J.B. Peptidomic profiling of human milk with LC–MS/MS reveals pH-specific proteolysis of milk proteins. *Food Chem.* **2019**, *274*, 766–774. [CrossRef]
10. Zou, D.; Luo, X.; Han, C.; Li, J.; Yang, P.; Li, Q.; Huang, N. Preparation of a biomimetic ECM surface on cardiovascular biomaterials via a novel layer-by-layer decellularization for better biocompatibility. *Mater. Sci. Eng. C* **2019**, *96*, 509–521. [CrossRef] [PubMed]
11. El-Ghaffar, M.A.A.; Hashem, M.S.; El-Awady, M.K.; Rabie, A.M. pH-sensitive sodium alginate hydrogels for riboflavin controlled release. *Carbohydr. Polym.* **2012**, *89*, 667–675. [CrossRef] [PubMed]
12. Zhang, K.; Shi, Z.; Zhou, J.; Xing, Q.; Ma, S.; Li, Q.; Zhang, Y.; Yao, M.; Wang, X.; Li, Q.; et al. Potential application of an injectable hydrogel scaffold loaded with mesenchymal stem cells for treating traumatic brain injury. *J. Mater. Chem.* **2018**, *6*, 2982–2992. [CrossRef]
13. Li, L.; Xu, Y.; Zhou, Z.; Chen, J.; Yang, P.; Yang, Y.; Li, J.; Huang, N. The effects of Cu-doped TiO$_2$ thin films on hyperplasia, inflammation and bacteria infection. *Appl. Sci.* **2015**, *5*, 1016–1032. [CrossRef]
14. Li, J.; Zhang, K.; Wu, F.; He, Z.; Yang, P.; Huang, N. Constructing bio-functional layers of hyaluronan and type IV collagen on titanium surface for improving endothelialization. *J. Mater. Sci.* **2015**, *50*, 3226–3236. [CrossRef]
15. Bai, Y.; Zhang, K.; Xu, R.; Liu, H.; Guan, F.; Liu, H.; Chen, Y.; Li, J. Surface Modification of Esophageal Stent Materials by a Drug-Eluting Layer for Better Anti-Restenosis Function. *Coatings* **2018**, *8*, 215. [CrossRef]
16. Zhang, K.; Bai, Y.; Wang, X.; Li, Q.; Guan, F.; Li, J. Surface modification of esophageal stent materials by a polyethylenimine layer aiming at anti-cancer function. *J. Mater. Sci. Mater. Med.* **2017**, *28*, 125. [CrossRef] [PubMed]
17. Yao, H.; Li, J.; Li, N.; Wang, K.; Li, X.; Wang, J. Surface modification of cardiovascular stent material-316L SS with estradiol loaded poly (trimethylene carbonate) film for better biocompatibility. *Polymers* **2017**, *9*, 598. [CrossRef]
18. Han, C.; Li, J.; Zou, D.; Luo, X.; Yang, P.; Zhao, A.; Huang, N. Mechanical Property of TiO$_2$ Micro/nano Surfaces based on the Investigation of Residual Stress, Tensile Force and Fluid Flow Shear Stress: For Potential Application of Cardiovascular Devices. *J. Nano Res.* **2017**, *49*, 190–201. [CrossRef]
19. Chen, L.; Li, J.; Chang, J.; Jin, S.; Wu, D.; Yan, H.; Wang, X.; Guan, S. Mg-Zn-Y-Nd coated with citric acid and dopamine by layer-by-layer self-assembly to improve surface biocompatibility. *Sci. Chin. Technol. Sci.* **2018**, *61*, 1228–1237. [CrossRef]
20. Zeeshan, M.; Ali, H.; Khan, S.; Khan, S.A.; Weigmann, B. Advances in orally-delivered pH-sensitive nanocarrier systems; an optimistic approach for the treatment of inflammatory bowel disease. *Int. J. Pharm.* **2019**, *558*, 201–214. [CrossRef]

21. Zou, D.; Luo, X.; Li, J.; Wang, S.; Zhang, K.; Sun, J.; Yang, P.; Zheng, Q.; Zhang, C. Investigating blood compatibility and tissue compatibility of a biomimetic ECM layer on cardiovascular biomaterials. *J. Biomater. Tissue Eng.* **2018**, *8*, 640–646. [CrossRef]
22. Yan, Y.; An, Q.; Xiao, Z.; Zheng, W.; Zhai, S. Flexible core-shell/bead-like alginate@PEI with exceptional adsorption capacity, recycling performance toward batch and column sorption of Cr(VI). *Chem. Eng. J.* **2017**, *313*, 475–486. [CrossRef]
23. Chen, L.; Li, J.; Wang, S.; Zhu, S.; Zhu, C.; Zheng, B.; Yang, G.; Guan, S. Surface modification of the biodegradable cardiovascular stent material Mg-Zn-Y-Nd alloy via conjugating REDV peptide for better endothelialization. *J. Mater. Res.* **2018**, *33*, 4123–4133. [CrossRef]
24. Li, J.G.; Wu, F.; Zhang, K.; He, Z.K.; Zou, D.; Luo, X.; Fan, Y.H.; Yang, P.; Zhao, A.S.; Huang, N. Controlling Molecular Weight of Hyaluronic Acid Conjugated on Amine-rich Surface: Toward Better Multifunctional Biomaterials for Cardiovascular Implants. *ACS Appl. Mater. Interfaces* **2017**, *9*, 30343–30358. [CrossRef] [PubMed]
25. Zhou, J.; Zhang, K.; Ma, S.; Liu, T.; Yao, M.; Li, J.; Wang, X.; Guan, F. Preparing an injectable hydrogel with sodium alginate and Type I collagen to create better MSCs growth microenvironment. *e-Polymers* **2019**, *19*, 95–100.
26. Wang, S.; Li, J.; Zhou, Z.; Zhou, S.; Hu, Z. Micro/nano scales direct cell behavior on biomaterials surface. *Molecules* **2019**, *24*, 75. [CrossRef]
27. Li, J.; Zhang, K.; Chen, H.; Liu, T.; Yang, P.; Zhao, Y.; Huang, N. A novel coating of type IV collagen and hyaluronic acid on stent material-titanium for promoting smooth muscle cells contractile phenotype. *Mater. Sci. Eng. C* **2014**, *38*, 235–243. [CrossRef]
28. Naeem, M.; Oshi, M.A.; Kim, J.; Lee, J.; Cao, J.; Nurhasni, H.; Im, E.; Jung, Y.; Yoo, J.W. pH-triggered surface charge-reversal nanoparticles alleviate experimental murine colitis via selective accumulation in inflamed colon regions. *Nanomed. Nanotechnol. Biol. Med.* **2018**, *14*, 823–834. [CrossRef]
29. Li, J.; Zhang, K.; Huang, N. Engineering Cardiovascular Implant Surfaces to Create a Vascular Endothelial Growth Microenvironment. *Biotechnol. J.* **2017**, *12*, 1600401. [CrossRef]
30. Birkett, R.T.; Mary, M.A.J.; O'Donnell, T.; Epstein, A.J.; Saur, N.M.; Bleier, J.I.S.; Paulson, E.C. Elective colon resection without curative intent in stage IV colon cancer. *Surg. Oncol.* **2019**, *28*, 110–115. [CrossRef]

 *materials* 

Article

Fabrication and Biocompatibility Evaluation of Nanodiamonds-Gelatin Electrospun Materials Designed for Prospective Tissue Regeneration Applications

Aida Şelaru [1,†], Diana-Maria Drăgușin [2,†], Elena Olăreț [2], Andrada Serafim [2], Doris Steinmüller-Nethl [3], Eugeniu Vasile [4], Horia Iovu [2], Izabela-Cristina Stancu [2], Marieta Costache [1,5,*] and Sorina Dinescu [1,5,†]

1   Department of Biochemistry and Molecular Biology, University of Bucharest, 050095 Bucharest, Romania; aida.selaru@bio.unibuc.ro (A.Ş.); sorina.dinescu@bio.unibuc.ro (S.D.)
2   Advanced Polymer Materials Group, University Politehnica of Bucharest, 011061 Bucharest, Romania; diana.dragusin@upb.ro (D.-M.D.); elena.olaret@stud.fim.upb.ro (E.O.); andrada.serafim@gmail.com (A.S.); horia.iovu@upb.ro (H.I.); izabela.stancu@upb.ro (I.-C.S.)
3   DiaCoating. GmbH, 6112 Wattens, Austria; doris.steinmueller@diacoating.com
4   Department of Science and Engineering of Oxide Materials and Nanomaterials, University Politehnica of Bucharest, 011061 Bucharest, Romania; eugeniu.vasile@upb.ro
5   Research Institute of University of Bucharest, 050107 Bucharest, Romania
*   Correspondence: marieta.costache@bio.unibuc.ro; Tel.: +40-21-3181575
†   These authors have equally contributed to this study.

Received: 5 August 2019; Accepted: 9 September 2019; Published: 11 September 2019

Abstract: Due to the reduced ability of most harmed tissues to self-regenerate, new strategies are being developed in order to promote self-repair assisted or not by biomaterials, among these tissue engineering (TE). Human adipose-derived mesenchymal stem cells (hASCs) currently represent a promising tool for tissue reconstruction, due to their low immunogenicity, high differentiation potential to multiple cell types and easy harvesting. Gelatin is a natural biocompatible polymer used for regenerative applications, while nanodiamond particles (NDs) are used as reinforcing nanomaterial that might modulate cell behavior, namely cell adhesion, viability, and proliferation. The development of electrospun microfibers loaded with NDs is expected to allow nanomechanical sensing due to local modifications of both nanostructure and stiffness. Two aqueous suspensions with 0.5 and 1% w/v NDs in gelatin from cold water fish skin (FG) were used to generate electrospun meshes. Advanced morpho- and micro-structural characterization revealed homogeneous microfibers. Nanoindentation tests confirmed the reinforcing effect of NDs. Biocompatibility assays showed an increased viability and proliferation profile of hASCs in contact with FG_NDs, correlated with very low cytotoxic effects of the materials. Moreover, hASCs developed an elongated cytoskeleton, suggesting that NDs addition to FG materials encouraged cell adhesion. This study showed the FG_NDs fibrous scaffolds potential for advanced TE applications.

Keywords: tissue engineering; diamond nanoparticles; fish gelatin; adipose-derived stem cells; biocompatibility

---

1. Introduction

The field of regenerative medicine and tissue engineering (TE) has emerged as a new approach for serious trauma injuries to essential organs of the body. Due to the considering disadvantages, namely insufficient donors, incompatibility between donors, and immune system issues that come along tissue grafting and organ transplants, TE holds a great promise towards regenerating the injured organ and, more importantly, gaining back the essential body functions towards normal life. Although serious steps are made in the field of microsurgical interventions, these still do not deliver the expected result in the setting of organ trauma [1]. Therefore, TE as a new and modern biological approach is being developed in order to overcome these challenges. TE combines the principles of engineering and life science in order to generate biological substitutes that will assist the injured tissue in regenerating and achieving normal functions. The key players in the frame of TE are cells, scaffolds and growth factors, molecules or nanoparticles which serve to stimulate a better interaction between the cells and the scaffold [2].

Since this field emerged, substantial efforts have been made to generate natural based polymers for biomedical applications, drug delivery, and regenerative medicine. In comparison to synthetic polymers, natural ones generate lower immune responses and better interact with the cellular components in terms of viability and toxicity [3]. Among them, fish gelatin (FG) is becoming one of the most promising biopolymer not only for medical applications, but also widely used for cosmetic and pharmaceutical purposes [4], due to a low risk of transmitting diseases as in case of mammalian gelatins [5]. Gelatin is a collagen derivate, obtained by partial hydrolysis of native collagen, therefore becoming attractive for scaffold preparation as it can deliver a similar structure to the extracellular matrix (ECM) found in most animal and human tissues [6,7], that can ensure appropriate chemical and biological cues for strong interactions with a large variety of cells [8]. In the past, several studies demonstrated that gelatin-based scaffolds are bio-friendly and interact well with different cell types, such as fibroblast [6,9], murine pre-osteoblasts [10], rat corneal keratocytes [11], mature adipocytes [12] and neonatal mouse cerebellum stem cells [13]. It has a wide use in the field of TE also due to its capability to deliver growth factors and to enhance the vascularization process within newly engineered tissue [14]. Thus, this polypeptide plays a key role in designing functional scaffolds, providing finely tuned templates which can facilitate cell adhesion, growth, proliferation, and differentiation. Therefore, gelatin has a great potential to serve as a substrate for multiple TE applications, including bone, adipose, corneal and nervous tissue regeneration.

Modern trends in scaffolding techniques include the functionalization and reinforcement at nano and micro scale of materials with different types of nanoparticles, such as gold nanoparticles, carbon nanotubes (CNT), halloysite nanotubes (HNT), diamond-like carbon (DLC) and diamond nanoparticles (NDs) [15], since cells seem to sense and respond to nanoscale or microscale in terms of adherence, growth, and proliferation [16]. Diamond, a material composed of carbon atoms arranged in a cubic crystal structure, has become a potential candidate for biotechnological and life sciences applications, especially due to its natural provenance. Mostly, it became attractive for a large range of TE applications under its nanostructured form, specifically NDs [17]. Moreover, NDs turned out to exhibit low or no cytotoxicity when brought in contact with cells [18]. This type of nanoparticles has been mostly explored for hard tissue regeneration, such as bone [19–21], yet carbon-based structures' potential for supporting regeneration is also evaluated for softer tissues, such as the skin [15,22] or the nervous tissue [23].

Stem cells are widely used as a tool in regenerative medicine and TE, due to their proliferation abilities and high potential to differentiate to multiple cell types. While embryonic stem cells (ESCs) and induced pluripotent stem cells (iPSCs) can follow multiple cell lineages belonging to all three embryonic sheets, adult stem cells maintain their versatility, but with a more reduced ability of differentiation. Among the adult stem cells types, adipose-derived mesenchymal stem cells (ASCs) can be easily isolated from fatty tissue due to minimal invasive procedures, such as liposuction. ASCs are able to undergo differentiation towards adipogenic [24], osteogenic [25] and chondrogenic [26] lineages. Interestingly, ASCs seem to be capable of differentiating also to several cell types originating from the ectoderm [27]. In this respect, very recent work demonstrated that these kind of cells are able to form neurospheres after 6 days of culture in specific culture media, being positive for microtubule-associated protein 2 (MAP2), glial fibrillary acidic protein (GFAP) and sex-determining region Y-box 2 (SOX2) neural markers [28]. All these characteristics proved that ASCs deliver expected behavior in engineered scaffolds and could be the solution for clinical applications.

It has been reported that the performance of stem cells behavior could be controlled by the presence of a nanostructure within the scaffold [29]. It seems that once cells interact with the nano-components, their adhesion, growth, and migration is modulated. Reinforcing natural based materials with these structures is a modern strategy for stem cell guidance and brings a great contribution to the vast field of TE [30]. The premise of our study is that cells respond differently to mechanical stimuli. As it was shown in previous studies, when in contact with an implantable device or scaffold material, the cellular behavior is influenced by the nanoscale or submicroscale characteristics of the surface [31]. In their study, Darling et al. [32] have shown a close connection between the cell morphology and the young modulus of the surface with which they interacted. Thus, hASCs exhibited a spread morphology when in contact with a stiffer surface than when in contact with a more elastic surface, the second substrate leading to a spherical cell morphology.

In this study, we sought to better understand the effect of the NDs on the nanomechanical properties of fibrous gelatin-based nanocomposite hydrogels and to investigate the mechanisms of the physical factors that could influence the cell response regarding adhesion, proliferation and cell morphology. Therefore, we expected that the incorporation of the nanoparticles within the polypeptide fibrous structure to contribute to the guidance of hASCs through mechanical sensing, leading to enhanced cellular activity. In this context, the aim of our study was to fabricate and characterize electrospun fibrous gelatin nanocomposite meshes containing NDs in terms of mechano-structural properties and the ability to support hASCs adhesion, growth, and proliferation, as a potential platform for tissue regeneration.

2. Materials and Methods

2.1. Preparation of Electrospun NDs-loaded FG Fibrous Scaffolds

Surface functionalized (COOH and OH) NDs suspension in ultrapure water was kindly provided by DiaCoating (Wattens, Austria). The precursors for the fibrous nanocomposite scaffolds were obtained by dissolution of FG (Sigma-Aldrich Co, Oakville, ON, Canada) into double distilled water (ddw) or NDs suspensions until a final protein concentration of 50% w/v and various ratios of NDs (0%, 0.5% and 1% w/v). The gelatin was left to dissolve at 40 °C under vigorous magnetic stirring for 4 h. For a homogenous distribution of the NDs all the mixtures were sonicated in an ultrasound bath (Elmasonic S 30/H, Elma Schmidbauer GmbH, Singen, Germany), for additional 4 h.

To fabricate the electrospun fibrous scaffolds (further denoted FG, FG_NDs 0.5% and FG_NDs 1%, respectively), climate-controlled electrospinning equipment was used (EC-CLI, IME Medical Electrospinning, Waalre, The Netherlands). Accordingly, each solution was loaded in a 5 mL syringe and placed on a syringe pump to precisely controlled flow rate. A constant volume of 350 µl of each precursor was injected through a G22 nozzle, using a flow rate of 12.5 µl min^{-1} at a constant temperature of 25 °C and a relative humidity of 45%. Randomly oriented fibers were collected on a cylindrical collector (Ø 20 mm, rotating at 100 rpm), at a distance between the tip of the nozzle and the collector of 12 cm and a speed of the nozzle lengthways of the collector of 5 mm s^{-1} (to ensure a homogenous deposition of the fibers on the entire length of the collector). The applied voltage depended on the composition of the injected precursors ranging from 21 kV for FG and FG_NDs 0.5% to 23 kV for FG_NDs 1%. Cross-linking was performed on fibers detached from the collector, using an 0.5% w/v ethanolic glutaraldehyde (GA) solution for 4 days, at RT (GA, 50% w/v aqueous solution, Sigma-Aldrich Co, St. Louis, MO, USA). Forwards, the crosslinked fibrous scaffolds were extensively washed with ethanol (analytical grade, Chimopar, Bucharest, Romania) for two days, and finally with double distilled water.

2.2. Characterization of the Fibrous Scaffolds

Rheological behavior of all precursors was assessed under isothermal conditions using a Kinexus Pro rheometer (Malvern Panalytical Ltd, Malvern, UK) with parallel plate geometry. The 20 mm diameter disposable plates were used and the gap between the plates was fixed at 1 mm. Steady state flow tests to measure the viscosity of sample formulations were performed at room temperature at different shear rates from $0.01\ s^{-1}$ to $1000\ s^{-1}$.

Attenuated total reflectance Fourier transform infrared (ATR-FTIR) spectrometry was performed using a Jasco 4200 spectrometer (JASCO Deutschland GmbH, Pfungstadt, Germany) equipped with a Specac Golden Gate ATR device (Specac Ltd, Orpington, UK), at a resolution of $4\ cm^{-1}$, in the wavenumber region of 4000–$600\ cm^{-1}$ and recording 200 scans/sample.

The morpho-structural features of the gold-sputtered fibrous scaffolds were examined through scanning electron microscopy (SEM) and high-resolution electron microscopy (HRSEM) using a Quanta Inspect F SEM device equipped with a field emission gun (FEG) (Fei Company, Hillsboro, OR, USA) with 1.2 nm resolution and with an X-ray energy dispersive spectrometer (EDX) (Fei Company, Hillsboro, OR, USA). Transmission electron microscopy (TEM) was also used in order provided information regarding the distribution of the NDs. The micrographs were registered using a TECNAI F30 G2 STWIN microscope (Fei Company, Hillsboro, OR, USA) operated at 300 kV with EDX and EELS facilities. Thus, a small piece of fibrous scaffold was deposited on a TEM copper grid and covered with a thin amorphous carbon film with holes. Single fibers were observed at the edge of the scaffolds.

Micro-computed tomography (microCT) investigation has explored the overall architectures of both mesh and tubular structures obtained by rolling the mesh around a plastic support. A SkyScan 1272, high-resolution X-Ray microtomograph (Bruker MicroCT, Kontich, Belgium) has been used. The samples were fixed with dental wax and placed in the scanning chamber, and the analyses were performed using an accelerating voltage of 45 kV and a beam current of 200 μA, with no filter present during scanning. The rotation step was set at 0.3 degrees. Images were processed using NRecon (Version 1.7.1.6, Bruker MicroCT, Kontich, Belgium), CTVox (Version 3.3.0r1403, Bruker MicroCT, Kontich, Belgium) and Data Viewer (Version 1.5.4.6, Bruker MicroCT, Kontich, Belgium) softwares.

Substrate surface contact angle was evaluated with a Drop Shape Analyzer 100 (DSA 100, KRÜSS GmbH, Hamburg, Germany), equipped with a high-resolution camera for capturing images during the measurements and evaluated using ADVANCE software (Version 1.7, KRÜSS GmbH, Hamburg, Germany). To avoid user-related errors and to assure repeatability an automatic program was defined in the software for the dosing system. A volume of 2 μl of double distilled water was placed on each surface, at room temperature (25 °C). Measurements were registered for 3 s (10 fps) and the drop's shape was analyzed using a sessile drop technique. For each sample, four measurements were performed on different areas and the values were averaged.

Mechanical properties at nanoscale (the Young's modulus, (E) and hardness (H)) were investigated by nanoindentation using a Nano Indenter G200 (Keysight Technologies, Santa Rosa, CA, USA). The samples consisted in aerogel films with composition corresponding to samples FG, FG_NDs 0.5% and FG_NDs 1%, air-dried in an oven, at 40 °C. They were glued on the sample stage. The continuous stiffness measurement (CSM) was performed using a Berkovich tip, with 25 indentations with 50 μm distance between them (to prevent interactions between indentations) for each sample. Indentations were set to a maximum penetration depth of 2000 nm with a strain rate target of 0.05/s. This was achieved by superimposing a 2 nm of amplitude and 45 Hz of frequency small oscillating force during the loading cycle. In brief, an indentation test started with the indenter tip approaching the surface with an approach velocity of 10 nm/sec. When the indenter touched the surface, it started to penetrate the surface at a rate of $0.05\ s^{-1}$. At the maximum penetration depth of 2000 nm, the load on the indenter was held constant for 10 s, then it was withdrawn from the sample. When the load on the sample reaches 10% of the maximum load on the sample, it was held constant for 100 s. The indenter was totally withdrawn, and the sample is moved for the next test.

2.3. Biocompatibility Evaluation of the Fibrous Scaffolds

To monitor the influence of the nanoparticles content, the FG_NDs nanocomposite meshes were tested for their biocompatibility in contact with human adipose-derived stem cells (hASCs). Cells were isolated from subcutaneous adipose tissue resulted after liposuction only after obtaining the informed consent of the patient. All performed studies that included hASCs were in compliance with the Helsinki Declaration, with the approval of the University of Bucharest Ethics Committee. In order to isolate stem cells, previously described protocol was applied [33,34]. Afterwards, the primary cell culture was maintained in Dulbecco's Modified Eagle Medium (DMEM, Sigma-Aldrich Co, Steinheim, Germany) supplemented with 1% antibiotic and 10% fetal bovine serum (FBS) and grown until passage 4. The scaffolds were sterilized by exposure to UV light, cells were seeded at 2×10^4 cells/cm^2 and incubated for one week in standard conditions (37 °C, 5% CO_2 and humidity). Biocompatibility assays were performed at 2, 4, and 7 days post-seeding and after 48 h actin F filaments were stained in order to evaluate cell adhesion. During quantitative biocompatibility studies, FG_NDs_0.5% and 1% composites were compared to pure FG considered as control and to the reference tissue culture plate (TCP) control.

To assess if hASCs maintain their metabolic activity when put in contact with FG_NDs, methylthiazolyldiphenyl tetrazolium bromide (MTT, Sigma-Aldrich Co, Steinheim, Germany) assay was performed. First, culture media was discharged and then MTT solution was put over the composites and incubated for 4h at dark in standard conditions. MTT solution was prepared at the recommended concentration of 1 mg/mL in DMEM lacking FBS. Then, the formed violet formazan crystals were solved with isopropanol and the obtained solution was measured at 550 nm using FlexStation3 Spectrophotometer (Molecular Devices, San Jose, CA, USA).

LDH assay was performed in order to assess if the materials exhibit significant toxicity on hASCs. Therefore, "In vitro toxicology assay kit lactate dehydrogenase based" TOX7 kit (Sigma Aldrich Co, Steinheim, Germany) was used and the test was performed following the manufacturer's instructions and the final solution was measured at 490 nm using FlexStation3 Spectrophotometer (Molecular Devices, USA).

Live/Dead staining was accomplished by using Live/Dead kit (ThermoFisher Scientific, Foster City, CA, USA). The solution was prepared following the manufacturer's instructions, after discharging the culture media, the solution was put in contact with the bioconstructs and incubated for 20 min in dark. Laser-scanning confocal microscope (Carl Zeiss LSM 710 system, Zeiss, Germany) was used for visualization and the obtained images were analyzed using corresponding Zeiss Zen 2010 software.

For cell adhesion investigation, F-actin filaments of hASCs were evidenced 48 h post-seeding. For fixation of the cells, a 4% paraformaldehyde solution (Sigma Aldrich Co, Steinheim, Germany) was used for 1 h. Then cell membrane was permeabilized with 0,1% Triton X-100 (Sigma Aldrich Co, Steinheim, Germany) solution in bovine serum albumin (BSA) for 45 min. Last, cell-scaffold systems were incubated for 1 h with phalloidin-FITC (Sigma Aldrich Co, Steinheim, Germany) and for 5 min with Hoechst 33342, in order to stain cell nuclei (ThermoFisher Scientific, Foster City, CA, USA). Laser-scanning confocal microscope (Carl Zeiss LSM 710 system, Zeiss, Germany) was used for visualization and the obtained images were analyzed using corresponding Zeiss Zen 2010 software. The quantification of the area covered by phalloidin-FITC (%) was made using Image J software (NIH, Bethesda, MD, USA, public software) as an average of several images for each composite (n = 10) and plots were obtained using GraphPad Prism version 3.

All experiments were performed in triplicate (n = 3) and the generated results were expressed as means ± standard deviation using GraphPad Prism 3.0 Software (GraphPad Software Inc., San Diego, CA, USA). The statistical relevance was assessed using this software, by performing one-way ANOVA and Bonferroni post-test, considering a statistical difference for $p < 0.05$.

3. Results

3.1. Rheological Evaluation of the Precursors

To assess the composition effect on the viscosity of the precursors, steady state flow tests were performed at different shear rates at room temperature; the resulted dynamic viscosities of the solutions are shown in Figure 1.

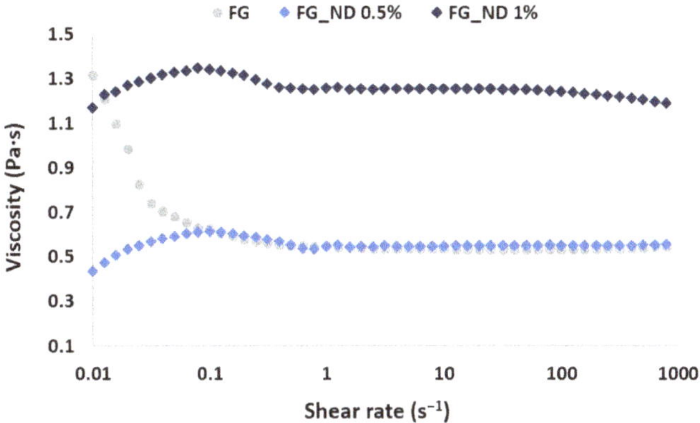

Figure 1. Graphical representation of the influence of precursors' composition on the viscosity at different shear rates and room temperature.

Gelatin solution exhibited a shear thinning behavior at low shear rates when compared with the NDs loaded compositions which presented a slight shear thickening behavior for the same values of the shear rate. Furthermore, increasing shear rate above 1 s^{-1} a linear viscosity was recorded for FG and FG_ND 0.5%, the measured viscosity becoming independent of the shear rate. The FG_ND 1% composition presented higher viscosity compared to the samples FG and FG_ND 0.5%. One could explain this behavior as a consequence of both the increased concentration of the nanospecies within the solution, but also as a consequence of the higher amount of hydrogen bonds formed between the functional groups at the surface of the nanoparticles with the functional groups from the backbone of gelatin (NH_2, COOH, OH) [35]. Moreover, another interesting aspect resulted from the measurements for FG_ND 1% is represented by the small tendency of decrease of the viscosity with increasing the shear rate (values over 80 s^{-1}). This pseudoplastic behaviour may appear due to a more efficient dispersion of NDs through the polymer chains. However, the higher NDs content lead to a 2.3 times increase of the viscosity (from 0.54 Pa·s to 1.24 Pa·s). This behavior suggests the reinforcing potential of the nanoparticles on the fibrous scaffolds at 1% loading ratio. Such rheologycal comportment of the precursors were in agreement with previous results of our group [36].

3.2. Wettability

The nanocomposite meshes are obtained after FG cross-linking in the presence of NDs. The formation of cross-links occurs with shrinkage of the biopolymer due to the reduction of the distance following generation of new chemical bonds. Figure 2a. is representative with respect to the network shrinkage during the cross-linking process. The contact angle experiments revealed a slight modification in the average values, increasing from a value of approximately 66° for FG to 70° for the nanocomposite loaded with 1% NDs. This is also visible in Figure 3b. No significant difference has been noticed between the wettability of pristine FG and the nanocomposite containing 0.5% NDs (Figure 2a,b).

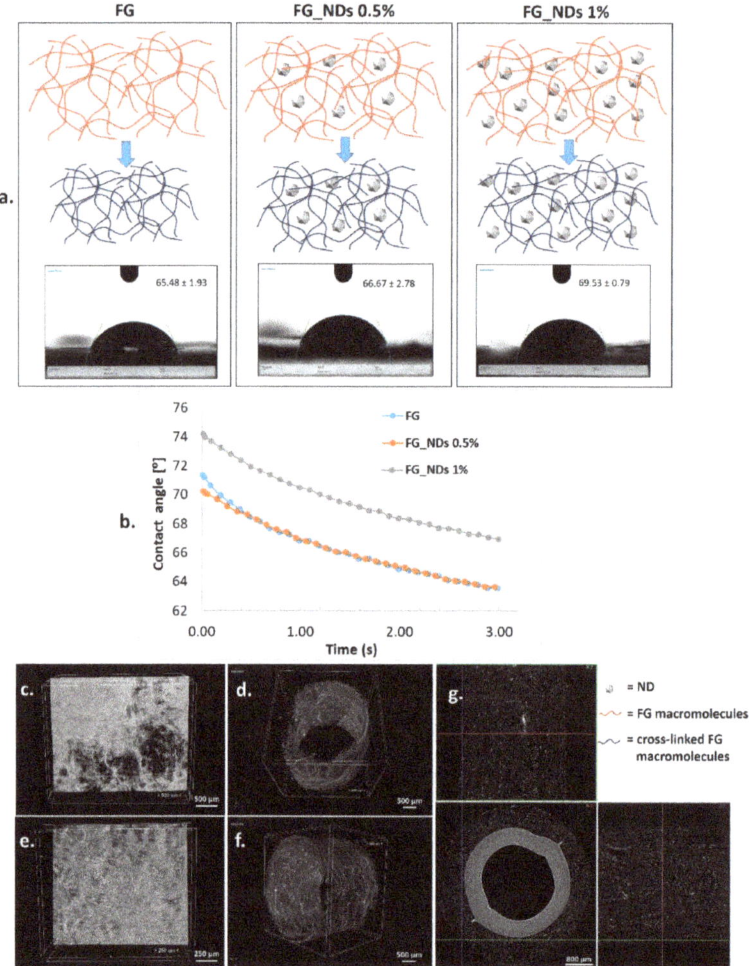

Figure 2. (**a**) Schematic representation of nanodiamond particles (NDs) nanocomposite synthesis from precursors, through fish gelatin (FG) cross-linking (red—FG macromolecules; blue—cross-linked FG macromolecules) (shrinkage during cross-linking is suggested; influence of the NDs content on the wettability expressed as contact angle results; micro-CT images of: (**b**) wettability evolution monitored during 3 s; (**c**) FG_NDs 0.5% mesh; (**d**) FG_NDs 0.5% rolled mesh; (**e**) FG_NDs 1% mesh; (**f**) FG_NDs 1% rolled mesh; (**g**) Data Viewer images of the microfibrous FG_NDs 1% rolled mesh; inner plastic support is visible as a compact ring.

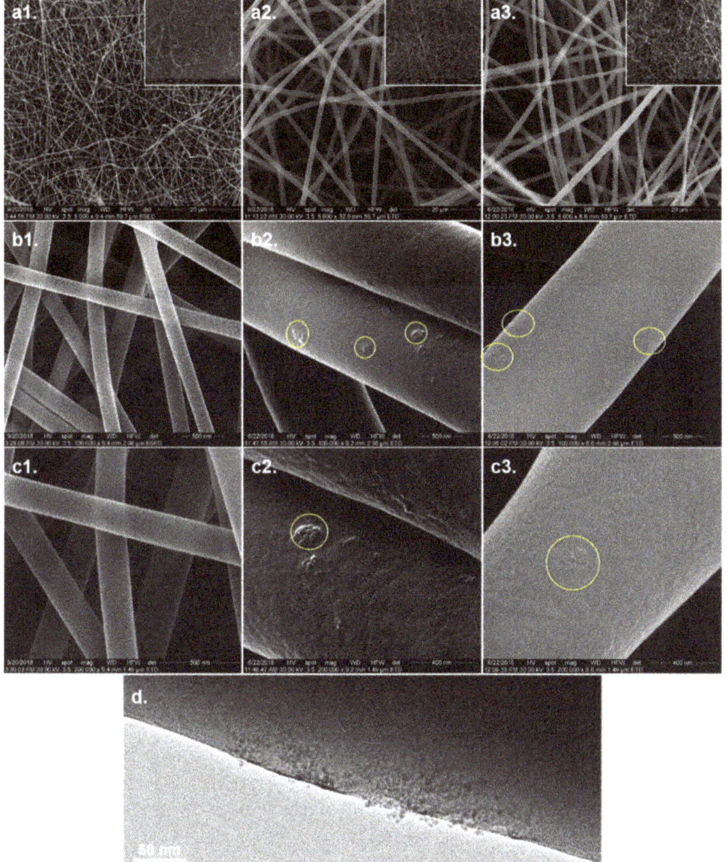

Figure 3. SEM images of electrospun fibers: (**a1–c1**) FG at different magnifications, (**a2–c2**) FG_ND 0.5% at different magnifications, (**a3–c3**) FG_ND 1% at different magnifications. (**d**) TEM micrograph of FG_ND 1%—functionalized NDs clusters in a fiber.

3.3. Microstructural Analysis

MicroCT imaging provided an overview of the microfibrous structure of the cross-linked electrospun scaffolds and of the tubular constructs that can be generated after rolling the meshes on polypropylene collectors. Representative images are given in Figure 2c–g, presenting the potential of this simple method to obtain tubular scaffolds which could be used for nerve guidance channels.

The microstructure of the scaffolds has been further investigated by SEM, HRSEM, and TEM. The electrospun meshes did not exhibited significant differences regarding the microstructure. All compositions led to fabrication of homogeneous fibers, continuous, with smooth surface, without defects along their length, entangled into mats with interconnected porosity. The average diameter of the fibers is of approximately 0.2 µm for FG, ~1.1 µm for FG_ND 0.5%, and ~1 µm for FG_ND 1% (with a larger diameter distribution) as one can observe in Figure 3(a1–a3). Figure 3(b2,b3) presented a rather uniform distribution of the nanoparticles into the polymeric fibers along their axes, in the form of nanoclusters. A higher tendency of cluster formation could be observed for the FG_ND 0.5% composition (as it can be seen in Figure 3(c2,c3)). TEM micrograph provided information on the distribution of the nanoparticles into the fibers. It can be observed that the nanoparticles are immobilized into the polymeric matrix, with a protruded aspect at the surface of the fiber. However, the

NDs are not homogeneously dispersed into the fiber, rather presenting a tendency to form nanoparticles agglomerates along the axes of the fibers (Figure 3d). The crystalline phase distinguishes from the amorphous glassy continuous matrix of the FG fibers; it is obviously different from the point of view of the mechanical properties. Such a distribution of the nanoparticles was expected due to the low content of nanoparticles used in this work and it is in agreement with previous results of our group [19].

3.4. Nanomechanical Investigation

Nanomechanical sensing influences the behavior of cells, therefore it became appealing to investigate how the mechanical properties at the nanoscale are influenced by the NDs content, at such low loadings. Nanoindentation can discriminate between similar, low-modulus samples when the composition is modified, and therefore we selected it as a method to explore the potential reinforcement effect of NDs nanoparticles, at low concentrations such as 0.5% and 1%. Preliminary characterization of dried FG_NDs scaffolds has been performed using the CSM based method as previously reported [37–40], and results were calculated according to Oliver and Pharr's method [41]. This method allows a continuous measuring of the stiffness throughout the indentation's loading cycle, allowing the calculation of E and H as a function of the indentation depth. Load is monitored as function of the displacement (Figure 4a), contributing to a better understanding of the nanomechanical behavior. This analysis revealed that the maximum load required to reach the same penetration depth increases with increasing NDs content. It has been noticed that during the 10 s peak hold time, a creep deformation appears regardless the sample's composition. The plastic deformation is also shown by the curves in Figure 4a.

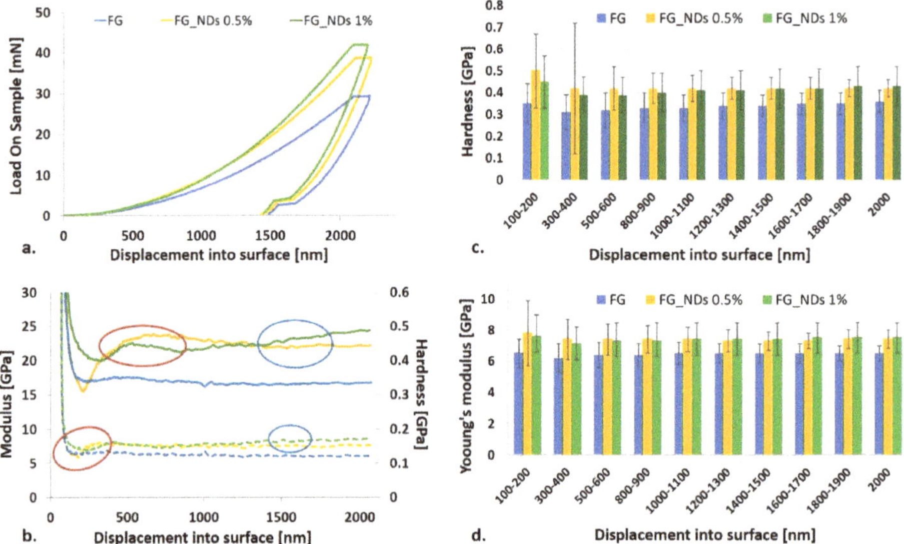

Figure 4. Influence of NDs loading on the nanomechanical properties of FG_NDs (**a**) load as function of penetration depth; (**b**) representative variation of modulus (top—dotted lines) and hardness (bottom—plain lines) up to a maximum depth of 2000 nm; red circles—the effect on FG-ND 0.5% which could be attributed to the presence of nanoparticles aggregates; blue circles—the expected behavior which appear with indentation depth increase; (**c**) hardness as a function of the indentation depth; (**d**) Young's modulus as a function of the indentation depth.

The variation of modulus and hardness with the penetration depth are displayed in Figure 4b–d. One can observe that results are more homogeneous with increasing the penetration depth, most probably due to the decrease of the nanomechanical heterogeneity in the dried samples from surface to bottom. The solvent evaporation during the samples' preparation leads to a shrinkage of the polymer material with a significant change in the resulting elasticity. If in the hydrated compositions, the reinforcing effect of the NDs has been emphasized by a significant increase in the viscosity as measured by rheology, it was expected that the dried samples to present only minor differences in terms of stiffness since they correspond to a total solid content of 50.5% for FG_NDs 0.5% and respectively 51% for FG_NDs 1%, respectively. The big standard deviations (Figure 4c,d) for the displacement range from 100 nm to 200 nm are due to the surface anomalies which, once with the increase of indentation depth have a lower influence. Table 1 summarizes the increase of the values for Young's modulus and hardness calculated as percentage related to the values obtained for simple FG. These values clearly suggest the decrease of both Young's modulus and hardness with increasing the indentation depth.

Table 1. The percent of mechanical properties increase with addition of NDs in respect to the values obtained for FG.

Indentation Depth [nm]	Young's Modulus (%)		Hardness (%)	
	FG_NDs 0.5%	FG_NDs 1%	FG_NDs 0.5%	FG_NDs 1%
100–200	20.00	16.92	42.86	28.57
300–400	19.35	14.52	35.48	25.81
500–600	15.63	14.06	31.25	21.88
800–900	15.63	14.06	27.27	21.21
1000–1100	13.85	13.85	27.27	24.24
1200–1300	12.31	13.85	23.53	20.59
1400–1500	12.31	13.85	23.53	23.53
1600–1700	12.31	15.38	20.00	20.00
1800–1900	13.85	15.38	20.00	22.86
2000	13.85	15.38	16.67	19.44

The values of the Young's modulus recorded for pure FG are ranging between 6.2 and 6.5 GPa on the entire indentation depth interval, when compared with FG_ND 0.5% for which the Young's modulus values are ranging between 7.3–7.8 GPa, and for FG_ND 1% the values of the Young's modulus are ranging between 7.1–7.6 GPa.

This observation is supported by the SEM and TEM micrographs (Figure 3) identifying agglomerated nanoparticles dispersed into the amorphous polymer fibers. While NDs are crystalline materials, the dried FG matrix is in a glassy state and accordingly, the resulting local mechanical heterogeneity can be discriminated by nanoindentation. Increasing the NDs content from 0.5% to 1% enhances the homogeneity of the nanoparticles' dispersion in the FG_NDs 1% nanocomposite. In the case of FG_NDs 0.5%, the higher values obtained for both modulus and hardness with significantly high standard deviations than for FG_ND1% at lower indentation depth, could be explained by the presence of nanoparticles aggregates closer to the surface.

3.5. FT-IR Analysis

FT-IR analysis confirmed the presence of characteristic functional groups from the two individual components of the fibrous scaffolds. Functionalized NDs (Figure 5(1)) present a large band between 1700 and 1800 cm^{-1}, attributed to C=O stretching in carboxylic groups. A peak related to C-O stretching was also visible at 1066 cm^{-1}, attributed to the presence hydroxyl group (confirmed by the large band between 3000 and 3600 cm^{-1} due to O-H stretching) and/or ether groups on the surface of the nanoparticles [19,42]. C–H stretching was also present between 2800 and 3000 cm^{-1}, linked to a very limited amount of amorphous carbon lying on the surface of the NDs. FG (Figure 5(4)) presents specific vibrations such as broad peak at 3289 cm^{-1}, which is common signal for O–H and N–H stretching, a small peak at 3082 cm^{-1} attributed to N–H, at 2938 cm^{-1} is present the specific vibration for saturated C–H stretch, and amide I peak at 1637 cm^{-1} and amide II peak at 1531 cm^{-1}. For the nanocomposite fibrous scaffolds, all the specific vibrations of the component can be noticed in the correspondent spectrum (Figure 5(2) depicts the IR spectrum for FG_ND 1%, and Figure 5(3) depicts the FT-IR spectrum for FG_ND 0.5%).

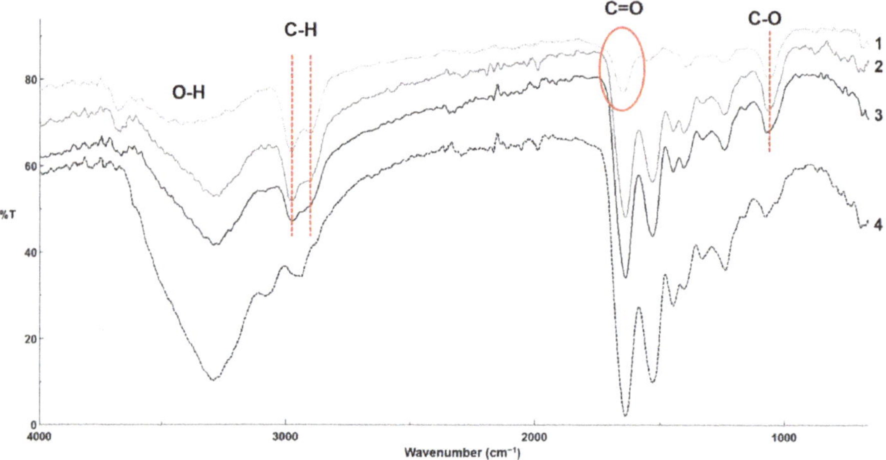

Figure 5. FT-IR spectra recorded on control samples: 1—NDs, and 4—FG and on the fibrous scaffolds 2—FG_ND 1%, 3—FG_ND 0.5% (increasing amount of NDs intensifies the specific vibrations characteristic for this component).

3.6. Biocompatibility Assessment

MTT profile (Figure 6a) indicateds, at 2 days post-seeding, that cell viability was the same on the two nanocomposites containing NDs as compared to the FG control and TCP control. On the other hand, at 4 days a significant ($p < 0.05$) difference was found between FG_NDs nanocomposites and FG, and no difference in terms of viability between FG_NDs 0.5% and FG_NDs 1%. After 7 days of culture, most significant ($p < 0.001$) proportion of viable cells were found in contact with FG_NDs 1% as compared to the FG control. Moreover, a relevant ($p < 0.05$) difference was also found on FG_NDs 0.5% as compared to FG_NDs 1%. No significant proliferation occurred from 2 to 4 days post-seeding on the tested composites. In contrast, proliferation rate was found to be higher for 4 to 7 days of culture, with a significant ($p < 0.001$) increase on the material loaded with 0.5% NDs. Also, a big increase in proliferation can be observed from 2 to 7 days on all tested composites, but with a significant ($p < 0.001$) relevance for the nanocomposite scaffold containing 1% NDs. This behavior can be correlated with the previously described nanomechanical properties. Therefore, it can be concluded that the addition of NDs in the fibrous hydrogel scaffolds supports cell viability and increases the proliferation rate.

LDH assay (Figure 6b) revealed overall small amounts of toxicity during 7 days of culture. Moreover, it can be observed how these remained constant from 2 days until 7 days post-seeding. Also, levels of LDH tend to be lower (but not significantly) on composites enriched with NDs as compared to the tested FG control and similar to the levels registered for TCP control. Thus, the presence of NDs in these materials did not induce significant cytotoxicity on hASCs.

By simultaneously labeling live (green) and dead (red) cells (Figure 6c), it can be noticed that there is a significant proportion between live and dead cells on all tested composites. Images obtained at the confocal microscopy revealed that cells proliferated and maintained their viability throughout one week of culture, thus confirming MTT and LDH results. Within 2 days of culture, more viable cells are to be observed on FG_NDs 1% as compared to the control and FG_NDs 0.5%. Moreover, hASCs in contact with FG_NDs 1% present a different morphology, they are elongated as compared to FG_NDs 0.5% where most of them are still in their round shape. After 7 days, a significant amount of cells is to be evidenced on all nanocomposites, but when it comes to composites enriched with NDs, hASCs have formed groups, this meaning that these particular scaffolds offered them optimal conditions for proliferation.

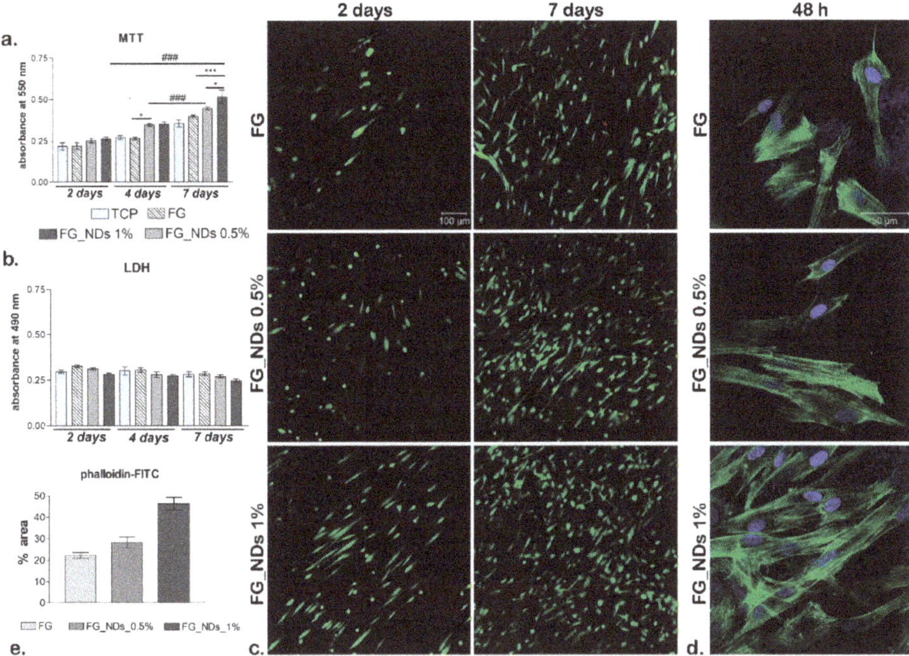

Figure 6. Biocompatibility and adhesion assays performed for hASCs/FG_NDs bioconstructs (**a**) Cell viability profile obtained after one week of culture by MTT test. Statistical significance: * $p < 0.05$; *** and ### $p < 0.001$. (**b**) Cytotoxicity levels exerted by FG_NDs on hASCs during 7 days of culture (**c**) Cell viability and proliferation qualitative analysis obtained after performing Live/Dead staining on the tested composites; live—green labeled with calcein AM and cell nuclei of dead cells—red labeled with ethidium bromide (**d**) F-actin filaments developed by hASCs in contact with FG_NDs after 48h of culture; filaments (green) are stained in phalloidin-FITC, while nuclei (blue) are stained with Hoechst 33342. (**e**) quantification of phalloidin-FITC levels in all controls.

After cell cytoskeleton was evidenced by staining with phalloidin-FITC (Figure 6d), a good cell adhesion was observed on the investigated nanocomposites. There are no significant changes in cell phenotype from a material to another. Long actin fibers were found on all composition, confirming that hASCs were able to develop cytoskeleton in contact with FG_NDs, similar to the one formed in contact with FG. However, F-actin filaments are clearly more visible on FG_NDs 1% than in case of FG_0.5% and the control, this meaning that after 48 h hASCs had a better interaction with this particular scaffold. Hereby this is another confirmation that incorporation of NDs in the structure of biomaterials has a positive impact on behavior of mesenchymal stem cells.

4. Discussion

In the last years, the scientists tried to better understand the interaction processes between a biomaterial and the living cells through nanomechanics responses induced by nano-cues loaded in commonly used biopolymers for TE applications. A deeper understanding of the small scale phenomena that have a great influence on the cells behavior (either positive [19,43] in stimulating cellular adhesion, proliferation, differentiation, or negative [44] through the alteration of the cells' properties which eventually leads to important conditions), could provide suitable solutions in case of TE and cellular therapy [16]. Due to the continuous need for improvement, natural polymers, nanotechnologies, and stem cells seem to hold a great promise for generating substitutes that could guide an injured tissue towards regeneration.

Based on our group previous studies [19,36] we synthesized randomly oriented multifunctional electrospun composites scaffolds based on FG and NDs and we thoroughly characterized them regarding physico-chemical and biocompatibility properties. In order to avoid the common tendency of the NDs to create aggregates, we have used a mild sonication treatment of the NDs dispersions in aqueous FG solutions during the obtaining of the fibers' precursors. Consistent with previous researches [35,36], the rheological characterization of the precursors provided information regarding the viscosity of the mixture. As expected, no significant differences were observed between the simple FG and FG_ND 0.5% solutions, but for FG_ND 1% precursor 2.3 times increase of the viscosity was recorded. The behavior could be explained due to interactions between the functional groups from the surface of NDs and the functional groups of the gelation macromolecules, in accordance with other studies [35,36,45]. The effect of the NDs can be slightly observed on the increased hydrophobicity of FG_ND 1% material. In their study, Mahdavi et al., 2016 [22], fabricated through electrospinning composite scaffolds based on chitosan, bacterial cellulose, and NDs. They observed a similar behavior regarding the wettability of their fibers with the behavior of our mats. They suggested that even though the NDs present many hydrophilic functional groups on their surface, a bigger influence in the final wettability properties it is given by the surface energy and the surface geometrical features of the material. One could explain the behavior of our materials through the same mechanisms, meaning that an increase of the NDs loading is leading to a decrease of the wettability [46,47].

An important repercussion of the viscosity of the precursors was further reflected in the influenced over the morphology of the fibers as shown in SEM micrographs. The introduction of the NDs increased considerably the diameter of the fibers. It is possible that the effect of the NDs is related to the enhanced viscosity and thus, changes of the conductivity could appear due to the formation of weak physical cross-linking bonds between the NDs and the FG. In a previous study of Li et al., 2016 [48], it was shown that increased viscosity of the solution translates in a weaker stretching of the liquid jet during the electrospinning process, which can lead to an increase of the diameter of the fabricated fibers. Also, Pacelli et al. 2017, have shown that the injectability of chitosan hydrogels loaded with NDs is affected because higher NDs concentration leads to a high number of polar functional groups in the gel [49]. This behavior, associated with the tendency of the nanoparticles to form clusters, could be an explanation of the bigger diameters of the nanocomposite fibers. Also, since the NDs can act as cross-linking points, the macromolecular coil of gelatin might not stretch enough to permit the fabrication of thin fibers, as in the case of the simple FG solution. These results presented in the literature can be correlated with the rheological characterization of our precursors and the diameter of the resulting fibers, namely, the superficial cross-liking between NDs and gelatin led to thicker but less homogeneous dimensions of the fibers. Regarding the distribution of the NDs within the fibers, our findings agree with previous studies of our group [19,36]. Thus, the NDs do not present a homogeneous distribution along the fibers, but they are rather forming clusters, present predominately at the surface of the fibers. We anticipated that the superficial nanostructuring of the surface of the fibers will have an important influence on the cellular behavior, as will further be discussed.

In our experiment, the bioconstruct consisting of natural polymer material—FG, loaded with NDs, and cultured with hASCs exhibited an overall good biocompatibility and promising potential as a support platform for TE and regenerative medicine. The main finding of this study is the ability of hASCs to respond differently to the variable composition of the FG_NDs composites, particularly depending on the NDs concentration within the materials. In other words, the inclusion of NDs in these fibrous gelatin electrospun meshes appears to be a useful tool for the modulation of the cellular component behavior in terms of cell growth, proliferation and adhesion.

As previously stated, gelatin has a great potential in serving as a substrate for cell growth within TE applications. Its potential for various directions in the field of TE has been proven in a recent study, where cells were encapsulated within the material. It seems that alginate/gelatin microspheres which contained hASCs have supported cell viability, proliferation and adipogenic differentiation [50], therefore demonstrating gelatin's favorable effect on cell behavior. As a consequence, in our study gelatin represented the base of the scaffold. Particularly, fish gelatin was used because it has low gelation temperature, which permitted the production of the fibers at room temperature from very concentrated aqueous solution, comparing with mammalian gelatins, thus, providing more similarities with the ECM [51]. As MTT profile revealed, there is no significant difference in cell proliferation from day 2 to day 7 post-seeding, but at the same time, no significant toxicity levels were found, therefore proving that gelatin is biocompatible with hASCs.

In order to enhance a materials' ability to promote cell proliferation, recent biomaterial development includes addition of nanoparticles in order to obtain optimal cell-scaffold interaction [17]. NDs are known for their exquisite properties and good interaction with cells and this has been proven several times in different studies. Previous findings indicated that treating lung epithelial cells and normal fibroblasts with a suspension of carboxylated NDs did not affect cell viability and cells did not undergo apoptosis, whereas nanotubes induced cytotoxicity in the same cell types, thus demonstrating that NDs are less toxic than other carbon-based structures used in the field of TE [52]. Moreover, it has been shown that these nanoparticles would not affect stem cells, the key players in regeneration. It has been demonstrated that including NDs to poly (L-lactide-co-ε-caprolactone) (PCL) scaffolds promoted cell proliferation and differentiation of human bone marrow-derived mesenchymal stem cells (hBM-MSCs) towards osteogenic lineage [21]. Moreover, data indicated that presence of NDs did not affect viability of human mesenchymal stem cells and stimulated their migration on the materials surface [53]. Therefore, NDs have a non-toxic effect on stem cells and highly promote their growth and proliferation on scaffolds. Similar to these observations, our biocompatibility results demonstrated that incorporation of NDs in the material structure had a benefic impact on hASCs cell culture in terms of viability, proliferation and adhesion. MTT profile revealed that cells maintained their viability and started to proliferate during 7 days of culture in contact with these composites. It was observed how the addition of NDs significantly increased the proliferation rate of hASCs from 4 to 7 days post-seeding ($p < 0.001$). LDH assay revealed no significant levels of cytotoxicity, thus demonstrating that NDs presence in the scaffold did not exhibit any negative impact upon hASCs. Interestingly, the staining of live cells by Live/Dead assay revealed that these stem cells already presented an equal distribution at 2 days post-seeding on FG_ND 1% surfaces, whereas, for the control FG and FG_ND 0.5%, smaller cell groups could be observed. The abundant groups of cells formed after 7 days of culture on both materials containing NDs demonstrating that the material continued to support cell viability and proliferation on the long term. This proves that incorporation of these nanoparticles enhances a good interaction between stem cells and nanocomposite fibrous scaffolds. These findings together with the microstructural and nanomechanical data suggest that it is likely that the nanocomposite clusters dispersed at smaller distance within the FG_NDs 1% fibers with respect to the 0.5% samples contribute through mechanical sensing to the guidance of cells.

The results of the nanoindentation tests come to support and explain the cell behavior. Thus, the results obtained for displacement range from 100 nm to 200 nm presented big standard deviations, phenomenon that can be explained by the heterogenous aspect of the surface. Higher indentation depth, not only that led to lower standard deviation but also to lower values for both Young modulus and hardness. The observation, supported both by SEM and TEM analysis, indicate the tendency of NDs to locate and form agglomerates at the surface of the fibers. These clusters formed at the surface of the fibers can act as guidance cues for cells in order to better adhere, migrate and further proliferate at the interface with the biomaterial.

Cell adhesion is a critical step in the evaluation of cell-scaffold interaction since cell surface receptors interact directly with the components and surface of the substrate. Therefore, it is important for the material to own specific proprieties that allow cells to easily attach to its surface. Recent work has established that gelatin generates a quite similar structure to the natural ECM, therefore cells adhere with ease to these surfaces. Fibrous structures usually tend to provide bigger areas and to promote cell adhesion and growth than other materials with a smooth surface. Guarino et al. [54] showed that human mesenchymal stem cells had a superior interaction with PCL membranes when gelatin was added to their structure. Besides, NDs have also shown to enhance cell adhesion due to their favorable physical proprieties [17,18]. Pereia et al. [55] fabricated a poly (lactic acid) scaffold which contained NDs and investigated biological and bioactive proprieties of this electrospun fiber. The conclusion of their study was that incorporation of these particles to the materials structure delivered satisfying cell-scaffold interaction. In our study, marking F-actin filaments after 48 h of culture, exposed an interesting behavior of hASCs in contact with these meshes, namely the elongation of the cytoskeleton increasing with NDs concentration in the material. We have concluded that the presence of NDs delivers a good impact on cell behavior, in terms of stimulating hASCs attachment to the scaffold and allowing cells to grow and multiply. This study confirmed that most favorable hASCs response was registered in contact with FG_NDs 1%, suggesting once more that a better cell-scaffold interaction occurs dependent on NDs' ratio in the final composite.

Even though lots of studies involving NDs are focusing on bone and cartilage tissue engineering, recent work has demonstrated NDs potential to support neuron attachment and neurite outgrowth, therefore making these particles potential candidates for nervous tissue engineering. Hopper et al. [56] studied the effect of NDs on NG108-15 cell line and on primary dorsal root ganglion (DRG) neurons, concluding that NDs promoted neuronal cell adhesion, migration and neurite outgrowth. Moreover, due to gelatin's hydrophilic surface and high water content, it has been demonstrated that this kind of material supported hASCs differentiation towards neural-*like* cell types, expressing specific markers as nestin and Tuj-1 [57]. All these findings together with our results confirm that functionalizing gelatin meshes with NDs and together with hASCs hold a great promise for more sophisticated applications in TE, such as peripheric nerve regeneration. Therefore, we envisage developing further studies where to analyze hASCs potential to differentiate towards neural or glial cell types, assisted by FG_NDs platforms that show ability to support this TE direction. Preliminary data shows that these substrates are capable to promote survival and growth of primary DRG neurons, therefore expanding the advantages of NDs to this new field of nerve regeneration.

5. Conclusions

An overall enhanced interaction between hASCs and FG_NDs revealed that cell viability and proliferation were not diminished by the incorporation of NDs up to 1% and confirmed low cytotoxicity of the obtained nanocomposite meshes. The main hypothesis of this study consisted in the potential modulation of cell response by NDs incorporation into the fibrous gelatin meshes. One of the most important monitored effects was the modification of the nano, micro-structure of the fibers surface and, accordingly, of the stiffness due to the increasing yet low nanoparticle content. Nanoindentation data combined with the microstructural analyses suggested a local increase of the scaffolds' stiffness more homogenous in the samples loaded with 1% NDs, due to a better distribution of small clusters along the fibers' longitudinal axis. Such behavior can be considered a proof for a nanomechanical sensing that can explain the modification of cell morphology, adhesion, and proliferation with the best results for the highest nanoparticle loading. Overall, hASCs displayed different behavior in response to increasing NDs loading in the FG_NDs composites, thus confirming that their use for different regenerative applications can be modulated by such tools. These findings and NDs potential have to be explored in the near-future to better understand the interaction between hASCs and the nano-components, in order to extend hASCs applications in other fields than the already confirmed ones - bone, cartilage, and adipose TE. Considering this idea and NDs potential for nerve regeneration, FG_NDs capability to support hASCs differentiation to neuronal cell types and consequent peripheric nerve reconstruction post-injury should be further investigated.

Materials **2019**, *12*, 2933

Author Contributions: Conceptualization, D.-M.D., I.-C.S., M.C. and S.D.; Funding acquisition, H.I. and M.C.; Investigation, A.Ş., D.-M.D., E.O., A.S., E.V. and S.D.; Methodology, A.Ş. and D.-M.D.; Project administration, I.-C.S. and M.C.; Resources, D.-M.D., D.S.-N., I.-C.S., M.C. and S.D.; Software, E.O.; Supervision, D.-M.D., I.-C.S. and M.C.; Validation, D.-M.D. and S.D.; Visualization, A.Ş., D.-M.D., E.O., I.-C.S. and S.D.; Writing—original draft, A.Ş., D.-M.D., E.O., I.-C.S. and S.D.; Writing—review & editing, D.-M.D., I.-C.S. and S.D.. A.Ş., S.D. and D.-M.D. have equally contributed to this study.

Funding: This research was supported by a grant of the Romanian Ministry of Research and Innovation, CCCDI-UEFISCDI, project number PN-III-P1-1.2-PCCDI-2017-0782/REGMED, within PNCDI III. The fabrication through electrospinning, nanoindentation, contact angle, and micro-CT analyses were possible due to European Regional Development Fund through Competitiveness Operational Program 2014-2020, Priority axis 1, Project No. P_36_611, MySMIS code 107066, Innovative Technologies for Materials Quality Assurance in Health, Energy and Environmental – Center for Innovative Manufacturing Solutions of Smart Biomaterials and Biomedical Surfaces—INOVABIOMED.

Conflicts of Interest: The authors declare no conflict of interest.

References

1. Dzobo, K.; Thomford, N.E.; Senthebane, D.A.; Shipanga, H.; Rowe, A.; Dandara, C.; Pillay, M.; Motaung, K.S.C.M. Advances in regenerative medicine and tissue engineering: Innovation and transformation of medicine. *Stem Cells Int.* **2018**. [CrossRef] [PubMed]

2. Salgado, A.J.; Oliveira, J.M.; Martins, A.; Teixeira, F.G.; Silva, N.A.; Neves, N.M.; Sousa, N.; Reis, R.L. Tissue engineering and regenerative medicine: Past, present, and future. In *Title of International Review of Neurobiology*; Stefano, G., Isabelle, P., Pierluigi, T., Bruno, B., Eds.; Academic Press: Boston, MA, USA, 2013; Volume 108, pp. 1–33.

3. Sell, S.A.; Wolfe, P.S.; Garg, K.; McCool, J.M.; Rodriguez, I.A.; Bowlin, G.L. The use of natural polymers in tissue engineering: A focus on electrospun extracellular matrix analogues. *Polymers* **2010**, *2*, 522–553. [CrossRef]

4. Karim, A.A.; Bhat, R. Fish gelatin: Properties, challenges, and prospects as an alternative to mammalian gelatins. *Food Hydrocolloid* **2009**, *23*, 563–576. [CrossRef]

5. Shakila, R.J.; Jeevithan, E.; Varatharajakumar, A.; Jeyasekaran, G.; Sukumar, D. Comparison of the properties of multi-composite fish gelatin films with that of mammalian gelatin films. *Food Chem.* **2012**, *135*, 2260–2267. [CrossRef] [PubMed]

6. Yoon, H.J.; Shin, S.R.; Cha, J.M.; Lee, S.H.; Kim, J.H.; Do, J.T.; Song, H.; Bae, H. Cold water fish gelatin methacryloyl hydrogel for tissue engineering application. *PloS ONE* **2016**, *11*, e0163902. [CrossRef] [PubMed]

7. Lee, K.Y.; Mooney, D.J. Hydrogels for tissue engineering. *Chem. Rev.* **2001**, *101*, 1869–1880. [CrossRef] [PubMed]

8. Afewerki, S.; Sheikhi, A.; Kannan, S.; Ahadian, S.; Khademhosseini, A. Gelatin-polysaccharide composite scaffolds for 3D cell culture and tissue engineering: Towards natural therapeutics. *Bioeng. Transl. Med.* **2019**, *4*, 96–115. [CrossRef]

9. Manikandan, A.; Thirupathi Kumara Raja, S.; Thiruselvi, T.; Gnanamani, A. Engineered fish scale gelatin: An alternative and suitable biomaterial for tissue engineering. *J. Bioact. Compat. Polym.* **2018**, *33*, 332–346. [CrossRef]

10. Fu, C.; Bai, H.; Hu, Q.; Gao, T.; Bai, Y. Enhanced proliferation and osteogenic differentiation of MC3T3-E1 pre-osteoblasts on graphene oxide-impregnated PLGA–gelatin nanocomposite fibrous membranes. *Rsc Adv.* **2017**, *7*, 8886–8897. [CrossRef]

11. Lai, J.Y.; Li, Y.T.; Cho, C.H.; Yu, T.C. Nanoscale modification of porous gelatin scaffolds with chondroitin sulfate for corneal stromal tissue engineering. *Int. J. Naonomed.* **2012**, *7*, 1101. [CrossRef]

12. Huber, B.; Borchers, K.; Tovar, G.E.; Kluger, P.J. Methacrylated gelatin and mature adipocytes are promising components for adipose tissue engineering. *J. Biomater. Appl.* **2016**, *30*, 699–710. [CrossRef] [PubMed]

13. Ghasemi-Mobarakeh, L.; Prabhakaran, M.P.; Morshed, M.; Nasr-Esfahani, M.H.; Ramakrishna, S. Electrospun poly (ε-caprolactone)/gelatin nanofibrous scaffolds for nerve tissue engineering. *Biomaterials* **2008**, *29*, 4532–4539. [CrossRef] [PubMed]

14. Yamamoto, M.; Tabata, Y.; Ikada, Y. Growth factor release from gelatin hydrogel for tissue engineering. *J. Bioact. Compat. Polym.* **1999**, *14*, 474–489. [CrossRef]

15. Houshyar, S.; Kumar, S.; Rifai, A.; Tran, N.; Nayak, R.; Shanks, R.A.; Padhye, R.; Fox, K.; Bhattacharyya, A. Nanodiamond/poly-ε-caprolactone nanofibrous scaffold for wound management. *Mat. Sci. Eng. C-Mater.* **2019**, *100*, 378–387. [CrossRef] [PubMed]

16. Chen, J. Nanobiomechanics of living cells: A review. *Interface Focus* **2014**, *4*, 20130055. [CrossRef] [PubMed]

17. Bacakova, L.; Broz, A.; Liskova, J.; Stankova, L.; Potocky, S.; Kromka, A. The application of nanodiamond in biotechnology and tissue engineering. In *Title of Diamond and Carbon Composites and Nanocomposites*; Mahmood, A., Ed.; IntechOpen: London, UK, 2016; pp. 59–88.

18. Bacakova, L.; Grausova, L.; Vandrovcova, M.; Vacik, J.; Frazcek, A.; Blazewicz, S.; Kromka, A.; Rezek, B.; Vanecek, M.; Nesladek, M.; et al. Carbon nanoparticles as substrates for cell adhesion and growth. In *Title of Nanopartiles: New Research*; Simone, L.L., Ed.; Nova Science Publishes, Inc.: New York, NY, USA, 2008; pp. 39–107.

19. Serafim, A.; Cecoltan, S.; Lungu, A.; Vasile, E.; Iovu, H.; Stancu, I.C. Electrospun fish gelatin fibrous scaffolds with improved biointeractions due to carboxylated nanodiamond loading. *Rsc Adv.* **2015**, *5*, 95467–95477. [CrossRef]

20. Eivazzadeh-Keihan, R.; Maleki, A.; de la Guardia, M.; Bani, M.S.; Chenab, K.K.; Pashazadeh-Panahi, P.; Baradaran, B.; Mokhtarzadeh, A.; Hamblin, M.R. Carbon based nanomaterials for tissue engineering of bone: Building new bone on small black scaffolds: A review. *J. Adv. Res.* **2019**, *18*, 185–201. [CrossRef] [PubMed]

21. Xing, Z.; Pedersen, T.O.; Wu, X.; Xue, Y.; Sun, Y.; Finne-Wistrand, A.; Kloss, F.R.; Waag, R.; Krueger, A.; Steinmuller-Nethl, D.; et al. Biological effects of functionalizing copolymer scaffolds with nanodiamond particles. *Tissue Eng. Part. A* **2013**, *19*, 1783–1791. [CrossRef]

22. Mahdavi, M.; Mahmoudi, N.; Anaran, F.R.; Simchi, A. Electrospinning of Nanodiamond-Modified Polysaccharide Nanofibers with Physico-Mechanical Properties Close to Natural Skins. *Mar. Drugs* **2016**, *14*, 128. [CrossRef]

23. Thalhammer, A.; Edgington, R.J.; Cingolani, L.A.; Schoepfer, R.; Jackman, R.B. The use of nanodiamond monolayer coatings to promote the formation of functional neuronal networks. *Biomaterials* **2010**, *31*, 2097–2104. [CrossRef]

24. Yu, G.; Floyd, Z.E.; Wu, X.; Hebert, T.; Halvorsen, Y.D.C.; Buehrer, B.M.; Gimble, J.M. Adipogenic Differentiation of Adipose-Derived Stem Cells. In *Title of Adipose-Derived Stem Cells. Methods in Molecular Biology (Methods and Protocols)*; Gimble, J., Bunnell, B., Eds.; Humana Press: Totowa, NJ, USA, 2011; Volume 702, pp. 193–200.

25. Grottkau, B.E.; Lin, Y. Osteogenesis of adipose-derived stem cells. *Bone Res.* 2013; 2, 133–145.

26. Oh, S.J.; Park, H.Y.; Choi, K.U.; Choi, S.W.; Kim, S.D.; Kong, S.K.; Cho, K.S. Auricular Cartilage Regeneration with Adipose-Derived Stem Cells in Rabbits. *Mediat. Inflamm.* **2018**, *2018*, 1–8. [CrossRef] [PubMed]

27. Ferroni, L.; Gardin, C.; Tussardi, I.T. Potential for Neural Differentiation of Mesenchymal Stem Cells. *Adv. Biochem. Eng. Biotechnol.* **2012**, *129*, 89–115.

28. Jahan-Abad, A.J.; Morteza-Zadeh, P.; Negah, S.S.; Gorji, A. Curcumin attenuates harmful effects of arsenic on neural stem/progenitor cells. *Avicenna J. Phytomed.* **2017**, *7*, 376.

29. Krishna, L.; Dhamodaran, K.; Jayadev, C.; Chatterjee, K.; Shetty, R.; Khora, S.S.; Das, D. Nanostructured scaffold as a determinant of stem cell fate. *Stem Cell Res.* **2016**, *7*, 188. [CrossRef] [PubMed]

30. Wang, Z.; Ruan, J.; Cui, D. Advances and prospect of nanotechnology in stem cells. *Nanoscale Res. Lett.* **2009**, *4*, 593. [CrossRef] [PubMed]

31. Huang, T.; He, D.; Kleiner, G.; Kuluz, J. Neuron-like Differentiation of Adipose-Derived Stem Cells From Infant Piglets in Vitro. *J. Spinal Cord Med.* **2007**, *30*, 535–540. [CrossRef] [PubMed]

32. Darling, E.M.; Topel, M.; Zauscher, S.; Vail, T.P.; Guilak, F. Viscoelastic properties of human mesenchymally-derived stem cells and primary osteoblasts, chondrocytes, and adipocytes. *J. Biomech.* **2008**, *41*, 454–464. [CrossRef]

33. Galateanu, B.; Dinescu, S.; Cimpean, A.; Dinischiotu, A.; Costache, M. Modulation of Adipogenic Conditions for Prospective Use of hADSCs in Adipose Tissue Engineering. *Int. J. Mol. Sci.* **2012**, *13*, 15881–15900. [CrossRef]

34. Dinescu, S.; Galateanu, B.; Albu, M.; Cimpean, A.; Dinischiotu, A.; Costache, M. Sericin enhances the bioperformance of collagen-based matrices preseeded with hADSCs. *Int. J. Mol. Sci.* **2013**, *14*, 1870–1889. [CrossRef]

35. Zhang, Y.; Hua, Q.; Zhang, J.M.; Zhao, Y.; Yin, H.; Dai, Z.; Zheng, L.; Tang, J. Enhanced thermal and mechanical properties by cost-effective carboxylated nanodiamonds in poly (vinyl alcohol). *Nanocomposites* **2018**, *4*, 58–67. [CrossRef]

36. Cecoltan, S.; Serafim, A.; Dragusin, D.-M.; Lungu, A.; Lagazzo, A.; Barberis, F.; Stancu, I.-C. The potential of NDPs-loaded fish gelatin fibers as reinforcing agent for fish gelatin hydrogels. *Key Eng. Mater.* **2016**, *695*, 278–283. [CrossRef]

37. Woo, D.J.; Sneed, B.; Peerally, F.; Heer, F.C.; Brewer, L.N.; Hooper, J.P.; Osswald, S. Synthesis of nanodiamond-reinforced aluminum metal composite powders and coatings using high-energy ball milling and cold spray. *Carbon* **2013**, *63*, 404–415. [CrossRef]

38. Hardiman, M.; Vaughan, T.J.; McCarthy, C.T. Fibrous composite matrix characterisation using nanoindentation: The effect of fibre constraint and the evolution from bulk to in-situ matrix properties. *Compos. Part. A Appl. Sci. Manuf.* **2015**, *68*, 296–303. [CrossRef]

39. Behler, K.D.; Stravato, A.; Mochalin, V.; Korneva, G.; Yushin, G. Nanodiamond-Polymer Composite Fibers and Coatings. *Asc Nano* **2009**, *3*, 363–369. [CrossRef] [PubMed]

40. Lee, S.; Teramoto, Y.; Wang, S.; Pharr, G.M.; Rials, T.G. Nanoindentation of Biodegradable Cellulose Diacetate-graft -Poly (L -lactide) Copolymers: Effect of Molecular Composition and Thermal Aging on Mechanical Properties. *J. Polym. Sci. Polym. Phys.* **2006**, *45*, 1114–1121. [CrossRef]

41. Oliver, W.C.; Pharr, G.M. An improved technique for determining hardness and elastic modulus using load and displacement sensing indentation experiments. *J. Mater. Res.* **1992**, *7*, 1564–1583. [CrossRef]

42. Kim, Y.; Lee, D.; Kim, S.Y.; Kang, E.; Kim, C.K. Nanocomposite Synthesis of Nanodiamond and Molybdenum Disulfide. *Nanomaterials* **2019**, *9*, 927. [CrossRef] [PubMed]

43. Huang, Y.A.; Kao, C.W.; Liu, K.K.; Huang, H.S.; Chiang, M.H.; Soo, C.R.; Chang, H.C.; Chiu, T.W.; Chao, J.I.; Hwang, E. The effect of fluorescent nanodiamonds on neuronal survival and morphogenesis. *Sci. Rep.* **2014**, *4*, 6919–6928. [CrossRef]

44. Lee, G.Y.H.; Lim, C.T. Biomechanics approaches to studying human diseases. *Trends Biotechnol.* **2007**, *25*, 111–118. [CrossRef]

45. Sun, Y.; Yang, Q.; Wang, H. Synthesis and Characterization of Nanodiamond Reinforced Chitosan for Bone Tissue Engineering. *J. Funct. Biomater.* **2016**, *7*, 27–42.

46. Ma, M.; Mao, Y.; Gupta, M.; Gleason, K.K.; Rutledge, G.C. Superhydrophobic fabrics produced by electrospinning and chemical vapor deposition. *Macromolecules* **2005**, *38*, 9742–9748. [CrossRef]

47. Wang, X.; Yu, J.; Sun, G.; Ding, B. Electrospun nanofibrous materials: A versatile medium for effective oil/water separation. *Mater. Today* **2015**, *19*, 403–414. [CrossRef]

48. Li, X.; Bian, F.; Lin, J.; Zeng, Y. Effect of electric field on the morphology and mechanical properties of electrospun fibers. *Rsc Adv.* **2016**, *6*, 50666–50672. [CrossRef]

49. Pacelli, S.; Acosta, F.; Chakravarti, A.R.; Samanta, S.G.; Whitlow, J.; Modaresi, S.; Ahmed, R.P.H.; Rajasingh, J.; Paul, A. Nanodiamond-based injectable hydrogel for sustained growth factor release: Preparation, characterization and in vitro analysis. *Acta Biomater.* **2017**, *58*, 479–491. [CrossRef] [PubMed]

50. Yao, R.; Zhang, R.; Luan, J.; Lin, F. Alginate and alginate/gelatin microspheres for human adipose-derived stem cell encapsulation and differentiation. *Biofabrication* **2012**, *4*, 025007. [CrossRef]

51. Chiou, B.S.; Avena-Bustillos, R.J.; Shey, J.; Yee, E.; Bechtel, P.J.; Imam, S.H.; Glenn, G.M.; Orts, W.J. Rheological and mechanical properties of cross-linked fish gelatins. *Polymer* **2006**, *47*, 6379–6386. [CrossRef]

52. Liu, K.K.; Cheng, C.L.; Chang, C.C.; Chao, J.I. Biocompatible and detectable carboxylated nanodiamond on human cell. *Nanotechnology* **2007**, *18*, 325102. [CrossRef]

53. Pacelli, S.; Maloney, R.; Chakravarti, A.R.; Whitlow, J.; Basu, S.; Modaresi, S.; Paul, A. Controlling adult stem cell behavior using nanodiamond-reinforced hydrogel: Implication in bone regeneration therapy. *Sci. Rep.* **2018**, *7*, 6577. [CrossRef]

54. Guarino, V.; Alvarez-Perez, M.; Cirillo, V.; Ambrosio, L. hMSC interaction with PCL and PCL/gelatin platforms: A comparative study on films and electrospun membranes. *J. Bioact. Compat. Polym.* **2011**, *26*, 144–160. [CrossRef]

55. Pereira, F.A.S.; Salles, G.N.; Rodrigues, B.V.M.; Marciano, F.R.; Pacheco-Soares, C.; Lobo, A.O. Diamond nanoparticles into poly (lactic acid) electrospun fibers: Cytocompatible and bioactive scaffolds with enhanced wettability and cell adhesion. *Mater. Lett.* **2016**, *183*, 420–424. [CrossRef]

56. Hopper, A.P.; Dugan, J.M.; Gill, A.A.; Fox, O.J.L.; May, P.W.; Haycock, J.W.; Claeyssens, F. Amine functionalized nanodiamond promotes cellular adhesion, proliferation and neurite outgrowth. *Biomed. Mater.* **2014**, *9*, 045009. [CrossRef] [PubMed]

57. Tsai, C.Y.; Lin, C.L.; Cheng, N.C.; Yu, J. Effects of nano-grooved gelatin films on neural induction of human adipose-derived stem cells. *Rsc Adv.* **2017**, *7*, 53537–53544. [CrossRef]

 *materials*

Review

Bioactive Glass-Based Endodontic Sealer as a Promising Root Canal Filling Material Without Semisolid Core Materials

Ayako Washio, Takahiko Morotomi, Shinji Yoshii and Chiaki Kitamura *

Division of Endodontics and Restorative Dentistry, Department of Oral Functions, Kyushu Dental University, Kitakyushu 803-8580, Japan; r05washio@fa.kyu-dent.ac.jp (A.W.); r13morotomi@fa.kyu-dent.ac.jp (T.M.); r08yoshii@fa.kyu-dent.ac.jp (S.Y.)
* Correspondence: r06kitamura@fa.kyu-dent.ac.jp; Tel.: +81-93-582-1131

Received: 30 October 2019; Accepted: 26 November 2019; Published: 29 November 2019

Abstract: Endodontic treatment for a tooth with damaged dental pulp aims to both prevent and cure apical periodontitis. If the tooth is re-infected as a result of a poorly obturated root canal, periapical periodontitis may set-in due to invading bacteria. To both avoid any re-infection and improve the success rate of endodontic retreatment, a treated root canal should be three-dimensionally obturated with a biocompatible filling material. Recently, bioactive glass, one of the bioceramics, is focused on the research area of biocompatible biomaterials for endodontics. Root canal sealers derived from bioactive glass-based have been developed and applied in clinical endodontic treatments. However, at present, there is little evidence about the patient outcomes, sealing mechanism, sealing ability, and removability of the sealers. Herein, we have developed a bioactive glass-based root canal sealer and provided evidence concerning its physicochemical properties, biocompatibility, sealing ability, and removability. We also review the classification of bioceramics and characteristics of bioactive glass. Additionally, we describe the application of bioactive glass to facilitate the development of a new root canal sealer. Furthermore, this review shows the potential application of bioactive glass-based cement as a root canal filling material in the absence of semisolid core material.

Keywords: bioceramics; bioactive glass; hydroxyapatite; root canal sealer

---

1. Introduction

Endodontic treatment for a tooth with damaged dental pulp aims to both prevent and cure apical periodontitis. After a root canal preparation and irrigation in order to both remove bacteria and suppress the inflammation of the periodontal ligament around root apex, dentists obturate the treated root canal with filling materials [1]. If the post-treatment tooth is re-infected due to a poor root canal obturation, periapical periodontitis sets-in due to an invasion of bacteria into the canal. It is well known that the success rate of endodontic retreatment on periapical periodontitis is no higher than that of the initial treatment [2–5]. Three-dimensional obturation of the treated root canal with biocompatible filling materials is vital to avoid re-infection as well as the root canal preparation and irrigation steps, thereby increasing the success rate of retreatment [6,7].

The primary functions of any root canal filling material are to seal the bacterial in-growth so as to prevent fluid influx from providing nutrients to the trapped bacterium [8]. Endodontic treatment techniques have been changing due to technological advances, and advances in root canal filling material have significantly contributed to increased rates in the successful treatment of patients. Root canal sealer, one of the many filling materials, has been shown to be essential for successful obturation, as the sealer should bond to the dentin of the canal walls and close-off the periapical area of the root canal system. However, conventional root canal sealer typified by Grossman's formula is hardly ideal as it is neither adhesive nor does it have a bonding effect with dentin.

Mineral trioxide aggregate (MTA)-based root canal sealers, such as EndoSequence BC Sealer (Brasseler USA, Savannah, GA, USA), have been developed and are now commercially available; these MTA-based sealers provide ideal performance as a root canal sealer. MTA is formulated from commercial Portland cement (tricalcium silicate, dicalcium silicate, tricalcium aluminate, tetracalcium aluminoferrite, calcium sulfate) [9,10], combined with bismuth oxide powder for radiopacity. MTA-based root canal sealer is generally believed to be a bioceramic-based sealer. However, MTA is not bioceramic as its crystals are non-vitreous. It has been reported that some MTA-based sealers show good physical and biological properties [11–13], as well as the ability to produce hydroxyapatite on its surface in the presence of phosphate-buffered saline [14,15]. A hypothesized mechanism for the formation of hydroxyapatite is initiated by the release of calcium hydroxide from MTA, which interacts with a phosphate-containing solution to produce a calcium-deficient apatite achieved via an amorphous calcium phosphate phase [16]. These characteristics indicate that MTA-based sealers may display bioactivity. However, several studies report that some MTA-based root canal sealers show non-biocompatibility due to the presence of arsenic, a low ability to seal, long setting time, and non-retreatability [17,18].

Recently, bioactive glass, one of confirmed bioceramics, has been the focus of a great deal of research in biomaterials for Endodontics. Furthermore, bioactive glass-based root canal sealers have been developed and applied within clinical endodontic treatments. Herein we will review the appropriate classification of materials as bioceramics and the specific characteristics of bioactive glass. Additionally, we describe the possible application of bioactive glass as a newly developed root canal sealer is described. Furthermore, this review promises the potential of bioactive glass-based cement as a root canal filling material.

1.1. Bioceramics

Biomaterials are defined as synthetic or natural materials that are capable of either replacing parts of a living system or functioning while in intimate contact with living tissues [19]. Biomaterial-based implants and medical devices are widely used to replace or to restore the functionality of traumatized or degenerated tissues. The foremost requirement when selecting a biomaterial is its biological acceptability as a long-term non-rejected implant within the body. To achieve this acceptability, applicable biomaterials must be non-toxic, non-carcinogenic, chemically inert, stable, and mechanically strong. The most common biomaterial classes are metals, polymers, and ceramics. These three classes are used either solely or in combination to form the most presently available implantation devices.

Ceramics, a class of biomaterial, are polycrystalline materials that display characteristic hardness, brittleness, strength, stiffness, resistance to corrosion and wear, and low density. Bioceramics are utilized to restore functionality to diseased or damaged hard tissues and are used in several different fields such as dentistry, orthopedics, and medical sensors. Presently available bioceramics come in three basic types: bioinert, bioactive, and bioresorbable ceramics [20]. The first generation of bioceramics was comprised of alumina and zirconia [21]. The main features of first-generation bioceramics were their good mechanical properties, especially their wear resistance. The second generation of bioceramics was comprised of bioactive glass (BG), hydroxyapatite, and calcium phosphate-based cement. Second generation bioceramics bond to and integrate with the living bone of the body without forming a fibrous tissue around them and without promoting either inflammation or toxicity [22]. Unique among the second-generation bioceramics, BG has instigated a revolution in healthcare appliances and has paved the way for modern biomaterial-driven medicine [23,24].

1.2. Bioactive Glass

BG contains the glass type of Na_2O-CaO-SiO_2-P_2O_5 in specific proportions [25], as a component of silica (SiO_2) is ≤ 50 mol%. The compositional phase diagram for BG, also highlighting at what mixture-levels particular biomaterial properties arise, is provided as Figure 1 [25,26]. BG has been applied in clinical settings for orthopedic surgery for several decades. When BG is implanted in a defect area close to bone, reactions on BG surfaces lead to the release of critical concentrations of soluble Si, Ca, P and Na ions, which induce favorable intracellular and extracellular responses leading to rapid bone formation [27]; this bone formation is then followed by the formation of silica-rich gel on its surface. Silica-rich gel reacts with ions present in bodily fluids, resulting in the formation of hydroxyapatite (HAp)-like on the surface of BG. Furthermore, osteoblasts produce new bone in the silica-rich gel, allowing BG to bond with the bone through both the formation of bone-like hydroxyapatite layers and biological interactions with collagen (Figure 2) [22,28]. Additionally, BG is able to stimulate bone cells to regenerate and self-repair, thus significantly accelerating tissue healing kinetics [27]. These properties are termed osteoconductivity and osteoinductivity [23,29]. BG has been mainly used for applications where it will contact bone tissue, yet BG has recently shown promise in inducing the repair of soft tissues, too [30,31]. BG has attracted the interests of many researchers, as the ionic dissolution products of BG were found to stimulate angiogenesis. Furthermore, there now exist other BG-based products for applications in wound healing and peripheral nerve regeneration [32]. These applications suggest that BG shows suitability and biocompatibility as a biomaterial capable of being applied both to hard tissues such as dentin or cementum—as these materials are similar to the bone—and to soft tissues such as dental pulp and periapical tissue [33].

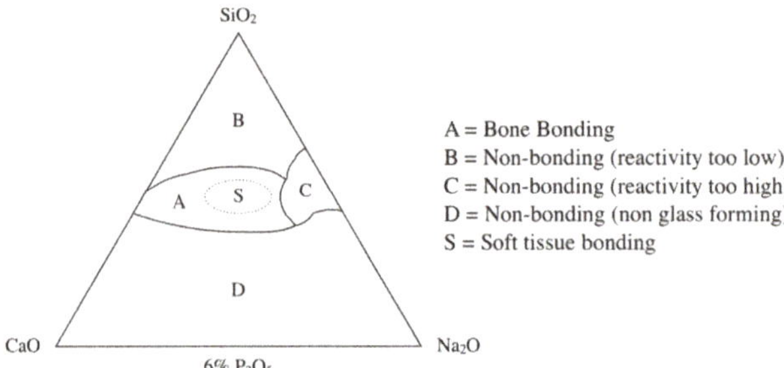

Figure 1. Compositional phase diagram of bioactive glasses with a focus on bone-bonding. Region S is the region of Class A bioactivity where bioactive glasses bond to both bone and soft tissues and display gene activating characteristics [25].

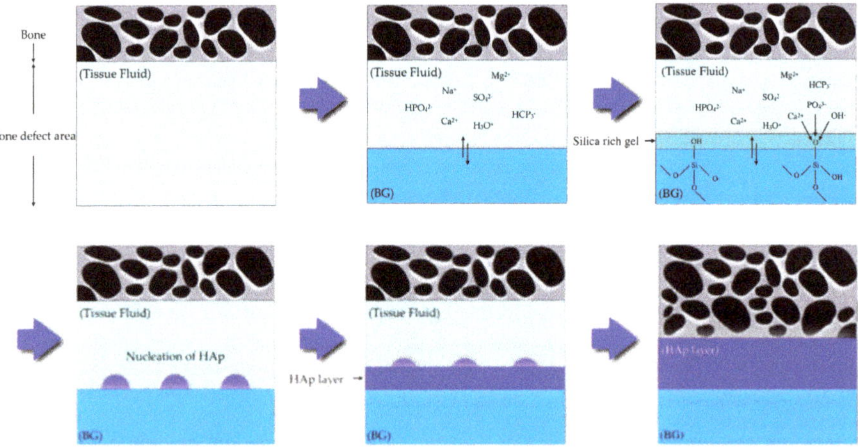

Figure 2. Scheme showing a proposed bonding mechanism of bioactive glass with bone [22,28].

1.3. Bioceramic-Based Root Canal Sealer

General practitioners desire a root canal sealer capable of strongly bonding to root canal walls with high sealing properties, high biocompatibility, as well as removability to accommodate retreatment. Researchers have found promising results in the application of bioceramics to solve these issues. Bioceramic-based materials have recently been introduced as endodontics materials as both repairing cement [34,35] and root canal sealer [13,36–39]. Bioceramic-based materials show an alkaline pH, antibacterial activity, radiopacity, biocompatible, nontoxic, non-shrinking, and are chemically stable within the biological environment. A further advantage of bioceramic materials is that they promote the formation of hydroxyapatite, ultimately facilitating a bond between dentin and the filling material during the setting process [11,38]. However, conventional bioceramic-based sealers show clinical disadvantages such as difficulty in handling, higher cytotoxicity in its freshly mixed state, a high pH during setting, long setting times, and that hardening requires sufficient moisture [18,40–43]. An additional disadvantage is that bioceramic-based sealers are difficult to remove when facilitating retreatment [44]. To overcome these disadvantages, we developed a next-generation bioceramic-based root canal sealer based on previous medically reliable BG-based materials, Nishika Canal Sealer BG (Nippon Shika Yakuhin, Yamaguchi, Japan).

1.4. Bioactive Glass-Based Root Canal Sealer

There are two well-known commercialized root canal sealers that include BG. One is GuttaFlow Bioseal (GFB) (Coltène/Whaledent AG, Altstätten, Switzerland), which is composed of gutta-percha, polydimethylsiloxane, platinum catalyzer, zirconium dioxide, and BG. GFB has shown a low solubility, low porosity, alkalization capacity [45], dentin penetrability [46], and cytocompatibility [47,48]. At present, only limited evidence is available concerning either the mechanism of GFB hardening or its ability to seal the canal and be removed for retreatment. The second product is Nishika Canal Sealer BG (CS-BG), shown in Figure 3; presently there exists compelling evidence concerning, with evidences about its physicochemical properties, biocompatibility, sealing ability, and removability. CS-BG was developed from BG-based biomaterials and originally intended for both dental pulp and bone regeneration therapies. CS-BG is a two-phased paste; Paste A consists of fatty acids, bismuth subcarbonate, and silica dioxide, whereas Paste B consists of magnesium oxide, calcium silicate glass (a type of BG), and silica dioxide, etc. By pushing the plunger of a double syringe, the two-phase paste can be dispensed at a 1:1 ratio. The dispensed paste can be mixed easily and quickly; this procedure is captured in Figure 4. A stainless-steel spatula may be corroded by the ingredients of the paste, we recommend the use of a plastic spatula to avoid contamination of metal implements. CS-BG paste tends to get hardened when exposed to heat or moisture. Therefore, it is recommended to store the syringes in the resealable aluminum foil bag, then placing the bag in a cold storage location (1–10 °C) without freezing.

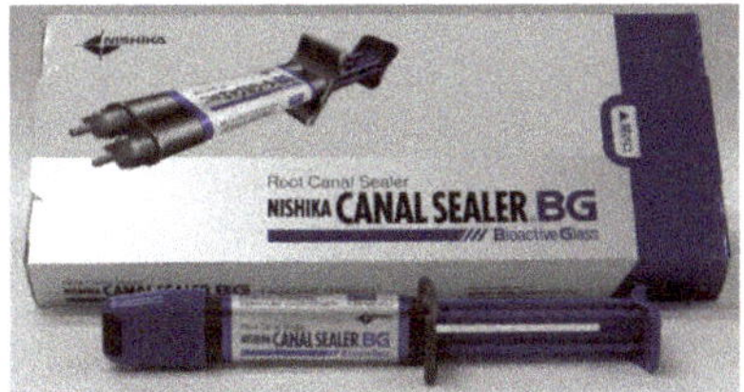

Figure 3. Nishika Canal Sealer BG.

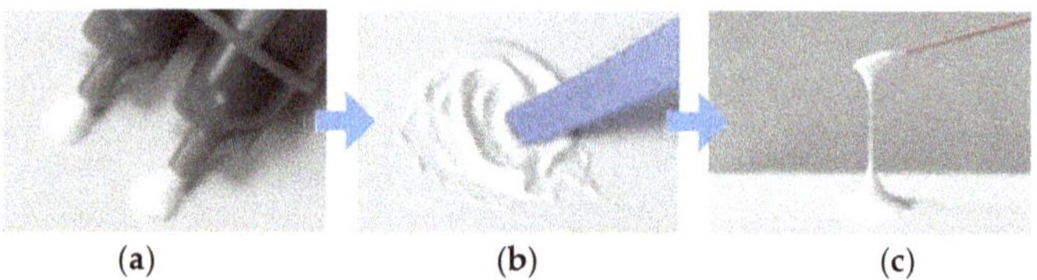

(a)   (b)   (c)

Figure 4. Mixing procedure of CS-BG. (**a**) Dispense the amount required. (**b**) Mix them gently. Mixing time: five seconds or more. (**c**) Ideal paste consistency for root canal obturation.

2. Physicochemical Properties

2.1. Physical Properties

Physical properties of CS-BG were analyzed according to the International Organization for Standardization (ISO) standards of root canal sealing materials (ISO 6876:2012 Requirement), and it is found that the properties of CS-BG were suitable for use as an endodontic sealer (Table 1).

Table 1. Physical properties of CS-BG.

Flow	28.7 mm	Solubility	0.5%
Working time	15 mm	Disintegration	None
Setting time	180 min	Radiopacity	5 mmAl.
Film thickenss	27.9 μm		

2.2. pH Change

The pH of a CS-BG sample that was hardened in simulated body fluid (SBF) was measured in the purified water. The pH gradually decreased during periodic immersion and stabilized at around pH = 10 (Figure 5); this pH is optimal for the formation of HAp on the BG surface [49–51]. This alkaline pH is maintained by ions evolving from non-BG components.

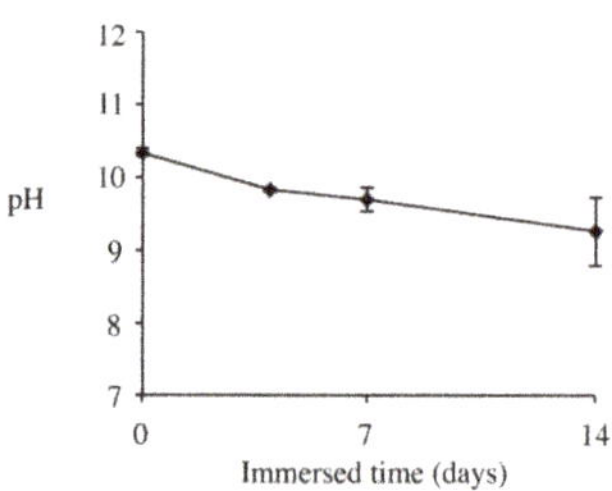

Figure 5. pH of purified water after immersion of hardened CS-BG [51].

2.3. HAp Formation on the Surface of CS-BG in SBF

The surface-structures of CS-BG discs (diameter 3.5 mm, height 6 mm) hardened in either SBF or purified water were analyzed using Field emission scanning electron microscope analysis (FE-SEM). The surface structure of CS-BG after immersion in SBF showed typical spherules of petal-like crystals (Figure 6a). X-ray diffraction analysis (XRD) showed that petal-like crystals were HAp. In SBF, HAp crystallization on the surface increased in a time-dependent manner (Figure 6c) [51]. On the other hand, the surface structure of CS-BG after immersion in purified water showed no petal-like crystals (Figure 6b,d).

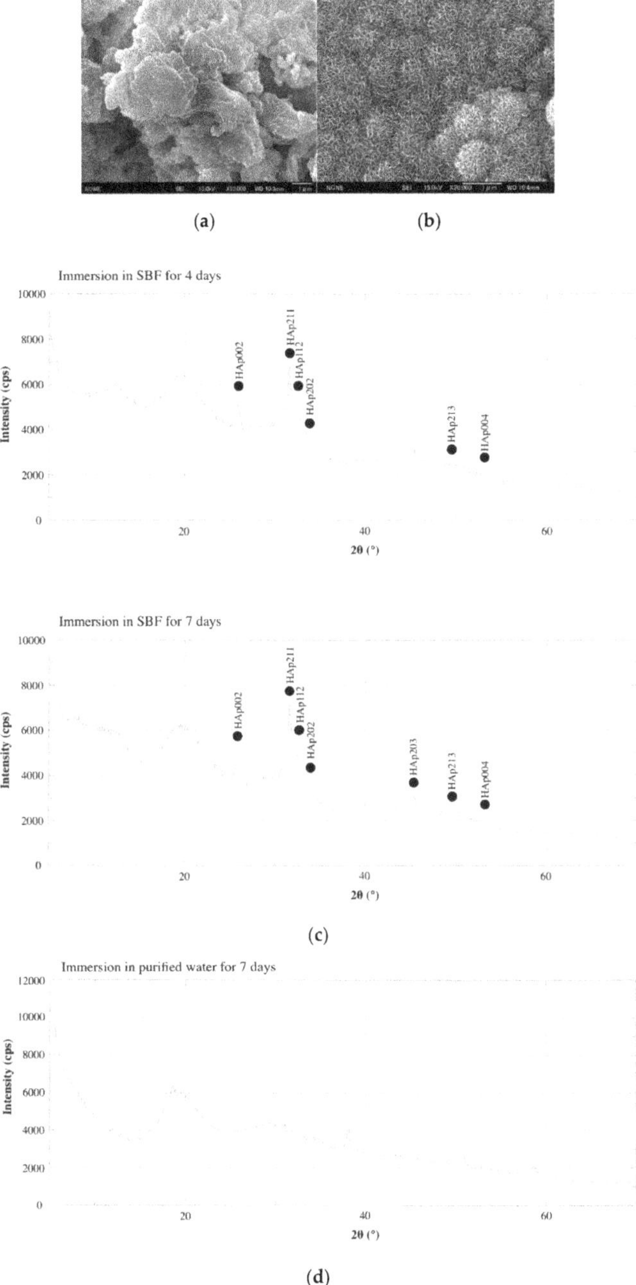

Figure 6. FE-SEM images and XRD patterns of the CS-BG surface. (**a**) FE-SEM images of CS-BG immersed in SBF. (**b**) FE-SEM images of CS-BG immersed in purified water. Scale bar 1 μm. (**c**) XRD patterns of CS-BG after immersion in SBF for four and seven days. (**d**) XRD patterns of CS-BG after immersion in purified water for seven days [51].

3. Biocompatibility

Biocompatibility is an essential property of any root filling material that is in direct contact with both hard tissues (e.g., dentin or cementum) and soft tissues (e.g., periodontal ligament) [52,53]. Reiterating, biocompatibility is the ability of a material to achieve a stable and advantageous host response during application [54]. Biocompatibility is typically assessed by cytotoxic tests in most studies [55]. The cytotoxicity of bioceramic-based sealers has been evaluated in vitro using mouse osteoblast cells, human osteoblast cells [38,56,57], and human periodontal ligament cells [58–60].

The in vitro biocompatibility of CS-BG was demonstrated by cell migration and viability assays using human periodontal ligament cells (HPDLC) and osteoblast-like cells. Migration and survival of both HPDLC and osteoblast-like cells under CS-BG showed no significant difference when compared to control (Figures 7 and 8) [61]. HPDLC and osteoblast-like cells proliferated and migrated in direct contact with the surface of hardened CS-BG (Figure 9) [62]. The in vivo biocompatibility of CS-BG was analyzed by both rat pulpectomy and root canal obturation models. These in vivo tests indicate that CS-BG does not inhibit the wound healing process of periapical tissue around the root apex of a canal filled with CS-BG (Figure 10) [63]. These in vitro and in vivo studies show that CS-BG has excellent biocompatibility for periapical tissue.

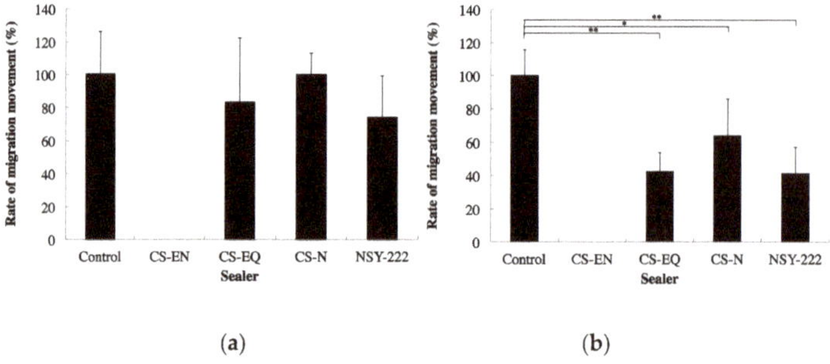

(a)   (b)

Figure 7. Effects of root canal filling sealers on cell migration ability. (**a**) HPDLCs. (**b**) osteoblast-like cells. The percentage of cell migration after scratching. Control: no sealer, CS-EN: eugenol-based sealer, CS-EQ: eugenol-based sealer quick, CS-N: non-eugenol-based sealer, CS-BG: BG based sealer. Each bar represents a mean ± SD. *, **: significant differences with $p < 0.05$ and $p < 0.01$, respectively [61].

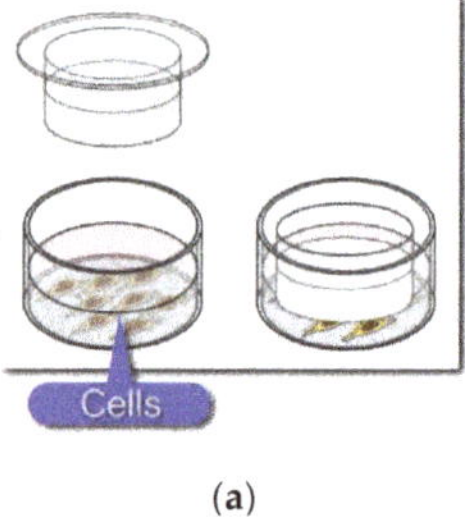

(a)

Figure 8. *Cont.*

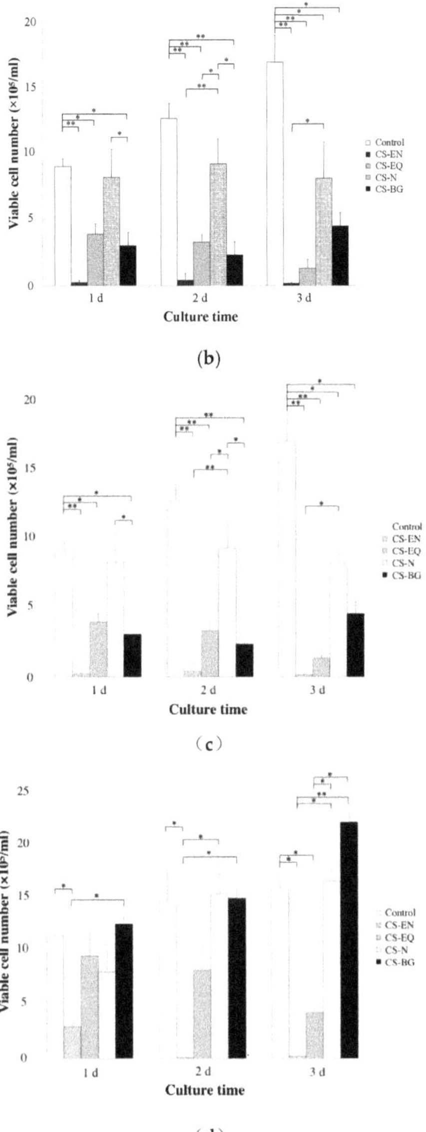

(**b**)

(**c**)

(**d**)

Figure 8. *Cont.*

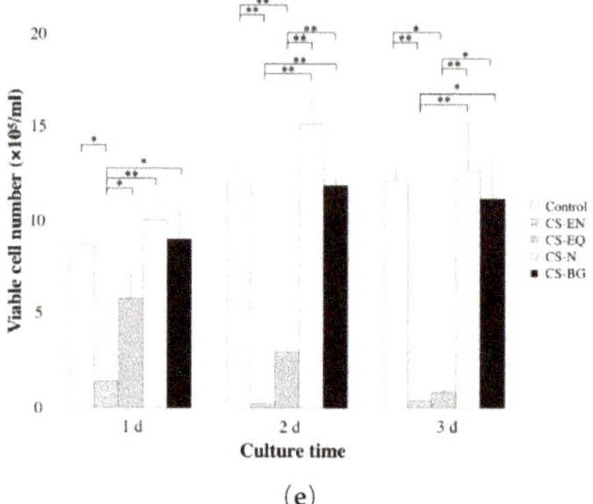

(e)

Figure 8. Effects of root canal filling sealers on cell viability. (**a**) Schematic of culture method. The cells (1 × 105/ well) were separately subcultured in 24-well Transwell plates. Transwell filter inserts including fresh and hardened sealers were inserted into the wells. (**b,c**) fresh sealer. (**d,e**) hardened sealer. (**b,d**) HPDLCs. (**c,e**) osteoblast-like cells. Control: no sealer, CS-EN: eugenol-based sealer, CS-EQ: eugenol-based sealer quick, CS-N: non-eugenol-based sealer, CS-BG: BG-based sealer. Each bar represents a mean ± SD. *, **: significant differences with p < 0.05 and p < 0.01, respectively [61].

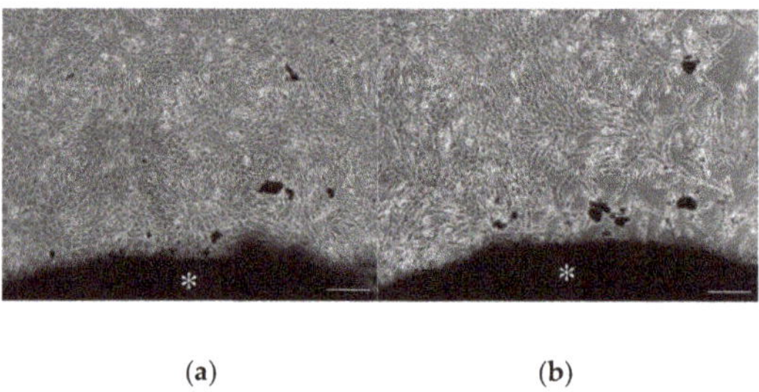

(a)   (b)

Figure 9. Phase-contrast microscopic photographs showed the attachment of cells to CS-BG (*). (a) HPDLCs. (b) osteoblast-like cells [62]. Scale bar 200 μm.

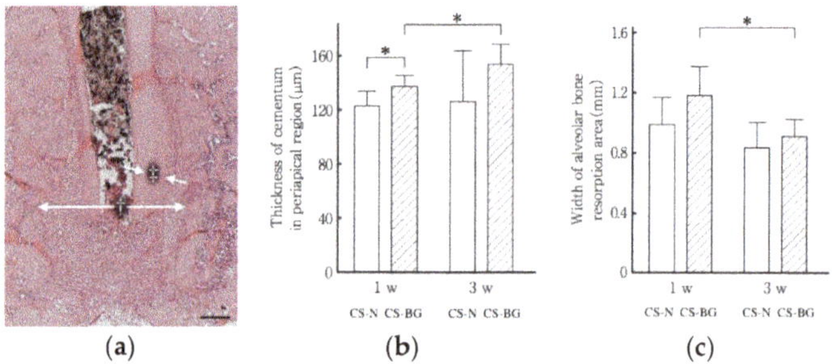

(a)   (b)   (c)

Figure 10. Semi-quantitative analysis of the tissue responses on periapical tissue after root canal obturation with CS-BG. (**a**) Each site of measurement for the width of periapical bone alveolar resorption (†) and the thickness of cementum (‡). Scale bar 200 μm. (**b**) Width of periapical alveolar bone resorption area. (**c**) The thickness of cementum in the periapical region. CS-N: non-eugenol-based sealer. Each bar represents a mean ± SD. *: significant differences with p < 0.05 [63].

4. Sealing Ability

The invasion of microorganisms into the interfacial region between filling materials and the root canal dentinal wall should be prevented to avoid re-infection [64–66]. Properly sealing this interface is dependent on the ability of the filling material to bind to the dentinal wall. There exists no standard method for measuring the sealing ability of a root canal sealer [67–70]. To assess the sealing ability of CS-BG, a dye leakage test was used to simulate the seepage of nutrient fluid into the sealed cavity. An amount of dye was sealed within the root canal and sealed with a combination of gutta-percha point and CS-BG by the lateral condensation technique. The total amount of leakage was approximately half in comparison with conventional root canal sealers (eugenol-based and non-eugenol-based sealer), and the leakage gradually decreased over time (Figure 11a,b) [71]. When a root canal was filled with CS-BG by the single-cone technique, the leakage was less than that observed for the CS-GB material applied by the lateral condensation technique (Figure 11a,b) [71]. These results showed the excellent sealing ability of CS-BG, especially when applied by the single cone method.

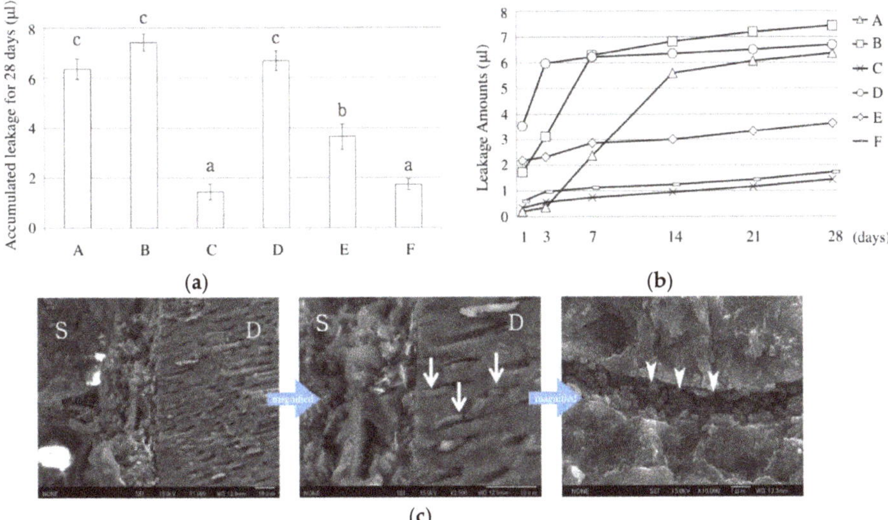

Figure 11. Sealing ability of root canal sealers after root canal obturation. (**a**) Time-dependent leakage evaluation of six different root canal obturations using dye penetration test. (**b**) The total leakage amount measured for 28 days. A: Eugenol-based root canal sealer (lateral condensation technique), B: Non-eugenol-based root canal sealer (lateral condensation technique), C: Bioceramics-based root canal sealer (lateral condensation technique), D: Bioceramics-based root canal sealer (single cone technique), E: CS-BG (lateral condensation technique), F: CS-BG (single cone technique). (**c**) FE-SEM images of the interface between the filled sealer and root canal wall. S: CS-BG, D: dentin. Arrows: the formation of tag-like structures in dentinal tubules, arrowheads: hydroxyapatite-like crystals in dentinal tubules [71]. Scale bar 10 μm.

The characteristics of the interface between the hardened CS-BG and the root canal wall was analyzed by both FE-SEM and Energy-dispersive X-ray spectrometry (EDX). The FE-SEM showed the formation of a tag-like structure comprised of typical spherules of petal-like crystals embedded into dentinal tubules and at the entrance of the tubules (Figure 11c) [71]. These crystals were identified as HAp by EDX analysis [71].

Figure 12 shows a proposed mechanistic scheme for the bonding of CS-BG to the dentin-based root canal wall. After the obturation of a root canal with CS-BG, the CS-BG makes contact with a small amount of dental fluid on the dentin. During hardening, the CS-BG releases ions from its matrix, these components consist of the non-BG component of the sealing paste. The evolution of these ions maintains a pH of approximately 10 in the surrounding dentinal fluid, which is the optimal pH for the formation of HAp on the surface of BG. The BG within the sealer mixture then reacts to the dental fluid, resulting in the release of critical concentrations of soluble Si, Ca, P, and Na ions. This causes the formation of a silica-rich gel on the BG surface that reacts with the ions now present in the dentinal fluid. As a result, HAp-like crystal layers are formed on the surface of BGs. Finally, HAp-like crystal tags interstitially grow into dentin tubules. The overall CS-BG bonding with the dentin wall is formed through the formation of these HAp layers and tags within the dentin tubules (Figure 12).

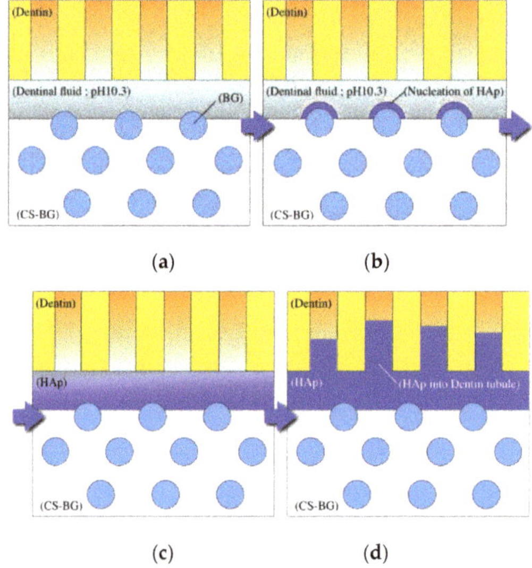

Figure 12. Schematic of the proposed mechanism for CS-BG to bond to dentin. (**a**) CS-BG matrix generates a pH of 10.3 in the dental fluid at the sealer-dentin interface. (**b**,**c**) CS-BG matrix displays an amphiphilic property and facilitates the growth of HAp. (**d**) After bonding with dentin, HAp crystals grow into the dentinal tubule.

5. Removability

When re-infection occurs at the periapical tissue of a treated tooth, dentists must first remove any present root canal sealer from the canal before proceeding with endodontic retreatment [72–74]. In vitro studies demonstrated that CS-BG is capable of being removed entirely by the standard methods of re-preparation and irrigation with an EDTA solution. Furthermore, the dentinal tubules of the dentin wall were reopened upon removal of CS-BG, shown by FE-SEM images in Figure 13 [75]. The easily removed nature of the sealer, coupled with the reopening of the dentin cavities, indicates that CS-BG does not inhibit retreatment.

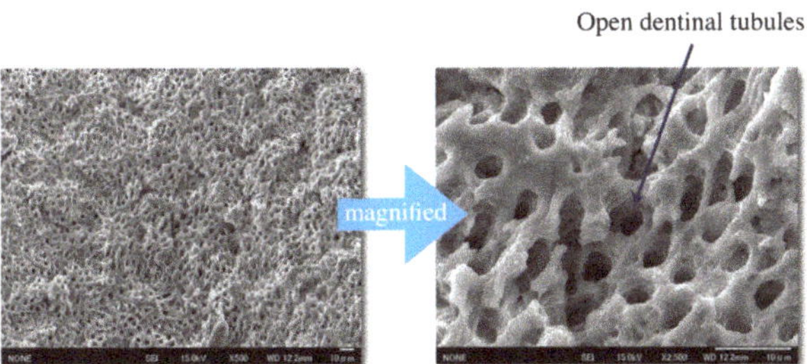

Figure 13. Root canal wall after the removal of hardened CS-BG and irrigation with a solution of EDTA. Dentinal tubules were observed [75]. Scale bar 10 μm.

6. Clinical Performance of Bioactive Glass-Based Root Canal Sealer

CS-BG is now available for use in root canal obturation and has been shown to induce good wound healing of periapical tissues. Figure 14 shows a clinical case (40-year-old female) upon whom CS-BG was applied during a root canal obturation. The radiographic image taken during pre-endodontic treatment (Figure 14a) shows an apparent radiolucency at the periradicular area of the maxillary left canine; this radiographic translucency was diagnosed as symptomatic apical periodontitis. After standard endodontic treatment, the canal was obturated using CS-BG and gutta-percha by a non-compaction technique (Figure 14b). Wound healing and bone formation of the periapical tissues were observed at both six and 14 months after the obturation (Figure 14c,d). From now on, we will follow the cases obturated using CS-BG for longer-term clinical efficacy.

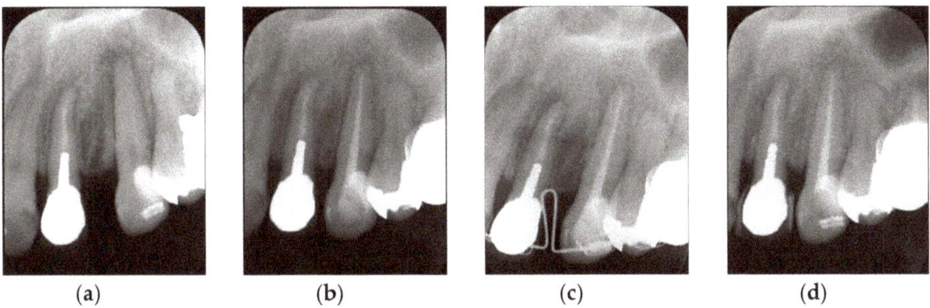

(a) (b) (c) (d)

Figure 14. Clinical performance of CS-BG. (**a**) Pre-treatment radiograph. (**b**) Post-obturation radiograph. (**c**) 6 months follow-up radiograph. (**d**) 14 months follow-up radiograph.

7. Potential of Bioactive Glass-Based Sealer as Root Canal Filling Material Without Semisolid Core Materials

At present, there are various techniques to obturate root canal systems, including sealer, sealer plus a single-core material, sealer coating combined with cold compaction of core materials, sealer coating combined with warm compaction of core materials, and sealer coating combined with a carrier-based core material [76,77]. It has also been reported that a significant number of general practitioners have chosen cold lateral or warm gutta-percha compaction with either eugenol-based or non-eugenol-based sealer [78–82]. However as revealed by several reviews, bioceramic-based root canal sealers applied by the single cone technique are more appropriate sealers to obtain good sealing of an obturated root canal [83,84]. When undertaking the single cone technique, the sealer delivery method is important. The root canal sealer should be delivered to the root apex before the insertion of the single-core material. Slight pressurization occurs during the insertion of core material; this pressure can cause the sealer within the canal to flow into lateral branches of the root canal. Within this review, it is indicated that CS-BG has the high sealing ability, suggestive of its potential as a sealer for root canal obturation even without semisolid core material such as a gutta-percha point. In hopes to achieve the best possible patient outcome after obturation with CS-BG—with or without a semisolid core material—the delivering device of the bioceramic-based sealer should be one of the next targets for translational research in Endodontics.

8. Conclusions

The next-generation bioactive glass-based root canal sealer examined in this work has displayed capabilities to form hydroxyapatite-like precipitations, biocompatibility, sealing ability, and removability. CS-BG is now available for use in root canal obturation and has been shown to induce good wound healing of periapical tissues in clinical cases. This accumulation of characteristics is suggested that CS-BG has its potential as a sealer for root canal obturation even without a semisolid core material.

Author Contributions: Conceptualization of the study was performed by C.K. and A.W.; methodology, investigation, analysis, writing, and draft preparation were conducted by T.M., A.W., and S.Y.; writing, reviewing, and editing was performed by A.W. and C.K.

Funding: This research was supported by Grants-in-Aid for Scientific Research (Kitamura, Morotomi, Washio, Yoshii) from the Ministry of Education, Science, and Culture of Japan.

Conflicts of Interest: The authors declare no conflict of interest. The funders had no role in the design of the study; in the collection, analyses, or interpretation of data; in the writing of the manuscript, or in the decision to publish the results.

References

1. Sjogren, U.; Figdor, D.; Persson, S.; Sundqvist, G. Influence of infection at the time of root filling on the outcome of endodontic treatment of teeth with apical periodontitis. *Int. Endod. J.* **2012**, *30*, 297–306. [CrossRef]
2. Van Nieuwenhuysen, J.P.; Aouar, M.; D'Hoore, W. Retreatment or radiographic monitoring in endodontics. *Int. Endod. J.* **1994**, *27*, 75–81. [CrossRef]
3. Rotstein, I.; Salehrabi, R.; Forrest, J.L. Endodontic treatment outcome: Survey of oral health care professionals. *J. Endod.* **2006**, *32*, 399–403. [CrossRef] [PubMed]
4. Imura, N.; Pinheiro, E.T.; Gomes, B.P.; Zaia, A.A.; Ferraz, C.C.; Souza-Filho, F.J. The outcome of endodontic treatment: A retrospective study of 2000 cases performed by a specialist. *J. Endod.* **2007**, *33*, 1278–1282. [CrossRef] [PubMed]
5. Ng, Y.L.; Mann, V.; Gulabivala, K. Outcome of secondary root canal treatment: A systematic review of the literature. *Int. Endod. J.* **2008**, *41*, 1026–1046. [CrossRef] [PubMed]
6. Ng, Y.L.; Mann, V.; Rahbaran, S.; Lewsey, J.; Gulabivala, K. Outcome of primary root canal treatment: Systematic review of the literature. Part 2. Influence of clinical factors. *Int. Endod. J.* **2008**, *41*, 6–31. [CrossRef] [PubMed]
7. Santos, J.M.; Palma, P.J.; Ramos, J.C.; Cabrita, A.S.; Friedman, S. Periapical inflammation subsequent to coronal inoculation of dog teeth root filled with resilon/epiphany in 1 or 2 treatment sessions with chlorhexidine medication. *J. Endod.* **2014**, *40*, 837–841. [CrossRef] [PubMed]
8. Emmanuel, S.; Shantaram, K.; Sushil, K.C.; Manoj, L. An in vitro evaluation and comparision of apical sealing ability of three different obturation technique-lateral condensation, Obtura II, and ThermaFil. *J. Int. Oral Health* **2013**, *5*, 35–43.
9. Sarkar, N.K.; Caicedo, R.; Ritwik, P.; Moiseyeva, R.; Kawashima, I. Physicochemical basis of the biologic properties of mineral trioxide aggregate. *J. Endod.* **2005**, *31*, 97–100. [CrossRef]
10. Dammaschke, T.; Gerth, H.U.; Züchner, H.; Schäfer, E. Chemical and physical surface and bulk material characterization of white ProRoot MTA and two Portland cements. *Dent. Mater.* **2005**, *21*, 731–738. [CrossRef]
11. Zhang, H.; Shen, Y.; Ruse, N.D.; Haapasalo, M. Antibacterial activity of endodontic sealers by modified direct contact test against *Enterococcus faecalis*. *J. Endod.* **2009**, *35*, 1051–1055. [CrossRef] [PubMed]
12. Zhou, H.M.; Shen, Y.; Zheng, W.; Li, L.; Zheng, Y.F.; Haapasalo, M. Physical properties of 5 root canal sealers. *J. Endod.* **2013**, *39*, 1281–1286. [CrossRef] [PubMed]

13. Candeiro, G.T.; Correia, F.C.; Duarte, M.A.; Ribeiro-Siqueira, D.C.; Gavini, G. Evaluation of radiopacity, pH, release of calcium ions, and flow of a bioceramic root canal sealer. *J. Endod.* **2012**, *38*, 842–845. [CrossRef] [PubMed]

14. Bozeman, T.B.; Lemon, R.R.; Eleazer, P.D. Elemental analysis of crystal precipitate from gray and white MTA. *J. Endod.* **2006**, *32*, 425–428. [CrossRef]

15. Reyes-Carmona, J.F.; Felippe, M.S.; Felippe, W.T. Biomineralization ability and interaction of mineral trioxide aggregate and white Portland cement with dentin in a phosphate-containing fluid. *J. Endod.* **2009**, *35*, 731–736. [CrossRef]

16. Tay, F.R.; Pashley, D.H.; Rueggeberg, F.A.; Loushine, R.J.; Weller, R.N. Calcium phosphate phase transformation produced by the interaction of the Portland cement component of white mineral trioxide aggregate with a phosphate-containing fluid. *J. Endod.* **2007**, *33*, 1347–1351. [CrossRef]

17. Bramante, C.M.; Demarchi, A.C.; de MOraes, I.G.; Bernadineli, N.; Garcia, R.B.; Spangberg, L.S.; Duarte, M.A. Presence of arsenic in different types of MTA and white and gray Portland cement. *Oral. Surg. Oral Med. Oral Pathol. Oral Radiol. Endod.* **2008**, *106*, 909–913. [CrossRef]

18. Torabinejad, M.; Chivian, N. Clinical applications of mineral trioxide aggregate. *J. Endod.* **1999**, *25*, 197–205. [CrossRef]

19. Williams, D.F. *The Williams Dictionary of Biomaterials*; Liverpool University Press: Liverpool, UK, 1999.

20. Dubok, V.A. Bioceramics—Yesterday, Today, Tomorrow. *Powder Metall. Met. Ceram.* **2000**, *39*, 381–394. [CrossRef]

21. Hulbert, S.F.; Hench, L.L.; Wilson, J. *An Introduction to Bioceramics*; World Scientific: Singapore, 1993; pp. 25–40.

22. Hench, L.L. Bioceramics: From Concept to Clinic. *J. Am. Ceram. Soc.* **1991**, *74*, 1487–1510. [CrossRef]

23. Hench, L.L. Genetic design of bioactive glass. *J. Eur. Ceram. Soc.* **2009**, *29*, 1257–1265. [CrossRef]

24. Jones, J.R.; Brauer, D.S.; Hupa, L.; Greenspan, D.C. Bioglass and bioactive glasses and their impact on healthcare. *Int. J. Appl. Glass Sci.* **2016**, *7*, 423–434. [CrossRef]

25. Hench, L.L. The story of Bioglass®. *J. Mater. Sci.* **2006**, *17*, 967–978. [CrossRef] [PubMed]

26. Jell, G.; Stevens, M.M. Gene activation by bioactive glasses. *J. Mater. Sci. Mater. Med.* **2006**, *17*, 997–1002. [CrossRef]

27. Xynos, I.D.; Edgar, A.J.; Buttery, L.D.K.; Hench, L.L.; Polak, J.M. Gene expression profiling of human osteoblasts following treatment with the ionic products of Bioglass 45S5 dissolution. *J. Biomed. Mater. Res.* **2001**, *55*, 151–157. [CrossRef]

28. Hoeland, W.; Vogel, W.; Waumann, K.; Gummel, J. Interface reactions between machinable bioactive glass-ceramics and bone. *J. Biomed. Mater. Res.* **1985**, *19*, 303–312. [CrossRef]

29. Hench, L.L.; Polack, J.M. Third-generation biomedical materials. *Science* **2002**, *295*, 1014–1017. [CrossRef]

30. Baino, F.; Novajra, G.; Miguez-Pacheco, V.; Boccaccini, A.R.; Vitale-Brovarone, C. Bioactive glasses: Special applications outside the skeletal system. *J. Non-Cryst. Solids* **2016**, *432*, 15–30. [CrossRef]

31. Kokubo, T.; Takadama, H. How useful is SBF in predicting in vivo bone bioactivity? *Biomaterials* **2006**, *27*, 2907–2915. [CrossRef]

32. Gilchrist, T.; Glasby, M.; Healy, D.; Kelly, G.; Lenihan, D.; McDowall, K.; Miller, I.; Myles, L. In vitro nerve repair-In vivo. The reconstruction of peripheral nerves by entubulation with biodegradable glass tubes—A preliminary report. *Br. J. Plast. Surg.* **1998**, *51*, 231–237. [CrossRef]

33. Oguntebi, B.; Clark, A.; Wilsin, J. Pulp capping with bioglass and autologous demineralized dentin in miniature swine. *J. Dent. Res.* **1993**, *72*, 484–489. [CrossRef] [PubMed]

34. Damas, B.A.; Wheater, M.A.; Bringas, J.S.; Hoen, M.M. Cytotoxicity comparison of mineral trioxide aggregates and EndoSequence bioceramic root repair materials. *J. Endod.* **2011**, *37*, 372–375. [CrossRef] [PubMed]

35. Leal, F.; De-Deus, G.; Brandão, C.; Luna, A.S.; Fidel, S.R.; Souza, E.M. Comparison of the root-end seal provided by bioceramic repair cements and White MTA. *Int. Endod. J.* **2011**, *44*, 662–668. [CrossRef] [PubMed]

36. Koch, K.; Brave, D. Bioceramic technology: The game changer in endodontics. *Endod. Pract.* **2009**, *2*, 17–21.

37. Hess, D.; Solomon, E.; Spears, R.; He, J. Retreatability of a bioceramic root canal sealing material. *J. Endod.* **2011**, *37*, 1547–1549. [CrossRef] [PubMed]

38. Loushine, B.A.; Bryan, T.E.; Looney, S.W.; Gillen, B.M.; Loushine, R.J.; Weller, R.N.; Pashley, D.H.; Tay, F.R. Setting properties and cytotoxicity evaluation of a premixed bioceramic root canal sealer. *J. Endod.* **2011**, *37*, 673–677. [CrossRef]

39. Massi, S.; Tanomaru-Filho, M.; Silva, G.F.; Duarte, M.A.; Grizzo, L.T.; Buzalaf, M.A.; Guerreiro-Tanomaru, J.M. pH, calcium ion release, and setting time of an experimental mineral trioxide aggregate-based root canal sealer. *J. Endod.* **2011**, *37*, 844–846. [CrossRef]

40. Walker, M.P.; Diliberto, A.; Lee, C. Effect of setting conditions on mineral trioxide aggregate flexural strength. *J. Endod.* **2006**, *32*, 334–336. [CrossRef]

41. Chen, L.; Suh, B.I. Cytotoxicity and biocompatibility of resin-free and resin modified direct pulp capping materials: A state-of-the-art review. *Dent. Mater. J.* **2017**, *36*, 1–7. [CrossRef]

42. Jafari, F.; Aghazadeh, M.; Jafari, S.; Khaki, F.; Kabiri, F. In vitro Cytotoxicity Comparison of MTA Fillapex, AH-26 and Apatite Root Canal Sealer at Different Setting Times. *Iran. Endod. J.* **2017**, *12*, 162–167.

43. Silva, E.J.; Rosa, T.P.; Herrera, D.R.; Jacinto, R.C.; Gomes, B.P.; Zaia, A.A. Evaluation of cytotoxicity and physicochemical properties of calcium silicate-based endodontic sealer MTA Fillapex. *J. Endod.* **2013**, *39*, 274–277. [CrossRef] [PubMed]

44. Oltra, E.; Cox, T.C.; LaCourse, M.R.; Johnson, J.D.; Paranjpe, A. Retreatability of two endodontic sealers, EndoSequence BC Sealer and AH Plus: A micro-computed tomographic comparison. *Restor. Dent. Endod.* **2017**, *42*, 19–26. [CrossRef]

45. Gandolfi, M.G.; Siboni, F.; Prati, C. Properties of a novel polysiloxane-guttapercha calcium silicate-bioglass-containing root canal sealer. *Dent. Mater.* **2016**, *32*, 113–126. [CrossRef] [PubMed]

46. Akcay, M.; Arslan, H.; Durmus, N.; Mese, M.; Capar, I.D. Dentinal tubule penetration of AH Plus, iRoot SP, MTA Fillapex, and GuttaFlow Bioseal root canal sealers after different final irrigation procedures: A confocal microscopic study. *Lasers. Surg. Med.* **2016**, *48*, 70–76. [CrossRef]

47. Collado-Gonzalez, M.; Tomas-Catala, C.J.; Onate-Sanchez, R.E.; Moraleda, J.M.; Rodríguez-Lozano, F.J. Cytotoxicity of GuttaFlow Bioseal, GuttaFlow2, MTA Fillapex, and AH Plus on human periodontal ligament stem cells. *J. Endod.* **2017**, *43*, 816–822. [CrossRef] [PubMed]

48. Rodríguez-Lozano, F.J.; Collado-González, M.; Tomás-Catalá, C.J.; García-Bernal, D.; López, S.; Oñate-Sánchez, R.E.; Moraleda, J.M.; Murcia, L. GuttaFlow Bioseal promotes spontaneous differentiation of human periodontal ligament stem cells into cementoblast-like cells. *Dent. Mater.* **2019**, *35*, 114–124. [CrossRef]

49. Kokubo, T.; Kushitani, H.; Sakka, S.; Kitsugi, T.; Yamamuro, T. Solutions able to reproduce in vivo surface-structure changes in bioactive glass-ceramic A-W. *J. Biomed. Mater. Res.* **1990**, *24*, 721–734. [CrossRef]

50. Kokubo, T.; Ito, S.; Huang, Z.T.; Hayashi, T.; Sakka, S.; Kitsugi, T.; Yamamuro, T. Ca, P-rich layer formed on high-strength bioactive glass-ceramic A-W. *J. Biomed. Mater. Res.* **1990**, *24*, 331–343. [CrossRef]

51. Washio, A.; Nakagawa, A.; Nishihara, T.; Maeda, H.; Kitamura, C. Physicochemical properties of newly developed bioactive glass cement and its effects on various cells. *J. Biomed. Mater. Res. B Appl. Biomater.* **2015**, *103*, 373–380. [CrossRef]

52. Orstavik, D. Materials used for root canal obturation: Technical, biological and clinical testing. *Endod. Top.* **2005**, *12*, 25–38. [CrossRef]

53. Al-Haddad, A.; Ab Aziz, C.; Zeti, A. Bioceramic-Based Root Canal Sealers: A Review. *Int. J. Biomater.* **2016**. [CrossRef] [PubMed]

54. Williams, D.F. *Definitions in Biomaterials: Proceedings of a Consensus Conference of the European Society for Biomaterials, Chester, UK, 3–5 March 1986*; Elsevier: Amsterdam, The Netherlands; New York, NY, USA, 1987; Volume 4.

55. Schmalz, G. Use of cell cultures for toxicity testing of dental materials—Advantages and limitations. *J. Dent.* **1994**, *22*, S6–S11. [CrossRef]

56. Salles, L.P.; Gomes-Cornélio, A.L.; Guimarães, F.C.; Herrera, B.S.; Bao, S.N.; Rossa-Junior, C. Guerreiro-Tanomaru, J.M.; Tanomaru-Filho, M. Mineral trioxide aggregate-based endodontic sealer stimulates hydroxyapatite nucleation in human osteoblast-like cell culture. *J. Endod.* **2012**, *38*, 971–976. [CrossRef]

57. Jung, S.; Sielker, S.; Hanisch, M.R.; Libricht, V.; Schäfer, E.; Dammaschke, T. Cytotoxic effects of four different root canal sealers on human osteoblasts. *PLoS ONE* **2018**, *13*, e0194467. [CrossRef] [PubMed]

58. Bae, W.J.; Chang, S.W.; Lee, S.I.; Kum, K.Y.; Bae, K.S.; Kim, E.C. Human periodontal ligament cell response to a newly developed calcium phosphate-based root canal sealer. *J. Endod.* **2010**, *36*, 1658–1663. [CrossRef]

59. Collado-González, M.; García-Bernal, D.; Oñate-Sánchez, R.E.; Ortolani-Seltenerich, P.S.; Lozano, A.; Forner, L.; Llena, C.; Rodríguez-Lozano, F.J. Biocompatibility of three new calcium silicate-based endodontic sealers on human periodontal ligament stem cells. *Int. Endod. J.* **2017**, *50*, 875–884. [CrossRef]

60. Lee, J.K.; Kim, S.; Lee, S.; Kim, H.C.; Kim, E. In Vitro Comparison of Biocompatibility of Calcium Silicate-Based Root Canal Sealers. *Materials* **2019**, *12*, 2411. [CrossRef]

61. Washio, A.; Yoshii, S.; Morotomi, T.; Maeda, H.; Kitamura, C. Effects of bioactive glass based sealer on cell migration ability and viability of periodontal ligament cells and osteoblast-like cells. *Jpn. J. Conserv. Dent.* **2017**, *60*, 96–104.

62. Washio, A.; Kitamura, C. The next-generation bioacvite glass-based root canal sealer inducing the ideal wound healing environment of periapical tissue ~Nishika Canal Sealer BG~. *Dent. Diam.* **2017**, *42*, 178–183.

63. Morotomi, T.; Hanada, K.; Washio, A.; Yoshii, S.; Matsuo, K.; Kitamura, C. Effect of newly-developed bioactive glass root canal sealer on periapical tissue of rat's molar. *Jpn. J. Conserv. Dent.* **2017**, *60*, 120–127.

64. Wilcox, L.R. Endodontic retreatment: Ultrasonics and chloroformas the final step in reinstrumentation. *J. Endod.* **1989**, *15*, 125–128. [CrossRef]

65. Schirrmeister, J.F.; Wrbas, K.T.; Meyer, K.M.; Altenburger, M.J.; Hellwig, E. Efficacy of different rotary instruments for gutta-percha removal in root canal retreatment. *J. Endod.* **2006**, *32*, 469–472. [CrossRef] [PubMed]

66. Gomes-Filho, J.E.; Watanabe, S.; Cintra, L.T.; Nery, M.J.; Dezan-Júnior, E.; Queiroz, I.O.; Lodi, C.S.; Basso, M.D. Effect of MTA-based sealer on the healing of periapical lesions. *J. Appl. Oral Sci.* **2013**, *21*, 235–242. [CrossRef] [PubMed]

67. Schwartz, R.S. Adhesive dentistry and endodontics. Part 2: Bonding in the root canal system—The promise and the problems: A review. *J. Endod.* **2016**, *32*, 1125–1134. [CrossRef]

68. Nagas, E.; Altundasar, E.; Serper, A. The effect of master point taper on bond strength and apical sealing ability of different root canal sealers. *Oral. Surg. Oral Med. Oral Pathol. Oral Radiol. Endod.* **2009**, *107*, 61–64. [CrossRef]

69. Lin, Z.; Ling, J.; Fang, J.; Liu, F.; He, J. Physicochemical properties, sealing ability, bond strength and cytotoxicity of a new dimethacrylate-based root canal sealer. *J. Formos. Med. Assoc.* **2010**, *109*, 819–827. [CrossRef]

70. Fang, J.; Mai, S.; Ling, J.; Lin, Z.; Huang, X. In vitro evaluation of bond strength and sealing ability of a new low-shrinkage, methacrylate resin-based root canal sealer. *J. Formos. Med. Assoc.* **2012**, *111*, 340–346. [CrossRef]

71. Yoshii, S.; Washio, A.; Morotomi, T.; Kitamura, C. Root canal sealing ability of bioactive glass-based sealer and its effects on dentin. *Jpn. J. Conserv. Dent.* **2016**, *59*, 463–471.

72. Wilcox, L.R.; Krell, K.V.; Madison, S.; Rittman, B. Endodontic retreatment: Evaluation of gutta-percha and sealer removal and canal reinstrumentation. *J. Endod.* **1987**, *9*, 453–457. [CrossRef]

73. Reddy, S.; Neelakantan, P.; Saghiri, M.A.; Lotfi, M.; Subbarao, C.V.; Garcia-Godoy, F.; Gutmann, J.L. Removal of gutta-percha/zinc-oxide-eugenol sealer or gutta-percha/epoxy resin sealer from severely curved canals: An in vitro study. *Int. J. Dent.* **2011**, *2011*, 541831. [CrossRef]

74. Uzunoglu, E.; Yilmaz, Z.; Sungur, D.D.; Altundasar, E. Retreatability of Root Canals Obturated Using Gutta-Percha with Bioceramic, MTA and Resin-Based Sealers. *Iran. Endod. J.* **2015**, *10*, 93–98.

75. Washio, A.; Yoshii, S.; Morotomi, T.; Kitamura, C. Evaluation of removability of root canals filled using bioactive glass based sealer. *Jpn. J. Conserv. Dent.* **2017**, *60*, 14–21.

76. Endodontics: Colleagues for Excellence. Canal preparation and obturation: An updated view of the two pillars of nonsurgical endodontics. In *American Association of Endodontists*; Dental Professional Community: Chicago, IL, USA, 2016; pp. 1–8.

77. Treatment standards. In *American Association of Endodontists*; Dental Professional Community: Chicago, IL, USA, 2018; pp. 1–20.

78. Michelle, L.; Johnathon, W.; Gary, H.; Jeffrey, S.; Rufus, C. Current Trends in Endodontic Practice: Emergency Treatments and Technological Armamentarium. *J. Endod.* **2009**, *35*, 35–39.

79. Gina, M.S.; Wael, S.; Christine, M.S.; Brian, W. Current Trends in Endodontic Treatment by General Dental Practitioners: Report of a United States National Survey. *J. Endod.* **2014**, *40*, 618–624.

80. Raoof, M.; Zeini, N.; Haghani, J.; Sadr, S.; Mohammadalizadeh, S. Preferred materials and methods employed for endodontic treatment by Iranian general practitioners. *Iran. Endod. J.* **2015**, *10*, 112–116.

81. Al-Omari, W.M. Survey of attitudes, materials and methods employed in endodontic treatment by general dental practitioners in North Jordan. *BMC Oral Health* **2004**, *4*, 1. [CrossRef]

82. Gupta, R.; Rai, R. The adoption of new endodontic technology by Indian dental practitioners: A questionnaire survey. *J. Clin. Diagn. Res.* **2013**, *7*, 2610–2614. [CrossRef]

83. Germain, S.; Meetu, K.; Issam, K.; Alfred, N.; Carla, Z. Impact of the Root Canal Taper on the Apical Adaptability of Sealers used in a Single-cone Technique: A Micro-Computed Tomography Study. *J. Contemp. Dent. Pract.* **2018**, *19*, 808–815.

84. Chybowski, E.A.; Glickman, G.N.; Patel, Y.; Fleury, A.; Solomon, E.; He, J. Clinical Outcome of Non-Surgical Root Canal Treatment Using a Single-cone Technique with EndoSequence Bioceramic Sealer: A Retrospective Analysis. *J. Endod.* **2018**, *44*, 941–945. [CrossRef]

Article

Human Mesenchymal Stem Cell Response to Lactoferrin-based Composite Coatings

Madalina Icriverzi [1,2], Anca Bonciu [3,4], Laurentiu Rusen [3], Livia Elena Sima [1], Simona Brajnicov [3], Anisoara Cimpean [2], Robert W. Evans [5], Valentina Dinca [3,*] and Anca Roseanu [1,*]

[1]   Institute of Biochemistry of the Romanian Academy, 060031 Bucharest, Romania; radu_mada@yahoo.co.uk (M.I.); livia_e_sima@yahoo.com (L.E.S.)
[2]   Department of Biochemistry and Molecular Biology, University of Bucharest, Faculty of Biology, 91–95 Splaiul Independentei, 050095 Bucharest, Romania; anisoara.cimpean@bio.unibuc.ro
[3]   National Institute for Laser, Plasma and Radiation Physics, 409 Atomistilor, 077125 Magurele, Romania; anca.bonciu@inflpr.ro (A.B.); laurentiu.rusen@inflpr.ro (L.R.); brajnicov.simona@inflpr.ro (S.B.)
[4]   Faculty of Physics, University of Bucharest, RO-077125 Magurele, Romania
[5]   School of Engineering and Design, Brunel University, London UB8 3PH, UK; robertwevans49@gmail.com
*   Correspondence: valentina.dinca@inflpr.ro (V.D.); roseanu@biochim.ro (A.R.)

Received: 27 August 2019; Accepted: 16 October 2019; Published: 18 October 2019

Abstract: The potential of mesenchymal stem cells (MSCs) for implantology and cell-based therapy represents one of the major ongoing research subjects within the last decades. In bone regeneration applications, the various environmental factors including bioactive compounds such as growth factors, chemicals and physical characteristics of biointerfaces are the key factors in controlling and regulating osteogenic differentiation from MSCs. In our study, we have investigated the influence of Lactoferrin (Lf) and Hydroxyapatite (HA) embedded within a biodegradable PEG-PCL copolymer on the osteogenic fate of MSCs, previous studies revealing an anti-inflammatory potential of the coating and osteogenic differentiation of murine pre-osteoblast cells. The copolymer matrix was obtained by the Matrix Assisted Pulsed Laser Evaporation technique (MAPLE) and the composite layers containing the bioactive compounds (Lf, HA, and Lf-HA) were characterised by Scanning Electron Microscopy and Atomic Force Microscopy. Energy-dispersive X-ray spectroscopy contact angle and surface energy of the analysed coatings were also measured. The characteristics of the composite surfaces were correlated with the viability, proliferation, and morphology of human MSCs (hMSCs) cultured on the developed coatings. All surfaces were found not to exhibit toxicity, as confirmed by the LIVE/DEAD assay. The Lf-HA composite exhibited an increase in osteogenic differentiation of hMSCs, results supported by alkaline phosphatase and mineralisation assays. This is the first report of the capacity of biodegradable composite layers containing Lf to induce osteogenic differentiation from hMSCs, a property revealing its potential for application in bone regeneration.

Keywords: mesenchymal stem cells; osteogenic differentiation; lactoferrin; polymer composite

---

1. Introduction

The behaviour of cells mediated by bioresponsive substrates and interfaces represents an important and ever-growing area in tissue engineering. There is an interest in finding innovative strategies for enhancing the efficacy of the biomedical devices used in the treatment of bone diseases such as osteoporosis.

The approaches used are based on either injecting bioactive compounds (i.e., natural protein) in a direct manner or attaching them to a biomaterial surface. The advantage of specific functionalising and addressing/controlling the locally attached proteins over the injecting approach is related to avoiding a rapid clearance of the interest biocompound from the body and side effects.

Lactoferrin (Lf) is a multifunctional protein that, apart from its antimicrobial, anti-inflammatory and anti-tumoral effects [1–5], plays a positive role in modulating osteoblast differentiation and inhibits osteoclastogenesis [6–9]. Different formulations or constructs containing Lf were designed and tested in vitro and in vivo for bone tissue regeneration enhancement. These included Lf incorporation into the collagen membrane [10], hydrogels [11], nanofiber loading [12], or into microspheres [13], or coupled with compounds [14], in order to improve the effectiveness of biomaterials used in bone regeneration. All these data suggest that the Lf-based coatings could be a useful strategy for controlling aspects related to inflammatory or osteogenic responses.

Hydroxyapatite (HA), which is highly biocompatible, osteoconductive and biodegradable and facilitates binding in a functional way to biomolecules, is already an established material for biomedical devices related to bone tissue engineering [15–17]. Recent studies demonstrated an effect of Lf-HA nanocrystals on mesenchymal stem cells (MSCs) and pre-osteoblast cells revealing the improved response of the combination of Lf and HA on osteogenic differentiation and an inhibition of osteoclast activity [18,19].

Our previous work demonstrated the synergetic effect of coupling Lf with either antitumoral drugs or osteogenic HA within a biodegradable polymeric matrix [20–22]. The hybrid coating of the PEG-PCL-Me-HA-Lf promoted matrix mineralization and osteogenic differentiation of the murine osteoblast cell model in osteoinductive conditions [21] and exhibited low levels of pro-inflammatory TNF-α, high levels of anti-inflammatory IL-10 cytokine, and increased polarization of human THP-1 macrophage cells towards M2 pro-reparative phenotype [22].

The need to use a biodegradable polymeric matrix arises from the long-term biological assays and use, as well as from the necessity to have a controllable and uniform distribution of the biocompounds as coating on the medical device. The future of implant surfaces lies in the design and development of surfaces that interact in a specific way to promote desired processes and minimise detrimental side effects. Biocompatible PEG-PCL copolymers are widely used as controlled drug delivery systems in different therapies [23] and biomedical applications [24,25]. Promising studies on PEG-PCL copolymer surfaces have demonstrated the ability of these materials to maintain the viability and functionality of human MSCs (hMSCs) and to induce morphology changes adapted to the structure and composition of the copolymer for better interaction with the surface [26–29].

Matrix Assisted Pulsed Laser Evaporation (MAPLE) is an efficient method to obtain multifunctional single or multiple compound biocoatings-polymers, proteins, graphene, nanoparticles and even larger compounds, such as microspheres, bacteria [20–22,30–33].

We recently demonstrated the use of MAPLE as a single-step method for embedding multiple bioactive factors into a biodegradable synthetic polymeric coating without losing the functionality of proteins or drugs [20,22]. It was shown that by entrapping the osteoconductive factor HA and low quantities of Lf within a biodegradable copolymer matrix, both the inflammatory response and osteoblasts' response were dictated by the coating composition and characteristics [20,22].

Given the potential and importance of hMSCs for implantology and cell-based therapy, an understanding of how the characteristics of biointerfaces can be used in controlling and regulating osteogenic differentiation from MSCs is essential. An assessment of results across our previous work has demonstrated that the characteristics of biomimetic interfaces can be used to instigate a specific tissue response. Therefore, we have investigated in this work the effect of Lf and HA embedded within a biodegradable copolymeric matrix on the osteogenic fate of MSCs.

2. Materials and Methods

2.1. Materials and Solution for Target Preparation

The materials used to obtain the single component and the lactoferrin-based composite coatings were all purchased from Sigma Aldrich: Lf (L0520 SIGMA, Aldrich, Saint Louis, MO, USA), HA nanoparticle powder (677418 Aldrich, Aldrich, Saint Louis, MO, USA), and Poly (ethylene glycol)-*block*-poly (ε-caprolactone) methyl ether (570303 Aldrich, Saint Louis, MO, USA) (PEG-block-PCL Me-average Mn~5000, PCL average Mn~5000). Double distilled water was used to prepare Lf (2%), and HA nanoparticles (1%). PEG-PCL Me solution (1%) was prepared in chloroform.

2.2. Coatings Deposition

The single element and the composite coatings were obtained by the MAPLE method using a single, double and triple module target system and a "Surelite II" pulsed Nd: YAG laser system (Continuum Company, Pessac, France) (266 nm, 6 ns pulse duration, 10 Hz repetition rate) as previously described [20,21] The modular target consisted of frozen solutions of PEG-block-PCL Me copolymer (Co), Lf, HA. In the case of single-component coatings, the solutions were individually prepared and maintained frozen in a copper container using liquid nitrogen. For the composite layers, a modular target system with one or two Teflon concentric rings, depending on the number of components, was used and the materials were frozen separately. The Teflon rings are removed after freezing to avoid the interaction of the laser beam and the exposed Teflon ring. The laser beam was scanned over the target surface and rotated to prevent overheating and drilling due to the laser irradiation. In this way, the laser beam energy is mainly absorbed by the frozen solvent, leading to the vaporization and transfer of the target molecules towards and onto the glass placed parallel, at a distance of 3 cm from the target in the vacuum chamber.

The Ca/P ratio reported for bone is within 1–2 range, depending on the type of bone, age, etc. For the chosen deposition parameters (450 mJ/cm^2, 0.01 cm^2 laser spot size, 120 kpulses for Lf (100 µg) 60 kpulses for PEG-PCL-Me and 60 kpulses for HA (134 µg) and according to EDAX measurements, the calculated Ca/P ratio was within the range of 1.3–1.84. The value is close to that previously reported by us for HA-based composite coatings. [20,22]. PEG-PCL-Me assures a gradual degrading period and the number of pulses for Co was chosen to provide a full coverage and/or an entrapping matrix for HA, LF or both.

2.3. Surface Characterization

Atomic Force Microscopy (AFM) (XE 100 AFM, Park Systems company, Suwon, Korea) measurements were performed in non-contact mode. Samples were sputter-coated with gold and observed by the FEI Inspect-S scanning electron microscope at an accelerating voltage between 5 and 20 kV in order to analyse the topography of the samples. The elements analysis of the coatings was conducted by the same Scanning Electron Microscopy (SEM) instrument equipped with energy-dispersive X-ray spectroscopy (EDX) system.

The contact angle measurements were performed by the sessile drop method using an optical measuring system (CAM101, KSV, Biolin Surface, Finland) with deionised water. Three drops of the liquid (9 µL) were examined on each substratum, and the contact angle was measured 3 s after the positioning of the drop.

Surface free energy (SFE) was determined based on the contact angle measurement of two wetting agents: water and di-iodomethane. This calculation was conducted using the concept of polar and dispersion components using the Owens, Wendt, Rabel, and Kaelble (OWRK) method for calculation [34–36].

2.4. Cell Culture

Biological studies were performed using human mesenchymal stem cells (hMSCs) obtained as previously described [37]. Bone marrow was harvested from one healthy patient undergoing surgery for the orthopaedic implant procedure, with the approval of the Ethics Committee of the University of Medicine and Pharmacy of Craiova (reference No. 68/11.07.2016). The phenotype of hMSC and their osteogenic capacity are presented in Supplementary data (Figures S1 and S2). Cells were cultured in expansion medium-low glucose (1 g/L D-glucose) DMEM + GlutaMax medium (Dulbecco's Minimal Essential Medium), supplemented with 10% (*v/v*) foetal bovine serum (FBS) and 1% (*v/v*) streptomycin/penicillin (all from Gibco) and kept at 37 °C with 5% CO_2. For osteogenic differentiation, hMSCs were cultured in osteoinductive conditions: α-MEM medium (Biochrom AG) supplemented with 82 µg/mL ascorbic acid, 100 nM dexamethasone (Sigma-Aldrich, Saint Louis, MO, USA), and 10 mM β-glycerophosphate (Calbiochem). The medium was changed twice a week during 28 days of culture.

2.5. Cell Proliferation Assay

Before in vitro assays, all samples were sterilised by immersion in 1% Penicillin-Streptomycin solution for 15 min.

Cell viability was assessed using the MTS method (CellTiter 96® Aqueous Non-Radioactive Cell Proliferation Assay, Promega, Fitchburg, WI, USA). hMSCs were cultured on surfaces in a 24-well plate (Costar flat bottom with lid, tissue culture treated) at a density of 5×10^3 cells/cm^2 in DMEM medium. After 72 h of incubation, the supernatant was removed and replaced with 360 µL of MTS solution (tetrazolium compound in cell culture media) for each well. The plate was incubated at 37 °C, in a humidified atmosphere of 5% CO_2 in the dark, for 60 min. After the incubation period, 100 µL of the culture solution was transferred to a 96-well clear bottom plate (Nunc, Thermo Scientific) and optical density at 450 nm was measured by a microplate reader (Mithras LB 940 DLReady, Berthold Technologies GmbH &Co. KG, Wildbad, Germany). The absorbance increases in proportion with cell density.

2.6. Viability/Cytotoxicity Cell Imaging Assay

The viability of hMSCs after 72 h of cultivation on the surfaces was evaluated using a LIVE/DEAD® Viability/Cytotoxicity kit (Molecular Probes, Eugene, OR). The cells (5×10^3 cells/cm^2) were incubated for 30 min. at 37 °C with 10 µM of Calcein AM and 4 µM Ethidium homodimer-1 mixture in DMEM medium and then fixed with 4% paraformaldehyde (PFA) for 15 min. Live-cell control is represented by hMSCs plated on the coverslip. Dead-cell control was obtained by the treatment of MSC cells with 70% ethanol for 5 min. ProLong Gold antifade reagent was used to mount the samples which were immediately examined with a Zeiss Axiocam ERc5s with ApoTome.2 slider module using 10× lens, equipped with AxioCam MRm, camera. The images were captured with the AxioVision Rel 4.8 program

2.7. Microscopic Evaluation of Cell Adhesion and Morphology

The effect of coatings on the adhesion and morphology of hMSCs was evaluated by fluorescence microscopy. Cells adhered on different surfaces were fixed with 4% p-formaldehyde for 15 min and permeabilised for 3 min at room temperature with 0.2% Triton-X-100. The samples were blocked for 1 h in 0.5% BSA in Phosphate-Buffered Saline (PBS) and then incubated with Alexa Fluor488-conjugated phalloidin (green) for actin filaments detection (1:50, A 12379 Invitrogen, Thermo Fisher Sci., CA, USA). Cell nuclei were labelled for 1 min at room temperature with Hoechst (blue, H 21492 Life Technologies, Molecular Probes, Eugene, OR, USA) at a dilution of 1:3000 in PBS. Before image acquisition, the specimens were mounted in ProlongGold Antifade Reagent (P 36934 Molecular Probes, Life Technologies, Eugene, OR, USA). Fluorescence images were acquired using a Zeiss Axiocam ERc5s Apotom microscope with a 20× lens. The images were analysed with the AxioVision Rel. 4.8 software (Zeiss).

For Scanning Electron Microscopy (SEM) studies, hMSCs seeded on analysed surfaces for 72 h, or 2 to 4 weeks with or without osteogenic factors were fixed with 2.5% glutaraldehyde in PBS for 20 min. The specimens were then subjected, as reported in [38], to dehydration and drying with 70%, 90%, and 100% ethanol solutions, for 15 min twice for each concentration and then 2 rounds of 3 min washing with 50% and 75% hexamethyldisilazane (HDMS, in ethanol) and finally with a 100% HDMS solution. The surfaces were air-dried overnight in a chemical Euroclone AURA 2000 M.A.C. fume hood.

2.8. Determination of Alkaline Phosphatase Activity

Alkaline phosphatise (ALP) activity of hMSCs cultured on different surfaces was determined using the Quantitative Alkaline Phosphatase ES Characterization Kit (SCR066 Merck, Millipore, Darmstadt, Germany). Briefly hMSCs at 14 days of incubation, with or without osteoinductive medium, were incubated with p-nitrophenylphosphate (pNPP) substrate for 20 min at room temperature. The method is based on the capacity of the cellular enzyme to hydrolyse p-NPP into phosphate and p-nitrophenol (yellow coloured). After the reaction is stopped, the absorbance of the coloured compound is measured at 405 nm with a spectrophotometer (Mithras, Berthold Technologies, Bad Wildbad, Germany). The amount of p-nitrophenol produced is proportional to that of ALP present within the reaction. The levels of ALP were extrapolated from a standard curve using recombinant ALP provided in the kit and expressed as ng/mL/sample.

2.9. Extracellular Matrix Mineralization Assessment

The samples were employed for detection and quantification of mineralisation occurred at 14 and 28 days after culturing hMSCs in the presence and absence of osteogenic factors on different surfaces. The cells were washed gently with PBS then fixed for 20 min with 4% PFA. After washing with distilled water and carefully removing the entire liquid, the samples were incubated with Alizarin Red S solution (ARS) (40 mM, pH 4.1–4.3) at room temperature for 10 min, repeatedly washed with PBS and microscopically visualised using a TissueFAXS imaging system (Tissue Gnostics, Vienna, Austria). For quantification, Alizarin red staining was extracted for 15 min at room temperature using a solution of 20% methanol and 10% acetic acid in water. Subsequently, the liquid was transferred to 96-well plates and the absorbance read at 450 nm using a Mithras LB940 Berthold spectrophotometer.

2.10. Statistical Analysis

Statistical analysis was performed with GraphPad Prism software using one-way ANOVA with Bonferroni's multiple comparison tests. Triplicate samples were used in all biological assays to ensure the reproducibility of the results. The data are presented as means ± SD (standard deviation). The *p* values < 0.05 were considered to be statistically significant.

3. Results and Discussions

3.1. Surface Characterisation of Composite Coatings

As surface characteristics are strongly correlated to cell response and behaviour, the Lf-based composite coatings characteristics were investigated in order to correlate those properties with the effect on the mesenchymal stem cells (MSCs). The energy-dispersive X-ray spectroscopy data show that the Ca/P ratio in HA containing samples was within 1.3–1.8, which is relatively close to the theoretical value in HA [39,40]. It was reported that HA with Ca/P ratio of 1.67 has good mechanical properties in terms of hardness and fracture toughness which make it a good candidate to serve as a standard for bone tissue regeneration [40]. The data presented in Figure 1A,B show that HA coatings have a Ca/P ratio of 1.61, suggesting a mineral composition very similar to that of bone [39]. The contents of Ca and P in the Lf based coatings were consistently higher than those of the copolymer, with the highest percentages of Ca and P being found for the Lf-HA coatings.

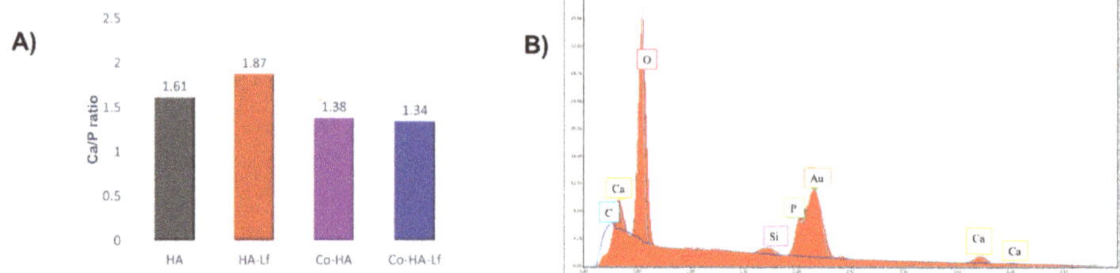

Figure 1. (**A**) Ca/P ratio of composite coatings obtained by Matrix Assisted Pulsed Laser Evaporation technique (MAPLE), (**B**) energy dispersive X-ray (EDX) spectra of hydroxyapatite (HA) coating obtained by MAPLE.

Furthermore, according to previous results obtained by FTIR spectroscopic evaluation, it was demonstrated that the functional groups (characteristic stretching and bending vibrations) of single element coatings, as well as a composite one, are maintained after the MAPLE process at 450 mJ/cm^2 [20,22].

The increase in the number of pulses for each component within the sample's composition resulted in roughness and hydrophilicity changes as compared to the previously reported samples. These changes in surface morphology and roughness observations are supported by the SEM and AFM results (Figures 2 and 3). For example, in the case of Co, the deposition of a large quantity of material led to its reorganisation on the substrate and wrinkle-like structures that led to an increased surface roughness (794 nm), while the HA, specifically HA-Lf based coatings, were similar to the previous ones [22], being characterised by HA nanoparticle accumulation, without forming cracks on its surface, but with an increased roughness, as shown in Figure 3.

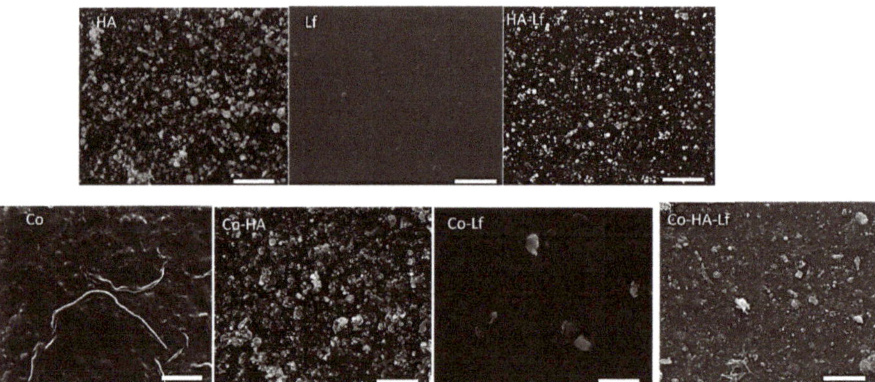

Figure 2. Scanning Electron Microscopy (SEM) images of the top morphology of the coatings obtained by MAPLE. Scale bar: 10 μm.

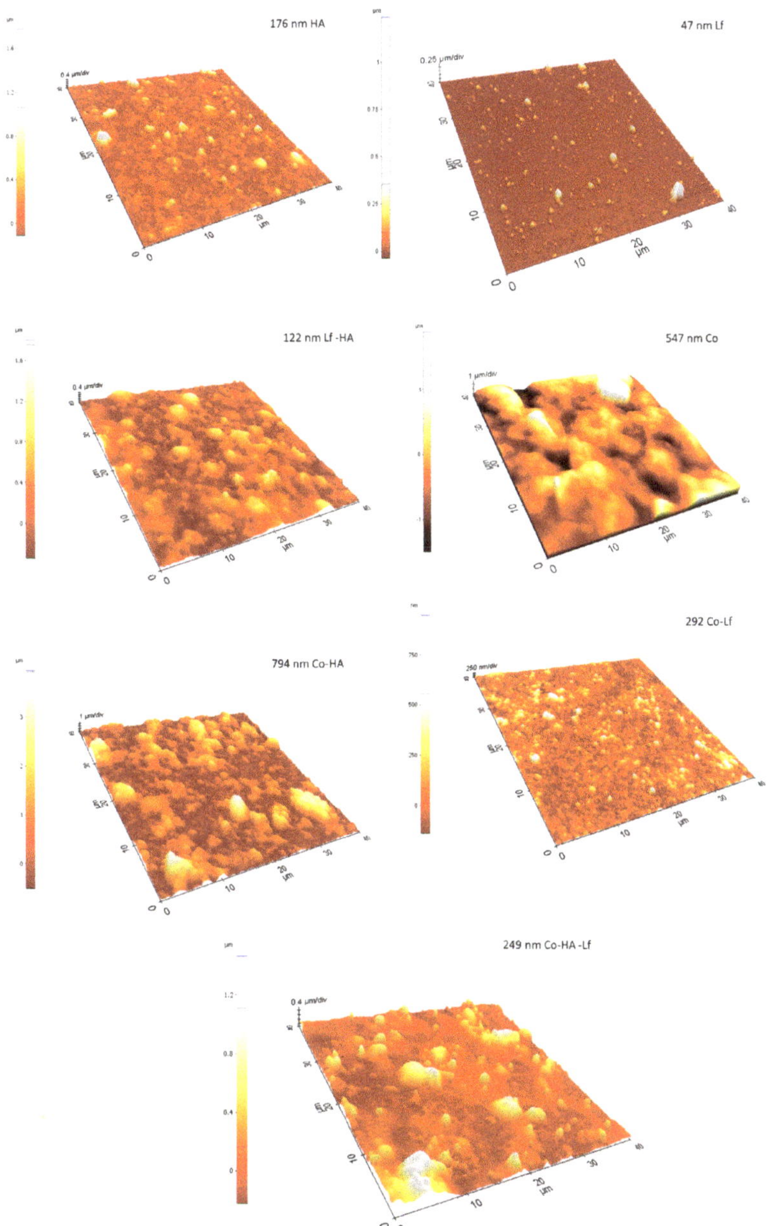

Figure 3. 3D Atomic Force Microscopy (AFM) images (40 µm × 40 µm) of single component and composite coatings obtained by MAPLE.

Therefore, the corresponding AFM images also show distinct features for the composite, with the surface root mean square roughness results (obtained from the AFM measurements) revealing increased roughness (47 nm Lf, 794 nm Co HA, 292 Co Lf,122 nm Lf HA, 249 nm Co Lf HA) (Figure 3), indicating clearly that the surface morphology and microstructure change depending on composition.

The surface charge and hydrophilicity of an implant have been known to influence osteointegration [41–43]. When the surface of the biomaterial comes in contact with a biological fluid, the molecules adsorbed create the conditions which will govern cell–surface interactions. One general observation is that protein adsorption is greater on hydrophobic surfaces compared to hydrophilic ones. In our case, the contact angle was significantly lower for Lf, HA-Lf and PEG-PCL-Me-HA (33.83° ± 2.04°, 45.26° ± 0.26° and 49.1° ± 0.65°) than for the HA, PEG-PCL-Me-Lf and PEG-PCL-Me-HA-Lf (64.74° ± 0.71°, 66.07° ± 0.24°, 57.85° ± 0.53°) respectively. A similar contact angle value for HA (62° ± 2°) was obtained by Siniscalco et al. [44]. It was shown by Li et al. that incorporation of PEG improves the hydrophilicity of the multiblock copolymers compared to the PCL homopolymer [45], which was also observed in our case, all of the samples showing hydrophilic character. The protein incorporation into the HA coatings led to a decrease of the contact angle (Figure 4) while its incorporation within the matrix of the copolymer did not induce major changes over the wettability of the composite coatings due to the ability of the PEG-PCL-Me to cover the nanoparticles just like a matrix (Figure 2). The variation in contact angle depending on the type of coatings can be related as well to the surface morphology, specifically roughness.

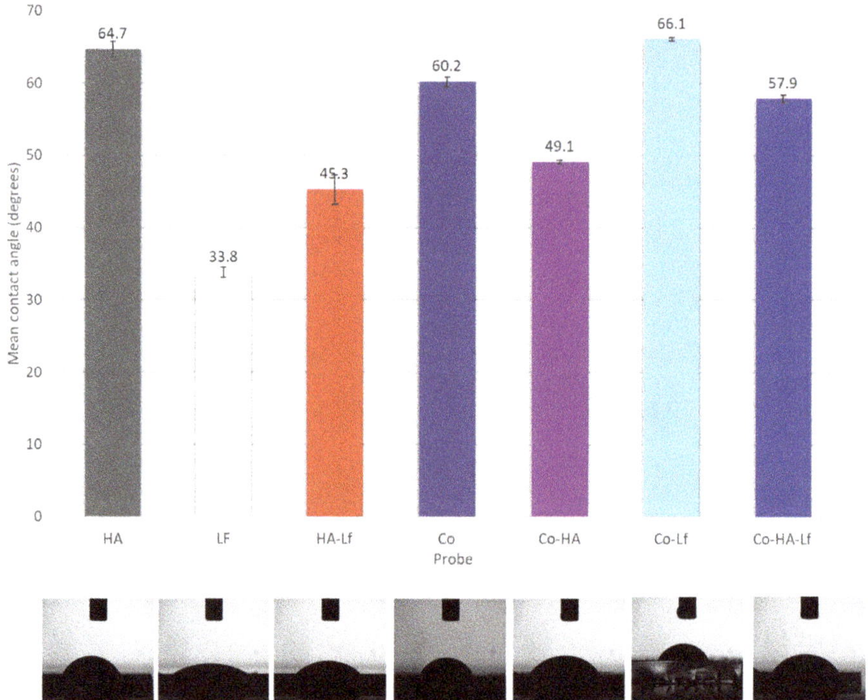

Figure 4. Histogram of contact angle measurements using water as a wetting agent.

Cell shape is strongly correlated with surface properties and generally increases in size with increases in hydrophilicity [46]. On hydrophilic surfaces, cells also show strong focal adhesions formation and stress fibre bundles within 3 h of plating. Therefore, MSCs on materials that permit cell spreading, would tend to adopt an osteogenic phenotype, while those whose spreading is limited would become adipocytes. Conversely, on hydrophobic surfaces, staining for actin is far more diffuse and vinculin staining for focal adhesions is lacking [47]. However, while cell attachment to a biomaterial surface is clearly important for good implant integration, the trend for improved cell behaviour with increasing adhesion is not perfect. Indeed, excessive adhesion may actually be detrimental. One report of high levels of MSC attachment on positively charged surfaces concomitantly showed reduced cell spreading and differentiation [48].

As previously mentioned, successful orthopaedic implant osteointegration relies on the quick and efficient formation of bone tissue at an implant surface. When biological fluids come in contact with an artificial material, water interactions, protein adsorption, and cell attachment are governed by the surface free energy of the material. Polymers are often considered to have low-energy surfaces due to their covalent and van der Waals bonding, therefore often leading to the surfaces being non-polar, and thus of a hydrophobic nature. As discussed above, cells generally adhere poorly to hydrophobic materials, and thus polymer surface modification is often necessary. In our case, using a laser evaporation technique, we have obtained for the copolymer a surface free energy of 57.64 mN/m with a polar component of 9.6 mN/m. Wei et al. showed that that hydrophilic surfaces strongly supported osteoblast attachment. [49]. The mean values of total surface free energy for HA (57.11 mN/m) is only slightly higher (54.6 mN/m) than the value obtained by Szczes et al. [50]. The total surface energy was equivalent on all the surfaces (Figure 5). The polar component was smaller than the dispersive one for all the coatings. Besides, Lf incorporation induced an increase of the polar component when compared with the original surfaces. Rapid hydration of the layers could facilitate the adhesion of biomolecules [51] and enhance bone apposition in the early healing phase [52,53]. Therefore, the hybrid coating layers with enhanced wettability produced in this study are expected to accelerate osteointegration.

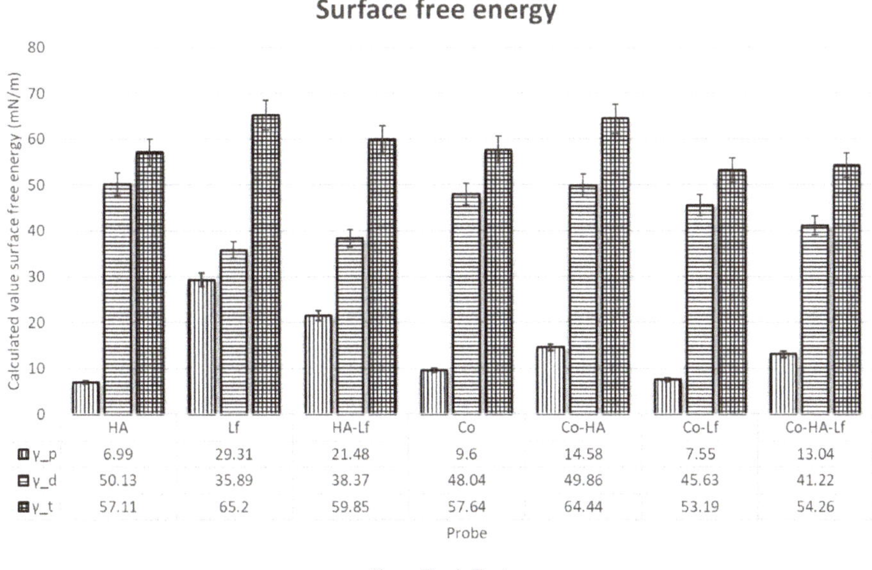

Probe	HA	Lf	HA-Lf	Co	Co-HA	Co-Lf	Co-HA-Lf
γ_p	6.99	29.31	21.48	9.6	14.58	7.55	13.04
γ_d	50.13	35.89	38.37	48.04	49.86	45.63	41.22
γ_t	57.11	65.2	59.85	57.64	64.44	53.19	54.26

Figure 5. Histogram presenting the total surface free energy, disperse and polar parts.

3.2. Lf Content in the Polymeric Coating Positively Affects the Cell Proliferation

The proliferation of hMSCs was determined after 72 h of culture in direct contact with the surfaces. The MTS assay revealed no toxicity of the surfaces and all samples supported attachment and proliferation of the cells. Cell proliferation after three days of culture was enhanced by addition of human lactoferrin (100 µg/surface) into the polymeric matrix, a significant increase ($p < 0.05$) compared with control being observed (Figure 6). A similar effect was seen on viability and proliferation of MC3T3 murine pre-osteoblast cells cultured on human lactoferrin loaded poly (ε-caprolactone) nanofibres [51]. Lactoferrin was reported to increase proliferation of a mouse pluripotent mesenchymal cell model in a dose-dependent manner [52]. Similar results were reported by Cornish et al. [6] where bovine lactoferrin treatment with concentrations similar to those found in vivo (1–100 µg/mL) on different cell types - rat and human osteoblast-like cells, primary human osteoblast cells, and bone marrow-derived stromal cells-stimulated proliferation. A significant increase in rabbit MSC proliferation was also observed in the case of in vitro Lf treatment for 7 days [18]. In addition, lactoferrin has a protective role against oxidative stress, senescence, and apoptosis of hMSCs, increasing the efficiency of some therapies involving the use of Lf-based biomaterials [53]. Different concentrations of Lf-functionalised biomimetic HA nanocrystals maintained the viability and proliferation of rabbit MSC cells [18] and murine pre-osteoblasts [19] for up to 14 days and Lf and HA in spongy-like hydrogels increased metabolic activity of human adipose-derived stem cells (hASCs) for up to 21 days [54].

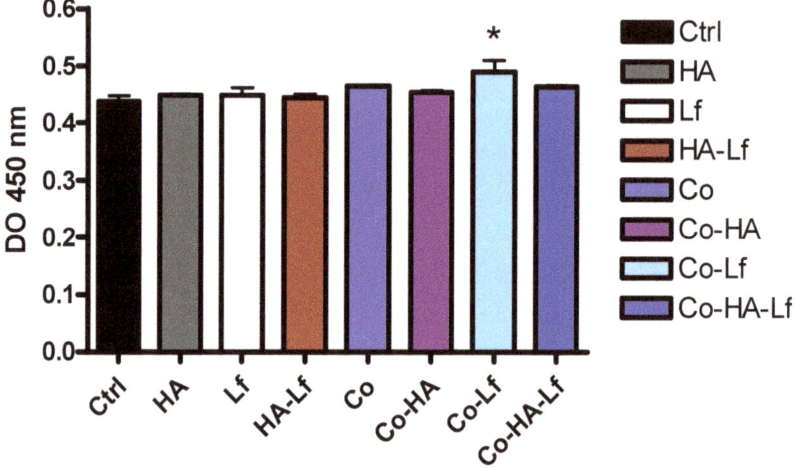

Figure 6. Viability/proliferation of human Mesenchymal stem cells (hMSCs) cultured in direct contact with analysed samples for three days as determined by MTS assay. Data analysis was based on mean ±SD (n = 3). * $p < 0.05$ versus control.

Recent studies with PEG-PCL polymeric surfaces have demonstrated biocompatibility of the material and the ability to support normal adhesion, proliferation, and morphology of murine pre-osteoblasts [30]. By incorporating HA and Lf within the polymeric substrate, the proliferation rate of the MC3T3-E1 murine pre-osteoblast line increased from day 1 to day 5. This suggests that the tested surfaces have increased biocompatibility, a high percentage of cells retaining their ability to be metabolically active in direct contact with hybrid substrates [31].

The proliferation observations are supported by the LIVE/DEAD assay results (Figure 7). No dead cells were observed on either type of surface suggesting that the composition did not impair the adherence and viability of the cells. Fluorescent microscopic inspection showed that all surfaces sustained attachment and proliferation of hMSCs after three days of culture in standard conditions. Different studies showed that Lf alone or in combination with other agents, deposited or incorporated in various supports, sustained the proliferation and viability of mesenchymal cells and bone cells. Also, based on the LIVE/DEAD test, it was demonstrated an increase of up to seven days of the proliferation of murine stromal cells derived from bone marrow encapsulated in a modified recombinant human Lf gel [55] and up to 20 days of MC3T3 murine pre-osteoblasts [56].

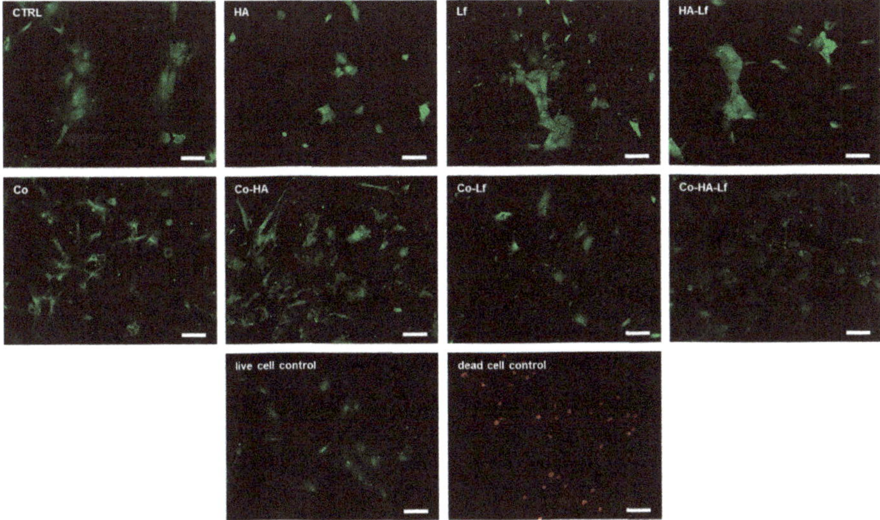

Figure 7. LIVE/DEAD assay. Green-fluorescent live cells and red-fluorescent dead cells are labelled with Calcein AM and ethidium homodimer-1 respectively. Fluorescence microscopy images of hMSCs cultured on different surfaces for 72 h (10×). Scale bar 100 μm.

3.3. Polymeric Coating Control hMSCs Behaviour and Spreading

We subsequently investigated the morphology and adhesion of hMSCs after three days of culture in standard conditions on the different substrates. Immunofluorescent investigation showed that all surfaces allowed monolayer cell attachment and normal morphology development. Phalloidin staining showed a characteristic cytoskeleton organisation, with elongated organised and well-defined actin filaments along the major axis of the cells, on all coatings. However, on surfaces with the copolymeric matrix, hMSCs showed a different behaviour with respect to surface coverage, compared with surfaces without the PEG-PCL component. Surfaces with one component displayed cells with a colony-like proliferation while composite materials seemed to promote the spreading of mesenchymal cells over the surface. Probably due to degradation of the copolymer and exposure of different components, biomaterial surface induced slight modification of the cell morphology with spindle-like extensions anchoring to the surface. Figure 8 shows morphological features and spreading of hMSCs cultured on the analysed surfaces. Additionally, the interaction of hMSCs with the substrate for 72 h was analysed by SEM investigation which confirmed attachment and specific spreading of mesenchymal stem cells on the different substrates. Cells on surfaces without the polymeric component appeared more flat and larger with a greater area and perimeter than hMSCs after 72 h of direct contact with substrates that contain the polymeric matrix. The latter surfaces induced some morphological modification regarding the cell anchorage on the substrates. Cells can be seen branching on the surface with cellular extensions that provide a better interaction with the composite materials. Studies with copolymeric surfaces with different concentrations of PEG-PCL components showed that the composition influences the attachment and spreading/aggregation of mesenchymal stem cells [26]. It seems that cell behaviour on these surfaces is influenced by the length and molar percentage of the polymeric units and by the characteristics of the components with cell-adhesive or cell-repellent effect with an important impact on cell-substrate and cell-cell interactions [27]. Observations regarding the ability to adjust the attachment and aggregation of human MSCs according to PEG-PCL composition were also made by Visan et al. [29]. It was also demonstrated that PEG-PCL copolymer films have the potential to reduce the senescence of hMSCs and maintain stem cell functionality [28]. Cell morphology is closely correlated with stem cell differentiation, and a branched morphology has been reported to be compatible with osteogenic differentiation. Kumar et al. consider that a branched form of cells can act as an osteocyte-like morphology to induce osteogenesis of human bone marrow stem cells in the absence of osteoinduction factors [57].

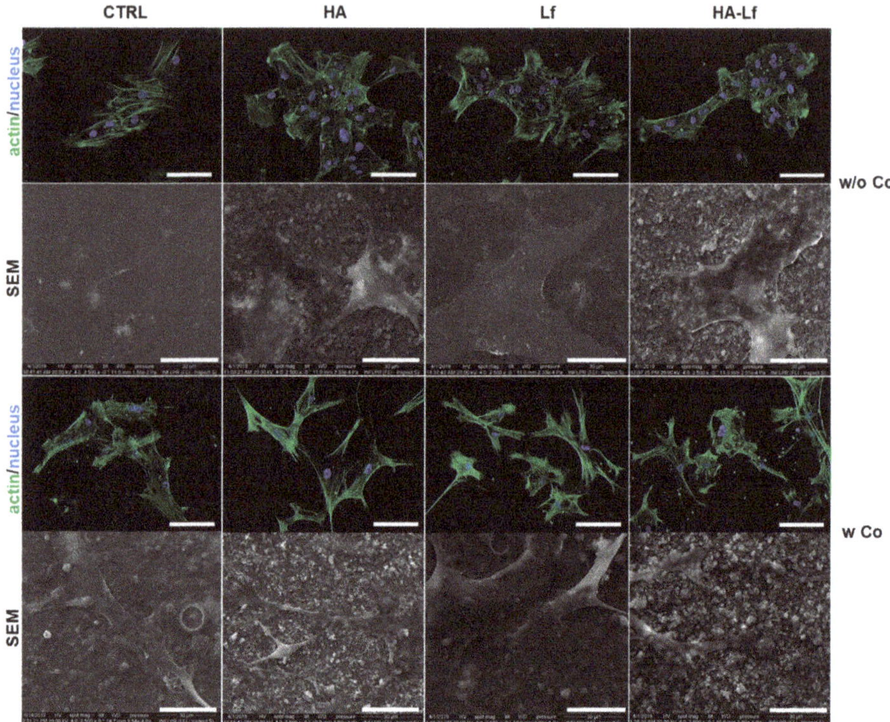

Figure 8. Representative fields of view of human mesenchymal stem cell adhesion and actin cytoskeleton organisation for each surface obtained by fluorescence microscopy. Cell F-actin (green) and nucleus (blue) were examined using a 20× lens. Scale bar is 100 μm. SEM micrographs were taken with 2500× objectives of hMSCs on substrates. Scale bar represents 30 μm.

3.4. Alkaline Phosphatase Activity in hMSCs Grown in Contact with Hybrid Surfaces

To investigate the effect of the surfaces on osteogenic differentiation of hMSCs, ALP activity, an early marker of bone formation was evaluated. The enzyme is expressed in many types of cells but its activity is increased in bone cells, having an important role in mineralisation [58]. The enzyme activity was quantified in the presence or absence of osteoinduction factors after two weeks of the culture of osteoprogenitor cells in direct contact with the surfaces. As shown in Figure 9, after 14 days of culture, ALP activity is statistically increased in osteoinductive conditions compared to standard conditions. Higher values of ALP compared to control (glass) are recorded for cells grown on HA-coated surfaces in osteoinductive conditions. Co-HA-Lf biomimetic polymeric interface exhibit only a slight increase in enzymatic activity compared to control, but higher compared to components alone.

Since Lf treatment led to increased ALP activity, it has been speculated that Lf has the ability to direct the development of mesenchymal stem cells [59] or immature osteoblasts to differentiated phenotypes [60]. Lf treatment also resulted in increased ALP activity in MC3T3-E1 cells, primary osteoblasts [61], but also in human osteosarcoma-derived MG63 cells [60]. Different structures that incorporate Lf are reported in the literature: structures that have been shown to increase the expression level or ALP activity [62–65]. Lf-functionalised HA nanoparticles [19] or Lf-coated

HA nanoparticles [66] led to increased ALP activity in murine pre-osteoblasts (MC3T3-E1) or stem cells derived from rabbit adipose tissue, demonstrating the synergistic effect of these compounds in inducing osteogenic differentiation.

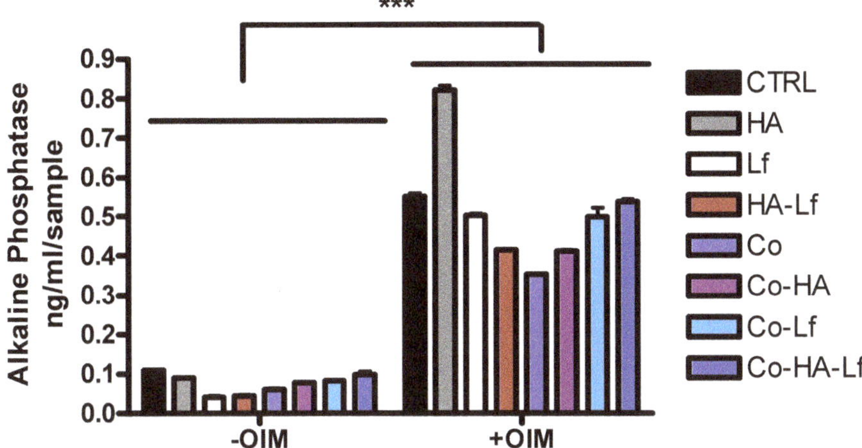

Figure 9. Differentiation of hMSCs to osteoblasts on different surfaces. Quantification of ALP activity at 14 days of hMSCs cultivation with and without osteogenic factors. Data analysis was based on mean ± SD (n = 3). The significance level between groups was *** $p < 0.001$.

3.5. Evaluation of Extracellular Matrix Mineralization

Mineralisation is an important late indicator of osteoblastic differentiation as well as an indicator of successful in vitro bone formation [67]. The potential of hybrid coatings to induce osteogenic differentiation of MSC was determined by evaluating mineralisation of the extracellular matrix. The ability of hMSCs to produce calcified matrix was analysed by staining with Alizarin Red S (ARS) solution. The extracellular calcium deposits (Figure 10A) and optical density values of staining solution after 14 and 28 days in the absence (Figure 10B) and respectively in the presence of osteoinduction factors (Figure 10C) were measured. It can be observed that the higher level of mineralisation, characterised by the presence of bone nodules, was detected in the case of hMSCs grown on substrates with bioactive HA and Lf components embedded in polymeric matrix for both short and long periods of time in the presence or absence of osteoinductive factors (Figure 10A).

All analysed surfaces have the ability to induce osteogenic differentiation of hMSCs but with different capacities depending on surface structure. Thus, in standard culture conditions (Figure 10B) low mineral deposition is observed for the surfaces studied, except those covered with HA, Co-HA, Co-Lf, and Co-HA-Lf. The highest level of mineralisation with a significant increase compared with the control ($p < 0.001$) was observed for both 14 and 28 days of culture of hMSCs on Co-HA-Lf surfaces.

The addition of osteoinductive factors led to an increase in mineralisation for all surfaces compared to control at both time points. Significant increases of calcium deposits ($p < 0.001$) were determined (Figure 10C) in the case of Co surface for 28 days and for HA, Co-HA, Co-Lf and Co-HA-Lf coated surfaces for both 14 and 28 days in osteogenic conditions.

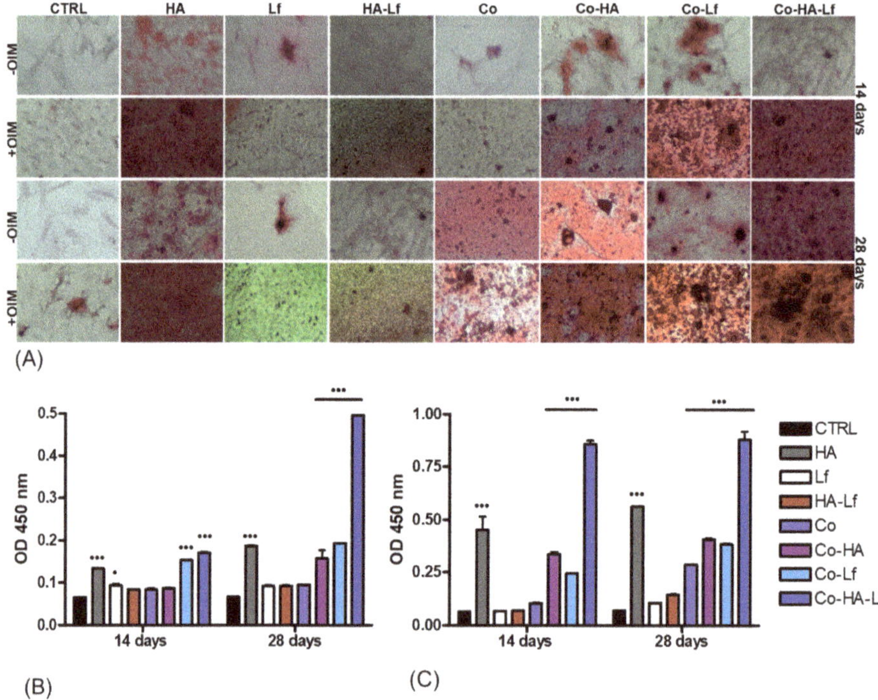

(A)

(B)

(C)

Figure 10. Alizarin Red labelling (**A**) of calcium deposits produced by hMSCs grown on different surfaces. Undifferentiated MSCs with no extracellular calcium deposits appear uncoloured or poorly coloured. The intense orange-red coloration represents the mineralised matrix. Colorimetric quantification of calcium production by hMSCs after 14 and 28 days of culture on analysed samples, in the absence (**B**), and in the presence of osteoinduction factors (**C**) Statistically significant values $p < 0.05$ and $p < 0.001$ vs control.

The HA-Lf-PEG-PCL surface induces a much higher matrix calcification than any other coating, both at 14 days and 28 days in the presence or absence of osteogenic factors. Such behaviour seems to indicate a possible synergistic effect of HA and Lf released from the polymeric substrate in promoting calcium deposition. An argument in favour of this assumption is the role in osteogenic differentiation and the support of mineralisation described in the literature for Lf, alone [6,61] or incorporated into different structures [56,63,64]. Furthermore, the combination of HA and Lf nanocrystals [18,19,66] or deposited in a polymeric matrix [21] led to increased mineralisation. The differentiation of hMSCs on this type of material could be also correlated with surface characteristics such as wettability, roughness, and surface free energy [67].

Our results are in agreement with that reported in the literature regarding the moderate hydrophilic character of surfaces for osteogenic differentiation. A water contact angle between 50° and 70° seems to favour initial steps of adhesion of pre-osteoblast cells [68] and promote spreading, proliferation and osteogenic differentiation of mouse and human MSCs [69]. Our results are supported by those obtained by SEM microscopy, the images in Figure 11 showing the presence of mineralization nodules on analysed surfaces at different time points in the absence or presence of osteoinduction factors.

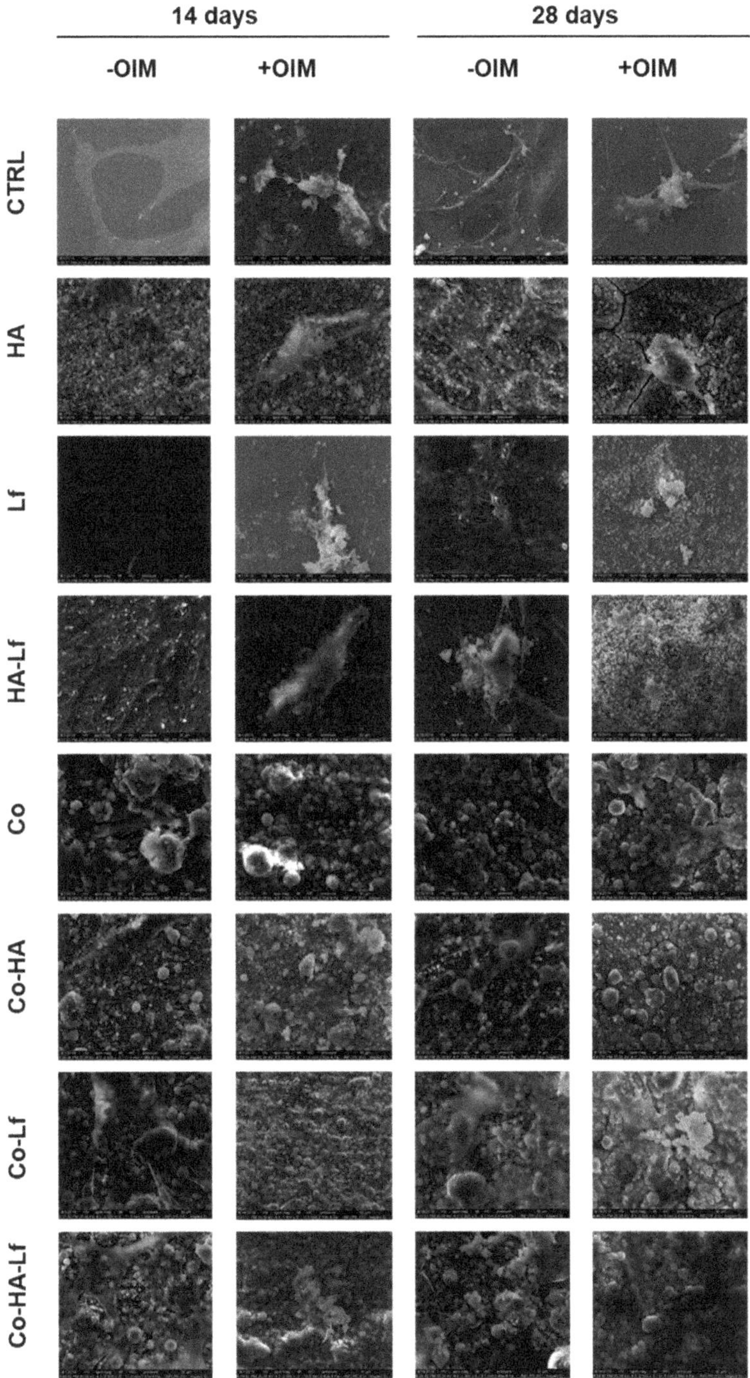

Figure 11. SEM images of hMSCs on each surface, after 14- and 28-days incubation in the presence and absence of the osteoinduction medium (magnification order 2500×—scale bar 30 μm).

4. Conclusions

In this work, we report the effect of hydrophilic Lf based composites coatings obtained by MAPLE onto MSCs.

The coatings were characterised by contents of Ca and P in the Lf-based coatings were close to a Ca/P ratio of 1.61, suggesting a mineral composition similar to that of bone. The distinct roughness and morphological features of the composite were shown by AFM and SEM, indicating the change in surface morphology and microstructure depending on composition. Contact angle and surface energy measurements showed that Lf incorporation into the HA coatings led to a decrease of the contact angle while its incorporation within the matrix of the copolymer did not induce major changes over the wettability of the composite coatings. It also induced an increase of the polar component when compared with the original surfaces.

The biocompatibility assays revealed the absence of a cytotoxic effect of the studied variants. Our results showed that HA and Lf incorporation into the PEG-PCL-Me polymeric layer promoted hMSCs adhesion and positively modulated morphology and cell spreading associated with an increase in the capacity of osteogenic differentiation, cells adapting to the surface characteristics. Co-HA-Lf surface up-modulated ALP activity and has been shown to be most effective in promoting bone regeneration. A significant improvement of the process of extracellular matrix mineralisation, both in osteoinductive conditions and in the absence of osteoinduction factors, was also demonstrated.

The incorporation of HA and Lf into the copolymer matrix proves to be a method of interest for the manufacture of bioactive surfaces with excellent biocompatibility and osteogenic promotion, properties useful for their application in bone regeneration.

Supplementary Materials: The following are available online at http://www.mdpi.com/1996-1944/12/20/3414/s1, Figure S1: hMSC immunophenotyping by flow cytometry, Figure S2: (A) Alkaline phosphatase enzymatic activity of primary cells grown in osteoblast differentiation media. (a) Dermal fibroblasts, (b) MSC and, (c) MSC P2 differentiated osteoblasts were tested in the NBT-BCIP substrate solution. (B) Osteoblast mineralisation assay. (a) Dermal fibroblasts, (b) MSC and, (c) MSC P2 differentiated osteoblasts were incubated with Alizarin Red, which specifically binds Ca^{2+} ions.

Author Contributions: Conceptualization, M.I., A.R. and V.D.; Methodology, M.I., L.R., L.E.S., S.B., A.B., A.R. and V.D.; Validation, M.I., L.R.; Formal Analysis, M.I. and V.D.; Investigation, M.I., L.R., S.B., A.B. and V.D.; Resources, V.D., M.I., L.E.S. and A.R.; Writing—Original Draft Preparation, M.I. and V.D.; Writing—Review and Editing, A.R., V.D., R.W.E. and A.C.; Supervision, A.R., V.D., R.W.E. and A.C.

Funding: This research received funding from the Romanian National Authority for Scientific Research (CNCS–UEFISCDI), under the projects TERAMED 63 PCCDI/2018, PN-III-P1-1.2-PCCDI-2017-072, and Nucleu 16N/2019, Structural and Functional Proteomics Research Program of the Institute of Biochemistry of the Romanian Academy, and by the University of Bucharest-Biology Doctoral School.

Conflicts of Interest: The authors declare no conflict of interest.

References

1. Farnaud, S.; Evans, R.W. Lactoferrin—A multifunctional protein with antimicrobial properties. *Mol. Immunol.* **2003**, *40*, 395–405. [CrossRef]
2. Legrand, D. Overview of Lactoferrin as a Natural Immune Modulator. *J. Pediatrics* **2016**, *173*, S10–S15. [CrossRef] [PubMed]
3. García-Montoya, I.A.; Cendón, T.S.; Arévalo-Gallegos, S.; Rascón-Cruz, Q. Lactoferrin a multiple bioactive protein: An overview. *Biochim. Biophys. Acta Gen. Subj.* **2012**, *1820*, 226–236. [CrossRef] [PubMed]
4. Lepanto, M.S.; Rosa, L.; Paesano, R.; Valenti, P.; Cutone, A. Lactoferrin in Aseptic and Septic Inflammation. *Molecules* **2019**, *24*, 1323. [CrossRef] [PubMed]
5. Chea, C.; Haing, S.; Miyauchi, M.; Shrestha, M.; Imanaka, H.; Takata, T. Molecular mechanisms underlying the inhibitory effects of bovine lactoferrin on osteosarcoma. *Biochem. Biophys. Res. Commun.* **2019**, *508*, 946–952. [CrossRef] [PubMed]
6. Cornish, J.; Callon, K.E.; Naot, D.; Palmano, K.P.; Banovic, T.; Bava, U.; Watson, M.; Lin, J.-M.; Tong, P.C.; Chen, Q.; et al. Lactoferrin Is a Potent Regulator of Bone Cell Activity and Increases Bone Formation In Vivo. *Endocrinology* **2004**, *145*, 4366–4374. [CrossRef]

7. Cornish, J.; Naot, D. Lactoferrin as an effector molecule in the skeleton. *BioMetals* **2010**, *23*, 425–430. [CrossRef]

8. Moreno-Expósito, L.; Illescas-Montes, R.; Melguizo-Rodríguez, L.; Ruiz, C.; Ramos-Torrecillas, J.; de Luna-Bertos, E. Multifunctional capacity and therapeutic potential of lactoferrin. *Life Sci.* **2018**, *195*, 61–64. [CrossRef]

9. Icriverzi, M.; Dinca, V.; Moisei, M.; Evans, R.W.; Trif, M.; Roseanu, A. Lactoferrin in Bone Tissue Regeneration. *Curr. Med. Chem.* **2019**, *26*. [CrossRef]

10. Takayama, Y.; Mizumachi, K. Effect of lactoferrin-embedded collagen membrane on osteogenic differentiation of human osteoblast-like cells. *J. Biosci. Bioeng.* **2009**, *107*, 191–195. [CrossRef]

11. Takaoka, R.; Hikasa, Y.; Hayashi, K.; Tabata, Y. Bone regeneration by lactoferrin released from a gelatin hydrogel. *J. Biomater. Sci. Polym. Ed.* **2011**, *22*, 1581–1589. [CrossRef] [PubMed]

12. Vandrovcova, M.; Douglas, T.E.; Heinemann, S.; Scharnweber, D.; Dubruel, P.; Bacakova, L. Collagen-lactoferrin fibrillar coatings enhance osteoblast proliferation and differentiation. *J. Biomed. Mater. Res. A* **2015**, *103*, 525–533. [CrossRef] [PubMed]

13. Görmez, U.; Kürkcü, M.; Benlidayi, M.E.; Ulubayram, K.; Sertdemir, Y.; Dağlioğlu, K. Effects of bovine lactoferrin in surgically created bone defects on bone regeneration around implants. *J. Oral Sci.* **2015**, *57*, 7–15. [CrossRef] [PubMed]

14. Shi, P.; Wang, Q.; Yu, C.; Fan, F.; Liu, M.; Tu, M.; Lu, W.; Du, M. Hydroxyapatite nanorod and microsphere functionalized with bioactive lactoferrin as a new biomaterial for enhancement bone regeneration. *Colloids Surf. B Biointerfaces* **2017**, *155*, 477–486. [CrossRef]

15. Kattimani, V.S.; Kondaka, S.; Krishna, P.L. Hydroxyapatite—Past, Present, and Future in Bone Regeneration. Bone Tissue Regen. *Insights* **2016**, *7*, 9–19.

16. Haider, A.; Haider, S.; Han, S.S.; Kang, I.-K. Recent advances in the synthesis, functionalization and biomedical applications of hydroxyapatite: A review. *RSC Adv.* **2017**, *7*, 7442–7458. [CrossRef]

17. Bigi, A.; Boanini, E. Functionalized biomimetic calcium phosphates for bone tissue repair. *J. Appl. Biomater. Funct. Mater.* **2017**, *15*, e313–e325. [CrossRef]

18. Montesi, M.; Panseri, S.; Iafisco, M.; Adamiano, A.; Tampieri, A. Effect of hydroxyapatite nanocrystals functionalized with lactoferrin in osteogenic differentiation of mesenchymal stem cells. *J. Biomed. Mater. Res. Part A* **2015**, *103*, 224–234. [CrossRef]

19. Montesi, M.; Panseri, S.; Iafisco, M.; Adamiano, A.; Tampieri, A. Coupling Hydroxyapatite Nanocrystals with Lactoferrin as a Promising Strategy to Fine Regulate Bone Homeostasis. *PLoS ONE* **2015**, *10*, e0132633. [CrossRef]

20. Dinca, V.; Florian, P.E.; Sima, L.E.; Rusen, L.; Constantinescu, C.; Evans, R.W.; Dinescu, M.; Roseanu, A. MAPLE-based method to obtain biodegradable hybrid polymeric thin films with embedded antitumoral agents. *Biomed. Microdevices* **2014**, *16*, 11–21. [CrossRef]

21. Rusen, L.; Brajnicov, S.; Neacsu, P.; Marascu, V.; Bonciu, A.; Dinescu, M.; Dinca, V.; Cimpean, A. Novel degradable biointerfacing nanocomposite coatings for modulating the osteoblast response. *Surf. Coat. Technol.* **2017**, *325*, 397–409. [CrossRef]

22. Icriverzi, M.; Rusen, L.; Brajnicov, S.; Bonciu, A.; Dinescu, M.; Cimpean, A.; Evans, R.W.; Dinca, V.; Roseanu, A. Macrophage in vitro Response on Hybrid Coatings Obtained by Matrix Assisted Pulsed Laser Evaporation. *Coatings* **2019**, *9*, 236. [CrossRef]

23. Grossen, P.; Witzigmann, D.; Sieber, S.; Huwyler, J. PEG-PCL-based nanomedicines: A biodegradable drug delivery system and its application. *J. Control. Release* **2017**, *260*, 46–60. [CrossRef] [PubMed]

24. Malikmammadov, E.; Endoğan, T.; Kiziltay, A.; Hasirci, V.; Hasirci, N. PCL and PCL-Based Materials in Biomedical Applications. *J. Biomater. Sci. Polym. Ed.* **2017**, *29*, 1–55. [CrossRef]

25. Deng, H.; Dong, A.; Song, J.; Chen, X. Injectable thermosensitive hydrogel systems based on functional PEG/PCL block polymer for local drug delivery. *J. Control. Release* **2019**, *297*, 60–70. [CrossRef]

26. Balikov, D.A.; Crowder, S.W.; Boire, T.C.; Lee, J.B.; Gupta, M.K.; Fenix, A.M.; Lewis, H.N.; Ambrose, C.M.; Short, P.A.; Kim, C.S.; et al. Tunable Surface Repellency Maintains Stemness and Redox Capacity of Human Mesenchymal Stem Cells. *ACS Appl. Mater. Interfaces* **2017**, *9*, 22994–23006. [CrossRef]

27. Crowder, S.W.; Balikov, D.A.; Boire, T.C.; McCormack, D.; Lee, J.B.; Gupta, M.K.; Skala, M.C.; Sung, H.-J. Copolymer-Mediated Cell Aggregation Promotes a Proangiogenic Stem Cell Phenotype In Vitro and In Vivo. *Adv. Healthc. Mater.* **2016**, *5*, 2866–2871. [CrossRef]

28. Balikov, D.A.; Crowder, S.W.; Lee, J.B.; Lee, Y.; Ko, U.H.; Kang, M.-L.; Kim, W.S.; Shin, J.H.; Sung, H.-J. Aging Donor-Derived Human Mesenchymal Stem Cells Exhibit Reduced Reactive Oxygen Species Loads and Increased Differentiation Potential Following Serial Expansion on a PEG-PCL Copolymer Substrate. *Int. J. Mol. Sci.* **2018**, *19*, 359. [CrossRef]

29. Visan, A.; Cristescu, R.; Stefan, N.; Miroiu, M.; Nita, C.; Socol, M.; Florica, C.; Rasoga, O.; Zgura, I.; Sima, L.E.; et al. Antimicrobial polycaprolactone/polyethylene glycol embedded lysozyme coatings of Ti implants for osteoblast functional properties in tissue engineering. *Appl. Surf. Sci.* **2017**, *417*, 234–243. [CrossRef]

30. Rusen, L.; Neacsu, P.; Cimpean, A.; Valentin, I.; Brajnicov, S.; Dumitrescu, L.N.; Banita, J.; Dinca, V.; Dinescu, M. In vitro evaluation of poly(ethylene glycol)-block-poly(ε-caprolactone) methyl ether copolymer coating effects on cells adhesion and proliferation. *Appl. Surf. Sci.* **2016**, *374*, 23–30. [CrossRef]

31. Caricato, A.P.; Arima, V.; Catalano, M.; Cesaria, M.; Cozzoli, P.D.; Martino, M.; Taurino, A.; Rella, R.; Scarfiello, R.; Tunno, T.; et al. MAPLE deposition of nanomaterials. *Appl. Surf. Sci.* **2014**, *30*, 92–98. [CrossRef]

32. Yang, S.; Zhang, J. Matrix-Assisted Pulsed Laser Evaporation (MAPLE) technique for deposition of hybrid nanostructures. *Front. Nanosci. Nanotech.* **2017**, *3*, 1–9. [CrossRef]

33. Stiff-Roberts, A.D.; Ge, W. Organic/hybrid thin films deposited by matrix-assisted pulsed laser evaporation (MAPLE). *Appl. Phys. Rev.* **2017**, *4*, 041303. [CrossRef]

34. Kaelble, D.H. Dispersion-Polar Surface Tension Properties of Organic Solids. *J. Adhes.* **1970**, *2*, 66–81. [CrossRef]

35. Owens, D.K.; Wendt, R.C. Estimation of the surface free energy of polymers. *J. Appl. Polym. Sci.* **1969**, *13*, 1741–1747. [CrossRef]

36. Rabel, W. Einige Aspekte der Benetzungstheorie und ihre Anwendung auf die Untersuchung und Veränderung der Oberflächeneigenschaften von Polymeren. *Farbe und Lack* **1971**, *77*, 997–1006.

37. Sima, L.E.; Stan, G.E.; Morosanu, C.O.; Melinescu, A.; Ianculescu, A.; Melinte, R.; Neamtu, J.; Petrescu, S.M. Differentiation of mesenchymal stem cells onto highly adherent radio frequency-sputtered carbonated hydroxylapatite thin films. *J. Biomed. Mater. Res. Part A* **2010**, *95A*, 1203–1214. [CrossRef]

38. Icriverzi, M.; Rusen, L.; Sima, L.E.; Moldovan, A.; Brajnicov, S.; Bonciu, A.; Mihailescu, N.; Dinescu, M.; Cimpean, A.; Roseanu, A. In vitro behavior of human mesenchymal stem cells on poly (N-isopropylacrylamide) based biointerfaces obtained by matrix assisted pulsed laser evaporation. *Appl. Surf. Sci.* **2018**, *440*, 712–724. [CrossRef]

39. Kourkoumelis, N.; Balatsoukas, I.; Tzaphlidou, M. Ca/P concentration ratio at different sites of normal and osteoporotic rabbit bones evaluated by Auger and energy dispersive X-ray spectroscopy. *J. Biol. Phys.* **2012**, *38*, 279–291. [CrossRef]

40. Trommer, R.M.; Santos, L.A.; Bergmann, C.P. Alternative technique for hydroxyapatite coatings. *Surf. Coat. Technol.* **2007**, *201*, 9587–9593. [CrossRef]

41. Rupp, F.; Scheideler, L.; Olshanska, N.; de Wild, M.; Wieland, M.; Geis-Gerstorfer, J. Enhancing surface free energy and hydrophilicity through chemical modification of microstructured titanium implant surfaces. *J. Biomed. Mater. Res. Part A* **2006**, *76A*, 323–334. [CrossRef] [PubMed]

42. Buser, D.; Broggini, N.; Wieland, M.; Schenk, R.K.; Denzer, A.J.; Cochran, D.L.; Hoffmann, B.; Lussi, A.; Steinemann, S.G. Enhanced Bone Apposition to a Chemically Modified SLA Titanium Surface. *J. Dent. Res.* **2004**, *83*, 529–533. [CrossRef] [PubMed]

43. Ferguson, S.J.; Broggini, N.; Wieland, M.; de Wild, M.; Rupp, F.; Geis-Gerstorfer, J.; Cochran, D.L.; Buser, D. Biomechanical evaluation of the interfacial strength of a chemically modified sandblasted and acid-etched titanium surface. *J. Biomed. Mater. Res. Part A* **2006**, *78A*, 291–297. [CrossRef] [PubMed]

44. Siniscalco, D.; Dutreilh-Colas, M.; Hjezi, Z.; Cornette, J.; El Felss, N.; Champion, E.; Damia, C. Functionalization of Hydroxyapatite Ceramics: Raman Mapping Investigation of Silanization. *Ceramics* **2019**, *2*, 372–384. [CrossRef]

45. Li, S.; Garreau, H.; Vert, M.; Petrova, T.; Manolova, N.; Rashkov, I. Hydrolytic degradation of poly(oxyethylene)–poly-(ε-caprolactone) multiblock copolymers. *J. Appl. Polym. Sci.* **1998**, *68*, 989–998. [CrossRef]

46. Lim, J.Y.; Taylor, A.F.; Li, Z.; Vogler, E.A.; Donahue, H.J. Integrin Expression and Osteopontin Regulation in Human Fetal Osteoblastic Cells Mediated by Substratum Surface Characteristics. *Tissue Eng.* **2005**, *11*, 19–29. [CrossRef]

47. Liu, X.; Lim, J.Y.; Donahue, H.J.; Dhurjati, R.; Mastro, A.M.; Vogler, E.A. Influence of substratum surface chemistry/energy and topography on the human fetal osteoblastic cell line hFOB 1.19: Phenotypic and genotypic responses observed in vitro. *Biomaterials* **2007**, *28*, 4535–4550. [CrossRef]

48. Qiu, Q.; Sayer, M.; Kawaja, M.; Shen, X.; Davies, J.E. Attachment, morphology, and protein expression of rat marrow stromal cells cultured on charged substrate surfaces. *J. Biomed. Mater. Res.* **1998**, *42*, 117–127. [CrossRef]

49. Wei, J.; Igarashi, T.; Okumori, N.; Igarashi, T.; Maetani, T.; Liu, B.; Yoshinari, M. Influence of surface wettability on competitive protein adsorption and initial attachment of osteoblasts. *Biomed. Mater.* **2009**, *4*, 045002. [CrossRef]

50. Szcześ, A.; Yan, Y.; Chibowski, E.; Hołysz, L.; Banach, M. Properties of natural and synthetic hydroxyapatite and their surface free energy determined by the thin-layer wicking method. *Appl. Surf. Sci.* **2018**, *434*, 1232–1238. [CrossRef]

51. James, E.N.; Nair, L.S. Development and Characterization of Lactoferrin Loaded Poly(ε-Caprolactone) Nanofibers. *J. Biomed. Nanotechnol.* **2014**, *10*, 500–507. [CrossRef] [PubMed]

52. Yagi, M.; Suzuki, N.; Takayama, T.; Arisue, M.; Kodama, T.; Yoda, Y.; Otsuka, K.; Ito, K. Effects of lactoferrin on the differentiation of pluripotent mesenchymal cells. *Cell Biol. Int.* **2009**, *33*, 283–289. [CrossRef] [PubMed]

53. Park, S.Y.; Jeong, A.-J.; Kim, G.-Y.; Jo, A.; Lee, J.E.; Leem, S.-H.; Yoon, J.-H.; Ye, S.K.; Chung, J.W. Lactoferrin Protects Human Mesenchymal Stem Cells from Oxidative Stress-Induced Senescence and Apoptosis. *J. Microbiol. Biotechnol.* **2017**, *27*, 1877–1884. [CrossRef] [PubMed]

54. Bastos, A.R.; da Silva, L.P.; Maia, F.R.; Pina, S.; Rodrigues, T.; Sousa, F.; Oliveira, J.M.; Cornish, J.; Correlo, V.M.; Reis, R.L. Lactoferrin-Hydroxyapatite Containing Spongy-Like Hydrogels for Bone Tissue Engineering. *Materials* **2019**, *12*, 2074. [CrossRef] [PubMed]

55. Amini, A.A.; Kan, H.-M.; Cui, Z.; Maye, P.; Nair, L.S. Enzymatically cross-linked bovine lactoferrin as injectable hydrogel for cell delivery. *Tissue Eng. Part A* **2014**, *20*, 2830–2839. [CrossRef] [PubMed]

56. Amini, A.A.; Nair, L.S. Recombinant human lactoferrin as a biomaterial for bone tissue engineering: Mechanism of antiapoptotic and osteogenic activity. *Adv. Healthc. Mater.* **2014**, *3*, 897–905. [CrossRef]

57. Kumar, G.; Tison, C.K.; Chatterjee, K.; Pine, P.S.; McDaniel, J.H.; Salit, M.L.; Young, M.F.; Simon, C.G., Jr. The determination of stem cell fate by 3D scaffold structures through the control of cell shape. *Biomaterials* **2011**, *32*, 9188–9196. [CrossRef]

58. Golub, E.E.; Harrison, G.; Taylor, A.G.; Camper, S.; Shapiro, I.M. The Role of Alkaline-Phosphatase in Cartilage Mineralization. *Bone Min.* **1992**, *17*, 273–278. [CrossRef]

59. Li, Y.; Zhang, W.; Ren, F.; Guo, H. Activation of TGF-β Canonical and Noncanonical Signaling in Bovine Lactoferrin-Induced Osteogenic Activity of C3H10T1/2 Mesenchymal Stem Cells. *Int. J. Mol. Sci.* **2019**, *20*, 2880. [CrossRef]

60. Takayama, Y.; Mizumachi, K. Effect of Bovine Lactoferrin on Extracellular Matrix Calcification by Human Osteoblast-Like Cells. *Biosci. Biotechnol. Biochem.* **2008**, *72*, 226–230. [CrossRef]

61. Zhang, W.; Guo, H.; Jing, H.; Li, Y.; Wang, X.; Zhang, H.; Jiang, L.; Ren, F. Lactoferrin Stimulates Osteoblast Differentiation Through PKA and p38 Pathways Independent of Lactoferrin's Receptor LRP1. *J. Bone Miner. Res.* **2014**, *29*, 1232–1243. [CrossRef] [PubMed]

62. Kim, S.E.; Yun, Y.-P.; Shim, K.-S.; Park, K.; Choi, S.-W.Q.; Suh, D.H. Effect of lactoferrin-impregnated porous poly(lactide-co-glycolide) (PLGA) microspheres on osteogenic differentiation of rabbit adipose-derived stem cells (rADSCs). *Colloids Surf. B Biointerfaces* **2014**, *122*, 457–464. [CrossRef] [PubMed]

63. Kim, S.E.; Yun, Y.-P.; Lee, J.Y.; Park, K.; Suh, D.H. Osteoblast activity of MG-63 cells is enhanced by growth on a lactoferrin-immobilized titanium substrate. *Colloids Surf. B Biointerfaces* **2014**, *123*, 191–198. [CrossRef] [PubMed]

64. Kim, S.E.; Lee, D.-W.; Yun, Y.-P.; Shim, K.-S.; Jeon, D.; Rhee, J.K.; Kim, H.-J.; Park, K. Heparin-immobilized hydroxyapatite nanoparticles as a lactoferrin delivery system for improving osteogenic differentiation of adipose-derived stem cells. *Biomed. Mater.* **2016**, *11*, 025004. [CrossRef]

65. Castillo Diaz, L.A.; Elsawy, M.; Saiani, A.; Gough, J.E.; Miller, A.F. Osteogenic differentiation of human mesenchymal stem cells promotes mineralization within a biodegradable peptide hydrogel. *J. Tissue Eng.* **2016**, *7*, 1–15. [CrossRef]

66. Blair, H.C.; Larrouture, Q.C.; Li, Y.; Lin, H.; Beer-Stoltz, D.; Liu, L.; Tuan, R.S.; Robinson, S.P.H.; Nelson, D.J. Osteoblast Differentiation and Bone Matrix Formation In Vivo and In Vitro. *Tissue Eng. Part B Rev.* **2017**, *23*, 268–280. [CrossRef]
67. Gentleman, M.; Gentleman, E. The role of surface free energy in osteoblast–biomaterial interactions. *Int. Mater. Rev.* **2014**, *59*, 417–429. [CrossRef]
68. Ardhaoui, M.; Naciri, M.; Mullen, T.; Brugha, C.; Keenan, A.K.; Al-Rubeai, M.; Dowling, D.P. Evaluation of Cell Behaviour on Atmospheric Plasma Deposited Siloxane and Fluorosiloxane Coatings. *J. Adhes. Sci. Technol.* **2010**, *24*, 889–903. [CrossRef]
69. Hao, L.; Yang, H.; Du, C.; Fu, X.; Zhao, N.; Xu, S.; Cui, F.; Mao, C.; Wang, Y. Directing the fate of human and mouse mesenchymal stem cells by hydroxyl–methyl mixed self-assembled monolayers with varying wettability. *J. Mater. Chem. B* **2014**, *2*, 4794–4801. [CrossRef]

Article

Protein-Polymer Matrices with Embedded Carbon Nanotubes for Tissue Engineering: Regularities of Formation and Features of Interaction with Cell Membranes

Michael M. Slepchenkov [1], Alexander Yu. Gerasimenko [2,3], Dmitry V. Telyshev [2,3] and Olga E. Glukhova [1,2,*]

[1] Department of Physics, Saratov State University, Astrakhanskaya street 83, Saratov 410012, Russia; slepchenkovm@mail.ru

[2] Laboratory of Biomedical Nanotechnology, I.M. Sechenov First Moscow State Medical University, Bolshaya Pirogovskaya street 2-4, Moscow 119991, Russia; gerasimenko@bms.zone (A.Y.G.); d_telyshev@mail.ru (D.V.T.)

[3] Institute of Biomedical Systems, National Research University of Electronic Technology MIET, Shokin Square 1, Zelenograd, Moscow 124498, Russia

* Correspondence: glukhovaoe@info.sgu.ru; Tel.: +7 8452 514562

Received: 22 August 2019; Accepted: 17 September 2019; Published: 21 September 2019

Abstract: This paper reveals the mechanism of nanowelding a branched network of single-walled carbon nanotubes (SWCNTs) used as a framework for the formation of protein–polymer matrices with albumin, collagen, and chitosan. It is shown that the introduction of certain point defects into the structure of SWCNTs (single vacancy, double vacancy, Stone–Wales defect, and a mixed defect) allows us to obtain strong heating in defective regions as compared to ideal SWCNTs. The wavelengths at which absorption reaches 50% are determined. Non-uniform absorption of laser radiation along with inefficient heat removal in defective regions determines the formation of hot spots, in which nanowelding of SWCNTs is observed even at 0.36 nm between contacting surfaces. The regularities of formation of layered protein–polymer matrices and the features of their interaction with cell membrane are revealed. All studies are carried out in silico using high-precision quantum approaches.

Keywords: protein–polymer matrices; nanowelding; single-walled carbon nanotubes; point defects; absorption; laser radiation; cell membrane

---

1. Introduction

At present, a 3D wireframe nanomaterial in the form of a branched network of multi-walled carbon nanotubes (MWCNTs) and SWCNTs is extremely popular in various fields, including biomedicine [1–6]. In recent years, 3D carbon nanotube (CNT) scaffolds have been especially demanded in the field of tissue engineering as a material for creating conductive scaffolds used for bone tissue regeneration [7]. The ability of 3D CNT scaffolds to stimulate the proliferation, maturation, and long-term survival of cardiomyocytes has already been proven [8]. An elastomeric scaffold developed based on the 3D CNT framework and polydimethylsiloxane stimulates the growth and electrophysiological maturation of cardiomyocytes, as well as the formation of functional syncytium [9]. 3D porous conductive CNT/polypyrrole scaffolds showed remarkable ability to regenerate astrocytes, which suggests their high potential as a neural prosthesis [10].

To create branched junctions of CNTs, one can use various methods and approaches, including direct growth [11–13], high-energy electron beam irradiation [14–16], forming junctions in strong electrical, optical, and thermal fields [17–19], as well as using atomic force microscopy [20]. The laser nanowelding technique seems to be the most promising method for obtaining high-quality 3D scaffolds from CNT junctions [21]. It was found that the key factors in nanowelding of MWCNTs are the degree of graphitization of MWCNTs, the irradiation time, the type of substrate, and others [21–24]. However, the degree of absorption of laser radiation by CNTs during welding at various wavelengths was not previously estimated. Also, the influence of the CNT structure, in particular CNT chirality, the presence of defects and their type on the nature of nanowelding and its quality was not taken into account. Knowledge of these regularities will allow a deeper understanding of the mechanism of splicing CNTs under the action of laser radiation in order to develop a controlled experimental method for obtaining 3D CNT scaffolds characterized by high conductive and strength properties.

Taking into account such a variety of factors influencing the welding of CNTs within a single experimental study is a difficult task. In addition, a detailed study of the CNT welding mechanism requires studies of the physical processes occurring in the structure of CNTs at the atomic and quantum levels. Such studies can only be carried out using computer simulation methods. The first works in this direction appeared at the beginning of the 21st century [25,26]. In these first works, the nanowelding of SWCNTs of various shapes under the influence of ion and electron irradiation was simulated using molecular dynamics. In the last decade, the process of nanowelding is studied using simulation methods from different angles. On the one hand, nanowelding is modeled as a result of simple Joule heating [18,27], and on the other hand, as a result of the formation of X, Y, and T-shaped SWCNT junctions due to the melting of silver nanoparticles [28–30]. In the second case, the nanoparticles play the role of a "bonding adhesive" at the junction of the SWCNTs. It has already been repeatedly shown that a welding of SWCNT framework occurs during strong heating to 1300–1700 K, which ensures a junction of the SWCNT open ends with each other and with individual SWCNTs [2,18]. When heated to 2500–3500 K, the SWCNTs do not just connect to each other, but weld seamlessly, forming a new X-shaped structure [17,31]. The main question in this case is the SWCNT nanowelding mechanism. As already mentioned, a strong heating of the SWCNTs is necessary to start the nanowelding mechanism. Today, several methods of heating are known. One of the promising methods is a welding using laser pulsed irradiation. Under the action of a laser beam, silver nanoparticles melt and connect the CNTs together. Another example is the method of laser nanowelding of bundles of double-walled CNTs, when fragments of CNTs resulting from the partial destruction of the outer shells of multilayer CNTs act as solder [32]. However, to date, there is no information about the dependence of the degree of absorption of electromagnetic waves by nanotubes on their chirality and the presence of defects, and also about the influence of these factors on the nanowelding process.

Another relevant direction of a study of the properties and applications of 3D nanomaterial in the form of a branched network of MWCNTs and SWCNTs is the use of these materials for the subsequent creation of protein and polymer matrices on their basis. Such matrices are already actively synthesized and in demand in various fields of biomedicine [33,34], including in the field of tissue engineering [34]. It has already been proven that these matrices are biocompatible, but it is not yet known how exactly such matrices contact cell membranes and what effect they have on them. One of the most promising candidates for creating protein and polymer matrices are albumin, collagen and chitosan. Albumin is already being used for neural tissue engineering applications [35], for lungs tissue engineering applications [36], for engineering functional cardiac tissues [37], and also to create a potential degradable tissue scaffold [38]. Collagen-based biomaterials function as cell scaffolds and replace the native extracellular matrix [39]. A collagen scaffold can be applied in tissue engineering, including nerve, bone, cartilage, tendon, ligament, blood vessel and skin [40,41]. The natural abundance, cost-effectiveness, biodegradability, biocompatibility and the ability to heal wounds have made chitosan a very popular polymer for tissue engineering and implantation [42,43].

In this work, we perform in silico study of the formation of a branched network of SWCNTs as a result of nanoscale welding of SWCNTs under the action of laser irradiation. The most effective frequency of laser irradiation for welding of SWCNTs is revealed. Modeling of the formation of layered protein–polymer matrices with a branched network of SWCNTs as a framework and albumin, collagen and chitosan as fillers is carried out. Some regularities of interaction of polymer–protein matrices with the cell membrane are also studied.

2. Computational Details

To understand the nature of CNT nanowelding, we considered the features of the interaction of the defective regions of SWCNTs and the open ends of SWCNTs with laser radiation in the ultraviolet (UV)–visible–infrared (IR) range. As was mentioned in the introduction, welding occurs during strong heating [2,18] precisely due to open ends, as well as due to the partial destruction of defective CNT regions [32]. It should also be noted that defective regions do not efficiently remove heat, so during general heating of the nanostructure, these regions overheat more noticeably and may eventually collapse.

Let us consider the absorption of energy in the case of normal incidence of a plane electromagnetic wave. To do this, we calculate the absorption coefficient using Maxwell's theory and quantum-mechanical approaches. The SWCNT array acts as the interface between two media, each of which is a vacuum. That is, we are considering the process of interaction of a SWCNT network with an incident wave, which passes from a vacuum through a SWCNT network again into a vacuum. We consider waves with the vector **E** directed along the CNT axis and perpendicular to it. The absorption coefficient is determined by the equation:

$$A = 1 - |R|^2 - |T|^2, \tag{1}$$

where $|R|^2$ and $|T|^2$ are the squares of the moduli of the complex reflection and transmission coefficients. For the case of normal wave incidence, the expressions for R and T coefficients are written as follows [44]:

$$R = \frac{-\sigma_{\alpha\beta}Z_0}{2 + \sigma_{\alpha\beta}Z_0}, T = \frac{2}{2 + \sigma_{\alpha\beta}Z_0}, \tag{2}$$

where Z_0 is the wave resistance in free space $Z_0 = 120\pi$ Ohm; $\sigma_{\alpha\beta}$ is an element of the complex optical conductivity tensor (α and β indices coincide in the direction of the axis of the nanotube or perpendicular to it). The elements of the complex optical conductivity tensor $\sigma_{\alpha\beta}(\Omega)$ were calculated using the Kubo–Greenwood formula that defines the conductivity as a function of photon energy Ω [45,46]:

$$\sigma_{\alpha\beta} = \frac{2e^2\hbar}{im_e^2 S_{cell}} \frac{1}{N_k} \sum_{k \in BZ}^{N_k} \sum_{m,n} \frac{\hat{P}_\alpha^{nm}(k) \cdot \hat{P}_\beta^{nm}(k)}{E_n(k) - E_m(k) + \Omega + i\eta} \times \frac{f_\beta[E_n(k) - \mu] - f_\beta[E_m(k) - \mu]}{E_n(k) - E_m(k)}, \tag{3}$$

where $f_\beta(E) = 1/(1 + \exp[\beta(E - \mu)])$ is the Fermi–Dirac function with chemical potential μ and the inverse of thermal energy $\beta = 1/k_B T$; S_{cell} is the area of the super-cell; N_k is the number of k-points needed to sample the Brillouin zone (BZ); $\hat{P}_\alpha^{nm}(k)$ are the matrix elements corresponding to the α-component of the momentum operator vector; $\hat{P}_\beta^{nm}(k)$ are the matrix elements corresponding to the β-component of the momentum operator vector; m_e and e are the free-electron mass and electron charge; $E_n(k)$ is the sub-band energy in the valence band, $E_m(k)$ is the sub-band energy in the conduction band. The spin degeneracy is already taken into account in the above equations by factor 2, η is a phenomenological parameter which characterizes the processes of electron scattering. To calculate the momentum matrix elements $\hat{P}_\alpha^{nm}(k)$ we used the well-known expression substitution $\hat{P}(k) \rightarrow (m_e/\hbar)\nabla_k \hat{H}(k)$, where $\hat{H}(k)$ is the Hamiltonian. The Hamiltonian was constructed as part of the self-consistent charge density functional tight-binding (SCC DFTB) method. All calculations were performed using open source Kvazar [47] and DFTB + package [48]. The use of the SCC DFTB method instead of ab initio methods is due to the polyatomicity of the supercells of the studied SWCNT. For example, supercells of chiral SWCNTs contain 500–1324 atoms. We also note that Equation (2) is valid for thin films whose thickness is much less than the wavelength. These equations were obtained previously for composite thin films [44] and well tested.

To simulate the nanotube nanowelding process, the nonequilibrium molecular dynamics method with the adaptive intermolecular reactive bond order (AIREBO) force field was used [49]. This method is implemented in open source Kvazar. The simulation time step was 0.1 fs. Polymer–protein matrices based on a SWCNT network were simulated by means of the coarse-grained approach using the MARTINI force field [50] and the GROMACS program [51].

3. Results and Discussion

In this work, we consider chiral and non-chiral SWCNTs with a diameter of 0.6–2 nm, which are most often synthesized: (i) non-chiral zigzag SWCNTs (m, 0) with m = 13, 14, 16, 20, 23, 32 and armchair SWCNTs (m, m) with m = 4, 12, 15, 20; (ii) chiral SWCNTs (11,10), (14,4), (12,6), (12,8), (23,6) and (9,4). That is, semiconductor SWCNTs account for ~69% of the total number of SWCNTs under consideration. This fact is completely consistent with the experimental data, according to which semiconductor SWCNTs are always composed of ~2/3 from an array of synthesized SWCNTs. Previously, the authors of this work showed that the nanowelding of SWCNTs occurs in those regions of SWCNTs where there is a defect or several defects. In particular, the formation of a tree-like structure from SWCNTs due to the formation of T-shaped contacts was shown using the molecular dynamics method [2]. Similar T-shaped SWCNT structures are formed due to the formation of covalent bonds between the open ends of one SWCNT and the atoms of the defective region of another SWCNT. We assume that the reason for the formation of covalent bonds between SWCNTs is the nonuniform absorption of laser radiation energy, which inevitably leads to nonuniform heating of SWCNTs and the appearance of "hot spots" in which nanowelding occurs. The basis of our assumptions is a series of works in which the strong influence of various defects on the temperature distribution in nanotubes is proved. For example, J. Park et al. have shown that the temperature can increase by several hundred degrees in the defect region, and the thermal conductivity of the defective regions drops sharply [52]. It was found that such defects as Stone–Wales and vacancies strongly affect the thermal conductivity of a CNT network. They prevent the free propagation of phonons reducing thermal conductivity in these local regions by 2–10 times [53,54]. Another paper by M. Chang et al. [55] demonstrated the presence of a large temperature gradient due to a drop in thermal conductivity in defective regions, which inevitably leads to the appearance of heat localization. It was also noted there that the presence of local regions of high heating is characteristic of all low-dimensional structures, which was previously observed for silicon and aluminum wires of submicron diameter [56,57]. The reason for this phenomenon is (i) the scattering of phonons, the mean free path of which decreases significantly due to topological defects, and (ii) the phonons are limited in the direction of propagation in low-dimensional structures. This leads to the localization of heat. In this work, the nonuniform absorption of laser energy is caused by the presence of defects in the atomic structure. To verify our assumption, we first investigated the regularities of absorption of electromagnetic wave energy by the defective SWCNTs in the UV–visible–IR range using Equations (1)–(3) and the SCC DFTB method.

3.1. Absorption of Electromagnetic Wave Energy by Defect-Free and Defective SWCNTs

We have constructed atomistic models of SWCNTs with point defects, such as single (1V) and double (2V) vacancies, Stone–Wales (SW) defects and an SW + 1V mixed defect. As an example, Figure 1a shows supercells of two SWCNTs with various point defects: SWCNT (15,15) with a mixed defect consisting of two SW defects and one 1V defect, SWCNT (20,0) with two 2V defects, SWCNT (20,0) with a single SW defect, SWCNT (15,15) with a single 2V defect. Heptagons are marked in yellow, atoms of adjacent pentagons are marked in green, and atoms in a region with 1V and 2V vacancies are marked in blue. The supercells of the defective SWCNTs under study were optimized and energetically favorable atomistic models of SWCNTs corresponding to the equilibrium structure were revealed. To calculate the optical conductivity and absorption coefficient, we constructed a model of a thin film of parallel SWCNTs located at a distance of ~0.5 nm. At this distance, the electron clouds of adjacent SWCNTs do not affect each other and, thus, we calculate the energy absorption of an individual SWCNT. Since the thickness of such a film of SWCNTs is no more than 2 nm, the range of the studied waves was 10–3000 nm. The wave vector is normal to the film. The calculated values of the absorption coefficient for SWCNTs of various diameters and chiralities are presented in Figure 1b. Since a lot of calculated data were obtained, this figure shows the results for only 12 SWCNTs, which are most often synthesized and have diameters in a wide range of 0.6–2 nm. Based on the calculation results, including data presented in Figure 1b, we can say that there are three wavelength regions in which the main number of absorption maxima is concentrated for all types of SWCNTs. These are wavelength ranges of 200–400 nm, 650–800 nm and 900–1150 nm. In the far infrared wavelength range of 1800–2400 nm, a certain number of absorption maxima is also observed, but only for certain types of defects and not for all types of SWCNTs. In order to evaluate exactly how defects in the atomic structure of SWCNTs affect the frequency distribution of absorption maxima, we calculated the absorption spectra of defect-free SWCNTs. Analyzing all the types of SWCNTs under study, one can conclude that the appearance of defects leads to the appearance of new absorption peaks, and at some wavelengths it also leads to an increase in the absorption coefficient. One such example is given for a SWCNT (12,8) in Figure 1c. Based on the calculated data, we can say that defective regions of SWCNTs are capable of absorbing more energy from external radiation than defect-free regions. And besides, defective regions are capable of absorbing energy in a wider range of wavelengths than defect-free ones.

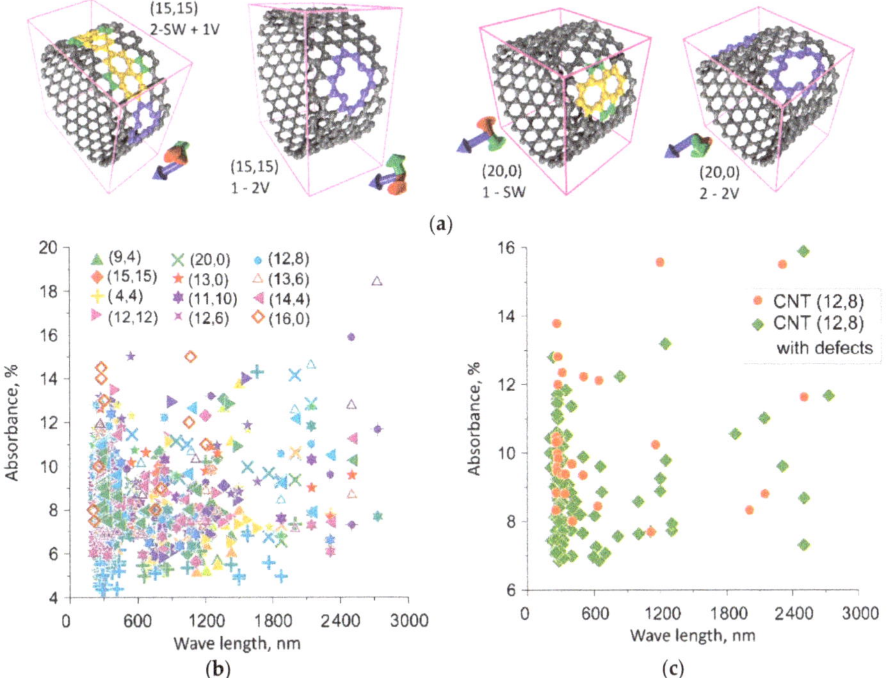

Figure 1. Absorption of the energy of electromagnetic waves by defective single-walled carbon nanotubes (SWCNTs): (**a**) Examples of atomistic models of defective SWCNTs (lilac box corresponds to a supercell); (**b**) distribution of the absorption coefficient maxima for the defective SWCNTs with a diameter of 0.6–2 nm; (c) distribution of the absorption coefficient maxima for defect-free and defective SWCNT (12,8).

The next step in studying the absorption capacity of SWCNTs was to study the regularities of absorption of electromagnetic wave energy by the open ends of SWCNTs of finite length. The above calculation data were obtained for SWCNTs of infinite length, since we applied periodic boundary conditions. In this connection, a logical question arises: How do short SWCNTs absorb energy and what is the role of the open ends of SWCNTs in this case? To answer this question, we investigated SWCNTs with a length of several supercells. The maximum length of the considered short SWCNTs was 38–40 nm. The choice of this length is due to the large number of atoms in chiral SWCNTs, namely, 6000–8000. Nanotubes with a large number of atoms cannot be investigated by quantum methods. To calculate the absorption coefficient of electromagnetic waves by the open ends of SWCNTs, we built a model of a film ~40 nm thick corresponding to the length of SWCNTs. SWCNTs were arranged in parallel, wherein the outer surface of SWCNTs was formed by the open ends. The wave vector of the incident wave was directed normal to the film, that is, along the axes of the SWCNTs. This model allows us to calculate the absorption properties of the open ends of the SWCNTs. However, since the film thickness was 40 nm, the considered wavelength range was 100–3000 nm. Figure 2 shows the distribution diagrams of the absorption maxima: Figure 2a—for SWCNTs of various chiralities and diameters, Figure 2b—for achiral semiconductor SWCNTs and metal SWCNTs (in the inset). In contrast to extended SWCNTs that absorb energy by the side surface, the open ends of short SWCNTs provide significantly greater energy absorption. Figure 2a shows that in the UV region, the absorption maxima are concentrated in the wavelength range of 200–400 nm; in the visible region, the absorption maxima are concentrated in the wavelength range of 600–700 nm. In the IR region, there is no such pronounced interval; the absorption maxima are distributed over the entire wavelength range of 800–3000 nm. The largest absorption maxima are observed at wavelengths of 1700, 2200 and 2300 nm. The inset in Figure 2b shows the distribution diagrams of absorption maxima above 10%. In the case of metal SWCNTs, the absorption reaches even 40% at wavelengths of 650–750 nm and 1050–1250 nm. In the case of semiconductor SWCNTs, the absorption reaches 36% only in the UV range at wavelengths of 250–350 nm.

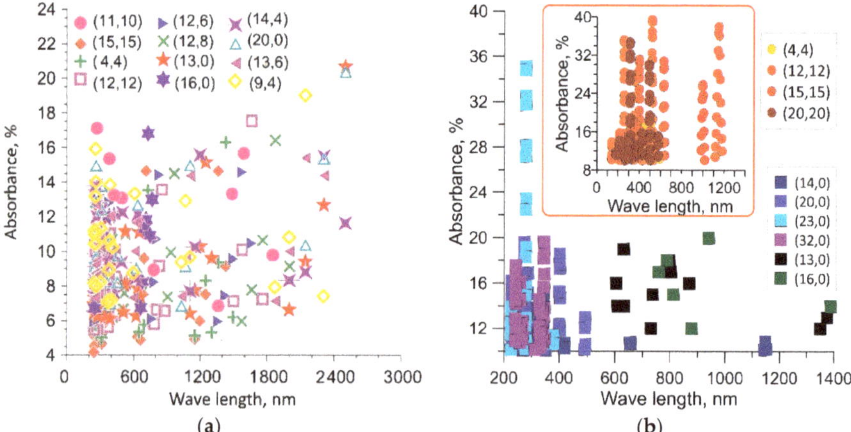

Figure 2. Distribution of absorption maxima for SWCNTs of different diameters: (**a**) SWCNTs of different chiralities; (**b**) achiral SWCNTs (circles correspond to armchair SWCNTs, squares—zigzag SWCNTs).

3.2. The formation of A Network of SWCNTs under The Action of Laser Radiation

Further, by the example of the SWCNTs (4,4), we studied the process of carbon network formation as a result of absorption of laser beam energy. A carbon network is a branched structure of SWCNTs. The choice of SWCNTs (4,4) is due to their small diameter, which makes it possible to introduce a larger number of SWCNTs into the model of SWCNT network with the same total number of atoms. Figure 3a shows a fragment of a network of SWCNTs (4,4). In total, the network model included 258 SWCNTs with a length of 5–20 nm. SWCNTs were located in a periodic box with size of 65 nm × 45 nm × 40 nm. The total number of atoms in the model was 174,776. SWCNTs were arranged randomly taking into account the van der Waals interaction between them, but without the formation of covalent bonds between the SWCNTs. The density of the created SWCNT network was 60 kg/m^3. After that, defects of various types were created in those places where the SWCNTs were located at a distance of 0.2–0.4 nm. The introduction of defects is necessary for the implementation of the nanowelding process. The nanowelding process was implemented by non-uniform temperature distribution. The interaction of an SWCNT array with a laser beam consists of non-uniform absorption of electromagnetic wave energy, that is, in non-uniform heating. According to the absorption coefficients calculated by us, in the defective regions and in the vicinity of the open ends of the SWCNTs, the temperature was 16%–20% higher than the temperature in defect-free regions. Further, a molecular dynamics simulation of the welding process was carried out. The red color in Figure 3a denotes the nanowelding points at the places of jointing open ends of the SWCNTs with defective SWCNTs, as well as at the junctions of the SWCNTs with each other due to the interaction between atoms of the defective regions. Figure 3b shows a plot of change in the carbon network energy (in units of energy per atom eV/atom, the violet axis and the violet curve) during the formation of covalent bonds between SWCNTs. The number of formed bonds is given as a percentage of the total possible number of covalent bonds for a given network. It can be seen that the energy decreases nonlinearly as the formed bonds increase. The plots of time of the formation of covalent bonds between the SWCNTs (red, green, blue, and dark blue curves) versus distance between them are also presented in Figure 3b. As expected, nanowelding most rapidly occurs at a small distance between the SWCNTs. The presented regularity of energy change (purple curve) is characteristic for all cases, regardless of the distance between the tubes, and is the result of averaging a large number of numerical experiments. The energy of the structure and the time of formation of covalent CNT–CNT bonds are directly linked. This can be seen from Figure 3b. The more CNT–CNT bonds are formed, the lower the energy, which means that the process of forming covalent bonds between the CNTs in the nanowelding process is energetically beneficial. According to our calculations, in general, a few picoseconds are enough to form bonds.

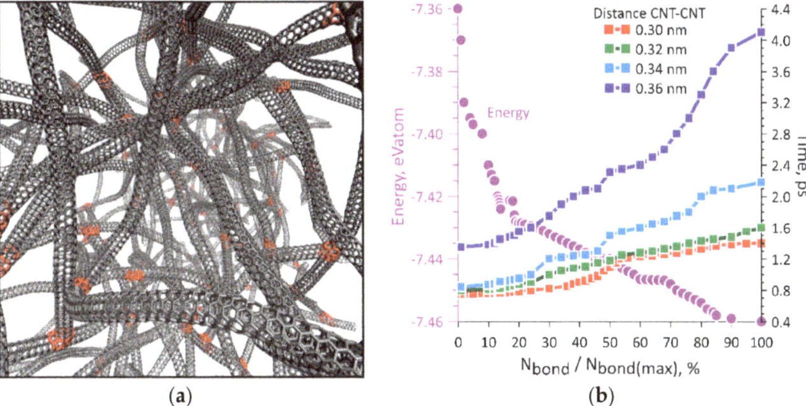

Figure 3. Formation of a carbon network: (**a**) A branched network of SWCNTs (4,4) with red regions corresponding to the nanowelding points; (**b**) plots of change in the energy of the network of SWCNTs (purple curve) and time of the formation of covalent bonds.

3.3. The Formation of Layered Protein–Polymer Matrices

Earlier, we studied the regularities of formation of protein–polymer matrices based on a network of SWCNTs, albumin, collagen, and chitosan [34]. In this paper, we study the regularities of formation of a layered protein–polymer matrix of three layers, including CNT-albumin, CNT-collagen, and CNT-chitosan layers. For this, as in the previous study [34], we used the coarse-grained modeling method. Coarse-grained models of individual matrix layers are shown in Figure 4. Figure 4a shows a periodic box of the SWCNT + albumin protein matrix, where SWCNTs with a diameter of 1.6 nm and albumin macromolecules are located. The size of the periodic box was 30 nm × 30 nm × 30 nm. The periodic box contains 19,827 grains of SWCNTs and 25,970 grains of five albumin molecules. Grains of water molecules are marked in blue. The box contains 200,000 grains of water. In total, the structure contains 245,797 grains. The periodic box of the SWCNT + collagen protein matrix is shown in Figure 4b, where 160,000 water grains, 19,827 SWCN grains and 56,498 collagen grains are present. The periodic box of the SWCNT + chitosan polymer matrix is shown in Figure 4c. Since chitosan chains are very numerous, SWCNTs are marked in bright red, and chitosan in green, pink, and blue. This box contains 200,000 water grains and 9,237 chitosan grains. All structures were optimized with a thermostat 310 K.

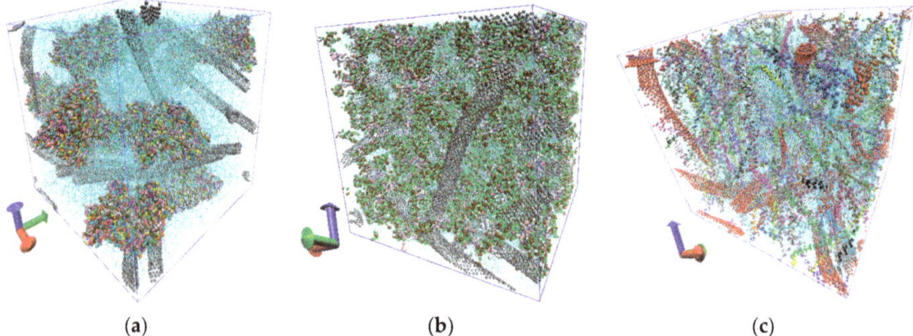

(a)   (b)   (c)

Figure 4. Coarse-grained models of individual layers of protein–polymer matrices: (**a**) SWCNT–albumin; (**b**) SWCNT–collagen; (**c**) SWCNT–chitosan.

Next, a layered structure of SWCNTs and natural polymers was built. The construction principle was as follows. Two different layers were taken and combined into a new periodic box, which is translated only in two directions along X and Y. The distance between adjacent layers was taken equal to the van der Waals distance between the components forming the layers. Molecular dynamic optimization was carried out at a temperature of 310 K. The box was also optimized in two directions. The result was a periodic box of a two-layer structure. During optimization, the components at the boundary of two layers mutually penetrated into the adjacent layer. An example of a layered protein matrix of SWCNTs, albumin, and collagen is shown in Figure 5a. Coarse-grained models of SWCNTs are shown in gray. Layers with different proteins are marked. The energy of such a structure changed during the formation and its plot is shown in Figure 5b. As the components of the layers at the boundary were optimized, the energy fell until it stabilized and stopped changing, which indicates the stabilization of the two-layer protein matrix. Similarly, a model of a two-layer protein–polymer matrix was constructed from SWCNTs, albumin, and chitosan. It is presented in Figure 5c. To make it easier to see both layers, all the chitosan threads are depicted in pink (bottom layer). SWCNTs are marked in gray. The plot of energy change is presented in Figure 5d, which shows the decrease in energy during the optimization of the structure. In contrast to the previous case, when the two-layer protein matrix stabilized in 4 ns, in this case the stabilization time was 6 ns. Another two-layer protein–polymer matrix is shown in Figure 5e, where the SWCNT–collagen layer acts as a protein layer. In this case, the structure optimization was already completed within the first 2 ns, as can be seen from the plot in

Figure 5f. This is explained by the fact that collagen contacts to SWCNTs rather tightly; therefore, large changes are not observed at the boundary of two layers. The thickness of each layer was ~11.5 nm in all cases; the size of the periodic box in the XY plane was ~19 nm × 19 nm in all cases.

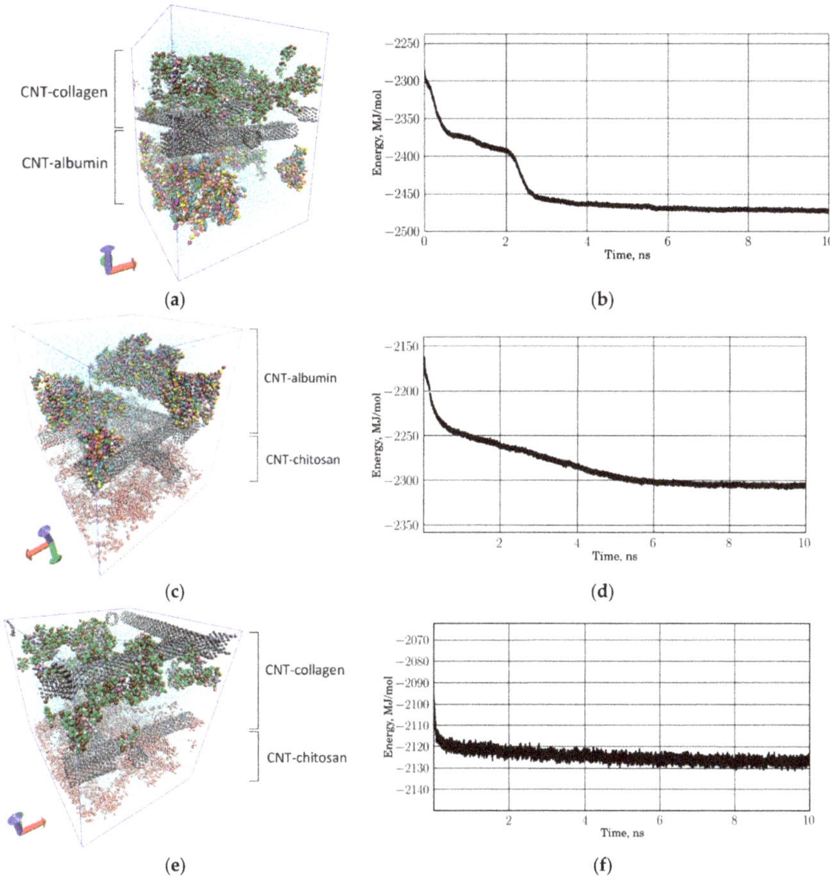

Figure 5. Layered 2D protein-polymer matrices: Coarse-grained model (**a**) and energy of albumin–collagen structure (**b**); coarse-grained model (**c**) and energy of albumin–chitosan structure (**d**); coarse-grained model (**e**) and energy of collagen–chitosan structure (**f**).

3.4. The Interaction of Layered Protein–Polymer Structures with the Cell Membrane

A layer of phospholipid molecules was taken as a cell membrane. The membrane is a layer of 8,450 coarse-grained molecules of DPPC (Dipalmitoylphosphatidylcholine). The total number of coarse grains in the membrane is 101,400. The length and width of the membrane are 50 nm, and the thickness is 4 nm. First, we studied the process of interaction of a CNT–albumin protein matrix with a membrane (see Figure 6). The model shown in Figure 4a was used. The process of contact of the membrane with the matrix was modeled as follows. The membrane moves to the matrix with an average blood flow velocity of 0.5 m/s, which simulates the collision of blood elements, for example, an erythrocyte with the surface of the CNT-albumin protein matrix. The surface of the matrix is sticking out SWCNTs with albumin molecules. SWCNTs are arranged at different angles in an arbitrary way. Additionally, part of the SWCNTs is strictly vertically oriented. When the membrane falls on the surface of the matrix, it is pierced by SWCNTs. Then the membrane rises, returning to its original position. This simulates the rolling of blood elements on the surface of a protein matrix. Figure 6a shows snapshots of the process of detachment of the membrane from the surface. The point in time 0 ps in the figure corresponds to the moment when the membrane begins to detach. The total time was 100 ps. In Figure 6a, the SWCNT network is presented in pink, the grains of albumin molecules have yellow and red colors, and the hydrophilic parts of the DPPC molecules are shown in green and hydrophobic heads in blue. For ease of viewing, water grains are removed. Molecular dynamic modeling of this process showed that three of the five albumin molecules are captured by the membrane and "stuck" to its surface. This is due to the fact that these molecules are directly on the surface. Two other albumin molecules located deep in the SWCNT framework remain in the matrix. Snapshots in Figure 6a also demonstrate that within 90–100 ps, the holes formed in the membrane are almost completely drawn out. Figure 6b,c show plots of the total energy of the matrix + membrane system during the falling membrane onto the surface of the matrix (see Figure 6b) and during the detachment of the membrane with its subsequent distancing (see Figure 6c). During contact and subsequent interaction, when some SWCNTs pierce the membrane, the energy increases sharply during the first ten picoseconds (see Figure 6b). With a further membrane falling, the energy increases, but more slowly. A similar behavior of energy is observed in the case of detachment and reverse movement of the membrane (see Figure 6c). In this case, a sharp increase in the energy is due to the beginning of the process of detachment of some albumin molecules from the carbon framework. Within 90 ps, the process of detachment of albumin molecules from SWCNTs finishes, therefore, an increase in energy replaced by its decrease. And further, the decrease in energy continues until the membrane structure is completely restored.

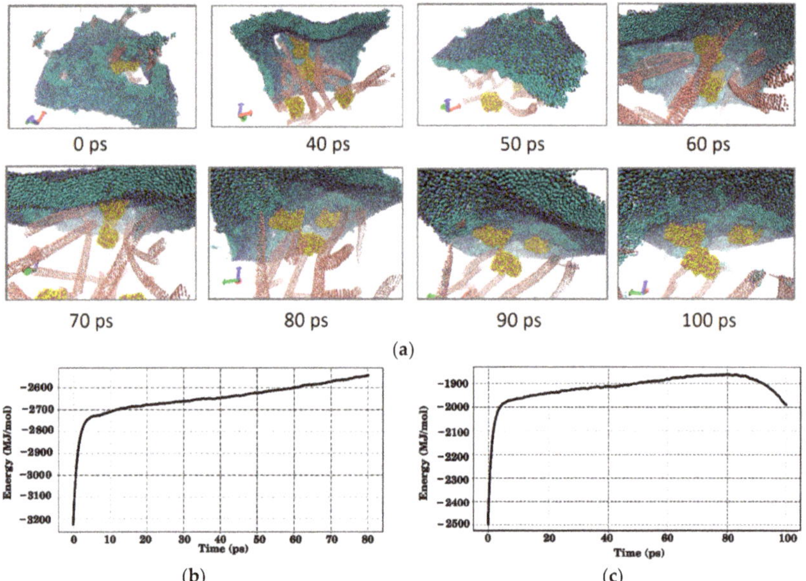

Figure 6. The interaction of the membrane with the SWCNT-albumin protein matrix: (**a**) Snapshots of the process of detachment of the membrane from the surface of the matrix; (**b**) a plot of change in the energy of the matrix + membrane system when the membrane falls on the matrix; (**c**) a plot of change in the energy of the matrix + membrane system when the membrane detaches from the matrix and moves away from it (SWCNTs are marked in pink, albumin molecules are marked in yellow and red, and the membrane is marked green and blue).

Similarly, the falling of the membrane onto the SWCNT–collagen protein matrix was studied. Based on the experience of studying the contact of the membrane with the SWCNT-albumin matrix, we took a fragment of the model imaged in Figure 4b, where the SWCNTs are vertically oriented relative to the membrane. This was done due to the fact that the most important is the piercing of the membrane by SWCNTs and the process of the membrane subsequent recovery. Collagen molecules are directly located on the SWCNTs. The simulation result of the process of membrane falling onto the matrix is shown in Figure 7a. The initial moment is the moment when the membrane is pierced by SWCNTs in contact with the matrix, and then the process of detachment of the membrane begins. Within 70 ps, a complete detachment of the membrane occurs, wherein some of the collagen molecules remain on the membrane. The membrane is deformed and bears traces of contacts with SWCNTs in the form of holes. Over the next 30 ps, the holes gradually tighten and by the point in time 100 ps they are almost gone. It can be concluded that the membrane is not damaged, since only a few DPPC molecules remain on the SWCNTs after contact with the membrane. All this process of interaction of the membrane with the matrix can be traced by change in the energy of the entire membrane + matrix system. A plot of energy in Figure 7b for a case of membrane falling onto the matrix clearly shows that when the membrane penetrates the SWCNTs, the energy of the entire system increases sharply. This occurs in the first 10 ps. Further, the energy slowly increases, which corresponds to the further threading of the membrane on the SWCNTs. The energy of the matrix + membrane system from the beginning of the membrane rise is shown in Figure 7c. It is completely identical to the visual snapshots of Figure 7a. During the first 70 ps after the beginning of detachment, the energy practically does not change, increasing very slowly. And then it decreases sharply when the membrane detached from SWCNTs. Further molecular dynamics modeling (after 100 ps) shows that the membrane completely restores its initial structure and shape.

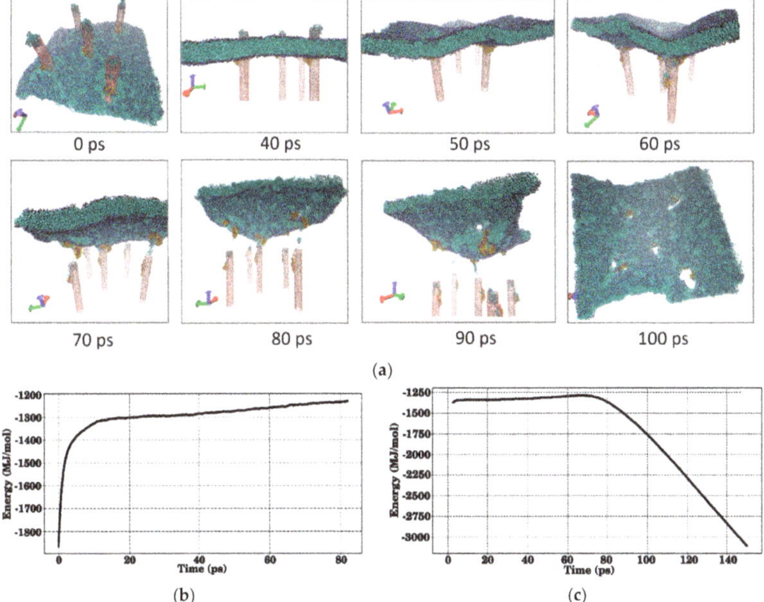

Figure 7. The interaction of the membrane with the SWCNT-collagen protein matrix: (**a**) Snapshots of the process of detachment of the membrane from the surface of the matrix; (**b**) a plot of change in the energy of the matrix + membrane system when the membrane falls on the matrix; (**c**) a plot of change in the energy of the matrix + membrane system when the membrane detaches from the matrix and moves away from it (SWCNTs are marked in pink, albumin molecules are marked in yellow and red, and the membrane is marked in green and blue).

And, finally, the process of membrane contact with a fragment of the SWCNT–chitosan matrix was studied. As in the previous case, a small fragment was taken with vertically oriented SWCNTs with respect to the membrane. The entire process of membrane contact with SWCNTs and chitosan threads between them is shown in Figure 8. The change in the SWCNT–chitosan structure is shown in Figure 8a, the change in energy during the membrane falling onto the matrix is shown in Figure 8b, and the change in energy during the detachment of the membrane is shown in Figure 8c. In general, the process is completely identical to the previous case. Some of the chitosan molecules "stick" to the membrane; some molecules remain in the matrix. Holes in the membrane from contact with nanotubes disappear within 100–200 ps.

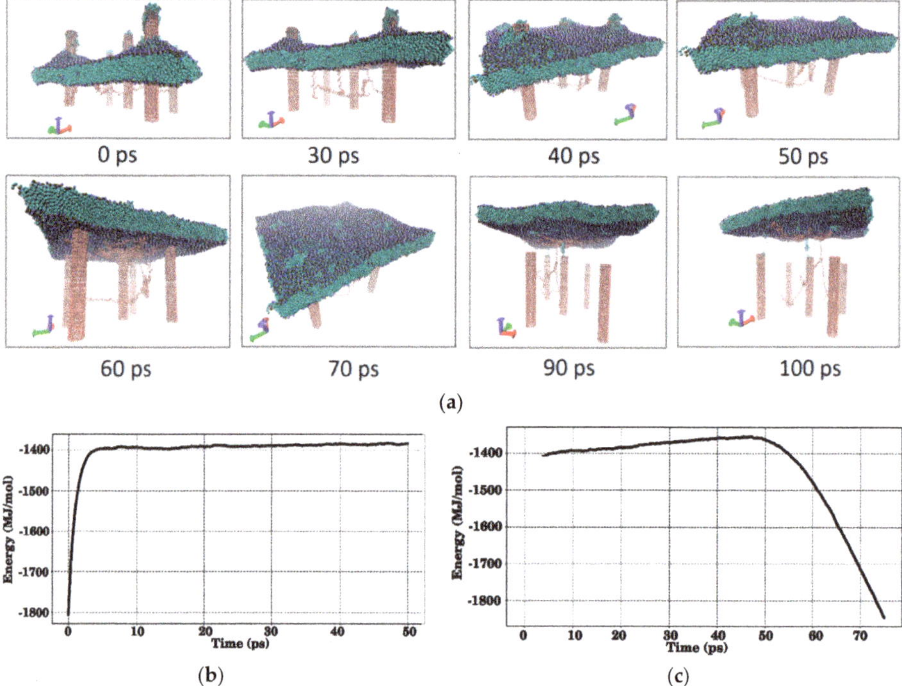

Figure 8. The interaction of the membrane with the CNT–chitosan polymer matrix: (**a**) Snapshots of the process of detachment of the membrane from the surface of the matrix; (**b**) a plot of change in the energy of the matrix + membrane system when the membrane falls on the matrix; (**c**) a plot of change in the energy of the matrix + membrane system when the membrane detaches from the matrix and its backward movement from the matrix (CNTs and chitosan filaments are marked in pink, the membrane is marked in green and blue).

4. Conclusions

The studied regularities of interaction of SWCNTs with various electromagnetic waves in the range of 10–3000 nm showed that defective regions of SWCNTs absorb more energy than defect-free ones. The same effect is observed for SWCNT regions near their open ends. It was found that the absorption of the incident wave energy can reach 36%–40%. This allows us to make an assumption that nanowelding of SWCNTs and the formation of a branched SWCNT network will be carried out precisely in these regions, which will act as a hot-spot. Molecular dynamic modeling of SWCNT nanowelding showed that this process occurs rather quickly in time for several tens of picoseconds. The time of nanowelding with the formation of covalent bonds between SWCNTs is determined primarily by the distance between the contacting SWCNTs. The shorter this distance, the faster new bonds form.

The modeling of the formation of layered protein and polymer matrices based on SWCNTs, albumin, collagen, and chitosan showed that individual layers of these matrices are able to interact with each other. This interaction is energetically beneficial. This shows that such a layered biocompatible material is very promising in biomedical applications, in particular in tissue engineering. In silico studies of the interaction of protein and polymer matrices with membranes showed that the matrices do no harm to the cell membrane. The resulting small holes from the contacts with nanotubes are rapidly tightened within 100–200 ps. Some molecules of albumin, collagen, or chitosan that adhere to the membrane do not harm the membrane because they are completely biocompatible. In the future, mechanical properties of layered protein and polymer matrices with embedded carbon nanotubes will be considered.

Author Contributions: Conceptualization, O.E.G.; Methodology, O.E.G.; Validation, A.Y.G. and D.V.T.; Funding acquisition, A.Y.G. and D.V.T.; Resources, D.V.T.; Funding acquisition, M.M.S.; Investigation, O.E.G. and M.M.S.; Writing-Original Draft Preparation, O.E.G. and M.M.S.; Writing-Review & Editing, O.E.G.; Supervision, O.E.G.

Funding: This research was funded by the grant of the President of the Russian Federation (project no. MK-2373.2019.2).

Conflicts of Interest: The authors declare no conflict of interest.

References

1. Marchesan, S.; Kostarelos, K.; Bianco, A.; Prato, M. The winding road for carbon nanotubes in nanomedicine. *Mater. Today* **2015**, *18*, 12–19. [CrossRef]
2. Gerasimenko, A.Y.; Glukhova, O.E.; Savostyanov, G.V.; Podgaetsky, V.M. Laser structuring of carbon nanotubes in the albumin matrix for the creation of composite biostructures. *J. Biomed. Opt.* **2017**, *22*, 065003. [CrossRef] [PubMed]
3. Scarselli, M.; Castrucci, P.; De Nicola, F.; Cacciotti, I.; Nanni, F.; Gatto, E.; Venanzi, M.; De Crescenzi, M. Applications of three-dimensional carbon nanotube networks. *Beilstein J. Nanotechnol.* **2015**, *6*, 792–798. [CrossRef] [PubMed]
4. Kobashi, K.; Ata, S.; Yamada, T.; Futaba, D.N.; Yumura, M.; Hata, K. A dispersion strategy: Dendritic carbon nanotube network dispersion for advanced composites. *Chem. Sci.* **2013**, *4*, 727–733. [CrossRef]
5. Xue, Y. Carbon nanotubes for biomedical applications. In *Industrial Applications of Carbon Nanotubes*, 1st ed.; Peng, H., Li, Q., Chen, T., Eds.; Elsevier: Oxford, UK, 2017; pp. 323–346.
6. Alshehri, R.; Ilyas, A.M.; Hasan, A.; Arnaout, A.; Ahmed, F.; Memic, A. Carbon nanotubes in biomedical applications: Factors, mechanisms, and remedies of toxicity. *J. Med. Chem.* **2016**, *59*, 8149–8167. [CrossRef] [PubMed]
7. Huang, B.; Vyas, C.; Roberts, I.; Poutrel, Q.A.; Chiang, W.H.; Blaker, J.J.; Huang, Z.; Bártolo, P. Fabrication and characterisation of 3D printed MWCNT composite porous scaffolds for bone regeneration. *Mater. Sci. Eng. C-Mater.* **2019**, *98*, 266–278. [CrossRef]
8. Peña, B.; Bosi, S.; Aguado, B.A.; Borin, D.; Farnsworth, N.L.; Dobrinskikh, E.; Rowland, T.J.; Martinelli, V.; Jeong, M.; Taylor, M.R.G.; et al. Injectable carbon nanotube-functionalized reverse thermal gel promotes cardiomyocytes survival and maturation. *ACS Appl. Mater. Interfaces* **2017**, *9*, 31645–31656. [CrossRef]
9. Nguyen, A.H.; Marsh, P.; Schmiess-Heine, L.; Burke, P.J.; Lee, A.; Lee, J.; Cao, H. Cardiac tissue engineering: State-of-the-art methods and outlook. *J. Biol. Eng.* **2019**, *13*, 57. [CrossRef]
10. Alegret, N.; Dominguez-Alfaro, A.; González-Domínguez, J.M.; Arnaiz, B.; Cossío, U.; Bosi, S.; Vázquez, E.; Ramos-Cabrer, P.; Mecerreyes, D.; Prato, M. Three-dimensional conductive scaffolds as neural prostheses based on carbon nanotubes and polypyrrole. *ACS Appl. Mater. Interfaces* **2018**, *10*, 43904–43914. [CrossRef]
11. Kia, K.K.; Bonabi, F.; Shokrzadeh, M. Electric field-oriented growth of carbon nanotubes and y-branches in a needle-pulsed arc discharge unit: Mechanism of the production. *J. Exp. Nanosci.* **2015**, *10*, 1093–1105.
12. Wei, Q.; Liu, Y.; Zhang, L.; Huang, S. Growth and formation mechanism of branched carbon nanotubes by pyrolysis of Iron(II) phthalocyanine. *Nano-Micro Lett.* **2013**, *5*, 124–128. [CrossRef]

13. Goswami, G.K.; Nandan, R.; Nanda, K.K. Growth of branched carbon nanotubes with doped/un-doped intratubular junctions by one-step co-pyrolysis. *Carbon* **2013**, *56*, 97–102. [CrossRef]

14. Wei, X.L.; Liu, Y.; Chen, Q.; Peng, L.M. Controlling electron-beam-induced carbon deposition on carbon nanotubes by Joule heating. *Nanotechnology* **2008**, *19*, 355304. [CrossRef] [PubMed]

15. Ni, Z.; Li, Q.; Yan, L.; Gong, J.; Zhu, D. Welding of multi-walled carbon nanotubes by ion beam irradiation. *Carbon* **2008**, *46*, 376–378. [CrossRef]

16. Dong, L.; Tao, X.; Zhang, L.; Zhang, X.; Nelson, B.J. Nanorobotic spot welding: Controlled metal deposition with attogram precision from copper-filled carbon nanotubes. *Nano Lett.* **2007**, *7*, 58–63. [CrossRef] [PubMed]

17. Yao, Y.; Fu, K.K.; Zhu, S.; Dai, J.; Wang, Y.; Pastel, G.; Chen, Y.; Li, T.; Wang, C.; Li, T.; et al. Carbon welding by ultrafast joule heating. *Nano Lett.* **2016**, *16*, 7282–7289. [CrossRef] [PubMed]

18. Ozden, S.; Brunetto, G.; Karthiselva, N.S.; Galvão, D.S.; Roy, A.; Bakshi, S.R.; Tiwary, C.S.; Ajayan, P.M. Controlled 3D carbon nanotube structures by plasma welding. *Adv. Mater. Interfaces* **2016**, *3*, 1500755. [CrossRef]

19. Yang, L.; Cui, J.; Wang, Y.; Hou, C.; Xie, H.; Mei, X.; Wang, W.; Wang, K. Nanospot welding of carbon nanotubes using near-field enhancement effect of AFM probe irradiated by optical fiber probe laser. *RSC Adv.* **2015**, *5*, 56677–56685. [CrossRef]

20. Guo, S. The creation of nanojunctions. *Nanoscale* **2010**, *2*, 2521–2529. [CrossRef]

21. Liu, Z.; Yuan, Y.; Shang, Y.; Han, W. Structural changes and electrical properties of nanowelded multiwalled carbon nanotube junctions. *Appl. Opt.* **2018**, *57*, 7435–7439. [CrossRef]

22. Yuan, Y.; Liu, Z.; Zhang, K.; Han, W.; Chen, J. Nanoscale welding of multi-walled carbon nanotubes by 1064 nm fiber laser. *Opt. Laser Technol.* **2018**, *103*, 327–329. [CrossRef]

23. Yuan, Y.; Chen, J. Nano-welding of multi-walled carbon nanotubes on silicon and silica surface by laser irradiation. *Nanomaterials* **2016**, *6*, 36. [CrossRef] [PubMed]

24. Yuan, Y.; Chen, J. Morphology adjustments of multi-walled carbon nanotubes by laser irradiation. *Laser Phys. Lett.* **2016**, *13*, 066001. [CrossRef]

25. Krasheninnikov, A.V.; Nordlund, K.; Keinonen, J.; Banhart, F. Ion-irradiation-induced welding of carbon nanotubes. *Phys. Rev. B* **2002**, *66*, 245403. [CrossRef]

26. Terrones, M.; Banhart, F.; Grobert, N.; Charlier, J.C.; Terrones, H.; Ajayan, P.M. Molecular junctions by joining single-walled carbon nanotubes. *Phys. Rev. Lett.* **2002**, *89*, 075505. [CrossRef] [PubMed]

27. Bullard, Z.; Meunier, V. Dynamical properties of carbon nanotube welding into X junctions. *Phys. Rev. B* **2013**, *88*, 035422. [CrossRef]

28. Cui, J.; Yang, L.; Wang, Y. Molecular dynamics study of the positioned single-walled carbon nanotubes with T-, X-, Y- junction during nanoscale soldering. *Appl. Surf. Sci.* **2013**, *284*, 392–396. [CrossRef]

29. Cui, J.; Yang, L.; Wang, Y. Nanowelding configuration between carbon nanotubes in axial direction. *Appl. Surf. Sci.* **2013**, *264*, 713–717. [CrossRef]

30. Cui, J.; Yang, L.; Zhou, L.; Wang, Y. Nanoscale soldering of axially positioned single-walled carbon nanotubes: A molecular dynamics simulation study. *ACS Appl. Mater. Interfaces* **2014**, *6*, 2044–2050. [CrossRef]

31. Kirca, M.; Yang, X.; To, A.C. A stochastic algorithm for modeling heat welded random carbon nanotube network. *Comput. Methods Appl. Mech. Eng.* **2013**, *259*, 1–9. [CrossRef]

32. Zhang, Y.; Gong, T.; Jia, Y.; Liu, W.; Wei, J.; Ma, M.; Wang, K.; Zhong, M.; Wu, D.; Cao, A. Tailoring the intrinsic metallic states of double-walled nanotube films by self-soldered laser welding. *Appl. Phys. Lett.* **2007**, *91*, 233109. [CrossRef]

33. Kroustalli, A.; Zisimopoulou, A.E.; Koch, S.; Rongen, L.; Deligianni, D.; Diamantouros, S.; Athanassiou, G.; Kokozidou, M.; Mavrilas, D.; Jockenhoevel, S. Carbon nanotubes reinforced chitosan films: Mechanical properties and cell response of a novel biomaterial for cardiovascular tissue engineering. *J. Mater. Sci. Mater. Med.* **2013**, *24*, 2889–2896. [CrossRef] [PubMed]

34. Savostyanov, G.V.; Slepchenkov, M.M.; Shmygin, D.S.; Glukhova, O.E. Specific features of structure, electrical conductivity and interlayer adhesion of the natural polymer matrix from the layers of branched carbon nanotube networks filled with albumin, collagen and chitosan. *Coatings* **2018**, *8*, 378. [CrossRef]

35. Hsu, C.C.; Serio, A.; Amdursky, N.; Besnard, C.; Stevens, M.M. Fabrication of hemin-doped serum albumin-based fibrous scaffolds for neural tissue engineering applications. *ACS Appl. Mater. Interfaces* **2018**, *10*, 5305–5317. [CrossRef] [PubMed]

36. Aiyelabegan, H.T.; Zaidi, S.S.Z.; Fanuel, S.; Eatemadi, A.; Ebadi, M.T.K.; Sadroddiny, E. Albumin-based biomaterial for lung tissue engineering applications. *Int. J. Polym. Mater. Po.* **2016**, *65*, 853–861. [CrossRef]

37. Fleischer, S.; Shapira, A.; Regev, O.; Nseir, N.; Zussman, E.; Dvir, T. Albumin fiber scaffolds for engineering functional cardiac tissues. *Biotechnol. Bioeng.* **2014**, *111*, 1246–1257. [CrossRef] [PubMed]

38. Overby, R.J.; Feldman, D.S. Influence of Poly(Ethylene Glycol) End groups on Poly(Ethylene Glycol)-albumin system properties as a potential degradable tissue scaffold. *J. Funct. Biomater.* **2019**, *10*, 1. [CrossRef] [PubMed]

39. Cen, L.; Liu, W.; Cui, L.; Zhang, W.; Cao, Y. Collagen tissue engineering: Development of novel biomaterials and applications. *Pediatr. Res.* **2008**, *63*, 492–496. [CrossRef]

40. Dong, C.; Lv, Y. Application of collagen scaffold in tissue engineering: Recent advances and new perspectives. *Polymers (Basel)* **2016**, *8*, 42. [CrossRef]

41. Irawan, V.; Sung, T.C.; Higuchi, A.; Ikoma, T. Collagen scaffolds in cartilage tissue engineering and relevant approaches for future development. *Tissue Eng. Regen Med.* **2018**, *15*, 673–697. [CrossRef]

42. Pandey, A.R.; Singh, U.S.; Momin, M.; Bhavsar, C. Chitosan: Application in tissue engineering and skin grafting. *J. Polym. Res.* **2017**, *24*, 125. [CrossRef]

43. Ahsan, S.M.; Thomas, M.; Reddy, K.K.; Sooraparaju, S.G.; Asthana, A.; Bhatnagar, I. Chitosan as biomaterial in drug delivery and tissue engineering. *Int. J. Biol. Macromol.* **2018**, *110*, 97–109. [CrossRef] [PubMed]

44. Glukhova, O.E.; Nefedov, I.S.; Shalin, A.S.; Slepchenkov, M.M. New 2D graphene hybrid composites as an effective base element of optical nanodevices. *Beilstein J. Nanotechnol.* **2018**, *9*, 1321–1327. [CrossRef] [PubMed]

45. Marder, M.P. *Condensed Matter Physics*, 2nd ed.; Wiley-VCH: Berlin, Germany, 2011; pp. 623–631.

46. Le, H.A.; Ho, S.T.; Nguyen, D.C.; Do, V.N. Optical Properties of Graphene Superlattices. *J. Phys. Condens. Matter* **2014**, *26*, 405304. [CrossRef]

47. Glukhova, O.E. Molecular dynamics as the tool for investigation of carbon nanostructures properties. In *Thermal Transport in Carbon-Based Nanomaterials*, 1st ed.; Zhang, G., Ed.; Elsevier: Oxford, UK, 2017; pp. 267–289.

48. Aradi, B.; Hourahine, B.; Frauenheim, T. DFTB+, a sparse matrix-based implementation of the DFTB method. *J. Phys. Chem. A* **2007**, *111*, 5678–5684. [CrossRef] [PubMed]

49. Nikitin, A.; Ogasawara, H.; Mann, D.; Denecke, R.; Zhang, Z.; Dai, H.; Cho, K.; Nilsson, A. Hydrogenation of Single-Walled Carbon Nanotubes. *Phys. Rev. Lett.* **2005**, *95*, 225507. [CrossRef] [PubMed]

50. de Jong, D.H.; Singh, G.; Bennett, W.F.D.; Arnarez, C.; Wassenaar, T.A.; Schäfer, L.V.; Periole, X.; Tieleman, D.P.; Marrink, S.J. Improved parameters for the martini coarse-grained protein force Field. *J. Chem. Th. Comp.* **2013**, *9*, 687–697. [CrossRef]

51. GROMACS. Available online: http://www.gromacs.org (accessed on 15 April 2019).

52. Park, J.; Bifano, M.F.P.; Prakash, V. Sensitivity of thermal conductivity of carbon nanotubes to defect concentrations and heat-treatment. *J. Appl. Phys.* **2013**, *113*, 034312. [CrossRef]

53. Kumanek, B.; Janas, D. Thermal conductivity of carbon nanotube networks: A review. *J. Mater. Sci.* **2019**, *54*, 7397–7427. [CrossRef]

54. Che, J.; Cagın, T.; Goddard, W.A., III. Thermal conductivity of carbon nanotubes. *Nanotechnology* **2000**, *11*, 65–69. [CrossRef]

55. Chang, M.; Fan, H.D.E.; Chowdhury, M.M.; Sawatzky, G.A.; Nojeh, A. Heat localization through reduced dimensionality. *Phys. Rev. B* **2018**, *98*, 155422. [CrossRef]

56. Bakan, G.; Khan, N.; Cywar, A.; Cil, K.; Akbulut, M.; Gokirmak, A.; Silva, H. Self-heating of silicon microwires: Crystallization and thermoelectric effects. *J. Mat. Res.* **2011**, *26*, 1061. [CrossRef]

57. Mecklenburg, M.; Hubbard, W.A.; White, E.R.; Dhall, R.; Cronin, S.B.; Aloni, S.; Regan, B.C. Nanoscale temperature mapping in operating microelectronic devices. *Science* **2015**, *347*, 629. [CrossRef] [PubMed]

Article

Lateral Spacing of TiO$_2$ Nanotubes Modulates Osteoblast Behavior

Madalina Georgiana Necula [1,†], **Anca Mazare** [2,†], **Raluca Nicoleta Ion** [1], **Selda Ozkan** [2], **Jung Park** [3], **Patrik Schmuki** [2] **and Anisoara Cimpean** [1,*]

[1] Department of Biochemistry and Molecular Biology, University of Bucharest, 050095 Bucharest, Romania; necula.madalina92@gmail.com (M.G.N.); rciubar@yahoo.com (R.N.I.)

[2] Department of Materials Science WW4-LKO, Friedrich-Alexander University, 91058 Erlangen, Germany; anca.mazare@fau.de (A.M.); selda.oezkan@ww.uni-erlangen.de (S.O.); schmuki@ww.uni-erlangen.de (P.S.)

[3] Division of Molecular Pediatrics, Department of Pediatrics, University Hospital Erlangen, 91054 Erlangen, Germany; jung.park@uk-erlangen.de

* Correspondence: anisoara.cimpean@bio.unibuc.ro

† Both authors contributed equally to this work.

Received: 14 August 2019; Accepted: 10 September 2019; Published: 12 September 2019

Abstract: Titanium dioxide (TiO$_2$) nanotube coated substrates have revolutionized the concept of implant in a number of ways, being endowed with superior osseointegration properties and local drug delivery capacity. While accumulating reports describe the influence of nanotube diameter on cell behavior, little is known about the effects of nanotube lateral spacing on cells involved in bone regeneration. In this context, in the present study the MC3T3-E1 murine pre-osteoblast cells behavior has been investigated by using TiO$_2$ nanotubes of ~78 nm diameter and lateral spacing of 18 nm and 80 nm, respectively. Both nanostructured surfaces supported cell viability and proliferation in approximately equal extent. However, obvious differences in the cell spreading areas, morphologies, the organization of the actin cytoskeleton and the pattern of the focal adhesions were noticed. Furthermore, investigation of the pre-osteoblast differentiation potential indicated a higher capacity of larger spacing nanostructure to enhance the expression of the alkaline phosphatase, osteopontin and osteocalcin osteoblast specific markers inducing osteogenic differentiation. These findings provide the proof that lateral spacing of the TiO$_2$ nanotube coated titanium (Ti) surfaces has to be considered in designing bone implants with improved biological performance.

Keywords: spaced TiO$_2$ nanotubes; osteoblast; cell adhesion and morphology; cell proliferation; osteogenic differentiation

---

1. Introduction

A key aspect for orthopedic implant integration is the ability to enhance the functional activity of osteoblasts at the tissue/implant interface without generating a fibrous tissue [1]. Despite its remarkable success as a bone tissue implant, accumulated experience in titanium (Ti) implantation has emphasized some aspects that need to be improved, such as the ability to stimulate osteointegration [2]. Therefore, research efforts have been focused on improving cell-material interactions, shown to depend on the surface physico-chemical properties of the implant and extensive studies have been made on evaluating these factors (wettability, roughness, topography) [3–5].

The addition of nanotopographic features to the surface of bone implants has become a growing area of research in the bone regenerative medicine, due to the nanometer scale structural hierarchy of natural bone [6]. Numerous studies emphasized the role of nanotopographical cues in directing osteoprogenitor cells behavior [7–11]. A simple, economical and easy method to modify the topography of Ti-based materials, widely used in orthopedics, is electrochemical anodization [12]. This process leads to the formation of self-ordered TiO_2 nanostructures directly on the surface of the Ti substrate and more importantly, the morphology of anodic TiO_2 nanotubes (TNTs), such as diameter, wall thickness and length, can be easily controlled through the anodization parameters (applied voltage, time, electrolyte compositions, temperature, etc.) [12,13]. Nowadays, a wide range of nanotubular structures can be grown such as smooth wall, stacks, single wall or spaced nanotubes [12,14].

Typically, anodic nanotubes grow in a closed packed hexagonal arrangement, showing very little or no spacing in between the nanotubes. Over the years, spaced or loose-packed nanotubes have been reported to grow in electrolytes containing diethylene glycol (DEG), dimethylsulfoxide (DMSO) or trietylene glycol electrolytes containing fluorine [14–17], or in very specific anodization condition (e.g., ethylene glycol, tri (tetra, poly)-ethylene glycol) [18,19].

Ti surfaces modified with anodic TNTs represent a highly biocompatible material that integrates well into the bone tissue, exhibits thermal and chemical stability, controllable dimensions, large contact surface, adjustable size pores, adjustable surface chemistry and high surface/volume ratio [4,12,20,21]. These particularities have proved to favor the proliferation and differentiation of bone cells [4,22].

Cell attachment and interaction with TNTs coated surfaces has been intensively studied, mostly in view of the effect of nanotube diameter on bone regeneration process [7,23–27], while other nanotube morphological parameters are less investigated, e.g., spacing. The impact of TNTs with different diameters varying from 15 nm to 100 nm on mesenchymal stem cells (MSCs) was initially reported by Park et al. [7,23], observing that 15 nm diameter nanotubes favor osteogenic differentiation by improving cell adhesion. At the same time, nanotubes with a diameter above 50 nm induced a sensitive alteration of stem cell behavior in terms of spreading and adhesion resulting in the reduction of cell proliferation, migration and differentiation and, finally, apoptosis. When the experimental conditions were optimized for cell-biomaterial interactions, additional cell proliferation and migration tests with osteoblasts, osteoclasts and endothelial cells showed similar results [12].

In case of the stem cells induced toward the committed status under high level of serum supplement or the already committed pre-osteoblast cell line, studies concluded that also a larger diameter from 70 nm to 100 nm can stimulate the growth and differentiation of bone cells [24,25,28,29]. Though the range of nanotubular diameters explored in various studies was confined up to 100 nm, there are also studies investigating osteoprogenitor cells interactions with large diameter TNTs (100 nm up to 470 nm) [30,31], indicating that 170 nm provides the optimal diameter to sustain osteoblast proliferation, differentiation and mineralization [31].

Regarding the spacing between the nanotubes, some studies hypothesized that this surface characteristic can contribute to the continuous flow of the culture medium as well as to the exchange of gases, nutrients and signaling molecules, thus stimulating the cellular metabolism [27,32] and mimicking better the in vivo conditions. Moreover, recent works have demonstrated that the cellular behavior can be positively or negatively regulated by the spacing between different nanostructures, such as nanorods [33,34] or nanopillars [35,36]. For example, Zhou et al. [33,34] synthesized strontium-doped hydroxyapatite (Sr_1-HA) nanorods with different interrod spacing (67.3 ± 3.8, 95.7 ± 4.2, and 136.8 ± 8.7 nm) on microporous TiO_2 and showed that the osteoblast adhesion, proliferation and differentiation can be regulated by the interrod spacing: the cellular response was significantly enhanced on the nanorods with spacing smaller than 96 nm while larger spacing exerted inhibitory effects.

Given the fact that the biological effect generated by the spacing of nanotubular structures is not well known, this study aims to test the influence of TNT with a spacing of 80 nm (TNT80) or 18 nm (TNT18), and both with similar nanotube diameter of ~78 nm, on the MC3T3 pre-osteoblast response (as reference material in these studies, flat Ti surface was used). The choice of ~78 nm diameter is motivated by previous studies showing an optimal osteogenic differentiation on similar diameter nanotubes [29–31], in vivo studies evaluating the osseointegration capacity of TNT modified Ti implants showing that the implants coated with ~70 nm diameter TNTs induced an accelerated bone formation [29,37]. Lastly, our previous in vitro studies revealed that ~78 nm diameter TNTs mitigate the macrophages inflammatory response [38,39].

2. Materials and Methods

2.1. TiO$_2$ Nanotube Growth and Characterization

2.1.1. TiO$_2$ Nanotube Growth

To grow nanotubes, 0.125 mm thick Ti foil (99.6% pure temper annealed, ADVENT, Oxford, UK) was used, cut in 2.5 × 2.5 cm. Ti foils were cleaned by ultrasonication (Emmi-H30, EMAG AG, Germany) in acetone (Emsure Merck, Darmstadt, Germany), in ethanol (>99.8% p.a. Roth, Karlsruhe, Germany), followed by rinsing with distilled water and drying in a N$_2$) stream. For anodization, a two-electrode configuration was used in which Pt (ADVENT, Oxford, UK) acts as a cathode, and the Ti as anode using an O-ring cell with a diameter of 2 cm.

For TNT18 (nanotubes with ~78 nm diameter and 18 nm spacing at the top), the Ti foils were used as such and anodization was performed in Glycerol (>99.7% p.a. Roth, Karlsruhe, Germany): H$_2$O (70:30 vol.%) + 0.5 wt.% Ammonium fluoride (NH$_4$F, >98% p.a. Roth, Karlsruhe, Germany) at 20 V for 2 h (room temperature). The distance between the electrodes was of 1.5 cm and around 50 mL of electrolyte were used.

The synthesis of spaced NTs with similar diameter and a spacing of 80 nm (TNT80) consists of a two-step anodization process. In the first step, the Ti foil was anodized in Ethylene glycol (>99.5% p.a. Roth, Karlsruhe, Germany) + 0.1 M NH$_4$F + 1 M H$_2$O at 53 V for 1 h, and then the nanotube layer was removed by ultrasonication in deionized water. The prepatterned substrate was used as substrate for the second anodization which was performed in Diethylene glycol (>99.5% p.a. Roth, Karlsruhe, Germany) + 4 wt.% Hydrofluoric acid (HF 40%, Sigma Aldrich, Germany) + 0.3 wt.% NH$_4$F + 7 wt.% H$_2$O at 27 V for 4h at 30 °C (using 60 mL electrolyte and 2 cm distance in between the electrodes). After anodization, spaced nanotubes were immersed in ethanol for 10 min, rinsed with distilled water and dried in an N$_2$ stream.

2.1.2. TiO$_2$ Nanotube Characterization

Samples morphology was characterized by scanning electron microscope (SEM, FE-SEM 4800SEM, Hitachi, Japan) coupled with an energy-dispersive X-ray Spectroscope (EDAX Genesis, fitted to the SEM chamber), while their chemical composition and chemical state was investigated by using X-ray photoelectron spectroscopy (XPS, PHI 5600, Physical Electronics, US)—peaks were shifted to C1s 284.8eV.

2.2. Cell Culture

Experiments were performed using MC3T3-E1 murine pre-osteoblast cell line (ATCC®, CRL-2593TM) which was grown in Dulbecco's Minimal Essential Medium (DMEM, Sigma-Aldrich Co.St. Louis, MO, USA) supplemented with 10% fetal bovine serum (FBS, Gibco, Grand Island, NY, USA) and 1% (*v/v*) penicillin/streptomycin (10,000 units mL^{-1} penicillin and 10 mg mL^{-1} streptomycin) (Sigma-Aldrich Co. St. Louis, MO, USA) in an incubator at 37 °C in humidified atmosphere with 5% CO$_2$. These cells were seeded on the Ti, TNT18 and TN80 surfaces at an initial cell density of 1×10^4 cells/cm^2 to assess the cellular survival/proliferation, adhesion and morphology and maintained in standard culture conditions for up to three days. For pre-osteoblast differentiation studies, the cell seeding density was 4×10^4 cells/cm^2. These studies were performed under osteogenic conditions by supplementing the standard culture medium with 50 µg/mL ascorbic acid (Sigma-Aldrich Co.St. Louis, MO, USA), 5 mM β-glycerophosphate (Sigma-Aldrich Co. St. Louis, MO, USA) and 10^{-8} M dexamethasone (Sigma-Aldrich Co. St. Louis, MO, USA). Prior to osteoblast seeding, the substrates were cleaned by three successive baths, 10-minute each, with 70% ethanol. Then, the samples were rinsed twice for 30 min in sterile-filtered Milli-Q water, air-dried and exposed to ultraviolet light in a sterile tissue culture hood, for 30 min on each side. All of the above experiments have been performed in triplicate.

2.3. Evaluation of Cellular Survival and Proliferation

To assess the survival capacity of MC3T3-E1 pre-osteoblasts seeded on the test materials, the LIVE/DEAD Cell Viability/Cytotoxicity Assay Kit (Molecular Probes, Eugene, OR, USA) was used. This kit is based on the simultaneous staining of viable cells which convert the non-fluorescent cell permeable dye, calcein acetoxymethyl (AM), to the green fluorescent calcein, and of the dead cells marked with ethidium homodimer-1 (EthD-1) which labels nucleic acids of membrane-compromised cells in red. This assay was performed at 1 day and 3 days post-seeding, as previously shown [40]. Briefly, the cell-populated substrates were washed with phosphate buffered saline (PBS, Life Technologies Corporation, Grand Island, NY, USA), incubated with a solution containing 2 µM calcein AM and 4 µM EthD-1 for 10 minutes in the dark, and washed again with PBS. Afterwards, they were visualized under an inverted fluorescence microscope Olympus IX71 (Olympus, Tokyo, Japan) and representative fields were captured using the Cell F image acquisition system (Version 5.0, Olympus Soft Imaging Solutions, Münster, Germany).

To quantify the viability and proliferation of the pre-osteoblasts grown in contact with the analyzed samples, the assays of the lactate dehydrogenase (LDH) release into the culture media and of the cells' potential to reduce the water-soluble tetrazolium salt WST-8 (2-(2-methoxy-4-nitrophenyl)-3-(4-nitrophenyl)-5-(2,4-disulfophenyl)-2H-tetrazolium, monosodium salt) were conducted. More specifically, the amount of LDH released from the cytosol of dead cells as a result of materials' cytotoxicity was measured by using "LDH-based In Vitro Toxicology Assay Kit" (Sigma-Aldrich Co. St. Louis, MO, USA) according to the manufacturer's protocol. The optical density (OD) of the reaction product was measured at 490 nm using an automatic plate reader (FlexStation 3 Multi-Mode Microplate Reader, Molecular Devices, San Jose, CA, USA). The potential of the MC3T3-E1 pre-osteoblasts to reduce WST-8 compound to a soluble formazan was quantified in cell culture medium by using the Cell Counting Kit-8 assay (CCK-8, Sigma-Aldrich Co.St. Louis, MO, USA). For this purpose, the culture medium was discarded and then replaced with fresh culture medium containing 10% CCK-8 reagent. After 2 hours incubation step at 37 °C in a humidified 5% CO_2 atmosphere, optical density (OD) was measured at 450 nm using an automatic plate reader (FlexStation 3 Multi-Mode Microplate Reader, Molecular Devices, San Jose, CA, USA).

2.4. Evaluation of Cell Adhesion and Morphology

To assess the cell adhesion and morphological features, MC3T3-E1 pre-osteoblasts in contact with the tested surfaces were fixed with a cold solution of 4% paraformaldehyde (Sigma-Aldrich Co., Steinheim, Germany) in PBS, blocked and permeabilized with a solution containing 2% bovine serum albumin (BSA, Sigma-Aldrich Co., Steinheim, Germany) and 0.1% Triton X-100 (Sigma-Aldrich Co., Steinheim, Germany) in PBS. This step was followed by an incubation with a mouse monoclonal anti-vinculin antibody (Santa Cruz Biotechnology, Dallas, TX, USA), dilution 1:50 in PBS containing 1.2% BSA, for 1 hour at room temperature. Afterwards, samples were incubated with a goat anti-mouse IgG antibody coupled with AlexaFluor 546, (Invitrogen, Eugene, OR, USA) diluted 1:200 in PBS containing 1.2% BSA, for 30 minutes in the dark. To stain the actin filaments, another incubation step with phalloidin coupled with AlexaFluor 488 (20 µg/mL, Invitrogen, Eugene, OR, USA) for 30 minutes in the dark, was performed. In a final step, the cells were incubated with 2 µg/mL 4'-6-diamidino-2-phenylindole (DAPI, Sigma-Aldrich Co., Steinheim, Germany), a specific dye for nuclei, for 10 minutes, in the dark. Each incubation step was followed by successive washes with PBS. The samples were visualized using a fluorescence microscope (Olympus IX71, Olympus, Tokyo, Japan). Representative images were captured using the Cell F acquisition system (Version 5.0, Olympus Soft Imaging Solutions, Münster, Germany).

The spread cell area and the number of focal adhesions per cell were analyzed with Image J software (Version 1.52d, National Institutes of Health, Bethesda, MD, USA). For measuring the spread area of each cell, the freehand selection tool was used. In order to quantify the number of focal adhesions, fluorescence images of vinculin staining, taken at 40×, were analyzed. In a first step, representative captured fields were transformed into grayscale images and the background was subtracted. Thereafter, the threshold was adjusted and the focal adhesions were counted using analyze particles function.

In addition, the cellular morphology parameters such as nuclear area/cytoplasm area ratio, nuclear elongation factor and cell roundness were analyzed by following the contour of each cell manually (n = 30) using ImageJ software [41]. Thus, nuclear elongation was calculated as major/minor axis of nucleus while roundness was expressed as $4 \times \text{area}/(\pi \times \text{major\_axis}^2)$ of cytoplasm. By definition, roundness is equal to 1 for a completely round cell.

2.5. Assay of Pre-Osteoblast Cell Differentiation

2.5.1. Evaluation of the Intracellular Alkaline Phosphatase Activity

Alkaline phosphatase (ALP) activity, an early marker of osteoblast differentiation, was quantitatively determined in cell lysates by a method based on alkaline hydrolysis of p-nitrophenyl phosphate (pNPP), a colorless organic compound, to p-nitrophenol (pNP), a yellow compound, and inorganic phosphate. The activity of this enzyme was measured at 7 days and 14 days after cell seeding using Alkaline Phosphatase Activity Colorimetric Assay Kit (BV-K412-500, BioVision, Milpitas, CA, USA), as reported in a previous study [42]. Briefly, at the end of the incubation period, the MC3T3-E1 pre-osteoblasts grown in contact with the materials' surfaces were lysed and centrifuged to remove the cell debris. Then, the cell lysate was mixed with pNPP solution and incubated for 60 minutes at 25 °C in the dark. The OD values of the reaction products were measured at 405 nm using an automatic plate reader (FlexStation 3 Multi-Mode Microplate Reader, Molecular Devices, San Jose, CA, USA) and related to a standard curve for ALP activity calculation. To avoid variations due to different protein concentrations, ALP activity was normalized to the corresponding protein concentration, previously determined by Bradford reaction.

2.5.2. Quantitation of the Secreted Osteopontin

The secretion of osteopontin (OPN) was determined in the culture media after 7, 14 and 21 days of cell incubation. For this purpose, the culture media were harvested, centrifuged to remove the debris and the supernatant was used for analysis, as shown in a previous paper [42]. An enzyme-linked immunosorbent assay (ELISA) technique was performed in accordance with the manufacturer's instructions (R&D Systems, Minneapolis, MN, USA).

2.5.3. Immunofluorescence Detection of Osteocalcin Expression

The level of the osteocalcin (OCN) expression on the materials' surfaces was assessed at 4 weeks post-seeding by immunofluorescence microscopy. Thus, after fixing, permeabilizing and blocking the cells, as mentioned in Section 2.4, the cell-populated samples were incubated for 2 h with primary anti-osteocalcin antibody (Santa Cruz, 1:50 dilution), washed with PBS and, subsequently, incubated in the dark with goat anti-mouse IgG antibody coupled with AlexaFluor 488 (1:200, Invitrogen, Eugene, OR, USA), for 30 minutes. Finally, the nuclei were labeled with DAPI (Sigma-Aldrich Co., Steinheim, Germany) for 10 mins. The labeled samples were visualized under the fluorescence inverted microscope (Olympus IX71, Olympus, Tokyo, Japan). Representative microscopic fields were captured using the Cell F (Version 5.0, Olympus Soft Imaging Solutions, Münster, Germany) acquisition system and the mean fluorescence intensity was measured using ImageJ software (Version 1.52d, National Institutes of Health, Bethesda, MD, USA).

2.6. Statistical Analysis

Statistical analysis of data was performed with GraphPad Prism software (Version 6, GraphPad, San Diego, CA, USA) using one-way ANOVA/two-way ANOVA with Bonferroni's multiple comparison tests. All values are expressed as means ± standard deviation (SD) and differences at $p < 0.05$ were considered statistically significant.

3. Results and Discussions

3.1. Nanotube Morphology and Characterization

As previously mentioned, TiO_2 nanotubes grown by electrochemical anodization of Ti usually grow in a hexagonally close-packed configuration, and the tube to tube spacing observed in top view SEM images is only present at the top of the nanotubes [12,21]. Such is the case for TNT18, close packed nanotubes grown in a glycerol: water electrolyte containing NH_4F at 20 V for 2 h [38], which have a tube diameter of ~78 nm diameter and lateral spacing of 18 nm (see also SEM images in Figure 1a).

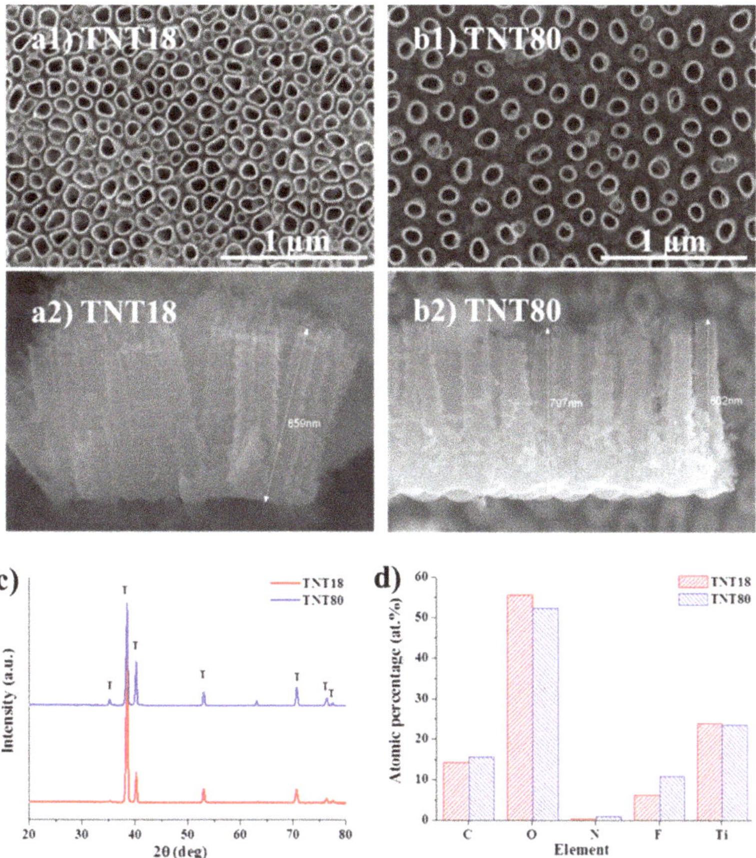

Figure 1. Top view and cross section SEM images of (**a1,a2**) TNT18, (**b1,b2**) TNT80. (**c**) XRD patterns of as-formed TiO_2 nanotubes (TNT18, TNT80); (**d**) Atomic percentage data computed from X-ray photoelectron spectroscopy (XPS) measurements for the two different nanotubular structures.

In our previous works we have shown that the growth of spaced tubes is based on self-organization on two scales and an investigation into the critical parameters affecting the spacing of tubes obtained in DEG based electrolytes revealed that the tube-spacing originates in the initial stages of tube growth [14,43]. This spacing and the spaced nanotube morphology is controlled by the anodization conditions, e.g., electrolyte composition (water content), applied voltage and temperature [14,43].

For the present work, the anodization conditions were optimized in order to reach a similar tube diameter with that of the close packed TNT and a spacing of ~80 nm (Figure 1b). We have previously shown [43] that controlling the temperature (of the substrate) significantly affects the morphology of spaced nanotubes, namely at 30 °C spaced nanotubes are uniformly spread on the Ti substrate (high uniformity) whereas without temperature control only a local tube formation (differences between regions) is achieved for 4 h anodization experiments. Moreover, the desired nanotubular morphology should be uniform on the surface and the amount of spongy oxide (small diameter nanotubes) in between the individual spaced tubes should be minimal, just at the bottom to achieve a true individual spacing but enough to ensure the presence of standing spaced nanotubes, i.e., from ion-milled cross-section it was observed that DEG spaced nanotubes are well-embedded in a fluoride-rich layer [44,45] while anodizing at higher temperatures of 50–60 °C leads to spongy oxide free spaced nanotubes which can collapse [43].

The above mentioned aspects led to the optimized anodization conditions established for the spaced nanotubes used in the present study, which consist of anodization at 27 V for 4 h at 30 °C in DEG + 4 wt.% HF + 0.3 wt.% NH_4F + 7 wt.% H_2O, using a double anodization procedure (for more detailed information, please see experimental part). From the cross-section SEM images, it is evident that in the case of close packed TNT (TNT18), the spacing is limited to the top of nanotubes while for the spaced tubes (TNT80), the spacing is visible from top to bottom (note that the TNT layers have similar lengths, ~0.85 µm).

Both nanotubular structures are amorphous, as only peaks arising from the Ti substrate are evident in the XRD patterns (Figure 1c). Additionally, by measuring the XPS spectra and computing the atomic percentage of elements (Figure 1d), we observed no significant difference between the samples—the slightly higher fluorine content in the spaced tubes (TNT80) is also due to the electrolyte composition (as HF is used as the main source of fluorine). As the XPS surface analysis can reach up to 5–10 nm of the top surface, we have also measured the EDX of samples, 3.8 at.% F for TNT18 and 5.0 at.% F for TNT80—the percentages obtained by both XPS and EDX are in line with literature data for nanotube surfaces [21,46]. Moreover, as it was previously shown that the nanotopography of the microenvironment is a dominant factor in comparison to the crystallinity or the fluorine content in the nanotubes, with regard to cells adhesion and proliferation (e.g., endothelial cells, mesenchymal stem cells) [47], the slightly higher fluorine content at the top layer of spaced nanotubes is not expected to influence the cell culture tests.

3.2. Cell Survival and Proliferation

The cellular response to materials developed for biomedical applications is strongly influenced by their surface properties such as roughness, oxide thickness, morphology, surface energy, chemical composition, nanostructure, etc. [48,49]. Considering the importance of surface nanotopography in guiding the cell fate and behavior in terms of cell viability, proliferation and/or differentiation [7–11,23–31], a first objective of our in vitro studies was to establish the survival rate of MC3T3-E1 pre-osteoblasts grown in contact with Ti, TNT18 and TNT80 surfaces by using the LIVE/DEAD Cell Viability/Cytotoxicity Assay (Figure 2a). The fluorescent images, shown in Figure 2a, denote the presence of a cellular monolayer represented by viable green-labeled cells. No red dead cells were detected on all analyzed materials. It is also important to note that cellular population was slightly non-homogeneous in distribution and morphology on the TNT80 substrate as compared to Ti and TNT18 surfaces after 1 day of culture. However, at 3 days post-seeding, the cells became confluent on all three surfaces suggesting their increased potential to induce cell proliferation.

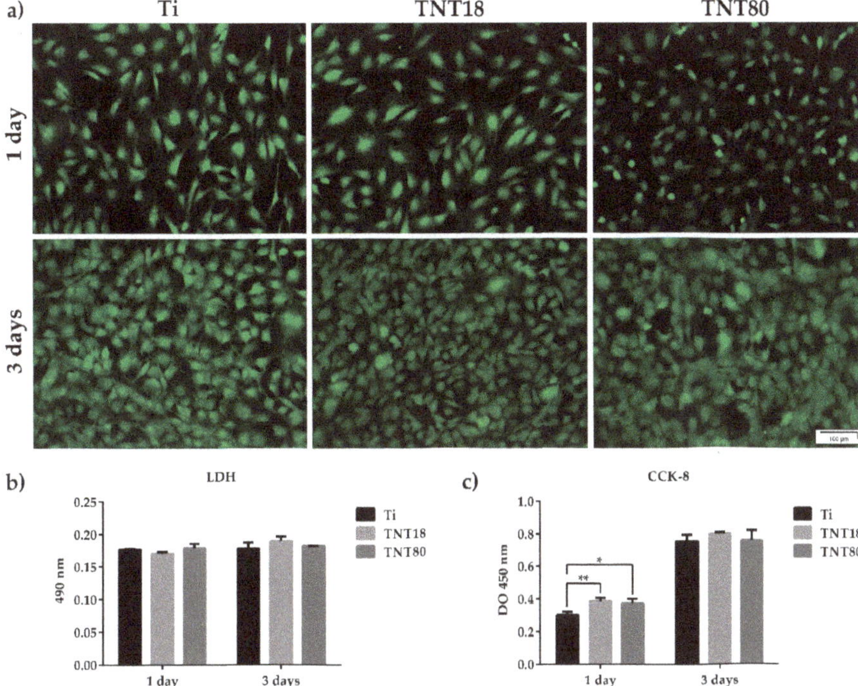

Figure 2. MC3T3-E1 pre-osteoblast survival and proliferation status. (**a**) Fluorescence images indicating the exclusive presence of green-labeled viable cells on all tested surfaces after staining with LIVE/DEAD Viability/Cytotoxicity kit; (**b**) The lactate dehydrogenase (LDH) levels released in the culture medium by cells grown in contact with Ti, TNT18 and TNT80 surfaces (n = 3, mean ± SD); (**c**) Pre-osteoblast proliferation capacity as evaluated by CCK-8 assay (n = 3, mean ± SD, * $p < 0.05$, ** $p < 0.01$ vs. Ti).

Further, the absence of cytotoxicity of TNT18 and TNT80 was confirmed by estimating the activity of cytoplasmic LDH released into the culture medium by cells that have lost membrane integrity. As shown in Figure 2b, both at 1 day and 3 days post-seeding, reduced and almost equal levels of LDH activity were detected in the culture media of the cells grown in contact with all three analyzed surfaces. Therefore, TNT spacing did not induce any cytotoxicity at the studied time points, creating a favorable environment for the MC3T3-E1 pre-osteoblast growth.

These observations are also supported by the results of the CCK-8 test—an assay used for quantifying the number of viable metabolically active cells. Thus, as noted in Figure 2c, the number of pre-osteoblasts grown in contact with Ti, TNT18 and TNT80 surfaces showed a time-dependent increase from 1 day to 3 days-incubation period. Moreover, after 1 day of culture the nanotubular surfaces exhibited a higher potential to sustain cell proliferation in comparison with the flat Ti surface. However, at 3 days post-seeding a similar number of cells was identified on all analyzed materials. These results are not surprising as a previous study approaching the behavior of MC3T3-E1 pre-osteoblasts in contact with large diameter anatase titania nanotubes (70–120 nm) exhibited stimulatory effects on the cell proliferation rates at early culture stage [28].

Overall, it might be inferred that both analyzed nanostructured surfaces have the potential to support cell survival and proliferation. They induced a differential stimulation of cell proliferation at 24 h post-seeding ($p < 0.05$ and $p < 0.01$ for TNT80 and TNT18, respectively) but, after 72 h of culture similar OD450 values were obtained for all studied surfaces.

3.3. Cell Adhesion and Morphological Features

Cellular adhesion to a surface is another decisive factor in determining the biocompatibility of a biomaterial. Information on the adhesion and morphology of the MC3T3-E1 pre-osteoblasts were obtained by fluorescence microscopy after labeling of actin cytoskeleton and vinculin (Figure 3a), the last one being a key protein in focal adhesions that can stabilize and modulate the dynamics of cell adhesions [50,51].

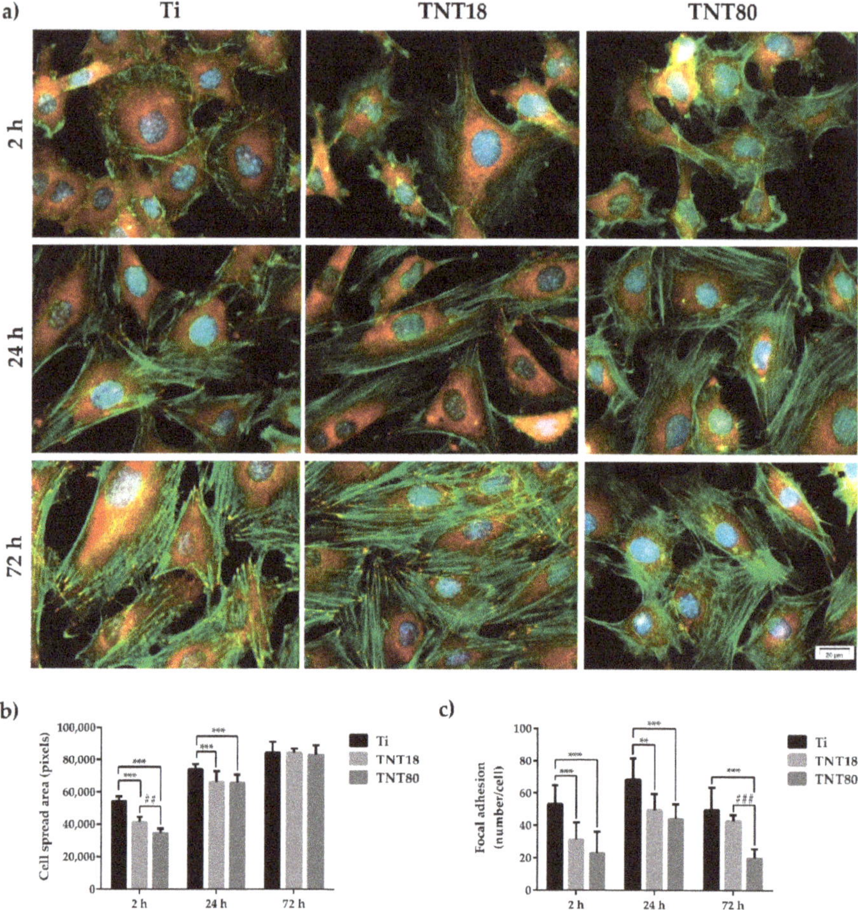

Figure 3. Cell adhesion, spreading and morphology at 2 h, 24 h and 72 h post-seeding. (**a**) Fluorescence micrographs used to assess adhesion and morphological characteristics of the MC3T3-E1 pre-osteoblasts maintained in contact with Ti, TNT18 and TNT80 surfaces. The cytoskeletal actin filaments were stained with phalloidin coupled with Alexa Fluor-488 (green); vinculin was labeled with anti-vinculin primary antibody and secondary antibody coupled with Alexa Fluor-546 (red); the nuclei were stained with DAPI (blue); (**b**) Quantitative analysis of cell spread area. Results are presented as means ± SD (n = 10 cells, *** $p < 0.001$ compared with Ti; ## $p < 0.01$ compared with TNT18); (**c**) Quantification of focal adhesions. Results are presented as means ± SD (n = 10 cells, ** $p < 0.01$, *** $p < 0.001$ compared with Ti; ### $p < 0.001$ compared with TNT18).

Accumulating data showed that cells can sense nanometer-scale variations in the average spacing of integrin ligands [52,53], and that the interactions mediated by these receptors are essential for providing information necessary for numerous adhesion-dependent cell functions, such as cell proliferation, differentiation and survival [54,55]. For example, ordered patterns of integrin ligands with a lateral spacing larger than 73 nm limited the focal adhesion (FA) formation and cell spreading while interdot distances of ≤58 nm allowed efficient FA formation and actin stress fibers assembly, and the cells adopted a well spread morphology [52,56]. In another study by Lee et al. [57] it was demonstrated that the variation in the nanoscale spacing of Arginine-Glycine-Aspartic Acid (RGD) ligands in alginate gels influenced adhesion, proliferation, and differentiation capacities of MC3T3-E1 pre-osteoblasts, where a decrease in the RGD island spacing from 78 to 36 nm induced an enhancement of cell proliferation rates and osteocalcin secretion. However, the threshold values mentioned above cannot be generalized owing to the strong dependence of cell behavior on the substrate properties [58].

In the present study, the images obtained after 2 h, 24 h and 72 h of cell culture revealed differences in pre-osteoblast morphology, actin cytoskeleton organization and distribution of vinculin between the analyzed surfaces (Figure 3a). Thus, at 2 h post-seeding on the flat Ti surface, the cells displayed spread morphology and larger dimensions (Figure 3b), and thin stress fibers throughout the cell body more numerous than in pre-osteoblasts grown in contact with the investigated nanotubular surfaces. Moreover, the presence of a higher number of vinculin positive signals on this surface, predominantly localized at the cell periphery, suggests the formation of multiple focal complexes (Figure 3c). On the contrary, the pre-osteoblasts grown on TNT18 and TNT80 surfaces adopted mostly a dendritic morphology with numerous cytoplasmic projections and significantly less vinculin-rich focal contact points at their extremities (Figure 3a) with an average of 31.2 and, respectively, 22.9 compared to 53.1 (Figure 3c). At the same time, the degree of cell spreading was lower than on the Ti surface (Figure 3b). This behavior could be ascribed to the limited surface area for cell attachment on the top wall surface owing to the large inner nanotube diameter and spacing gap between nanotubes. The more spacing gap between nanotubes, the lower pre-osteoblast spreading was noticed.

After an incubation period of 24 h, the MC3T3-E1 pre-osteoblasts still exhibited distinct cell morphologies on the different surfaces. For example, the cells grown on the nanotubular surfaces were more elongated and possessed more cellular protrusions as compared with the flat Ti surface. Furthermore, the most obvious and numerous focal adhesions were expressed on the Ti substrate, revealing a progressive decrease in the following order Ti > TNT18 > TNT80 (Figure 3a,c). This finding is in line with the results reported by Park et al. [7] showing the formation of less focal contacts on larger titania nanotubes (≥70 nm diameter) than on the flat Ti substrate while more focal contacts were visible on smaller nanotubes (≤30 nm diameter). Likewise, well defined bright green-labeled actin filaments, thinner on the nanotubular surfaces, mostly oriented in a parallel direction along the cell body and within the cellular protrusions are visible (Figure 3a). Compared with the flat control surface, the pre-osteoblast cells displayed smaller average cell areas on both TNT18 and TNT80 (Figure 3b) but the differences between the nanotubular surfaces were reduced with time.

Additionally, the scanning electron microscopy (SEM) micrographs of cells incubated for 2 h and 24 h on the analyzed surfaces, from which selected SEM images for the 24 h time incubation point are shown in Figure 4, confirmed the above morphological observations by fluorescence microscopy. Thus, besides the surface features of the three investigated materials, spread MC3T3-E1 pre-osteoblasts displaying different morphological features and cellular extensions, in the form of lamellipodia and filopodia, can be distinguished (Figure 4). Noticeable, more pronounced protrusion of filopodia with significantly longer and wider configuration, spread across the pores of both analyzed nanotubular arrays, is visible. An important characteristic of some of these filopodial extensions is their transparency, a feature that has previously been shown to be typical for the cells attached to the large diameter nanotubes [7].

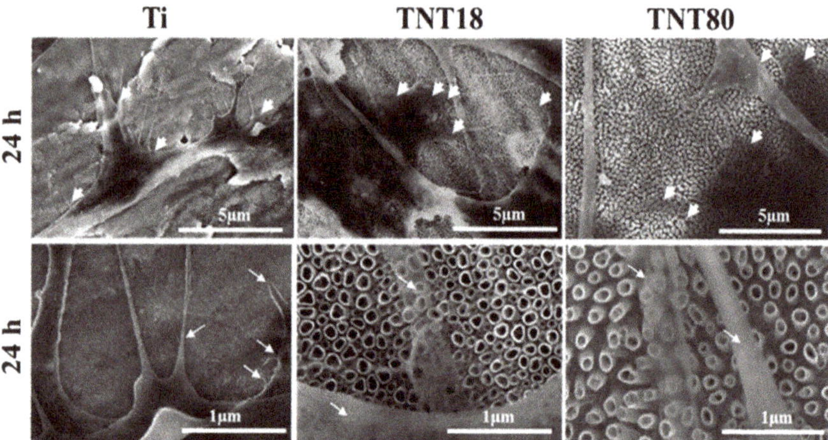

Figure 4. SEM images of the MC3T3-E1 pre-osteoblasts maintained in contact with Ti, TNT18 and TNT80 surfaces for 24 h. The white arrowheads show the location of cellular extensions; filopodia are indicated with white arrows.

Actin/vinculin immunofluorescence analysis has also been performed at 72 h post-seeding. As shown in Figure 3, at this point in time both nanotubular arrays influence the cells' shape and the organization of the actin cytoskeleton (Figure 3a), as well as the pattern and number of focal adhesions (Figure 3a,c). Thus, the adherent cells on the TNT18 surface exhibited polygonal or spindle-shaped osteoblast-like morphologies similar to the pre-osteoblasts grown in contact with the flat Ti surface but on the TNT80 substrate they adopted mixed shapes, either a less-broad cobblestone-like or spindle-shaped morphology (Figure 3a). Furthermore, well-defined actin stress fibers oriented in a parallel fashion along the main cellular axis and vinculin immunoreactive sites at their termini are visible on the flat Ti and TNT18 surfaces. On the contrary, MC3T3-E1 cells grown in direct contact with the TNT80 substrate showed mainly a branched actin filament network as well as more intercellular connections established at the level of the filopodial and lamellipodial protrusions. Likewise, they exhibited a significantly lower number of focal adhesions than on Ti and TNT18 surfaces (Figure 3c). We speculate that the larger spacing gap may provide less chance for cells to form integrin clustering leading to focal contact formation compared to less spaced, dense TNT with the same inner diameter (Figures 3 and 4). Importantly, the vinculin signals exhibited a punctiform pattern on the flat Ti surface whereas on both studied titania nanotubes they appeared to be elongated, suggesting acquisition by these cells of a more migratory phenotype [59,60]. This finding is consistent with previously reported studies showing that various human cell types display enhanced motility on nanotopographies compared with flat surfaces [7,61,62]. It is noteworthy that the vinculin recruitment to the focal adhesion sites has been shown to correlate directly with the traction force applied on the same focal adhesion [63]. It reinforces focal adhesions by crosslinking actin filaments to a large cytoskeletal molecule, talin, a critical step in cell mechanics connecting the cell to its substrate [64,65]. Actually, vinculin conveys force inside-out by increasing integrin–talin complexes via the head domain, while its tail domain is needed to propagate force to the cytoskeletal actin. To note that the assembly of focal adhesions is affected not only by internally-generated forces exerted on them [66], but also by the physical state and mechanical properties of the external cellular environment [56]. Of particular interest is the modulation of the number, arrangement, and size of focal adhesions, redistribution of cyto-and nucleo-skeletal components, as well as of cell and nuclear morphologies by nanotopography [67–70]. In fact, nanostructured surfaces evoke architectural rearrangements that activate, through focal adhesions, the signaling cascades leading to indirect downstream effects on gene expression and induce mechanical changes in the cell that involve physical pulling of the cytoskeleton on the nucleus. These induce changes in gene transcription by imposing mechanical forces on nuclear components. Considering these potential mechanotransductive effects elicited by nanotopographical surfaces on cell structural components and the above results showing the modulation of the pre-osteoblast behavior in terms of cell adhesion and morphology, cytoskeleton organization, cellular expansion, and focal adhesion patterns by lateral spacing of titania nanotubes, we further evaluated the next cellular spreading parameters: nuclear area/cytoplasm area ratio, nuclear elongation factor and cell roundness (Figure 5). These parameters can provide information about the extent of the cellular response to traction forces coming from the underlying substrates and of the cellular forces exerted on them [41]. In this way the cells can probe the rigidity of the extracellular environment, develop focal adhesions, trigger signaling, and remodel the extracellular matrix forces. As the results show, cell stretching and nuclear elongation were distinguishable only in initial phase of cultivation (2 h) between groups. This result is in agreement with our results showing cell spreading area. Considering that focal contacts on TNT80 were constantly less detectable throughout observation period (2 h–72 h), focal contact formation certainly via cell-substrate sensing mechanism including integrin signaling might be continuously controlled by TNT topographic differences. This result seems to be in good accordance with previous studies showing that larger lateral gap provides less focal contact formation.

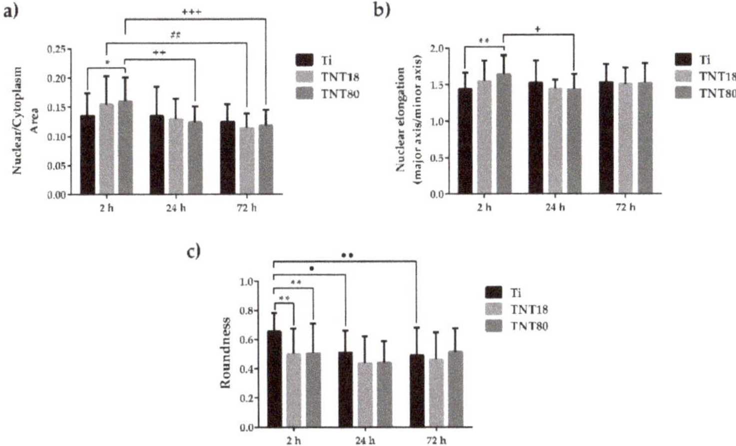

Figure 5. Cellular morphology parameters evaluated at 2 h, 24 h and 72 h post-seeding. (**a**) Quantification of Nuclear-Cytoplasmic area ratio. Results are presented as means ± SD (n = 30 cells, * $p < 0.05$ compared with Ti; ## $p < 0.01$ between TNT18 at 24 h vs. 72 h; ++ $p < 0.01$, +++ $p < 0.001$ between TNT80 at 2 h vs. 24 h, and 2h. vs. 72 h, respectively); (**b**) Nuclear elongation measurement calculated as the ratio between major axis and minor axis. Results are presented as means ± SD (n = 30 cells, ** $p < 0.01$ compared with Ti; + $p < 0.05$ between respective groups at 2 h vs. 24 h); (**c**) Quantification of cell roundness. Results are expressed as means ± SD (n = 30 cells, ** $p < 0.01$ compared with Ti; ● $p < 0.05$, ●● $p < 0.01$ comparison between Ti at 2 h vs. 24 h, and 2h vs. 72 h, respectively).

3.4. Pre-osteoblast Cell Differentiation

Generally, nanostructures have been reported to support the osteogenic differentiation of stem cells and osteoblasts [24,71,72]. In this study, the expression of the bone cell-specific markers such as ALP, OPN and OCN was measured.

ALP is a ubiquitous membrane-bound homodimeric metalloenzyme that catalyzes the hydrolysis of phospho-monoesters, releasing inorganic phosphate (Pi) and alcohol, and is one of the most commonly used biochemical markers to assess the osteoblast activity [73]. It appears that the mechanism of action of this enzyme consists both in increasing the local concentration of inorganic phosphate, a mineralization promoter, and in reducing the extracellular concentration of pyrophosphate, a mineralization inhibitor. In the present study, intracellular ALP activity was quantified at 7-days and 14-days post-seeding in order to estimate the ability of tested materials to induce bone mineralization. As it can be seen in Figure 6a, after 7 days of cell incubation under osteogenic conditions, the nanotube coated surfaces (TNT18 and TNT80) induced an increase in ALP activity by ~50% compared to the Ti surface. It is also noted that at 14 days post-seeding, ALP activity recorded increased values on all analyzed materials. Furthermore, at this time point, the differences in the expression of ALP activity by the pre-osteoblasts grown on all three surfaces were more obvious than at 7 days of cell incubation. Specifically, both nanotubular surfaces exhibited higher levels of ALP activity in comparison with the flat Ti surface. This finding is not surprising since, overall, nanotubular TiO_2 surfaces are well known for their ability to enhance ALP activity [27,31,74]. However, in the present study, this enhancement was significant in the case of the pre-osteoblasts grown on TNT80 substrate. Furthermore, ALP activity in the lysates of these cells recorded a significant increase when compared with intracellular ALP activity exhibited by the cells grown in contact with TNT18 surface.

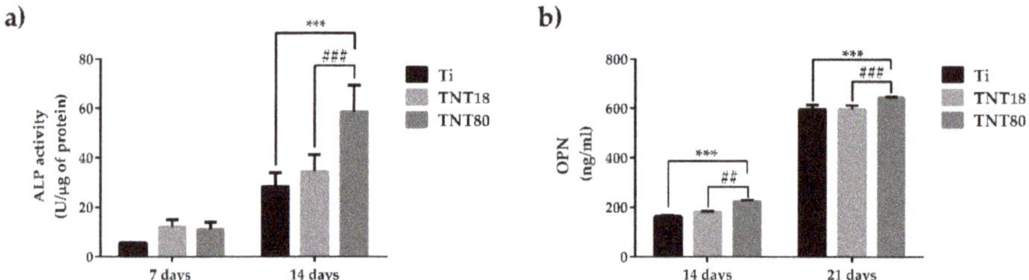

Figure 6. Evaluation of alkaline phosphatase (ALP) and osteopontin (OPN) expression. (**a**) The expression level of intracellular ALP activity of osteoblasts grown on Ti, TNT18 and TNT80 surface at 7 and 14 days in the presence of osteogenic differentiation media (n = 3, mean ± SD, *** $p < 0.001$ compared with Ti; ### $p < 0.001$ compared with TNT18); (**b**) The expression level of osteopontin released into the culture medium by osteoblasts cultivated on the tested surfaces for 14 and 21 days in the presence of osteogenic medium (n = 3, mean ± SD, *** $p < 0.001$ compared with Ti; ## $p < 0.01$, ### $p < 0.001$ compared with TNT18).

In order to get a more complete picture of the ability of the analyzed surfaces to induce the early cell differentiation, the concentration of OPN secreted in the culture medium by MC3T3-E1 cells grown on Ti, TNT18 and TNT80 was determined by ELISA technique at 14 days and 21 days post-seeding (Figure 6b). Osteopontin is a highly phosphorylated glycoprotein that strongly links to extracellular matrix non-collagen proteins, and exhibits multiple biological functions [75]. For instance, OPN in the osseous tissue is released from osteoblasts and osteoclasts eliciting three major functions during biomineralization phase of bone structuring including modulation of bone cells adhesion, modulation of osteoclast function, and modulation of matrix mineralization, as well. The results obtained in the present study showed a time-dependent increase in OPN synthesis and extracellular release in culture media maintained in contact with all analyzed substrates. It is noteworthy that TNT80 elicited a stronger effect in inducing OPN secretion and, implicitly, pre-osteoblast differentiation in comparison with TNT18 and Ti surfaces at both incubation time points.

The next objective of our study was to quantify the level of expression for the most abundant non-collagenous bone matrix protein, OCN, in MC3T3-E1 pre-osteoblasts grown in direct contact with the three analyzed surfaces. This protein is often studied as a late marker for bone formation, playing the role of a regulator in bone mineralization and bone turnover [76,77]. However, it can be stated that the exact role of OCN in bone is still incompletely understood, although several lines of evidence proved that OCN enhances bone formation. For example, it was shown that OCN increases the adhesion of osteoblast-like cells on biocement [78]. In addition, Rammelt et al. demonstrated its potential to enhance the appearance of active osteoblasts and bone healing around hydroxyapatite/collagen composites [79].

As shown in Figure 7a, immunofluorescence detection of OCN expression in MC3T3-E1 pre-osteoblasts grown for 4 weeks in contact with Ti, TNT18 and TNT80 materials, under osteogenic conditions, denotes quite a similar staining pattern of this protein on their surfaces. However, the quantification of the OCN fluorescence intensity (Figure 7b) indicates that both nanotubular surfaces induced a statistically significant increase in the expression of this osteoblast differentiation marker when compared to flat Ti surface.

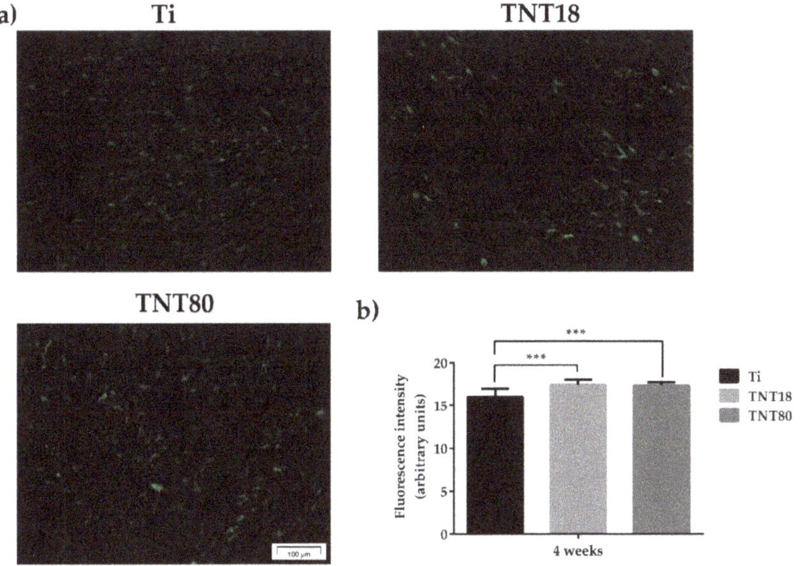

Figure 7. Evaluation of OCN expression. (**a**) Fluorescent images that highlight osteocalcin expression in MC3T3-E1 osteoblasts grown on Ti, TNT18, TNT80 surfaces for 4 weeks in the presence of osteogenic differentiation media using anti-osteocalcin (green) specific antibody; (**b**) Measurement of fluorescence intensity using Image J software (n = 20, mean ± SD, *** $p < 0.001$).

The above-mentioned results are in good agreement with previous studies on MSC [24] and pre-osteoblast [27,28,80] showing that large diameter (70–100 nm) TiO_2 nanotubes strongly induced osteogenic differentiation when compared to smaller diameter nanotubes. Hence, taken together, these experimental data emphasize the ability of the analyzed nanotubular surfaces, mainly TNT80, to enhance the induction of osteoblast differentiation. It is becoming increasingly clear that nanotopograhy represents a viable strategy to modulate cell differentiation and that the cell function is highly regulated by mechanotransduction [81,82]. Taking into account the results of this study, we assume that one of the mechanisms responsible for the differential osteogenic response of the MC3T3-E1 pre-osteoblasts is driven by the mechanotransductive signals induced by the analyzed surfaces. This assumption is supported by a recent study performed by Zhang et al. [80] who investigated the intracellular mechanisms involved in stimulation of the osteogenic differentiation of MC3T3-E1 cells by large diameter titania nanotubes (LTNTs; 90 nm inner diameter) in comparison to small diameter nanotubes (STNTs; 30 nm inner diameter) and flat Ti surface. The Real-time PCR analysis showed that LTNTs elicit increased gene expression of the bone differentiation markers, Runt-related transcription factor 2 (Runx2) and osterix (Osx), when compared with cells in contact with flat Ti surfaces. This finding has also been confirmed by histological analysis performed on the regeneration bone tissue after implantation into rat tibiae showing that titania nanotubes, mainly LTNTs, induced better implant osseointegration. To clarify the underlying mechanisms of this differential osteogenic response provoked by the analyzed surfaces, the expression levels of focal adhesion kinase, both total (FAK) and activated (pY397-FAK), as well as FAK recruitment to focal adhesions have also been investigated. The results demonstrated that when compared to flat Ti substrate, both nanotubular surfaces, more significantly LTNTs diminished FAK activity and its recruitment to focal adhesions. As a result, a reduction in the activity of the Ras homolog gene family member A (RhoA), a small GTPase able to sense and respond to mechanical cues, occurred. RhoA and FAK interact together in order to perceive the mechanical stimuli and regulate cell differentiation [83]. The alteration of the FAK/Rho signaling was followed by the export in cytosol of the Yes-associated protein (YAP), which has been shown to be implicated in transmission of mechanical signals to the nucleus, and activation of the bone differentiation marker Runt-related transcription factor 2 (Runx2). This export reduced the YAP/Runx2 binding probability leading to the Runx2 activation and initiation of the osteogenesis on nanotubes. Considering the stimulatory effects of these larger lateral gaps on osteogenic induction, our results may indicate that larger spacing gaps play a role in enhancing osteogenic induction of pre-osteoblastic cells in addition to the effect of large nanotube diameter.

We do not rule out other nanotopography-mediated signal transduction pathways or the above presented mechanism that could be responsible for the modulation of pre-osteoblast behavior by lateral nanotube spacing. Further research will be necessary to confirm these assumptions and elucidate the underlying mechanisms.

4. Conclusions

Our results provide some insight on the effects of titania nanotube spacing on osteoblast cell functions in vitro. Both nanotubular structures, closed packed nanotubes with tube to tube spacing at the top of 18 nm and spaced nanotubes with tube to tube spacing of 80 nm, were found to sustain cell viability and proliferation to almost a similar extent as the flat Ti substrate. Furthermore, the data obtained demonstrate that spaced nanotubes (tube to tube spacing of 80 nm) influence other aspects of cell behavior including cell adhesion and morphology, cytoskeleton organization and focal adhesion patterns, as well as osteogenic differentiation. They have been proved to elicit beneficial effects on the pre-osteoblast osteogenic differentiation, as demonstrated by the increased ALP activity, and osteopontin and osteocalcin protein expression. Overall, the study provides a new variable worthy of investigation in designing and optimizing titanium-based materials as platforms for bone regeneration and drug delivery.

Author Contributions: Conceptualization, A.C., A.M., R.N.I. and P.S; methodology, A.C., R.N.I., M.G.N. and A.M.; Software, M.G.N., R.N.I. and A.M.; validation, R.N.I., M.G.N., A.M., and S.O.; formal analysis, S.O., M.G.N., A.M. and R.N.I.; investigation, M.G.N, A.M., R.N.I. and S.O; resources, A.C. and P.S.; data curation, A.C., R.N.I., and J.P.; writing—original draft preparation, M.G.N, A.M., R.N.I. and A.C.; writing—review and editing, A.C., J.P. and P.S.; visualization, A.C. and M.G.N.; supervision, A.C. and P.S.; project administration, A.C.; funding acquisition, A.C., J.P. and P.S.

Funding: This research was funded by the Romanian Ministry of National Education, CNCS-UEFISCDI, grant PCE 55/2017 to A.C. and by DFG grant PA 2537/1-3 to J.P. and P.S.

Acknowledgments: Helga Hildebrand is acknowledged for the XPS measurements.

Conflicts of Interest: The authors declare no conflict of interest.

References

1. Mavrogenis, A.F.; Dimitriou, R.; Parvizi, J.; Babis, G.C. Biology of implant osseointegration. *J. Musculoskelet. Neuronal Interac.* **2009**, *9*, 61–71.
2. Civantos, A.; Martínez-Campos, E.; Ramos, V.; Elvira, C.; Gallardo, A.; Abarrategi, A. Titanium Coatings and Surface Modifications: Toward Clinically Useful Bioactive Implants. *ACS Biomater. Sci. Eng.* **2017**, *3*, 1245–1261. [CrossRef]

3.   Bauer, S.; Schmuki, P.; von der Mark, K.; Park, J. Engineering biocompatible implant surfaces. Part I: Materials and surfaces. *Prog. Mater. Sci.* **2013**, *58*, 261–326. [CrossRef]

4.   Kulkarni, M.; Mazare, A.; Gongadze, E.; Perutkova, Š.; Kralj-Iglič, V.; Milosev, I.; Schmuki, P.; Iglič, A.; Mozetič, M. Titanium nanostructures for biomedical applications. *Nanotechnol.* **2015**, *26*, 62002. [CrossRef] [PubMed]

5.   Cheng, Y.; Yang, H.; Yang, Y.; Huang, J.; Wu, K.; Chen, Z.; Wang, X.; Lin, C.; Lai, Y. Progress in TiO$_2$ nanotube coatings for biomedical applications: A review. *J. Mater. Chem. B* **2018**, *6*, 1862–1886. [CrossRef]

6.   Gong, T.; Xie, J.; Liao, J.; Zhang, T.; Lin, S.; Lin, Y. Nanomaterials and bone regeneration. *Bone Res.* **2015**, *3*, 15029. [CrossRef] [PubMed]

7.   Park, J.; Bauer, S.; Von Der Mark, K.; Schmuki, P. Nanosize and Vitality: TiO$_2$ Nanotube Diameter Directs Cell Fate. *Nano Lett.* **2007**, *7*, 1686–1691. [CrossRef] [PubMed]

8.   Lotz, E.M.; Olivares-Navarrete, R.; Berner, S.; Boyan, B.D.; Schwartz, Z. Osteogenic response of human MSCs and osteoblasts to hydrophilic and hydrophobic nanostructured titanium implant surfaces. *J. Biomed. Mater. Res. Part A* **2016**, *104*, 3137–3148. [CrossRef]

9.   Karazisis, D.; Ballo, A.M.; Petronis, S.; Agheli, H.; Emanuelsson, L.; Thomsen, P.; Omar, O. The role of well-defined nanotopography of titanium implants on osseointegration: Cellular and molecular events in vivo. *Int. J. Nanomed.* **2016**, *11*, 1367–1382.

10.   Castro-Raucci, L.M.S.; Francischini, M.S.; Teixeira, L.N.; Ferraz, E.P.; Lopes, H.B.; de Oliveira, P.T.; Hassan, M.Q.; Rosa, A.L.; Beloti, M.M. Titanium with nanotopography induces osteoblast differentiation by regulating endogenous bone morphogenetic protein expression and signaling pathway. *J. Cell. Biochem.* **2015**, *117*, 1718–1726. [CrossRef]

11.   Park, J.; Mazare, A.; Schneider, H.; Von Der Mark, K.; Fischer, M.J.; Schmuki, P. Electric field-induced osteogenic differentiation on TiO$_2$ nanotubular layer. *Tissue Eng. Part C Methods* **2016**, *22*, 809–821. [CrossRef] [PubMed]

12.   Lee, K.; Mazare, A.; Schmuki, P. One-dimensional titanium dioxide nanomaterials: Nanotubes. *Chem. Rev.* **2014**, *114*, 9385–9454. [CrossRef] [PubMed]

13.   Riboni, F.; Nguyen, N.T.; So, S.; Schmuki, P. Aligned metal oxide nanotube arrays: Key-aspects of anodic TiO$_2$ nanotube formation and properties. *Nanoscale Horiz.* **2016**, *1*, 445–466. [CrossRef]

14.   Ozkan, S.; Nguyen, N.T.; Mazare, A.; Cerri, I.; Schmuki, P. Controlled spacing of self-organized anodic TiO$_2$ nanotubes. *Electrochem. Commun.* **2016**, *69*, 76–79. [CrossRef]

15.   Albu, S.P.; Schmuki, P. TiO$_2$ nanotubes grown in different organic electrolytes: Two-size self-organization, single vs. double-walled tubes, and giant diameters. *Phys. Status solidi (RRL)-Rapid Res. Lett.* **2010**, *4*, 215–217. [CrossRef]

16.   Ozkan, S.; Nguyen, N.T.; Mazare, A.; Hahn, R.; Cerri, I.; Schmuki, P. Fast growth of TiO$_2$ nanotube arrays with controlled tube spacing based on a self-ordering process at two different scales. *Electrochem. Commun.* **2017**, *77*, 98–102. [CrossRef]

17.   Nguyen, N.T.; Ozkan, S.; Hwang, I.; Mazare, A.; Schmuki, P. TiO$_2$ nanotubes with laterally spaced ordering enable optimized hierarchical structures with significantly enhanced photocatalytic H$_2$ generation. *Nanoscale* **2016**, *8*, 16868–16873. [CrossRef]

18.   Lu, H.H.; Lu, S.S.; Chen, Y.; Wang, X.M. Large-scale sparse TiO$_2$ nanotube arrays by anodization. *J. Mater. Chem.* **2012**, *22*, 5921.

19.   Albu, S. Morphology and Growth of Titania Nanotubes: Nanostructuring and Applications. Ph.D. Thesis, Friedrich-Alexander-Universität Erlangen-Nürnberg (FAU), Erlange, Germany, October 2012.

20.   Losic, D.; Aw, M.S.; Santos, A.; Gulati, K.; Bariana, M. Titania nanotube arrays for local drug delivery: Recent advances and perspectives. *Expert Opin. Drug Deliv.* **2015**, *12*, 103–127. [CrossRef]

21.   Kulkarni, M.; Mazare, A.; Park, J.; Gongadze, E.; Killian, M.S.; Kralj, S.; Von Der Mark, K.; Iglič, A.; Schmuki, P. Protein interactions with layers of TiO$_2$ nanotube and nanopore arrays: Morphology and surface charge influence. *Acta Biomater.* **2016**, *45*, 357–366. [CrossRef]

22.   Su, E.P.; Justin, D.F.; Pratt, C.R.; Sarin, V.K.; Nguyen, V.S.; Oh, S.; Jin, S. Effects of titanium nanotubes on the osseointegration, cell differentiation, mineralisation and antibacterial properties of orthopaedic implant surfaces. *Bone Jt. J.* **2018**, *100B*, 9–16. [CrossRef] [PubMed]

23. Park, J.; Bauer, S.; Schlegel, K.A.; Neukam, F.W.; Von Der Mark, K.; Schmuki, P. TiO$_2$ nanotube surfaces: 15 nm-an optimal length scale of surface topography for cell adhesion and differentiation. *Small* **2009**, *5*, 666–671. [CrossRef] [PubMed]

24. Oh, S.; Brammer, K.S.; Li, Y.S.J.; Teng, D.; Engler, A.J.; Chien, S.; Jin, S. Stem cell fate dictated solely by altered nanotube dimension. *Proc. Natl. Acad. Sci.* **2009**, *106*, 2130–2135. [CrossRef] [PubMed]

25. Lv, L.; Liu, Y.; Zhang, P.; Zhang, X.; Liu, J.; Chen, T.; Su, P.; Li, H.; Zhou, Y. The nanoscale geometry of TiO$_2$ nanotubes influences the osteogenic differentiation of human adipose-derived stem cells by modulating H3K4 trimethylation. *Biomaterials* **2015**, *39*, 193–205. [CrossRef] [PubMed]

26. Mu, P.; Li, Y.; Zhang, Y.; Yang, Y.; Hu, R.; Zhao, X.; Huang, A.; Zhang, R.; Liu, X.Y.; Huang, Q.; et al. High-throughput screening of rat mesenchymal stem cell behavior on gradient TiO$_2$ nanotubes. *ACS Biomater. Sci. Eng.* **2018**, *4*, 2804–2814. [CrossRef]

27. Brammer, K.S.; Oh, S.; Cobb, C.J.; Bjursten, L.M.; Van Der Heyde, H.; Jin, S. Improved bone-forming functionality on diameter-controlled TiO$_2$ nanotube surface. *Acta Biomater.* **2009**, *5*, 3215–3223. [CrossRef] [PubMed]

28. Xu, L.; Yu, W.Q.; Jiang, X.Q.; Zhang, F.Q.; Yu, W.; Jiang, X.; Zhang, F. The effect of anatase TiO$_2$ nanotube layers on MC3T3-E1 preosteoblast adhesion, proliferation, and differentiation. *J. Biomed. Mater. Res. Part A* **2010**, *94*, 1012–1022.

29. Wang, N.; Li, H.; Lü, W.; Li, J.; Wang, J.; Zhang, Z.; Liu, Y. Effects of TiO$_2$ nanotubes with different diameters on gene expression and osseointegration of implants in minipigs. *Biomaterials* **2011**, *32*, 6900–6911. [CrossRef]

30. Zhang, R.; Wu, H.; Ni, J.; Zhao, C.; Chen, Y.; Zheng, C.; Zhang, X. Guided proliferation and bone-forming functionality on highly ordered large diameter TiO$_2$ nanotube arrays. *Mater. Sci. Eng. C* **2015**, *53*, 272–279. [CrossRef]

31. Zhang, Y.; Luo, R.; Tan, J.; Wang, J.; Lu, X.; Qu, S.; Weng, J.; Feng, B. Osteoblast behaviors on titania nanotube and mesopore layers. *Regen. Biomater.* **2017**, *4*, 81–87. [CrossRef]

32. Hamlekhan, A.; Takoudis, C.; Sukotjo, C.; Mathew, M.T.; Virdi, A.; Shahbazian-Yassar, R.; Shokuhfar, T. Recent progress toward surface modification of bone/Dental implants with titanium and zirconia dioxide nanotubes fabrication of TiO$_2$ nanotubes. *J. Nanotechnol. Smart Mater.* **2014**, *1*, 301–314.

33. Zhou, J.; Li, B.; Lu, S.; Zhang, L.; Han, Y. Regulation of osteoblast proliferation and differentiation by interrod spacing of Sr-HA nanorods on microporous titania coatings. *ACS Appl. Mater. Interfaces* **2013**, *5*, 5358–5365. [CrossRef] [PubMed]

34. Zhou, J.; Han, Y.; Lu, S. Direct role of interrod spacing in mediating cell adhesion on Sr-HA nanorod-patterned coatings. *Int. J. Nanomed.* **2014**, *9*, 1243–1260.

35. Hanson, L.; Lin, Z.C.; Xie, C.; Cui, Y.; Cui, B. Characterization of the cell–nanopillar interface by transmission electron microscopy. *Nano Lett.* **2012**, *12*, 5815–5820. [CrossRef] [PubMed]

36. Kim, H.S.; Yoo, H.S. Differentiation and focal adhesion of adipose-derived stem cells on nano-pillars arrays with different spacing. *RSC Adv.* **2015**, *5*, 49508–49512. [CrossRef]

37. Von Wilmowsky, C.; Bauer, S.; Roedl, S.; Neukam, F.W.; Schmuki, P.; Schlegel, K.A. The diameter of anodic TiO$_2$ nanotubes affects bone formation and correlates with the bone morphogenetic protein-2 expression in vivo. *Clin. Oral Implants Res.* **2012**, *23*, 359–366. [CrossRef]

38. Neacsu, P.; Mazare, A.; Cimpean, A.; Park, J.; Costache, M.; Schmuki, P.; Demetrescu, I. Reduced inflammatory activity of RAW 264.7 macrophages on titania nanotube modified Ti surface. *Int. J. Biochem. Cell Boil.* **2014**, *55*, 187–195. [CrossRef]

39. Neacsu, P.; Mazare, A.; Schmuki, P.; Cimpean, A. Attenuation of the macrophage inflammatory activity by TiO$_2$ nanotubes via inhibition of MAPK and NF-κB pathways. *Int. J. Nanomed.* **2015**, *10*, 6455–6467.

40. Ion, R.; Stoian, A.B.; Dumitriu, C.; Grigorescu, S.; Mazare, A.; Cimpean, A.; Demetrescu, I.; Schmuki, P. Nanochannels formed on TiZr alloy improve biological response. *Acta Biomater.* **2015**, *24*, 370–377. [CrossRef]

41. Domura, R.; Sasaki, R.; Okamoto, M.; Hirano, M.; Kohda, K.; Napiwocki, B.; Turng, L.S. Comprehensive study on cellular morphologies, proliferation, motility, and epithelial–mesenchymal transition of breast cancer cells incubated on electrospun polymeric fiber substrates. *J. Mater. Chem. B* **2017**, *5*, 2588–2600. [CrossRef]

42. Neacsu, P.; Staras, A.I.; Voicu, S.I.; Ionascu, I.; Soare, T.; Uzun, S.; Cojocaru, V.D.; Pandele, A.M.; Croitoru, S.M.; Miculescu, F.; et al. Characterization and in vitro and in vivo assessment of a novel cellulose acetate-coated mg-based alloy for orthopedic applications. *Materials* **2017**, *10*, 686. [CrossRef]

43. Ozkan, S.; Mazare, A.; Schmuki, P. Critical parameters and factors in the formation of spaced TiO_2 nanotubes by self-organizing anodization. *Electr. Acta* **2018**, *268*, 435–447. [CrossRef]

44. Ozkan, S.; Nguyen, N.T.; Mazare, A.; Schmuki, P. Optimized Spacing between TiO_2 nanotubes for enhanced light harvesting and charge transfer. *Chem. Electro. Chem.* **2018**, *5*, 3183–3190. [CrossRef]

45. Ozkan, S. Self-ordered arrays of "spaced" nanotubes and their applications in energy conversion. Ph.D. Thesis, Friedrich Alexander University, Erlange, Germany, October 2017.

46. Peng, Z.; Ni, J. Surface properties and bioactivity of TiO_2 nanotube array prepared by two-step anodic oxidation for biomedical applications. *R. Soc. Open Sci.* **2019**, *6*, 181948. [CrossRef]

47. Park, J.; Bauer, S.; Schmuki, P.; Von Der Mark, K. Narrow window in nanoscale dependent activation of endothelial cell growth and differentiation on TiO_2 nanotube surfaces. *Nano Lett.* **2009**, *9*, 3157–3164. [CrossRef]

48. Yang, L. Fundamentals of nanotechnology and orthopedic materials. In *Nanotechnology-Enhanced Orthopedic Materials*; Elsevier BV: Amsterdam, The Netherlands, 2015; pp. 1–25.

49. Damiati, L.; Eales, M.G.; Nobbs, A.H.; Su, B.; Tsimbouri, P.M.; Salmeron-Sanchez, M.; Dalby, M.J. Impact of surface topography and coating on osteogenesis and bacterial attachment on titanium implants. *J. Tissue Eng.* **2018**, *9*. [CrossRef]

50. Bays, J.L.; DeMali, K.A. Vinculin in cell–cell and cell–matrix adhesions. *Cell. Mol. Life Sci.* **2017**, *74*, 2999–3009. [CrossRef]

51. Burridge, K. Focal adhesions: A personal perspective on a half century of progress. *FEBS J.* **2017**, *284*, 3355–3361. [CrossRef]

52. Arnold, M.; Cavalcanti-Adam, E.; Glass, R.; Blümmel, J.; Eck, W.; Kantlehner, M.; Kessler, H.; Spatz, J.P.; Cavalcanti-Adam, E.A. Activation of integrin function by nanopatterned adhesive interfaces. *Chem. Phys. Chem.* **2004**, *5*, 383–388. [CrossRef]

53. Cavalcanti-Adam, E.A.; Volberg, T.; Micoulet, A.; Kessler, H.; Geiger, B.; Spatz, J.P. Cell spreading and focal adhesion dynamics are regulated by spacing of integrin ligands. *Biophys. J.* **2007**, *92*, 2964–2974. [CrossRef]

54. Cavalcanti-Adam, E.A.; Micoulet, A.; Blummel, J.; Auernheimer, J.; Kessler, H.; Spatz, J.P. Lateral spacing of integrin ligands influences cell spreading and focal adhesion assembly. *Eur. J. Cell Biol.* **2005**, *85*, 219–224. [CrossRef]

55. Arnold, M.; Jakubick, V.C.; Lohmu, T.; Heil, P.; Blummel, J.; Cavalcanti-Adam, E.A.; López-García, M.; Walther, P.; Kessler, H.; Geiger, B.; et al. Induction of cell polarization and migration by a gradient of nanoscale variations in adhesive ligand spacing. *Nano Lett.* **2008**, *8*, 2063–2069. [CrossRef]

56. Girard, P.P.; Cavalcanti-Adam, E.A.; Kemkemer, R.; Spatz, J.P. Cellular chemomechanics at interfaces: Sensing, integration and response. *Soft Matter.* **2007**, *3*, 307. [CrossRef]

57. Lee, K.Y.; Alsberg, E.; Hsiong, S.; Comisar, W.; Linderman, J.; Ziff, R.; Mooney, D. Nanoscale adhesion ligand organization regulates osteoblast proliferation and differentiation. *Nano Lett.* **2004**, *4*, 1501–1506. [CrossRef]

58. Ventre, M.; Causa, F.; Netti, P.A. Determinants of cell–material crosstalk at the interface: Towards engineering of cell instructive materials. *J. R. Soc. Interface* **2012**, *9*, 2017–2032. [CrossRef]

59. Kim, D.H.; Wirtz, D. Focal adhesion size uniquely predicts cell migration. *FASEB J.* **2013**, *27*, 1351–1361. [CrossRef]

60. Lehtimäki, J.; Hakala, M.; Lappalainen, P. Actin filament structures in migrating cells. *Bile Acids and Their Receptors* **2016**, 123–152.

61. Brammer, K.S.; Oh, S.; Gallagher, J.O.; Jin, S. Enhanced cellular mobility guided by TiO_2 nanotube surfaces. *Nano Lett.* **2008**, *8*, 786–793. [CrossRef]

62. Liliensiek, S.J.; Wood, J.A.; Yong, J.; Auerbach, R.; Nealey, P.F.; Murphy, C.J. Modulation of human vascular endothelial cell behaviors by nanotopographic cues. *Biomaterials* **2010**, *31*, 5418–5426. [CrossRef]

63. Dumbauld, D.W.; Lee, T.T.; Singh, A.; Scrimgeour, J.; Gersbach, C.A.; Zamir, E.A.; Fu, J.; Chen, C.S.; Curtis, J.E.; Craig, S.W.; et al. How vinculin regulates force transmission. *Proc. Natl. Acad. Sci.* **2013**, *110*, 9788–9793. [CrossRef]

64. Goldmann, W.H. Role of vinculin in cellular mechanotransduction. *Cell Boil. Int.* **2016**, *40*, 241–256. [CrossRef]

65. Grashoff, C.; Hoffman, B.D.; Brenner, M.D.; Zhou, R.; Parsons, M.; Yang, M.T.; McLean, M.A.; Sligar, S.G.; Chen, C.S.; Ha, T.; et al. Measuring mechanical tension across vinculin reveals regulation of focal adhesion dynamics. *Nature* **2010**, *466*, 263–266. [CrossRef]

66. Mullen, C.A.; Vaughan, T.J.; Voisin, M.C.; Brennan, M.A.; Layrolle, P.; McNamara, L.M. Cell morphology and focal adhesion location alters internal cell stress. *J. R. Soc. Interface* **2014**, *11*, 20140885. [CrossRef]

67. Dalby, M.J.; Biggs, M.J.; Gadegaard, N.; Kalna, G.; Wilkinson, C.D.; Curtis, A.S. Nanotopographical stimulation of mechanotransduction and changes in interphase centromere positioning. *J. Cell. Biochem.* **2007**, *100*, 326–338. [CrossRef]

68. Curtis, A.; Casey, B.; Gallagher, J.; Pasqui, D.; Wood, M.; Wilkinson, C. Substratum nanotopography and the adhesion of biological cells. Are symmetry or regularity of nanotopography important? *Biophys. Chem.* **2001**, *94*, 275–283. [CrossRef]

69. Biggs, M.J.; Richards, R.; Gadegaard, N.; Wilkinson, C.; Dalby, M.; Richards, R.; Dalby, M. Regulation of implant surface cell adhesion: Characterization and quantification of S-phase primary osteoblast adhesions on biomimetic nanoscale substrates. *J. Orthop. Res.* **2007**, *25*, 273–282. [CrossRef]

70. Biggs, M.J.P.; Richards, R.G.; Dalby, M.J. Nanotopographical modification: A regulator of cellular function through focal adhesions. *Nanomed. Nanotechnol. Boil. Med.* **2010**, *6*, 619–633. [CrossRef]

71. Le Guéhennec, L.; Martin, F.; Lopez-Heredia, M.A.; Louarn, G.; Amouriq, Y.; Cousty, J.; Layrolle, P. Osteoblastic cell behavior on nanostructured metal implants. *Nanomedicine* **2008**, *3*, 61–71. [CrossRef]

72. Shen, X.; Ma, P.; Hu, Y.; Xu, G.; Zhou, J.; Cai, K. Mesenchymal stem cell growth behavior on micro/nano hierarchical surfaces of titanium substrates. *Colloids Surf. B Biointerfaces* **2015**, *127*, 221–232. [CrossRef]

73. Golub, E.E.; Boesze-Battaglia, K. The role of alkaline phosphatase in mineralization. *Curr. Opin. Orthop.* **2007**, *18*, 444–448. [CrossRef]

74. Ding, X.; Wang, Y.; Xu, L.; Zhang, H.; Deng, Z.; Cai, L.; Wu, Z.; Yao, L.; Wu, X.; Liu, J.; et al. Stability and osteogenic potential evaluation of micro-patterned titania mesoporous-nanotube structures. *Int. J. Nanomed.* **2019**, *14*, 4133–4144. [CrossRef]

75. Içer, M.A.; Gezmen-Karadağ, M. The multiple functions and mechanisms of osteopontin. *Clin. Biochem.* **2018**, *59*, 17–24. [CrossRef]

76. Tsao, Y.T.; Huang, Y.J.; Wu, H.H.; Liu, Y.A.; Liu, Y.S.; Lee, O.K. Osteocalcin mediates biomineralization during osteogenic maturation in human mesenchymal stromal cells. *Int. J. Mol. Sci.* **2017**, *18*, 159. [CrossRef]

77. Neve, A.; Corrado, A.; Cantatore, F.P. Osteocalcin: Skeletal and extra-skeletal effects. *J. Cell. Physiol.* **2013**, *228*, 1149–1153. [CrossRef]

78. Knepper-Nicolai, B.; Reinstorf, A.; Hofinger, I.; Flade, K.; Wenz, R.; Pompe, W. Influence of osteocalcin and collagen I on the mechanical and biological properties of biocement D. *Biomol. Eng.* **2002**, *19*, 227–231. [CrossRef]

79. Rammelt, S.; Neumann, M.; Hanisch, U.; Reinstorf, A.; Pompe, W.; Zwipp, H.; Biewener, A. Osteocalcin enhances bone remodeling around hydroxyapatite/collagen composites. *J. Biomed. Mater. Res. Part A* **2005**, *73*, 284–294. [CrossRef]

80. Zhang, H.; Cooper, L.F.; Deng, F.; Song, J.; Yang, S. Titanium nanotubes induce osteogenic differentiation through the FAK/RhoA/YAP cascade. *RSC Adv.* **2016**, *6*, 44062–44069. [CrossRef]

81. McNamara, L.E.; McMurray, R.J.; Biggs, M.J.P.; Kantawong, F.; Oreffo, R.O.C.; Dalby, M.J. Nanotopographical control of stem cell differentiation. *J. Tissue Eng.* **2010**, *2010*, 120623. [CrossRef]

82. Dobbenga, S.; Fratila-Apachitei, L.E.; Zadpoor, A.A. Nanopattern-induced osteogenic differentiation of stem cells-A systematic review. *Acta Biomater.* **2016**, *46*, 3–14. [CrossRef]

83. Xu, B.; Song, G.; Ju, Y.; Li, X.; Song, Y.; Watanabe, S. RhoA/ROCK, cytoskeletal dynamics, and focal adhesion kinase are required for mechanical stretch-induced tenogenic differentiation of human mesenchymal stem cells. *J. Cell. Physiol.* **2012**, *227*, 2722–2729. [CrossRef]

 *materials*

Article

Co-Culture of Osteoblasts and Endothelial Cells on a Microfiber Scaffold to Construct Bone-Like Tissue with Vascular Networks

Kouki Inomata and Michiyo Honda *

Department of Applied Chemistry, School of Science and Technology, Meiji University, 1-1-1 Higashimita, Tama-ku, Kawasaki, Kanagawa 214-8571, Japan
* Correspondence: michiyoh@meiji.ac.jp; Tel.: +81-44-934-7210; Fax: +81-44-934-7906

Received: 28 August 2019; Accepted: 4 September 2019; Published: 5 September 2019

Abstract: Bone is based on an elaborate system of mineralization and vascularization. In hard tissue engineering, diverse biomaterials compatible with osteogenesis and angiogenesis have been developed. In the present study, to examine the processes of osteogenesis and angiogenesis, osteoblast-like MG-63 cells were co-cultured with human umbilical vein endothelial cells (HUVECs) on a microfiber scaffold. The percentage of adherent cells on the scaffold was more than 60% compared to the culture plate, regardless of the cell type and culture conditions. Cell viability under both monoculture and co-culture conditions was constantly sustained. During the culture periods, the cells were spread along the fibers and extended pseudopodium-like structures on the microfibers three-dimensionally. Compared to the monoculture results, the alkaline phosphatase activity of the co-culture increased 3–6 fold, whereas the vascular endothelial cell growth factor secretion significantly decreased. Immunofluorescent staining of CD31 showed that HUVECs were well spread along the fibers and formed microcapillary-structures. These results suggest that the activation of HUVECs by co-culture with MG-63 could enhance osteoblastic differentiation in the microfiber scaffold, which mimics the microenvironment of the extracellular matrix. This approach can be effective for the construction of tissue-engineered bone with vascular networks.

Keywords: bone tissue engineering; three-dimensional co-culture; osteoblast; endothelial cell; microfiber scaffold; osteogenesis; angiogenesis

---

1. Introduction

The reconstruction of structural and functional tissues is highly challenging. The ability of excessively damaged organs and tissues to regenerate themselves is low. Since organs and tissues have different structures and functions, treatment should be selected appropriately in consideration of their biochemical properties and anatomical features. Bone has the potential to regenerate. For instance, an iliac crest, which is used for autologous bone grafting to treat bone fractures, could induce vascularization around the implanted area to promote immediate bone formation [1,2]. However, this method has many limitations, so new approaches should be developed.

Tissue engineering to improve or replace damaged tissue has attracted attention. Typical tissue engineering strategies combine three elements (e.g., cells, growth factors, and scaffolds) to construct three-dimensional tissues to enhance the regenerating capability to the original state [3–5]. It is essential for the three-dimensional tissue to maintain its structural functions so that blood vessels supply oxygen and nutrients and remove metabolic wastes. These absences cause low cell viability and cell death around the defects [6,7].

Bone tissue is based on a sophisticated system of mineralization and vascularization. In bone, blood vessels are indispensable for the maintenance of bone homeostasis, i.e., the balance between bone formation and bone resorption [8–10]. Extensive vascular networks in bone play a significant role in not only providing oxygen and nutrients and draining wastes but also in transporting inorganic ions (e.g., calcium and phosphate ions) necessary for calcification. In the current study, Grüneboom and their colleagues found the existence of an effective communication between the bone marrow vascular system and external circulation. The blood vessels in the periosteum and on the surface of bone connected periosteal circulation, mediated the recruitment of immune cells to the circulation [11].

In this context, diverse biomaterials compatible with osteogenesis and angiogenesis have been proposed and developed, such as porous ceramics, hydroxyapatite-based nanocomposites, and bioactive glasses [12–14]. However, all materials have not been necessarily designed for the cell culture. For this reason, it is essential to consider cellular characteristics when constructing three-dimensional cell culture scaffolds.

In bone tissue engineering, two interactions are key in the construction of tissue-engineered bone. These interactions are between bone formation cells (osteoblasts) and blood vessel constitution cells (vascular endothelial cells) and between the above-mentioned cells and biomaterials. Previous studies have shown that co-cultures of osteoblastic cells with endothelial cells resulted in the stimulation of osteoblast differentiation and the formation of microcapillary-like structures [15,16]. The adequate adhesion of the cells to the scaffold also has an important role in the expression of normal cell functions, such as cell survival, proliferation, and differentiation [17,18]. Above all, porous scaffolds have attracted attention due to their potential to support vascular tube formation [19,20]. However, those studies have not revealed the interaction between the co-cultured osteoblast, endothelial cells, and the porous scaffolds.

In the present study, osteoblasts were co-cultured with endothelial cells on a microfiber scaffold to examine the characteristics of the co-culture system, such as cell attachment, survival, osteoblast differentiation, and endothelial cell tube formation, which are regarded as important in bone tissue engineering. This model has shown that the three-dimensional co-culture system could be effective in regulating cell survival, adhesion, osteoblastic differentiation, and vascular tube formation. Therefore, the approach would allow for the production of tissue-engineered bone with vascular networks *in vitro*.

2. Materials and Methods

2.1. Materials

Chemicals and reagents were purchased from the following manufacturers. Eagle's minimum essential medium (EMEM), non-essential amino acid solution, trypsin-EDTA, monoclonal mouse anti-vinculin antibody (hVIN-1), and CelLytic™ cell lysis reagent were purchased from Sigma-Aldrich (St. Louis, MO, USA). Endothelial cell growth medium 2 kit (ECGM2) was purchased from Takara Bio (Shiga, Japan). Fetal bovine serum (FBS), penicillin, streptomycin, phosphate-buffered saline (PBS, sodium chloride, potassium chloride, disodium hydrogenphosphate, and potassium dihydrogen phosphate), dimethyl sulfoxide (DMSO), 4% paraformaldehyde in PBS, Triton X-100, bovine serum albumin (BSA), 25% glutaraldehyde solution, LabAssay™ ALP, and protein assay Bradford reagent were purchased from FUJIFILM Wako Pure Chemical (Osaka, Japan). Alexa Flour® 488-labeled phalloidin, Alexa Flour® 594-labeled goat anti-mouse IgG$_1$, and alamarBlue® cell viability reagent were purchased from Invitrogen (Carlsbad, CA, USA). 3-(4,5-dimethyl-2-thiazolyl)-2,5-diphenyl-tetrazolium bromide (MTT) and 4′,6-diamino-2-phenylindole (DAPI) were purchased from Dojindo (Kumamoto, Japan). Monoclonal mouse anti-human CD31 antibody was purchased from Dako (Glostrup, Denmark). Human VEGF quantikine ELISA kit was purchased from R&D Systems (Minneapolis, MN, USA). MG-63 cells were obtained from ATCC® CRL-1427™ (Manassas, VA, USA). Human umbilical vein endothelial cells (HUVECs) were obtained from PromoCell (Heidelberg, Germany). The microfiber scaffold was kindly supplied by ORTHO ReBIRTH (Kanagawa, Japan).

2.2. Fabrication of the Microfiber Mesh Scaffolds

The microfiber mesh scaffolds consisted of 30 wt.% of poly (lactic-co-glycolic acid) (PLGA, LG855S, Evonik Japan, Tokyo, Japan), 40 wt.% of β-tricalcium phosphate (β-TCP, β-TCP-100, Taihei Chemical Industrial Co., LTD, Osaka, Japan), and 30 wt.% of silicon-doped vaterite (SiV) powders [21]. The above composites were dissolved in 8 wt.% of chloroform for electrospinning. Electrospinning was carried out using a nanofiber electrospinning system (NANON-03, MECC Co., LTD. Fukuoka, Japan). PLGA/β-TCP/SiV solution was put into a syringe (diameter: 15.8 mmφ, volume: 10 mL) and pumped out of the syringe at a rate of 4 mL h^{-1}. A voltage supplier was used to maintain the voltage at 24 kV. All experiments were carried out at room temperature and samples were also dried at room temperature for overnight. The microfiber mesh scaffolds for the cell culture were sterilized by gamma ray irradiation.

2.3. Cell Culture

In this study, human osteoblast-like MG-63 cells and HUVECs were used as models of osteoblasts and endothelial cells. The MG-63 cells were cultured in EMEM supplemented with 10% heat-inactivated FBS, 1% non-essential amino acid solution, 100 U/mL penicillin, and 100 µg/mL streptomycin. The HUVECs were cultured in ECGM2. Both the MG-63 cells and HUVECs were grown under a humidified atmosphere containing 5% CO_2 at 37 °C. All experiments were conducted using the MG-63 cells at passage 3–6 and the HUVECs at passage 1–3.

2.4. Co-Culture of the MG-63 Cells and HUVECs on the Microfiber Scaffold

The MG-63 cells and HUVECs were suspended in a ratio of 1:4 with ECGM2, according to a previously described method [14]. Before cell seeding, a microfiber scaffold was placed in each well of a 24-well plate and hydrophilized with growth medium for 30 min in a humidified atmosphere containing 5% CO_2 at 37 °C. The suspended cells (5.0×10^4 or 5.0×10^5 cells) were seeded on each scaffold. As a control, MG-63 cells were seeded alone in EMEM at the same density. The culture medium was refreshed every two days, and the culture was maintained for a maximum of 14 days.

2.5. Initial Cell Attachment

Initial cell attachment was determined using an MTT-based assay, according to the manufacturer's protocol. At 6 h after incubation, the MTT reagent was added to the medium (final concentration of 0.5 mg/mL) and incubated for 4 h. The stain was then eluted with DMSO and centrifuged. The absorbance was measured at 570 nm (measurement wavelength) and 650 nm (reference wavelength) using a microplate reader (Multiskan FC, Thermo Fisher Scientific, Waltham, MA, USA). Initial cell attachment was calculated by the ratio between the absorbance of the cells that adhered to the scaffolds and the absorbance of the cells cultured without the scaffold.

2.6. Cell Viability

Metabolic activity of the cells on the scaffolds was assessed as cell viability using a resazurin-based assay, according to the manufacturer's procedure. At 5, 7, and 14 days, the medium was exchanged for growth medium containing 10% alamarBlue® reagent and incubated for 4 h. The absorbance was measured at 570 nm (measurement wavelength) and 595 nm (reference wavelength) using a microplate reader. The percent of reduced alamarBlue® reagent was calculated as previously described [22].

2.7. Immunofluorescence Microscopy

The cells were washed twice with PBS and fixed with 4% paraformaldehyde in PBS for 15 min at room temperature. The cells were then permeabilized with 0.1% Triton X-100 in PBS for 15 min at room temperature. After rinsing with PBS twice, the cells were blocked with 3% BSA in PBS for 1 h at room temperature and incubated with a primary antibody, monoclonal mouse anti-vinculin antibody

(1:400), or monoclonal mouse anti-human CD31 antibody (1:250), which was diluted in PBS containing 3% BSA overnight at 4 °C. The cells were washed with PBS twice and then stained with Alexa Flour® 488-labeled Phalloidin (1:250) for F-actin, Alexa Flour® 594-labeled goat anti-mouse IgG$_1$ (1:500) for vinculin, or CD31 and DAPI (1:500) for the nucleus diluted in PBS for 1 h at room temperature in the dark. The cells were washed with PBS and then examined by florescence microscopy (BZ X-710, Keyence, Osaka, Japan).

2.8. Scanning Electron Microscopy

The cells were washed twice with PBS and fixed with 2.5% glutaraldehyde in PBS overnight at 4 °C. After rinsing with PBS twice, the cells were freeze-dried for three days. The specimens were coated with gold using sputtering and then examined by scanning electron microscopy (SEM, VE-9800, Keyence, Osaka, Japan) at an accelerating voltage of 5 kV.

2.9. Measurements of Alkaline Phosphatase and Vascular Endothelial Growth Factor

Alkaline phosphatase (ALP) activity and vascular endothelial growth factor (VEGF) secretion were measured similarly to the method described in previous studies [15,23]. At 5, 7, and 14 days, the scaffolds and supernatants were collected and stored at −80 °C. The cells were lysed in CelLytic™ M and homogenized by sonication. The cell lysates were centrifuged and assayed for ALP activity using a colorimetric analysis using *p*-nitrophenyl phosphate as the substrate following the protocol of LabAssay™ ALP. One unit was defined as the activity that produced one nanomole of *p*-nitrophenol after 15 min. The supernatants were assessed for VEGF secretion using the sandwich enzyme-linked immunosorbent assay (ELISA), according to each manufacturer's instructions. Total protein concentrations were determined by the Bradford standard method.

2.10. Statistical Analysis

The data were statistically analyzed for determination of the mean and the standard deviation (SD) of the mean. The Student's *t*-test was carried out with a significance level of $p < 0.05$.

3. Results

3.1. Initial Cell Attachment and Cell Morphology

In the present study, we used a microfiber scaffold, which was composed of a three-dimensional porous matrix (Figure 1). The scaffold consisted of random fibers with an average fiber diameter of 10 to 30 μm. The three-dimensional microfiber structure in the scaffold might be a suitable geometry for cell growth and formation of vascular networks.

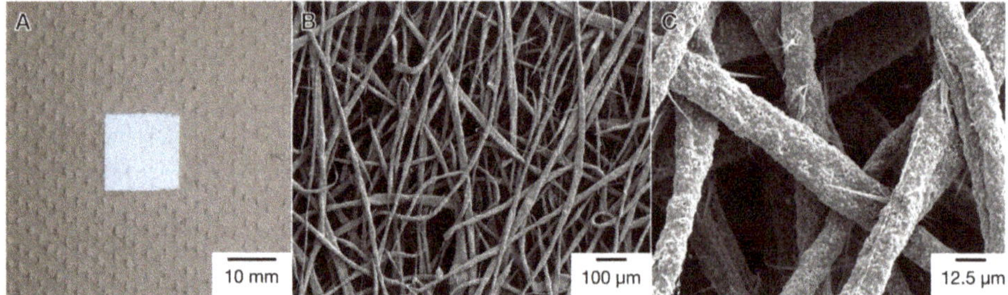

Figure 1. Light photomicrograph (**A**) and SEM images (**B,C**) of a microfiber mesh scaffold. Bars indicate 10 mm (**A**), 100 μm (**B**), and 12.5 μm (**C**).

Initial cell attachment is a key index for evaluating the biocompatibility of materials. The interaction between cells and biomaterials contributes to cell activity on the scaffold, such as cell survival, proliferation, and differentiation. To examine the initial attachment on the porous microfibers, the cells were seeded on the scaffold and incubated for 6 h. The percentage of adherent cells was then determined using an MTT assay (Figure 2). The results showed that more than 60% of the cells adhered to the scaffold were compared to the culture plate, regardless of the cell type and culture conditions. Osteoblasts showed the highest attachment among all the cell types. However, there were no significant differences depending on cellular types and culture conditions. The geometry and surface properties of the scaffold might induce higher initial attachment.

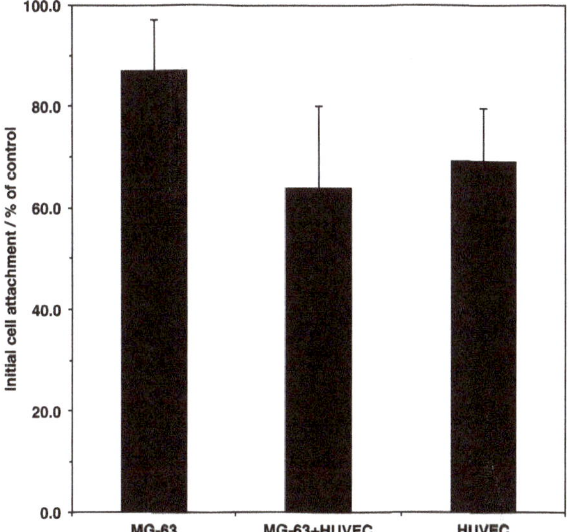

Figure 2. Initial cell attachment of MG-63 cells and/or HUVECs on a microfiber mesh scaffold. Cells were seeded at 5.0×10^5 cells cm^{-3} on the scaffold placed in each well of a 24-well plate and cultured for 6 h. Initial cell attachment was assessed using an MTT assay described in Materials and Methods and calculated by the ratio between the absorbance of the cells that adhered to the scaffold and the absorbance of the cells cultured without the scaffold. Data were determined from three replicate samples and are shown as mean ± SD. There were no significant differences among them.

Next, to investigate the cell adhesion and cell spreading on the scaffold, fluorescence microscopy was used to observe the actin cytoskeleton and vinculin of the cells (Figure 3). These cell behaviors also affect cell activity on the scaffold. Vinculin is an adaptor protein connecting actin filaments with integrin and is then expressed in spreading cells adherent to the extracellular matrix via integrin. Therefore, the immunofluorescent staining of vinculin can be used to evaluate cell adhesion on the scaffold. On the first day after seeding, the cells had exhibited both spindle-shaped and shrinkage-rounded morphology (Arrowheads and asterisks in Figure 3A,B). As the culture time passed, the cells were well spread, and they expressed vinculin. In addition, to confirm the micro-surrounding of the cells on the scaffold, the cells were observed by SEM (Figure 4). Under each culture condition, we observed two types of cells that were elongating along the fibers and extending pseudopodium-like structures between the fibers (Arrowheads in Figure 4A,B). These results suggested that cells could adhere to the scaffold via integrin. This was supported by previous reports [21,24,25].

A

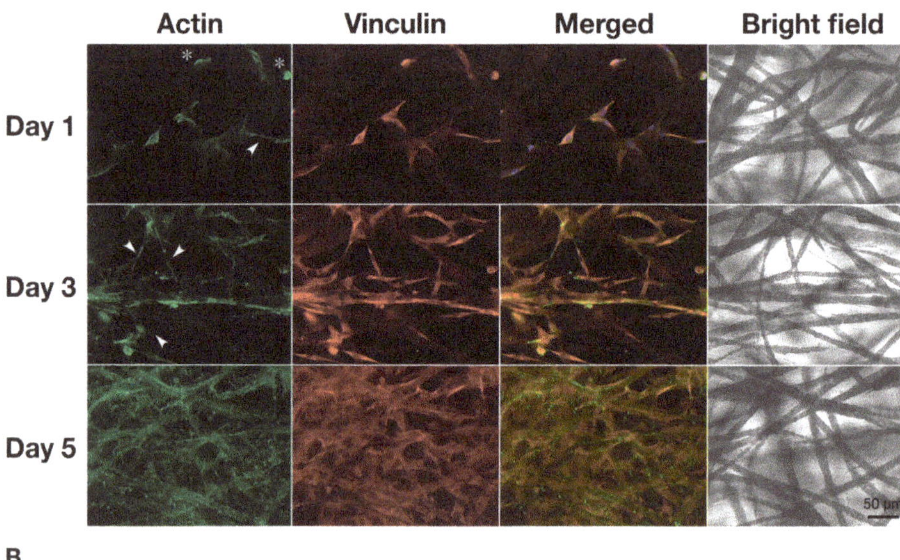

B

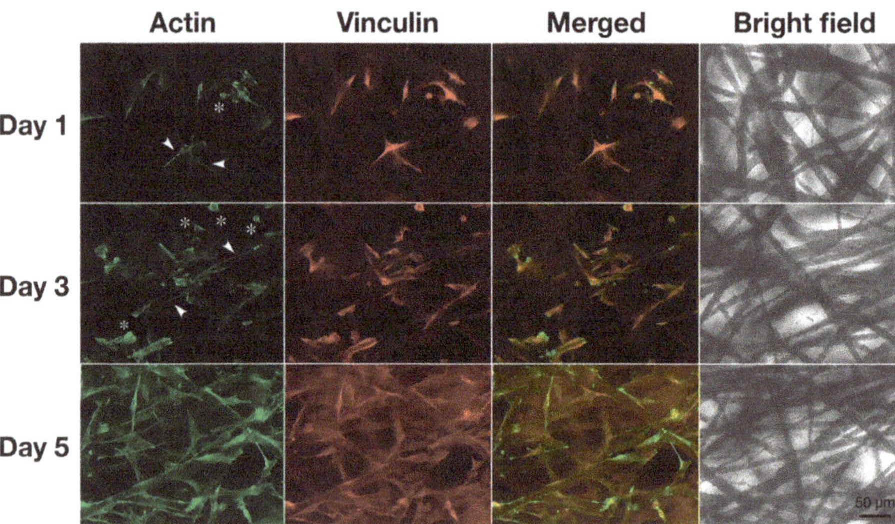

Figure 3. Morphological observation of (**A**) monoculture and (**B**) co-culture cells by immunofluorescence microscopy. Cells were seeded at 5.0×10^4 cells cm^{-3} on a microfiber mesh scaffold placed in each well of a 24-well plate and cultured for 1, 3, and 5 days. At the culture time, the cells were fixed and stained with Alexa Flour® 488-labeled phalloidin for actin (green) and anti-vinculin for vinculin (red). They were viewed through a fluorescence phase-contrast microscope at 20× magnifications (scale bars: 50 µm). Arrowheads show the spindle-shaped cells and asterisks indicate shrinkage-rounded cells.

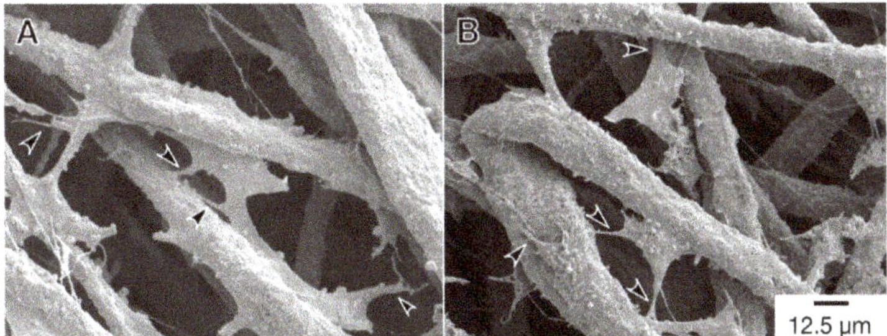

Figure 4. Morphological observation of (**A**) monoculture and (**B**) co-culture cells by SEM. Cells were seeded at 5.0×10^4 cells cm^{-3} on a microfiber mesh scaffold placed in each well of a 24-well plate and cultured for five days. At the culture time, the cells were fixed and viewed with SEM at 800× magnifications (scale bar: 12.5 µm). Arrowheads show pseudopodium-like structures.

3.2. Cell Viability

Three-dimensional cell culture scaffolds are designed for the construction of engineered tissue with biomimetic environments ex vivo. Cell survival on biomaterials is important for supposing cell behavior under similar situations in vivo. In the present study, the metabolic activity of the cells cultured on the scaffold was measured using a resazurin-based assay (Figure 5). Cell viability in the monoculture of the MG-63 cells slightly decreased from day 5 to day 7, even though there was no significant difference between the culture periods. On the other hand, the viability in co-culture of MG-63 cells HUVECs was constantly sustained. Comparing the monoculture and co-culture cells, no significant differences between culture conditions were observed. These results showed that the porous and fibrous scaffold could support cell survival by serving as a pathway for nutrients and oxygen [26,27].

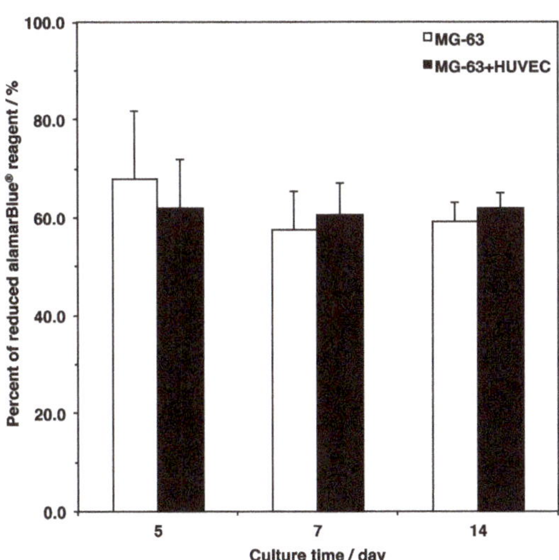

Figure 5. Cell viability of monoculture and co-culture on a microfiber mesh scaffold. Cells were seeded at 5.0×10^5 cells cm^{-3} on the scaffold placed in each well of a 24-well plate and cultured for 5, 7, and 14 days. Cell viability was assessed using an alamarBlue® assay described in Materials and Methods. Data were determined from three replicate samples and are shown as mean ± SD. There were no significant differences among them.

3.3. Osteogenic Differentiation

Previous studies have shown that co-culturing osteoblasts with endothelial cells enhances osteogenesis [15,16]. To investigate the effects on osteoblast differentiation by co-culture of osteoblasts and endothelial cells on the scaffold, bone-specific protein alkaline phosphatase (ALP) activity was quantitated using a *p*-nitrophenyl phosphate substrate assay (Figure 6). The ALP activity in the MG-63 monoculture was constantly sustained from day 5 to day 7 and slightly increased from day 7 to day 14. However, there were no significant differences between the culture times. In contrast, the ALP activity in the co-culture of the MG-63 cells and HUVECs was remarkably high during all culture periods. Comparing the two, the ALP activity under the co-culture conditions was three to six times higher than that under the monoculture conditions. These results indicated that endothelial cells could also enhance osteoblast differentiation on the fibrous scaffold.

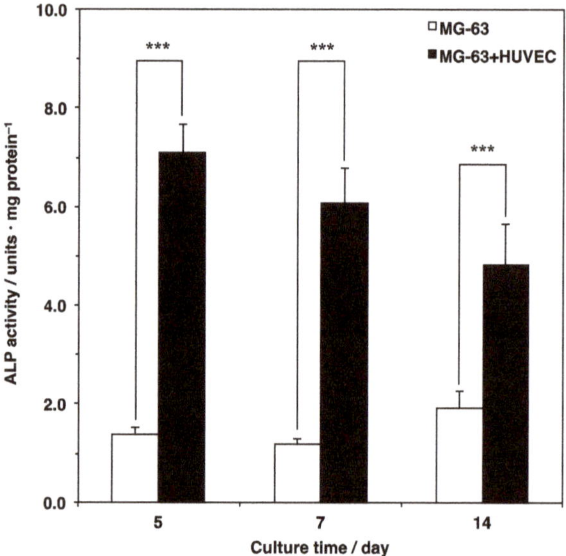

Figure 6. Comparison of ALP activity in a monoculture and a co-culture on a microfiber mesh scaffold. Cells were seeded at 5.0×10^5 cells cm^{-3} on the scaffold placed in each well of a 24-well plate and cultured for 5, 7, and 14 days. ALP activity was assessed using a *p*-nitrophenyl phosphate substrate assay described in Materials and Methods. Data were determined from three replicate samples, which are shown as mean ± SD. *** $p < 0.001$ compared with the monoculture (MG-63 cells alone).

3.4. Angiogenic Properties

In bone tissue engineering, the lack of functional microvasculature induces the deficient supply of oxygen and nutrients and decreases the removal of metabolic wastes. This leads to low cell viability and cell apoptosis, which results in clinical problems [7,28,29]. To examine the vascularization in the co-culture system on the microfiber scaffold, the actin cytoskeleton and CD31 of the cells were observed by fluorescence microscopy (Figure 7). CD31 is well known as the platelet endothelial cell adhesion molecule-1 (PECAM-1) and a specific marker of endothelial cells that can discriminate between MG-63 cells (CD31-negative cells) and HUVECs (CD31-positive cells). On the other hand, F-actin, which is a cytoskeleton protein, is expressed in both osteoblasts and endothelial cells. Five days after seeding, the HUVECs were distributed along the fibers. After seven days, the HUVECs were well spread between the fibers and formed microcapillary-like structures (Arrowheads in Figure 7A). At 14 days of culture, the cells filled in the spaces where the fibers were intertwined and vascular networks were formed among the HUVECs (Figure 7B).

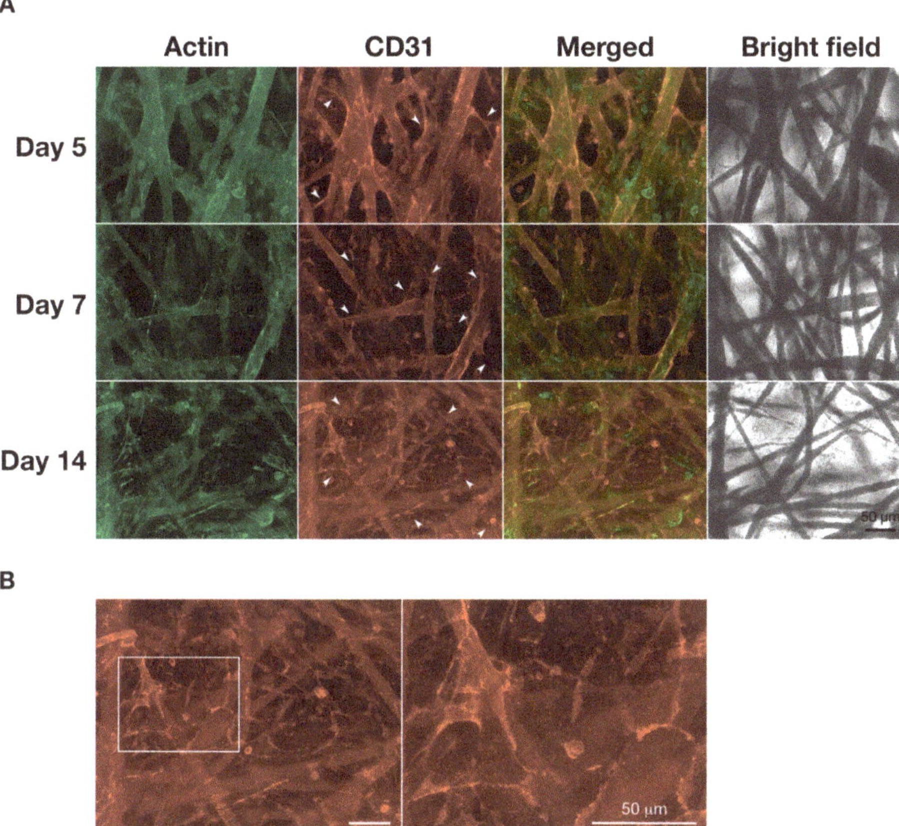

Figure 7. Morphological observation of monoculture and co-culture cells by immunofluorescence microscopy. Cells were seeded at 5.0×10^5 cells cm^{-3} on a microfiber mesh scaffold placed in each well of a 24-well plate and cultured for five, seven, and 14 days. At the culture time, the cells were fixed and stained with Alexa Flour® 488-labeled phalloidin for actin (green) and anti-human CD31 for CD31 (red). They were viewed through a fluorescence phase-contract microscope at 20× magnifications (scale bars: 50 µm). Arrowheads (**A**) show microcapillary-like structures and square enclosure (**B**) indicates a vascular network area.

Next, to confirm the process of angiogenesis in detail, VEGF, which is a growth factor that positively contributes to vascularization, was quantified using a sandwich ELISA (Figure 8). In the MG-63 monoculture, high VEGF secretion was detected at all culture periods. On the other hand, the concentration of VEGF in the co-culture of the MG-63 cells and HUVECs significantly decreased compared with the monoculture. These results suggested that the HUVECs might consume the VEGF, which is mainly produced by MG-63 cells to form the vascular lumens.

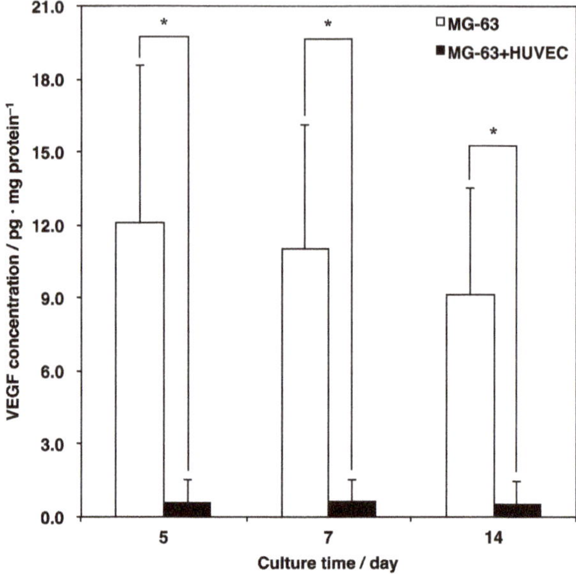

Figure 8. Comparison of vascular endothelial growth factor (VEGF) secretion in a monoculture and a co-culture on a microfiber scaffold. Cells were seeded at 5.0×10^5 cells cm^{-3} on the scaffold placed in each well of a 24-well plate and cultured for five, seven, and 14 days. VEGF production was assessed using a sandwich ELISA described in Materials and Methods. Data were determined from three replicate samples, which are shown as mean ± SD. * $p < 0.05$ compared with monoculture (MG-63 cells alone).

4. Discussion

The effective introduction of blood vessels into engineered tissue is a common challenge in tissue engineering. To date, diverse biomaterials have been developed in consideration of cellular distribution and penetration [20,30]. When pores are inappropriately filled with excess cells, decreased cell survival and necrosis can be induced. However, co-culturing target cells with endothelial cells could be employed to avoid such situations. Additionally, endothelial cells can form vascular networks to supply oxygen and nutrients and remove metabolic wastes in engineered tissue. In bone tissue engineering, a co-culture system of osteoblasts with endothelial cells on porous scaffolds has already been proposed [8,26]. However, the interaction between the co-culture cells and the scaffold was unclear.

In the present study, the processes of osteogenesis and angiogenesis in a co-culture of osteoblasts and endothelial cells on a microfiber scaffold was investigated, and cell attachment, extension, survival, osteoblastic differentiation, and endothelial cell tube formation were the measured variables. A porous and fibrous scaffold was adopted. The scaffolds were prepared using PLGA, β-TCP, and SiV composites, and they had *in vivo* biological properties such as biocompatibility, bio-resorbability, and osteo-conduction [21,24,31]. In particular, PLGA has a property that complements hydrophilicity and hydrophobicity to support cell attachment. The fibrous structure composed of PLGA contributes to cell survival by serving as the pathways for nutrients and oxygen [20,26,27]. Our results showed that the percentage of adherent cells was more than 60%, and their cell viabilities were constantly sustained regardless of the cell types and culture conditions. It is probable that these cell activities were supported by the geometry and surface properties of the scaffold. As previously reported, electro-spun scaffolds could provide the three-dimensional interconnectivities that allow integration between cells and fibers [32,33]. In random fiber networks with large pores, cells could easily integrate inside the scaffold. In this study, a microfiber scaffold with random networks could provide an appropriate environment for cell proliferation and the circularity of oxygen and nutrients. Additionally, a large surface area, which has higher quantity of protein adsorption, could enhance the cell adhesion [34].

Bone-specific ALP is a marker of osteoblast differentiation *in vitro*, which is known to be expressed until maturation. The ALP activity in the co-culture of the MG-63 cells and HUVECs was approximately three to six times higher than that in the monoculture of the MG-63 cells. These results were consistent with previous studies [15,16,35]. In addition, some studies have suggested that the up-regulation of ALP activity could differ between direct and indirect cell cultures [36–38]. This would mean that endothelial cells support osteoblasts via different pathways in proximal and remote conditions. Since the ALP activity on the scaffold was between the effects of direct and indirect contact in the two-dimensional culture, the co-culture on the fibrous scaffold can potentially reflect proximal and remote conditions. However, the mechanisms lack detail.

Previous studies demonstrated that co-culturing osteoblasts and endothelial cells enhances angiogenesis as well as osteogenesis [15,16]. In the present study, we also observed that vascular lumens were formed among the HUVECs over time. Some studies have suggested that osteoblasts could be a main source of VEGF in co-culture systems of osteoblasts and endothelial cells [36,39,40]. In fact, the concentration of VEGF under the co-culture condition significantly decreased when compared with the MG-63 cell monoculture. These results indicate that HUVECs might consume MG-63 cell-derived VEGF to drive cell activity such as proliferation, survival, migration, and tube formation.

The co-culture of osteoblasts and endothelial cells on a scaffold is a dynamic system based on two interactions, which are between osteoblasts and endothelial cells as well as between these cells and the scaffold. Osteoblasts and endothelial cells support and enhance each other's cell behavior via three main elements (e.g., humoral factors, cell junctions, and extracellular matrixes) [41–44]. The importance of cell communication is divided between direct and indirect culture. For instance, direct contact between osteoblasts and endothelial cells can be a starting point for gap junction formation to up-regulate osteogenesis and angiogenesis. However, plostanoids negatively modulates the VEGF-mediated crosstalk between osteoblasts and endothelial cells in direct contact [36,37]. The co-culture of the MG-63 cells and HUVECs on the microfiber scaffold can reflect both remote effects (via humoral factors) and proximal effects (via cell junctions) from the perspective of ALP activity, vascular lumen formation, and VEGF secretion. Therefore, these results suggest that the VEGF produced mainly by the MG-63 cells could be consumed by binding to the VEGF receptor on the HUVECs to contribute to signaling pathways. As a result, the activation of HUVECs can contribute to the behavior of the MG-63 cells, which results in the enhancement of osteoblastic differentiation. However, the pathways have not yet been identified.

It has been reported that cell adhesion to biomaterials could be involved in osteoblast proliferation and differentiation as well as endothelial cell tube formation [45,46]. The cells in this study exhibited spindle-shaped and shrinkage-rounded morphologies at 1 day after seeding, even though the cells exhibited the spread form and pseudopodium-like structures over time. Furthermore, in the case of large cell numbers, the HUVECs distributed themselves along and between the fibers, which resulted in the formation of microcapillary-like structures in three dimensions. On the other hand, it is known that insufficient cell attachment and penetration cannot effectively induce osteogenesis and angiogenesis in the case of dense ceramics and metallics [46]. Therefore, it is considered that the fibrous scaffolds can regulate the vascularization topologically. However, additional studies are warranted to elucidate the interactions between osteoblasts and endothelial cells and between these cells and scaffolds. Furthermore, histological analyses of the bone repair process using a microfiber scaffold in an animal model would make it possible to provide new findings of osteogenesis and angiogenesis. In particular, the histomorphometric analyses of bone structure and remodeling (e.g., bone volume, trabecular thickness, bone mineral density, and bone formation rate) would indicate valuable information about bone metabolism [47].

In summary, the co-culture of osteoblasts with endothelial cells on a microfiber scaffold stimulated cell activity including cell adhesion, survival, osteoblastic differentiation, and endothelial cell tube formation via cell–cell communication and cell–scaffold interaction. Therefore, this approach highlights tissue-engineered bone with vascular networks by crosstalk between osteoblasts and endothelial cells and the interaction between these cells and the scaffold.

5. Conclusions

The reconstruction of three-dimensional tissue requires consideration of the biochemical properties and anatomical features of objective tissue. In tissue engineering, it is essential to investigate the biological properties of the scaffold such as cell attachment, adhesion, extension, survival, proliferation, and differentiation. The induction of vascularization into engineered tissue is also key to avoid clinical issues caused by implanted tissue. In the present study, osteoblasts were co-cultured with endothelial cells on a microfiber biomaterial to examine the processes of osteogenesis and angiogenesis on the scaffold. Our results showed that the cells could attach and spread with the formation of pseudopodium-like structures and that the cells constantly sustained their own viability. Furthermore, osteoblasts and endothelial cells could enhance and improve each other's functions such as osteoblastic differentiation and endothelial cell tube formation. These results were supported by the properties of the scaffold such as fibrous and topological structures. However, (i) cell-to-cell signal transduction on the microfiber scaffold and (ii) the potential of bone repair after implantation of the scaffold are still unclear. Further studies will provide new insights into angiogenesis in bone remodeling and bone metabolism. Taken together, this model has shown that the three-dimensional co-culture system could regulate both osteogenesis and angiogenesis to effectively construct bone-like tissue with vascular networks in vitro.

Author Contributions: Conceptualization, K.I. and M.H. Investigation, K.I. Methodology, K.I. and M.H. Writing—K.I. and M.H. Writing—review and editing, M.H. Visualization, K.I. Supervision, M.H. Project administration, M.H.

Funding: A grant-in-aid from the Japan Society for the Promotion of Science Grant Number 17K01396 partly supported this work.

Conflicts of Interest: The authors declare no conflict of interest.

Abbreviations

EMEM	Eagle's minimum essential medium
ECGM2	endothelial cell growth medium 2 kit
FBS	fetal bovine serum
PBS	phosphate-buffered saline
DMSO	dimethyl sulfoxide
BSA	bovine serum albumin
MTT	3-(4,5-dimethyl-2-thiazolyl)-2,5-diphenyl-tetrazolium bromide
DAPI	4′,6-diamino-2-phenylindole
HUVEC	human umbilical vein endothelial cell
SEM	scanning electron microscopy
ALP	alkaline phosphatase
VEGF	vascular endothelial growth factor
ELISA	enzyme-linked immunosorbent assay
PECAM-1	platelet endothelial cell adhesion molecule-1
β-TCP	β-tricalcium phosphate
PLGA	poly (lactic-co-glycolic acid)
SiV	siloxane-doped vaterite

References

1. Wang, W.; Yeung, K.W.K. Bone grafts and biomaterials substitutes for bone defect repair: A review. *Bioact. Mater.* **2017**, *2*, 224–247. [CrossRef]

2. Pape, H.C.; Evans, A.; Kobbe, P. Autologous bone graft: properties and techniques. *J. Orthop. Trauma* **2010**, *24*, S36–S40. [CrossRef]

3. Dzobo, K.; Thomford, N.E.; Senthebane, D.A.; Shipanga, H.; Rowe, A.; Dandara, C.; Pillay, M.; Motaung, K. Advances in regenerative medicine and tissue engineering: innovation and transformation of medicine. *Stem Cells Int.* **2018**, *2018*, 2495848. [CrossRef]

4. Langer, R.; Vacanti, J. Advances in tissue engineering. *J. Pediatr. Surg.* **2016**, *51*, 8–12. [CrossRef]

5. Lanza, R.; Langer, R.; Vacanti, J. Tissue engineering. *Science* **1993**, *260*, 920–926.

6. Laranjeira, M.S.; Fernandes, M.H.; Monteiro, F.J. Reciprocal induction of human dermal microvascular endothelial cells and human mesenchymal stem cells: time-dependent profile in a co-culture system. *Cell Prolif.* **2012**, *45*, 320–334. [CrossRef]

7. Rouwkema, J.; Rivron, N.C.; van Blitterswijk, C.A. Vascularization in tissue engineering. *Trends Biotechnol.* **2008**, *26*, 434–441. [CrossRef]

8. Grosso, A.; Burger, M.G.; Lunger, A.; Schaefer, D.J.; Banfi, A.; Di Maggio, N. It takes two to tango: coupling of angiogenesis and osteogenesis for bone regeneration. *Front. Bioeng. Biotechnol.* **2017**, *5*, 68. [CrossRef]

9. Black, C.R.; Goriainov, V.; Gibbs, D.; Kanczler, J.; Tare, R.S.; Oreffo, R.O. Bone tissue engineering. *Curr. Mol. Biol. Rep.* **2015**, *1*, 132–140. [CrossRef]

10. Pirraco, R.P.; Marques, A.P.; Reis, R.L. Cell interactions in bone tissue engineering. *J. Cell. Mol. Med.* **2010**, *14*, 93–102. [CrossRef]

11. Grüneboom, A.; Hawwari, I.; Weidner, D.; Culemann, S.; Müller, S.; Henneberg, S.; Brenzel, A.; Merz, S.; Bornemann, L.; Zec, K.; et al. A network of trans-cortical capillaries as mainstay for blood circulation in long bones. *Nat. Metab.* **2019**, *1*, 236–250. [CrossRef]

12. Autefage, H.; Allen, F.; Tang, H.M.; Kallepitis, C.; Gentleman, E.; Reznikov, N.; Nitiputri, K.; Nommeots-Nomm, A.; O'Donnell, M.D.; Lange, C.; et al. Multiscale analyses reveal native-like lamellar bone repair and near perfect bone-contact with porous strontium-loaded bioactive glass. *Biomaterials* **2019**, *209*, 152–162. [CrossRef]

13. Hiratsuka, T.; Uezono, M.; Takakuda, K.; Kikuchi, M.; Oshima, S.; Sato, T.; Suzuki, S.; Moriyama, K. Enhanced bone formation onto the bone surface using a hydroxyapatite/collagen bone-like nanocomposite. *J. Biomed. Mater. Res. Part B, Appl. Biomater.* **2019**. [CrossRef]

14. Fiocco, L.; Li, S.; Stevens, M.M.; Bernardo, E.; Jones, J.R. Biocompatibility and bioactivity of porous polymer-derived Ca-Mg silicate ceramics. *Acta Biomater.* **2017**, *50*, 56–67. [CrossRef]

15. Honda, M.; Aizawa, M. Preliminary study for co-culture of osteoblasts and endothelial cells to construct the regenerative bone. *Key Eng. Mater.* **2017**, *758*, 269–272. [CrossRef]

16. Herzog, D.P.; Dohle, E.; Bischoff, I.; Kirkpatrick, C.J. Cell communication in a coculture system consisting of outgrowth endothelial cells and primary osteoblasts. *Biomed. Res. Int.* **2014**, *2014*, 320123. [CrossRef]

17. Filova, E.; Vandrovcova, M.; Jelinek, M.; Zemek, J.; Houdkova, J.; Jan, R.; Kocourek, T.; Stankova, L.; Bacakova, L. Adhesion and differentiation of Saos-2 osteoblast-like cells on chromium-doped diamond-like carbon coatings. *J. Mater. Sci. Mater. Med.* **2017**, *28*, 17. [CrossRef]

18. Honda, M.; Fujimi, T.J.; Izumi, S.; Izawa, K.; Aizawa, M.; Morisue, H.; Tsuchiya, T.; Kanzawa, N. Topographical analyses of proliferation and differentiation of osteoblasts in micro- and macropores of apatite-fiber scaffold. *J. Biomed. Mater. Res. A* **2010**, *94*, 937–944. [CrossRef]

19. Koduru, S.V.; Leberfinger, A.N.; Pasic, D.; Forghani, A.; Lince, S.; Hayes, D.J.; Ozbolat, I.T.; Ravnic, D.J. Cellular based strategies for microvascular engineering. *Stem Cell Rev.* **2019**, *15*, 218–240. [CrossRef]

20. Stevens, M.M. Biomaterials for bone tissue engineering. *Mater. Today* **2008**, *11*, 18–25. [CrossRef]

21. Obata, A.; Hotta, T.; Wakita, T.; Ota, Y.; Kasuga, T. Electrospun microfiber meshes of silicon-doped vaterite/poly (lactic acid) hybrid for guided bone regeneration. *Acta Biomater.* **2010**, *6*, 1248–1257. [CrossRef]

22. Goegan, P.; Johnson, G.; Vincent, R. Effects of serum protein and colloid on the alamarBlue assay in cell cultures. *Toxicol. In Vitro* **1995**, *9*, 257–266. [CrossRef]

23. Honda, M.; Kikushima, K.; Kawanobe, Y.; Konishi, T.; Mizumoto, M.; Aizawa, M. Enhanced early osteogenic differentiation by silicon-substituted hydroxyapatite ceramics fabricated via ultrasonic spray pyrolysis route. *J. Mater. Sci. Mater. Med.* **2012**, *23*, 2923–2932. [CrossRef]

24. Obata, A.; Ozasa, H.; Kasuga, T.; Jones, J.R. Cotton wool-like poly (lactic acid)/vaterite composite scaffolds releasing soluble silica for bone tissue engineering. *J. Mater. Sci. Mater. Med.* **2013**, *24*, 1649–1658. [CrossRef]

25. Marquis, M.E.; Lord, E.; Bergeron, E.; Drevelle, O.; Park, H.; Cabana, F.; Senta, H.; Faucheux, N. Bone cells-biomaterials interactions. *Front. Biosci. (Landmark Ed.)* **2009**, *14*, 1023–1067. [CrossRef]

26. Santos, M.I.; Reis, R.L. Vascularization in bone tissue engineering: physiology, current strategies, major hurdles and future challenges. *Macromol. Biosci.* **2010**, *10*, 12–27. [CrossRef]

27. Murphy, W.L.; Simmons, C.A.; Kaigler, D.; Mooney, D.J. Bone regeneration via a mineral substrate and induced angiogenesis. *J. Dent. Res.* **2004**, *83*, 204–210. [CrossRef]

28. Stegen, S.; van Gastel, N.; Carmeliet, G. Bringing new life to damaged bone: the importance of angiogenesis in bone repair and regeneration. *Bone* **2015**, *70*, 19–27. [CrossRef]

29. Santos, M.; Pashkuleva, I.; Alves, C.; Gomes, M.E.; Fuchs, S.; Unger, R.E.; Reis, R.; Kirkpatrick, C.J. Surface-modified 3D starch-based scaffold for improved endothelialization for bone tissue engineering. *J. Mater. Chem.* **2009**, *19*, 4091–4101. [CrossRef]

30. Bose, S.; Roy, M.; Bandyopadhyay, A. Recent advances in bone tissue engineering scaffolds. *Trends Biotechnol.* **2012**, *30*, 546–554. [CrossRef]

31. Osada, N.; Makita, M.; Nishikawa, Y.; Kasuga, T. Cotton-Wool-Like Resorbable Bone Void Fillers Containing β-TCP and Calcium Carbonate Particles. *Key Eng. Mater.* **2018**, *782*, 53–58. [CrossRef]

32. Stachewicz, U.; Szewczyk, P.K.; Kruk, A.; Barber, A.H.; Czyrska-Filemonowicz, A. Pore shape and size dependence on cell growth into electrospun fiber scaffolds for tissue engineering: 2D and 3D analyses using SEM and FIB-SEM tomography. *Mater. Sci. Eng. C Mater. Biol. Appl.* **2019**, *95*, 397–408. [CrossRef]

33. Wojak-Cwik, I.M.; Hintze, V.; Schnabelrauch, M.; Moeller, S.; Dobrzynski, P.; Pamula, E.; Scharnweber, D. Poly (L-lactide-co-glycolide) scaffolds coated with collagen and glycosaminoglycans: Impact on proliferation and osteogenic differentiation of human mesenchymal stem cells. *J. Biomed. Mater. Res. Part A* **2013**, *101*, 3109–3122. [CrossRef]

34. Son, S.R.; Linh, N.B.; Yang, H.M.; Lee, B.T. In vitro and in vivo evaluation of electrospun PCL/PMMA fibrous scaffolds for bone regeneration. *Sci. Technol. Adv. Mater.* **2013**, *14*, 015009. [CrossRef]

35. Guillotin, B.; Bourget, C.; Remy-Zolgadri, M.; Bareille, R.; Fernandez, P.; Conrad, V.; Amedee-Vilamitjana, J. Human primary endothelial cells stimulate human osteoprogenitor cell differentiation. *Cell. Physiol. Biochem.* **2004**, *14*, 325–332. [CrossRef]

36. Clarkin, C.E.; Emery, R.J.; Pitsillides, A.A.; Wheeler-Jones, C.P. Evaluation of VEGF-mediated signaling in primary human cells reveals a paracrine action for VEGF in osteoblast-mediated crosstalk to endothelial cells. *J. Cell. Physiol.* **2008**, *214*, 537–544. [CrossRef]

37. Clarkin, C.E.; Garonna, E.; Pitsillides, A.A.; Wheeler-Jones, C.P. Heterotypic contact reveals a COX-2-mediated suppression of osteoblast differentiation by endothelial cells: A negative modulatory role for prostanoids in VEGF-mediated cell: Cell communication? *Exp. Cell Res.* **2008**, *314*, 3152–3161. [CrossRef]

38. Guillotin, B.; Bareille, R.; Bourget, C.; Bordenave, L.; Amedee, J. Interaction between human umbilical vein endothelial cells and human osteoprogenitors triggers pleiotropic effect that may support osteoblastic function. *Bone* **2008**, *42*, 1080–1091. [CrossRef]

39. Furumatsu, T.; Shen, Z.N.; Kawai, A.; Nishida, K.; Manabe, H.; Oohashi, T.; Inoue, H.; Ninomiya, Y. Vascular endothelial growth factor principally acts as the main angiogenic factor in the early stage of human osteoblastogenesis. *J. Biochem.* **2003**, *133*, 633–639. [CrossRef]

40. Wang, D.S.; Miura, M.; Demura, H.; Sato, K. Anabolic effects of 1,25-dihydroxyvitamin D3 on osteoblasts are enhanced by vascular endothelial growth factor produced by osteoblasts and by growth factors produced by endothelial cells. *Endocrinology* **1997**, *138*, 2953–2962. [CrossRef]

41. Simunovic, F.; Winninger, O.; Strassburg, S.; Koch, H.G.; Finkenzeller, G.; Stark, G.B.; Lampert, F.M. Increased differentiation and production of extracellular matrix components of primary human osteoblasts after cocultivation with endothelial cells: A quantitative proteomics approach. *J. Cell. Biochem.* **2019**, *120*, 396–404. [CrossRef]

42. Maes, C.; Clemens, T.L. Angiogenic-osteogenic coupling: the endothelial perspective. *Bonekey Rep.* **2014**, *3*, 578. [CrossRef]

43. Hager, S.; Lampert, F.M.; Orimo, H.; Stark, G.B.; Finkenzeller, G. Up-regulation of alkaline phosphatase expression in human primary osteoblasts by cocultivation with primary endothelial cells is mediated by p38 mitogen-activated protein kinase-dependent mRNA stabilization. *Tissue Eng. Part A* **2009**, *15*, 3437–3447. [CrossRef]

44. Villars, F.; Guillotin, B.; Amedee, T.; Dutoya, S.; Bordenave, L.; Bareille, R.; Amedee, J. Effect of HUVEC on human osteoprogenitor cell differentiation needs heterotypic gap junction communication. *Am. J. Physiol.-Cell Physiol.* **2002**, *282*, C775–C785. [CrossRef]

45. Santos, M.I.; Unger, R.E.; Sousa, R.A.; Reis, R.L.; Kirkpatrick, C.J. Crosstalk between osteoblasts and endothelial cells co-cultured on a polycaprolactone-starch scaffold and the in vitro development of vascularization. *Biomaterials* **2009**, *30*, 4407–4415. [CrossRef]

46. Unger, R.E.; Sartoris, A.; Peters, K.; Motta, A.; Migliaresi, C.; Kunkel, M.; Bulnheim, U.; Rychly, J.; Kirkpatrick, C.J. Tissue-like self-assembly in cocultures of endothelial cells and osteoblasts and the formation of microcapillary-like structures on three-dimensional porous biomaterials. *Biomaterials* **2007**, *28*, 3965–3976. [CrossRef]

47. Parfitt, A.M.; Drezner, M.K.; Glorieux, F.H.; Kanis, J.A.; Malluche, H.; Meunier, P.J.; Ott, S.M.; Recker, R.R. Bone histomorphometry-standardization of nomenclature, symbols, and units. *J. Bone Miner. Res.* **1987**, *2*, 595–610. [CrossRef]

Article

Real-Time Live-Cell Imaging Technology Enables High-Throughput Screening to Verify in Vitro Biocompatibility of 3D Printed Materials

Ina G. Siller, Anton Enders, Tobias Steinwedel, Niklas-Maximilian Epping, Marline Kirsch, Antonina Lavrentieva, Thomas Scheper and Janina Bahnemann *

Institute of Technical Chemistry, Leibniz University Hannover, Callinstraße 5, 30167 Hannover, Germany
* Correspondence: Jbahnemann@iftc.uni-hannover.de; Tel.: +49-511-762-2568

Received: 11 June 2019; Accepted: 28 June 2019; Published: 2 July 2019

Abstract: With growing advances in three-dimensional (3D) printing technology, the availability and diversity of printing materials has rapidly increased over the last years. 3D printing has quickly become a useful tool for biomedical and various laboratory applications, offering a tremendous potential for efficiently fabricating complex devices in a short period of time. However, there still remains a lack of information regarding the impact of printing materials and post-processing techniques on cell behavior. This study introduces real-time live-cell imaging technology as a fast, user-friendly, and high-throughput screening strategy to verify the in vitro biocompatibility of 3D printed materials. Polyacrylate-based photopolymer material was printed using high-resolution 3D printing techniques, post-processed using three different procedures, and then analyzed with respect to its effects on cell viability, apoptosis, and necrosis of adipogenic mesenchymal stem cells (MSCs). When using ethanol for the post-processing procedure and disinfection, no significant effects on MSCs could be detected. For the analyses a novel image-based live-cell analysis system was compared against a biochemical-based standard plate reader assay and traditional flow cytometry. This comparison illustrates the superiority of using image-based detection of in vitro biocompatibility with respect to analysis time, usability, and scientific outcome.

Keywords: real-time live-cell imaging technology; in vitro study; biocompatibility; 3D printing; flow cytometry; adipogenic mesenchymal stem cells

---

1. Introduction

3D printing has become a highly attractive tool with numerous different applications in the last decade. Already established technologies in the realm of rapid prototyping, 3D printing techniques are now increasingly being used to fabricate individually-designed devices in a comparatively easy, time and cost-effective way. Several 3D printing technologies are now available on the market, diverging mainly in the printing process and/or the physical state of the material bases utilized. The most established of these technologies create devices by melting and extruding thermoplastic filaments, fusing small particles of polymer powder together, or curing liquid resins via photopolymerization [1]. There are some fundamental similarities, however, for example, all 3D printing techniques make use of a "layer by layer" fabrication process, which facilitates almost unlimited complexity with respect to the final printed product.

Facilitated by 3D printing, the rise of rapid prototyping has great potential to accelerate the progression of biomedicine, biotechnology, and tissue engineering. Put differently, 3D printing permits the rapid fabrication of customized medical products and equipment, which can enable more individualized medical application [1–3]. The generation of implantable, highly porous 3D scaffolds has become an increasingly important concept within the field of tissue engineering [3,4]. Such porous, personalized scaffolds provide a suitable surface for patient-specific cells to proliferate under ideal conditions. Nevertheless, despite these recent advances in medical applications involving 3D printing, the in vivo use of 3D printed materials should still be treated with some degree of caution, given the tremendous complexity of interactions within the human body. One major challenge associated with introducing foreign material into an organic system is the concept of "biological compatibility" or "biocompatibility" [1]. A generally accepted definition of this concept was given by D. F. Williams in 1987, "Biocompatibility is the ability of a material to perform with an appropriate host response in a specific application" [5].

An "appropriate host response" includes a normal healing process, resistances to bacterial colonization or biofilm formation and the prevention of blood clotting [6]. The biocompatibility of all materials being considered for use in real-world biomedical applications must first therefore be carefully assessed and confirmed via in vivo and in vitro studies [6,7]. If leachables or extractables show a negative impact on mammalian cells in vitro, then a material cannot be characterized as biologically compatible [8]. International standards—such as ISO 10993—provide extensive information, which can be used to develop appropriate assays and otherwise inform about biocompatibility testing methods [9].

A variety of cell culture-based in vitro assays are available for investigation of cytotoxicity of materials. These methods vary widely, from analysis and counting of viable/dead cells via microscope to biochemical-based assays and flow cytometric analyses. Microscopic observations of changes in cell morphology and counting of viable/dead cells form the basis [7]. As vital dyes such as Trypan blue can only enter—and thereby mark out—cells with disrupted cell membranes, use of these dyes in tissue cultures allows researchers to visually distinguish living and dead cells [10]. Biochemical-based assays provide a more reliable and specific outcome [7,11]. Numerous commercial assays are available, each one dealing with a different process of cellular metabolism [7,12]. For example, the CellTiter-Blue® assay (CTB assay) used within this study relies on the conversion of resazurin to the fluorescent product resorufin, which highlights the intracellular reduction potential of living cells [13]. However, although assays like this are widely used, they can only highlight the fundamental distinction between living and dead cells—they do not allow any further nuanced analysis into the different mechanisms by which cell death may occur [7]. Analyses of apoptosis and necrosis provide more detailed information on this front. The apoptotic pathway describes the mechanism of an internally "programmed" cell death [14,15]. By contrast, necrosis is a cellular death mechanism that has been triggered by external factors, such as injury or infection [14]. Both of these pathways display distinct morphological and biochemical features that can be observed and analyzed using specific fluorescence detection markers.

Due to the versatility of 3D printing technologies, a wide variety of printing materials—as well as post-processing procedures and surface finishing steps—are now being utilized. The materials can differ (for example) in their physical state, melting temperature, strength, and/or durability [1]. And the potential fields of application for any given method—as well as associated necessary post-processing or sterilization steps—are ultimately dependent on the properties of the underlying materials [7,16]. For example, materials with a high heat distortion temperature can be sterilized by thermal sterilization (autoclaving) for subsequent use in biological applications, while materials with a lower heat distortion tolerance require alternative sterilization procedures. Support materials such as wax are used by many 3D printing technologies to provide a scaffold with which to stabilize the building material. Following the printing process, this support material must be removed. The post-processing and removal of support material residues is also material-dependent, and can cause difficulties—especially with respect to detailed 3D structures and small channels (for example in microfluidic systems [17]. Depending on the post-processing and sterilization procedure, different end products with different properties can be obtained from the same material formulation. For applications in cell culture, every material formulation and post-processed product needs an individual investigation for biocompatibility. Accordingly, there is an apparent need for high-throughput screening assays.

This study seeks to help to fill in that gap by introducing real-time live-cell imaging technology as a fast, cost-effective and easy to use screening method to examine the in vitro biocompatibility of materials. To that end, a translucent clear, solid polyacrylate was printed via a high-resolution MultiJet 3D printing process, and was then post-processed to remove the supporting material. Following this post-processing procedure, three different disinfection and sterilization methods were examined, using ultra violet (UV) light as a physical sterilization method, as well as ethanol (70%, *v/v*) and sodium hypochlorite (2%, *v/v*) as chemical reagents. Afterwards, all of the post-processed objects were analyzed and screened for their suitability in cell culture applications by comparison of different in vitro biocompatibility methods. For biocompatibility evaluation, extraction media were obtained in accordance to ISO 10993:2012 standards and its impact on adipogenic mesenchymal stem cells (MSCs) was observed. Metabolic activity (representing cell viability) was assessed using a CellTiter-Blue® (CTB) cell viability assay. Analyses of apoptotic and necrotic responses as a measure of biocompatibility were also performed in a comparative study, using both modern image-based live-cell analysis technology and traditional flow cytometry.

2. Materials and Methods

2.1. Experimental Procedure

After 3D printing of a translucent polyacrylate material was completed, the printed parts are then cleaned in post-processing steps and sterilized or disinfected, respectively, using three different approaches. According to ISO 10993:12, extraction media are obtained for studying the influence of the 3D printed material on MSCs in in vitro biocompatibility assays. Three different methods to assess in vitro biocompatibility were compared in this study: (1) A biochemical-based viability assay (CTB assay) in a standard plate reader; (2) traditional flow cytometry; and (3) novel image-based live-cell analysis. A schematic overview of the experimental procedure is shown in Figure 1.

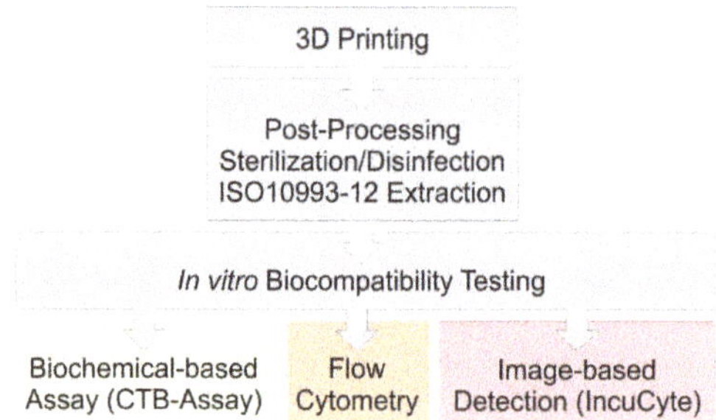

Figure 1. Flow chart of performed experiments. The in vitro biocompatibility of 3D printed material was evaluated using three different approaches.

2.2. 3D Printing

3D printed constructs were manufactured using the high-resolution MultiJet 3D printer ProJet® MJP 2500 Plus (3D Systems, Rock Hill, SC, USA). The 3D printing material analyzed in this study is VisiJet® M2R-CL (3D Systems, Rock Hill, SC, USA). It appears as a translucent clear, solid polyacrylate following a UV-curing process, and it is printed with a resolution of 800×900 dots per inch and a layer resolution of 32 μm [18,19]. As support material for the printing process, VisiJet® M2-SUP is used. For studying the success of the post-processing and the potential influence of leachables, $5 \times 5 \times 5$ mm cubes were printed—representing a total surface area of $1.5 \text{ cm}^2 \cdot \text{ml}^{-1}$. The known hazardous components in the liquid state of the present acrylic photopolymer material are 3-hydroxy-2,2-dimethylpropyl, 3-hydroxy-2,2-dimethylpropionate, the polymerization initiator diphenyl(2,4,6-trimethylbenzoyl) phosphine oxide, and monofunctional urethane acrylate. Together with the constituent tricyclodecane dimethanol diacrylate, these components are all listed as being potentially harmful to aquatic organisms and/or as otherwise potentially causing adverse effects on aquatic environments in their liquid state (i.e., before polymerization) [19]. During the printing process, the polyacrylic material is polymerized by UV light—after which it can be declare harmless. No additional information regarding the material was provided by the manufacturer.

2.3. Post-Processing of 3D Printed Objects

All steps of the post-processing are shown in Figure 2, and they include freezing the printing plate for 15 min at −18 °C, and placing 3D printed objects in a heat steam bath of the EasyClean unit (3D systems, Rock Hill, SC, USA) for 45 min at 65 °C and incubation in an ultrasonic bath (Bandelin electronic, Berlin, Germany) with detergent (Fairy Ultra Plus, Procter and Gamble, CT, USA) for 30 min at 65 °C. Deionized water, provided by Arium® (Sartorius Stedim Biotech GmbH, Göttingen, Germany), was used in all experiments.

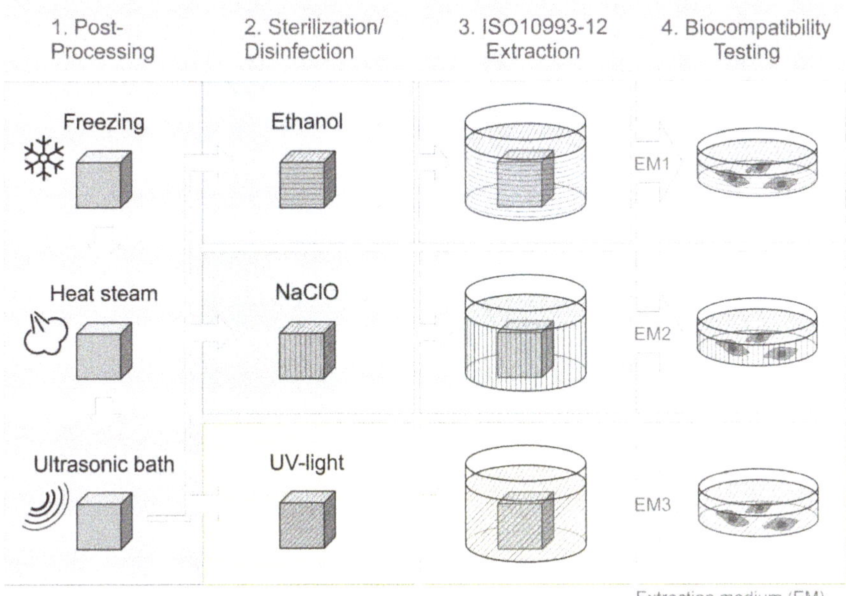

Figure 2. Schematic process of 3D printing, post-processing and extraction medium preparation. Cleaning steps (**1**): freezing of 3D printed objects (15 min, −18 °C), heat steam in a water bath (45 min, 65 °C), ultrasonic bath with detergent (30 min, 65 °C). Sterilization steps (**2**): disinfection in ethanol (70%, *v/v*, 1 h, RT) or sodium hypochlorite (2%, *v/v*, 1 h, RT) or UV light exposure (1 h, RT). Biocompatibility testing steps (**4**) then followed an incubation of 3D printed objects in cell culture medium according to EN ISO 10993-12 (2012) (**3**). (EM = extraction medium). EM 1: EM obtained by incubation of 3D printed material treated with ethanol (70%, *v/v*) in a disinfection process. EM 2: EM obtained by incubation of 3D printed material treated with sodium hypochlorite (2%, *v/v*) and EM 3: EM obtained by incubation of 3D printed material sterilized by UV light.

2.4. Sterilization/Disinfection of 3D Printed Objects

One disadvantage of many 3D printed materials is their relatively low heat distortion temperature and their corresponding incompatibility with thermal sterilization approaches [19–21]. However, a guaranteed sterile and disinfected product is necessary for the use in biomedical applications [22,23]. The polyacrylic material used in this study has a heat distortion temperature around 80 °C, as a result the most common sterilization method (autoclaving) is not a possibility [19,24,25]. But physical and chemical procedures can also be used to sterilize and disinfect materials [24]. In this study, two different methods for chemical disinfection were used: The product was subjected to incubation in either ethanol (Carl Roth GmbH und Co. KG, Karlsruhe, Germany), 70%, *v/v*, or sodium hypochlorite, 2%, *v/v*, (Carl Roth GmbH und Co. KG, Karlsruhe, Germany), for 1 h at room temperature. In addition, UV irradiation (UV Sterilization Cabinet KT-09DC, Alexnld, Tiberias, Israel, 6 W, λ = 266 nm) was also used as a physical sterilization method. In order to cover every side of the 3D printed cubes with UV light, the cubes are turned around within a total of 1 h of UV light exposure at room temperature. After sterilization or disinfection procedure all cubes were washed thoroughly with sterile phosphate-buffered saline (PBS).

2.5. Preparation of Extraction Media (EM) for Biocompatibility Studies

Potential leaching properties of the 3D printed material, or remaining support material, were evaluated by obtaining extraction medium (EM) according to EN ISO 10993-12:2012 (Biological evaluation of medical devices—art 12: Sample preparation and reference materials). After post-processing, the aforementioned 3D printed cubes were incubated in cell culture medium Minimum Essential Medium Eagle, with alpha modification (α-MEM) (Thermo Fisher Scientific Inc., Waltham, MA, USA) containing 10% human serum (c.c.pro GmbH, Oberdorla, Germany) and 1% Gentamicin (PAA Laboratories GmbH, Pasching, Austria), for 72 h at 37 °C with a surface area/volume ratio of 3 $cm^2 \cdot ml^{-1}$. The obtained medium is referred to as extraction medium (EM). EM obtained by incubation of 3D printed material treated with ethanol (70%, *v/v*) in post-processing process is referred to as "EM 1." EM obtained by incubation of 3D printed material treated with sodium hypochlorite (2%, *v/v*) is hereafter referred to as "EM 2", and EM obtained by incubation of 3D printed material sterilized by UV light is referred to as "EM 3." Cell culture medium incubated for 72 h at 37 °C without 3D printed objects served as a control for all biocompatibility experiments.

2.6. Cell Line and Cell Culture Conditions

For all experiments, human adipogenic tissue-derived mesenchymal stem cells (MSCs) were used. After obtaining the donor's informed written consent, as approved by the Institutional Review Board (Hannover Medical School) with the reference number 3475-2017, adipose tissue was received following abdominoplasty surgery. After isolation, MSCs have been extensively characterized by surface marker analysis and functional properties as described earlier [26]. Cultivation of MSCs was performed in cell culture medium in a 5% CO_2, 21% O_2, humidified atmosphere at 37 °C (Heracell 150i incubator, Thermo Fisher Scientific Inc., Waltham, USA). The MSCs were routinely maintained in 75 cm^2 cell culture flasks (Corning, CellBind Surface, Corning, NY, USA), and then harvested at about 85% confluency by accutase treatment (Merck KGaA, Darmstadt, Germany) for detachment [26]. 24 h prior to the start of an experiment, cells were seeded in 6-, and 96-well plates (at a density of 18,000 cells·cm^{-2} and 1100 cells·cm^{-2}, respectively) (Sarstedt AG and Co. KG, Nürnbrecht, Germany). Experiments were performed with cells of passages two to six.

2.7. CellTiter Blue® (CTB) Viability Assay in Fluorescence Plate Reader

For indirect evaluation of cell viability using a standard method, a CellTiter-Blue® cell viability assay (Promega, GmbH, Mannheim, Germany) was performed using the background and standard controls specified in the accompanying manual. Metabolically active cells are able to reduce blue resazurin into a purple, fluorescent resorufin via action of numerous redox enzymes in different intracellular compartments [13,27,28]. The fluorescence intensity produced by this reaction is therefore indicative of the number of viable cells. The product formation is monitored at an extinction wavelength of 544 nm, and an emission wavelength of 590 nm, using a fluorescence plate reader (Fluoroskan Acent, Thermo Fisher Scientific Inc., Waltham, MA, USA). MSCs were seeded in 96-well plates at a density of 8000 cells/well in 100 µl cell culture medium and incubated for 24 h at 37 °C in a humid atmosphere supplemented with 5% CO_2. Subsequently, the MSCs were cultivated in the related extraction medium (see Section 2.5) or control medium for 24 h. After 24 h, extraction or control medium was removed,

100 µl fresh culture medium containing 10% CTB stock solution was added to each well and the MSCs were incubated for 1.5 h before measuring the fluorescence in a plate reader. Each experiment was repeated 13 times ($n = 13$).

2.8. Cell Viability Analysis by Flow Cytometry

Flow cytometry represents the traditional method used to monitor and quantitatively examine cell apoptosis and necrosis [29]. The BD FACSAria™ Fusion (Becton Dickinson, Franklin Lakes, NJ, USA) used in this study contains four lasers with numerous filters, which allow for a combination of multiple fluorescence markers within one sample. The basic principle of a flow cytometer is the analyses of hydrodynamically focused single cells that pass orthogonally through a bundled laser beam of a suitable wavelength. As they pass through the laser beam, the cells can be identified and classified by their physical characteristics (i.e., according to cell size, granularity, or specific fluorescence labeling) [30].

2.8.1. Sample Preparation

MSCs were seeded at a density of 18,000 cells·cm^{-2} in 6-well plates and then incubated for 24 h at 37 °C under 5% CO_2. Before related extraction media or control medium was used (as described below in Section 2.5), MSCs were washed once with PBS to remove non-adherent cells. MSCs were then cultivated in correspondent media for another 24 h. Cell samples for cell counting and flow cytometry experiments were obtained by detachment of adherent cells using accutase treatment. Before dyeing and analysis, the detached cells were sedimented by centrifugation for 5 min at 200× g and then resuspended in fresh culture medium [31,32]. The cell number and viability was estimated viw cell counting using a 0.4% Trypan blue stain ($n = 4$) in a haemocytometer (Brand GmbH + Co. KG, Wertheim, Germany) [10]. Trypan blue can be used to visually identify cells with disrupted cell membranes since dead or damaged cells possess a compromised membrane integrity which allows the dye to enter the cell and visibly mark it as distinct from a healthy living surrounding.

2.8.2. Measurement and Quantification of Apoptosis and Necrosis

MSCs were centrifuged for 5 min at 200× g, resuspended, and then washed with PBS twice. Necrotic cells were marked and identified using the SYTOX® AADvanced™ Dead Cell Stain, which is provided in the CellEvent™ Caspase-3/7 Green Flow Cytometry Assay Kit (Thermo Fisher Scientific Inc., Waltham, MA, USA). These cells were stained as instructed in the manual. Necrotic cells possess disrupted cell membranes which allow the Dead Cell Stain to enter the cell and intercalate in DNA structures, thereby visually marking out the cell. Necrosis can be measured at an excitation maximum of 546 nm and an emission maximum of 647 nm. Apoptotic cells express and activate the enzymes caspase-3 and caspase-7 [33]. Hence, apoptosis can be evaluated by the detection of active caspase-3/7 using the CellEvent™ Caspase-3/7 Green Stain, provided in the same assay kit. The corresponding green fluorescence signal has an excitation maximum of 511 nm and an emission maximum of 533 nm, and was captured with appropriate laser and filter settings using a BD FACSAria™ Fusion flow cytometer. The same number of cells were stained in each sample, in order to maintain an equal distribution of fluorescence reagents to cells. To represent a positive control for apoptosis, 50 µm cisplatin (cisplatin-induced apoptosis) was also added to the cells (control experiments were performed in triplicate). Cisplatin is a platin derivative that blocks DNA synthesis, induces apoptosis via p53-dependent and independent signaling mechanisms, and activates caspase-3. It is a well-known DNA-alkylating antitumor agent which is used as a chemotherapeutic drug [34,35]. MSCs cultivated in normal cell culture medium, without contact to 3D printing material, served as a negative control. The MSCs were cultivated for a period of 30 h, with cell samples taken every 4–6 h ($n = 6$). Cell samples were handled and counted via the Trypan blue exclusion method (described in Section 2.8.1). The BD FACS Diva™ Software v8.0 (Becton Dickinson, Franklin Lakes, NJ, USA) was used for analysis. Flow cytometry analysis is predicated on the principle of "gating", by placing gates around cell populations with common characteristics, different cell populations can be segregated and selected for further investigation. Here, a uniform gating strategy was used for all experiments in order to separately analyze and quantify apoptotic, necrotic and living cells. Necrotic and apoptotic cells, respectively, possess higher red and green fluorescence signal intensities compared with living cells. Gates were determined based on both positive and negative cell controls. At least 10,000 events per sample were analyzed with an "event" being defined as a single particle detected by the system. The experiment was performed with three biological replicates.

2.9. Cell Viability Analysis by Real-Time Live-Cell Imaging System

The IncuCyte® Live-Cell Analysis System (Sartorius Stedim Biotech GmbH, Göttingen, Germany) is an image-based real-time system that allows for an automatic acquisition and analysis of cell images. With the use of two lasers, both phase contrast as well as fluorescence images can be captured. The entire system is placed inside a cell culture incubator in order to guarantee controlled cultivation conditions during real-time monitoring. Phase contrast and fluorescence images are automatically recorded and analyzed using customized software tools in the IncuCyte® S3 image analysis software (Sartorius Stedim Biotech GmbH, Göttingen, Germany). With pre-defined imaging masks, fluorescence signals of the recorded images are then analyzed and counted. Parameters such as minimum fluorescence signal intensity are considered and defined in advance (e.g., to exclude diffuse background noise from the evaluation). The same imaging masks are applied to all acquired images. The data is exported as Counts/Image, which represents the counted fluorescence signals with respect to a single image. The applied dynamic image processing and analysis enables quantitative real-time analyses of fluorescence signals in an imaging field. In addition, by using pre-defined cell-specific imaging masks containing information on cell size and shape, cell growth can be monitored in real-time, by analyzing the occupied area of an imaging field in phase contrast images. Accordingly, this system provides both quantitative and kinetic data. A schematic workflow of the real-time live-cell imaging system is shown in Figure 3.

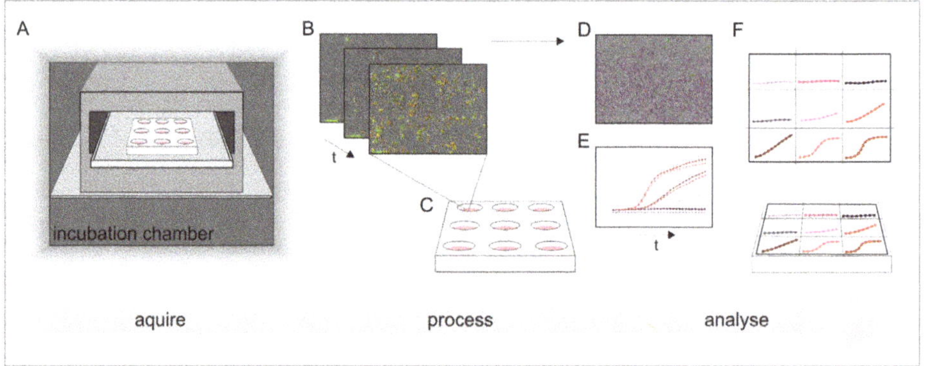

Figure 3. Schematic illustration of the working process of real-time live-cell analysis. (**A**) Placing of the real-time live-cell imaging system inside a cell culture incubator; (**B**) automatically acquire images over time; (**C**) receive images of all locations in the culture vessel at once; (**D**) imaging masks identify regions of interest; and (**E**) the results can be monitored in real-time and (**F**) display quantitative and kinetic analyses of all culture vessels at once.

2.9.1. Sample Preparation

MSCs were seeded in 96-well plates at a density of 8000 cells/well in 100 µl cell culture medium and then incubated for 24 h at 37 °C in a humid atmosphere supplemented with 5% CO_2. Staining reagents for quantification of apoptosis and necrosis were diluted in respective cell culture medium obtained as described in Section 2.5. Before the staining reagents containing media were added to these cultivation wells, the old medium was first discarded and all non-adherent cells were removed by a washing step with PBS.

2.9.2. Measurement and Quantification of Apoptosis and Necrosis

A quantitative analysis of apoptosis and necrosis of MSCs over time was ascertained during the cultivation in extraction medium 1 (EM 1) and extraction medium 2 (EM 2), with regular cell culture medium serving as a control. Staining of the cells was performed using the IncuCyte® Cytotoxicity and Apoptosis Detection Kits (Sartorius Stedim Biotech GmbH, Göttingen, Germany), according to the manufacturer's protocols. Real time measurement of necrosis is based on the cell membrane integrity—i.e., the same principle as used for necrosis detection in flow cytometry experiments. In the case of a damaged cell membrane, the IncuCyte® Cytotox dye enters the cell, intercalates into DNA, and thereby marks out the nuclei. Red fluorescence of the cytotoxicity dye can be measured at an excitation maximum of 612 nm and an emission maximum of 631 nm. As in flow cytometry experiments, apoptotic cells were analyzed by detection of active caspase-3/7. IncuCyte® Caspase-3/7 reagent was used for the detection of active caspase-3/7, which is expressed and activated in apoptotic cells. Apoptotic cells can be identified by measuring green fluorescence at an excitation maximum of 500 nm and an emission maximum of 530 nm. The apoptosis inducer cisplatin was added in three wells to a final concentration of 50 μm to regular cell culture medium, in order to represent a positive control for apoptosis. MSCs cultivated in regular cell culture medium, without contact to 3D printing material, served as a control. As soon as the staining reagents with the corresponding medium were added to the cell culture wells, the monitoring was started using the IncuCyte® S3 Live-Cell Analysis System. Phase contrast and fluorescence images were automatically captured every hour for a duration of 30 h. The experiment was performed with six biological replicates, every measurement in triplicates. Quantitative analyses of caspase-3/7 and cytotoxicity signals, as well as of cell proliferation were performed with pre-defined cell-specific masks in the IncuCyte® S3 image analysis software.

3. Results

3D printed polyacrylic material was post-processed using three different sterilization or disinfection methods. To evaluate the efficiency of each post-processing and disinfection method as well as to investigate potential leaching properties of the 3D printed polyacrylic material itself, a comparative study using a biochemical-based standard plate reader assay (CTB Assay), standard flow cytometry, and an image-based live-cell analysis system was conducted. The leaching of acrylate monomers, degradation products, or other components from polymer-based materials is a well-known problem that often has negative effects on the biological environment [7,36–38]. Leachables can lead to cytotoxic effects on cells (which can manifest as irritations and/or allergic reactions within the human body) [8,36,39].

3.1. Biochemical-Based CTB Cell Viability Assay

Metabolic activity as an indicator of cell viability of MSCs is analyzed by performing biochemical-based CTB cell viability assays during cultivation in extraction medium, which is prepared according to EN ISO 10993-12 (2012) (see Section 2.5). This CTB assay presents a biochemical-based method for assessing the cytotoxicity of a material. These results are summarized in Figure 4, where the cell viability observed during MSC cultivation in different extraction media is plotted. The cell viability is normalized to the control cultivation. Here, the use of ethanol (70%, *v/v*) (EM 1) as disinfectant did not show a significant difference in metabolic capacity and cell viability compared to control cultures. By contrast, both chemical disinfection methods of the 3D printed objects with sodium hypochlorite (2%, *v/v*) (EM 2), and irradiation sterilization (EM 3), caused a significant decrease in metabolic activity—resulting in only 35.5 ± 13,0% and 25.4 ± 17.0% viable cells, respectively, when compared to the control culture. From these results, the following conclusions could be drawn: (1) cleaning and disinfection of the 3D printed parts using ethanol 70% was successful, and (2) EM 1 did not contain any toxic leachables.

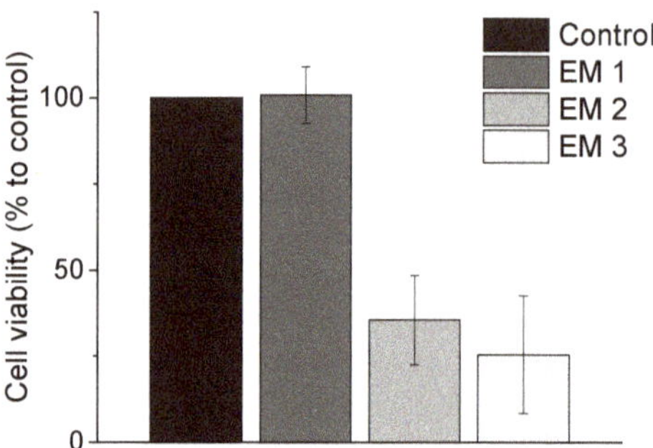

Figure 4. Results of CellTiter-Blue® cell viability assay (CTB assay) to analyze the metabolic capacity (shown as cell viability in %) of MSCs. (EM = extraction medium). EM 1: EM obtained by incubation of 3D printed material treated with ethanol (70%, *v/v*) in a disinfection process. EM 2: EM obtained by incubation of 3D printed material treated with sodium hypochlorite (2%, *v/v*). EM 3: EM obtained by incubation of 3D printed material sterilized by UV light. All experiments were repeated several times ($n = 13$) and compared to MSC cultivation in regular cell culture medium (Control).

It can further be concluded that UV light is not a suitable sterilization method for the 3D printed material used in this study. The negative effects of EM 3 on cell viability may be due to several factors. UV light can have an adverse effect on both the optical and mechanical properties of polymer materials [40,41]. In our experiments, a slight change in color and translucency, as well as an increased brittleness of the surface of the material, was noticed after only 1 h of UV light exposure. Applications involving polymers are restricted due to the capability of photo-degradation, particularly under exposure of UV light [41,42]. Photooxidative reactions aroused by UV light are also associated with the formation of free radicals, which can lead to a radical chain mechanism and ultimately result in the rupture of a polymer structure. The degree of impact depends on the UV light intensity and duration—but this process initially manifests as a change in the color and an increased degree of "mistiness" observed in the polymer material [41,43]. These reactions may also lead to a release of leachables, which can have cytotoxic effects on cells. It should also be noted that the UV sterilization method was also rather impractical in this instance, because the 3D printed parts had to be rotated permanently in order to ensure uniform UV exposure. Since it would be difficult to maintain uniform UV irradiation across all surfaces of complex 3D printed structures—such as embedded channels in microfluidic systems—they would therefore be difficult to sterilize using this procedure.

Similarly, although sodium hypochlorite is the most widely used disinfectant in the food industry and a commonly used irritant in endodontic practice, a significant decrease in cell viability of MSCs was observed in our CTB assays after cultivation in EM 2 using sodium hypochlorite as a disinfection agent in the post-processing process [10,44–46]. This is perhaps not surprising; a study on mesenchymal stem cells of the human bone marrow from Alkahtani et al. has previously shown that even low concentrations of sodium hypochlorite exhibit cytotoxicity [47]. Treatment of sodium hypochlorite can thus damage cell membrane proteins and lead to cell lysis [48]. Such damage might have been responsible for the decreased metabolic activity observed in our MSCs. In contrast, the use of ethanol (70%, *v/v*) as a disinfection agent in the post-processing process of the 3D printed polyacrylic material has no negative impact on metabolic capacity of MSCs. Ethanol functioned as an effective disinfectant here without impacting either the optical or mechanical properties of the material. In addition, ethanol (70%, *v/v*) is also already a commonly used disinfectant in the health services field [23,49].

The CTB assay can score with its fast and user-friendly implementation while also allowing for high-throughput screenings. As a method performed in a standard plate reader, there is no need of sophisticated instruments. However, there is one important limitation on the CTB assay: it only provides information about the count of viable cells, and it is not sensitive to measuring the different mechanisms that can lead to cellular death, which present important information about the material formulation under investigation. Accordingly, to more precisely consider the impact of the post-processed 3D printed material on cell behavior, further studies aimed at measuring the rate of specific death mechanisms (i.e., apoptosis and necrosis) were also necessary. The use of specific dyes which mark out particular apoptotic and necrotic intracellular signals allowed for more detailed evaluations of cellular behavior and cytotoxicity mechanisms to assess in vitro biocompatibility. The standard plate reader used for CTB assays is not capable of detecting multiple fluorescence signals simultaneously. The follow section therefore considers the practicability of performing apoptosis and necrosis staining and analyses in a flow cytometry study vs. using a novel high-throughput image-based analysis system.

3.2. Analysis of Cell Death Responses via Flow Cytometry

Flow cytometry is a standard method used to monitor and quantitatively examine cell death via apoptosis and necrosis [29]. As cells undergoing necrosis experience a disruption of the cell membrane, the use of a red fluorescence dye that enters and labels the DNA of damaged cells with disrupted cell membranes is an elegant and effective way to visibly mark out such cells [33]. Specific fluorescence labeling can also be used to visually detect apoptotic cells, which express and activate the enzymes caspase-3 and caspase-7 [33]. Here, a green fluorescence dye that is sensitive to active caspase-3/7 was used to identify apoptosis (see Section 2.8.2). The relative percentage of necrotic vs. apoptotic MSCs within a sample can then be assessed and used to analyze the biocompatibility of the 3D printed material after post-processing and disinfection (see Section 2.5). Figure 5 shows the flow cytometric analysis of MSCs cultivated over a period of 30 h.

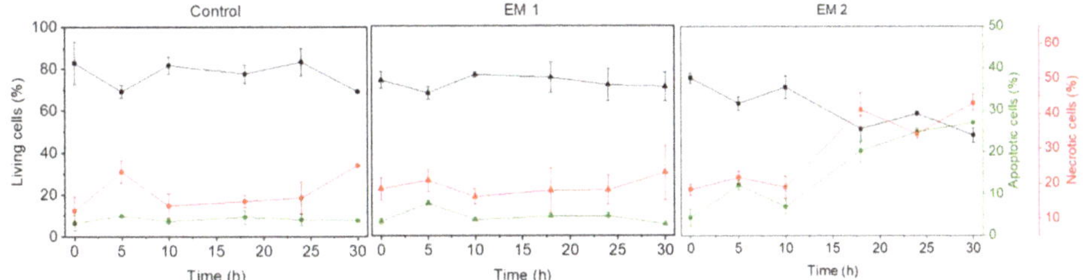

Figure 5. Results of flow cytometric studies on apoptosis and necrosis of MSCs over a period of 30 h. The percentage of living, apoptotic and necrotic cells are analyzed per cultivation. A caspase 3/7 signal (green) represents apoptotic cells; the cytotox-signal (red) is correlated to necrotic cells. (EM = extraction medium). EM 1: EM obtained by incubation of 3D printed material treated with ethanol (70%, *v/v*) in a disinfection process. EM 2: EM obtained by incubation of 3D printed material treated with sodium hypochlorite (2%, *v/v*). The experiments are compared to MSC cultivation in regular cell culture medium (Control) and were performed three times (*n* = 3).

As explained above (see Section 3.1), UV light is not a suitable sterilization method for the 3D printed material used in this study. Therefore, as shown in Figure 5, UV light as sterilization method was no longer analyzed. MSCs that were cultivated in extraction medium 1 (EM 1), obtained by incubation of 3D printed material disinfected by ethanol (2%, *v/v*), showed no significant difference with respect to the relative percentages of living, apoptotic, and necrotic cells when compared to control cultures; in both cases, the percentage of apoptotic cells was about 4%, the percentage of necrotic cells was about 16%, and the balance were living cells. Since the same number of cells was stained and used for each measurement, the data does not show any increase in the count of living cells due to cell growth. In contrast to the MSCs in EM 1 and the control cultures, the cultivation of MSCs in extraction medium 2 (EM 2)—obtained by incubation of 3D printed material disinfected by sodium hypochlorite (2%, *v/v*)—resulted in a strong increase in both apoptotic and necrotic cells. In this medium, the percentage of apoptotic and necrotic cells increased over time from 4% and 18%, respectively, to approximately 30% and 45%, while the percentage of living cells correspondingly decreased from 80% to 50%. Each experiment showed a slight increase in the percentage of apoptotic and necrotic cells, as well as a simultaneous decrease in the count of living cells (after 5 h). This occurrence may be related to the change of cell culture medium to relevant extraction or control medium, and adaption of the cells to their new environment—which is associated with cellular stress [50].

Figure 6 illustrates the calculated cell growth over a cultivation period of 30 h. For MSCs cultivated in EM 1, no significant difference in cell growth was observed when compared to control cultures. Over the cultivation period, the number of living cells increased by a factor of approximately 2, both for cultivation in control medium and in EM 1. By contrast, cultivation in EM 2 leads to a strong decrease in cell viability, which resulted in a significant decrease in the number of living cells (by more than half) within 30 h.

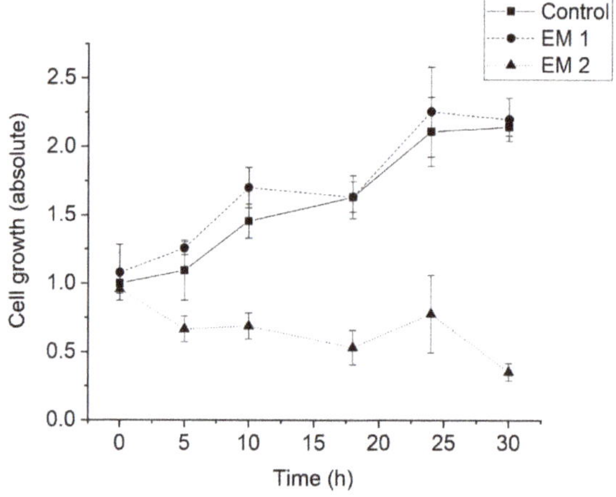

Figure 6. Cell growth of MSCs determined by cell counting using Trypan blue staining. (EM = extraction medium). EM 1: EM obtained by incubation of 3D printed material treated with ethanol (70%, *v/v*) in a disinfection process. EM 2: EM obtained by incubation of 3D printed material treated with sodium hypochlorite (2%, *v/v*). The experiments are compared to MSC cultivation in regular cell culture medium (Control) and were performed three times (*n* = 3).

In summary, then, apoptosis/necrosis analyses over 30 h reveal no evidence of any behavior in MSCs cultivated in EM 1 that could be attributed to potential toxic leachables in the 3D printed material. And a post-processing procedure that included disinfection with ethanol (70%, *v/v*) proved to be the most advisable approach tested for handling this high-resolution polyacrylic 3D printed material. In general, the flow cytometry results confirm the results of the CTB assay, but it provides more detailed information about the mechanism of cell death that was observed.

3.3. Analysis of Cell Death Responses via Image-Based Live-Cell Analysis System

Another approach for analyzing apoptotic and necrotic responses of cells in order to assess in vitro biocompatibility of a material is represented by comparatively novel image-based live-cell analysis systems. The IncuCyte® Live-Cell Analysis System used in this study is an image-based real-time system that allows the automatic acquisition and analysis of phase contrast and fluorescence images of cells using customized software tools.

Using this system, MSCs cultivated either in extraction or in a control medium (see Section 2.5) were monitored and analyzed automatically over a period of 30 h. Phase contrast, as well as fluorescence images, were captured every 1 h following the addition of fluorescence reagents for the purpose of highlighting apoptosis and necrosis. A contrasting juxtaposition—representing the cell phenotype data of individual cell populations cultivated in extraction or control medium—is shown in Figure 7 Green fluorescence signals show apoptotic cells; red fluorescence signals show necrotic cells. As a positive apoptosis control, MSCs were cultivated with the addition of the apoptosis inducer cisplatin. In keeping with previous investigations (see Sections 3.1 and 3.2 above), no differences in cell morphology, cell growth, or layer formation was observed for MSCs cultivated in EM 1 compared with control cultures. By contrast, MSCs cultivated in EM 2 show similar characteristics compared to the cultivation of MSCs with cisplatin (positive apoptosis control). After 15 h of incubation in EM 2 or cisplatin, large gaps in cell layer, less connected cells, and cell rounding as well as shrinkage were all observed. These are common characteristics associated with cell apoptosis [51]. After 30 h of MSC cultivation in EM 2 and the positive apoptosis control, a high increase in apoptotic and necrotic signals was observed via measurements of corresponding fluorescence signals.

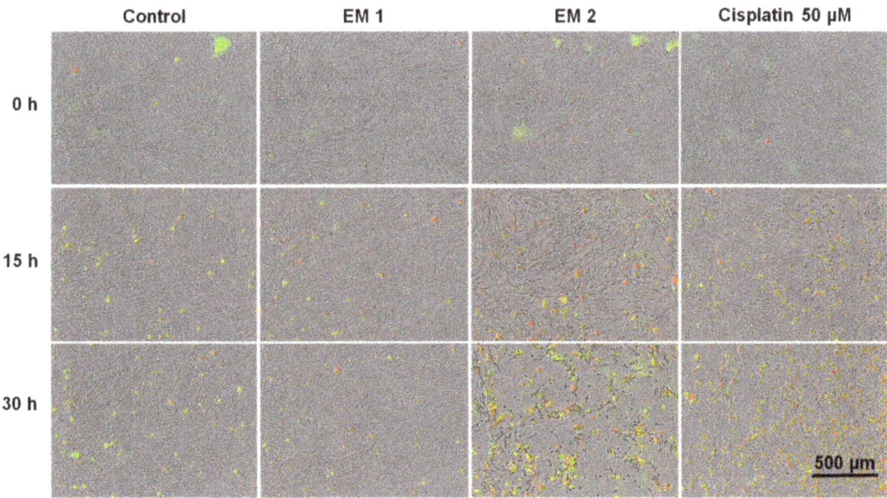

Figure 7. Fluorescence images of MSCs over time by image-based live-cell analysis system (IncuCyte). Green fluorescence is related to apoptotic cells; red fluorescence shows necrotic cells. (EM = extraction medium). EM 1: EM obtained by incubation of 3D printed material treated with ethanol (70%, *v/v*) in a disinfection process. EM 2: EM obtained by incubation of 3D printed material treated with sodium hypochlorite (2%, *v/v*). The experiments are compared to MSC cultivation in regular cell culture medium (Control) and were performed three times (*n* = 3). Cisplatin 50 µm: Positive control for apoptosis.

Figure 8 shows kinetic analyses of MSC growth, as well as apoptotic and necrotic signals obtained by dynamic image processing of phase contrast and fluorescence images, as described in Section 2.9. In this Figure, the unit Counts/Image was based on fluorescence signals provoked by apoptotic or necrotic cells in a specific imaging field. MSCs were cultivated in corresponding extraction media or control medium (see Section 2.5, above). As was to be expected from the previous investigations, there was no relevant difference observed in the cell behavior of MSCs cultured with EM 1 compared to the control cell culture medium. Over the duration of the experiment, cell confluency (representing the cell growth) increased. Living cells grow, expand, and divide. Furthermore, the number of apoptotic and necrotic cells per image field during MSC cultivation in EM 1 and control medium remained minimal. By contrast, MSC cultivation in EM 2 stagnated, and a strong relative increase in apoptotic and cytotoxic signals was also observed. A subsequent decline in cell proliferation after 10 h in EM 2 was likely related to the changes in cell morphology (e.g., cell rounding, shrinkage) and detachment of dead cells from the surface as a result of increased apoptosis and necrosis [14]. Detached dead cells might migrate into the supernatant, beyond the focal point of the laser, where they cannot be recognized and counted adequately.

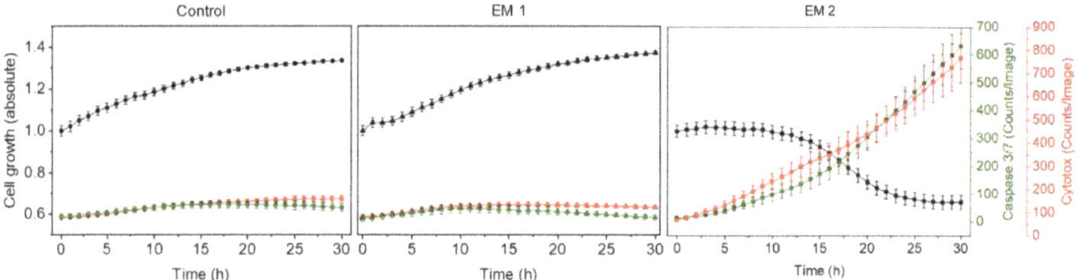

Figure 8. Analysis of cytotoxicity of the 3D printing polyacrylic material and apoptosis of MSCs by an image-based live-cell analysis system (IncuCyte). Cell growth, apoptosis and necrosis of MSCs are analyzed per cultivation. A caspase 3/7 signal (green) represents apoptotic cells; the cytotox-signal (red) is correlated to necrotic cells. (EM = extraction medium). EM 1: EM obtained by incubation of 3D printed material treated with ethanol (70%, *v/v*) in a disinfection process. EM 2: EM obtained by incubation of 3D printed material treated with sodium hypochlorite (2%, *v/v*). The experiments are compared to MSC cultivation in regular cell culture medium (Control) and were performed eighteen times (*n* = 18).

4. Discussion

The results obtained from the image-based evaluation conducted via live-cell analysis system were in full agreement with the results obtained via both the CTB assay and flow cytometry method—and all three methods confirmed that EM 1 had no significant influence on MSCs. It can therefore be assumed that the post-processing procedure including disinfection with ethanol (70%, *v/v*) was successful, and no critical amount of cytotoxic substances leached out of the 3D printed polyacrylic material. Since the 3D printed polyacrylic material had no negative impact on cell behavior or cell morphology of MSCs, it can be considered in vitro biocompatible. These findings collectively mark out a solid starting point for further investigations, and open the door for potential biological and biomedical applications using the analyzed 3D printed high-resolution polyacrylic material, which is promising not only for micro-scale and microfluidic applications, but also for rapid prototyping of various devices for cell culture and lab scale experiments [17].

Comparing the three methodologies used to evaluate in vitro biocompatibility here reveals some major disadvantages of both the flow cytometry and CTB assay methods. In both oh those cases, cell sample preparation and analysis must take place outside the cell culture incubator, which is designed to ensure a constant temperature and high humidity to facilitate cell growth under a CO_2 atmosphere. Such handling of the cells outside the incubator disrupts these optimal conditions, which may impose cellular stress and could also potentially impact cell growth, apoptosis, and/or necrosis [52,53]. Furthermore, as noted above, analysis of biocompatibility via biochemical-based CTB assay only provides information about cell viability in general. As a result, it can at best be considered a first analysis assay useful to obtaining a general sense of the cytotoxicity potential of a material, before continuing with further considerations. Flow cytometry and the image-based analysis system both allow for more detailed and specific analyses of cellular behavior and reactions on potential cytotoxic material constituents in assessing biocompatibility. For flow cytometry experiments, the cells of an individual cultivation well were harvested and examined for each measuring point. This means that in flow cytometry analyses, different cell populations are compared with each other, and therefore temporal investigation based only on a single cell population is not possible (this applies to adherent growing cells). Attempting to track dynamic functional cellular processes and morphology over the whole time frame of an experiment accordingly becomes an arduous task; and flow cytometry is particularly ill-suited to the task of monitoring rapid cellular changes (e.g., in response to external influences). Due to well-to-well variations and differences in cell treatment and seeding, the comparability of the obtained data cannot be guaranteed [54]. In addition, sample preparation for flow cytometry studies is laborious, requiring substantial time expenditure and good cell culture practices [54]. The extensive sample handling also results in a substantial delay from cell detachment to analyses. Additionally, the multiple centrifugation steps required during sample preparation and dyeing procedures expose the cells to mechanical stress [54]. Disruption and damage of the cell membrane triggered by stress factors can lead to apoptotic or necrotic responses and thus to false-positive results.

In contrast, image-based live-cell analysis gives the ability to visualize cellular phenotypes images as well as to perform kinetic analyses and quantifications of apoptotic and necrotic cell responses simultaneously in high-throughput. Based on microscopic data, numerous cell specific analyses can be performed directly, using customized tools and software. Live-cell imaging technology offers the possibility to monitor and study the same cell population for an indefinite period of time by analyzing the same imaging field. Since the imaging and analysis is realized fully automated inside a cell culture incubator, there is no need to physically move cells and risk exposing them to lower temperatures and potential cellular stress. Culture perturbations in performing assays with traditional methods such as flow cytometry and CTB viability assays are affecting cellular behavior and provoke cellular stress [52–54]. That includes the physical movement of cultures by removing cell culture flasks or plates from the laboratory cell culture incubator as well as changes in temperature and atmospheric conditions while performing the experiment. The real-time analysis system does not have to take into account any of the aforementioned disturbances.

5. Conclusions

This study presents a comprehensive comparison of three different methodologies for the in vitro evaluation of biocompatibility of 3D printed polyacrylic material. The superiority of an image-based live-cell analysis system with respect to time, usability, and scientific outcome was shown. Image-based real-time analyses allow for simultaneous observations of changes in cell morphology via microscopic imaging as well as kinetic analyses and quantifications of apoptotic and necrotic cell responses. Conventional methods for testing in vitro biocompatibility—such as microscopy or biochemically based assays—were comparatively outshone. The fast and simple handling; the potential of performing screenings in high throughput; and the high quantity and informative value of cellular data all make real-time live-cell imaging technology an ideal tool not only for the study of biocompatibility, but also for the usage in numerous cell culture applications on a daily basis. With the possibility of integrating up to two fluorescence channels in addition to phase contrast, and the choice between three different objective nosepieces, countless image-based cell assays can potentially be performed and monitored in real-time. Long term assays for studying chemotaxis, angiogenesis or stem cell differentiation are just as simple to realize as measurements of cellular health in drug screenings.

At the same time, this study also highlighted the importance of analyzing and comparing different post-processing procedures of 3D printed materials considered for biological applications Even though the tested material itself is in vitro biocompatible, remaining support material or contaminations due to insufficient post-processing methods could still potentially lead to adverse effects on surrounding cellular environment. 3D printing materials produced for a specific printer system are often not considered for use in cell culture or biomedical applications where biocompatibility is a central demand [3]. Manufacturers often give no suggestions for a proper disinfection and sterilization of their numerous material formulations. It is accordingly up to the researcher to investigate the materials in terms of biocompatibility and appropriate post-processing and sterilization protocols. On that account, high-throughput screening methods as the image-based live-cell analysis system are critical for both finding biocompatible material formulations, and also finding the best solution of post-processing for one given material.

Author Contributions: I.G.S., A.E. and J.B. designed the experiments; I.G.S. conducted the experimental work, drafted and revised the manuscript; T.S. (Tobias Steinwedel) performed sterilization/disinfection experiments; N.-M.E. and M.K. assisted with the FACS analyses and cell cultivation; A.L., T.S. (Thomas Scheper) and J.B. supervised the work, revised the manuscript, and provided helpful ideas for the present work.

Funding: This research was funded by the German Research Foundation (DFG) via the Emmy Noether programme, project ID 346772917 and the publication of this article was funded by the Open Access fund of Leibniz Universität Hannover.

Acknowledgments: The authors would like to thank Natalie Rotermund for her support in creating the illustrations and acknowledge the corporation with Sarah Strauss and Peter Vogt, Hannover Medical School (Germany), who provided tissue material for hAD-MSCs isolation.

Conflicts of Interest: The authors declare no conflict of interest.

Ethics Approval and Consent to Participate: The use of donated tissues and cells is approved by the local ethics committee of Hannover Medical School (reference number 3475-2017). Patients gave their written consent for tissue donations. Consents were archived within the patients' charts. All donations were performed anonymously and were not traceable by the scientists. The set of information for the scientists contained only age and gender. Patients with severe co-morbidities were not included in the study at hand.

References

1. Gross, B.C.; Erkal, J.L.; Lockwood, S.Y.; Chen, C.; Spence, D.M. Evaluation of 3D Printing and Its Potential Impact on Biotechnology and the Chemical Sciences. *Anal. Chem.* **2014**, *86*, 3240–3253. [CrossRef] [PubMed]
2. Ventola, C.L. Medical Applications for 3D Printing: Current and Projected Uses. *Pharm. Ther.* **2014**, *39*, 704–711.
3. Chia, H.N.; Wu, B.M. Recent advances in 3D printing of biomaterials. *J. Biol. Eng.* **2015**, *9*, 4. [CrossRef] [PubMed]
4. Derby, B. Printing and Prototyping of Tissues and Scaffolds. *Sci.* **2012**, *338*, 921. [CrossRef] [PubMed]
5. Williams, D.F. Definitions in biomaterials. In Proceedings of the Consensus of the European Society for Biomaterials Conference, Chester, UK, 3–5 March 1986.
6. Ratner, B.D.; Hoffman, A.S.; Schoen, F.J.; Lemons, J.E. *Biomaterials Science. An Introduction to Materials in Medicine*, 3rd ed.; Academic Press: San Diego, CA, USA.
7. Bernard, M.; Jubeli, E.; Pungente, M.D.; Yagoubi, N. Biocompatibility of polymer-based biomaterials and medical devices—regulations, in vitro screening and risk-management. *Biomater. Sci.* **2018**, *6*, 2025–2053. [CrossRef] [PubMed]
8. Badylak, S. The Impact of Host Response on Biomaterial Selection. In *Host Response to Biomaterials*; Academic Press: San Diego, CA, USA.
9. NSAI Standards. *I. S. EN ISO 10993-12:2012 Biological Evaluation of Medical Devices*; National Standards Authority of Ireland: Dublin, Ireland, 2012.
10. Strober, W. Trypan Blue Exclusion Test of Cell Viability. *Curr. Protoc. Immunol.* **2001**, *21*, A.3B.1–A.3B.2.
11. Eisenbrand, G.; Pool-Zobel, B.; Baker, V.; Balls, M.; Blaauboer, B.J.; Boobis, A.; Carere, A.; Kevekordes, S.; Lhuguenot, J.C.; Pieters, R.; et al. Methods of in vitro toxicology. *Food Chem. Toxicol.* **2002**, *40*, 193–236. [CrossRef]
12. Single, A.; Beetham, H.; Telford, B.J.; Guilford, P.; Chen, A. A Comparison of Real-Time and Endpoint Cell Viability Assays for Improved Synthetic Lethal Drug Validation. *J. Biomol. Screen.* **2015**, *20*, 1286–1293. [CrossRef] [PubMed]
13. O'Brien, J.; Wilson, I.; Orton, T.; Pognan, F. Investigation of the Alamar Blue (resazurin) fluorescent dye for the assessment of mammalian cell cytotoxicity. *Eur. J. Biochem.* **2000**, *267*, 5421–5426. [CrossRef]
14. Fiers, W.; Beyaert, R.; Declercq, W.; Vandenabeele, P. More than one way to die: apoptosis, necrosis and reactive oxygen damage. *Oncogene* **1999**, *18*, 7719–7730. [CrossRef]
15. Norbury, C.J.; Hickson, I.D. Cellular responses to DNA damage. *Annu. Rev. Pharmacol. Toxicol.* **2001**, *41*, 367–401. [CrossRef]

16. Walczak, R. Inkjet 3D printing—towards new micromachining tool for MEMS fabrication. *Bull. Pol. Acad. Sci. Tech. Sci.* **2018**, *66*, 179–186.

17. Enders, A.; Siller, I.G.; Urmann, K.; Hoffmann, M.R.; Bahnemann, J. 3D Printed Microfluidic Mixers—A Comparative Study on Mixing Unit Performances. *Small* **2019**, *15*, 1804326. [CrossRef] [PubMed]

18. 3D Systems. Projet MJP 2500 MultiJet Plastic Printers Tech. Specs. Available online: https://de.3dsystems.com/3d-printers/projet-mjp-2500-series/specifications (accessed on 6 June 2019).

19. 3D Systems. Safety Data Sheet: VisiJet M2R-CL. Available online: http://infocenter.3dsystems.com/materials/mjp/visijet-m2r-cl (accessed on 6 June 2019).

20. 3D Systems. Safety Data Sheet: VisiJet M3 Crystal. Available online: http://infocenter.3dsystems.com/materials/mjp/visijet-m3-crystal (accessed on 6 June 2019).

21. 3D Systems. Safety Data Sheet: VisiJet M2G-CL. Available online: http://infocenter.3dsystems.com/materials/mjp/visijet-m2g-cl?_ga=2.257034957.758772822.1561980611-943946801.1561980611 (accessed on 6 June 2019).

22. Block, S.S. *Disinfection, Sterilization, and Preservation*, 5th ed.; Lippincott Williams & Wilkins: Philadelphia, PA, USA, 2001.

23. Smith, P.N.; Palenik, C.J.; Blanchard, S.B. Microbial contamination and the sterilization/disinfection of surgical guides used in the placement of endosteal implants. *Int. J. Oral Maxillofac. Implants* **2011**, *26*, 274–281. [PubMed]

24. Rutala, W.A.; Weber, D.J. Disinfection and sterilization: An overview. *Am. J. Infect. Control* **2013**, *41*, S2–S5. [CrossRef]

25. Rutala, W.A.; Weber, D.J.; HICPAC. *Guideline for Disinfection and Sterilization in Healthcare Facilities*; Centers for Disease Control and Prevention: Chapel Hill, NC, USA, 2008.

26. Pepelanova, I.; Kruppa, K.; Scheper, T.; Lavrentieva, A. Gelatin-Methacryloyl (GelMA) Hydrogels with Defined Degree of Functionalization as a Versatile Toolkit for 3D Cell Culture and Extrusion Bioprinting. *Bioengineering (Basel)* **2018**, *5*, 55. [CrossRef] [PubMed]

27. Niles, A.L.; Moravec, R.A.; Riss, T.L. In vitro viability and cytotoxicity testing and same-well multi-parametric combinations for high throughput screening. *Curr. Chem. Genom.* **2009**, *3*, 33–41. [CrossRef]

28. Gonzalez, R.J.; Tarloff, J.B. Evaluation of hepatic subcellular fractions for Alamar blue and MTT reductase activity. *Toxicol. In Vitro* **2001**, *15*, 257–259. [CrossRef]

29. Wlodkowic, D.; Skommer, J.; Darzynkiewicz, Z. Flow cytometry-based apoptosis detection. *Methods Mol. Boil.* **2009**, *559*, 19–32.

30. Adan, A.; Alizada, G.; Kiraz, Y.; Baran, Y.; Nalbant, A. Flow cytometry: basic principles and applications. *Crit. Rev. Biotechnol.* **2017**, *37*, 163–176. [CrossRef]

31. Bajpai, R.; Lesperance, J.; Kim, M.; Terskikh, A.V. Efficient propagation of single cells accutase-dissociated human embryonic stem cells. *Mol. Reprod. Dev.* **2007**, *75*, 818–827. [CrossRef] [PubMed]

32. Salzig, D.; Leber, J.; Merkewitz, K.; Lange, M.C.; Köster, N.; Czermak, P. Attachment, Growth, and Detachment of Human Mesenchymal Stem Cells in a Chemically Defined Medium. *Stem Cells Int.* **2016**, *2016*, 1–10. [CrossRef] [PubMed]

33. Elmore, S. Apoptosis: A review of programmed cell death. *Toxicol. Pathol.* **2007**, *35*, 495–516. [CrossRef] [PubMed]

34. Dasari, S.; Tchounwou, P.B. Cisplatin in cancer therapy: molecular mechanisms of action. *Eur. J. Pharmacol.* **2014**, *740*, 364–378. [CrossRef] [PubMed]

35. Florea, A.M.; Büsselberg, D. Cisplatin as an Anti-Tumor Drug: Cellular Mechanisms of Activity, Drug Resistance and Induced Side Effects. *Cancers* **2011**, *3*, 1351–1371. [CrossRef] [PubMed]

36. Kopperud, H.M.; Kleven, I.S.; Wellendorf, H. Identification and quantification of leachable substances from polymer-based orthodontic base-plate materials. *Eur. J. Orthod.* **2011**, *33*, 26–31. [CrossRef] [PubMed]

37. Oesterreicher, A.; Wiener, J.; Roth, M.; Moser, A.; Gmeiner, R.; Edler, M.; Pinter, G.; Griesser, T. Tough and degradable photopolymers derived from alkyne monomers for 3D printing of biomedical materials. *Polym. Chem.* **2016**, *7*, 5169–5180. [CrossRef]

38. Amato, S.F.; Ezzell, R.M. *Regulatory Affairs for Biomaterials and Medical Devices: Woodhead Publishing Series in Biomaterials*; Woodhead Publishing: Sawston, UK, 2015.

39. Rashid, H.; Sheikh, Z.; Vohra, F. Allergic effects of the residual monomer used in denture base acrylic resins. *Eur. J. Dent.* **2015**, *9*, 614–619. [CrossRef]

40. Goddard, J.M.; Hotchkiss, J.H. Polymer surface modification for the attachment of bioactive compounds. *Prog. Polym. Sci.* **2007**, *32*, 698–725. [CrossRef]

41. Rudko, G.; Kovalchuk, A.; Fediv, V.; Chen, W.M.; Buyanova, I.A. Enhancement of polymer endurance to UV light by incorporation of semiconductor nanoparticles. *Nanoscale Res. Lett.* **2015**, *10*, 81. [CrossRef]

42. Vijayalakshmi, S.P.; Madras, G. Photodegradation of poly (vinyl alcohol) under UV and pulsed-laser irradiation in aqueous solution. *J. Appl. Polym. Sci.* **2006**, *102*, 958–966. [CrossRef]

43. Gogotov, I.N.; Barazov, S.K. The Effect of Ultraviolet Light and Temperature on the Degradation of Composite Polypropylene. *Int. Polym. Sci. Technol.* **2014**, *41*, 55–58. [CrossRef]

44. Fukuzaki, S. Mechanisms of actions of sodium hypochlorite in cleaning and disinfection processes. *Biocontrol Sci.* **2006**, *11*, 147–157. [CrossRef] [PubMed]

45. Rutala, W.A.; Weber, D.J. Uses of inorganic hypochlorite (bleach) in health-care facilities. *Clin. Microbiol. Rev.* **1997**, *10*, 597–610. [CrossRef] [PubMed]

46. Spencer, H.R.; Ike, V.; Brennan, P.A. Review: the use of sodium hypochlorite in endodontics—potential complications and their management. *Bdj open* **2007**, *202*, 555. [CrossRef]

47. Alkahtani, A.; Alkahtany, S.M.; Anil, S. An in vitro evaluation of the cytotoxicity of varying concentrations of sodium hypochlorite on human mesenchymal stem cells. *J. Contemp. Dent. Pract.* **2014**, *15*, 473–481. [CrossRef]

48. Hidalgo, E.; Bartolome, R.; Dominguez, C. Cytotoxicity mechanisms of sodium hypochlorite in cultured human dermal fibroblasts and its bactericidal effectiveness. *Chem. Biol. Interact.* **2002**, *139*, 265–282. [CrossRef]

49. Graziano, M.U.; Graziano, K.U.; Pinto, F.M.G.; Bruna, C.Q.M.; Souza, R.Q.S.; Lascala, C.A. Effectiveness of disinfection with alcohol 70% (w/v) of contaminated surfaces not previously cleaned. *Rev. Lat. Am. Enfermagem* **2013**, *21*, 618–623. [CrossRef]

50. Garcia-Montero, A.; Vasseur, S.; Mallo, G.V.; Soubeyran, P.; Dagorn, J.C.; Iovanna, J.L. Expression of the stress-induced p8 mRNA is transiently activated after culture medium change. *Eur. J. Cell Biol.* **2001**, *80*, 720–725. [CrossRef]

51. Rello, S.; Stockert, J.C.; Moreno, V.; Gámez, A.; Pacheco, M.; Juarranz, A.; Cañete, M.; Villanueva, A. Morphological criteria to distinguish cell death induced by apoptotic and necrotic treatments. *Apoptosis* **2005**, *10*, 201–208. [CrossRef]

52. Watanabe, I.; Okada, S. Effects of temperature on growth rate of cultured mammalian cells (L5178Y). *J. Cell Biol.* **1967**, *32*, 309–323. [CrossRef] [PubMed]

53. Tchao, R. Fluid shear force and turbulence-induced cell death in plastic tissue culture flasks. *In Vitro Toxicol.* **1996**, *9*, 93–100.

54. Gelles, J.D.; Chipuk, J.E. Robust high-throughput kinetic analysis of apoptosis with real-time high-content live-cell imaging. *Cell Death Dis.* **2016**, *7*, e2493. [CrossRef] [PubMed]

 *materials*

Article

Bactericidal and Biocompatible Properties of Plasma Chemical Oxidized Titanium (TiOB®) with Antimicrobial Surface Functionalization

Stefan Kranz [1,*], André Guellmar [1], Andrea Voelpel [1], Tobias Lesser [1], Silke Tonndorf-Martini [1], Juergen Schmidt [2], Christian Schrader [2], Mathilde Faucon [3], Ulrich Finger [3], Wolfgang Pfister [4], Michael Diefenbeck [5] and Bernd Sigusch [1]

[1] Department of Conservative Dentistry and Periodontology, University Hospital Jena, An der Alten Post 4, 07743 Jena, Germany; Andre.Guellmar@med.uni-jena.de (A.G.); Andrea.Voelpel@med.uni-jena.de (A.V.); Tobias.Lesser@med.uni-jena.de (T.L.); Silke.Tonndorf-Martini@med.uni-jena.de (S.T.-M.); bernd.w.sigusch@med.uni-jena.de (B.S.)

[2] Innovent Technology Development, Prüssingstraße 27 B, 07745 Jena, Germany; js@innovent-jena.de (J.S.); cs@innovent-jena.de (C.S.)

[3] Königsee Implantate GmbH, OT-Aschau—Am Sand 4, 07426 Allendorf, Germany; Mathilde.Faucon@koenigsee-implantate.de (M.F.); Ulrich.Finger@koenigsee-implantate.de (U.F.)

[4] Department of Microbiology, University Hospital Jena, Erlanger Allee 101, 07747 Jena, Germany; Wolfgang.Pfister@med.uni-jena.de

[5] Department of Trauma-, Hand- and Reconstructive Surgery, University Hospital Jena, Erlanger Allee 101, 07747 Jena, Germany; Michael.Diefenbeck@bonesupport.com

[*] Correspondence: Stefan.Kranz@med.uni-jena.de

Received: 14 February 2019; Accepted: 11 March 2019; Published: 15 March 2019

Abstract: Coating of plasma chemical oxidized titanium (TiOB®) with gentamicin-tannic acid (TiOB® gta) has proven to be efficient in preventing bacterial colonization of implants. However, in times of increasing antibiotic resistance, the development of alternative antimicrobial functionalization strategies is of major interest. Therefore, the aim of the present study is to evaluate the antibacterial and biocompatible properties of TiOB® functionalized with silver nanoparticles (TiOB® SiOx Ag) and ionic zinc (TiOB® Zn). Antibacterial efficiency was determined by agar diffusion and proliferation test on Staphylocuccus aureus. Cytocompatibility was analyzed by direct cultivation of MC3T3-E1 cells on top of the functionalized surfaces for 2 and 4 d. All functionalized surfaces showed significant bactericidal effects expressed by extended lag phases (TiOB® gta for 5 h, TiOB® SiOx Ag for 8 h, TiOB® Zn for 10 h). While TiOB® gta (positive control) and TiOB® Zn remained bactericidal for 48 h, TiOB® SiOx Ag was active for only 4 h. After direct cultivation for 4 d, viable MC3T3-E1 cells were found on all surfaces tested with the highest biocompatibility recorded for TiOB® SiOx Ag. The present study revealed that functionalization of TiOB® with ionic zinc shows bactericidal properties that are comparable to those of a gentamicin-containing coating.

Keywords: titanium implants; dental implants; antibacterial coating; gentamicin; silver; zinc; cytotoxicity; MC3T3-E1; Staphylococcus aureus; plasma chemical oxidation

---

1. Introduction

Surface functionalization mainly aims at increasing the osseointegration of orthopedic and dental implants by supporting the adherence of endogenous cells [1–3]. Besides commonly used techniques such as coating of the implant surface with hydroxyapatite [4], growth factors [5], or bisphosphonates [6], our group recently showed that plasma chemical oxidation (PCO) also is of great potential [7,8]. In the electrochemical process of PCO, the naturally occurring oxidation layer on titanium is converted into a ceramic-like surface of high porosity with up to four micrometers in thickness (so called bioactive TiOB® surface). As observed by Diefenbeck et al., titanium implants modified by PCO are especially well osseointegrated with high bone-to-implant contact rates [7].

On the one hand, functionalization is applied to improve healing and incorporation of implants, but on the other, it can also be used to render the surface antimicrobial self-active. Studies have proven that implants with antimicrobial surface activity are capable of preventing microbial adherence efficiently, without affecting the viability of endogenous cells like osteoblasts [1,2]. Antimicrobial surface functionalization mainly aims at reducing the risk of infection-associated implant failure [9]. However, in cases of implant-associated infections, Staphylococcus spp. are considered the predominant bacteria, with Staphylococcus aureus often being the main causative agent [10,11].

In this context, our group clearly showed that a gentamicin-containing coating on TiOB® implants sufficiently prevents Staphylococcus aureus from surface colonization in an osteomyelitis study in rats [12].

In times of increasing drug resistance and antibiotic-related events, the development of alternative and novel antibacterial approaches that are not dependent on the action of traditional antibiotics is of major concern [13]. Therefore, the present in vitro study focuses on the preparation and the antibacterial as well as biocompatible testing of bioactive TiOB® surfaces functionalized with silver nanoparticles and ionic zinc.

The antimicrobial action of silver has been known since historic times, and silver particles have already been applied in many biomedical applications such as wound dressings [14], catheter systems [15], bone cements [16], and implant coatings [17] due to their bactericidal properties.

Application of zinc is another very common strategy to prevent microbial colonization and biofilm formation [18] and has already been evaluated in association with bioactive glass and bioceramics [19,20]. Currently, Zn is intensively studied in biodegradable metal alloys with self-active antimicrobial characteristics as well [21].

The focus of the present study is to analyze the bactericidal and biocompatible properties of bioactive TiOB® functionalized with silver nanoparticles and ionic zinc. Since functionalization of TiOB® implants with coatings of gentamicin-tannic acid (TiOB® gta) already proved its antimicrobial efficiency in vivo [12], it was used as control in the present study.

2. Materials and Methods

2.1. Sample Preparation

Samples were manufactured from surgical grade titanium TiAI6V4 alloy rods (DIN ISO 5832-3) by Königsee Implantate GmbH, Aschau, Germany. Two different sample geometries were prepared: Cylindrical pins (active area: 12 mm long, 1.5 mm in diameter, n = 48 per TiOB® surface functionalization) were used for the microbiological tests, and disk-shaped samples (2 mm thick, 15 mm in diameter, n = 6 per TiOB® surface modification) for the cytotoxicity tests (Figure 1a,b). All samples were uniformly shot-blasted with aluminum oxide abrasives and subsequently with RKSP 120 ceramic particles (Rösler Oberflächentechnik GmbH, Untermerzbach, Germany).

TiOB® surfaces were established by plasma chemical oxidation (PCO) as described earlier [7,8]. In brief, PCO is an anodic oxidation-based modification process of the naturally occurring oxidation layer found on titanium causing the formation of a ceramic-like, macro-porous, bioactive surface called TiOB®. The quality of the produced TiOB® surface is dependent upon the applied electrolyte, the provided anode material, and the current and voltage curves used. PCO was performed by INNOVENT e.V., Jena, Germany.

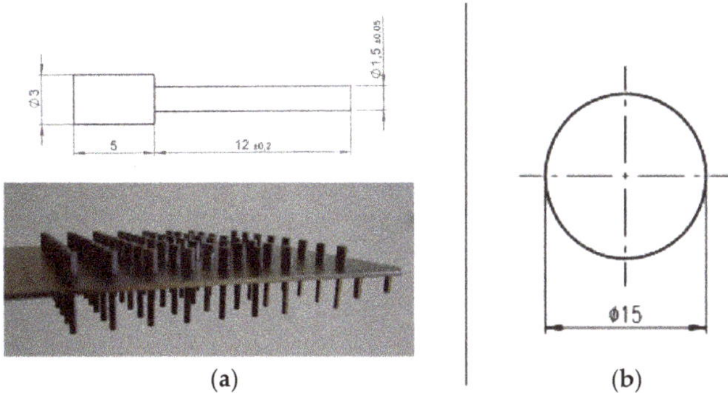

Figure 1. (**a**) Customized holding device and geometrical dimensions of the cylindrical samples used in the proliferation assay; (**b**) design of the disc shape samples used in the cytotoxicity tests.

2.2. Antibacterial Functionalization of TiOB®

Antibacterial functionalization of TiOB® with gentamicin-tannic acid (TiOB® gta), silver nanoparticles (TiOB® SiOx Ag), and ionic zinc (TiOB® Zn) was performed by INNOVENT e.V., Jena, Germany.

For visualization, all functionalized TiOB® surfaces were characterized by scanning electron microscopy (SEM). Therefore, samples were subjected to carbon vapor coating and afterwards imaged using a ZEISS Supra 55 VP (Carl Zeiss Microscopy GmbH, Jena, Germany) equipped with InLens-SE-detector (Carl Zeiss Microscopy GmbH, Jena, Germany) and Everhard-Thornley-SE-detector (Carl Zeiss Microscopy GmbH, Jena, Germany), both driven in HV-mode.

2.2.1. TiOB® Gta

Functionalization of bioactive TiOB® with gentamicin-tannic acid was performed exactly as described by Diefenbeck et al. [12]. In brief, tannic acid of pharmaceutical purity (Ph. Eur.) was used and bound to gentamicin by neutralization of a gentamicin-2 H_2SO_4 solution with NaOH causing precipitation of gentamicin-tannin complexes. Subsequently, 500 mg of the precipitated gentamicin-tannin complex was dissolved in formic acid (Ph. Eur.) and applied onto TiOB® samples by dip coating. Immediately afterwards, samples were dried by volatilization of the solvent, leading to the formation of a crystalline gentamicin-tannic acid layer on top of the TiOB® surface with 5 mg of gentamicin in total.

Because TiOB® gta already proved an efficient bacteriostatic effect in vivo, it served as a control in the present study.

2.2.2. TiOB® SiOx Ag

For functionalization of bioactive TiOB® with silver nanoparticles, all samples were primarily embedded into silicon and afterwards subjected to atmospheric-pressure plasma-assisted chemical vapor deposition (APCVD), as described by Beier et al. [22].

In brief, silver nanoparticles were deposited on TiOB® in the presence of vaporized hexamethyldisiloxane (HMDSO) (Sigma-Aldrich Chemie GmbH, Taufkirchen, Germany) at room temperature under atmospheric pressure (2–6 bar). As the plasma source, a commercially available BLASTER MEF system (TIGRES, Rellingen, Germany) was used. Via an atomizer, vaporized HMDSO was sprayed into cold low-pressure plasmas in the presence of water-free air, leading to the deposition of silicon oxide (SiOx) on the TiOB® surface. Simultaneously, a 5% silver nitrate solution (purity 99.8%, Merck, Darmstadt, Germany) was additionally sprayed into the plasma, causing the incorporation of silver nanoparticles into the growing SiOx layer. In each run, a SiOx Ag layer of 25 nm in thickness was formed on the macroporous TiOB® surface.

2.2.3. TiOB® Zn

TiOB® Zn was formed by two consecutive PCO steps. First, primary TiOB® coating was conducted in a calcium phosphate electrolyte up to an end-point voltage of 200 V. Subsequently, the samples were rinsed with deionized water and dried with compressed air.

In a second step, PCO was carried out in a electrolyte containing 0.14 mmol/L ammonium hydrogenphosphate, 13.4 mmol ammonia solution (25%), and 0.5 mmol/L zinc ions ($Zn(CH_3CO_2)_2$) stabilized by chelating agents. During the second PCO treatment, zinc ions were incorporated into the TiOB® surface at an end-point voltage of 350 V.

2.3. X-ray Photoelectron Spectroscopy (XPS)

The presence of zinc and silver in the respective sample surface was verified by surface-sensitive XPS (Theta Probe, Thermo VG Scientific, Paisley, UK). The detection depth of the XPS measurement was approximately 10 nm. A depth profile analysis was established by local sputtering of the surface with an ion beam and successive measurements. The spectrometer used was a Theta Probe (Thermo VG Scientific, Paisley, UK) with monochromatic Al Kα radiation (1486.6 eV). Excitation (100 W) was performed at a voltage of 15 kV and an emission current of 6.7 mA in a raster area of 400 µm in diameter at 10^{-9}–10^{-8} mbar. The spectrometer was equipped with an ion gun, and sputtering was performed with argon gas at approximately 1.5×10^{-7} mbar. The ions produced were then accelerated to 3 keV in a drift section at an ion current of approximately 1 µA. The focal spot of the ion beam was 4 mm^2 and, with a sputter yield of 1.39, led to a Ti removal of approximately 0.1 nm/s For sputtering of non-conducting samples, an electron gun was additionally used to prevent electrostatic charging during sputtering and measurement. For this purpose, electrons were accelerated onto the samples by applying 6 eV and 15 µA. Before analysis of a spectrum, the energy axis was calibrated on the basis of the O1s peak (EB(O 1s) = 531.0 eV). The quantitative analysis was performed using the peak areas and considering a background correction according to Shirley.

2.4. Experimental Section

For the following in vitro tests, the samples were assigned to four experimental groups (Table 1). All samples were sterilized by autoclaving (35 min at 134–138 °C and 2.16 × 105 Pa) before use.

Table 1. Group assignments and technical information.

Group Assignment	Technical Information
TiOB control	Ti6Al4V, PCO (280 V), TiOB surface
TiOB gentamicin-tannic acid (TiOB gta)	Ti6Al4V, PCO (280 V), TiOB surface, dip coating
TiOB SiOx Ag	Ti6Al4V, PCO (280 V), TiOB surface, APCVD
TiOB Zn	Ti6Al4V, 1st PCO (200 V), TiOB surface, 2nd PCO (350 V), Zn electrolyte

2.4.1. Bactericidal Properties of Functionalized TiOB®

For all antibacterial experiments the gram-positive species Staphylococcus aureus subsp. Rosenbach (ATCC 49230) was used. The species was cultivated aerobically on TS agar (tryptose–soybean agar, Oxoid, Wesel, Germany) at 37 °C. Test batches were arranged by suspending 2–3 colonies in TS broth (tryptose–soybean broth, Oxoid, Germany) followed by cultivation for approximately 4 h at 37 °C. Bacteria were harvested in the exponential phase of growth.

2.4.1.1. Agar Diffusion Test

In this subsection the release of antibacterial active components from the functionalized TiOB® surfaces was observed by agar diffusion test. Therefore, functionalized TiOB® samples of cylindrical shape were placed in microplates (Figure 1a) and incubated in 300 μL PBS at 37 °C. After 2, 4, 6, 12, 24, and 48 h of incubation, eluates from 8 samples per TiOB® surface were collected and stored at −20 °C until use.

For testing, TS agar was supplemented with proliferating Staph. aureus and poured into sterile petri dishes. After solidification of the agar, 100 μL of the stored eluates were pipetted into punched holes of 9 mm in diameter. The correspondent inhibition zones were analyzed after incubation for 24 h at 37 °C.

2.4.1.2. Proliferation Test

The antibacterial efficiency of the functionalized TiOB® surfaces was assessed by a proliferation assay adopted from Bechert et al. [23]. In brief, the test is based upon the capability of adherent Staph. aureus to proliferate.

Therefore, 8 samples of each TiOB® surface were placed in a 96-well microplate using a customized holding device (Figure 1a). Each sample was then incubated with 300 μL bacterial solution (OD$_{546}$ 0.5—equivalent to approximately 10^6–10^7 bacterial cells/mL) for one hour under slight shaking at 37 °C. Subsequently, the samples were removed, and non-adherent bacterial cells were discharged by washing twice with 100 μL PBS. Afterwards the rinsed samples were placed into 300 μL/well freshly prepared nutrient-reduced medium (PBS, 1% sterile TS broth, 0.25% glucose, 0.2% $(NH_4)2SO_4$)) and incubated for another 18 h. Subsequently, all samples were removed and fresh TS broth at a ratio of 1:3 was applied. Optical density (OD540 nm) was measured at baseline and after incubation for 1, 2, 3, 4, 5, 6, 7, 8, and 24 h at 37 °C.

In a second part, all functionalized TiOB® samples were primarily aged (pre-incubated) for 2, 4, 6, 12, 24, and 48 h in PBS. The proliferation test was then performed in the same way as described above.

2.4.2. Biocompatible Properties of Functionalized TiOB®

In this part of the experiment, the viability of MC3T3-E1 cells (DSMZ ACC 210) was determined after direct cultivation on top of the functionalized TiOB® surfaces for 2 and 4 d.

Cells were cultivated in alpha-MEM (minimal essential medium) supplemented with 10% FBS (fetal calf serum) and 1% PenStrep (penicillin-streptomycin) (all GIBCO; Karlsruhe, Germany) at 37 °C with 5% of CO_2. Confluent cells at a density of 4.500 cells/cm^2 were transferred to 12-well plates and cultivated on the top of disk-shaped TiOB® samples in 2 mL α-MEM (37 °C, 5% CO_2) for 2 d or 4 d.

Afterwards, the samples were washed twice with 2 mL PBS and each labeled with freshly prepared staining solution (12 μL fluorescein diacetate (FDA)/16 μL ethidium bromide in 2 mL PBS). Labeled cells were imaged using a fluorescence microscope (Nikon, Labophot, Japan, 10 × phase-contrast objective, λex = 455–495 nm). Viable cells appeared in green whereas non-viable cells showed red fluorescence. Nine fields of view per sample were examined with regard to live/dead cells, with three repetitions per TiOB® surface.

2.5. Statistics

The data were analyzed using SPSS 19.0 for Windows (IBM, Armonk, NY, USA). Statistical significance in the proliferation test was determined by one-way ANOVA, addressed by Bonferoni correction to diminish accumulation of alpha errors. The level of significance was set to $p < 0.05$.

The results of the cytotoxicity test were graphically displayed by boxplots. Significant differences in the cell counts were checked by t-test.

3. Results

3.1. Surface Characterization by Scanning Electron Microscopy

SEM observation revealed different surface configurations (Figure 2a–d). In Figure 2a, the surface topography of non-functionalized bioactive TiOB® (control) is shown. Bioactive TiOB® without antibacterial functionalization is characterized by a specific macroporous surface and described elsewhere in detail [7,8]. In contrast, functionalization of TiOB® with gentamicin-tannic acid (TiOB® gta, Figure 2b) resulted in a rather smooth surface topography. In the case of TiOB® gta, the entire macroporous TiOB® structure is covered by a thick layer of gentamicin-tannic acid. A similar appearance is found for samples functionalized with ionic zinc (TiOB® Zn, Figure 2d). TiOB® Zn is characterized by smooth surface areas interrupted by some large pores. In Figure 2c, the surface of TiOB® functionalized with silver nanoparticles (TiOB® SiOx Ag) is shown. Here, the sample surface is similar to non-functionalized bioactive TiOB® and also of macroporous appearance.

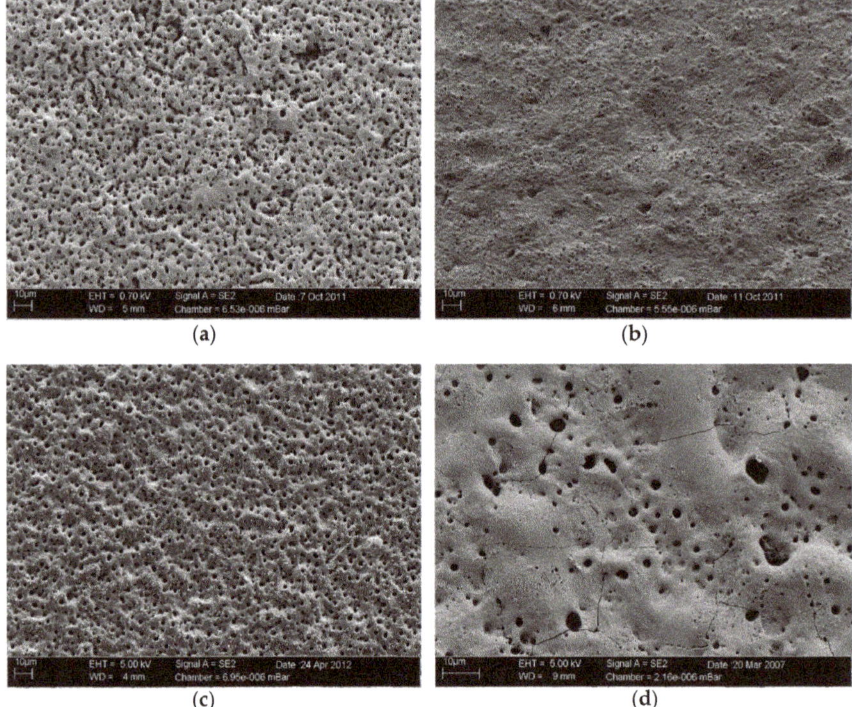

Figure 2. Surface characterization by scanning electron microscopy: (**a**) bioactive macroporous TiOB® (TiOB® control); (**b**) TiOB® functionalized with gentamicin-tannic acid (TiOB® gta); (**c**) TiOB® functionalized with silver nanoparticles (TiOB® SiOx Ag); (**d**) TiOB® functionalized with ionic zinc (TiOB® Zn).

3.2. XPS Analysis

Samples functionalized with gentamicin-tannic acid by dip coating showed a brownish crystalline cover layer with 5 mg gentamicin per sample.

The TiOB® SiOx Ag samples revealed a layer of approximately 100 nm in thicknesses with a content of 5% silver incorporated into the silicon oxide matrix (Figure 3a).

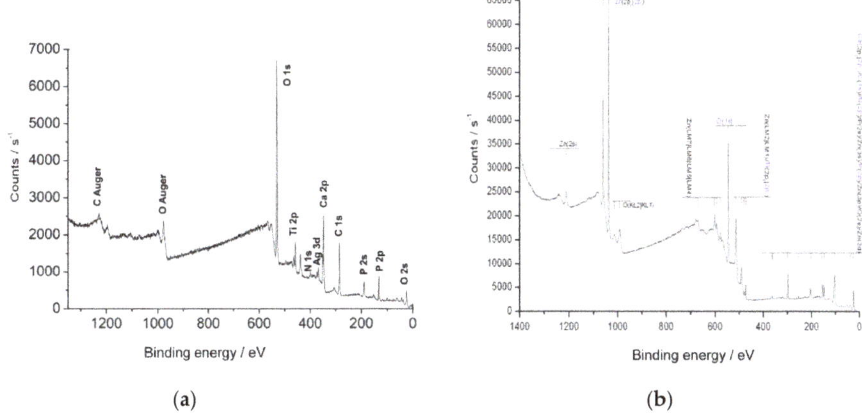

Figure 3. X-ray photoelectron spectroscopic analysis: (**a**) TiOB® SiOx Ag; (**b**) TiOB® Zn.

In the case of the TiOB® Zn samples, a matrix layer of 13 µm in thickness was produced with high concentrations of zinc in the top layers. In detail, the content of Zn varied from 20% in the top layer to 7% at a depth of 160 nm (Figure 3b).

3.3. Bactericidal Properties of Functionalized TiOB®

The release of gentamicin and active Ag and Zn from the respective functionalized TiOB® surfaces was analyzed by means of an agar diffusion test.

It was shown that only eluates obtained from TiOB® functionalized with gentamicin-tannic acid were able to inhibit the growth of Staph. aureus (Table 2). In case of TiOB® gta, inhibition zones were observed during the entire study period (48 h) with the largest zones recorded at baseline.

Table 2. Inhibition of *Staphylococcus aureus* in an agar diffusion test by eluates collected from the functionalized surfaces after 0, 2, 4, 6, 12, 24, and 48 h.

	Agar Diffusion Test- -							
	Eluates Collecting Time							
	0 h	**2 h**	**4 h**	**6 h**	**12 h**	**24 h**	**48 h**	
TiOB® control	0	0	0	0	0	0	0	
TiOB® gta	0	1.94	1.88	0.50	0.94	1.50	1.25	mean inhibition
TiOB8® SiOx Ag	0	0	0	0	0	0	0	zones [mm]
TiOB® Zn	0	0	0	0	0	0	0	

In contrast, eluates collected from the surfaces functionalized with Ag NPs or ionic Zn completely failed to induce any inhibition zones in the agar diffusion test. TiOB® without any antibacterial functionalization (TiOB® control) also showed no signs of bacterial inhibition (Table 2).

In the second part of the examination, the capability of Staph. aureus to proliferate on the functionalized surfaces was observed (Figure 4). It was shown that all functionalized TiOB® surfaces caused reduced rates of bacterial proliferation accompanied by extended lag phases (Figure 4). For TiOB® gta, a lag phase of 5 h was detected, whereas TiOB® SiOx Ag and TiOB® Zn showed lag phases of 8 and 10 h. Further, the bacterial solutions obtained from Staph. aureus grown for 24 h in direct contact with the antibacterial surfaces were also of less optical density compared to those arranged from the TiOB® controls (blue graphs in Figure 4).

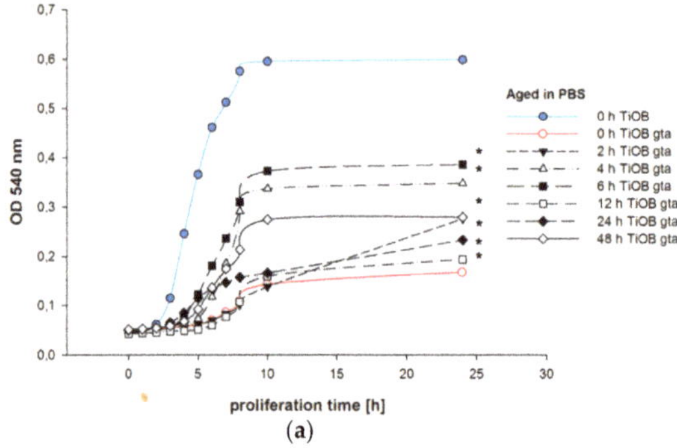

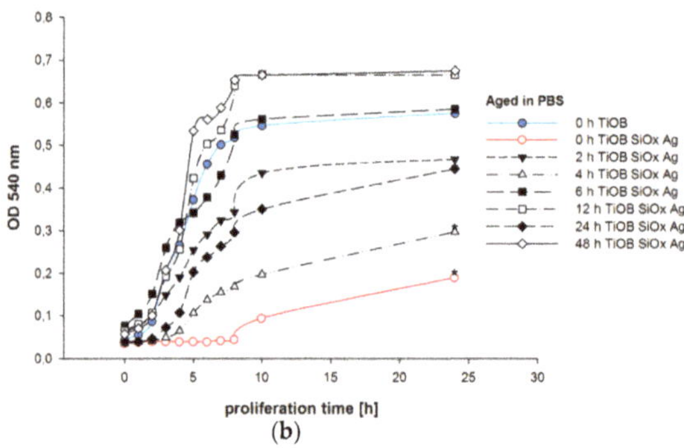

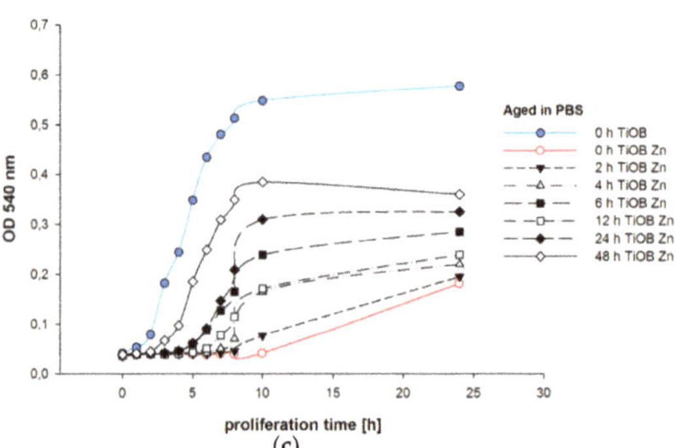

Figure 4. Proliferation of Staphylococcus aureus on functionalized TiOB®: (**a**) TiOB® gta; (**b**) TiOB® SiOx Ag; (**c**) TiOB® Zn. Samples were pre-incubated (aged) in PBS for 0, 2, 4, 6, 12, 24, and 48 h. Non-functionalized TiOB® (TiOB® control) is presented as a blue graph. Non-aged but antibacterial functionalized samples are shown in red. Significances to non-functionalized TiOB® are marked by stars ($p < 0.05$).

Moreover, it was found that the antibacterial activity of all functionalized TiOB® surfaces was dependent upon the time of pre-incubation in PBS (aging).

Among all surfaces examined, the strongest antibacterial effect was observed for samples that were not incubated in PBS prior to examination (presented by red graphs in Figure 4). In the cases of TiOB® gta and TiOB® Zn, significant antibacterial activity was still detected after pre-incubation in PBS for 24 h (Figure 4a,c).

However, in the case of the TiOB® SiOx Ag surfaces, significant bacterial suppression was detected only for samples that were incubated in PBS for 2 and 4 h (Figure 4b).

3.4. Biocompatible Properties of Functionalized TiOB®

In Figure 5, the results of the LSM observation are shown. As expected, non-functionalized TiOB® (TiOB® control) was colonized by viable MC3T3-E1 cells to the highest extent (Figures 5 and 6). Even after only 2 days of cultivation, the majority of cells exhibited a typical spread-like cell shape (Figure 5). In the case of the functionalized surfaces, best cytocompatibility was observed for TiOB® SiOx Ag and TiOB® Zn. In detail, after 2 days of cultivation, the largest number of viable cells, with an equal share in non-viable cells, was observed for TiOB® SiOx Ag (Figure 6a). Similar to the TiOB® control, viable cells grown on TiOB® SiOx Ag showed a healthy spread-like shape (Figure 5). However, after 2 days of cultivation, the number of healthy cells grown on TiOB® gta and TiOB® Zn was rather low when compared to the TiOB® control (Figure 5). On these surfaces, the majority of viable cells were of round or spheroidal shape (Figure 5).

After 4 days of cultivation, a significant increase in viable cells was observed for TiOB® SiOx Ag and also TiOB® Zn, with healthy cells grown on the entire surface (Figures 5 and 6b). Unfortunately, no significant increase in viable cells was witnessed for TiOB® gta (Figure 6b). Even after 4 days of cultivation, the majority of cells remained of round and spheroidal shape (Figure 5).

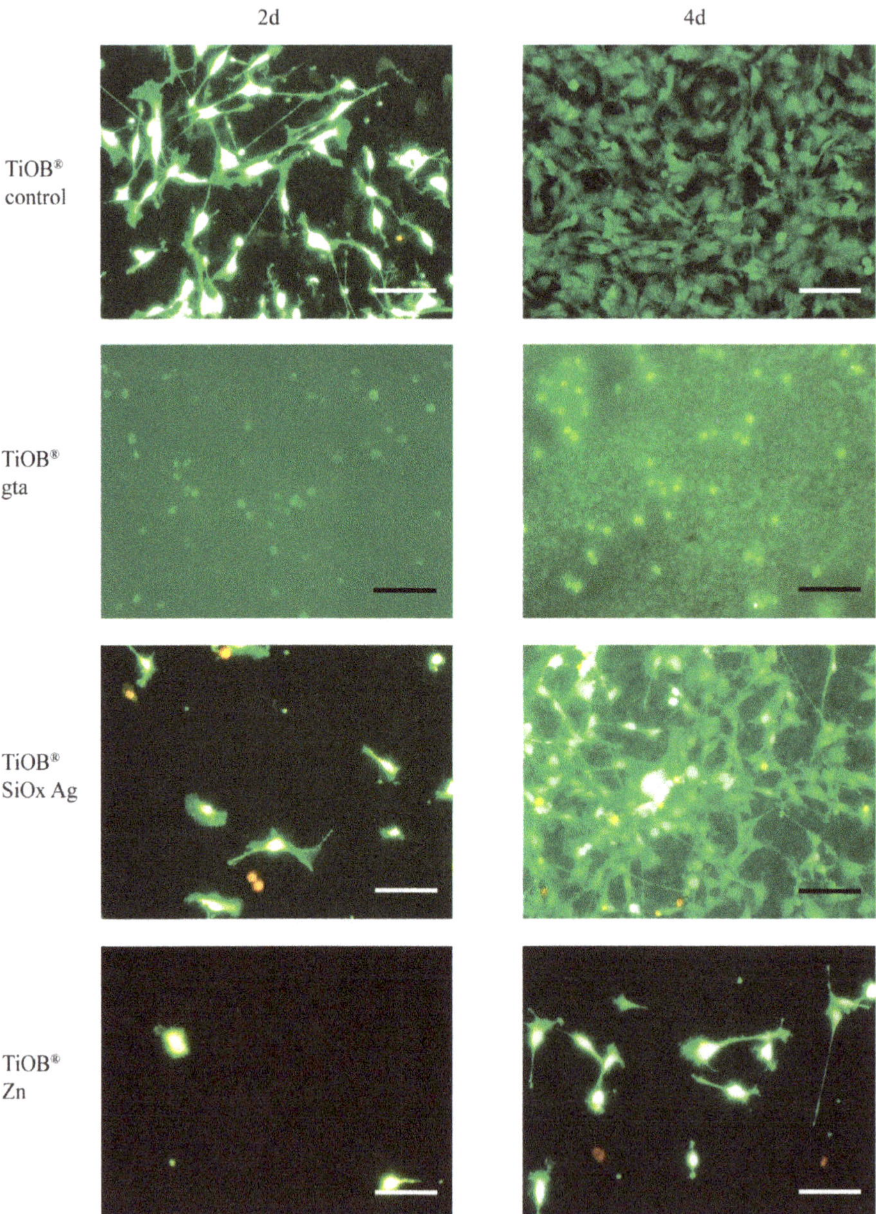

Figure 5. Laser scanning microscopy images of MC3T3-E1 cells cultivated in direct contact with non-functionalized TiOB® (TiOB® control), TiOB® gta, TiOB® SiOx Ag, and TiOB® Zn for 2 and 4 days. Cells were stained by FDA/Ethidium bromide. Viable cells appear in green whereas nuclei of non-viable cells are shown in red.

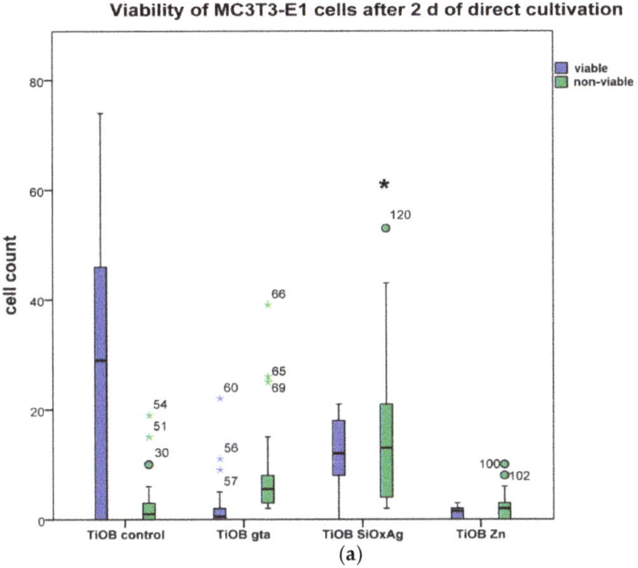

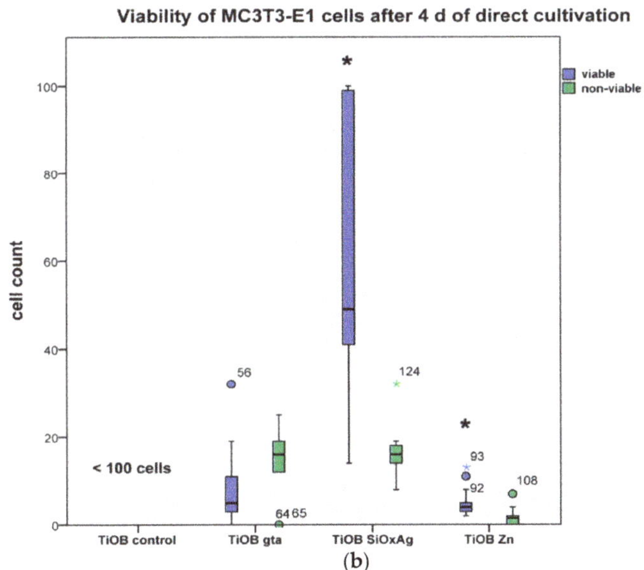

Figure 6. Share of viable and non-viable cells grown in direct contact with non-functionalized TiOB® (TiOB® control), TiOB® gta, TiOB® SiOx Ag, and TiOB® Zn after 2 (**a**) and 4 (**b**) days of cultivation. Significant differences are marked with a star ($p < 0.05$).

4. Discussion

In dental and orthopedic implant-based surgery, postoperative infections with pathogenic bacteria often cause serious complications [10]. Various studies have shown that implants with antibacterial surface activity are capable of preventing microbial colonization [24]. However, the major challenge is still to combine significant bacterial growth inhibition with sufficient biocompatibility [25].

The present study aims to determine the bactericidal and biocompatible properties of plasma chemical oxidized titanium functionalized with coatings of gentamicin-tannic acid, silver nanoparticles, and ionic zinc.

As shown by our group already, colonization of TiOB® surfaces by *Staph. aureus* can sufficiently be prevented by a gentamicin-tannic acid coating. Early cell response towards TiOB® gta has not yet been evaluated in detail.

However, the results of the present study revealed a reduced biocompatibility of TiOB® gta. In contrast to uncoated TiOB® (TiOB® control), which was sufficiently colonized by MC3T3-E1, the majority of cells grown in direct contact with TiOB® gta remained of round and spheroidal cell shape even after cultivation for 4 d.

Usually, gentamicin eluted from implant coatings is of good local and systemic biocompatibility [26]. This is also supported by findings of Popat et al. who observed no cytotoxic effect of gentamicin eluted from functionalized implant surfaces on MC3T3-E1 cells. Moreover, the authors realized that the characteristics of the surface seem to play a distinct role in early cell response [27]. The influence of cell attachment and proliferation behavior by topography features has already been studied by many other authors [28].

In the present study, SEM observations revealed a rather smooth surface of TiOB® gta with complete coverage of the entire bioactive macro-porous TiOB® structure, which might be one of the reasons for the reduced biocompatibility. Besides the unfavorable microstructure of the TiOB® gta surface, viability might also be negatively influenced by the eluted tannic acid. In this context, Sahiner et al. clearly showed that an increase in tannic acid concentration is directly accompanied with a loss in viability of A549 cancerous cells and L929 fibroblasts [29].

However, as previously shown by our group, the applied gentamicin-tannic acid coating is dissolvable in water with partial exposure of the native TiOB® surface after elution for 72 h [12].

We conclude that the rather smooth surface topography in combination with a concentration-dependent cytotoxic effect of the eluted tannic acid are responsible for the reduced biocompatibility observed for TiOB® gta during the first four days of direct cultivation.

On the other hand, the applied gentamicin-tannic acid coating most efficiently suppressed Staph. aureus in the present study. Especially during the initial phase (<4 h), large inhibition zones were observed in the agar diffusion assay. Vester et al. also observed a rapid release of gentamicin from titanium implants coated with poly(D,L-lactide) [30].

Other groups have shown that an initial burst release of gentamicin by 40% within the first hour, 70% within the first 24 h, and 80% within the first 48 h is associated with good clinical, laboratory, and radiological outcomes [26]. More specific information regarding the active release of gentamicin from TiOB® gta has been published by Diefenbeck et al. [12].

In general, biocompatibility of materials and drugs used in dentistry and surgery is different and highly dependent upon the type of substance or composition applied [31–33].

One major aim of the present study was to propose different antimicrobial functionalization strategies that are not dependent on the action of traditional antibiotics combined with surface characteristics better tolerated by endogenous cells. As shown by SEM observation, plasma-assisted chemical vapor deposition (PACVD) of silver nanoparticles (Ag NPs) resulted in surface topography features similar to those of the TiOB® control. In comparison to TiOB® gta or TiOB® Zn, strong colonization by viable MC3T3-E1 cells was already observed after 2 days of direct cultivation.

So far, implants functionalized by silver nanoparticles have often shown reduced biocompatible properties which are caused by either the release of silver ions in high concentrations or by the small size of the particles. A moderate cytotoxic effect was confirmed by Kheur and coworkers, who reported a biocompatible response of only 69% in viable cells when primary human gingival fibroblasts were incubated in the presence of titanium sputtered with Ag for 5 min [34]. Further, Smeets et al. proved that Ag/SiOxCy-coated titanium implants were significantly less osseointegrated compared to grit-blasted or acid-etched implants [35].

In the present study, a significant increase in viable and spread-out cells was observed after 4 d of direct cultivation on TiOB® SiOx Ag. This was directly accompanied by a decrease in bactericidal activity. As shown in the proliferation assay, significant bacterial inhibition of TiOB® SiOx Ag only lasted for a maximum of 4 h. Unlike TiOB® gta, no inhibition zones were observed in the agar diffusion test either. These results are in line with findings of Lisher et al., who also observed strong antibacterial effects of minimal Ag-containing plasma polymer coatings (Ag/amino-hydrocarbon) during the first day of immersion in deionized water [36].

When Ag NPs are oxidized, ionic Ag will be released with strong bactericidal activity [37]. Although the mechanisms of action are not fully understood, Ag has been applied as an antimicrobial agent since historical times. One reason for the antibacterial effect is found in the ability of silver ions to interact with essential thiol (sulfhydryl) groups causing a loss in bacterial enzyme function [38,39]. Further, Jaiswal et al. [39] and Feng et al. [38] discussed an inhibitory effect of Ag ions on microbial DNA replication and ATP synthesis. In addition, silver ions and Ag NPs are also likely to interact with phospholipids, leading to disturbance of the bacterial cell wall integrity [40,41]. Further, Ag NPs might trigger the formation of highly reactive oxygen radicals which are known to cause oxidative damage to DNA and proteins [42].

A recent study reported that the minimal inhibitory concentration (MICs) of Ag NPs on Staph. aureus is 300 µg/mL and the minimum bactericidal concentration (MBC) is 600 µg/mL, whereas 30 mg/mL are required for complete bacterial killing [43]. In order to increase the antibacterial efficiency of TiOB® SiOx Ag, the content in Ag NPs has to be increased which then will certainly be accompanied by an undesired decrease in biocompatibility.

Bioactive TiOB® was also functionalized by ionic zinc applied in two consecutive PCO steps. Similar to TiOB® gta, the macroporous surface of non-functionalized TiOB® was entirely covered. SEM observation revealed an implant surface with smooth areas interrupted by some large pores colonized by viable MC3T3-E1 cells with spread-like shape.

Zn is currently intensively studied in biodegradable metal alloys and, overall, presents good biocompatible properties. Investigations by Zhu et al. have shown that adhesion and proliferation of human bone marrow mesenchymal stem cells on biomedical metals can be increased by the additional application of Zn. Obviously, Zn plays a distinct role in the activation of several different intracellular pathways associated with gene activation, regulation, cell growth, differentiation, extracellular matrix mineralization, and osteogenesis [44]. Further, Yu and co-workers recently studied the biological properties of Zn incorporated in micro/nano-textured titanium surfaces by high-current anodization and found a positive effect on the proliferation behavior and alkaline phosphatase activity of osteoblasts [45]. In addition, Shen et al. also confirmed that Zn incorporated into coatings positively affects the proliferation and differentiation behavior of osteoblasts, resulting in enhanced peri-implant bone formation [46]. To the contrary, Pagano et al. showed a decrease in the cell number when human gingival fibroblasts and human keratinocytes were incubated with extracts obtained from glass ionomer cements modified with 6 wt% zinc L-carnosine [47]. In our study, the Zn coating showed high bactericidal activity towards Staph. aureus, which lasted for up to 48 h. Comparable to TiOB® SiOx Ag, no inhibition zones were discovered in the agar diffusion test. Although the antibacterial mechanisms are still not fully understood, a damage by direct or electrostatic interaction with the cell surface or by oxidation due to the formation of highly reactive oxygen species such as $1O_2$, HO·, or H_2O_2 have been discussed [48,49]. Similar to the gentamicin-tannic acid coating, the Zn-functionalization still showed an inhibitory effect, even after pre-incubation in PBS for 48 h.

5. Conclusions

All antibacterial self-active TiOB® surfaces observed in the present study proved a significant antibacterial effect on Staph. aureus with different biocompatible properties. Since gentamicin is able to elute out of TiOB® gta, inhibition of bacterial growth in the peri-implant region is possible, whereas the antibacterial effect of TiOB® SiOx Ag and TiOB® Zn is restricted to only the surface. The highest initial biocompatibility of all functionalized surfaces was observed for TiOB® SiOx Ag, which also proved the best-structured surface topography. TiOB® Zn showed an antibacterial efficiency comparable to that of TiOB® gta, with good biocompatible aspects after four days of direct cultivation. Further examinations are needed to also observe the performance of these functionalized TiOB® surfaces in vivo.

Author Contributions: Conceptualization: S.K., A.V., A.G., J.S., U.F., and M.D.; Methodology: A.V., A.G., S.T.-M., W.P. and M.D.; Software: S.K., A.V. and A.G.; Validation: S.K., A.V., A.G. and M.D.; Formal analysis: S.K., A.V., A.G. and T.L.; Investigation: S.K., A.V., A.G., S.T.-M. and M.D.; Resources: S.K., A.G. and S.T.-M.; Data curation: A.V., T.L., S.T.-M.; Writing—original draft preparation: S.K., A.G. and A.V.; Writing—review and editing: S.K., A.G., B.S., M.F., U.F., J.S., C.S., and M.D.; Visualization: S.K., A.G. and J.S.; Supervision: A.V., A.G. and B.S.; Project administration: B.S.

Funding: This research received no external funding.

Acknowledgments: This manuscript is dedicated to the memory of our colleague and friend Andrea Voelpel. We acknowledge all employees of Innovent Technology Development, Jena, Germany and Königsee Implantate GmbH, Allendorf, Germany for their technical support and provision of materials.

Conflicts of Interest: The authors declare no conflict of interest.

References

1. Jager, M.; Jennissen, H.P.; Dittrich, F.; Fischer, A.; Kohling, H.L. Antimicrobial and osseointegration properties of nanostructured titanium orthopaedic implants. *Materials* **2017**, *10*, 1302. [CrossRef] [PubMed]
2. Spriano, S.; Yamaguchi, S.; Baino, F.; Ferraris, S. A critical review of multifunctional titanium surfaces: New frontiers for improving osseointegration and host response, avoiding bacteria contamination. *Acta Biomater.* **2018**, *79*, 1–22. [CrossRef]
3. Su, Y.; Luo, C.; Zhang, Z.; Hermawan, H.; Zhu, D.; Huang, J.; Liang, Y.; Li, G.; Ren, L. Bioinspired surface functionalization of metallic biomaterials. *J. Mech. Behav. Biomed. Mater.* **2018**, *77*, 90–105. [CrossRef] [PubMed]
4. Lugovskoy, A.; Lugovskoy, S. Production of hydroxyapatite layers on the plasma electrolytically oxidized surface of titanium alloys. *Mater. Sci. Eng. C Mater. Biol. Appl.* **2014**, *43*, 527–532. [CrossRef] [PubMed]
5. Fiorellini, J.P.; Glindmann, S.; Salcedo, J.; Weber, H.P.; Park, C.J.; Sarmiento, H.L. The effect of osteopontin and an osteopontin-derived synthetic peptide coating on osseointegration of implants in a canine model. *Int. J. Periodontics Restor. Dent.* **2016**, *36*, e88–e94. [CrossRef]
6. Gao, Y.; Zou, S.; Liu, X.; Bao, C.; Hu, J. The effect of surface immobilized bisphosphonates on the fixation of hydroxyapatite-coated titanium implants in ovariectomized rats. *Biomaterials* **2009**, *30*, 1790–1796. [CrossRef] [PubMed]
7. Diefenbeck, M.; Muckley, T.; Schrader, C.; Schmidt, J.; Zankovych, S.; Bossert, J.; Jandt, K.D.; Faucon, M.; Finger, U. The effect of plasma chemical oxidation of titanium alloy on bone-implant contact in rats. *Biomaterials* **2011**, *32*, 8041–8047. [CrossRef] [PubMed]
8. Schrader, C.; Schmidt, J.; Diefenbeck, M.; Muckley, T.; Zankovych, S.; Bossert, J.; Jandt, K.D.; Faucon, M.; Finger, U. Bioactive tiob-coating on titanium alloy implants enhances osseointegration in a rat model. *Adv. Eng. Mater.* **2012**, *14*, B21–B27. [CrossRef]
9. De Avila, E.D.; de Molon, R.S.; Vergani, C.E.; de Assis Mollo, F., Jr.; Salih, V. The relationship between biofilm and physical-chemical properties of implant abutment materials for successful dental implants. *Materials* **2014**, *7*, 3651–3662. [CrossRef]
10. Dapunt, U.; Radzuweit-Mihaljevic, S.; Lehner, B.; Haensch, G.M.; Ewerbeck, V. Bacterial infection and implant loosening in hip and knee arthroplasty: Evaluation of 209 cases. *Materials* **2016**, *9*, 871. [CrossRef] [PubMed]
11. Laffer, R.R.; Graber, P.; Ochsner, P.E.; Zimmerli, W. Outcome of prosthetic knee-associated infection: Evaluation of 40 consecutive episodes at a single centre. *Clin. Microbio. Infect. Off. Publ. Eur. Soc. Clin. Microbio. Infect. Dis.* **2006**, *12*, 433–439. [CrossRef]
12. Diefenbeck, M.; Schrader, C.; Gras, F.; Muckley, T.; Schmidt, J.; Zankovych, S.; Bossert, J.; Jandt, K.D.; Volpel, A.; Sigusch, B.W.; et al. Gentamicin coating of plasma chemical oxidized titanium alloy prevents implant-related osteomyelitis in rats. *Biomaterials* **2016**, *101*, 156–164. [CrossRef] [PubMed]

13. Li, B.; Webster, T.J. Bacteria antibiotic resistance: New challenges and opportunities for implant-associated orthopedic infections. *J. Orthop. Res. Off. Publ. Orthop. Res. Soc.* **2018**, *36*, 22–32. [CrossRef]

14. Radulescu, M.; Andronescu, E.; Dolete, G.; Popescu, R.C.; Fufa, O.; Chifiriuc, M.C.; Mogoanta, L.; Balseanu, T.A.; Mogosanu, G.D.; Grumezescu, A.M.; et al. Silver nanocoatings for reducing the exogenous microbial colonization of wound dressings. *Materials* **2016**, *9*, 345. [CrossRef] [PubMed]

15. Bechert, T.; Boswald, M.; Lugauer, S.; Regenfus, A.; Greil, J.; Guggenbichler, J.P. The erlanger silver catheter: In vitro results for antimicrobial activity. *Infection* **1999**, *27* (Suppl. 1), S24–S29. [CrossRef]

16. Alt, V.; Bechert, T.; Steinrucke, P.; Wagener, M.; Seidel, P.; Dingeldein, E.; Domann, E.; Schnettler, R. An in vitro assessment of the antibacterial properties and cytotoxicity of nanoparticulate silver bone cement. *Biomaterials* **2004**, *25*, 4383–4391. [CrossRef]

17. Mei, S.; Wang, H.; Wang, W.; Tong, L.; Pan, H.; Ruan, C.; Ma, Q.; Liu, M.; Yang, H.; Zhang, L.; et al. Antibacterial effects and biocompatibility of titanium surfaces with graded silver incorporation in titania nanotubes. *Biomaterials* **2014**, *35*, 4255–4265. [CrossRef]

18. Toledano-Osorio, M.; Babu, J.P.; Osorio, R.; Medina-Castillo, A.L.; Garcia-Godoy, F.; Toledano, M. Modified polymeric nanoparticles exert in vitro antimicrobial activity against oral bacteria. *Materials* **2018**, *11*, 1013. [CrossRef]

19. Liu, L.; Pushalkar, S.; Saxena, D.; LeGeros, R.Z.; Zhang, Y. Antibacterial property expressed by a novel calcium phosphate glass. *J. Biomed. Mater. Res. Part B Appl. Biomater.* **2014**, *102*, 423–429. [CrossRef]

20. Yu, J.; Li, K.; Zheng, X.; He, D.; Ye, X.; Wang, M. In vitro and in vivo evaluation of zinc-modified ca-si-based ceramic coating for bone implants. *PLoS ONE* **2013**, *8*, e57564. [CrossRef]

21. Bakhsheshi-Rad, H.R.; Hamzah, E.; Low, H.T.; Kasiri-Asgarani, M.; Farahany, S.; Akbari, E.; Cho, M.H. Fabrication of biodegradable zn-al-mg alloy: Mechanical properties, corrosion behavior, cytotoxicity and antibacterial activities. *Mater. Sci. Eng. C Mater. Biol. Appl.* **2017**, *73*, 215–219. [CrossRef]

22. Beier, O.; Pfuch, A.; Horn, K.; Weisser, J.; Schnabelrauch, M.; Schimanski, A. Low temperature deposition of antibacterially active silicon oxide layers containing silver nanoparticles, prepared by atmospheric pressure plasma chemical vapor deposition. *Plasma Process. Polym.* **2013**, *10*, 77–87. [CrossRef]

23. Bechert, T.; Steinrucke, P.; Guggenbichler, J.P. A new method for screening anti-infective biomaterials. *Nat. Med.* **2000**, *6*, 1053–1056. [CrossRef]

24. Lin, W.T.; Zhang, Y.Y.; Tan, H.L.; Ao, H.Y.; Duan, Z.L.; He, G.; Tang, T.T. Inhibited bacterial adhesion and biofilm formation on quaternized chitosan-loaded titania nanotubes with various diameters. *Materials* **2016**, *9*, 155. [CrossRef]

25. Dapunt, U.; Hansch, G.M.; Arciola, C.R. Innate immune response in implant-associated infections: Neutrophils against biofilms. *Materials* **2016**, *9*, 387. [CrossRef]

26. Alt, V. Antimicrobial coated implants in trauma and orthopaedics-a clinical review and risk-benefit analysis. *Injury* **2017**, *48*, 599–607. [CrossRef]

27. Popat, K.C.; Eltgroth, M.; Latempa, T.J.; Grimes, C.A.; Desai, T.A. Decreased staphylococcus epidermis adhesion and increased osteoblast functionality on antibiotic-loaded titania nanotubes. *Biomaterials* **2007**, *28*, 4880–4888. [CrossRef]

28. Babuska, V.; Palan, J.; Kolaja Dobra, J.; Kulda, V.; Duchek, M.; Cerny, J.; Hrusak, D. Proliferation of osteoblasts on laser-modified nanostructured titanium surfaces. *Materials* **2018**, *11*, 1827. [CrossRef]

29. Sahiner, N.; Sagbas, S.; Aktas, N. Single step natural poly(tannic acid) particle preparation as multitalented biomaterial. *Mater. Sci. Eng. C Mater. Biol. Appl.* **2015**, *49*, 824–834. [CrossRef]

30. Vester, H.; Wildemann, B.; Schmidmaier, G.; Stockle, U.; Lucke, M. Gentamycin delivered from a pdlla coating of metallic implants: In vivo and in vitro characterisation for local prophylaxis of implant-related osteomyelitis. *Injury* **2010**, *41*, 1053–1059. [CrossRef]

31. Caldas, I.P.; Alves, G.G.; Barbosa, I.B.; Scelza, P.; de Noronha, F.; Scelza, M.Z. In vitro cytotoxicity of dental adhesives: A systematic review. *Dent. Mater. Off. Publ. Acad. Dent. Mater.* **2019**, *35*, 195–205. [CrossRef] [PubMed]

32. Marinucci, L.; Balloni, S.; Bodo, M.; Carinci, F.; Pezzetti, F.; Stabellini, G.; Conte, C.; Lumare, E. Patterns of some extracellular matrix gene expression are similar in cells from cleft lip-palate patients and in human palatal fibroblasts exposed to diazepam in culture. *Toxicology* **2009**, *257*, 10–16. [CrossRef] [PubMed]

33. Wiesli, M.G.; Ozcan, M. High-performance polymers and their potential application as medical and oral implant materials: A review. *Implant Dent.* **2015**, *24*, 448–457. [CrossRef]

34. Kheur, S.; Singh, N.; Bodas, D.; Rauch, J.Y.; Jambhekar, S.; Kheur, M.; Rajwade, J. Nanoscale silver depositions inhibit microbial colonization and improve biocompatibility of titanium abutments. *Colloids Surf. B Biointerfaces* **2017**, *159*, 151–158. [CrossRef] [PubMed]

35. Smeets, R.; Precht, C.; Hahn, M.; Jung, O.; Hartjen, P.; Heiland, M.; Grobe, A.; Holthaus, M.G.; Hanken, H. Biocompatibility and osseointegration of titanium implants with a silver-doped polysiloxane coating: An in vivo pig model. *Int. J. Oral Maxillofac. Implant.* **2017**, *32*, 1338–1345. [CrossRef] [PubMed]

36. Lischer, S.; Korner, E.; Balazs, D.J.; Shen, D.; Wick, P.; Grieder, K.; Haas, D.; Heuberger, M.; Hegemann, D. Antibacterial burst-release from minimal ag-containing plasma polymer coatings. *J. R. Soc. Interface* **2011**, *8*, 1019–1030. [CrossRef]

37. Van Hengel, I.A.J.; Riool, M.; Fratila-Apachitei, L.E.; Witte-Bouma, J.; Farrell, E.; Zadpoor, A.A.; Zaat, S.A.J.; Apachitei, I. Selective laser melting porous metallic implants with immobilized silver nanoparticles kill and prevent biofilm formation by methicillin-resistant staphylococcus aureus. *Biomaterials* **2017**, *140*, 1–15. [CrossRef]

38. Feng, Q.L.; Wu, J.; Chen, G.Q.; Cui, F.Z.; Kim, T.N.; Kim, J.O. A mechanistic study of the antibacterial effect of silver ions on escherichia coli and staphylococcus aureus. *J. Biomed. Mater. Res.* **2000**, *52*, 662–668. [CrossRef]

39. Jaiswal, S.; Duffy, B.; Jaiswal, A.K.; Stobie, N.; McHale, P. Enhancement of the antibacterial properties of silver nanoparticles using beta-cyclodextrin as a capping agent. *Int. J. Antimicrob. Agent.* **2010**, *36*, 280–283. [CrossRef]

40. Dror-Ehre, A.; Mamane, H.; Belenkova, T.; Markovich, G.; Adin, A. Silver nanoparticle-e. Coli colloidal interaction in water and effect on e. Coli survival. *J. Colloid Interface Sci.* **2009**, *339*, 521–526. [CrossRef]

41. Sondi, I.; Salopek-Sondi, B. Silver nanoparticles as antimicrobial agent: A case study on e. Coli as a model for gram-negative bacteria. *J. Colloid Interface Sci.* **2004**, *275*, 177–182. [CrossRef]

42. Li, J.; Xie, B.; Xia, K.; Li, Y.; Han, J.; Zhao, C. Enhanced antibacterial activity of silver doped titanium dioxide-chitosan composites under visible light. *Materials* **2018**, *11*, 1403. [CrossRef]

43. Deshmukh, S.P.; Mullani, S.B.; Koli, V.B.; Patil, S.M.; Kasabe, P.J.; Dandge, P.B.; Pawar, S.A.; Delekar, S.D. Ag nanoparticles connected to the surface of TiO_2 electrostatically for antibacterial photoinactivation studies. *Photochem. Photobiol.* **2018**, *94*, 1249–1262. [CrossRef]

44. Zhu, D.; Su, Y.; Young, M.L.; Ma, J.; Zheng, Y.; Tang, L. Biological responses and mechanisms of human bone marrow mesenchymal stem cells to zn and mg biomaterials. *ACS Appl. Mater. Interfaces* **2017**, *9*, 27453–27461. [CrossRef]

45. Yu, H.; Huang, X.; Yang, X.; Liu, H.; Zhang, M.; Zhang, X.; Hang, R.; Tang, B. Synthesis and biological properties of zn-incorporated micro/nano-textured surface on ti by high current anodization. *Mater. Sci. Eng. C Mater. Biol. Appl.* **2017**, *78*, 175–184. [CrossRef]

46. Shen, X.; Hu, Y.; Xu, G.; Chen, W.; Xu, K.; Ran, Q.; Ma, P.; Zhang, Y.; Li, J.; Cai, K. Regulation of the biological functions of osteoblasts and bone formation by zn-incorporated coating on microrough titanium. *ACS Appl. Mater. Interfaces* **2014**, *6*, 16426–16440. [CrossRef]

47. Pagano, S.; Chieruzzi, M.; Balloni, S.; Lombardo, G.; Torre, L.; Bodo, M.; Cianetti, S.; Marinucci, L. Biological, thermal and mechanical characterization of modified glass ionomer cements: The role of nanohydroxyapatite, ciprofloxacin and zinc l-carnosine. *Mater. Sci. Eng. C Mater. Biol. Appl.* **2019**, *94*, 76–85. [CrossRef]

48. Fiedot-Tobola, M.; Ciesielska, M.; Maliszewska, I.; Rac-Rumijowska, O.; Suchorska-Wozniak, P.; Teterycz, H.; Bryjak, M. Deposition of zinc oxide on different polymer textiles and their antibacterial properties. *Materials* **2018**, *11*, 707. [CrossRef]

49. Xie, Y.; He, Y.; Irwin, P.L.; Jin, T.; Shi, X. Antibacterial activity and mechanism of action of zinc oxide nanoparticles against campylobacter jejuni. *Appl. Environ. Microbiol.* **2011**, *77*, 2325–2331. [CrossRef]

Article

Particulated, Extracted Human Teeth Characterization by SEM–EDX Evaluation as a Biomaterial for Socket Preservation: An In Vitro Study

José Luis Calvo-Guirado [1], Alvaro Ballester Montilla [2], Piedad N De Aza [3], Manuel Fernández-Domínguez [4,*], Sergio Alexandre Gehrke [5], Pilar Cegarra-Del Pino [2], Lanka Mahesh [6], André Antonio Pelegrine [7], Juan Manuel Aragoneses [8] and José Eduardo Maté-Sánchez de Val [9]

[1] Department of Oral and Implant Surgery, Faculty of Health Sciences, Universidad Católica San Antonio de Murcia, 30002 Murcia, Spain; jlcalvo@ucam.edu

[2] International Dentistry Research Cathedra, Universidad Católica San Antonio de Murcia, 30002 Murcia, Spain; aballestermontilla@hotmail.com (A.B.M.); pilarcedepi@gmail.com (P.C.-D.P.)

[3] Institute of Bioengineering, Miguel Hernández University, 03203 Elche-Alicante, Spain; piedad@umh.es

[4] Department of Translational Medicine, CEU San Pablo University, 28223 Madrid, Spain

[5] Biotecnos Research Center—Tecnología e Ciencia, Ltd., Montevideo 11800, Uruguay; sergio.gehrke@hotmail.com

[6] Implantologist, Private Practice, New Delhi 110002, India; drlanka.mahesh@gmail.com

[7] Faculdade São Leopoldo Mandic, Instituto de Pesquisas São Leopoldo Mandic, São Paulo 13024-530, Brazil; andre.pelegrine@slmandic.edu.br

[8] Department of Dental Research in Universidad Francisco Henríquez y Carvajal, 10107 Santo Domingo, Dominican Republic; jmaragoneses@gmail.com

[9] Department of Oral and Implant Surgery, Faculty of Health Sciences, Universidad Católica San Antonio de Murcia, 30107 Murcia, Spain; jemate@ucam.edu

* Correspondence: clinferfun@yahoo.es

Received: 4 December 2018; Accepted: 23 January 2019; Published: 25 January 2019

Abstract: The aim of the study was to evaluate the chemical composition of crushed, extracted human teeth and the quantity of biomaterial that can be obtained from this process. A total of 100 human teeth, extracted due to trauma, decay, or periodontal disease, were analyzed. After extraction, all the teeth were classified, measured, and weighed on a microscale. The human teeth were crushed immediately using the Smart Dentin Grinder machine (KometaBio Inc., Cresskill, NJ, USA), a device specially designed for this procedure. The human tooth particles obtained were of 300–1200 microns, obtained by sieving through a special sorting filter, which divided the material into two compartments. The crushed teeth were weighed on a microscale, and scanning electron microscopy (SEM) evaluation was performed. After processing, 0.25 gr of human teeth produced 1.0 cc of biomaterial. Significant differences in tooth weight were found between the first and second upper molars compared with the lower molars. The chemical composition of the particulate was clearly similar to natural bone. Scanning electron microscopy–energy dispersive X-ray (SEM–EDX) analysis of the tooth particles obtained mean results of Ca% 23.42 $\pm$ 0.34 and P% 9.51 $\pm$ 0.11. Pore size distribution curves expressed the interparticle pore range as one small peak at 0.0053 µm. This result is in accordance with helium gas pycnometer findings; the augmented porosity corresponded to interparticle spaces and only 2.533% corresponded to intraparticle porosity. Autogenous tooth particulate biomaterial made from human extracted teeth may be considered a potential material for bone regeneration due to its chemical composition and the quantity obtained. After grinding the teeth, the resulting material increases in quantity by up to three times its original volume, such that two extracted mandibular lateral incisors teeth will provide a sufficient amount of material to fill four empty mandibular alveoli. The tooth particles present intra and extra pores up to 44.48% after pycnometer evaluation in order to increase the blood supply and support slow resorption of the grafted material, which supports healing and replacement resorption to achieve lamellar bone. After SEM–EDX evaluation, it appears that calcium and phosphates are still present within the collagen components even after the particle cleaning procedures that are conducted before use.

Keywords: smart dentin grinder; autogenous particulate dentin graft; tooth graft; ground teeth; human teeth; bone grafts; autologous graft

---

1. Introduction

In recent years, the physicochemical properties of biomaterials have been analyzed extensively to identify characteristics that will maximize the clinical outcomes of bone defect repair. In this context, two characteristics—grain size and the biomaterial's composition—directly influence the biomaterial's resorption activity and the speed of resorption [1].

A bone replacement material must have "bimodal" behavior, which, in the early stages of differentiation, allows osteoblasts to build bridges between grains of different sizes and integrate with other osteoblasts, supporting both proliferation and differentiation. New bone formation is stimulated by the activation of mesenchymal stem cells on the rough surfaces of biomaterials [2–4]. The ultimate goal is the union of completely differentiated osteoblasts, which will support the production of the bone matrix. This requires a bone replacement material with a porous structure including macropores, micropores, and nanopores [5,6]. In terms of roughness and external porosity, the surface of the bone replacement material's particles will directly influence the attachment of solvents to the surface of the biomaterial, allowing advanced cell colonization and the process of biomaterial remodeling to commence.

The presence of macropores and micropores in the particles of the graft biomaterial has been shown to be a very important criterion, allowing blood vessels to enter and favoring bone growth through osteoconduction within the pores. The structural properties and the physical and chemical characteristics of composite ceramics have been seen to affect their behavior *in vivo*, whether dependently or independently, whereby the outcome will depend on the case's individual bone repair parameters. Synthetic scaffolds can be used in regenerative and reconstructive surgery to treat bone defects. Biomaterials consisting of collagen and ceramic material are typically evaluated in terms of the proportions of liquid, collagen, and hydroxyapatite they contain. Porcine hydroxyapatite (HA) has lower crystallinity due to the presence of collagen in its composition. Changing the size, porosity, and crystallinity of each HA-based bone substitute material will influence the integration of the biomaterial within the implantation site and new bone formation [7,8]. To allow tissue penetration into the pores (and thus bone repair), they must be greater than 100 μm [9–13].

The most commonly used biomaterials are bioceramics based on calcium phosphate (Ca-P). The Ca-Ps have a composition and structure highly similar to the bone mineral phase, which presents osteoconductive properties and thus stimulates bone formation. Among the various materials assayed in recent years, tricalcium phosphate (TCP) has shown promising results in animal experiments and clinical studies [14–16].

At least one case series and several animal studies have reported promising results from a technique in which extraction sockets were augmented with autologous, particulate, mineralized dentin placed immediately after tooth extraction [17–19]. Although the supply of human teeth is, in fact, limited, when an extraction takes place, the tooth is naturally available and should be used to correct the damage caused by the extraction and subsequent lack of function, which leads to extensive resorption. To perform this procedure, the "Smart Dentin Grinder" [TM] machine was specially designed to crush, grind, and classify extracted teeth into different size particles. A special Dentin chemical cleanser is applied for 5 min to eliminate bacteria from the tooth, and right after, the tooth is washed with PBS two times. This novel procedure can be performed with any extracted teeth. Technically speaking, an autologous material can be returned to its donor without any treatment. However, in the case of the protocol that we have followed, there are multiple steps where strong disinfectant agents are used that are very effective in removing any bacteria/virus and many other biohazards that might be present.

Although this is true for allografts, this is not the case for autografts. Using our own biology in order to treat ourselves is not subject to ethical considerations.

The aim of this study was to determine the chemical composition and the amount of biomaterial obtained from crushed human teeth in order to fill empty alveolus with material from the manufacturer's protocol.

2. Materials and Methods

The study protocol was approved by the Catholic University of Murcia Ethics Committee (UCAM; registration number 6781; 21-07-2017).

Human teeth were extracted from 50 patients aged between 36 and 65 years, who received no financial compensation. All the patients signed informed consent forms to donate their teeth for use in the study. The teeth were extracted because of trauma, decay, or periodontal disease that had caused damage to one or two teeth in the upper maxilla and/or mandible. A total of 100 teeth were collected from 50 donors. The teeth were cleaned using straight fissure carbide burs, trimming the periodontal ligament, and dried with an air syringe. Each tooth was immediately classified, measured, and weighed. All the teeth were stored in separate sterile crystal containers at room temperature for 1–3 months, 1 per donor, labeling each container with the donor's details and the characteristics of the teeth (type, weight, dimensions).

After being cleaned and dried, the teeth were immediately crushed using the "Smart Dentin Grinder" device (KometaBio Inc., Cresskill, NJ, USA). The idea was to process an autologous dentin graft as a replacement for autologous bone harvesting. By doing so, we can preserve the tooth in the form of a particulate without diminishing the bioactive properties of dentin, a plethora of BMPs (bone morphogenic proteins) and growth factors, therefore leveraging it as a biocompatible, bioactive, bio-inert graft. Using dentin for non-autologous purposes or alternatively as an allograft, which requires extensive processing, is certainly not efficient and is not part of the current study's parameters. Autologous bone is still considered the gold standard for grafting. Autologous dentin not only has the same effects as autologous bone, we argue that, due to its inert and strong scaffold of dense HA, it is, in effect, better than autologous bone. The tooth particles were sized at 300–1200 microns, obtained by sieving the particles into two different compartments (Figure 1). The tooth particulate was then immersed in a basic alcohol cleanser in a sterile container for 10 min to dissolve all organic waste and bacteria. Afterward, the teeth particles were placed in ethylenediaminetetra-acetic acid (EDTA) for 2 min for partial demineralization and then washed with sterile saline for 3 min (Figure 2). Virus and fungi are all eliminated using the dentin cleanser that is part of the protocol. The dentin cleanser is a strong alkali (sodium hydroxide and ethanol combination) that is very effective in removing all bacteria, virus, and fungi. As for prions, we are not sure whether the dentin cleanser is able to remove all prions, but, again, these are the patient's own prions, because this is an autologous graft.

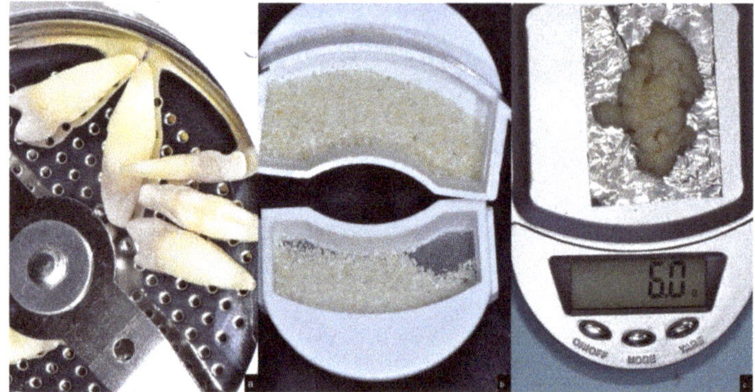

Figure 1. (**a**) Human teeth inside the Smart Dentin Grinder chamber; (**b**) Upper and lower compartment of different sized particles ranging from 300 to 1200 microns; (**c**) grounded teeth being weighed.

Figure 2. The manufacturer's protocol for grinding teeth.

The ground tooth material was analyzed by scanning electron microscopy (SEM) to evaluate its characteristics (Figure 3). For the SEM study, the particulate samples were placed in liquid nitrogen for approximately 2 min. The particles were coated with a carbon film (BalTec CED 030; BalTec, Balzers, Liechtenstein) for SEM analysis at ×10 magnification. The resolution was 0.8nm @ 15KV; 1.4nm @ 1 KV; 0.6 nm @ 30KV (STEM mode); 3.0@ 20 kV at 10 nA; and WD 8.5 nm using a Gemini II Electron Optics (Carl Zeiss Microscopy Gmbh, Jena, Germany), which is fitted with detectors for secondary electrons and backscattered electrons in order to allow for exploration of the different biological processes involved in tissue healing and to identify morphological changes in the cellular components of different materials. Mineralogical analysis of the material was performed by X-ray diffraction (XRD). XRD patterns were obtained using a Bruker AXS D8-ADVANCE X-ray Diffractometer (Karlsruhe, Germany) applying CuK1 radiation (0.15418 nm) and a second curved graphite monochromator. Diffractograms of the samples were compared with data from the Joint Committee on Powder Diffraction Standards (JCPDS) database (Figure 4).

Figure 3. (**a**) Scanning electron microscopy of teeth particles at 1-mm magnification; (**b**) augmented evaluation of collagenized tooth particles at 200 microns; (**c**) particle measurements at 200 microns; and (**d**) dentin tube measurements at 10 microns.

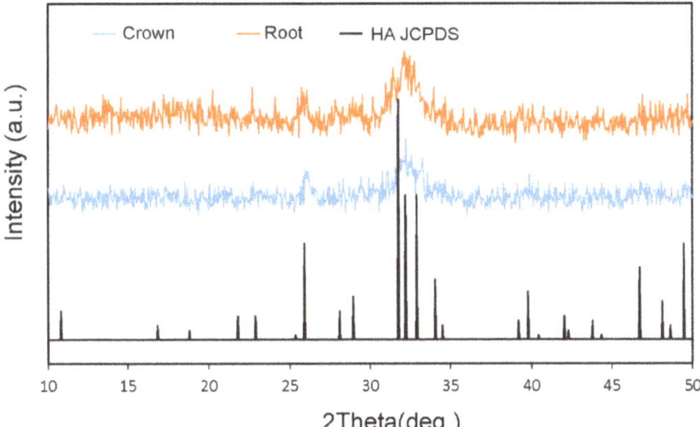

Figure 4. The diffractograms of the samples were compared with data from the Joint Committee on Powder Diffraction Standards (JCPDS) database.

The samples' porosity and pore size distribution were analyzed by mercury porosimetry using an automatic pore size analyzer (Poremaster-60 GT, Quantachrome Instruments, Boyton Beach, FL, USA) within a 6.215–411,475.500 KPa pressure range, corresponding to a pore diameter range of 236,641.05–3.57 nm. A total of 3 particulate samples (~0.47 g) were analyzed using this technique. An additional sample was also used in every case if the measured values for porosity differed by more than 5%. Helium gas pycnometry (Quantachrome Instruments, Boyton Beach, FL, USA) was used to determine the particle's real density (sample mass/volume of the solid), excluding empty spaces.

Statistical Analysis

Statistical analysis was performed using PASW Statistics v.18.0.0 software (SPSS). A descriptive test of a mean and standard deviation of each human tooth length, width, and weight was conducted. One-way ANOVA was applied for the comparison of the means, assuming a level of significance of 95% ($p < 0.05$).

3. Results

Human upper central incisors measured 6.5 ± 0.2 mm in length, 1.2 ± 0.6 mm in width, and weighed 1.3 ± 0.9 gr, while first mandibular molars measured 6.9 ± 0.2 mm in length, 2.1 ± 0.7 mm in width, and weighed 2.2 ± 1.1. These data show the significant differences between central incisors and first molars, which presented twice the width and weight of the incisors. Table 1 shows the mean tooth dimensions obtained for each type of tooth.

Table 1. Descriptive test of a mean and standard deviation of each human tooth length, width, and weight of 100 teeth.

Human Teeth	Mean length ± SD (mm)	Mean width ± SD (mm)	Mean weight ± SD (gr)
Upper central incisor	6.5 ± 0.2	1.2 ± 0.6	1.3 ± 0.9
Upper lateral incisor	5.9 ± 0.4	0.9 ± 0.1	0.9 ± 0.5
Upper canine	7.1 ± 1.2	1.3 ± 0.3	1.4 ± 1.1
Upper premolar	5.6 ± 0.6	0.9 ± 0.4	1.4 ± 0.2
Upper molar	7.8 ± 0.9	1.5 ± 0.3	1.9 ± 1.1
Lower central incisor	5.2 ± 0.8	1.2 ± 0.1	0.7 ± 0.2
Lower lateral incisor	5.1 ± 0.4	1.1 ± 0.2	0.6 ± 0.7
Lower canine	6.9 ± 0.5	1.2 ± 0.7	$1,2 \pm 0.6$
Lower premolar	6.1 ± 0.7	1.3 ± 0.6	1.4 ± 0.2
Lower molar	6.9 ± 0.2	2.1 ± 0.7	2.2 ± 1.1

Figure 3 shows the X-ray diffraction (XRD) patterns of central incisor tooth particles. XRD patterns are associated with the biomaterial's chemical composition. The crushed tooth particles presented high crystallinity (Figure 4).

A human extracted tooth weighing 0.25 gr produced at least 1.0 cc of particulate (Table 2).

Table 2. Comparison of the weight and volume of the human extracted teeth after grinding.

	Mineralized Human Particulated Dentin Graft								
Weight after extraction	0.25 gr	0.50 gr	1.0 gr	2.0 gr	3 gr	4 gr	5 gr	6.gr	7gr
Volume after grinding	0.75 cc	1.51 cc	3.10 cc	6.11 cc	9.12 cc	12.7 cc	15.62 cc	18.21 cc	21.74 cc

Analyzing the material by mercury porosimetry, two kinds of spaces were identified: those that correspond to empty spaces between particles (commonly designated as "interstices" or "interparticle" spaces) and those that correspond to the spaces within the particles themselves (known as "pores" or "intraparticle" spaces). The results obtained for the granules of human teeth particles showed that with increasing pressure, mercury penetrated into the increasingly amorphous pores.

Pore size distribution curves must be interpreted, a technique in which it is important to specify the size range of the measured pores. The size of these spaces depends on particle size, number, and shape, as well as the distribution of particle sizes. A big peak is related to a big particle (47.2 μm), corresponding to the intrusion of mercury into the interparticle spaces. The cumulative curve denoted an intrusion into the pores of between 219 μm and 38.2 μm, followed by a plateau after 38.2 μm, where no intrusion was detected. The initial rise of the curve mostly corresponded to the filling of the spaces between the particles, whereas the later stage of rising was related to the pores within the individual particles. The intraparticle pore range was more obvious, in which one small peak at 0.0053 μm was clearly visible.

These results were in accordance with the helium gas pycnometer (Table 3), in which 44.48% porosity corresponded to interparticle spaces and 2.533% corresponded to intraparticle porosity (Figure 5).

Table 3. Mercury-intruded volume, mode (most frequent diameter) of intraparticle pores, total porosity, and interparticle porosity. (a) 1 μm < pores < 220 μm; (b) pores < 1 um.

Human Teeth	Intruded Volume (cc/g)	Total Porosity (%)	Intraparticle Porosity (%) a	Interparticle Porosity (%) b
Upper central incisor	0.321	48.31	32.13	45.78
Upper lateral incisor	0.236	44.89	33.29	44.27
Upper canine	0.456	59.87	38.78	47.81
Upper premolar	0.562	58.20	33.29	39.76
Upper molar	0.786	67.98	36.87	45.71
Lower central incisor	0.145	42.17	31.89	45.99
Lower lateral incisor	0.164	41.74	31.78	42.29
Lower canine	0.472	61.87	33.34	46.32
Lower premolar	0.501	56.98	37.65	47.22
Lower molar	0.672	66.67	38.42	48.24
Mean ± Sd	0.431 ± 0.213	54.868 ± 9.871	34.745 ± 2.841	45.339 ± 2.610

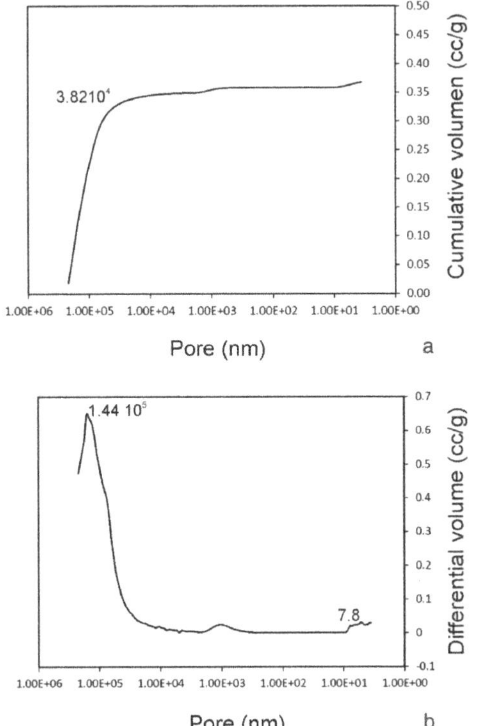

Figure 5. Results obtained by the helium gas pycnometer evaluating the interparticle and intraparticle porosity of the teeth grafts. (**a**) pore volume; (**b**) comparative volume pores.

The results of the SEM–EDX evaluation are shown in Table 4 (mean and standard deviation): O (%) 58.91 ± 1.1; Ca (%) 22.41 ± 0.28; C (%) 13.56 ± 0.44; P (%) 11.76 ± 0.45; N (%) 7.97 ± 0.21; Mg (%) 1.36 ± 0.18; and Na (%) 0.74 ± 0.45.

Table 4. Scanning electron microscopy–energy dispersive X-ray (SEM–EDX) evaluation of each crushed tooth's chemical composition.

Human Teeth	O (%)	Ca (%)	C (%)	P (%)	N (%)	Mg (%)	Na (%)
Upper central incisor	57.39 ± 0.11	23.78 ± 0.31	15.48 ± 0.12	9.53 ± 0.12	4.89 ± 0.11	0.96 ± 0.11	0.56 ± 0.13
Upper lateral incisor	51.38 ± 0.42	22.41 ± 0.28	14.29 ± 0.22	8.42 ± 0.11	4.07 ± 0.44	0.72 ± 0.17	0.44 ± 0.35
Upper canine	58.91 ± 1.1	24.89 ± 0.46	16.75 ± 0.23	10.23 ± 0.52	5.08 ± 0.32	0.98 ± 0.82	0.67 ± 1.8
Upper premolar	57.99 ± 0.22	24.56 ± 0.11	16.98 ± 1.87	10.55 ± 0.14	6.87 ± 0.24	1.36 ± 0.18	0.71 ± 0.23
Upper molar	61.27 ± 0.28	25.87 ± 0.67	17.39 ± 0.26	11.76 ± 0.45	7.97 ± 0.21	1.79 ± 0.22	0.74 ± 0.45
Lower central incisor	49.87 ± 0.33	21.11 ± 0.72	13.56 ± 0.44	7.82 ± 0.12	4.01 ± 0.66	0.77 ± 0.14	0.88 ± 0.56
Lower lateral incisor	48.66 ± 0.26	20.78 ± 0.65	13.11 ± 0.27	7.43 ± 0.54	3.99 ± 0.81	0.69 ± 0.36	0.48 ± 0.12
Lower canine	52.19 ± 0.15	24.56 ± 0.77	16.21 ± 0.98	9.68 ± 0.78	4.67 ± 0.81	0.97 ± 0.26	0.66 ± 0.24
Lower premolar	53.46 ± 0.23	24.82 ± 0.12	16.34 ± 0.29	10.23 ± 0.56	5.47 ± 0.54	1.06 ± 0.31	0.79 ± 0.33
Lower molar	57.82 ± 0.45	25.65 ± 0.38	17.13 ± 0.31	10.98 ± 0.33	6.03 ± 0.16	1.45 ± 0.24	0.82 ± 0.12

4. Discussion

Bone graft materials derived from teeth with an absence of antigenicity improve bone formation and bone remodeling capabilities. A wide range of bone graft materials are available, and choosing the right one presents a challenging decision that will be dictated by the bone substitute material's physicochemical properties in relation to the type of defect and the main purpose of the procedure [20–22].

Bone grafts derived from teeth can be considered to be an attractive option due to their autogenous origin and favorable clinical results, which have shown that these materials offer good osteoinductive capacities. Nevertheless, they pose some risk of viral infection and are limited in quantity, while most of the synthetic materials offer osteoconductive competence and can be supplied in unlimited quantities [23–25].

The SEM micrographs provided information about the morphology of the crushed tooth particulate, which presented no critical defect and a homogeneous microstructure with aggregates of high density.

In mercury porosimetry analysis, the inter and intraparticle pore distinction is not always clear. The information provided by pore size distribution curves must be interpreted, a technique in which the size range of the measured pores is of fundamental importance. In the present study, the tooth particles consisted of a highly porous network with an average pore size of 0.431 ± 0.213 μm. The total porosity of the samples analyzed had an average of 54.868%, which is comparable to replacement biomaterials of different origins and with the most useful ones, which are around 60%. As research has demonstrated, the degree of porosity and its disposition directly influences the biological behavior of biomaterial grafts. In addition, there is a direct relationship between these parameters and resorption rates [26,27].

EDX was used to determine the elemental composition of the dentin particulate, obtaining a Ca/P ratio of 1.67 ± 0.09, which is similar to that of synthetic HA. The presence of traces of magnesium was also observed, known to be the impurity in calcium phosphate as a raw material. The composition of the samples determined by quantitative analysis at different points of the sample surfaces showed the presence of Ca, P, and O.

Although demineralized dentin exhibits matrix-derived growth and differentiating factors for effective osteogenesis, the newly formed bone that is generated and the residual demineralized dentin is too weak to allow adequate implant anchorage. However, the use of the Smart Dentin Grinder Machine enables us to prepare a natural biomaterial from freshly extracted autologous teeth in the form of a bacteria-free particulate for immediate use as an autogenous graft biomaterial in a single surgical session. Teeth and mandibular/maxillary bone have a high level of similarity with dentin, both presenting similar chemical structures and composition in organic, protein, and mineral phases.

For this reason, our research team (in light of our own findings and those of other investigations) proposes that non-functional extracted teeth or periodontally involved teeth should no longer be discarded [28]. Extracted teeth can be ground to produce an autogenous dentin particulate within 15 min of extraction and can then be grafted into the post-extraction alveoli. In this way, the patient's own extracted tooth acts as a clinically useful bone graft material that offers all the advantages of autogenous bone due to the similarity of composition between bone and dentin. The particulate tooth material provides excellent biocompatibility without eliciting an immune response or a foreign material reaction or infection after it is used. In addition, it has osteoinduction, osteoconduction, and progressive substitution capabilities, and it can be worked into various sizes and shapes [28]. Moreover, some patients refuse allografts or xenografts on the basis of their origins—a problem that this technique overcomes.

5. Conclusions

Autogenous tooth particulate biomaterial made from human extracted teeth may be considered a potential material for bone regeneration due to its chemical composition and the high quantity of material obtained from each tooth. After grinding the teeth, the resulting material increases in volume by up to three times, so that two extracted mandibular lateral incisors teeth will provide a sufficient amount of material to fill four empty mandibular alveoli. The tooth particles present intra- and extra-porosity up to 44.48% after pycnometer evaluation in order to increase the blood supply and support slow resorption of the grafted material, which will support healing and replacement resorption to achieve lamellar bone. After SEM–EDX evaluation, it appears that calcium and phosphates are still present within the collagen components even after the particle cleaning procedures that are conducted before use.

Author Contributions: Conceptualization, J.L.C.-G.; methodology, J.E.M.-S.d.V. and M.F.-D.; software, A.A.P., P.N.D.A.; validation, J.L.C.-G. and P.C.-D.P.; formal analysis, L.M. and J.M.A.; investigation, A.B.M., J.L.C.-G., and J.E.M.-S.d.V.; resources, S.A.G.; data curation, P.C.-D.P.; original draft preparation, M.F.-D. and J.L.C.-G.; review and editing of the manuscript, L.M. and J.L.C.-G.; validation, J.M.A.; visualization, M.F.-D.; supervision, A.B.M. and J.E.M.-S.d.V.; project administration, J.L.C.-G.; and funding acquisition, J.L.C.-G.

Funding: This research received no external funding.

Acknowledgments: The authors would like to thank Nuria García Carrillo, Veterinarian, University of Murcia, Spain.

Conflicts of Interest: The authors declare no conflicts of interest.

References

1. Werner, J.; Linner-Kremar, B.; Friess, W.; Greil, P. Mechanical properties and in vitro cell compatibility of hydroxyapatite ceramics with graded pore structure. *Biomaterials* **2002**, *23*, 4285–4294. [CrossRef]
2. Fan, H.; Ikoma, T.; Tanaka, J.; Zhang, X. Surface structural biomimetics and the osteoinduction of calcium phosphate biomaterials. *Nanosci. Nanotechnol.* **2007**, *7*, 808–813. [CrossRef]
3. Zyman, Z.Z.; Tkachenko, M.V.; Polevodin, D.V. Preparation and characterization of biphasic calcium phosphate ceramics of desired composition. *J. Mater. Sci. Mater. Med.* **2008**, *19*, 2819–2825. [CrossRef] [PubMed]
4. Saldana, L.; Sanchez-Salcedo, S.; Izquierdo-Barba, I.; Bensiamar, F.; Munuera, L.; Vallet-Regi, M.; Vilaboa, N. Calcium phosphate-based particles influence osteogenic maturation of human mesenchymal stem cells. *Acta Biomater.* **2009**, *5*, 1294–1305. [CrossRef] [PubMed]
5. Gauthier, O.; Bouler, J.M.; Aguado, E.; Legeros, R.Z.; Pilet, P.; Daculsi, G. Elaboration conditions influence physicochemical properties and in vivo bioactivity of macroporous biphasic calcium phosphate ceramics. *J. Mater. Sci. Mater. Med.* **1999**, *10*, 199–204. [CrossRef] [PubMed]
6. Rivera-Munoz, E.; Diaz, J.R.; Rogelio Rodriguez, J.; Brostow, W.; Castano, V.M. Hydroxyapatite spheres with controlled porosity for eyeball prosthesis: Processing and characterization. *J. Mater. Sci. Mater. Med.* **2001**, *12*, 305–311. [CrossRef]

7. Calvo-Guirado, J.L.; Aguilar-Salvatierra, A.; Ramírez-Fernández, M.P.; Maté Sánchez de Val, J.E.; Delgado-Ruiz, R.A.; Gómez-Moreno, G. Bone response to collagenized xenografts of porcine origin (mp3®) and a bovine bone mineral grafting (4BONE(™) XBM) grafts in tibia defects: Experimental study in rabbits. *Clin. Oral Implants Res.* **2016**, *27*, 1039–1046. [CrossRef] [PubMed]

8. Maté Sánchez de Val, J.E.; Calvo-Guirado, J.L.; Gómez-Moreno, G.; Pérez-Albacete Martínez, C.; Mazón, P.; De Aza, P.N. Influence of hydroxyapatite granule size, porosity, and crystallinity on tissue reaction in vivo. Part A: Synthesis, characterization of the materials, and SEM analysis. *Clin. Oral Implants Res.* **2016**, *27*, 1331–1338. [CrossRef]

9. Carotenuto, G.; Spagnuolo, G.; Ambrosio, L.; Nicolais, L. Macroporous hydroxyapatite as alloplastic material for dental applications. *J. Mater. Sci. Mater. Med.* **1999**, *10*, 671–676. [CrossRef]

10. Tampieri, A.; Celotti, G.; Sprio, S.; Delcogliano, A.; Franzese, S. Porosity-graded hydroxyapatite ceramics to replace natural bone. *Biomaterials* **2001**, *22*, 1365–1370. [CrossRef]

11. Parrilla-Almansa, A.; García-Carrillo, N.; Ros-Tárraga, P.; Martínez, C.M.; Martínez-Martínez, F.; Meseguer-Olmo, L.; De Aza, P.N. Demineralized Bone Matrix Coating Si-Ca-P Ceramic Does Not Improve the Osseointegration of the Scaffold. *Materials* **2018**, *11*, 1580. [CrossRef] [PubMed]

12. Zuleta, F.; Murciano, A.; Gehrke, S.A.; Maté-Sánchez de Val, J.E.; Calvo-Guirado, J.L.; De Aza, P.N. A New Biphasic Dicalcium Silicate Bone Cement Implant. *Materials* **2017**, *10*, 758. [CrossRef] [PubMed]

13. Maté-Sánchez de Val, J.E.; Calvo-Guirado, J.L.; Delgado-Ruiz, R.A.; Ramírez-Fernández, M.P.; Negri, B.; Abboud, M.; Martínez, I.M.; de Aza, P.N. Physical properties, mechanical behavior, and electron microscopy study of a new α-TCP block graft with silicon in an animal model. *J. Biomed. Mater. Res. A* **2012**, *100*, 3446–3454. [CrossRef] [PubMed]

14. Krekmanov, L. The efficacy of various bone augmentation procedures for dental implants: A Cochrane systematic review of randomized controlled clinical trials. *Int. J. Oral Maxillofac. Implants* **2006**, *21*, 696–710.

15. Esposito, M.; Grusovin, M.G.; Kwan, S.; Worthington, H.V.; Coulthard, P. Interventions for replacing missing teeth: Bone augmentation techniques for dental implant treatment. *Cochrane Database Syst. Rev.* **2008**, *3*, CD003607.

16. Chopra, P.M.; Johnson, M.; Nagy, T.R.; Lemons, J.E. Micro-computed tomographic analysis of bone healing subsequent to graft placement. *J. Biomed. Mater. Res. B Appl. Biomater.* **2009**, *88*, 611–618. [CrossRef] [PubMed]

17. Valdec, S.; Pasic, P.; Soltermann, A.; Thoma, D.; Stadlinger, B.; Rücker, M. Alveolar ridge preservation with autologous particulated dentin—A case series. *Int. J. Implant Dent.* **2017**, *3*, 12. [CrossRef]

18. Binderman, I.; Hallel, G.; Nardy, C.; Yaffe, A.; Sapoznikov, L. A Novel Procedure to Process Extracted Teeth for Immediate Grafting of Autogenous Dentin. *J. Interdiscipl. Med. Dent. Sci.* **2014**, *2*, 154.

19. Calvo-Guirado, J.L.; Cegarra Del Pino, P.; Sapoznikov, L.; Delgado-Ruiz Fernández-Domínguez, M.; Gehrke, S.A. A new procedure for processing extracted teeth for immediate grafting in post-extraction sockets. An experimental study in American Fox Hound dogs. *Ann. Anat.* **2018**, *217*, 14–23. [CrossRef]

20. Ramírez Fernández, M.P.; Gehrke, S.A.; Mazón, P.; Calvo-Guirado, J.L.; De Aza, P.N. Implant Stability of Biological Hydroxyapatites Used in Dentistry. *Materials* **2017**, *10*, 644. [CrossRef]

21. Ramírez Fernández, M.P.; Mazón, P.; Gehrke, S.A.; Calvo-Guirado, J.L.; De Aza, P.N. Comparison of Two Xenograft Materials Used in Sinus Lift Procedures: Material Characterization and In Vivo Behavior. *Materials* **2017**, *10*, 623. [CrossRef]

22. Ramírez Fernández, M.P.; Gehrke, S.A.; Pérez Albacete Martinez, C.; Calvo Guirado, J.L.; de Aza, P.N. SEM-EDX Study of the Degradation Process of Two Xenograft Materials Used in Sinus Lift Procedures. *Materials* **2017**, *10*, 542. [CrossRef]

23. Kim, Y.K.; Kim, S.G.; Byeon, J.H.; Lee, H.J.; Um, I.U.; Lim, S.C. Development of a novel bone grafting material using autogenous teeth. *Oral Surg. Oral Med. Oral Pathol. Oral Radiol. Endod.* **2010**, *109*, 496503. [CrossRef]

24. Kim, Y.K.; Lee, J.; Um, I.W.; Kim, K.W.; Murata, M.; Akazawa, T.; Mitsugi, M. Tooth-derived bone graft material. *J. Korean Assoc. Oral Maxillofac. Surg.* **2013**, *39*, 103–111. [CrossRef]

25. Kim, Y.K.; Kim, S.G.; Yun, P.Y.; Yeo, I.S.; Jin, S.C.; Oh, J.S.; Kim, H.-J.; Yu, S.-K.; Lee, S.-Y.; Kim, J.-S.; et al. Autogenous teeth used for bone grafting: A comparison with traditional grafting materials. *Oral Surg. Oral Med. Oral Pathol. Oral Radiol.* **2014**, *117*, e39–e45. [CrossRef]

26. De Val, J.E.M.-S.; Mazon, P.; Calvo-Guirado, J.L.; Ruiz, R.A.D.; Fernandez, M.P.R.; Negri, B.; Abboud, M.; De Aza, P.N. Comparison of three hydroxyapatite/b-tricalcium phosphate/collagen ceramic scaffolds: An in vivo study. *J. Biomed. Mater. Res. Part A* **2014**, *102*, 1037–1046. [CrossRef]

27. Maté Sánchez de Val, J.E.; Mazón, P.; Piattelli, A.; Calvo-Guirado, J.L.; Mareque-Bueno, J.; Granero-Marín, J.M.; De Aza, P. Comparison among the physical properties of calcium phosphate-based bone substitutes of natural or synthetic origin. *Int. J. Appl. Ceram. Technol.* 2018. [CrossRef]

28. Kim, S.G.; Chung, C.H.; Kim, Y.K.; Park, J.C.; Lim, S.C. The use of particulate dentin–plaster of Paris combination with/without platelet-rich plasma in the treatment of bone defects around implants. *Int. J. Oral Maxillofac. Implants* **2002**, *17*, 86–94.

 *materials*

Article

In Vitro Comparison of the Efficacy of Peri-Implantitis Treatments on the Removal and Recolonization of *Streptococcus gordonii* Biofilm on Titanium Disks

Selena Toma [1,2,*], Catherine Behets [2], Michel C. Brecx [1] and Jerome F. Lasserre [1]

[1] Department of Periodontology, Université Catholique de Louvain (UCL)—Cliniques Universitaires Saint Luc, 1200 Brussels, Belgium; michel.brecx@uclouvain.be (M.C.B.); jerome.lasserre@uclouvain.be (J.F.L.)

[2] Institut de Recherche Expérimentale et Clinique (IREC), Pôle de Morphologie, Université Catholique de Louvain (UCL), 1200 Brussels, Belgium; catherine.behets@uclouvain.be

* Correspondence: selena.toma@uclouvain.be; Tel.: +32–2–764-57-14

Received: 19 November 2018; Accepted: 1 December 2018; Published: 6 December 2018

Abstract: Objective: To compare the efficacy of four commonly used clinical procedures in removing *Streptococcus gordonii* biofilms from titanium disks, and the recolonization of the treated surfaces. **Background:** Successful peri-implantitis treatment depends on the removal of the dental biofilm. Biofilm that forms after implant debridement may threaten the success of the treatment and the long-term stability of the implants. **Methods:** *S. gordonii* biofilms were grown on titanium disks for 48 h and removed using a plastic curette, air-abrasive device (Perio-Flow®), titanium brush (TiBrush®), or implantoplasty. The remaining biofilm and the recolonization of the treated disks were observed using scanning electron microscopy and quantified after staining with crystal violet. Surface roughness (Ra and Rz) was measured using a profilometer. **Results:** *S. gordonii* biofilm biomass was reduced after treatment with Perio-Flow®, TiBrush®, and implantoplasty (all $p < 0.05$), but not plastic curette ($p > 0.05$), compared to the control group. Recolonization of *S. gordonii* after treatment was lowest after Perio-Flow®, TiBrush®, and implantoplasty (all $p < 0.05$ vs. control), but there was no difference between the plastic curette and the control group ($p > 0.05$). Ra and Rz values ranged from 1–6 µm to 1–2 µm and did not differ statistically between the control, plastic curette, Perio-Flow, and TiBrush groups. However, the implantoplasty group showed a Ra value below 1 µm ($p < 0.01$, ANOVA, Tukey). **Conclusions:** Perio-Flow®, TiBrush®, and implantoplasty were more effective than the plastic curette at removing the *S. gordonii* biofilm and preventing recolonization. These results should influence the surgical management of peri-implantitis.

Keywords: peri-implantitis; biofilm; dental implants; in vitro model

1. Introduction

Dental implants are a treatment option for the replacement of missing teeth, restoring dental function, and esthetics. However, approximately 30% of patients with implants develop peri-implantitis, a major reason for implant failure [1–4]. Peri-implantitis is characterized by biofilm-related inflammation of the tissues surrounding dental implants. The term peri-implantitis was first used in 1987 to describe a periodontitis-like disease characterized by biofilm-related inflammation of the tissues surrounding dental implants [5]. Subsequently, alveolar bone loss, visible by X-ray analysis, allowed peri-implantitis to be distinguished from peri-implant mucositis [6].

The colonization of the implant surface by oral bacteria organized in biofilms, similarly to periodontitis, is considered as a primary etiological factor of peri-implantitis [7]. Bacterial adhesion and biofilm organization in dental plaque play a crucial role in the pathogenesis of peri-implantitis [1].

Smoking and a history of periodontitis are risk factors for developing the disease [8,9]. Other risk factors have been highlighted recently, including excess cement, genetic polymorphisms, diabetes, cardiovascular diseases, and the absence of keratinized tissue adjacent to the implant [10].

As with periodontitis treatment, dental plaque removal is a major approach of peri-implantitis treatment. Numerous mechanical procedures have been proposed to remove the dental biofilm and to improve peri-implant health. From the least to the most abrasive techniques, these include using plastic, metals or ultrasonic scalers, rubber polisher, an air-powder abrasive device, rotating brushes, and implantoplasty protocols [11,12]. However, recent reviews have highlighted the absence of reliable evidence for the most effective interventions in peri-implantitis treatment owing to their complexity [13], and standardized, evidence-based protocols are still lacking [14]. For these reasons, a focus on the mechanical elimination of the dental biofilm instead of a combination of treatment modalities was preferred. Based on the instrumentation used for the treatment of periodontitis, the plastic curette appeared as a potential cleaning tool to remove dental biofilm from the titanium implant surface. The use of a non-metallic scaler was first advised to avoid major surface modification associated with a risk of lower biocompatibility and re-osseointegration. The studies evaluating the cleaning efficacy of an air-powder abrasive demonstrated constant results. According to Tastepe et al., in vitro cleaning efficiency of the method is reported to be high [15]. Airflow devices using glycine powders seem to constitute an efficient therapeutic option in the debridement of implants with peri-implantitis [16]. Rotating titanium brushes have been proposed as an alternative for the mechanical treatment of peri-implantitis. Meager results are reported in the literature. According to John et al., a rotating titanium brush seems effective for mechanical cleansing of sand-blasted, large grit, acid etched (SLA) surfaces, while inducing no surface alteration [17]. Implantoplasty consists of the elimination of surface roughness together with the implant threads. This technique has been proposed to optimize maintenance, and facilitate oral hygiene when implant threads are exposed. Few clinical studies have evaluated the effects of this protocol. Romeo and coworkers obtained an implant survival rate of 100% after three years, with improvement in clinical and radiological parameters, as compared to those without implantoplasty [18].

Early bacterial colonizers such as *Streptococcus gordonii* (*S. gordonii*) are known to play a crucial role for bacterial adhesion of middle colonizers (*Fusobacterium nucleatum*, *F. nucleatum*) and late colonizers (*Porphyromonas gingivalis*, *P. gingivalis*) in the beginning of the formation of oral biofilm [19]. Therefore, elimination of early colonizers on the surface of dental implants could be decisive for long term implant success.

Therefore, the present in vitro study aims to compare the efficacy of four mechanical methods (a plastic curette, an air-abrasive device (Perio-Flow), titanium brush (TiBrush), and implantoplasty) on the removal of *Streptococcus gordonii* from titanium disks, and the bacterial recolonization of disks previously treated.

2. Materials and Methods

Titanium disks

Two hundred and fifty sterile, microrough titanium disks (diameter, 5 mm; thickness, 2 mm) that had been sand-blasted with aluminum oxide beads (75–170 μm) and treated with solvents by the manufacturer (Southern Implants, Irene, South Africa) were used. The disks were handled by their circumference to avoid contact with the surface to be treated and analyzed.

2.1. Decontamination Assay

Saliva coating of the disks

As already described by Ota-Tsuzuki et al. [20], unstimulated saliva was collected from three healthy donors (aged 24–26 years) for 30 min per day for 7 days. After collection of 300 mL, the saliva samples were frozen at −20 °C. Then, the saliva samples were pooled and centrifuged (6000× g rpm for 30 min at 4 °C), and the supernatant was filtered (5 µm and 0.22 µm). The sterile disks were placed in a sterile 24-well polystyrene cell-culture plate containing 500 µL saliva per well for 30 min at 37 °C to allow salivary pellicle formation.

Biofilm formation

The disks were placed in a new 24-well polystyrene plate after aspiration of the saliva. Standard reference-strain *S. gordonii* (ATCC 10558) was used to prepare inoculum. Inoculation, and incubation was done under anaerobic conditions (80% N_2, 10% H_2, and 10% CO_2) for 24 h at 37 °C. The bacterial cells were suspended in BHI agar, adjusting the turbidity to an optical density (OD) of 0.15 at 630 nm with 106 colony-forming units/mL, and 500 µL of this suspension added to the wells and incubated for 48 h under anaerobic conditions [20]. After formation of the *S. gordonii* biofilms, unattached cells were removed by washing with sterile saline solution, placed in a new sterile 24-well plate, and randomly allocated to the different treatment groups.

Titanium surface treatment

In all groups, experiments were carried out using 10 disks/group and performed five times (n = 5). As already described in Toma et al., [21], the disks were treated with plastic curette, Perio-Flow, titanium brush (Ti-Brush), and implantoplasty. Non-treated disks were used as controls.

Plastic curette—The entire surface of the disk was scaled with a plastic curette (*Implacare, Hu-Friedy, Chicago, IL, USA) at an angle of 70° for 30 s. The tip of the curette was made from high-grade resin. Each side of the curette was used for five disks.

Perio-Flow—The disks were treated using an air-abrasive system ([†] Perio-Flow, Perio-Flow nozzle, EMS, Nyon, Switzerland) using tap water and an air-power setting with glycine powder (25 µm) (Air-Flow Perio Powder, EMS). The specially designed nozzle, consisting of a thin flexible plastic tube (length 1.7 cm; diameter 0.8 mm at the tip), was fixed on a handpiece (Air-Flow EL-308/A, EMS, Nyon, Switzerland). Perio-Flow was applied in a circular, non-contact mode, parallel to the disk surface for 30 s. After, glycine powder was removed by irrigation with sterile saline (20 mL, 20 s).

Titanium brush—The Ti-Brush® ([‡] Straumann®, Basel, Switzerland) is made of titanium bristles with a stainless steel shaft. Disks were processed using a Ti-Brush® fixed on a surgical handpiece ([§]Bien-Air Medical Technologies, Bienne, Switzerland) oscillating in a clockwise/counterclockwise direction at low speed (maximum of 900 oscillations per minute, 30 s). Sterile saline solution (NaCl 0.9%) was used for irrigation and cooling of the treatment site. Each brush was used for five disks and then replaced.

Implantoplasty—Disks were polished with a diamond round shaped bur (30 µm particle size egg-shaped bur) ([‖] Komet, Gerb. Brasseler GmbH, Lemgo, Germany) and assembled on a handpiece ([¶]KaVo Dental GmbH, Biberach, Germany) working at 15,000 rpm. The disks were treated for 30 s and rinsed with sterile saline to remove any titanium particles.

Crystal violet assay—Crystal violet was used to evaluate the total amount of biofilm. After treatment, the disks were dried at 45 °C for 60 min, then immersed in a 1% crystal violet solution in the dark, at room temperature for 15 min. After three rinses in phosphate-buffered saline, 1 mL acetic acid (33%) was added to each well and left in the dark at room temperature for 15 min. The absorbance was measured at 450 nm using a microplate spectrophotometer ([#]Bio-Rad Laboratories, CA, USA) and reported as OD.

2.2. Recolonization Assay

Treated disks (10/group, repeated five times) were assigned to their experimental groups, coated with saliva, and cultured with *S. gordonii* to obtain a biofilm, as described above. This biofilm was also assessed using crystal violet staining and OD measurement after 48 h.

Scanning electron microscopy (SEM)

SEM was used to examine the remaining *S. gordonii* biofilm (48 h culture, n = 3/group, in duplicate) after each of the four treatments, as well as the biofilm recolonization of previously treated titanium disks. Bacterial biofilm was fixed in 2.5% glutaraldehyde in 0.05 mol at pH 7.4 for 1 h, post-fixed with 1% osmium tetroxide at pH 7.4 for 1 h, and dehydrated through an ethanol series (30%, 50%, 70%, 90%, and 100%; 20 min per concentration). Finally, the disks were sputter-coated with gold ([††] Emitech K550; Houston, TX, USA) and examined under a scanning electron microscope (JEOL 7200, Tokyo, Japan) at 15 kV.

Measurement of surface roughness

The surface roughness of the control and treatment groups was measured using a contact profilometer (DektakXT Bruker Stylus Profiler; Billerica, MA, USA) equipped with a diamond microneedle (diameter, 7 μm). The profilometer scanned each disk along a length of 2 mm. The horizontal movements of the tip, generated by surface irregularities, were transferred to a transducer that created an electric stimulus. Three measurements per disk (in duplicate) were taken. All measurements were carried out in the same direction. Arithmetical mean roughness (Ra) and ten-point mean roughness (Rz) were recorded.

Statistical analysis

Data are expressed as mean $\pm$ SD. To compare data between the groups, those following a normal distribution and homogeneity of variance were analyzed using a one-way analysis of variance (ANOVA) and a Tukey *post-hoc* test. Those following a non-normal distribution were analyzed using a non-parametric Kruskall–Wallis test with a Nemenyi–Damico–Wolfe–Dunn post-hoc test (GraphPad InStat version 3; GraphPad Software, La Jolla, CA, USA). Differences were considered statistically significant when $p < 0.05$.

3. Results

3.1. Bacterial Elimination After Surface Decontamination

Decontamination of the titanium surfaces by the different procedures was quantified by measuring the in vitro biofilm biomass using OD after crystal violet staining and SEM. Biofilm OD was significantly lower in the Perio-Flow®, TiBrush®, and implantoplasty groups than in the control and plastic curette groups ($p < 0.05$), demonstrating elimination of a greater part of the biofilm (Figure 1). No statistical difference was observed between the biofilm biomass of the control and plastic curette groups ($p > 0.05$).

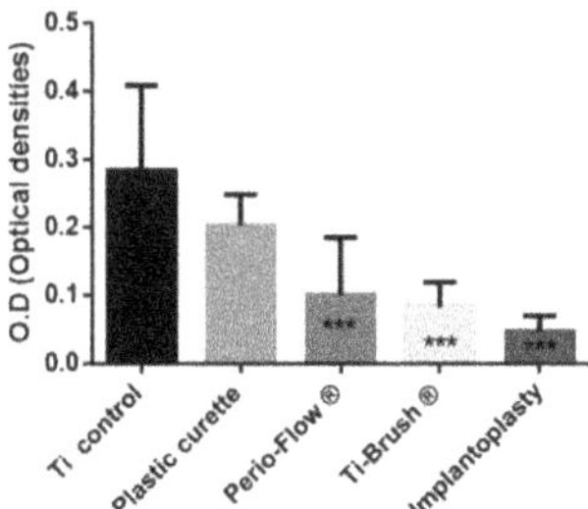

Figure 1. *Streptococcus Gordonii* Biofilm elimination after treatment. Results expressed in optical densities (O.D) (Crystal violet staining). Ti control vs. Plastic curette (ns $p > 0.05$), Ti control vs. Perio-Flow ® ** p <0.001, Ti control vs. Ti-Brush ® ** $p < 0.001$, Ti control vs. Implantoplasty *** $p < 0.001$, Plastic curette vs. Perio-Flow ® * $p < 0.05$, Plastic curette vs. Ti-Brush ® *** $p < 0.001$,Plastic curette vs. Implantoplasty *** $p < 0.001$, Perio-Flow ® vs. Ti-Brush ® ns $p > 0.05$,Perio-Flow ® vs. Implantoplasty ns $p > 0.05$, Ti-Brush ® vs. Implantoplasty ns $p > 0.05$ (*Kruskal Wallis, Dunn's*).

SEM revealed that biofilm colonies were more abundant on the untreated disks and those treated with the plastic curette than on other disks (Figure 2). The metallic surface of the disks was hardly visible in the control, plastic curette, Perio-Flow®, and TiBrush® groups, as they were covered in biofilm, whereas few bacteria were visible on the implantoplasty-treated disks, allowing the smooth titanium surface to be observed.

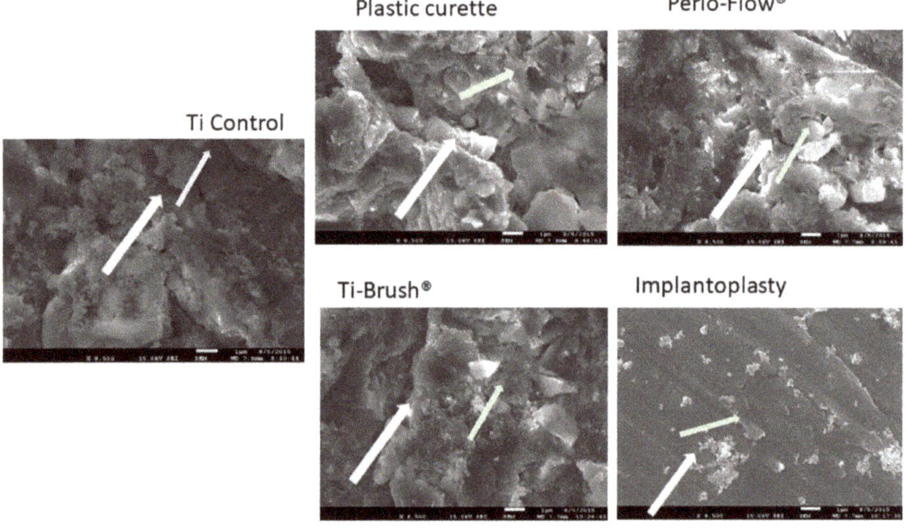

Figure 2. Scanning electron micrographs of Streptococcus Gordonii biofilm (48 h) elimination on titanium disks after treatments (8500 ×). Biofilm colonies were more abundant on the untreated disks, the plastic curette group than on other disks. Few bacteria were visible on the implantoplasty-treated disks, allowing the smooth titanium surface to be observed.

3.2. Bacterial Recolonization After Surface Treatment

The OD of the *S. gordonii* bacterial biomass was quantified after 48 h of culture and crystal violet staining of previously treated titanium disks (Figure 3). OD was significantly lower on the disks treated with Perio-Flow®, TiBrush®, and implantoplasty than on the control disks ($p < 0.05$). There was no difference in OD between the control and plastic curette groups ($p > 0.05$). SEM revealed a multilayered *S. gordonii* biofilm on all titanium surfaces, with more pronounced colonization of the control and plastic curette groups and more titanium visible on the implantoplasty group (Figure 4).

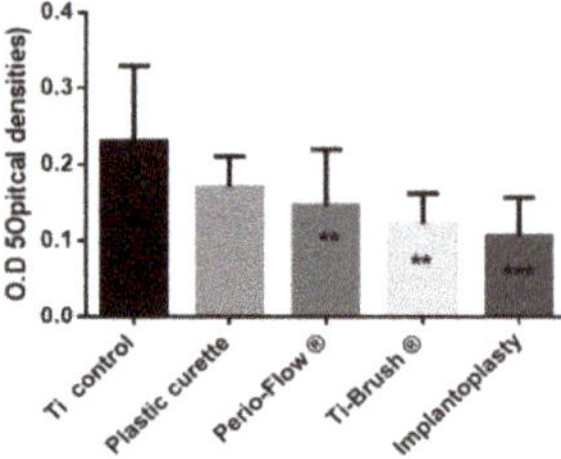

Figure 3. *Streptococcus Gordonii* biofilm recolonisation on treated disks. Results expressed in optical densities (O.D) after Crystal violet staining. Ti control vs. Plastic curette ns $p > 0.05$, Ti control vs. Perio-Flow ® ** $p < 0.01$, Ti control vs. Ti-Brush ® *** $p < 0.001$, Ti control vs. Implantoplasty * $p < 0.05$, Plastic curette vs. Implantoplasty *** $p < 0.001$, Perio-Flow ® vs. Ti-Brush ® $p > 0.05$, Perio-Flow ® vs. Implantoplasty ns $p > 0.05$, Ti-Brush ® vs. Implantoplasty ns $p > 0.05$ (*Kruskal Wallis, Dunn's*).

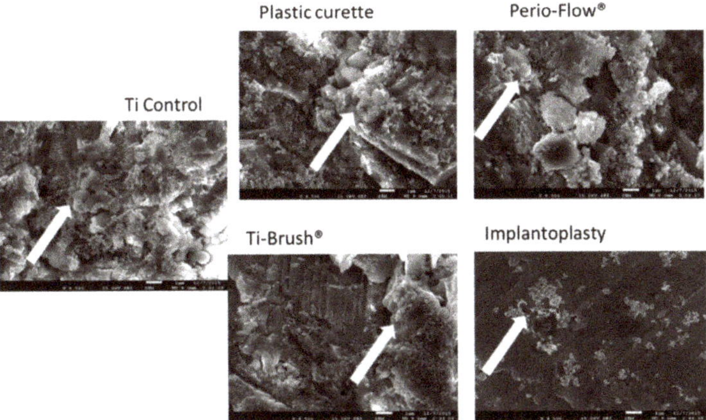

Figure 4. Scanning electron micrographs of Streptococcus Gordonii biofilm colonization (48 h) of previously treated titanium disks (8500 ×). A multilayered *S. gordonii* biofilm was present on all titanium surfaces, with a more pronounced colonization of the control and plastic curette groups. Few bacteria were visible on the implantoplasty-treated disks, allowing the smooth titanium surface to be observed.

3.3. Surface Roughness

All descriptive data for Ra and Rz is presented in Table 1. The surface of the implantoplasty group was smoother (Ra < 1 μm) than that of the control group ($p < 0.01$, ANOVA, Tukey). There was no statistically significant difference between the control, plastic curette, Perio-Flow, and TiBrush groups ($p > 0.05$, ANOVA, Tukey).

Table 1. The surface roughness of the control group and the four treatment groups were measured using a contact profilometer (DektakXT, Bruker, Billerica, MA, USA). Ra and Rz are expressed in μm. Three measurements per disks (in duplicate), repeated twice, were performed in parallel directions (Anova, Tukeys, ns $p > 0.05$).

Treatment	Ra (Mean ± SD)	Rz (Mean ± SD)	p Value
Control group	1.65 ± 0.107	9.79 ± 0.34	ns
Plastic curette	1.61 ± 0.17	9.98 ± 0.24	ns
Perio-Flow®	1.31 ± 0.14	9.22 ± 0.27	ns
Ti-Brush®	1.22 ± 0.31	8.76 ± 0.15	ns
Implantoplasty	0.98 ± 0.12	7.26 ± 0.022	** $p < 0.01$

4. Discussion

The aim of the present study was to compare the efficacy of four mechanical methods—plastic curette, an air abrasive device (Perio-Flow®), TiBrush®, and implantoplasty—in decontaminating and preventing the recolonization of titanium.

Within the limitations of this in vitro study, the plastic curette was the least effective method for removing *S. gordonii*. In contrast, the Perio-Flow®, TiBrush®, and implantoplasty methods eliminated significantly more of the biofilm. This study also showed that biofilm growth was significantly lower on the disks treated using Perio-Flow®, TiBrush®, and implantoplasty than on the untreated disks and those treated using the plastic curette.

If preservation of surface integrity is the primary therapeutic objective, the plastic curette may be preferable in the case of mucositis; nevertheless, its ability to effectively remove calculus and biofilm from smooth and rough surfaces has been widely questioned [22]. Previous studies showed no marked alterations of smooth or rough surfaces after treatment with plastic curettes [22–24]. The results obtained in the present study are consistent with this and suggest that this technique should not be considered as a treatment for peri-implantitis.

Air-powder abrasive systems were introduced to decontaminate implant surfaces following the failure of non-metallic instruments. Glycine powder, which does not markedly change implant surfaces or damage fibroblast attachment, is recommended for the treatment of rough implant surfaces [12,22]. Light titanium surface changes, such as rounding of the angles and edges of rough surfaces and occasional surface pitting, are visible using SEM after treatment with the air-powder system [25]. In contrast, no quantifiable change in roughness is detected on smooth or rough surfaces using a profilometer [26], which is in agreement with present results. It is important to note that surface changes may vary according to titanium hardness, duration of exposure, air pressure, size and hardness of the abrasive particles, and distance and angulation of the tip [27]. According to Menini et al., air polishing using either glycine or sodium bicarbonate powder for 5 or 20 s of machined and acid-etched titanium surfaces does not damage surface morphology [28].

In this study, TiBrush® treatment produced no significant changes in Ra on the titanium surface, although slight changes were observed in the SEM images (Figure 2) as previously demonstrated [21]. Ra averages all peaks and valleys of the roughness profile and thus could be considered too general. Nevertheless, it remains one of the most widely used parameters of roughness and is considered a good general indicator of the surface texture [29,30]. In accordance with previous results, TiBrush® seemed to be an effective instrument for mechanical cleansing while being gentle to the implant surface [17]. Combining this technique with a non-mechanical treatment such as photodynamic therapy increases its effectiveness on smooth and SLA surfaces [31].

Implantoplasty has been increasingly studied over the past decade [18,32–34] and now appears to be a viable alternative treatment for controlling peri-implantitis. The aim of this method is to polish the implant surface, then to decontaminate the implant surface, and finally to create a new surface less prone to plaque and bacterial adhesion [35]. The data obtained in the present study confirm that this method reduces surface roughness (Ra). However, the Ra we obtained is high compared to the results from a previous study [35]. This could be explained by the fact that we used only one diamond bur in this in vitro study. It seems that a combination of different burs associated with a polishing stone is recommended in order to reduce the Ra value considerably [35]. A recent study validated this technique for the management of exposed SLA implant surfaces regarding biocompatibility [36].

Instruments used to decontaminate implants should also attempt to reduce *de novo* bacterial recolonization and biofilm formation, as previously described by Duarte et al. [26]. In the present study, titanium disks treated with Perio-Flow®, TiBrush®, and implantoplasty demonstrated less *S. gordonii* adhesion than untreated control surfaces and plastic curette-treated surfaces, probably due to the difference of surface profiles observed using SEM (Figure 4). Indeed, the texture produced by TiBrush® and implantoplasty is characterized by flattened edges and a smoother surface (Figure 4). The low bacterial adhesion on the surfaces treated with the air-powder abrasive system could be explained by the presence of deposits of glycine powder. Moreover, our results suggest that the type of instrument plays an important role in *de novo* biofilm formation on rough surfaces.

Surface roughness is considered one of the most important factors influencing oral biofilm formation [37]. Rougher surfaces are conducive to bacterial adhesion [37,38], plaque formation, and plaque adherence [39–41]. According to Quirynen et al., roughness provides a large surface area and niches for microbial adhesion, reduces shear forces, and reduces bacterial desorption during initial adhesion [42]. Free surface energy determines biofilm development and plaque formation. The contact angle of each material influences surface energy. In the present study, surface roughness appeared to be associated with low bacterial adhesion. In previously published data, implantoplasty-treated titanium disks appeared smoother than control disks [21]. The difference in bacterial elimination and adhesion between the plastic curette and implantoplasty groups could be explained by this change in surface topography.

Furthermore, recent data confirmed the hydrophilic character of the implantoplasty-treated titanium disks [21]. Bacterial adhesion is promoted on hydrophobic surfaces via proteins acting as specific binding sites for bacteria. Bacterial adhesion is thus precipitate and facilitate [43]. Hydrophilization of surfaces has been shown to inhibit biofilm development [44] and could explain the low adhesion of bacteria on implantoplasty-treated disks. This method, performed during surgical debridement, could become the preferred treatment for peri-implantitis [45]. The modification of wettability observed in the implantoplasty group, in association with its smooth aspect and its particular chemical composition, may represent an advantage in terms of biocompatibility [21].

One limitation of this study is the fact that the oral cavity environment cannot be identically reproduced due to its mixed microbiota, shearing forces, and the antimicrobial effects of the saliva [46]. Biofilms formed in vivo and in vitro are not easily comparable [46]. In vivo, questions remain on the role of the salivary acquired pellicle on bacterial adhesion [47]. Another limitation is the fact that we investigated the effect of treatments on the decontamination and adhesion of only one bacterial species (*S. gordonii*). Although they are associated with healthy implant sites, streptococci—particularly *S. gordonii*—are considered early colonizers and accessory pathogens that facilitate the attachment of organisms normally incapable of binding to host surfaces and, therefore, can lead to biofilm development [48]. Many secondary colonizers, adherent to early colonizers previously present in the biofilm, are known to be implicated in peri-implant diseases like *Fusobacterium*, *Capnocytophaga*, *Porphyromonas*, and *Prevotella spp.* [49]. The use of this monospecies *S. gordonii* biofilm allowed us to reproduce early surface contamination. Furthermore, *Streptococcus gordonii* cells are known to be hydrophobic. Adhesion of microorganisms on different types of surfaces can be influenced by their hydrophobicity [49]. Hydrophobic surfaces attract more hydrophobic cells, while hydrophilic surfaces are more attractive for hydrophilic cells [50,51]. Since implants are made of hydrophobic materials, hydrophobic microorganisms easily adhere to them. The use of another bacterial species may have led to different results. Furthermore, a longer culture time (> 48 h) (more similar to what happens in vivo) might lead to more pronounced differences between groups. Roughness could emerge as a disadvantage as it renders the surface more prone to bacterial adhesion [37].

A decrease in O.D values can be observed between decontamination and recolonization (Figures 1 and 3) in the control group and plastic curette group, explaining this difference.

In this study, titanium disks were used instead of entire dental implants. The macrostructure of entire dental implants differs considerably from that of titanium disks, but the disks have been used in several previous in vitro studies to test experimental conditions [52,53]. The configuration of the peri-implant lesion, as well as implant surface accessibility, should also be considered prior to treating peri-implantitis. Whenever possible, pre-operative removal of the superstructure is recommended to ensure accessibility. Because of these morphological particularities, large rotary instruments, such as an implantoplasty diamond bur, might not be appropriate in some clinical situations.

5. Conclusions

This study showed that Perio-Flow®, TiBrush®, and implantoplasty protocols were more efficient than the plastic curette to remove in vitro *S. gordonii* biofilm from titanium disks. The effects of implantoplasty on the surface properties of the disks highlight it as a promising treatment for peri-implantitis. Further ex vivo microbiological studies should be performed to confirm these results.

Author Contributions: Conceptualization, S.T. and J.F.L.; Methodology, S.T.; Validation, S.T., C.B., and M.C.B.; Formal Analysis, S.T.; Investigation, S.T.; Writing—Original Draft Preparation, S.T.; Writing—Review and Editing, S.T., C.B., M.C.B., and J.F.L.; Supervision, J.F.L.

Funding: This research received no external funding.

Acknowledgments: The authors would like to thank Southern Implants for providing the titanium disks. The authors thank Delphine Magnin and Colette Douchamps for their assistance in SEM, and Estelle Marichal, Ana Dos Santos for her assistance in biofilm culture and crystal violet technique.

Conflicts of Interest: The authors declare no conflict of interest.

References

1. Berglundh, T.; Persson, L.; Klinge, B. A systematic review of the incidence of biological and technical complications in implant dentistry reported in prospective longitudinal studies of at least 5 years. *J. Clin. Periodontol.* **2002**, *29*, 197–212. [CrossRef]
2. Mombelli, A.; Muller, N.; Cionca, N. The epidemiology of peri-implantitis. *Clin. Oral Implants Res.* **2012**, *23*, 67–76. [CrossRef]
3. Roos-Jansaker, A.M.; Lindahl, C.; Renvert, H.; Renvert, S. Nine- to fourteen-year follow-up of implant treatment. Part I: Implant loss and associations to various factors. *J. Clin. Periodontol.* **2006**, *33*, 283–289. [CrossRef] [PubMed]
4. Marrone, A.; Lasserre, J.; Bercy, P.; Brecx, M.C. Prevalence and risk factors for peri-implant disease in Belgian adults. *Clin. Oral Implants Res.* **2013**, *24*, 934–940. [CrossRef] [PubMed]
5. Mombelli, A.; van Oosten, M.A.; Schurch, E., Jr.; Land, N.P. The microbiota associated with successful or failing osseointegrated titanium implants. *Oral Microbiol. Immunol.* **1987**, *2*, 145–151. [CrossRef] [PubMed]
6. Zitzmann, N.U.; Berglundh, T. Definition and prevalence of peri-implant diseases. *J. Clin. Periodontol.* **2008**, *35*, 286–291. [CrossRef] [PubMed]
7. Heitz-Mayfield, L.J.; Lang, N.P. Comparative biology of chronic and aggressive periodontitis vs. peri-implantitis. *Periodontology* **2010**, *53*, 167–181. [CrossRef] [PubMed]
8. Roccuzzo, M.; Bonino, F.; Aglietta, M.; Dalmasso, P. Ten-year results of a three arms prospective cohort study on implants in periodontally compromised patients. Part 2: Clinical results. *Clin. Oral Implants Res.* **2012**, *23*, 389–395. [CrossRef] [PubMed]
9. Roccuzzo, M.; De Angelis, N.; Bonino, L.; Aglietta, M. Ten-year results of a three-arm prospective cohort study on implants in periodontally compromised patients. Part 1: Implant loss and radiographic bone loss. *Clin. Oral Implants Res.* **2010**, *21*, 490–496. [CrossRef] [PubMed]
10. Renvert, S.; Quirynen, M. Risk indicators for peri-implantitis. A narrative review. *Clin. Oral Implants Res.* **2015**, *26*, 15–44. [CrossRef] [PubMed]
11. Louropoulou, A.; Slot, D.E.; Van der Weijden, F. The effects of mechanical instruments on contaminated titanium dental implant surfaces: A systematic review. *Clin. Oral Implants Res.* **2014**, *25*, 1149–1160. [CrossRef] [PubMed]
12. Louropoulou, A.; Slot, D.E.; Van der Weijden, F.A. Titanium surface alterations following the use of different mechanical instruments: A systematic review. *Clin. Oral Implants Res.* **2012**, *23*, 643–658. [CrossRef] [PubMed]
13. Esposito, M.; Grusovin, M.G.; Worthington, H.V. Interventions for replacing missing teeth: Treatment of peri-implantitis. *Cochrane Database Syst. Rev.* **2012**, *1*, Cd004970. [CrossRef] [PubMed]
14. Renvert, S.; Polyzois, I.; Claffey, N. Surgical therapy for the control of peri-implantitis. *Clin. Oral Implants Res.* **2012**, *23*, 84–94. [CrossRef] [PubMed]
15. Tastepe, C.S.; van Waas, R.; Liu, Y.; Wismeijer, D. Air powder abrasive treatment as an implant surface cleaning method: A literature review. *Int. J. Oral Maxillofac. Implants* **2012**, *27*, 1461–1473. [PubMed]
16. Sahrmann, P.; Ronay, V.; Hofer, D.; Attin, T.; Jung, R.E.; Schmidlin, P.R. In vitro cleaning potential of three different implant debridement methods. *Clin. Oral Implants Res.* **2015**, *26*, 314–319. [CrossRef] [PubMed]
17. John, G.; Becker, J.; Schwarz, F. Rotating titanium brush for plaque removal from rough titanium surfaces—An in vitro study. *Clin. Oral Implants Res.* **2014**, *25*, 838–842. [CrossRef]
18. Romeo, E.; Ghisolfi, M.; Murgolo, N.; Chiapasco, M.; Lops, D.; Vogel, G. Therapy of peri-implantitis with resective surgery. A 3-year clinical trial on rough screw-shaped oral implants. Part I: Clinical outcome. *Clin. Oral Implants Res.* **2005**, *16*, 9–18. [CrossRef]
19. Geng, H.; Yuan, Y.; Adayi, A.; Zhang, X.; Song, X.; Gong, L.; Zhang, X.; Gao, P. Engineered chimeric peptides with antimicrobial and titanium-binding functions to inhibit biofilm formation on Ti implants. *Mater. Sci. Eng. C Mater. Biol. Appl.* **2018**, *82*, 141–154. [CrossRef]
20. Ota-Tsuzuki, C.; Martins, F.L.; Giorgetti, A.P.; de Freitas, P.M.; Duarte, P.M. In vitro adhesion of Streptococcus sanguinis to dentine root surface after treatment with Er:YAG laser, ultrasonic system, or manual curette. *Photomed. Laser Surg.* **2009**, *27*, 735–741. [CrossRef]

21. Toma, S.; Lasserre, J.; Brecx, M.C.; Nyssen-Behets, C. In vitro evaluation of peri-implantitis treatment modalities on Saos-2osteoblasts. *Clin. Oral Implants Res.* **2016**, *27*, 1085–1092. [CrossRef] [PubMed]

22. Augthun, M.; Tinschert, J.; Huber, A. In vitro studies on the effect of cleaning methods on different implant surfaces. *J. Periodontol.* **1998**, *69*, 857–864. [CrossRef] [PubMed]

23. Mengel, R.; Buns, C.E.; Mengel, C.; Flores-de-Jacoby, L. An in vitro study of the treatment of implant surfaces with different instruments. *Int. J. Oral Maxillofac. Implants* **1998**, *13*, 91–96. [PubMed]

24. Ruhling, A.; Kocher, T.; Kreusch, J.; Plagmann, H.C. Treatment of subgingival implant surfaces with Teflon-coated sonic and ultrasonic scaler tips and various implant curettes. An in vitro study. *Clin. Oral Implants Res.* **1994**, *5*, 19–29. [CrossRef] [PubMed]

25. Parham, P.L., Jr.; Cobb, C.M.; French, A.A.; Love, J.W.; Drisko, C.L.; Killoy, W.J. Effects of an air-powder abrasive system on plasma-sprayed titanium implant surfaces: An in vitro evaluation. *J. Oral Implantol.* **1989**, *15*, 78–86. [PubMed]

26. Duarte, P.M.; Reis, A.F.; de Freitas, P.M.; Ota-Tsuzuki, C. Bacterial adhesion on smooth and rough titanium surfaces after treatment with different instruments. *J. Periodontol.* **2009**, *80*, 1824–1832. [CrossRef] [PubMed]

27. Chairay, J.P.; Boulekbache, H.; Jean, A.; Soyer, A.; Bouchard, P. Scanning electron microscopic evaluation of the effects of an air-abrasive system on dental implants: A comparative in vitro study between machined and plasma-sprayed titanium surfaces. *J. Periodontol.* **1997**, *68*, 1215–1222. [CrossRef]

28. Menini, M.; Piccardo, P.; Baldi, D.; Dellepiane, E.; Pera, P. Morphological and chemical characteristics of different titanium surfaces treated by bicarbonate and glycine powder air abrasive systems. *Implant Dent.* **2015**, *24*, 47–56. [CrossRef]

29. Park, J.B.; Jang, Y.J.; Koh, M.; Choi, B.K.; Kim, K.K.; Ko, Y. In vitro analysis of the efficacy of ultrasonic scalers and a toothbrush for removing bacteria from resorbable blast material titanium disks. *J. Periodontol.* **2013**, *84*, 1191–1198. [CrossRef]

30. Park, J.B.; Kim, N.; Ko, Y. Effects of ultrasonic scaler tips and toothbrush on titanium disc surfaces evaluated with confocal microscopy. *J. Craniofac. Surg.* **2012**, *23*, 1552–1558. [CrossRef]

31. Widodo, A.; Spratt, D.; Sousa, V.; Petrie, A.; Donos, N. An in vitro study on disinfection of titanium surfaces. *Clin. Oral Implants Res.* **2016**, *27*, 1227–1232. [CrossRef] [PubMed]

32. Matarasso, S.; Iorio Siciliano, V.; Aglietta, M.; Andreuccetti, G.; Salvi, G.E. Clinical and radiographic outcomes of a combined resective and regenerative approach in the treatment of peri-implantitis: A prospective case series. *Clin. Oral Implants Res.* **2014**, *25*, 761–767. [CrossRef] [PubMed]

33. Romeo, E.; Lops, D.; Storelli, S.; Ghisolfi, M. Clinical peri-implant sounding accuracy in the presence of chronic inflammation of peri-implant tissues. Clinical observation study. *Miner. Stomatol.* **2009**, *58*, 81–91.

34. Schwarz, F.; John, G.; Schmucker, A.; Sahm, N.; Becker, J. Combined surgical therapy of advanced peri-implantitis evaluating two methods of surface decontamination: A 7-year follow-up observation. *J. Clin. Periodontol.* **2017**, *44*, 337–342. [CrossRef]

35. Ramel, C.F.; Lussi, A.; Ozcan, M.; Jung, R.E.; Hammerle, C.H.; Thoma, D.S. Surface roughness of dental implants and treatment time using six different implantoplasty procedures. *Clin. Oral Implants Res.* **2016**, *27*, 776–781. [CrossRef]

36. Schwarz, F.; John, G.; Becker, J. The influence of implantoplasty on the diameter, chemical surface composition, and biocompatibility of titanium implants. *Clin. Oral Investig.* **2017**, *21*, 2355–2361. [CrossRef] [PubMed]

37. Teughels, W.; Van Assche, N.; Sliepen, I.; Quirynen, M. Effect of material characteristics and/or surface topography on biofilm development. *Clin. Oral Implants Res.* **2006**, *17*, 68–81. [CrossRef]

38. Subramani, K.; Wismeijer, D. Decontamination of titanium implant surface and re-osseointegration to treat peri-implantitis: A literature review. *Int. J. Oral Maxillofac. Implants* **2012**, *27*, 1043–1054.

39. Busscher, H.J.; van der Mei, H.C. Physico-chemical interactions in initial microbial adhesion and relevance for biofilm formation. *Adv. Dent. Res.* **1997**, *11*, 24–32. [CrossRef]

40. Carlen, A.; Nikdel, K.; Wennerberg, A.; Holmberg, K.; Olsson, J. Surface characteristics and in vitro biofilm formation on glass ionomer and composite resin. *Biomaterials* **2001**, *22*, 481–487. [CrossRef]

41. Rimondini, L.; Fare, S.; Brambilla, E.; Felloni, A.; Consonni, C.; Brossa, F.; Carrassi, A. The effect of surface roughness on early in vivo plaque colonization on titanium. *J. Periodontol.* **1997**, *68*, 556–562. [CrossRef]

42. Quirynen, M.; Bollen, C.M. The influence of surface roughness and surface-free energy on supra- and subgingival plaque formation in man. A review of the literature. *J. Clin. Periodontol.* **1995**, *22*, 1–14. [CrossRef]

43. Ge, J.; Catt, D.M.; Gregory, R.L. Streptococcus mutans surface alpha-enolase binds salivary mucin MG2 and human plasminogen. *Infect. Immun.* **2004**, *72*, 6748–6752. [CrossRef]

44. Okada, A.; Nikaido, T.; Ikeda, M.; Okada, K.; Yamauchi, J.; Foxton, R.M.; Sawada, H.; Tagami, J.; Matin, K. Inhibition of biofilm formation using newly developed coating materials with self-cleaning properties. *Dent. Mater. J.* **2008**, *27*, 565–572. [CrossRef]

45. Romeo, E.; Lops, D.; Chiapasco, M.; Ghisolfi, M.; Vogel, G. Therapy of peri-implantitis with resective surgery. A 3-year clinical trial on rough screw-shaped oral implants. Part II: Radiographic outcome. *Clin. Oral Implants Res.* **2007**, *18*, 179–187. [CrossRef]

46. Al-Ahmad, A.; Wunder, A.; Auschill, T.M.; Follo, M.; Braun, G.; Hellwig, E.; Arweiler, N.B. The in vivo dynamics of Streptococcus spp., Actinomyces naeslundii, Fusobacterium nucleatum and Veillonella spp. in dental plaque biofilm as analysed by five-colour multiplex fluorescence in situ hybridization. *J. Med. Microbiol.* **2007**, *56*, 681–687. [CrossRef]

47. Hannig, C.; Hannig, M. The oral cavity—A key system to understand substratum-dependent bioadhesion on solid surfaces in man. *Clin. Oral Investig.* **2009**, *13*, 123–139. [CrossRef]

48. Kolenbrander, P.E.; London, J. Adhere today, here tomorrow: Oral bacterial adherence. *J. Bacteriol.* **1993**, *175*, 3247–3252. [CrossRef]

49. Van Loosdrecht, M.C.; Norde, W.; Zehnder, A.J. Physical chemical description of bacterial adhesion. *J. Biomater. Appl.* **1990**, *5*, 91–106. [CrossRef]

50. Krasowska, A.; Sigler, K. How microorganisms use hydrophobicity and what does this mean for human needs? *Front. Cell. Infect. Microbiol.* **2014**, *4*, 112. [CrossRef]

51. Giaouris, E.; Chapot-Chartier, M.P.; Briandet, R. Surface physicochemical analysis of natural Lactococcus lactis strains reveals the existence of hydrophobic and low charged strains with altered adhesive properties. *Int. J. Food Microbiol.* **2009**, *131*, 2–9. [CrossRef]

52. Conserva, E.; Generali, L.; Bandieri, A.; Cavani, F.; Borghi, F.; Consolo, U. Plaque accumulation on titanium disks with different surface treatments: An in vivo investigation. *Odontology* **2018**, *106*, 145–153. [CrossRef]

53. Canullo, L.; Genova, T.; Wang, H.L.; Carossa, S.; Mussano, F. Plasma of Argon Increases Cell Attachment and Bacterial Decontamination on Different Implant Surfaces. *Int. J. Oral Maxillofac. Implants* **2017**, *32*, 1315–1323. [CrossRef]

 *materials*

Article

Physical and Histological Comparison of Hydroxyapatite, Carbonate Apatite, and β-Tricalcium Phosphate Bone Substitutes

Kunio Ishikawa [1],*, Youji Miyamoto [2], Akira Tsuchiya [1], Koichiro Hayashi [1], Kanji Tsuru [1,3] and Go Ohe [2]

[1] Department of Biomaterials, Faculty of Dental Sciences, Kyushu University, 3-1-1 Maidashi, Higashi-ku, Fukuoka 812-8582, Japan; tsuchiya@dent.kyushu-u.ac.jp (A.T.); khayashi@dent.kyushu-u.ac.jp (K.H.); tsuruk@college.fdcnet.ac.jp (K.T.)

[2] Department of Oral Surgery, Institute of Biomedical Sciences, Tokushima University Graduate School, 3-18-15 Kuramotocho, Tokushima 770-8504, Japan; miyamoto@tokushima-u.ac.jp (Y.M.); go.ohe@tokushima-u.ac.jp (G.O.)

[3] Section of Bioengineering, Department of Dental Engineering, Fukuoka Dental College, Fukuoka 814-0193, Japan

* Correspondence: ishikawa@dent.kyushu-u.ac.jp; Tel.: +81-92-642-6344

Received: 28 September 2018; Accepted: 11 October 2018; Published: 16 October 2018

Abstract: Three commercially available artificial bone substitutes with different compositions, hydroxyapatite (HAp; Neobone®), carbonate apatite (CO_3Ap; Cytrans®), and β-tricalcium phosphate (β-TCP; Cerasorb®), were compared with respect to their physical properties and tissue response to bone, using hybrid dogs. Both Neobone® (HAp) and Cerasorb® (β-TCP) were porous, whereas Cytrans® (CO_3Ap) was dense. Crystallite size and specific surface area (SSA) of Neobone® (HAp), Cytrans® (CO_3Ap), and Cerasorb® (β-TCP) were 75.4 ± 0.9 nm, 30.8 ± 0.8 nm, and 78.5 ± 7.5 nm, and 0.06 m^2/g, 18.2 m^2/g, and 1.0 m^2/g, respectively. These values are consistent with the fact that both Neobone® (HAp) and Cerasorb® (β-TCP) are sintered ceramics, whereas Cytrans® (CO_3Ap) is fabricated in aqueous solution. Dissolution in pH 5.3 solution mimicking Howship's lacunae was fastest in CO_3Ap (Cytrans®), whereas dissolution in pH 7.3 physiological solution was fastest in β-TCP (Cerasorb®). These results indicated that CO_3Ap is stable under physiological conditions and is resorbed at Howship's lacunae. Histological evaluation using hybrid dog mandible bone defect model revealed that new bone was formed from existing bone to the center of the bone defect when reconstructed with CO_3Ap (Cytrans®) at week 4. The amount of bone increased at week 12, and resorption of the CO_3Ap (Cytrans®) was confirmed. β-TCP (Cerasorb®) showed limited bone formation at week 4. However, a larger amount of bone was observed at week 12. Among these three bone substitutes, CO_3Ap (Cytrans®) demonstrated the highest level of new bone formation. These results indicate the possibility that bone substitutes with compositions similar to that of bone may have properties similar to those of bone.

Keywords: carbonate apatite; hydroxyapatite; β-tricalcium phosphate; artificial bone substitute; crystallite size; dissolution rate; hybrid dog

---

1. Introduction

Bone apatite is known as carbonate apatite [CO_3Ap: $Ca_{10-a}(CO_3)_b(PO_4)_{6-c}$], which contains 6–9 wt % carbonate in an apatitic structure. However, sintered hydroxyapatite [HAp; $Ca_{10}(PO_4)_6(OH)_2$] and sintered β-tricalcium phosphate [β-TCP: $Ca_3(PO_4)_2$] have been used in clinics as typical synthetic bone substitutes, while CO_3Ap has not been used [1–11].

The lack of chemically pure artificial CO_3Ap bone substitute is due to the thermal decomposition status of CO_3Ap powder. Preparation of chemically pure CO_3Ap powder is possible, however, CO_3Ap powder begins to decompose at around 400 °C. Thus, CO_3Ap blocks or granules cannot be fabricated by sintering without thermal decomposition or the release of CO_2. This led to the invention of HAp and β-TCP, which do not contain carbonate and, thus, can be sintered.

Sintered HAp demonstrated excellent tissue response and good osteoconductivity. Therefore, production of sintered HAp proves useful for patients. However, osteoconductivity of sintered HAp is poor when compared to that of autografts. Moreover, in contrast to autografts, which are replaced by new bone through a bone remodeling process, sintered HAp would not be replaced by new bone and remains at the bone defect. In contrast to HAp, β-TCP is resorbed at the bone defect and, thus, could be replaced by new bone. However, the dissolution rate of β-TCP is sometimes faster than the rate of bone formation, leading to unsatisfactory bone formation, or a bone defect filled with fibrous tissue.

Autograft still remains the gold standard for reconstruction of bone defects for most of the countries, even though there are serious drawbacks, including intervention at healthy sites to collect autografts. Since autograft exhibits properties superior to those of HAp and β-TCP, it is natural to develop artificial bone substitutes with compositions similar to those of bone or CO_3Ap.

Recently, CO_3Ap block was fabricated in aqueous solution via a dissolution–precipitation reaction, using a precursor block [12–24]. CO_3Ap block upregulated differentiation of osteoblastic cells, even when compared to Hap [25]. Further, CO_3Ap showed higher osteoconductivity than Hap [12]. CO_3Ap block was resorbed by osteoclasts similar to bone. Thus, similar to autografts, CO_3Ap block can be replaced by new bone [12].

Fortunately, chemically pure CO_3Ap granules (Cytrans®) have become commercially available. Since CO_3Ap block is not fabricated via sintering but through a dissolution–precipitation reaction in aqueous solution using a calcite block, the physical and histological behaviors of CO_3Ap may be expected to be different from those of sintered HAp and sintered β-TCP. However, no comparison has been made between the commercially available bone substitutes.

In the present study, physical and histological behaviors of artificial bone substitutes were compared using three commercially available artificial bone substitutes, Neobone® (HAp), Cytrans® (CO_3Ap), and Cerasorb® (β-TCP).

2. Materials and Methods

2.1. Artificial Bone Substitutes

Neobone® (HAp, CoorsTec, Tokyo, Japan), Cytrans® (CO_3Ap, GC, Tokyo, Japan), and Cerasorb® (β-TCP, Hakuho, Tokyo, Japan), three artificial bone substitutes commercially available in Japan, were chosen for this study.

2.2. Scanning Electron Microscopic Observation

Surface morphology of the three artificial bone substitutes were observed using a scanning electron microscope (SEM: S-3400N; Hitachi High Technologies Co., Tokyo, Japan) under an accelerating voltage of 15 kV, after being coated with gold–palladium. Coating was performed using a magnetron sputtering machine (MSP-1S: Vacuum Device Co., Ibaraki, Japan).

2.3. Compositional Analysis

Compositions of the three artificial bone substitutes were analyzed using a powder X-ray diffractometer, a Fourier transform infrared (FT-IR) spectrometer, and elemental analysis.

For X-ray diffraction (XRD) analysis, the samples were ground to fine powder, and the XRD patterns were recorded using a diffractometer (D8 Advance, Bruker AXS GmbH, Karlsruhe, Germany), generating CuKα radiation at 40 kV and 40 mA. Samples were scanned from 2θ of 10° to 40° (where θ is the Bragg angle) in a continuous mode. Crystallite size was also calculated from the XRD pattern, using Scherrer's equation.

For FT-IR analysis, FT-IR spectra were measured using a KBr disc method, with a spectrometer (SPECTRUM 2000LX; Perkin Elmer Co. Ltd., Kanagawa, Japan).

Carbonate content was measured using elemental analysis or a CHN coder (MT-6; Yanako Analytical Instruments, Kyoto, Japan).

2.4. Specific Surface Area Measurement

Specific surface areas were evaluated with a multiple Brunauer–Emmett–Teller (BET) method which evaluates nitrogen adsorption–desorption isotherms at 77 K using a BELSORP-mini II (MicrotracBEL, Osaka, Japan).

2.5. Porosity Measurement

Porosity of the bone substitutes was evaluated using the formula

$$\text{Porosity } (\%) = \frac{V_p}{V_p + V_s} \times 100, \tag{1}$$

where V_p and V_s are the pore volume and the volume of the specimens, respectively. The pore volume of the bone substitutes was measured using a specific surface area analyzer. The volume of the specimens was measured by the pycnometer method, where ethanol was used as the immersion liquid.

2.6. Dissolution Behavior

Dissolution behavior of the three artificial bone substitutes were measured using a method defined in JIS T0330-3; Bioceramics—Part 3: Testing method of measuring dissolution rate of calcium phosphate ceramics. In short, the dissolution rate of calcium phosphate ceramics was measured by immersing in 0.08 mol/L pH 5.50 acetic acid–sodium acetate solution, and 0.05 mol/L pH 7.30 tris(hydroxymethyl)aminomethane–HCl buffer solution at 25 ± 3 °C. Sample (100 ± 1 mg) was lifted in the solution using a thread with continuous stirring, and the Ca concentration in the solution was measured using a Ca ion electrode (n = 5).

According to the testing method, a Ca ion-selective combination electrode (6583-10C, Horiba, Ltd., Kyoto, Japan) connected to a pH/mV-meter (Type F-72, Horiba, Ltd., Kyoto, Japan) was employed in this study.

2.7. Animal Experiments

Two adult male hybrid dogs (weight; 3.0–3.5 kg) were purchased from Kitayama Rabesu (Nagano, Japan). All hybrid dogs were housed in the Animal Unit of Hamri Co., Ltd. (Ibaragi, Japan) Animal protocols were reviewed and approved by the Hamri Animal Studies Committee (13-H065). The animals maintained in individual cages were provided with water and animal specific pellet-type laboratory animal food. After premedication with diethyl ether by inhalation, the animals were anesthetized with ketamine (10 mg/kg) and xylazine (3 mg/kg) via intravenous injection. Three months following the extraction of premolars and molars, a bone defect of 3.6 mm in diameter and 8 mm in depth was made in alveolar bone using an Aadva twist drill, φ3.6 mm (GC). The defect was reconstructed with Neobone® (HAp), Cerasorb® (β-TCP), and Cytrans® (CO_3Ap). The animals were sacrificed using intravenous pentobarbital (120 mg/kg), 4 and 12 weeks after sample implantation.

2.8. Histological Analysis

The specimens with surrounding tissue were carefully harvested from the mandibular bone and fixed in 10% buffered formalin for 3 days, and then dehydrated by immersion in a 70% to 100% ethanol gradient with immersion in each grade for 3 days. Finally, the bone samples were embedded in methyl methacrylate. Then, sections were prepared for pathological analysis using a modified interlocked diamond saw (Exakt, Hamburg, Germany). Then, all sections were stained with Villanueva Goldner and observed using a light microscope (DM6000B, Leica Microsystems, Heerbrugg, Switzerland). The new bone area in the defect was measured using Image J ver1.51 software (US National Institutes of Health, Bethesda, MD, USA).

2.9. Statistical Analysis

For statistical analysis, one-way factorial analysis of variance (ANOVA) and Fisher's least significant difference (LSD) post hoc test were performed using Kaleida Graph 4. Values are expressed as means $\pm$ SD. A $p < 0.05$ value was considered to indicate statistically significant differences.

3. Results

Typical SEM images of the (a–c) Neobone® (HAp), (d–f) Cytrans® (CO_3Ap), and (g–i) Cerasorb® (β-TCP) are summarized in Figure 1.

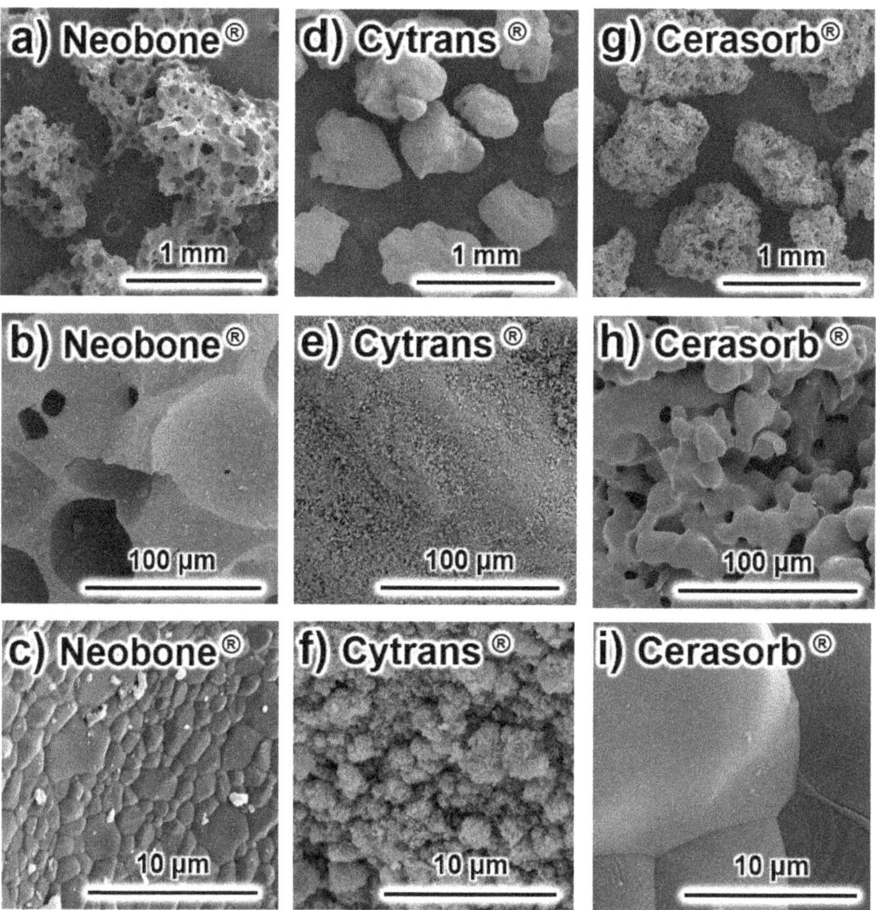

Figure 1. Typical scanning electron microscopy images of Neobone® (HAp) (**a–c**), Cytrans® (CO_3Ap) (**d–f**), and Cerasorb® (β-TCP) (**g–i**).

Both Neobone® and Cerasorb® displayed a porous structure, where Cerasorb® had much smaller pores compared to Neobone®. By contrast, Cytrans® was dense, and had no pores in the granules. Higher magnification showed that both Neobone® and Cerasorb® had smooth surfaces typical for sintering. By contrast, the surface of Cytrans® was rough and consisted of small precipitated crystals.

The XRD patterns of (a) Neobone® (b) Cytrans®, and (c) Cerasorb® are summarized in Figure 2. The XRD pattern of the standard (d) HAp and (e) β-TCP are also shown to facilitate comparison.

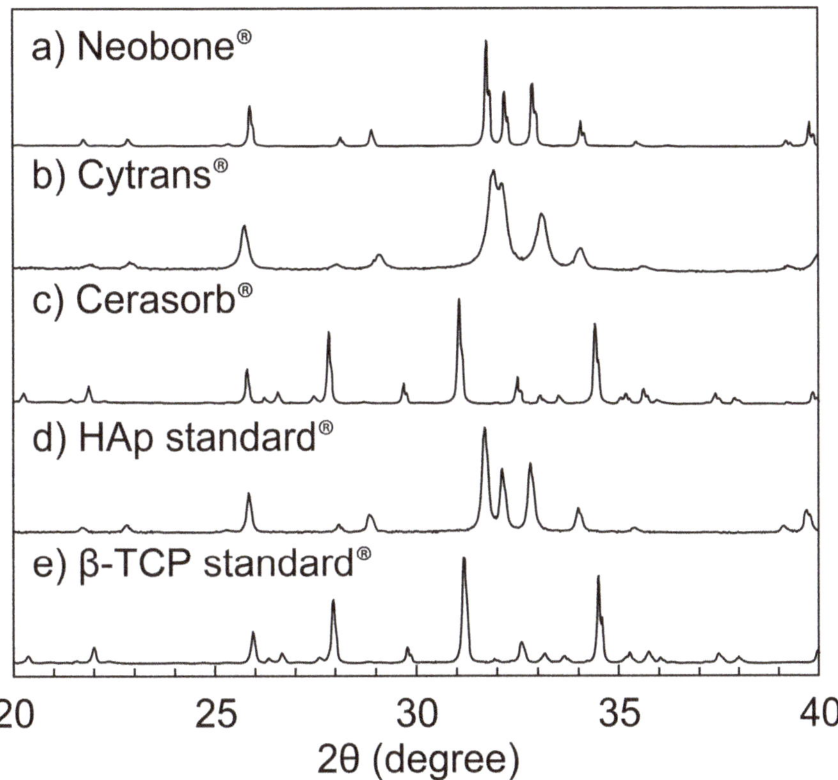

Figure 2. X-ray diffraction patterns of (**a**) Neobone® (HAp), (**b**) Cytrans® (CO$_3$Ap), (**c**) Cerasorb® (β-TCP), (**d**) standard Hap, and (**e**) standard β-TCP.

As shown, both Neobone® and Cytrans® showed typical apatitic patterns with no other peaks, whereas Cerasorb® showed an XRD pattern typical of β-TCP. Peaks of the Neobone® and Cerasorb® were sharp, indicating high crystallinity, whereas those of Cytrans® were broad, indicating low crystallinity.

The FT-IR spectra of (a) Neobone® (b) Cytrans®, and (c) Cerasorb® are summarized in Figure 3. The FT-IR spectra of standard (d) HAp and (e) β-TCP are also shown to facilitate comparison.

Both Neobone® and Cytrans® showed typical apatitic patterns, similar to the XRD pattern. In Cytrans®, an absorption peak typical for CO$_3$ was observed at 1450 cm^{-1}, 878 cm^{-1}, and 871 cm^{-1} [26,27]. No absorption peak assigned to OH was observed at 3573 cm^{-1} (ν_sOH) and 638 cm^{-1} (ν_LOH). By contrast, no absorption peak ascribed to CO$_3$ was observed in Neobone®. Instead, absorption peaks ascribed to OH were observed at 3573 cm^{-1} and 638 cm^{-1}. Cerasorb® showed an FT-IR pattern typical of β-TCP.

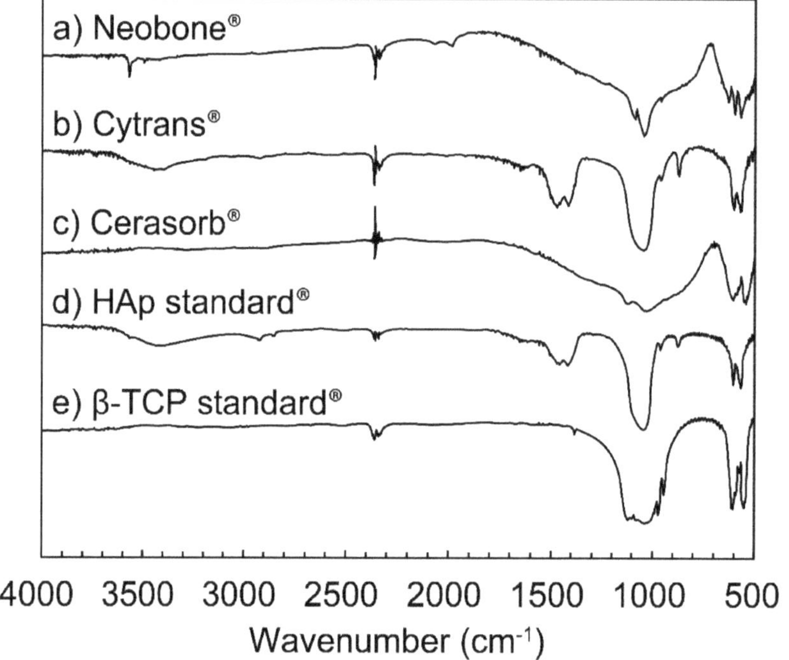

Figure 3. Fourier-transform infrared spectroscopy spectra of (**a**) Neobone® (HAp), (**b**) Cytrans® (CO$_3$Ap), (**c**) Cerasorb® (β-TCP), (**d**) standard Hap, and (**e**) standard β-TCP.

Crystallite sizes were calculated using Scherrer's equation, and specific surface area (SSA), carbonate content, bulk density, and porosity are summarized (Table 1). Crystallite sizes of Neobone® (HAp) and Cerasorb® (β-TCP) were approximately 80 nm, whereas the crystallite size of Cytrans® (CO$_3$Ap) was approximately 30 nm, which was smaller than half of Neobone® and Neobone®. Specific surface area was different by two orders of magnitude. SSA of Neobone® (HAp) was 1.0 m^2/g. On the other hand, SSA of Cerasorb® (β-TCP) was one order of magnitude smaller than Neobone® (HAp) and was 0.06 m^2/g. By contrast, SSA of Cytrans® (CO$_3$Ap) was one order of magnitude larger than Neobone® (HAp) and was 18.2 m^2/g. No carbonate was found in Neobone® (HAp) and Cerasorb® (β-TCP), whereas Cytrans® (CO$_3$Ap) had 11.9 wt % of carbonate content. Cytrans® (CO$_3$Ap) had the largest bulk density and Neobone® (HAp) had the smallest bulk density.

Table 1. Physical properties of Neobone® (HAp), Cytrans® (CO$_3$Ap), and Cerasorb® (β-TCP).

Bone Substitute	Crystallite Size (nm)	SSA (m^3/g)	CO$_3$ Content (%)	Bulk Density (g/cm^3)	Porosity (%)
Neobone® (HAp)	75.4 ± 0.9	1.0	-	0.47 ± 0.02	85.1 ± 0.5
Cytrans® (CO$_3$Ap)	30.8 ± 0.8	18.2	11.9	0.99 ± 0.03	68.7 ± 0.9
Cerasorb® (β-TCP)	78.5 ± 7.5	0.06	-	0.72 ± 0.03	76.4 ± 0.8

As a result of bulk density and similar theoretical density, packing porosity was largest in Neobone® (HAp) followed by Cerasorb® (β-TCP) and Cytrans® (CO$_3$Ap).

The dissolution behaviors of Neobone® (HAp), Cytrans® (CO$_3$Ap), and Cerasorb® (β-TCP) in pH 7.3 CaPO$_4$-free buffer solution, are summarized in Figure 4. In the neutral solution, simulating physiological condition, Cerasorb® (β-TCP) showed the fastest dissolution rate followed by Cytrans® (CO$_3$Ap) and Neobone® (HAp).

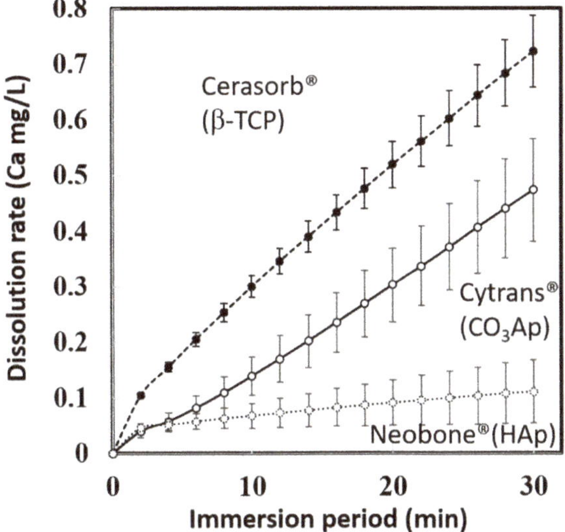

Figure 4. Dissolution behavior of Neobone® (HAp), Cytrans® (CO₃Ap), and Cerasorb® (β-TCP) in CaPO₄-free pH 7.3 buffer solution.

The dissolution behaviors of Neobone® (HAp), Cytrans® (CO₃Ap), and Cerasorb® (β-TCP) in pH 5.5 buffer solution are summarized in Figure 5. In the weak acidic solution, simulating the inside of the ruffle border of osteoclasts, Cytrans® (CO₃Ap) showed the fastest dissolution rate followed by Cerasorb® (β-TCP) and Neobone® (HAp), instead of Cerasorb® (β-TCP), which showed fastest dissolution at pH 7.3.

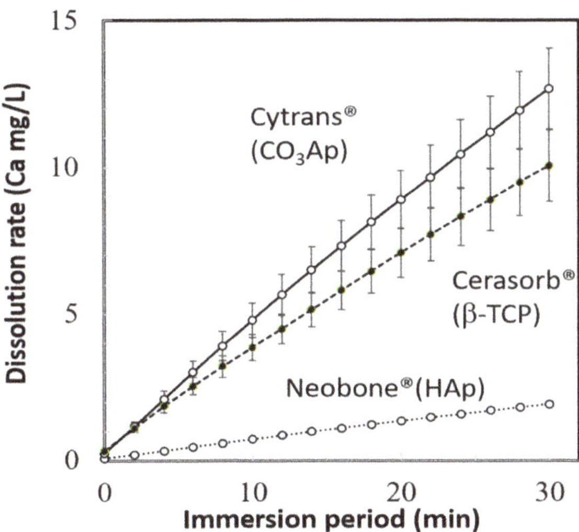

Figure 5. Dissolution behavior of Neobone® (HAp), Cytrans® (CO₃Ap), and Cerasorb® (β-TCP) in pH 5.5 buffer solution.

The Villanueva Goldner-stained histological images of the three artificial bone substitutes, 4 weeks and 12 weeks after implantation, are shown in Figures 6 and 7.

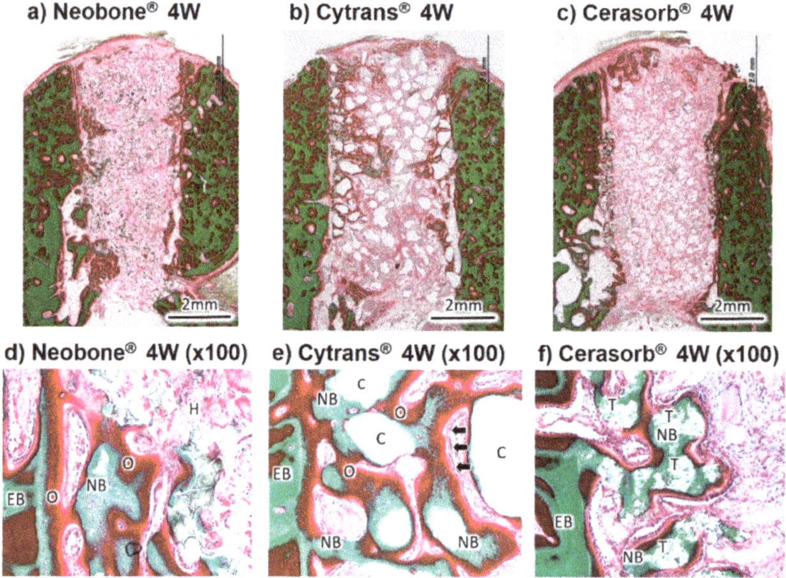

Figure 6. Histological findings of Neobone® (HAp), Cytrans® (CO$_3$Ap), and Cerasorb® (β-TCP) implanted into dog mandibular bone defect at 4 weeks after implantation (Villanueva Goldner staining). Green area, bone; red area, osteoid. Arrows indicate aligned cuboidal osteoblasts around the abundantly formed osteoid on the surface of the new bone. EB, existing bone; NB, new bone; O, osteoid; H, HAp (Neobone®); C, CO$_3$Ap (Cytrans®); T, β-TCP (Cerasorb®).

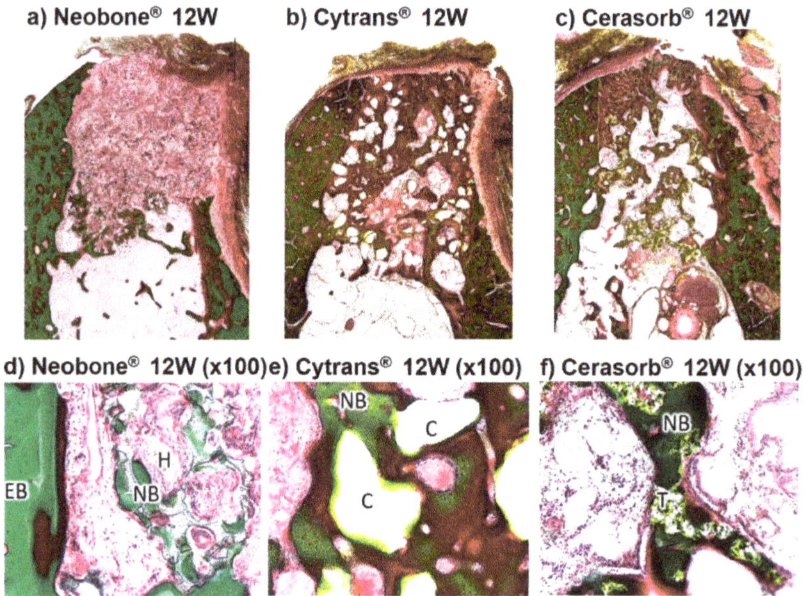

Figure 7. Histological findings of Neobone® (HAp), Cytrans® (CO$_3$Ap), and Cerasorb® implanted into dog mandibular bone defect at 12 weeks after implantation (Villanueva Goldner staining). Green area, bone; red area, osteoid. EB, existing bone; NB, new bone; O, osteoid; H, HAp (Neobone®); C, CO$_3$Ap (Cytrans®); T, β-TCP (Cerasorb®).

At 4 weeks after implantation, new bone formation was observed from existing bone to the center of the bone defect in the images of Neobone® (HAp), Cytrans® (CO$_3$Ap), and Cerasorb® (β-TCP) (Figure 6). New bone was formed on the surfaces and inside the pores of Cerasorb® (β-TCP) (Figure 6f). In the central area of the defect, fibrous soft tissue had infiltrated into the interconnected porous structure of Neobone® (HAp) and Cerasorb® (β-TCP). The structure of Cerasorb® (β-TCP) was apparently resorbed and became obscure. The new bone in Neobone® (HAp) and Cerasorb® (β-TCP) (Figure 6d,f) was only observed in the neighborhood of existing bone (Figure 6a). On the other hand, new bone formation in Cytrans® (CO$_3$Ap) had reached more central areas of the defects compared to those in Neobone® (HAp) and Cerasorb® (β-TCP) (Figure 6b). New bone and Cytrans® (CO$_3$Ap) granules were in direct contact with each other without any intermediate fibrous tissue in between (Figure 6e). Aligned cuboidal osteoblasts (arrows) were observed around the abundantly formed osteoid on the surface of the new bone, suggesting active bone formation by osteoblasts (Figure 6e). Slight chronic inflammatory cell infiltration- composed lymphocytes and plasma cells was observed in three materials (Figure 6d–f), and the strongest inflammation was recognized in Neobone® (HAp) and the weakest in Cytrans® (CO$_3$Ap). More granulation tissue was also observed in Neobone® (HAp) than Cerasorb® (β-TCP) and Cytrans® (CO$_3$Ap).

At 12 weeks after implantation, osteoid, which was observed at 4 weeks in the image of Neobone® (HAp), Cytrans® (CO$_3$Ap), and Cerasorb® (β-TCP), was gradually replaced by mature bone. The amount of new bone in Cerasorb® (β-TCP) and Cytrans® (CO$_3$Ap) was larger than those in Neobone® (HAp) (Figure 7d–f). The granulation tissue seen at 4 weeks was replaced by fibrous tissue around Neobone® (HAp), Cytrans® (CO$_3$Ap), and Cerasorb® (β-TCP), and the inflammatory reaction had become slighter at 12 weeks compared to that at 4 weeks. Therefore, all showed good biocompatibility as evidenced by extremely slight inflammatory reaction. The size of Cytrans® (CO$_3$Ap) granules at 12 weeks was smaller than that at 4 weeks.

The amount of bone at the bone defect 4 and 12 weeks after implantation is summarized (Figure 8). As indicated in Figures 6 and 7, the largest bone amount was observed at both 4 weeks and 12 weeks in the defects reconstructed with Cytrans® (CO$_3$Ap). Larger amounts of bone were observed at 12 weeks compared to those at 4 weeks, regardless of the type of bone substitute.

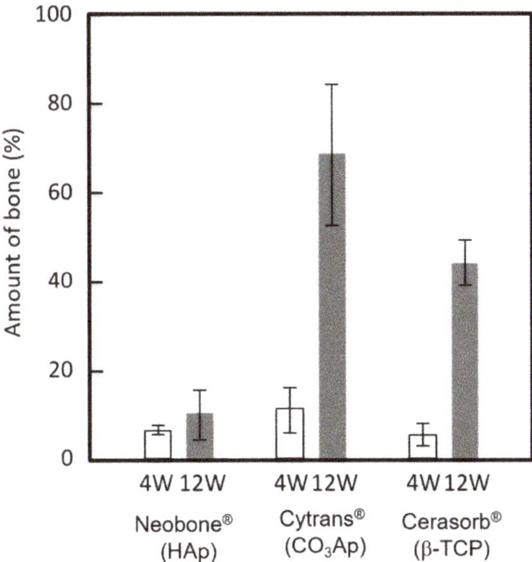

Figure 8. Amount of new bone formed in bone defect area at 4 weeks and 12 weeks after implantation. (*n* = 2).

4. Discussion

In this study, three commercially available bone substitutes with different composition, HAp (Neobone[®]), CO_3Ap (Cytrans[®]), and β-TCP (Cerasorb[®]), were compared with respect to their physical character and tissue response. Sintered HAp (Neobone[®]), and sintered β-TCP (Cerasorb[®]) showed physical properties typical of ceramics fabricated by sintering. Crystallite size was large and specific surface areas were limited. When HAp (Neobone[®]) was compared with β-TCP (Cerasorb[®]), β-TCP (Cerasorb[®]) had a much smaller SSA of 0.06 m^2/g than the 1.0 m^2/g of HAp (Neobone[®]). Also, β-TCP (Cerasorb[®]) had smaller porosity, 76.4 ± 0.8%, compared to HAp (Neobone[®]), 85.1 ± 0.5%.

These differences appear to be reasonable for obtaining proper dissolution behavior under physiological and osteoclastic condition. β-TCP is known to have much higher solubility than HAp, which was also confirmed in the current study using commercially available β-TCP (Cerasorb[®]) and HAp (Neobone[®]). Although the higher solubility of β-TCP is the key reason why β-TCP is replaced by new bone, β-TCP has a degree of solubility which is too high for physiological and osteoclastic pH. Therefore, Cerasorb[®] (β-TCP) may be fabricated as it has very small SSA and small porosity, which minimizes the dissolution rate.

CO_3Ap (Cytrans[®]) behaves differently when compared to HAp (Neobone[®]) and β-TCP (Cerasorb[®]). It has a smaller crystallite size and a much higher SSA value. This is thought to be caused by the fabrication method, which is via a dissolution–precipitation reaction using $CaCO_3$ block as a precursor in aqueous solution. Thus, crystallinity is low and crystallite size calculated from XRD pattern is smaller (30.8 ± 0.8 nm) when compared to HAp (Neobone[®]) or β-TCP (Cerasorb[®]) with crystallite sizes of 75.4 ± 0.9 nm and 78.5 ± 7.5 nm, respectively.

Dissolution of β-TCP (Cerasorb[®]) in pH 7.3 solutions was the fastest, even though it had the smallest SSA (0.06 m^2/g), followed by CO_3Ap (Cytrans[®]) that had the largest SSA (18.2 m^2/g), and HAp (Neobone[®]) that had a medium SSA (1.0 m^2/g). It should be noted that the dissolution test was done using a buffer solution containing no Ca-PO_4, according to standards. Body fluid was supersaturated with respect to apatite. Therefore, the least dissolution of HAp (Neobone[®]) and CO_3Ap (Cytrans[®]) should be lower than that seen in these results.

Interestingly, dissolution in the pH 5.5 solution was the fastest in CO_3Ap (Cytrans[®]), followed by β-TCP (Cerasorb[®]) and HAp (Neobone[®]). This may due to the release of CO_2 during dissolution in pH 5.5 solution in the case of CO_3Ap (Cytrans[®]). Although behavior of bone substitute at the bone defect may be more complex, dissolution behavior in pH 5.5 is thought to be related to the replacement by new bone. Both CO_3Ap (Cytrans[®]) and β-TCP (Cerasorb[®]) are known to be replaced by new bone, whereas HAp (Neobone[®]) is stable at the bone defect, and would not be replaced with bone.

Histological results, summarized in Figures 6–8, revealed that bone formation and resorption of bone substitute is different according to the type of bone substitute. In HAp (Neobone[®]), new bone formation was limited in the neighborhood of existing bone 4 weeks after implantation, and the amount of new bone was very small. At 12 weeks, the amount of bone was almost the same as that at 4 weeks.

In contrast to HAp (Neobone[®]), much larger bone was formed at both 4 weeks and 12 weeks in the case of CO_3Ap (Cytrans[®]). Most of the CO_3Ap granules were partially replaced by new bone at 12 weeks. Cell-to-cell signaling and material-to-cell signaling may be the cause of larger new bone formation in the case of CO_3Ap (Cytrans[®]). In other words, osteoclasts activated by resorption may release clastokines, including TRAcP, sphingosine-1-phosphate (S1P), BMP6, wingless-type 10b (Wnt10b), hepatocyte growth factor (HGF), collagen triple helix repeat-containing protein 1 (CTHRC1), where these factors activate preosteoblasts and osteoblasts [28,29]. For material-to-cell signaling, Nagai et al. reported that CO_3Ap upregulates osteoblastic differentiation of the human bone marrow compared to HAp [12].

β-TCP (Cerasorb[®]) showed interesting bone formation. At 4 weeks, almost no bone formation was observed. At this stage, materials remained at the bone defect. However, at 12 weeks after implantation, a larger amount of bone was formed, and most of the materials were resorbed at this stage. Although histological studies were performed only at 4 and 12 weeks, dissolution of the materials may be a key factor for β-TCP (Cerasorb®).

5. Conclusions

The CO_3Ap (Cytrans®) bone substitute, which is fabricated through dissolution–precipitation reaction using a precursor block in aqueous solution, showed a much higher specific surface area and a low crystallite size when compared to sintered HAp (Neobone®) and sintered β-TCP (Cerasorb®). In acidic pH 5.3 solution mimicking Howship's lacunae, CO_3Ap (Cytrans®) dissolved most rapidly, but showed limited dissolution in physiological pH 7.3 solution, indicating stability under physiological conditions. Further, CO_3Ap (Cytrans®) elicited the largest amount of new bone compared to HAp (Neobone®) and β-TCP (Cerasorb®). Composition similar to that of bone may be an advantageous factor for artificial bone substitutes.

Author Contributions: Conceptualization, K.I. and Y.M.; Investigation, A.T., K.H., K.T., G.O.; Writing—Review & Editing, K.I. and Y.M.

Acknowledgments: This research was supported, in part, by AMED under grant number JP18im0502004.

Conflicts of Interest: The authors declare no conflict of interest.

References

1. Giannoudis, P.V.; Dinopoulos, H.; Tsiridis, E. Bone substitutes: An update. *Injury* **2005**, *36*, S20–S27. [CrossRef] [PubMed]
2. Tamai, N.; Myoui, A.; Tomita, T.; Nakase, T.; Tanaka, J.; Ochi, T.; Yoshikawa, H. Novel hydroxyapatite ceramics with an interconnective porous structure exhibit superior osteoconduction in vivo. *J. Biomed. Mater. Res.* **2002**, *59*, 110–117. [CrossRef] [PubMed]
3. Matsumine, A.; Myoui, A.; Kusuzaki, K.; Araki, N.; Seto, M.; Yoshikawa, H.; Uchida, A. Calcium hydroxyapatite ceramic implants in bone tumour surgery. A long-term follow-up study. *J. Bone Jt. Surg. Br.* **2004**, *86*, 719–725. [CrossRef] [PubMed]
4. Bauer, T.W.; Smith, S.T. Bioactive materials in orthopaedic surgery: Overview and regulatory considerations. *Clin. Orthop. Relat. Res.* **2002**, *395*, 11–22. [CrossRef]
5. de Groot, K. Bioceramics consisting of calcium phosphate salts. *Biomaterials* **1980**, *1*, 47–50. [CrossRef]
6. Bohner, M. Calcium orthophosphates in medicine: from ceramics to calcium phosphate cements. *Injury* **2000**, *31*, 37–47. [CrossRef]
7. Hak, D.J. The use of osteoconductive bone graft substitutes in orthopaedic trauma. *J. Am. Acad. Orthop. Surg.* **2007**, *15*, 525–536. [CrossRef] [PubMed]
8. Yamasaki, N.; Hirao, M.; Nanno, K.; Sugiyasu, K.; Tamai, N.; Hashimoto, N.; Yoshikawa, H.; Myoui, A. A comparative assessment of synthetic ceramic bone substitutes with different composition and microstructure in rabbit femoral condyle model. *J. Biomed. Mater. Res. B Appl. Biomater.* **2009**, *91*, 788–798. [CrossRef] [PubMed]
9. Bauer, T.W.; Muschler, G.F. Bone graft materials. An overview of the basic science. *Clin. Orthop. Relat. Res.* **2000**, *371*, 10–27. [CrossRef]
10. Moore, W.R.; Graves, S.E.; Bain, G.I. Synthetic bone graft substitutes. *ANZ J. Surg.* **2001**, *71*, 354–361. [CrossRef] [PubMed]
11. Bohner, M.; Galea, L.; Doebelin, N. Calcium phosphate bone graft substitutes: failures and hopes. *J. Eur. Ceramic Soc.* **2012**, *32*, 2663–2671. [CrossRef]
12. Ishikawa, K. Bone substitute fabrication based on dissolution-precipitation reaction. *Materials* **2010**, *3*, 1138. [CrossRef]
13. Ishikawa, K.; Matsuya, S.; Lin, X.; Zhang, L.; Yuasa, T.; Miyamoto, Y. Fabrication of low crystalline B-type carbonate apatite block from low crystalline calcite block. *J. Ceram. Soc. Jpn.* **2010**, *118*, 341. [CrossRef]

14. Lee, Y.; Hahm, Y.M.; Matsuya, S.; Nakagawa, M.; Ishikawa, K. Characterization of macroporous carbonate-substituted hydroxyapatite bodies prepared in different phosphate solutions. *J. Mater. Sci.* **2007**, *42*, 7843. [CrossRef]

15. Zaman, C.T.; Takeuchi, A.; Matsuya, S.; Zaman, Q.H.; Ishikawa, K. Fabrication of B-type carbonate apatite blocks by the phosphorization of free-molding gypsum-calcite composite. *Dent. Mater. J.* **2008**, *27*, 710. [CrossRef] [PubMed]

16. Daitou, F.; Maruta, M.; Kawachi, G.; Tsuru, K.; Matsuya, S.; Terada, Y.; Ishikawa, K. Fabrication of carbonate apatite block based on internal dissolution-precipitation reaction of dicalcium phosphate and calcium carbonate. *Dent. Mater. J.* **2010**, *29*, 303. [CrossRef] [PubMed]

17. Maruta, M.; Matsuya, S.; Nakamura, S.; Ishikawa, K. Fabrication of low-crystalline carbonate apatite foam bone replacement based on phase transformation of calcite foam. *Dent. Mater. J.* **2011**, *30*, 14. [CrossRef] [PubMed]

18. Sunouchi, K.; Tsuru, K.; Maruta, M.; Kawachi, G.; Matsuya, S.; Terada, Y.; Ishikawa, K. Fabrication of solid and hollow carbonate apatite microspheres as bone substitutes using calcite microspheres as a precursor. *Dent. Mater. J.* **2012**, *31*, 549. [CrossRef] [PubMed]

19. Wakae, H.; Takeuchi, A.; Udoh, K.; Matsuya, S.; Munar, M.; LeGeros, R.Z.; Nakasima, A.; Ishikawa, K. Fabrication of macroporous carbonate apatite foam by hydrothermal conversion of α-tricalcium phosphate in carbonate solutions. *J. Biomed. Mater. Res. A* **2008**, *87*, 957. [CrossRef] [PubMed]

20. Takeuchi, A.; Munar, M.L.; Wakae, H.; Maruta, M.; Matsuya, S.; Tsuru, K.; Ishikawa, K. Effect of temperature on crystallinity of carbonate apatite foam prepared from α-tricalcium phosphate by hydrothermal treatment. *Bio-Med. Mater. Eng.* **2009**, *19*, 205.

21. Karashima, S.; Takeuchi, A.; Matsuya, S.; Udoh, K.; Koyano, K.; Ishikawa, K. Fabrication of low-crystallinity hydroxyapatite foam based on the setting reaction of alpha-tricalcium phosphate foam. *J. Biomed. Mater. Res. A* **2009**, *88*, 628.

22. Sugiura, Y.; Tsuru, K.; Ishikawa, K. Fabrication of carbonate apatite foam based on the setting reaction of α-tricalcium phosphate foam granules. *Ceram. Int.* **2016**, *42*, 204. [CrossRef]

23. Tsuru, K.; Kanazawa, M.; Yoshimoto, A.; Nakashima, Y.; Ishikawa, K. Fabrication of carbonate apatite block through a dissolution–precipitation reaction using calcium hydrogen phosphate dihydrate block as a precursor. *Materials* **2017**, *10*, 374. [CrossRef] [PubMed]

24. Kanazawa, M.; Tsuru, K.; Fukuda, N.; Sakemi, Y.; Nakashima, Y.; Ishikawa, K. Evaluation of carbonate apatite blocks fabricated from dicalcium phosphate dihydrate blocks for reconstruction of rabbit femoral and tibial defects. *J. Mater. Sci: Mater. Med.* **2017**, *28*, 85–96. [CrossRef] [PubMed]

25. Nagai, H.; Fujioka-Kobayashi, M.; Fujisawa, K.; Ohe, G.; Takamaru, N.; Hara, K.; Uchida, D.; Tamatani, T.; Ishikawa, K.; Miyamoto, Y. Effects of low crystalline carbonate apatite on proliferation and osteoblastic differentiation of human bone marrow cells. *J. Mater. Sci. Mater. Med.* **2015**, *26*, 99–107. [CrossRef] [PubMed]

26. Rei, C.; Collins, B.; Goehal, T.; Dickson, I.R.; Glimcher, M.J. The carbonate environment in bone mineral: A resolution-enhanced fourier transform infrared spectroscopy study. *Calcif Tissue Int.* **1989**, *45*, 157–164. [CrossRef]

27. Vignoles, M.; Bonel, G.; Young, R.A. Occurrence of nitrogenous species in precipitated B type carbonated hydroxyapatites. *Calcif. Tissue Int.* **1987**, *40*, 64–70. [CrossRef] [PubMed]

28. Charles, J.F.; Aliprantis, A.O. Osteoclasts: More than 'bone eaters'. *Trends Mol. Med.* **2014**, *20*, 449–459. [CrossRef] [PubMed]

29. Cappariello, A.; Maurizi, A.; Veeriah, V.; Teti, A. The Great Beauty of the osteoclast. *Arch. Biochem. Biophys.* **2014**, *558*, 70–78. [CrossRef] [PubMed]

Review

Comprehensive In Vitro Testing of Calcium Phosphate-Based Bioceramics with Orthopedic and Dentistry Applications

Radu Albulescu [1,2], **Adrian-Claudiu Popa** [3,4], **Ana-Maria Enciu** [1,5], **Lucian Albulescu** [1], **Maria Dudau** [1,5], **Ionela Daniela Popescu** [1], **Simona Mihai** [1], **Elena Codrici** [1], **Sevinci Pop** [1], **Andreea-Roxana Lupu** [1,6], **George E. Stan** [3], **Gina Manda** [1] and **Cristiana Tanase** [1,7,*]

[1] Victor Babes National Institute of Pathology, Biochemistry-Proteomics Department, 050096 Bucharest, Romania; radu_a1@yahoo.com (R.A.); ana.enciu@ivb.ro (A.-M.E.); albulescu.l@gmail.com (L.A.); dudau_maria2002@yahoo.com (M.D.); danabini@yahoo.com (I.D.P.); simona.mihai21@gmail.com (S.M.); raducan.elena@gmail.com (E.C.); sevincipop@yahoo.com (S.P.); ldreea@gmail.com (A.-R.L.); gina.manda@gmail.com (G.M.)

[2] Department Pharmaceutical Biotechnology, National Institute for Chemical-Pharmaceutical R&D, 031299 Bucharest, Romania

[3] National Institute of Materials Physics, 077125 Magurele, Romania; adrian_claudiu_popa@yahoo.com (A.-C.P.); george_stan@infim.ro (G.E.S.)

[4] Army Centre for Medical Research, 010195 Bucharest, Romania

[5] Department of Cellular and Molecular Biology and Histology, Carol Davila University of Medicine and Pharmacy, 050047 Bucharest, Romania

[6] Cantacuzino National Medico-Military Institute for Research and Development, 050096 Bucharest, Romania

[7] Cajal Institute, Titu Maiorescu University, 004051 Bucharest, Romania

* Correspondence: bioch@vbabes.ro

Received: 30 September 2019; Accepted: 5 November 2019; Published: 10 November 2019

Abstract: Recently, a large spectrum of biomaterials emerged, with emphasis on various pure, blended, or doped calcium phosphates (CaPs). Although basic cytocompatibility testing protocols are referred by International Organization for Standardization (ISO) 10993 (parts 1–22), rigorous in vitro testing using cutting-edge technologies should be carried out in order to fully understand the behavior of various biomaterials (whether in bulk or low-dimensional object form) and to better gauge their outcome when implanted. In this review, current molecular techniques are assessed for the in-depth characterization of angiogenic potential, osteogenic capability, and the modulation of oxidative stress and inflammation properties of CaPs and their cation- and/or anion-substituted derivatives. Using such techniques, mechanisms of action of these compounds can be deciphered, highlighting the signaling pathway activation, cross-talk, and modulation by microRNA expression, which in turn can safely pave the road toward a better filtering of the truly functional, application-ready innovative therapeutic bioceramic-based solutions.

Keywords: bioceramics; in vitro testing; hydroxyapatite; angiogenesis; osteogenesis; signaling pathways; microRNA

---

1. Introduction

Due to their physical–chemical similarity with bone mineral, calcium phosphate (CaP) bioceramics, with prominent exponents hydroxyapatite (HA, $Ca_{10}(PO_4)_6(OH)_2$) and β-tricalcium phosphate (β-TCP, $Ca_3(PO_4)_2$), are already used or envisaged to be employed, in pure or blended form, in a continuously increasing number of biomedical applications with the main focus on tissular regeneration within the skeletal system (bones, joints, and teeth) [1,2]. If used for orthopedic and dental transplants, CaPs may interact directly with the surrounding tissue, either supporting tissue growth or inducing tissue regeneration, while presenting a good compatibility with the biological systems [3–5]. Synthetic ceramics such as HA and β-TCP possess excellent biocompatibility and can interact directly with surrounding tissues through the formation of chemical and biochemical bonds [6,7]. Although many different type of materials were investigated and engineered (e.g., metallic, polymeric, or ceramic; natural or synthetic; bioinert, bioresorbable, or bioactive) to fabricate bone regeneration scaffolds or implant coatings, bioactive ceramics prominently arose as materials of choice due to their remarkable ability to create a strong bond with hard tissues, as well as prevent their encapsulation in fibrillary connective tissue, in contrast to renown bioceramics such as alumina or yttria, polymers, and metals [8–11]. Their bioactivity could also enhance gene activation for osteogenesis and angiogenesis. These materials evolved into an integral and vital segment of the modern healthcare system, and they can be integrated into the human body as permanent biomedical devices, due to their improved biocompatibility [12]. However, pure HA has limited usage in biomedicine because of its fragility and overall weak mechanical properties, unsuitable for developing load-bearing biomedical applications [13]. Nevertheless, the designed doping of CaP bioceramics with various cations (e.g., Na^+, Mg^{2+}, Sr^{2+}, and Zn^{2+}) or anions (e.g., $(CO_3)^{2-}$, $(SiO_4)^{4-}$, F^-, and Cl^-) have now the potential to transform them into major candidates for the future development of "smart materials", due to their capability to combine biocompatibility and mechanical performance with specific effects such as antimicrobial activity, angiogenesis induction, and drug delivery capacity [8–10]. When designed as resorbable biomaterials with various resorption kinetics (spanning from days to months), their ion dissolution products (Ca^{2+}, $(PO_4)^{3-}$, Na^+, $Si4^+$, Mg^{2+}, and Sr^{2+} ions) can usually be processed via normal metabolism [14] or can even be exploited to exert desired therapeutic effects, such as the promotion of angiogenesis or osteogenesis properties, and antimicrobial activity [11,15]. This new generation of CaP-based materials is envisaged to be employed in healthcare in various shapes and forms: bulk (especially for bone graft substitutes, e.g., porous scaffolds) [16–18], highly crystalline or nano-structured implant coatings [19], and dispersed nanoparticles (e.g., as antimicrobials or carriers in biological systems for drug delivery, transfection, gene silencing, or imaging) [20–22] or nano-objects (e.g., nano-rods, nano-wires, nano-tubes, and nano-needles) [23–26].

Capitalizing on their osteoconductivity and biocompatibility [27], recent studies reported that nano-sized pure and cation- and/or anion-substituted HAs could represent promising candidates for bone regeneration, as they closely mimic the structural and compositional features of the inorganic component of native bone matrix [1,28,29]. Bone tissue engineering emerged as a rapidly developing strategy for bone regeneration due to the increased clinical demand for biocompatible bone scaffolds and novel biomaterials, in orthopedic and dental medicine [30–32].

Bone healing requires a plethora of biological intricately linked events, such as angiogenesis, osteogenesis, and inflammatory reactions, in order to stimulate the complex regeneration processes [33]. A comprehensive in vitro testing of newly developed bioactive materials should inquire at least these properties. The advanced testing should also assess gene modulation or gene toxicity since, in some cases, implants based on bioactive ceramics should function properly for long periods of time (sometimes exceeding one decade).

Currently, the biomaterials scientific community is quite hyperactive in this respect, producing an immense quantity of information, especially in the realm of cation- and/or anion-substituted CaPs. Although extensively characterized from compositional, morphological, and structural points of view, the pure and substituted CaPs were evaluated to a lesser extent from a biological point of view. Regularly, such studies only tackled the topic superficially or incompletely, limited to the evaluation of the biomineralization capacity [34–36] (https://www.iso.org/standard/65054.html) and cytotoxicity/cytocompatibility assays which do not provide in-depth understanding of biological processes and hinder, therefore, a rapid transition from bench to bedside. Unfortunately, only few such materials reach an advanced in vitro biological assessment stage, and even fewer get to the stage of in vivo testing as a prerequisite for biomaterial transfer in medicine [1].

The scope of this article is to review the actual state of the in vitro safety assessment methods, applicable for bioceramics (with a focus on CaP-based compounds), stressing the critical aspects of commonly used procedures/regulations and recommending the ways to improve the selection algorithm of such biomaterials for the best biological outcome when implanted in vivo. Although progress was recently recorded in the accuracy, complexity, and fastness of biological testing of bioceramics, no complete review on the achieved progress was published to the best of our knowledge. Such a breviary study could popularize the forefront technologies (which become increasingly available)/biological protocols, as well as their judicious coupling, and reorient the focus of the biomaterials community toward the insightful and comprehensive understanding of the biological mechanisms of their materials, rather than toward a simple screening of seldom functionalities. The application of in vitro methods to evaluate the eventual deleterious effects of materials (such as cytotoxicity, genotoxicity, or production of reactive oxygen species) is also highlighted. Certain attention was dedicated to the biofunctional analysis of bioceramics, offering information on the state-of-the-art methods for the evaluation of two key factors: angiogenesis and osteogenesis. Moreover, it is suggested that deciphering the mechanisms of action of these bioceramic CaP compounds can be accomplished by involving the specific modulation of microRNAs (miRNAs) and cell signaling within osteogenesis and dentinogenesis.

2. Biocompatibility Assessment of CaP-Based Bioceramics

2.1. Regulatory Aspects

There is a complex set of International Organization for Standardization (ISO) standards governing the evaluation of biocompatibility. According to ISO 10993-1:2018 "Evaluation and Testing within a Risk Management Process", a set of mandatory tests must be selected, depending on the nature and way of contact of the biomaterial/nanomaterial with the body [37]. In connection with ISO 10993-1:2018, a set of assays should be considered when checking for the biocompatibility of biomaterials, for both types of materials of interest, i.e., external communicating devices coming in contact with tissue, bone or dentin and internal communicating ones of similar uses. The ISO:10993-1:2018 recommends running in vitro assays regarding cytotoxicity, genotoxicity, immunotoxicity, and material-mediated pyrogenicity, as well as a series of in vivo assays regarding irritation, subacute/sub-chronic toxicity, and implantation. A general workflow for material testing, also applicable to CaP bioceramics, with emphasis on in vitro testing, is presented in Figure 1.

While these recommendations regard all kind of materials (in bulk or in thin-film form), some specific provisions are made for nano-powders and nano-objects (e.g., (i) particulate bioceramics of various designed nano-shapes, or (ii) nano-debris as a result of in situ wear/degradation of the biomedical device). Such nano-debris can be generated during the life cycle of a medical device; therefore, the evaluation of possible adverse effects caused by the implantation or generation of nano-objects, whether during preparation or in situ use, wear, or degradation of medical devices, needs to be addressed. This applies to medical devices having the potential to generate nanoscale wear and/or degradation particles. For the biological evaluation of medical devices, knowledge on the potential generation and/or release of nano-objects from such materials and on their effects is essential [3,8,38]. Such an example of release of nanoparticles (NPs) is provided, for instance, for titanium implants, discussed by Kim et al. [39], who reviewed the literature and found that particles and ions from titanium alloys can deposit in surrounding tissue, mainly because of the corrosion and wear of implants, further causing bone loss, and failure of osteointegration.

The procedures for the biological evaluation of medical devices, described in the ISO 10993 series of standards, can also be applied for the biological evaluation of medical devices containing nano-objects (e.g., nano-powders, nano-rods, nano-wires, nano-tubes, and nano-needles) as an integrated part of the device (e.g., HA nanowire/collagen [24], or HA nano-needle/poly(L-lactic acid [26] composite scaffolds, and CaP nanoparticles with an intrinsic antimicrobial effect [20]). However, when the release of nano-objects from the medical device is possible, a safety evaluation should also be performed on these released nanoscale entities. Furthermore, in addition to evaluating a medical device as a whole, its nanoscale components or constituents should also be separately assessed.

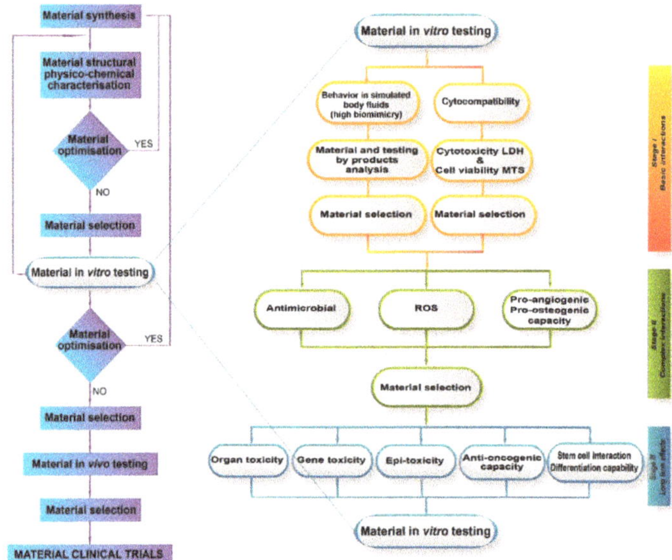

Figure 1. Workflow for selection of calcium phosphate (CaP)-based bioceramics suitable for patient applications. The left diagram is a general workflow diagram, from material synthesis to clinical application. In vitro testing is detailed in the right panel.

With regard to in vitro cytotoxicity assays, which are fast and reliable filters for bioceramics with undesired effects, several approaches are available. The main advantage is that such assays are not limited to certain cell lines; thus, a large variety of non-human mammalian and human cells can be employed. Moreover, the testing is not limited to "immortal" or "immortalized" cells; in certain situations, it can also be achieved using primary cells. These last two variants (immortalized and primary cells) were recently enhanced by quite a wide offer of such products, mainly from the American Type Culture Collection (ATCC). ISO 10993-5:2009 "Tests for Cytotoxicity—In Vitro Methods" stands as a regulatory document for these assays [40].

The ISO standard provides a large set of assays that can be applied in order to assess the potential acute adverse (toxicological) effects of extractables from medical device materials on mammalian cells, using the settings outlined in ISO 10993-5. While there is no explicit limitation with regard to the mammalian cell line to be used, the most frequent models are based on mouse or human cells. The standard testing procedures involve cell monolayers, grown near confluence in adequate recipients, exposed to extracts or eventually to biomaterials per se, under standard conditions, for various times (usually 24–72 h). Different techniques are employed for examination, starting from visual inspection under microscope (which allows for the evaluation of change in size and number of cell organelles, or disruption), viability assessments, such as lactate dehydrogenase (LDH), 3-(4,5-dimethylthiazol-2-yl)-2,5-diphenyltetrazolium bromide (MTT), 3-(4,5-dimethylthiazol-2-yl)-5-(3-carboxymethoxyphenyl)-2-(4-sulfophenyl)-2*H*-tetrazolium (MTS), etc., quantitation, and estimation of cell apoptosis or necrosis using specific assay methods. Usually, ISO 10993-5:2009 advises the use of cells from recognized repositories. Various culture types can be used, such as primary cell cultures, cell lines, or organotypic cultures, provided that accuracy and reproducibility of the response can be demonstrated [40].

There were several studies that demonstrated that HA and its related dissolution products could be quite harmless for use in devices, such as, for instance, bone implants [41–43]. However, some other studies still considered the potential risks of such materials, mainly due to the addition of other "doping" elements, such as Sr, Zn, etc. While in low concentrations they seem to be harmless, the increase of their content can result in strong modifications of the crystal lattice, leading to undesired effects (e.g., rapid dissolution rates and decomposition). For instance, the cytotoxicity of strontium-doped CaP coatings deposited onto AZ31 degradable magnesium alloy was evaluated with MC3T3-E1 mouse preosteoblasts and human mesenchymal stem cells (hMSCs), for contact times of 24–72 h [29]. For both cell types, the proliferation decreased upon increasing the Sr concentration. However, both osteogenic gene and protein expression significantly increased upon increasing Sr concentration. These results suggest that Sr-doped coatings are capable of promoting osteogenesis, in comparison to the undoped CaP coatings [44].

Generally, the safety assessment is clearly outlined by regulatory documents (such as the 10,993 series), while the efficacy assessment is more prone to the application of different testing methods of choice, involving both in vitro and in vivo assays similar to the situation for efficacy testing of medicines.

The most relevant colorimetric method for evaluating cytotoxicity is the neutral red (NRU) assay, based on the ability of viable cells to incorporate and retain neutral red dye in lysosomes [45]. Comparing NRU results obtained for HA, natural coral, and polyhydroxybutarate on CRL-1543 cells, Shamsuria et al. concluded that all materials are non-cytotoxic even after 72 h of treatment [46]. Moreover, HA induced osteoblast cell proliferation (123% vs. control) which could be interpreted as biofunctionality [46]. However, NRU and MTT tests showed that HA-based cements can exert cytotoxicity by changing the concentration of ions (calcium, magnesium, and phosphate) in cell culture media [47]. Quantitative cell viability measurement using the NRU test allows identifying materials able to promote cell growth in regeneration studies.

2.2. Cell Viability and Cytotoxicity

In most cases, HA materials are considered highly biocompatible and, therefore, suitable for bone tissue applications (e.g., medical implants and bone regeneration) [48–50]. However, the response of biological tissues and cell lines to HA-based materials (including HA nanoparticles and nano-objects) shows great variability [51–54].

Colorimetric viability tests are used for in vitro biocompatibility evaluation, where the absorbance of sample presumably is directly proportional to the number of viable cells [55]. Colorimetric viability tests using formazan salts (MTT, MTS, 2,3-bis-(2-methoxy-4-nitro-5-sulfophenyl)-2H-tetrazolium-5-carboxanilide (XTT), and Cell Counting Kit-8 (WST-8)) are based on the fact that mitochondrial activity and cell viability are correlated. Most common formazan salts used to assess cell viability are MTT (3-(4,5-dimethylthiazol-2-yl)-2,5-diphenyltetrazolium bromide), MTS (3-(4,5-dimethylthiazol-2-yl)-5-(3-carboxymethoxyphenyl)-2-(4-sulfophenyl)-2H-tetrazolium), XTT (2,3-bis-(2-methoxy-4-nitro-5-sulfophenyl)-2H-tetrazolium-5-carboxanilide), WST-1, and WST-8 reagents [56–60]. The MTT/MTS test is widely used to assess the biocompatibility of CaPs [52,61–63]. Using the MTS method, Santos et al. [52] showed that both chemically and hydrothermally synthesized HA did not affect the viability of MG63 cells treated for three and six days with HA concentrations lower than 500 µg/mL. Their results are important in the context of sterilization of HA-containing medical devices. Applying the MTT test on the same cell line (MG63 osteoblast-like cells), Aghaei et al. [64] demonstrated the biocompatibility of a silica mesoporous (MCM-48)/HA composite and its potential use as a drug delivery agent. The potential cytotoxicity of nanostructured HA was evaluated (MTS and LDH method), aiming to obtain highly biocompatible materials for bone surgery and dentistry applications [65,66].

In interpreting the data, it must be taken into account that an increase in the absorbance of samples may be due to an increased mitochondrial activity (with no significant change in the number of cells), rather than cellular proliferation. In addition, the number and activity of mitochondria in the tested cells are important, especially when discussing the comparative effect of materials on different cell lines or depending on the stage of cell differentiation [1,67]. In methods in which liquid biological samples containing the compounds of interest are subjected to optical absorbance determinations, the materials possessing a high refractive index, as well as fluorescence and photocatalytic properties, may induce altered results [68–72]. During optical density determinations, by absorbing or reflecting the incident light, suspended nanoparticles may increase the measured absorbance (turbidity/opacity). When exposed to ultraviolet (UV)–visible light, materials with photocatalytic properties may lead to the generation of reactive oxygen species and redox reactions on their surface, processes which may alter the molecules of interest in the biological sample or generate unknown or unwanted reaction products. Fluorescent nanomaterials may emit light at unwanted wavelengths which may induce erroneous results in optical determinations or induce undesirable physicochemical processes in the tested samples.

Given that colorimetric viability tests (using both tetrazolium salts and LDH activity) are based on comparing the absorbance of the samples with the absorbance of the controls, and the optical density of the solution depends on its clarity, it is necessary to perform parallel non-cellular tests (i.e., by applying the exact same working protocol, but in the absence of cells). These should follow the exact same working protocol, but in the absence of cells. If the absorbance of the samples differs from that of the control samples, the values of the optical densities obtained in the cell test are subtracted from the values corresponding to the cell test.

Moreover, the protein corona effect has to be considered. Adsorption on the surface of bioceramic samples of molecules from the biological environment can have different biological effects compared to the initial materials, due to modified colloidal stability, changes in shape and volume, hiding of some functional groups, and exposure of cryptic peptide epitopes [73–75]. In vitro and in vivo protein corona-dependent changes in viability were already shown for HA and magnetic HA scaffolds [76,77]. Considering all error sources mentioned above, for the accurate evaluation of biocompatibility, it is recommended to use at least two different techniques (with respect to cellular mechanisms and/or principle of the technique). Biocompatibility studies require a good characterization of the tested materials concerning all aspects potentially involved in biological effects; the choice of testing methods should be done according to these specific characteristics of the bioceramic samples. The cellular viability and cytotoxicity in the presence of CaP-based bioceramics on several types of cells are summarized in Table 1.

Table 1. Cellular viability and cytotoxicity tests. CaP—calcium phosphate; HA—hydroxyapatite; MTS—3-(4,5-dimethylthiazol-2-yl)-5-(3-carboxymethoxyphenyl)-2-(4-sulfophenyl)-2*H*-tetrazolium; MTT—3-(4,5-dimethylthiazol-2-yl)-2,5-diphenyltetrazolium bromide; LDH—lactate dehydrogenase.

Type of CaP	Type of Cells	Methodological Approach	Main Effects	References
		Cell Viability		
HA nanoparticles produced via wet chemical synthesis (37 °C) and hydrothermal synthesis (180 °C)	MG63 osteoblast-like cells	MTS cell proliferation assay	Neither particle, in doses lower than 0.5 mg/mL, affected cell viability and proliferation. For concentrations between 0.5 and 2 mg/mL, the inhibition of cell proliferation was time-dependent, with slightly higher values corresponding to chemically synthesized HA when compared with hydrothermally synthesized HA.	[52]

Table 1. *Cont.*

Type of CaP	Type of Cells	Methodological Approach	Main Effects	References
Cell Viability				
Nano-HA–silica-incorporated glass ionomer cement (HA–SiO$_2$–GIC)	human Dental Pulp Stem Cells (DPSC)	MTT assay	HA–SiO$_2$–GIC showed cytotoxic effects for all tested concentrations (3.125–200 mg/mL).	[61]
HA coatings prepared by a sol–gel method on Ti6Al4V	human fetal osteoblasts, subcultures 4–6	MTT assay	HA sol–gel-derived coatings showed low toxicity in osteoblast cell culture after 3 days (due to poor adhesion of the cells). Subsequently, cell viability increased in cells treated for 7 and 14 days with HA.	[62]
HA nanoparticles (HA NPs)	Reconstructed human gingival epithelium (HGE)	MTT test; LDH assay	3.1% HA NP solution did not induce cell death after 10 min, 1 h, and 3 h of incubation.	[63]
HA composite with the mesoporous silicate MCM-48	MG68 cells	MTT assay	MTT results showed the biocompatibility of the new material and supported its possible use as drug carrier.	[64]
HA–Au nanoparticles	Human mesenchymal stem cells	MTS test; LDH assay	When compared with controls, the MTS assay showed no significant differences in the cell viability of cells exposed to 1–100 µg/mL HA–Au nanoparticles. LDH results indicated minimal damage to the cell membranes.	[65]
High-temperature annealed nano-HA obtained via wet chemistry at 800 °C, 900 °C, and 1000 °C	L929 (NCTC clone 929) mouse fibroblast cells	MTT assay	All tested samples slightly decreased the viability of cells treated with 2.5, 5, 10, and 20 g/mL nanoparticle suspensions.	[66]

2.3. Genotoxicity

Genotoxicity assessment represents an inevitable assay for permanent implantable bioceramics. It is done observing the regulation of ISO-10993-3:2014 "Test for Genotoxicity, Carcinogenicity, and Reproductive Toxicity" [78].

Both in vivo and in vitro methods can be used, but the recommended and largely applied genotoxicity assays rely on in vitro methods, due to the practical advantages (low amounts of compounds, ease of set-up, and compatibility to automation) and the circumvention of laboratory animal use. Several gene effects can be produced by genotoxins—gene mutations, chromosomal aberrations, and other DNA effects. No in vitro assay covers all possible effects. According to ISO 10993-3:2014, the use of in vivo methods is required only if in vitro testing indicates a need for further testing, due to possible pharmacokinetic mechanisms or complex metabolic effects leading to bio-activation of the compounds. Generally, the testing methods indicated in ISO 10993-3:2014 are further detailed in several Organization for Economic Co-operation and Development (OECD) guidelines [79]. Some modifications of the protocols are required, since the original methods were designed for soluble compounds (OECD guidelines); for biomaterials, modifications were made, such as accommodating the evaluation of fluid extracts. This is usually achieved by the use of fluids able to extract polar and non-polar chemicals, and the methods are described in ISO 10993-12:2012 "Sample Preparation and Reference Materials". Depending on the test model used, the incorporation of nutrient media can also occur in the extractive formulation [80].

Gene Mutation Tests

Such point mutations affect small portions of DNA molecule and include frameshifts and base-pair substitutions. The most common assay for the detection of such mutations is the Ames bacterial reverse mutation assay, using histidine-dependent *Salmonella typhimurium* strains as the test organisms. S9 active rat liver microsomes are also incorporated in the assay, to provide simulation of whole-animal exposure. There are several distinct strains (3–5), eliciting distinct mechanisms of DNA damage. Following exposure, the cells are reverted and regain the ability to grow without histidine, thus allowing them to be counted on the plates.

A mammalian system used to detect gene mutation is the mouse lymphoma assay, using L5178Y cells [81]; these are exposed to extracts, with or without metabolic activation. After incubation, cultures are cloned in restrictive media for mutant phenotypes, and then assessed at the thymidine kinase (TK) locus to detect base-pair mutations, frameshift mutations, and small deletions. Cells that underwent mutations in the TK locus become resistant to growth in the presence of trifluorothymidine (TFT), unlike the parental cells, which cannot grow. Since mutant colonies exhibit a characteristic size distribution frequency, colony measurements can be used to distinguish the type of genetic effect.

Chromosomal aberration tests are used to detect chromosomal damage induced after one cellular division. The in vitro model employs Chinese hamster ovary cells. The assay is performed in the presence and absence of exogenous metabolic activation. Most aberrations can be identified as either chromosomal or chromatid type. Gaps, breaks, and exchanges are other examples of observable aberrations.

More recently, a relatively rapid test, the Comet assay, which detects the amount of broken DNA (the tail length), was proposed. The assay can be achieved on any cell line, and it is relatively fast and reliable [82].

By using the Ames test and the Comet assay, Wahab et al. [83] evaluated the genotoxic risks following the exposure of dental pulp cells to biphasic calcium phosphate (BCP). The study revealed that the average number of revertant colonies in the Ames test was about half of the number of revertant colonies in the negative control plate, meaning that the compound did not display any genotoxic effect.

Using a model of cultivated hepatocytes, Sonmez et al. [84] evaluated the several potential toxic and genotoxic effects of HA nanoparticles (NPs). With regard to genotoxicity, they evaluated the rate of the liver and measured the levels of 8-oxo-2-deoxyguanosine (8-OH-dG). Using increasing doses of NPs, they found increases in the number of micronucleated hepatocytes and 8-OH-dG levels compared to the control culture; however, these occurred only at high doses (1000 $\mu g/cm^2$).

Coelho et al. [41] investigated both cytotoxic and genotoxic effects of a bacterial cellulose membrane functionalized with HA and bone morphogenetic protein (BMP). Genotoxicity was evaluated by applying the in vitro Comet and micronucleus (cytokinesis-blocked micronucleus) assays on C3T3-E1 cells. The findings demonstrated that bacterial cellulose–HA was not genotoxic compared with the negative control, in both testing models.

Seyedmadiji et al. [85] investigated the functionality of HA/bioactive glass (BG) and fluorapatite (FA)/BG materials. They also employed the Comet assay to investigate potential genotoxic effects on Saos-2 cells and found a dose-dependent increase in DNA degradation, but within the limits of safety (therefore, below any threshold of genotoxicity). Kido et al. [86] used the Comet assay as a final assessment for genotoxicity on tissue samples obtained from rats that were exposed to a ceramic scaffold covered with HA and bioglass; their assays demonstrated the lack of genotoxic effects of the investigated material. Oledzka et al. [87] investigated the cyto- and geno-toxicity of a new multifunctional composite based on HA porous granules doped with selenite ions ($SeO_3)^{2-}$, and their study proved that the investigated materials were non-gentotoxic, as demonstrated by the Umu test (carried out on *S. typhimurium* TA1535/pSK1002). Yamamura et al. used in vivo models for the investigation of biocompatibility, and the lack of cyto- and geno-toxicity in blood, liver, kidney, and lung was noted 30 days after HA implantation [88].

Several studies [86–88] generally demonstrated, using various models, the lack of genotoxicity of HA-based materials.

2.4. Oxidative Stress

A critical issue in regenerative medicine and tissue engineering (bone engineering included) is related to oxidative stress and altered redox signaling, which may be inflicted on normal and pathologic cells by the implant itself as an unwanted side-effect [89]. Oxygen is critical for aerobic organisms but it is also the source of reactive oxygen species (ROS), such as superoxide anion, hydrogen peroxide, and hydroxyl radical [90,91]. Low levels of ROS are necessary for physiological cell functioning by modulating cell survival and differentiation through tightly regulated redox signaling pathways [92]. Meanwhile, if ROS levels overcome a cell-dependent threshold through increased ROS production and/or downregulation of the endogenous antioxidant system [93], a deleterious oxidative stress is generated, which can seriously alter proteins, lipids, carbohydrates, and nucleic acids, thereby profoundly disturbing cellular homeostasis and even cell survival [90]. Persistent deregulation of the redox balance (antioxidants versus oxidants) may have long-term consequences on tissue physiology [94], thus raising concerns regarding the safety of biomaterials, including bioceramics.

Considering the critical involvement of ROS in many physiologic and pathological processes, a large panel of methods was developed for precisely detecting various types of ROS [95]. Current technical challenges are mainly related to the short life of highly reactive species. Conventional methods for detecting ROS rely on their ability to change the ROS indicator and to shift it to a more stable oxidized form. Nevertheless, the specificity of the currently available indicators for particular types of ROS is poor, and sometimes the chemistry behind the detection method is not very well characterized and, therefore, data may be misinterpreted. A more precise method to detect ROS with a radical nature (e.g., superoxide anion and hydroxyl radical) is electron paramagnetic resonance (EPR) with specific spin traps such as cyclic nitrones that form relatively stable spin adducts with radicals with a longer half-life to allow detection [96]. The EPR method is not easily accessible for most biomedical laboratories due to the high cost of the equipment and the required expertise to process EPR data. Assessment of oxidative stress markers in biologic fluids and tissues might be more informative for an initial screening aimed at evaluating the magnitude of the oxidative stress [97]. Only afterward is it worth attempting to define the ROS profile and dynamics for in-depth mechanistic studies. The in vitro studies performed so far regarding the impact of HA on ROS generation by various types of cells were performed by flow cytometry using the general intracellular ROS indicator dichloro-dihydro-fluorescein diacetate (DCFH-DA) [98] or by luminol-enhanced chemiluminescence for detecting released ROS [99], as described below.

Considering that specific groups on the surface of functionalized HA-based implants may be highly reactive sites that interact with molecular dioxygen and generate uncontrolled oxidative stress, ROS-induced cytotoxicity is becoming an important component of the screening panel for assessing biocompatibility. Almost all studies developed so far for evaluating in vitro the impact of HA on oxidative stress were generally performed using HA NPs, and this was mainly due to methodological drawbacks in assessing the oxidative activity of cells cultivated on discs that better mimic the implant surface than NPs. Nevertheless, HA NPs covering orthopedic, spinal, and dental implants made of metals, ceramics, and polymers are under investigation, aiming to improve their osseointegration through enhanced biomimicry of host structures [100]. Moreover, HA scaffolds, alone or combined with bioactive molecules or genes and cells, are now being tested as promising bone grafts in hard-tissue engineering [101,102].

Oxidative stress and inflammation represent the best developed paradigm to explain many of the toxic effects of NPs in general [103]. The main mechanisms through which NPs can trigger enhanced oxidative stress comprise [104] (a) pro-oxidant or active redox cycling functional groups on the surface of NPs, (b) particle localization in cellular compartments involved in ROS generation, such as the electron transport chain in mitochondria [105] or activation of reduced membrane nicotinamide adenine dinucleotide phosphate (NADPH) oxidases [106], resulting in increased superoxide formation and further generation of more aggressive types of ROS (hydrogen peroxide and hydroxyl radical) through enzymatic reactions mediated by superoxide dismutases (SOD) or chemical reactions (Haber–Weiss and Fenton-type reactions) [107], and (c) indirect generation of ROS through inflammatory responses elicited by NPs, that are mediated by upregulation of NF-κB (nuclear factor κB), mitogen-activated protein kinase (MAPK), and phosphoinositide 3-kinase (PI3K) signaling pathways [108,109].

Due to its chemical and structural similarity with bone mineral, HA exhibits good biocompatibility, non-immunogenicity, and high osteoinductivity [110]. Early investigations on the biocompatibility of HA particles of different sizes, performed using bone-relevant cells (primary osteoblastic/osteoclastic cells), evidenced that larger-sized microparticles (500–841 μm) presented a better biocompatibility profile with respect to smaller-sized ones (<53 μm), which were shown to promote osteoclast activity and to restrain osteoblast activity [111]. As such, damaging effects are expected in the case of implant fracture or abrasion/erosion. It was also found that the uptake of nano-sized HA (20–80 nm in diameter) by bone marrow mesenchymal stem cells (MSCs) and osteosarcoma cells (U2OS) was dependent on the size of NPs and the type of cells [112]. Thus, 20-nm-sized HA particles greatly sustained the proliferation of beneficent MSCs, while the multiplication of tumor cells was inhibited. It is worth mentioning that human adipose-derived stem cells (hADSCs) which exhibit higher capacity of proliferation and multi-lineage differentiation in vitro are considered the most attractive MSCs due to the ease of their withdrawal and large availability. hADSCs can be charged in HA scaffolds, which promote in vitro osteogenic differentiation [113] and ectopic bone formation using HA scaffolds subcutaneously implanted in mice [114]. An increase in mitochondrial metabolism and consequent ROS generation underline the adipogenic differentiation of MSCs, as demonstrated by specific inhibition of the mitochondrial respiratory pathways [115].

Various in vitro studies highlighted that the interaction of cells with HA NPs can be modulated not only by size, but also through their charge and shape, thereby influencing their biocompatibility and the intended medical use. It was demonstrated that the uptake of nano-sized HA with a positive charge by MC3T3-E1 mouse pre-osteoblastic cells was higher compared to HA NPs bearing a negative charge, possibly due to better interaction with the negatively charged cell membrane [116]. The shape of HA nanostructures was proven to influence cellular viability and proliferation [117]. Thus, the study of Xu et al. [118], using the MTT assay and primary cultured rat osteoblasts, demonstrated that HA nanostructures with smaller specific surface areas induced lower apoptosis rates in the concentration range of 20–100 μg/mL. Moreover, needle-shaped particles (10–20 nm in diameter and 30–50 nm in length) inflicted greater cellular injury than spherical (10–30 nm in diameter) or rod-like particles (20–40 nm in diameter and 200–400 nm in length; 20–40 nm in diameter and 70–90 nm in length). At higher concentrations (100 μg/mL), the mentioned nano-sized HA particles were shown to induce enhanced ROS generation, as assessed by flow cytometry with DCFH-DA. The most active were the needle-shaped and spherical NPs which also induced higher apoptosis rates, thus indicating that oxidative stress underlined at least partially the observed cell death. Jin et al. [119] evidenced that HA nano-rods (~20 nm in diameter and ~80 nm in length) in the concentration range of 10–40 μg/mL were taken up into MC3T3-E1 mouse pre-osteoblasts via micropinocytosis and induced apoptosis through mitochondria- and lysosome-dependent damage pathways (mitochondria: altered expressions of B-cell lymphoma 2 (Bcl-2) and Bcl-2-associated X protein (Bax), decrease in mitochondrial membrane potential, and activation of caspase 3; lysosomes: lysosomal membrane permeabilization and increased the release of cathepsins B). Apoptosis was mainly correlated with oxidative stress ensuing from increased ROS formation (assessed by flow cytometry with DCFH-DA) and a decrease in endogenous antioxidant mechanisms related to SOD and glutathione peroxidase (GPx) activities. The involvement of this redox shift in inducing cell death was demonstrated using the ROS scavenger N-acetylcysteine (NAC) which was able to significantly protect osteoblasts against the apoptotic signals delivered by oxidative stress.

The clinical failure or success of an implanted bioceramic depends not only on the bone remodeling cellular system comprising osteoblasts, osteocytes, and osteoclasts, but also on the interaction of blood leukocytes with the implant surface and the released ionic components. Phagocytes (monocytes, macrophages, and neutrophils) mediate key events for tissue repair at the interface with the implant surface, and inflammation triggered by these cells is essential for promoting wound healing and for restoring local homeostasis [120]. The pro-inflammatory activity of monocytes/macrophages can be modulated by the surface topography of HA-based implants [121]. For instance, monocytes cultivated on HA discs with plate-like surface morphology (micrometric size) were shown to release higher levels of ROS (detected using a luminol-amplified luminescence assay) as compared to monocytes cultivated on discs made of needle-like agglomerates (nano-sized); thus better inducing bone healing [122]. Therefore, a smart design of implant surfaces is needed for improving the clinical performance of HA-based implants through modulation of inflammatory and oxidative processes which greatly impact bone remodeling.

HA NPs also impact angiogenesis, an important mechanism involved in tissue repair and regeneration, aimed at fulfilling the increased local need for nutrients, oxygen, and growth factors [123]. As comprehensively demonstrated by Shi et al. using umbilical vein endothelial cells (HUVECs) [28], HA NPs (~20 nm and ~80 nm) were taken up by endothelial cells via energy-dependent endocytosis pathways involving clathrin and caveolin, while micro-sized NPs (~12 μm) were incorporated by macrocytosis. HA NPs localized and interacted mainly with lysosomes, and induced a decline in nitric oxide (NO) production in a concentration-dependent manner, in line with the trends seen in cell migration and tube formation. The major impact of HA NPs on endothelial cells seems to be underlined by the downregulation of reactive nitrogen species (RNS), and not by enhanced ROS production. The reduction of NO levels was attributable to a dose-dependent decrease in the phosphorylation of endothelial nitric oxide synthase (eNOS), which was associated with the downregulation of phosphorylated protein kinase B (Akt). These alterations in the PI3K/Akt/eNOS signaling pathway trigger a decrease in the viability of endothelial cells and tube formation ability, thereby limiting angiogenesis and tissue repair. Moreover, PI3K/Akt inhibition is expected to hinder the activation of NADPH oxidase and, hence, limit superoxide-driven oxidative stress in endothelial cells [124].

HA was also investigated in relation to tumor cells considering its potential applications in tumor-associated bone segmental defects [125]. Han et al. [126] demonstrated using cancer (MGC-803, Os-732, Bel-7402) and normal cell lines (L-02 hepatocytes, MRC-5 lung fibroblasts and HaCaT keratinocytes) that HA NPs, especially those with lower dimensions of 60 nm, inhibited the proliferation of tumor cells more than normal ones, partly due to enhanced incorporation. This effect was also attributed to a decrease in protein synthesis following the interaction of ribosomes with NPs, and was apparently not related to increased levels of ROS. Other studies also emphasized that HA NPs may specifically kill or inhibit the proliferation of tumor cells, such as osteosarcoma cells [127]. The cytotoxic effect exerted by HA NPs in cancer cells was not underlined by an enhanced oxidative stress, as stated by the authors, considering that the activities of succinate dehydrogenase and SOD, key enzymes responsible for ROS generation and scavenging, respectively, were significantly decreased by HA NPs in both cancer and normal cells [126]. Different results were communicated in the study of Xu et al. [128], showing that the observed decrease in the number of human gastric C6 cancer cells exposed to HA NPs was due to apoptosis associated with increased ROS generation (detected by flow cytometry with DCFH-DA) and decreased antioxidant defense mechanisms related to SOD activity. Pre-treatment of cells with the ROS scavenger N-(2-mercaptopropionyl)-glycine was able to protect glioma cells against the HA-triggered apoptosis, thus demonstrating that increased ROS levels indeed contributed to cell death. The conflicting results obtained in different studies might be attributed to distinct reactivity to HA treatment exhibited by different types of cells with various origins and status (normal or tumor), or might be related to dosage, size, or shape of the investigated NPs. For instance, Zhao et al. [129] showed that rod-shaped HA NPs with different surface areas, in the concentration range of 10–300 μg/mL, did not affect the viability human epithelial virus-transformed lung cells (BEAS-2B), murine macrophages (RAW264.7) and human hepatocellular carcinoma cells (HepG2). Nevertheless, a significant, but not cytotoxic, increase in ROS generation was registered at 4 h post exposure, as detected by flow cytometry with DCFH-DA. The highest levels of ROS were generated by those NPs having the largest specific surface area, which increased the cell–particle interaction.

Altogether, there is experimental evidence that HA NPs, in particular situations, may induce oxidative stress in cells involved in osseointegration of implants and in bone remodeling. Accordingly, there is an urgent need to identify and develop therapeutic strategies aimed at limiting such alterations of the redox balance that may compromise the clinical outcomes of implantation.

Doping of HA with cerium recently emerged as a promising approach in bone implantation, providing a complex control of oxidative stress and redox signaling through the redox cycling of cerium ions [130,131]. Cerium NPs (ceria) are used for their remarkable antioxidant activities [132], deriving from the changes in the oxidation state between Ce^{4+} (fully oxidized) and Ce^{3+} (fully reduced) and from the exquisite ability of cerium to adjust the electronic configuration for best fitting the close environment and for recycling [133]. Cerium is a potent ROS scavenger which significantly decreases the levels of ROS, especially of the toxic hydroxyl radical [134]. Moreover, it exhibits excellent SOD and catalase-mimetic properties, thereby reinforcing the cellular tools for antioxidant protection against the bursts of superoxide anion and hydrogen peroxide at the cell–implant interface. It was found that higher levels of Ce^{3+} were efficient scavengers of superoxide [135], while higher levels of Ce^{4+} favored the catalase-mimetic activity [136]. The Ce^{3+}/Ce^{4+} oxidation state ratio was shown to be dependent on the primary particle size, with smaller cerium NPs having increased concentrations of Ce^{3+} as compared to their larger counterparts [137]. Moreover, cerium oxide proved good oxidative-stress anti-microbial activity [138], especially against Gram-negative bacteria. Therefore, cerium provides an additional advantage by counteracting potential low-grade infections at the site of implantation, thus reducing the need for prophylactic antibiotic treatment.

From the redox perspective, another option for doping HA is to use selenium, an essential micronutrient involved in various metabolic processes [139] as part of selenoenzymes, which are able to prevent and reverse even severe oxidative damages [140]. Selenium was also shown to play an important role in bone development [141]. The selenite ion represents a physiologic ionic doping agent for HA due to its almost identical size with respect to the orthophosphate tetrahedron. As demonstrated by Uskoković et al. [142], nano-sized selenite-HA has osteoinductive effects by eliciting an overall higher metabolic activity of cells. It is worth noting that, although selenium has antioxidant functions, selenite itself may trigger ROS production at levels that sustain osteogenesis [143]. Only if ROS levels increase above a threshold, for instance, in the case of cells exposed to NPs containing 3% selenite, do the apoptotic pathways get activated and cell death occurs [142]. The cytotoxic effect of selenite is of paramount importance for inhibiting osteoclastic bone resorption [144], as well as for decreasing the progression of certain bone tumors [144]. Selenite-induced oxidative stress and consequent apoptosis were found to be closely related to the intracellular level of reduced glutathione (GSH), a major cellular antioxidant molecule, with a key role in bone remodeling [145]. GSH proved to have a dual role in the effects of selenium on cancer versus normal cells. Thus, GSH may act as a pro-oxidant in cancer cells, facilitating selenium-induced oxidative stress, and as an antioxidant in normal cells, protecting them against oxidative stress and apoptosis. In addition to these effects, selenite-doped HA, like cerium-doped HA, exhibits anti-microbial activity [142] and could, therefore, reduce the post-implantation use of antibiotics. Altogether, selenite–HA materials show characteristics that make them suitable for improved osseointegration of bone implants and also for counteracting potential side-effects related to oxidative stress and infection.

A more general therapeutic approach to control the redox balance might be the pharmacologic activation of the cytoprotective transcription factor nuclear factor erythroid 2-related factor 2 (NRF2), which boosts the endogenous antioxidant system and controls the transcription of more than

250 cytoprotective genes [146,147]. Moreover, NRF2 was shown to contribute directly to bone remodeling by maintaining the equilibrium between osteoblast and osteoclast activities [148].

Concluding, potential alterations of the redox balance by HA and/or by the doping agents has to be taken in consideration when designing materials for bone implants, being aware that the levels of ROS and of endogenous antioxidant activity are critically involved in maintaining bone homeostasis and in sustaining regenerative processes.

2.5. Methods to Assess Oxidative Stress

Considering the critical involvement of ROS in various physiologic and pathological processes, a large panel of methods was developed and is still under construction for precisely detecting various types of ROS and their end result. More complex investigations regarding oxidative stress and redox signaling are needed for in-depth evaluation of the involvement of oxidative stress and redox signaling in bone regeneration and engineering using HA-based implants in order to identify therapeutic targets for improving osseointegration and the long-term outcome of implantation.

3. Efficacy Evaluation of CaP-Based Bioceramics

CaP-based bioceramics have a demonstrated impact on various biological processes, such as redox balance, cell signaling, or epigenetic control of cellular activity, and various basic research findings of these interactions are discussed below. Figure 2 summarizes this interactions and cellular outputs, with a focus on doped HA.

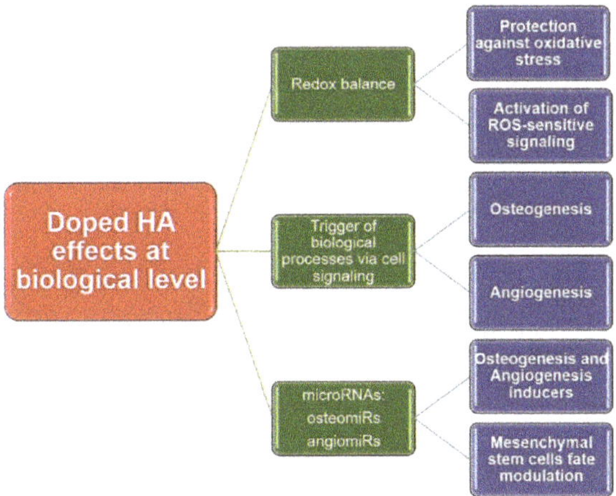

Figure 2. Interactions of doped hydroxyapatite (HA) with several main biological processes, and their cellular outputs.

3.1. Assessment of Osteogenic Effects

HA and its cation- and/or anion-substituted varieties are mainly envisioned for the development of bone regeneration applications. For osteogenesis studies, HA is used as implant coating, granules, or block bulk structures which act as non-resorbable materials in the short term, but still subjected to partial degradation and metabolization over time [149]. However, in addition to having poor mechanical properties, scaffolds of pure HA do not significantly promote vascularization and osteoinductivity [17]. Hence, additional doping is usually performed to augment osteoregeneration. HA can be modified by cationic and anionic substitution [1,19,150], or incorporated in a large variety of composite materials (together with metals, polymers, or proteins) [151]. The HA-based materials are evaluated on target cell populations and tissues (bone tissue and osteoblasts, alveolar bone, and tooth and mesenchymal cells), followed by animal testing. Although HA is a biocompatible material, each new composite or material containing HA must be tested again for cytotoxicity and biocompatibility, according to ISO 10993-6:2016 "Tests for Local Effects after Implantation". For in vivo testing, an implantation test is the recommended method to evaluate the biocompatibility of a material with the surrounding tissue. Usually, intramuscular or subcutaneous implantation is routinely performed, but special sites, such as bone, can be used, if justified accordingly. Rabbits or small rodents are recommended, but larger mammals can be used for long-term testing, especially if the period of use exceeds the lifespan of a small rodent. For non-absorbable materials, such as HA, the short-term responses are normally assessed from one week up to 12 weeks, and the long-term responses are tested for periods longer than 12 weeks. Implantation in bone tissue may require longer observation periods before a steady state is reached. At the end of the testing period, histological sampling is used to assess the local tissue response to the implant [152].

Section 3.2 presents the most common in vitro and in vivo models for osteoregeneration studies of HA-based bioceramics.

3.2. In Vitro Models of Osteogenesis

In vitro models of osteogenesis rely on using osteogenic cell cultures, namely, osteoblasts, osteoblast-like cells (e.g., MG63 cells, which are derived from osteosarcoma, and MC3T3-E1), or MSC-derived osteoblasts. MSC-derived osteoblasts are obtained through special cell culture conditions, during which specific osteoblast differentiating medium is used, for a variable time length, depending on the protocol. MSCs can be derived from different sources, such as adipose [153], bone marrow [153–156], and umbilical cord blood [157]. Despite exhibiting significant similarities, MSCs from different sources are not identical in terms of osteogenic properties and response to osteogenic stimulation [153], which highlights the importance of selecting the appropriate combination of MSC population and stimulation for performing osteogenesis studies.

The effects of HA on osteogenesis are assessed by a number of methods. The most basic approaches measure the cell proliferation, which can be achieved through different techniques, from cell counting to double-stranded DNA (dsDNA) quantification, mitochondrial activity assessment, and real-time cell analysis [158]. These methods overlap with those used for assessing cytocompatibility/cytotoxicity and are covered more extensively in Section 2.2. However, osteoblast proliferation is merely a first step in studying osteogenesis in vitro. Analyzing cell proliferation alone is not sufficient for assessing osteogenesis, as, during this process, osteoblasts reach complete differentiation and stop proliferating, making functional readouts compulsory.

Calcium deposition is indicative of complete differentiation of osteoblasts, and the Alizarin red S assay, based on a dye that binds calcium salts, was used to assess osteogenesis for decades [159]. However, this approach has limitations when it comes to studying osteogenesis using cells grown on an opaque substrate, as it does not allow for bright-field microscopy observation and subsequent quantification. Another method to assess the formation of new bone matrix that was used extensively is the von Kossa staining, a method that quantifies mineralization in cell culture, as well as tissue sections. However, this method is not specific to calcium phosphate; von Kossa staining alone was shown not to be appropriate for the identification and quantitation of bone-like mineral depositions, and it should only be used in combination with other methods to verify calcium phosphate presence and quality [160].

Alkaline phosphatase (ALP) is considered an early marker of osteoblast differentiation and represents a key player in mineralization [161–163]. ALP activity is frequently determined to assess functional osteogenesis using a *p*-nitrophenol phosphate disodium solution or commercially available enzymatic assays.

During osteogenesis, following the MSC proliferation phase, there is a significant expression of Runt-related transcription factor 2 (RUNX2), and its regulation is essential in bone formation [164–166]. The transcription factor RUNX2 has the capacity to upregulate the expression of collagen (Col-I), alkaline phosphatase (ALP), and osteocalcin (OCN) genes [167]. Moreover, the morphological changes and transformation of preosteoblasts into mature osteoblasts requires increased expression of (Osx) and secretion of bone matrix proteins such as OCN, bone sialoprotein (BSP) I/II, and Col-I [166,168–171]. Transcriptomic, proteomic, and metabolomic approaches, as well as immunodetection techniques (immunofluorescence - IFA, Western blot, enzyme-linked immunosorbent assay - ELISA), can be used to assess the expression levels of these osteogenic markers. Klontzas et al. compared the efficiency of different osteogenic agents on umbilical cord blood MSCs to promote the osteogenic process by quantifying osteonectin, OCN, ALP, and RUNX2 through qRT-PCR and reported differences between two commonly used osteoinductive agents, dexamethasone and BMP-2 [157]. Kulanthaivel et al. reported the improved osteogenic properties of Co^{2+}- and Mg^{2+}-doped HA, as confocal micrographs revealed increased expression of RUNX2 in MG-63 cells [172].

Transcriptomic, proteomic, and metabolomic approaches offer the ability to perform high-throughput studies; HTP proteomics includes both micro-arrays and mass spectrometry (MS) instruments; both are very sensitive, and both are able to generate a large amount of data. Microarrays are fast, with a large coverage of known proteins, and they could be useful if you want to "enlarge" the set of molecules to investigate. The same goes with MS instruments; the recent versions are faster and more sensitive, and they could be eventually used to both discover new markers and confirm the presence of some.

Trabecular bone histological organization is characterized by a three-dimensional (3D) network of osseous trabeculae of calcified matrix and isolated mature osteocytes, creating a spongy scaffold. The cavities of this scaffold are lined by osteoblasts and filled with blood and blood cell precursors.

The compact bone is a highly 3D organized structure composed of osteons. An osteon is the structural and functional unit, comprising concentric bone cylinders, all centered by a canal outlined by osteoblasts and filled with blood vessels and nerves.

Existing 3D culture models for osteogenesis and/or bone tissue regeneration rely on scaffolds made of collagen-derived gelatin [173], collagen–HA [174], ceramic materials, inorganic HA [175], chitosan, and alginate [176,177]. These scaffolds are a better model for bone tissue. Curtin et al. showed that collagen–HA scaffolds have better osteoinductive properties when compared to collagen-only scaffolds, as revealed by Alizarin red and von Kossa stainings, results that corroborated well with osteocalcin expression, as shown by immunofluorescence staining [174].

While 3D models allow in vitro testing using structures that resemble bone tissue, reduced effects in 3D cultured MSCs in comparison to results in monolayers were noted in previous reports [156,178]. One study reported that the expression of OCN and ALP, as well as the activity of ALP, was reduced in a 3D model to approximately half with respect to that detected in a two-dimensional (2D) model. However, this reduction from 2D to 3D did not prevent the treatment from enhancing bone repair when tested in vivo [179], highlighting the importance of moving from in vitro models to in vivo ones.

In vivo HA testing is usually performed in rabbits or little rodent species over short-term periods of time. There are several animal models widely used for the assessment of HA substitutes on bone reconstruction (as reviewed in Reference [180]). For flat bone reconstruction, the most common is the calvarial defect model, which is frequently tested in small rodents: rat [181,182] and rabbits [183]. Long bone reconstruction uses a distraction osteogenesis model—a surgical model that creates a gap in a long bone (usually tibia), using continuous traction on the osteotomy ends. This model was used with various rodent species (e.g., mouse [184–187], rat [188–191], and rabbit [192–194]) and even larger mammalians (e.g., sheep [195,196] or goats [197]). Endo-osseous implants are used for dental reconstruction or oromaxilofacial reconstruction after tumor resection [198], but their biocompatibility also needs testing [199]. The evaluation of biological effects requires imagistic evaluation (radiographs, micro computed tomography (CT)), histologic evaluation of newly formed osseous tissue, and biomechanical testing. Depending on the need, there are also animal models for specific bone-related pathologies, such as osteoporosis (as reviewed in References [200,201]), osteonecrosis [202], or rare bone diseases [203].

It is now common knowledge that HA bioceramics or even HA-based composites are a suitable solution for osteoregeneration, from both mechanical and osteoinductive points of view. Due to improved fracture toughness, Mg- [27,204–206] and Zn-doped [207–209] HAs are most frequently tested for osteoregenerative properties; however, the incorporation of other ions was also tested. Lithium-doped HA scaffolds, for example, demonstrated improved mechanical properties and increased osteogenesis potential over pure HA [17]. The rationale of Li-doping is supported by several in vitro studies showing its role in osteogenic progeny growth and development, as well as in shifting mesenchymal stem cell fate toward osteogenic differentiation (as reviewed in References [1,210]). The main drawback of Li–HA is the lack of angiogenic stimulation [17,211], which can be overcome by including it in a composite [212]. Strontium doping was also favorable for osteogenesis when compared to pure HA in calvarial bone defect models [213], as well as in osteoporosis models [214–216]. Beneficial in vivo results were also reported for manganese, cobalt, copper, and gallium. For a more detailed presentation of different varieties of cation- and anion-substituted HA scaffolds and their biological effects, some recent reviews can be addressed [1,210].

The same animal models are used for testing whether the scaffold architecture matters or not for the biological output. Data from implant testing and bone-specific animal models showed that, beyond chemical composition, the osteoinductive properties of bioceramic-based bone graft substitutes depend on gross architecture (e.g., 3D printing versus nanostructured scaffolds [217], the thickness and size of macropores [218], or the nanoarchitecture of the scaffold [219]). Osteoregeneration and biomechanical properties are improved by tailoring of pore sizes [220] or even adjusting the shape and distribution of HA crystals [221]. Furthermore, Cu-doping was shown to be a modifier of the micro/nano-structured surface on the HA scaffolds, with a positive impact on angiogenesis [222].

As novel and better models emerge from preclinical testing, translation toward clinical implementation is also in need of standardization. HA derivates are used in clinical trials as substitutes for autologous bone grafts (for systematic reviews, see References [223,224]), and included in guidelines as grade C (low quality of evidence) [225]; hence, more clinical trials are required. Also, for standardization purposes, in 2016, the Special Interest Group on 3D Printing (SIG) was created to fulfill two goals: "to provide recommendations toward the consistent and safe production of 3D printed models derived from medical images, and to describe a set of clinical scenarios for 3D printing appropriate for the intended use of caring for patients with those medical conditions" [226]. In terms of bone reconstruction, so far only craniomaxillofacial reconstruction following trauma, genetic diseases, or different types of tumors was addressed [226]. The investigation methods used to assess osteogenic effects are summarized in Table 2.

Table 2. Investigation methods used to assess osteogenic effects. MSC—mesenchymal stem cell; ALP—alkaline phosphatase; BSP—bone sialoprotein; IFA—immunofluorescence assay; OCN—osteocalcin; CT—computed tomography; 3D—three-dimensional; CCK-8—Cell Counting Kit-8.

Type of CaP	Biological Samples	Methodological Approach	Main Effects	References
Collagen/HA, HA, biphasic calcium phosphate	Rat MSCs	Cell proliferation (MTT) qRT-PCR	Rapid increase of osteogenic marker gene expression; increased expression of ALP	[154]
Sr-doped CaP	Human MSCs	Cell proliferation (LDH) ALP activity qRT-PCR of BSP II	Increased proliferation; enhanced ALP activity; increased expression of BSP II	[155]
Collagen–nano-HA scaffolds functionalized with microRNA (miRNA)	Human MSCs	qRT-PCR Mineral deposits quantification Calcium deposition IFA	Increased osteogenic markers; mineral deposition	[156]

Table 2. *Cont.*

Type of CaP	Biological Samples	Methodological Approach	Main Effects	References
Ag-doped hydroxyapatite/calcium silicate coating nano-Ti substrates	Mouse preosteobasts (MC3T3-E1 cells)	Cell proliferation (MTT) ALP activity ELISA	Enhanced proliferation; enhanced ALP activity; increased OCN expression	[162]
Co^{2+}- and Mg^{2+}-doped HA	MG-63 osteoblasts	Flow cytometry IFA	Similar cell-cycle profile as control cells; induced RUNX2 expression	[172]
Biphasic calcium phosphate ceramics	Animal tissue	Histological analysis	Mineralized bone formation	[197]
HA-coated implants	Animal tissue	Removal torque test Histological analysis	Higher removal torque value for HA group; new bone formation with increased density	[198]
HA-coated titanium implants	Animal tissue	Implant stability test	HA-favorable effect on osseointegration	[199]
Ca-doped MgP, HA	Animal tissue	Histological analysis	Bone healing results with complete osseointegration	[204]
Nano-to-submicron hydroxyapatite coatings	MSCs	Cell count and morphology analysis	Reduced cell adhesion	[205]
Sr-doped HA	MC3T3-E1 Animal tissue	Cell proliferation ALP activity Histological analysis	Enhanced proliferation and ALP activity; new bone formation	[213]
Sr-doped HA	Animal tissue	Histological analysis	Higher regeneration efficacy of Sr-doped HA compared to HA and control	[214]
Sr-doped HA	Animal tissue	Micro-CT assessment Histological analysis	Increased bone density around Sr-HA implants; improved trabecular microarchitecture compared to HA	[215,216]
Nanostructured HA scaffolds	Animal tissue	Histological analysis Micro-CT	Superior osteogenic capacity of foamed scaffolds compared to 3D-printed structures	[217]
β-TCP scaffolds	MSCs Animal tissue	Cell proliferation (CCK-8), Micro-CT Histological analysis	Smaller pore sizes; improved bone regeneration	[220]
Nano-HA	Animal tissue	Histological analysis	Bone regeneration similar to commercially available materials	[221]
Nano-HA scaffolds	MSCs Animal tissue	Cell proliferation (MTT) ALP activity qRT-PCR Western blot Micro-CT Histological analysis	Nanostructured HA surfaces promote cell attachment, proliferation, and osteogenic differentiation; enhanced osteo- and angiogenesis in vivo	[227]

3.3. Assessment of Angiogenic Effects

Mimicking the native microenvironment of bone extracellular matrix (ECM) by providing an optimal vascularization represents the major goal in bone regeneration processes, to be accomplished through a series of important characteristics: biocompatibility, optimal biodegradability, pre-vascularized structure, osteoconductivity, and less immunogenic responses. Unfortunately, currently available bioceramics do not entirely satisfy all these requirements [30–32]. The bone regeneration process is based on complex interconnected biological events including angiogenesis, osteogenesis, and inflammatory reactions [33].

Tissue regeneration is largely dependent on cell signaling that is mediated by cellular interactions with various key molecules, including growth factors. The angiogenic process involves a wide range of regulatory angiogenic proteins, such as vascular endothelial growth factor (VEGF), fibroblast growth factor-2 (FGF-2), epidermal growth factor (EGF), platelet-derived growth factor (PDGF), placental growth factor (PlGF), insulin-like growth factor (IGF), angiopoietins, matrix metalloproteinases, endostatin, thrombospondin-1, and insulin-like growth factor-binding protein-3, as well as pro-inflammatory cytokines (interleukin (IL)-1b, IL-6, tumor necrosis factor-α (TNFα)). The myriad of molecules involved in the regeneration processes are crucial players that trigger many signaling pathways, such as the VEGF signaling pathway, PI3K/Akt/eNOS, Ras/mitogen-activated protein kinase (MEK)/extracellular signal-regulated kinase (ERK), Sonic Hedgehog (SHH), Notch, and Wingless-related integration site (Wnt) [228].

Deficiencies in the vessel renewal or the paucity of neo-angiogenesis frequently result in delayed healing or tissue regeneration failure [229]. Notably, many studies reported in vivo implantation failures predominantly due to the lack of angiogenesis in the scaffold [230], delaying osteoid deposition and matrix development [229], consequently decreasing the bone healing rate [231]. Different strategies were designed to solve this issue, amongst which one can mention pre-vascularization by co-culturing of angiogenic and osteogenic cells [232].

To address these challenges, the HA-based bone scaffolds need to be designed for regulating cell behavior; therefore, generating the necessary angiogenesis is highly important for promoting a successful regenerating process.

A recent study pointed out that the expression of angiogenic genes (*eNOS, VEGF,* and *bFGF*) was upregulated in the conditioned media of HA micro/nanostructures, sustaining the high potential for inducing angiogenesis by regulating the immune microenvironment of macrophages. Based on these results, they pointed out that the basic regulation of macrophages influences the angiogenesis potential of endothelial cells (ECs), indicating that the biomaterial structure could regulate angiogenesis during tissue regeneration via a multi-pathway mechanism, either directly (endothelial cell stimulation) or indirectly (stimulation by activating macrophages) [233].

Aptamer-related technologies represent a revolutionary tool in bone tissue engineering. In this regard, a study conducted by Son et al. [234] developed an aptamer-conjugated HA (Apt-HA) that promotes angiogenesis, specifically targeting VEGF. In order to evaluate in vivo angiogenesis and bone regeneration, Apt-HA and HA were bilaterally implanted into a rabbit model and analyzed after eight weeks using micro-CT, histology, and histomorphometry. The results of this study demonstrated that Apt-HA showed significant increased blood vessel formation compared to simple HA, making the engineered Apt-HA an innovative candidate with great potential in promoting angiogenesis and bone regeneration [234].

Various potential strategies were explored to enhance the angiogenic capacity of the HA scaffolds. In this regard, several ions were observed to possess the potential to increase neovascularization (Mg, Cu, and Co), osteogenesis (Li, Zn, Sr, and Mn), or both (Si and B).

Magnesium is known to be a fundamental element in bone and tooth composition, and it was already reported that increased levels of Mg were correlated with a favorable endothelial function. It was shown that high concentrations of Mg modulate in vitro vascular endothelial cell behavior, giving new insights into the role of Mg in angiogenesis. In addition, it was shown that high concentrations of Mg enhanced the synthesis of nitric oxide, through the upregulation of endothelial nitric oxide synthase (eNOS). In such a supportive microenvironment, endothelial cells are induced to migrate and grow, thus accelerating the formation of new vessels through a VEGF-like mechanism of inducing angiogenesis [235].

In this scenario, Deng et al. [13] studied the angiogenic effects in a goat model using magnesium-doped porous HA (MgHA) combined with recombinant human bone morphogenetic protein-2 (rhBMP2). The in vitro studies revealed that the VEGF expression, assessed by ELISA, for the 5% MgHA/rhBMP2 group was significant different compared with other groups at days seven and 14. The outcome was higher for the 5% MgHA group at day 14 than that of the other two groups. The in vivo study showed a significantly higher expression of VEGF and collagen I messenger RNA (mRNA) expression (evaluated by PCR) at 12 weeks in the MgHA/rhBMP-2 group.

Sartori et al. tested two new bi-layered scaffolds, one for chondral regeneration/type C (containing type I collagen), and another for the regeneration/osteogenic/type O of the subchondral layer (containing bioactive Mg-doped HA crystals), both seeded and incubated with hMSCs [236]. At four weeks from implantation in a mouse model, immunohistochemistry analysis revealed that only inside the type O scaffold layer was new vessel formation observed, suggesting a neo-angiogenesis process. Furthermore, a large number of positive cells for anti-human VEGF were observed from the scaffold surface to the center, while a few positive cells for anti-mouse VEGF were found near the scaffold surface. At eight weeks from implantation, bone tissue formation was observed in the O scaffold layer, appearing smooth and pale stained in comparison to the surrounding tissue. Osteoblasts were present on newly formed trabecular tissue using O scaffold as a template [236].

This study is in accordance with a Yang et al. report [237], which noted that the hMSC engineered 3D scaffold of HA/TCP, polyurethane (PU) foam, poly(lactic-co-glycolic acid)/poly(ε-caprolactone) electrospun fibers (PLGA/PCL), and collagen I gel seemed to be very effective in stimulating blood vessel formation, thereby ensuring facilitated oxygen and nourishment diffusion inside the scaffold, as also shown by the positive expression of anti-human VEGF by immunohistochemistry.

Canullo et al. [238], in a double-blinded randomized controlled trial, attempted to histologically evaluate the complex connection between angiogenesis and osteogenesis in post-extraction sockets enhanced with Mg-enriched HA, via indirect immunohistochemistry, using alkaline phosphatase, cluster of differentiation 34 (CD34), and caveolin-1 antibodies. The histomorphometric analysis indicated early angiogenesis followed by early osteogenesis processes, generated by Mg-enriched HA, which suggested it as a suitable material for post-extraction ridge preservation in dental medicine [238]. Also, a strong expression of caveolin-1 in preosteoblasts, osteoblasts, and osteoclasts was found [238], in accordance with Frank et al. [239], who noted that caveolin-1 was strongly abundant in endothelial cells regulating functions such as angiogenesis, vascular permeability, and transcytosis [239]. In another study on caveolin-1-deficient mice, angiogenesis was found to be markedly reduced in comparison with control mice [240].

Sun et al. [241] explored the functions and properties of an HA nanowire/magnesium silicate core–shell (HANW@MS/CS) porous scaffold by SEM (scanning electron microscopy) and inductively coupled plasma (ICP) optical emission spectrometry (for the release behavior of ions), and pointed out that HANW@MS/CS promoted the attachment and growth of rat bone marrow-derived mesenchymal stem cells (rBMSCs), inducing the expression of osteogenic differentiation related genes and the *VEGF* gene of rBMSCs. Moreover, the HANW@MS/CS scaffold stimulated in vivo angiogenesis and bone regeneration, by enhancing the gene expression of *VEGF*, *RUNX2*, *OCN*, and *OPN* (osteopontin) compared with the HANW/CS and CS scaffolds (assessed by RT-qPCR). The experimental results indicated that the abilities of these scaffolds in stimulating angiogenic and osteogenic responses of rBMSCs followed the trend HANW@MS/CS scaffold > HANW/CS scaffold > CS scaffold [241].

An interesting study designed by Calabrese et al. [242] attempted to evaluate the osteoinductive and angiogenic potential of a cell-free collagen–MHA (magnesium-enriched HA) scaffold containing a bilayer scaffold made of collagen I alone (layer 1: Coll) and collagen–MHA (layer 2: Coll–MHA), using innovative biomaterials that closely mimic the bone characteristics. Using fluorescence molecular tomography (FMT) imaging, an increase in de novo formation of vessels was revealed, and, using AngioSense 680, the fluorescent signal appeared reasonably elevated until four weeks, before recording a decreased signal by about 40%, with no statistical significance. Hematoxylin and eosin (HE) staining also confirmed the vascularization, indicating the presence of structures resembling blood vessels, which were abundant at four and eight weeks post implantation mainly within the Coll–MHA layer.

The results revealed by HE staining were validated by the expression of CD31 endothelial marker, which recorded a weak signal at the first week, with an increasing trend up to week eight. These interesting findings bring novel insights into collagen–HA scaffolds designed using an innovative biological method, having the special capacity to recruit host cells by themselves, promoting bone regeneration. In addition, the spontaneous appearance of vascular structures within the innovative scaffold holds promise for successful bone regeneration using scaffolds alone, without supplementary growth factors or other in vitro manipulated cells, as many other studies confirmed [242].

Due to copper's known stimulatory effect on endothelial cells toward angiogenesis, many studies were conducted using Cu-doped HA for increasing the angiogenesis capacity [243]. Barralet et al. discovered that a CaP scaffold doped with low doses of Cu^{2+} led to the formation of new vessels along the macro-pore axis, as confirmed by immunohistochemistry [244]. In addition, scaffolds containing angiogenic factors, especially copper and a copper–VEGF combination, expressed a greater degree of tissue ingrowth than the control. Moreover, the addition of Cu ions strengthened the bioactivity [222,245], in which it was shown that Cu-assisted hydrothermal deposition techniques provide a reliable route toward engineering micro/nano-structured surfaces on Cu-doped HA scaffolds, with beneficial properties in terms of angiogenesis and bone regeneration. Based on the findings that the doped elements not only affect the apatite physical structure, but also strengthen its biological function, it was observed, by scanning electronic microscopy (SEM), that Cu concentration also affects the morphologies of CaP coatings that grow on HA scaffolds, significantly increasing cell proliferation [222].

Strontium is currently used in the treatment of osteoporosis. Sr-doped HA scaffolds enhanced the solubility and stimulated earlier bone formation, while also sustaining a better cell attachment and proliferation [246].

Different in vitro studies mentioned that Sr-doped HA supports osteoblast proliferation and differentiation processes by triggering calcium-sensing receptors, as well as stimulating angiogenesis and osteogenesis. It was observed that, in comparison with calcium polyphosphate (CPP) and HA scaffolds, the formation of a tube-like structure and the expression of platelet endothelial cell adhesion molecule (PECAM) in the co-cultured model was better in the Sr-doped CPP (SCPP) scaffold [247]. Also, a positive effect of Sr on angiogenesis is supported by in vivo studies which revealed the formation of new vessels, highlighted by positive staining for CD31, especially in 50% (molar ratio) Sr-doped HA (50Sr-HA), after four weeks of implantation compared to HA and 8Sr-HA [248].

Also, the capability of Sr-doped calcium polyphosphate (SCPP) to stimulate angiogenic and osteogenic processes was analyzed in vitro and in vivo by Gu et al. [247]. They used an in vitro co-cultured model of human umbilical vein endothelial cells (HUVECs) and osteoblasts and then cultured the cells with SCPP, calcium polyphosphate (CPP), and HA scaffolds. Subsequently, ELISA analysis demonstrated that PECAM-1 concentration in the SCPP group was significantly higher than in the CPP group and HA group, with a maximum at the 28th day, in accordance with immunofluorescence analysis. Strings of tube-like structured (TLS) HUVECs were detected spreading through the co-cultured model. The PECAM-1 expression of HUVEC and the formation of TLS were longer for the SCPP group in comparison with CPP, demonstrating that SCPP has a higher ability to induce angiogenesis in vitro. The in vivo model revealed a positive immunostaining for VEGF in newly regenerated tissue in both CPP and SCPP groups at weeks eight and 16 post operation with more formation of new bone and tube-like structures (TLSs) in the SCPP group. On the other hand, at 16 weeks, the HA group presented a mild positive result in VEGF expression and the formation of bone, while no TLSs were observed. The intensity of positive VEGF staining decreased at 16 weeks in the CPP and SCPP groups. In conclusion, the SCPP scaffold could represent a potential biomaterial that stimulates angiogenesis in bone tissue engineering [247].

Cobalt represents another promising essential element in the bone regeneration field due to its hypoxia-mediated angiogenesis capacity by hypoxia-inducible factor (HIF-1α) activation [246]. Based on histological and immunohistochemical analyses, it was observed that the substitution of Co^{2+} could improve the angiogenesis properties of HA, by mimicking hypoxia conditions, upregulating HIF-1a and VEGF expression [249].

Dual doping of bivalent Mg^{2+} and Co^{2+} ions was evaluated, and the in vitro analysis on bone cells (MG-63) showed that HAC (5% ($CoCl_2MgCl_2$)–HA) induced the highest expression of VEGF, followed by HAN (5% ($Co(NO_3)_2$–$Mg(NO_3)_2$)–HA), while HA showed the lowest expression, equivalent to the control. This finding highlighted that the high expression of HIF-1α in HAC was directly influenced the VEGF synthesis. In brief, dual doping improves the osteogenic and angiogenic properties of HA, resulting in an improved biomaterial for bone tissue engineering [172].

Zinc is another essential trace element, being important in the structure of various metalloenzymes, such as alkaline phosphatase (ALP), an extremely important molecule for the maturation of new bone formation.

Nano-HAs, with/without nano-zinc oxide (n-ZnO), were studied, and a chicken embryo chorioallantoic membrane (CAM) assay indicated the induction of angiogenesis for the scaffolds containing n-ZnO, as well as significant upregulation of angiogenic-related genes, confirmed by RT-PCR analysis. In consequence, scaffolds containing n-ZnO have substantial importance for inducing osteogenesis and angiogenesis processes in bone tissue engineering strategies [232].

Lithium is present in organisms as a trace metal, and various studies reported that Li could have effects in increasing bone density [250] and promoting osteogenic differentiation of bone marrow mesenchymal stem cells (BMMSCs) by activating the Wnt/glycogen synthase kinase 3β (GSK-3β) signaling pathway [251,252]. In a recent study, an innovative lithium-doped HA scaffold (Li-HA) was evaluated, seeded with hypoxia-preconditioned bone marrow mesenchymal stem cells (BMMSCs). When the seeded cells were preconditioned in hypoxia medium, the new bone formation was improved, with higher β-catenin and lower GSK-3β expression. Also, the HIF-1α, VEGF, and CD31 expression, evaluated by qPCR, was upregulated, exerting a positive effect on activating the Wnt and HIF-1α signaling pathway [211]. The investigation methods used to assess angiogenic effects are summarized in Table 3.

Table 3. Investigation methods used to assess the angiogenic effects. HUVEC—human umbilical vein endothelial cells; IHC—immunohistochemistry; TCP—tricalcium phosphate; PU—polyurethane; PLGA/PCL—poly(lactic-co-glycolic acid)/poly(ε-caprolactone); HANW—HA nanowire; MS/CS—magnesium silicate core–shell; BMMSC—bone marrow-derived mesenchymal stem cells; CAM—chorioallantoic membrane; IF—immunofluorescence; VEGF—vascular endothelial growth factor; SCPP—Sr-doped calcium polyphosphate; n-ZnO—nano-zinc oxide; HIF-1α—hypoxia-inducible factor.

Type of CaP	Biological Samples	Methodological Approach	Main Effects	References
Mg-doped HA	Co-culture model of HUVECs and MG63	ELISA PCR	Significant effects on bone formation and angiogenesis; Increasing VEGF	[13,238]
	-	IHC	Early angiogenesis followed by early osteogenesis	
Bi-layered scaffold (type I collagen and Mg/HA)	hMSCs	IHC	Stimulating proliferation and differentiation of hMSCs for tissue growth and neo-angiogenesis	[236]
3D scaffold (HA/TCP, PU, PLGA/PCL and collagen I gel.	-	IHC	Stimulating blood vessel formation	[237]
HANW@MS/CS (Magnesium Silicate)	rBMMSCs	SEM RT-qPCR analysis	Mg and Si elements contribute to angiogenic induction, bone formation, and blood vessel formation	[241]

Table 3. *Cont.*

Type of CaP	Biological Samples	Methodological Approach	Main Effects	References
Cu-doped HA	Animal tissue	IHC	CaP scaffold doped with low doses of copper sulfate led to the formation of micro-vessels	[244]
	Animal tissue	SEM	The micro/nano-structure of the Cu5–HA scaffold resulted in more angiogenesis, which formed the new blood vessels	[222]
Sr-doped CaP scaffold	Co-culture model of HUVEC and osteoblasts Animal tissue	Phase-contrast microscopy IHC	Formation of tube-like structure and the expression of platelet endothelial cell adhesion molecule in co-cultured model was better in SCPP scaffold Potential to promote the formation of angiogenesis	[247]
	Animal tissue	IHC	New vessel formation in Matrix-50Sr-HA explants, mainly after 4 weeks of implantations, suggested a positive effect of Sr on angiogenesis	[248]
Co-doped HA	Animal tissue	IHC	Enhanced vascularization in vivo; large blood vessels were predominantly found in Co-doped HA	[249]
Zn-doped HA	-	CAM assay	The number of vessel branches in the modified scaffolds with n-ZnO was significantly higher compared to the modified scaffolds without n-ZnO	[232]
Li-doped HA	BMMSCs (bone-marrow mesenchymal stem cells)	Western blot analysis IHC and IF	HIF-1α and VEGF immunohistochemistry indicated that the hypoxia BMMSCs group had significantly more positive cells than the other three groups	[211]

In conclusion, in bone tissue engineering, biological processes such as angiogenesis and osteogenesis are finely concerted during lifelong bone formation, and many studies established that the microenvironment could directly control the development of these processes.

3.4. Signaling Pathways Involved in Osteo- and Angio-genesis

New bone formation, as well as bone regeneration and bone healing, requires both diffusible signals and proper vascularization. Hence, angiogenesis is frequently investigated alongside with osteoinductivity of various cation- and/or anion-substituted-HA bioceramics. Furthermore, the same signals (see Figure 3) are responsible for inducing osteodifferentiation/proliferation and angiogenesis, depending on the receiver cell type.

Such extensive knowledge on signaling proteins involved in bone formation and regeneration was translated into clinical practice by clinical trials aiming at bone defect repair. To date, two members of the BMP family (BMP-2 and BMP-7) were approved as treatment in orthopedic and maxillofacial reconstruction. In selected pathologies, they were shown to outperform bone autograft, but potentially severe side effects called for caution in their clinical use [253–255]. Initially used as recombinant proteins, BMPs and growth factors are now delivered using various scaffolds, including HA substitutes [256–258]. HA by itself was shown to trigger "a specific intracellular signal transduction cascade during early osteoblast adhesion, activating proteins involved with cytoskeleton rearrangement, and induction of osteoblast differentiation" [259]. Additional growth factors such as VEGF, PDGF, IGF, or transforming growth factors (TGF-1 and TGF-2) can be adsorbed onto the bioceramic bone scaffold; however, to avoid the expensive costs, they were replaced with ions (e.g., Li, Co, Ni, Mg, Sr, and La) having similar effects. The incorporation of Au activates the Wnt/β-catenin signaling pathway, explaining the osteoinductive capability of HA–Au NPs [260]. The silk fibroin (SF)/HA/polyethyleneimine-functionalized graphene oxide (GO-PEI) scaffolds loaded with mir-214 inhibitor (SF/HA/GPM) increased the expression of activating transcription factor 4 (ATF4) and activated the Akt and ERK1/2 signaling pathways in mouse osteoblastic cells (MC3T3-E1) in vitro [261]. Boron-containing HA was shown to affect genes involved in Wnt, TGF-β, and response to stress signaling pathways [262]. An increase in CeO_2 content in HA coatings increased alkaline phosphatase (ALP) activity, calcium deposition activity, and the Wnt/β-catenin signaling pathway [263]. HA promoted the osteogenic differentiation of hBMSCs, possibly by increasing cell attachment and promoting the Yes-associated protein (YAP)/Tafazzin (TAZ) signaling pathway [264].

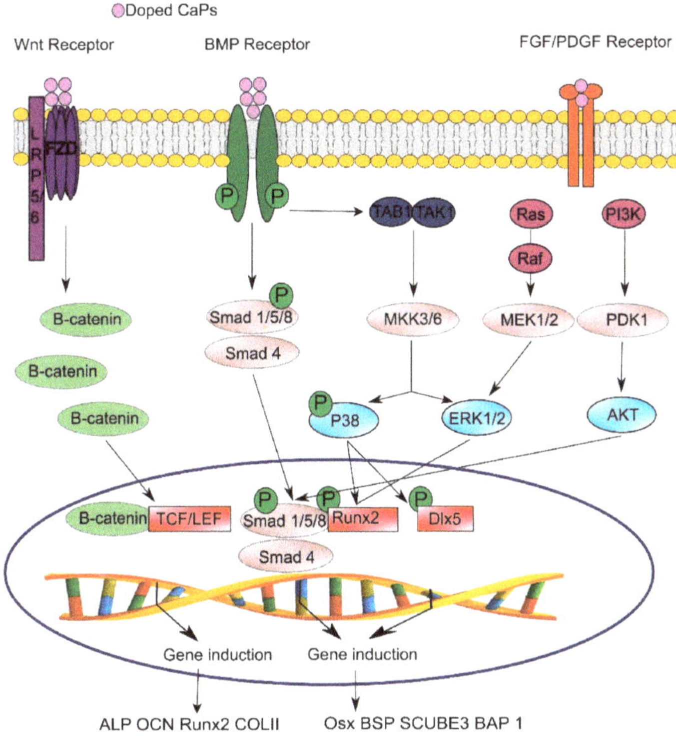

Figure 3. Cell signaling pathways activated in osteogenesis and angiogenesis models: Wingless-related integration site (Wnt) pathway, bone morphogenetic protein (BMP) pathway, and fibroblast growth factor (FGF)/platelet-derived growth factor (PDGF) pathway. Runt-related transcription factor 2 (RUNX2) is a major hub where all these pathways merge and cross-talk to guide the differentiation of bone and new vessels. Abbreviations: mitogen-activated protein kinase (MAPK), phosphatidylinositol 3-kinase (PI3K), osteocalcin (OCP), alkaline phosphatase (ALP), Osterix (Osx), BSP- Bone sialoprotein, BAP 1-BRCA1 associated protein-1, COL II- Collagen type II, SCUBE 3-Signal peptide-CUB-EGF-like domain-containing protein 3.

The development of functional HA bound to signaling peptides for the promotion of bone regeneration was studied actively. A stimulating effect of bone cell growth by capturing VEGF from Apt-HA in both in vitro and in vivo environments was observed [234]. Thrombin-peptide 508 (TP-508), erythropoietin (EPO), and blocking of thrombospondin-2 (TSP2) could also improve bone healing via angiogenesis mechanisms [265]. For example, using HA-based scaffold of Li-nHA/gelatin microsphere (GM)/rhEPO improved the viability of glucocorticoid-treated bone marrow mesenchymal stem cells and vascular endothelial cells and increased the expression of osteogenic and angiogenic factors. The Li-nHA/GM/rhEPO scaffold could upregulate the Wnt and HIF-1/VEGF pathways at the same time, with effects on improving osteogenesis and angiogenesis [212].

The VEGF-derived "QK" peptide was synthesized with a heptaglutamate (E7) domain, a motif that has strong affinity for CaP bone graft materials with greater activation of Akt and ERK1/2 in cells exposed to the E7–QK-coated discs. This angiogenic potential holds promise for augmenting the regenerative capacity of non-autologous bone grafts [266]. The connective tissue growth factor (CTGF)-loaded HA-based bioceramics could enhance cellular attachment through interaction with integrin, promoting actin cytoskeletal reorganization. CTGF-loaded HA also enhanced the differentiation of osteoblasts by integrin-mediated activation of the signaling pathways [267].

Improved vasculogenesis and bone matrix formation through a co-culture of endothelial cells and stem cells in tissue-specific methacryloyl gelatin-based hydrogels contributed to stimulate the interplay between osteogenesis and angiogenesis in vitro, as a basis for engineering vascularized bone [268].

Signals for osteo- and angio-genesis can also be delivered by means of the substrate micro-/nano-architecture. Bai et al. [269] demonstrated that nano-rod-decorated micro-surfaces better enable osteogenesis and angiogenesis, with respect to NP-decorated ones. Their study unraveled that the immune response of macrophages can be manipulated by the nano-/micro-surface, leading to a differential effect on osteointegration. The additional knowledge obtained from this study may provide a foundation and reference for the future design of coating materials for implantable materials [269].

Further understanding of cue signals that coordinate osteoinductive and pro-angiogenic effects will improve the generation of more performant HA-based bioceramic and biocomposite materials for orthopedic and dental applications.

3.5. MicroRNAs Involved in Osteo- and Angiogenesis

MicroRNAs (miRNAs) are evolutionarily conserved small non-coding RNAs, single-stranded molecules of about 22–25 nucleotides in length, which are involved in post-transcriptional gene regulation. MicroRNAs exert their function by partial or total binding to a specific mRNAs based on sequence complementary [270]. Series of miRNAs act in a complex functional network in which each miRNA might control hundreds of distinct target genes, and the expression of a single coding gene can be regulated by several miRNAs [271–273]. Up- or downregulation of the miRNA itself by stage- and tissue-specific expression patterns can lead to modified expression of its target genes and might be considered to act as fine-tuning of protein expression.

In the past several years [274], major progress was made in understanding the biological functions of miRNAs in bone formation and remodeling. The development of the next-generation high-throughput sequencing technologies [275,276] made possible the identification of classes of microRNAs involved in osteogenesis and angiogenesis. The availability of synthetic enhancers (mimics) or inhibitors (antagomiRs) triggered the investigation of the potential of miRNA for improved biomimetic composites for smart materials, mainly in combination with bioceramics. Therefore, specific miRNAs could be exploited to either induce stem-cell chondrogenic differentiation for articular cartilage regeneration or osteogenic differentiation for bone regeneration [277].

OsteomiRs were identified to regulate chondrocyte, osteoblast, and osteoclast differentiation by positively targeting principal osteogenic transcription factors such as RUNX2, Osterix (Osx), and ATF4 (activating transcription factor 4), and several signaling pathways including BMP, Notch, and Wnt, which control osteogenesis [278–280]. For example, miR-31 modulates osteogenic differentiation and mineralization of hBM-MSCs, by targeting the bone-specific transcription factor Osx [281], and miR-20a controls the expression of other important proteins involved in osteogenesis—BMP2, BMP4, and RUNX2 [282]. There are also several microRNAs with a specific role in processes of osteo- and angiogenesis. The highly conserved microRNA, mir-9 positively stimulates osteo- and endothelial progenitor cell formation. The miR-9 mimic-transfected HUVEC cells showed increased VEGF, VE-cadherin, and FGF protein expression levels, leading to increased EC migration and capillary tube formation in vitro. The activation of the AMP-activated protein kinase (AMPK) signaling pathway was the underlying molecular mechanism for the regulation of osteoblast differentiation and angiogenesis [283]. Nevertheless, the exact mechanisms of skeletal miRNAs governing the complex interactions and signaling pathways of different bone-forming cells are only beginning to be elucidated.

To accelerate bone regeneration, cytokines and growth factors could be delivered at the implantation site, but their use in clinical settings is constrained due to the poor stability of proteins, high cost, and short half-life [284]. Thus, more proper alternatives are needed to accelerate bone formation, and microRNA delivery using biocompatible systems seems to be more appropriate and less expensive.

For example, bone-specific miRNAs, such as miR-21 that promote osteogenesis in bone marrow stem cells, were delivered by biocompatible chitosan (CS)/hyaluronic acid NPs, thereby accelerating the osteogenesis process in human bone marrow mesenchymal stem cells (hBMMSCs) [285]. Also, the miR-21-functionalized microarc-oxidized (MAO) Ti surfaces demonstrated cell viability, cytotoxicity, and cell spreading comparable to that exhibited by naked MAO Ti surfaces and led to significantly higher expression of osteogenic genes. This novel miR-21-functionalized Ti implant may be used in the clinic to allow more effective and robust osteointegration [286].

Alternatively, the field of tissue engineering aims to regenerate damaged tissues, instead of replacing them, by developing biological substitutes that restore, maintain, or improve tissue function. The field relies extensively on the use of stem cells in combination with porous 3D scaffolds that house the cells and provide the appropriate environment for the regeneration of tissues and organs. The bioceramic-based HA NPs are potentially the main candidates as vectors, because of their major advantage of proven high biocompatibility. HA NPs offer additional advantageous properties for use in bone regeneration applications due to the chemical mimicry of the inorganic component of bones, as well as their demonstrated osteoconductive properties in vitro and in vivo [174].

In a recent in vitro study, a culture of stem cells derived from human periodontal ligament (hPDLSCs) was seeded on a scaffold made by fully deproteinated and sterilized HA bioceramic. The morphology, viability, osteogenic differentiation, VEGF release, and miR-210 expression of these cells were assessed. The promising results indicated that the 3D scaffold in contact with hPDLSCs showed good osteoconductive properties, evaluated through the adhesion and proliferation process, and presented the ability to stimulate VEGF secretion in hPDLSCs via miR-210 involvement. The induction of the production of this growth factor from hPDLSCs could represent a goal for tissue engineering, for the therapeutic growth of new blood vessels around the biomaterial in the first phase of osteointegration. Thus, the hPDLSC/glucose (G) construct could represent an interesting strategy to prefabricate a vascularized bone segment to be transplanted into the defect site [287].

Also, one study showed that the functionalization of porous collagen–nano-HA bone scaffolds with miR-133a-inhibiting complexes, delivered using non-viral HA NPs, enhanced human mesenchymal stem cell-mediated osteogenesis through the key activator of osteogenesis, RUNX2 [156]. The increased RUNX2 and osteocalcin expression, as well as higher ALP activity and calcium deposition, thus, demonstrated the further enhanced therapeutic potential of a biomaterial previously optimized for bone repair applications. In addition, miR-133a was identified as a direct negative regulator of the master transcription factor of osteogenesis, RUNX2; hence, the direct relationship between miR-133a levels and RUNX2 expression provides the possibility to target a central activator of osteogenesis. This nanoantagomiR-133a system also produced a rapid pro-osteogenic effect in hMSCs in 3D culture platforms [156]. The promising features of this platform offer the potential for applications beyond bone repair and tissue engineering, and constitute a new paradigm for microRNA-based therapeutics [156].

A study by Vimalraj et al. [288] demonstrated that a biocomposite scaffold based on carboxymethyl cellulose, zinc-doped nano-HA, and ascorbic acid (CMC/Zn-nHA/AC), along with microRNA-15b, transfected into mouse mesenchymal stem cells promoted osteoblast differentiation faster than control experiments. The early detection of alkaline phosphatase mRNA, which is an osteoblast differentiation marker gene [289], and the significantly increased expression of RUNX2 at the mRNA and protein level demonstrated the additive effect of the scaffold with bioactive molecule mirR-15b. This result demonstrated that biocompatible HA-based scaffolds might be functionalized with osteo-miRNAs in order to improve their response to osteogenic differentiation of mesenchymal stem cells. The increased effect of the bioengineered transfected cell-based scaffold suggested that there are different intracellular signaling pathways activated in cells, resulting in an enhanced osteogenic effect [288].

MicroRNAs are involved in several cellular mechanisms, but one distinct role of these non-coding RNA sequences is the modulation of the epigenetic mechanisms of gene expression. Epigenetic regulation is the biological mechanism whereby DNA, RNA, and proteins are chemically or structurally modified without changing their primary sequence. These epigenetic modifications play critical roles in the regulation of numerous cellular processes, including gene expression and DNA replication and recombination. Epigenetic regulatory mechanisms include, in addition to small (microRNAs) and long non-coding RNAs (lncRNAs), DNA methylation and hydroxymethylation, histone modification, chromatin remodeling, and RNA methylation. At present, there are a limited number of studies that investigated the impact of biomaterials on the epigenome, with most studies focusing on titanium and titanium dioxide (TiO_2), and a few on silica, glass, and graphene [290,291].

It should be noted that these studies considered just the biomaterials of nanometer dimensions that can be absorbed into the cellular environment and that might have an immediate effect at the molecular level. A recent study reported the biological effects of nano-HA (10 nm up to 100 nm) on the lineage commitment and differentiation of bone-forming osteoblasts and highlighted the impact of HA on the epigenome [292]. The nano-HA stimulated a strong dose-dependent suppression of the ALP, BSP, and OSC RNA levels, and this effect was sustained for weeks even in the absence of nano-HA. The study reported a 40% increase in DNA methylation at the promoter region of the osteoblast lineage commitment alkaline phosphatase gene (*ALPL*) in murine bone marrow stromal cells, following treatment with nano-HA. In general, the gene's promoter region hypermethylation is associated with gene silencing, and a less methylated promoter denotes a transcriptionally active gene. Furthermore, the exposure of osteoblasts to nano-HA resulted in dramatic and sustained changes in gene expression, whereas later-stage osteoblasts were much less responsive. These results suggested a potentially permanent alteration in the epigenome after HA exposure, with direct implication on osteogenic gene regulation. Collectively, this study identified for the first time that nano-HA is a potent regulator of the osteoblast lineage through changes in gene expression and identified methylation as a novel regulatory mechanism [292]. Although these results are interesting, future research analyzing single cytosine–phosphate–guanine (CpG) methylation at different regions of the *ALPL* gene is needed to determine the precise role of nanoparticle-dependent DNA methylation changes in gene expression and to determine the molecular mechanism through which nano-HA induces its effects [293].

Currently, there is limited knowledge on the epigenetic effects of biomaterials and the topography of 3D scaffolds on cellular activities. Greater investigations are necessary for a better understanding of the impact of biocompatible materials at the molecular level of the human epigenome. Finding the critical epigenetic mechanisms involved in stem-cell differentiation and studying the impact of bioactive material on the epigenome may be imperative for clinical translation into tissue engineering and bone regeneration.

4. Future Perspectives

Taken together, remarkable progress was made in unraveling the role of CaP-based bioceramics in stimulating angiogenic and osteogenic processes, which could open the path toward highly functional bone engineering medical devices with applicability in orthopedics and dentistry. Further studies are necessary for the in-depth evaluation of these complex processes by deciphering signaling pathways and miRNA involvement, using cutting-edge technologies for the assessment of the biological performance of such novel biomaterials and implants. The recent advancements in tissue engineering technologies, including three-dimensional (3D) printing, offers hitherto remarkable opportunities to develop a next generation of bone tissue substitute (grafts) with significant advantages over the conventional ones. Personalized implants can be produced with tailored characteristics better adapted to the patient-specific bone tissue regions/defects that are needed to be replaced/reinforced. To transfer these technologies to clinical practice, material science and tissue engineering need to be closely assisted by biomedical researchers in order to confer the safety risk assessment, as well as efficacy at high standards. A "systems biology" approach is needed for comprehensive analysis of the biological mechanisms of CaP-based bioceramics, with emphasis on the biocompatibility and biofunctional efficacy, joining critical processes such as oxidative stress angiogenesis, osteogenesis, etc. In the near future, the development of complex testing strategies will help to unveil the network of biological events elicited by CaP-based bioceramics in bulk, coating, or nanoparticle form, which are essential to ensure a longer and safer implant life in orthopedic and dentistry applications.

Author Contributions: All authors contributed equally to the manuscript, as follows: Conceptualization R.A., A.-C.P., A.-M.E, L.A., M.D., I.D.P., S.M., E.C., S.P., A.-R.L., G.E.S., G.M., C.T.; writing—original draft preparation R.A., A.-C.P., A.-M.E, L.A., M.D., I.D.P., S.M., E.C., S.P., A.-R.L., G.E.S., G.M., C.T.; writing—review and editing R.A., A.-C.P., A.-M.E, L.A., M.D., I.D.P., S.M., E.C., S.P., A.-R.L., G.E.S., G.M., C.T.

Funding: This article was funded by the Romanian Ministry of Research and Innovation, CCCDI-UEFISCDI, in the framework of: project PN-III-P1-1.2-PCCDI-2017-0062 (contract no. 58)/component project no. 2 and Program 1—The Improvement of the National System of Research and Development, Subprogram 1.2—Institutional Excellence—Projects of Excellence Funding in RDI, Contract No. 7PFE/16.10.2018.

Acknowledgments: The authors acknowledge Project PN-III-P1-1.2-PCCDI-2017-0062 (contract no. 58)/component project no. 2; Program 1—The Improvement of the National System of Research and Development, Subprogram 1.2—Institutional Excellence—Projects of Excellence Funding in RDI, Contract No. 7PFE/16.10.2018; Core Program PN 19.29.01.04; grant COP A 1.2.3., ID: P_40_197/2016 (Romanian Ministry of Research and Innovation). We would also like to thank Paula Antonia Moraru for support with artwork.

Conflicts of Interest: The authors declare no conflicts of interest.

References

1. Tite, T.; Popa, A.C.; Balescu, L.M.; Bogdan, I.M.; Pasuk, I.; Ferreira, J.M.F.; Stan, G.E. Cationic substitutions in hydroxyapatite: Current status of the derived biofunctional effects and their in vitro interrogation methods. *Materials* **2018**, *11*, 2081. [CrossRef] [PubMed]

2. Popa, A.C.; Stan, G.E.; Enculescu, M.; Tanase, C.; Tulyaganov, D.U.; Ferreira, J.M. Superior biofunctionality of dental implant fixtures uniformly coated with durable bioglass films by magnetron sputtering. *J. Mech. Behav. Biomed. Mater.* **2015**, *51*, 313–327. [CrossRef] [PubMed]

3. Miculescu, F.; Maidaniuc, A.; Voicu, S.I.; Thakur, V.K.; Stan, G.E.; Ciocan, L.T. Progress in hydroxyapatite–starch based sustainable biomaterials for biomedical bone substitution applications. *ACS Sustain. Chem. Eng.* **2017**, *5*, 8491–8512. [CrossRef]

4. Antoniac, I.V. *Handbook of Bioceramics and Biocomposites*; Springer: Berlin, Germany, 2016.

5. Mucalo, M. *Hydroxyapatite (HAp) for Biomedical Applications*; Elsevier: Amsterdam, The Netherlands, 2015.

6. Vichery, C.; Nedelec, J.M. Bioactive Glass Nanoparticles: From Synthesis to Materials Design for Biomedical Applications. *Materials* **2016**, *9*, 288. [CrossRef] [PubMed]

7. Combes, C.; Cazalbou, S.; Rey, C. Apatite Biominerals. *Minerals* **2016**, *6*, 34. [CrossRef]

8. Miculescu, F.; Mocanu, A.C.; Stan, G.E.; Miculescu, M.; Maidaniuc, A.; Cîmpean, A.; Mitran, V.; Voicu, S.I.; Machedon-Pisu, T.; Ciocan, L.T. Influence of the modulated two-step synthesis of biogenic hydroxyapatite on biomimetic products' surface. *Appl. Surf. Sci.* **2018**, *438*, 147–157. [CrossRef]

9. Io, O.; Og, A.; Og, O.; Ao, B.; Mo, P. Non-synthetic sources for the development of hydroxyapatite. *J. Appl. Biotechnol. Bioeng.* **2018**, *5*. [CrossRef]

10. Jaber, H.L.; Hammood, A.S.; Parvin, N. Synthesis and characterization of hydroxyapatite powder from natural Camelus bone. *J. Aust. Ceram. Soc.* **2017**, *54*, 1–10. [CrossRef]

11. Baino, F.; Novajra, G.; Vitale-Brovarone, C. Bioceramics and Scaffolds: A Winning Combination for Tissue Engineering. *Front. Bioeng. Biotechnol.* **2015**, *3*, 202. [CrossRef]

12. Šupová, M. Substituted hydroxyapatites for biomedical applications: A review. *Ceram. Int.* **2015**, *41*, 9203–9231. [CrossRef]

13. Deng, L.; Li, D.; Yang, Z.; Xie, X.; Kang, P. Repair of the calvarial defect in goat model using magnesium-doped porous hydroxyapatite combined with recombinant human bone morphogenetic protein-2. *Biomed. Mater. Eng.* **2017**, *28*, 361–377. [CrossRef] [PubMed]

14. Habibovic, P.; Barralet, J.E. Bioinorganics and biomaterials: Bone repair. *Acta Biomater.* **2011**, *7*, 3013–3026. [CrossRef] [PubMed]

15. Vargas, G.E.; Haro Durand, L.A.; Cadena, V.; Romero, M.; Mesones, R.V.; Mackovic, M.; Spallek, S.; Spiecker, E.; Boccaccini, A.R.; Gorustovich, A.A. Effect of nano-sized bioactive glass particles on the angiogenic properties of collagen based composites. *J. Mater. Sci. Mater. Med.* **2013**, *24*, 1261–1269. [CrossRef] [PubMed]

16. Yin, J.; Yu, J.; Ke, Q.; Yang, Q.; Zhu, D.; Gao, Y.; Guo, Y.; Zhang, C. La-Doped biomimetic scaffolds facilitate bone remodelling by synchronizing osteointegration and phagocytic activity of macrophages. *J. Mater. Chem. B* **2019**, *7*, 3066–3074. [CrossRef]

17. Luo, Y.; Li, D.; Zhao, J.; Yang, Z.; Kang, P. In vivo evaluation of porous lithium-doped hydroxyapatite scaffolds for the treatment of bone defect. *Bio-Med. Mater. Eng.* **2018**, *29*, 699–721. [CrossRef]

18. Pina, S.; Canadas, R.F.; Jimenez, G.; Peran, M.; Marchal, J.A.; Reis, R.L.; Oliveira, J.M. Biofunctional Ionic-Doped Calcium Phosphates: Silk Fibroin Composites for Bone Tissue Engineering Scaffolding. *Cells Tissues Organs* **2017**, *204*, 150–163. [CrossRef]

19. Graziani, G.; Boi, M.; Bianchi, M. A Review on ionic substitutions in hydroxyapatite thin films: Towards complete biomimetism. *Coatings* **2018**, *8*, 269. [CrossRef]

20. Wu, V.M.; Tang, S.; Uskokovic, V. Calcium phosphate nanoparticles as intrinsic inorganic antimicrobials: The antibacterial effect. *ACS Appl. Mater. Interfaces* **2018**, *10*, 34013–34028. [CrossRef]

21. Epple, M.; Ganesan, K.; Heumann, R.; Klesing, J.; Kovtun, A.; Neumann, S.; Sokolova, V. Application of calcium phosphate nanoparticles in biomedicine. *J. Mater. Chem.* **2010**, *20*, 18–23. [CrossRef]

22. Banik, M.; Basu, T. Calcium phosphate nanoparticles: A study of their synthesis, characterization and mode of interaction with salmon testis DNA. *Dalton Trans.* **2014**, *43*, 3244–3259. [CrossRef]

23. Sadat-Shojai, M.; Atai, M.; Nodehi, A.; Khanlar, L.N. Hydroxyapatite nanorods as novel fillers for improving the properties of dental adhesives: Synthesis and application. *Dent. Mater.* **2010**, *26*, 471–482. [CrossRef] [PubMed]

24. Sun, T.-W.; Zhu, Y.-J.; Chen, F.; Chen, F.-F.; Jiang, Y.-Y.; Zhang, Y.-G.; Wu, J. Ultralong hydroxyapatite nanowires/collagen scaffolds with hierarchical porous structure, enhanced mechanical properties and excellent cellular attachment. *Ceram. Int.* **2017**, *43*, 15747–15754. [CrossRef]

25. Chandanshive, B.B.; Rai, P.; Rossi, A.L.; Ersen, O.; Khushalani, D. Synthesis of hydroxyapatite nanotubes for biomedical applications. *Mater. Sci. Eng. C* **2013**, *33*, 2981–2986. [CrossRef] [PubMed]

26. Nejati, E.; Firouzdor, V.; Eslaminejad, M.B.; Bagheri, F. Needle-like nano hydroxyapatite/poly(l-lactide acid) composite scaffold for bone tissue engineering application. *Mater. Sci. Eng. C* **2009**, *29*, 942–949. [CrossRef]

27. Kozuma, W.; Kon, K.; Kawakami, S.; Bobothike, A.; Iijima, H.; Shiota, M.; Kasugai, S. Osteoconductive potential of a hydroxyapatite fiber material with magnesium: In vitro and in vivo studies. *Dent. Mater. J.* **2019**. [CrossRef]

28. Shi, X.; Zhou, K.; Huang, F.; Wang, C. Interaction of hydroxyapatite nanoparticles with endothelial cells: Internalization and inhibition of angiogenesis in vitro through the PI3K/Akt pathway. *Int. J. Nanomed.* **2017**, *12*, 5781–5795. [CrossRef]

29. Wang, P.; Zhao, L.; Liu, J.; Weir, M.D.; Zhou, X.; Xu, H.H. Bone tissue engineering via nanostructured calcium phosphate biomaterials and stem cells. *Bone Res.* **2014**, *2*, 14017. [CrossRef]

30. Dissanayaka, W.L.; Zhu, L.; Hargreaves, K.M.; Jin, L.; Zhang, C. Scaffold-free prevascularized microtissue spheroids for pulp regeneration. *J. Dent. Res.* **2014**, *93*, 1296–1303. [CrossRef]

31. Gomez, S.; Vlad, M.D.; Lopez, J.; Fernandez, E. Design and properties of 3D scaffolds for bone tissue engineering. *Acta Biomater.* **2016**, *42*, 341–350. [CrossRef]

32. Hosseinkhani, M.; Mehrabani, D.; Karimfar, M.H.; Bakhtiyari, S.; Manafi, A.; Shirazi, R. Tissue engineered scaffolds in regenerative medicine. *World J. Plastic Surg.* **2014**, *3*, 3–7.

33. Bose, S.; Fielding, G.; Tarafder, S.; Bandyopadhyay, A. Understanding of dopant-induced osteogenesis and angiogenesis in calcium phosphate ceramics. *Trends Biotechnol.* **2013**, *31*, 594–605. [CrossRef] [PubMed]

34. Kokubo, T.; Takadama, H. How useful is SBF in predicting in vivo bone bioactivity? *Biomaterials* **2006**, *27*, 2907–2915. [CrossRef] [PubMed]

35. Kokubo, T.; Yamaguchi, S. Simulated body fluid and the novel bioactive materials derived from it. *J. Biomed. Mater. Res. Part. A* **2019**, *107*, 968–977. [CrossRef] [PubMed]

36. Popa, A.-C.; Stan, G.; Husanu, M.-A.; Mercioniu, I.; Santos, L.; Fernandes, H.; Ferreira, J. Bioglass implant-coating interactions in synthetic physiological fluids with varying degrees of biomimicry. *Int. J. Nanomed.* **2017**, *12*, 683–707. [CrossRef]

37. ISO. *10993-1:2018—Biological Evaluation of Medical Devices*; ISO: Geneve, Switzerland, 2018.

38. Popa, A.C.; Marques, V.M.F.; Stan, G.E.; Husanu, M.A.; Galca, A.C.; Ghica, C.; Tulyaganov, D.U.; Lemos, A.F.; Ferreira, J.M.F. Nanomechanical characterization of bioglass films synthesized by magnetron sputtering. *Thin Solid Films* **2014**, *553*, 166–172. [CrossRef]

39. Kim, K.T.; Eo, M.Y.; Nguyen, T.T.H.; Kim, S.M. General review of titanium toxicity. *Int. J. Implant Denti.* **2019**, *5*, 10. [CrossRef]

40. ISO. *10993-5:2009 Biological Evaluation of Medical Devices—Part 5: Tests for In Vitro Cytotoxicity*; ISO: Geneve, Switzerland, 2009; p. 34.

41. Coelho, F.; Cavicchioli, M.; Specian, S.S.; Scarel-Caminaga, R.M.; Penteado, L.A.; Medeiros, A.I.; Ribeiro, S.J.L.; Capote, T.S.O. Bacterial cellulose membrane functionalized with hydroxiapatite and anti-bone morphogenetic protein 2: A promising material for bone regeneration. *PLoS ONE* **2019**, *14*, e0221286. [CrossRef]

42. Coelho, C.C.; Grenho, L.; Gomes, P.S.; Quadros, P.A.; Fernandes, M.H. Nano-hydroxyapatite in oral care cosmetics: Characterization and cytotoxicity assessment. *Sci. Rep.* **2019**, *9*, 11050. [CrossRef]

43. Guan, R.G.; Johnson, I.; Cui, T.; Zhao, T.; Zhao, Z.Y.; Li, X.; Liu, H. Electrodeposition of hydroxyapatite coating on Mg-4.0Zn-1.0Ca-0.6Zr alloy and in vitro evaluation of degradation, hemolysis, and cytotoxicity. *J. Biomed. Mater. Res. Part. A* **2012**, *100*, 999–1015. [CrossRef]

44. Singh, S.S.; Roy, A.; Lee, B.E.; Ohodnicki, J.; Loghmanian, A.; Banerjee, I.; Kumta, P.N. A study of strontium doped calcium phosphate coatings on AZ31. *Mater. Sci. Eng. C* **2014**, *40*, 357–365. [CrossRef]

45. Repetto, G.; del Peso, A.; Zurita, J.L. Neutral red uptake assay for the estimation of cell viability/cytotoxicity. *Nat. Protoc.* **2008**, *3*, 1125–1131. [CrossRef] [PubMed]

46. Shamsuria, O.; Fadilah, A.S.; Asiah, A.B.; Rodiah, M.R.; Suzina, A.H.; Samsudin, A.R. In vitro cytotoxicity evaluation of biomaterials on human osteoblast cells CRL-1543; hydroxyapatite, natural coral and polyhydroxybutarate. *Med. J. Malaysia* **2004**, *59* (Suppl. B), 174–175.

47. Przekora, A.; Czechowska, J.; Pijocha, D.; Ślósarczyk, A.; Ginalska, G. Do novel cement-type biomaterials reveal ion reactivity that affects cell viability in vitro? *Open Life Sci.* **2014**, *9*, s11535–s12013. [CrossRef]

48. Gradinaru, S.; Popescu, V.; Leasu, C.; Pricopie, S.; Yasin, S.; Ciuluvica, R.; Ungureanu, E. Hydroxyapatite ocular implant and non-integrated implants in eviscerated patients. *J. Med. Life* **2015**, *8*, 90–93.

49. Kattimani, V.S.; Kondaka, S.; Lingamaneni, K.P. Hydroxyapatite–Past, present, and future in bone regeneration. *Bone Tissue Reg. Insights* **2016**, *7*, BTRI.S36138. [CrossRef]

50. Harun, W.S.W.; Asri, R.I.M.; Sulong, A.B.; Ghani, S.A.C.; Ghazalli, Z. Hydroxyapatite-based coating on biomedical implant. *Mater. Sci.* **2018**. [CrossRef]

51. Hopkins, G.O. Multiple joint tuberculosis presenting as HLA-B27 disease. *Postgrad. Med. J.* **1983**, *59*, 113–115. [CrossRef]

52. Santos, C.; Gomes, P.S.; Duarte, J.A.; Franke, R.P.; Almeida, M.M.; Costa, M.E.; Fernandes, M.H. Relevance of the sterilization-induced effects on the properties of different hydroxyapatite nanoparticles and assessment of the osteoblastic cell response. *J. R. Soc. Interface* **2012**, *9*, 3397–3410. [CrossRef]

53. Ciobanu, C.S.; Iconaru, S.L.; Pasuk, I.; Vasile, B.S.; Lupu, A.R.; Hermenean, A.; Dinischiotu, A.; Predoi, D. Structural properties of silver doped hydroxyapatite and their biocompatibility. *Mater. Sci. Eng. C* **2013**, *33*, 1395–1402. [CrossRef]

54. Zhao, X.; Ng, S.; Heng, B.C.; Guo, J.; Ma, L.; Tan, T.T.; Ng, K.W.; Loo, S.C. Cytotoxicity of hydroxyapatite nanoparticles is shape and cell dependent. *Arch. Toxicol.* **2013**, *87*, 1037–1052. [CrossRef]

55. *Assay Guidance Manual*; Sittampalam, G.S.; Grossman, A.; Brimacombe, K.; Arkin, M.; Auld, D.; Austin, C.; Baell, J.; Bejcek, B.; Caaveiro, J.M.M.; Chung, T.D.Y.; et al. (Eds.) Bethesda: Rockville, MD, USA, 2004.

56. Altman, F.P. Tetrazolium salts and formazans. *Progress Histochem. Cytochem.* **1976**, *9*, III-51. [CrossRef]

57. Aslantürk, Ö.S. *In Vitro Cytotoxicity and Cell Viability Assays: Principles, Advantages, and Disadvantages*; IntechOpen: London, UK, 2018. [CrossRef]

58. Kim, K.M.; Lee, S.B.; Lee, S.H.; Lee, Y.K.; Kim, K.N. Comparison of validity between WST-1 and MTT test in bioceramic materials. *Key Eng. Mater.* **2005**, *284–286*, 585–588. [CrossRef]

59. Menyhart, O.; Harami-Papp, H.; Sukumar, S.; Schafer, R.; Magnani, L.; de Barrios, O.; Gyorffy, B. Guidelines for the selection of functional assays to evaluate the hallmarks of cancer. *Biochim. Biophys. Acta* **2016**, *1866*, 300–319. [CrossRef] [PubMed]

60. Chamchoy, K.; Pakotiprapha, D.; Pumirat, P.; Leartsakulpanich, U.; Boonyuen, U. Application of WST-8 based colorimetric NAD(P)H detection for quantitative dehydrogenase assays. *BMC Biochem.* **2019**, *20*, 4. [CrossRef]

61. Noorani, T.Y.; Luddin, N.; Rahman, I.A.; Masudi, S.M. In vitro cytotoxicity evaluation of novel nano-hydroxyapatite-silica incorporated glass ionomer cement. *J. Clin. Diagn. Res.* **2017**, *11*, ZC105–ZC109. [CrossRef]

62. El Hadad, A.A.; Peon, E.; Garcia-Galvan, F.R.; Barranco, V.; Parra, J.; Jimenez-Morales, A.; Galvan, J.C. Biocompatibility and corrosion protection behaviour of hydroxyapatite sol-gel-derived coatings on Ti6Al4V alloy. *Materials* **2017**, *10*, 94. [CrossRef]

63. Ramis, J.; Coelho, C.; Córdoba, A.; Quadros, P.; Monjo, M. Safety assessment of nano-hydroxyapatite as an oral care ingredient according to the EU cosmetics regulation. *Cosmetics* **2018**, *5*, 53. [CrossRef]

64. Aghaei, H.; Nourbakhsh, A.A.; Karbasi, S.; JavadKalbasi, R.; Rafienia, M.; Nourbakhsh, N.; Bonakdar, S.; Mackenzie, K.J.D. Investigation on bioactivity and cytotoxicity of mesoporous nano-composite MCM-48/hydroxyapatite for ibuprofen drug delivery. *Ceram. Int.* **2014**, *40*, 7355–7362. [CrossRef]

65. Ferreira dos Santos, C.; Gomes, P.S.; Almeida, M.M.; Willinger, M.-G.; Franke, R.-P.; Fernandes, M.H.; Costa, M.E. Gold-dotted hydroxyapatite nanoparticles as multifunctional platforms for medical applications. *RSC Adv.* **2015**, *5*, 69184–69195. [CrossRef]

66. Szymonowicz, M.; Korczynski, M.; Dobrzynski, M.; Zawisza, K.; Mikulewicz, M.; Karuga-Kuzniewska, E.; Zywickab, B.; Rybak, Z.; Wiglusz, R.J. Cytotoxicity evaluation of high-temperature annealed nanohydroxyapatite in contact with fibroblast cells. *Materials* **2017**, *10*, 590. [CrossRef]

67. Gao, J.; Feng, Z.; Wang, X.; Zeng, M.; Liu, J.; Han, S.; Xu, J.; Chen, L.; Cao, K.; Long, J.; et al. SIRT3/SOD2 maintains osteoblast differentiation and bone formation by regulating mitochondrial stress. *Cell Death Diff.* **2018**, *25*, 229–240. [CrossRef]

68. Guo, L.; Von Dem Bussche, A.; Buechner, M.; Yan, A.; Kane, A.B.; Hurt, R.H. Adsorption of essential micronutrients by carbon nanotubes and the implications for nanotoxicity testing. *Small* **2008**, *4*, 721–727. [CrossRef]

69. Wohlleben, W.; Kolle, S.N.; Hasenkamp, L.-C.; Böser, A.; Vogel, S.; Vacano, B.V.; Ravenzwaay, B.V.; Landsiedel, R. Artifacts by marker enzyme adsorption on nanomaterials in cytotoxicity assays with tissue cultures. *J. Phys. Conf. Ser.* **2011**, *304*, 012061. [CrossRef]

70. Pailleux, M.; Boudard, D.; Pourchez, J.; Forest, V.; Grosseau, P.; Cottier, M. New insight into artifactual phenomena during in vitro toxicity assessment of engineered nanoparticles: Study of TNF-alpha adsorption on alumina oxide nanoparticle. *Toxicol. In Vitro* **2013**, *27*, 1049–1056. [CrossRef]

71. Lupu, A.R.; Popescu, T. The noncellular reduction of MTT tetrazolium salt by TiO_2 nanoparticles and its implications for cytotoxicity assays. *Toxicol. In Vitro* **2013**, *27*, 1445–1450. [CrossRef]

72. Popescu, T.; Lupu, A.R.; Raditoiu, V.; Purcar, V.; Teodorescu, V.S. On the photocatalytic reduction of MTT tetrazolium salt on the surface of TiO_2 nanoparticles: Formazan production kinetics and mechanism. *J. Colloid Interface Sci.* **2015**, *457*, 108–120. [CrossRef]

73. Lee, Y.K.; Choi, E.J.; Webster, T.J.; Kim, S.H.; Khang, D. Effect of the protein corona on nanoparticles for modulating cytotoxicity and immunotoxicity. *Int. J. Nanomed.* **2015**, *10*, 97–113. [CrossRef]

74. Treuel, L.; Docter, D.; Maskos, M.; Stauber, R.H. Protein corona-from molecular adsorption to physiological complexity. *Beilstein J. Nanotechnol.* **2015**, *6*, 857–873. [CrossRef]

75. Nguyen, V.H.; Lee, B.J. Protein corona: A new approach for nanomedicine design. *Int. J. Nanomed.* **2017**, *12*, 3137–3151. [CrossRef]

76. Zhu, Y.; Yang, Q.; Yang, M.; Zhan, X.; Lan, F.; He, J.; Gu, Z.; Wu, Y. Protein corona of magnetic hydroxyapatite scaffold improves cell proliferation via activation of mitogen-activated protein kinase signaling pathway. *ACS Nano* **2017**, *11*, 3690–3704. [CrossRef]

77. Zhu, Y.; Jiang, P.; Luo, B.; Lan, F.; He, J.; Wu, Y. Dynamic protein corona influences immune-modulating osteogenesis in magnetic nanoparticle (MNP)-infiltrated bone regeneration scaffolds in vivo. *Nanoscale* **2019**, *11*, 6817–6827. [CrossRef]

78. ISO. *10993-3:2014 Biologicalevaluationof Medicaldevices -Part 3:Tests Forgenotoxicity,Carcinogenicityand Reproductivetoxicity*; ISO: Geneve, Switzerland, 2014.

79. OECD Guidelines. Available online: http://www.oecd.org (accessed on 24 October 2019).

80. ISO. *10993-12:2012 Biological Evaluation of Medical Devices—Part 12: Sample Preparation and Reference Materials*; ISO: Geneve, Switzerland, 2012; p. 20.

81. Lloyd, M.; Kidd, D. The mouse lymphoma assay. *Methods Mol. Biol.* **2012**, *817*, 35–54. [CrossRef]

82. Ab238544. Comet Assay Kit (3-well Slides). Available online: https://www.abcam.com/ps/products/238/ab238544/documents/ab238544%20Comet%20Assay%20Kit%20(3-well%20slides)_v2a%20(website).pdf (accessed on 24 October 2019).

83. Wahab, N.; Kannan, T.P.; Mahmood, Z.; Rahman, I.A.; Ismail, H. Genotoxity assessment of biphasic calcium phosphate of modified porosity on human dental pulp cells using Ames and Comet assays. *Toxicol. In Vitro* **2018**, *47*, 207–212. [CrossRef]

84. Sonmez, E.; Cacciatore, I.; Bakan, F.; Turkez, H.; Mohtar, Y.I.; Togar, B.; Stefano, A.D. Toxicity assessment of hydroxyapatite nanoparticles in rat liver cell model in vitro. *Hum. Exp. Toxicol.* **2016**, *35*, 1073–1083. [CrossRef]

85. Seyedmajidi, S.; Seyedmajidi, M.; Zabihi, E.; Hajian-Tilaki, K. A comparative study on cytotoxicity and genotoxicity of the hydroxyapatite-bioactive glass and fluorapatite-bioactive glass nanocomposite foams as tissue scaffold for bone repair. *J. Biomed. Mater. Res. Part A* **2018**, *106*, 2605–2612. [CrossRef]

86. Kido, H.W.; Ribeiro, D.A.; de Oliveira, P.; Parizotto, N.A.; Camilo, C.C.; Fortulan, C.A.; Marcantonio, E., Jr.; da Silva, V.H.; Renno, A.C. Biocompatibility of a porous alumina ceramic scaffold coated with hydroxyapatite and bioglass. *J. Biomed. Mater. Res. Part A* **2014**, *102*, 2072–2078. [CrossRef]

87. Oledzka, E.; Pachowska, D.; Orlowska, K.; Kolmas, J.; Drobniewska, A.; Figat, R.; Sobczak, M. Pamidronate-conjugated biodegradable branched copolyester carriers: synthesis and characterization. *Molecules* **2017**, *22*, 1063. [CrossRef]

88. Yamamura, H.; da Silva, V.H.P.; Ruiz, P.L.M.; Ussui, V.; Lazar, D.R.R.; Renno, A.C.M.; Ribeiro, D.A. Physico-chemical characterization and biocompatibility of hydroxyapatite derived from fish waste. *J. Mech. Behav. Biomed. Mater.* **2018**, *80*, 137–142. [CrossRef]

89. Sthijns, M.; van Blitterswijk, C.A.; LaPointe, V.L.S. Redox regulation in regenerative medicine and tissue engineering: The paradox of oxygen. *J. Tissue Eng. Regen. Med.* **2018**, *12*, 2013–2020. [CrossRef]

90. Yang, Y.; Karakhanova, S.; Werner, J.; Bazhin, A.V. Reactive oxygen species in cancer biology and anticancer therapy. *Curr. Med. Chem.* **2013**, *20*, 3677–3692. [CrossRef]

91. Sies, H. Oxidative stress: A concept in redox biology and medicine. *Redox Biol.* **2015**, *4*, 180–183. [CrossRef]

92. Mueller, C.F.; Laude, K.; McNally, J.S.; Harrison, D.G. ATVB in focus: Redox mechanisms in blood vessels. *Arterioscl. Thromb. Vasc. Biol.* **2005**, *25*, 274–278. [CrossRef]

93. Sthijns, M.M.; Weseler, A.R.; Bast, A.; Haenen, G.R. Time in redox adaptation processes: from evolution to hormesis. *Int. J. Mol. Sci.* **2016**, *17*, 1649. [CrossRef]

94. Forman, H.J.; Ursini, F.; Maiorino, M. An overview of mechanisms of redox signaling. *J. Mol. Cell. Cardiol.* **2014**, *73*, 2–9. [CrossRef]

95. The Molecular Probes Handbook—A Guide to Fluorescent Probes and Labeling Technologies. Available online: https://www.thermofisher.com/ro/en/home/references/molecular-probes-the-handbook.html (accessed on 24 October 2019).

96. Suzen, S.; Gurer-Orhan, H.; Saso, L. Detection of reactive oxygen and nitrogen species by electron paramagnetic resonance (EPR) technique. *Molecules* **2017**, *22*, 181. [CrossRef]

97. Marrocco, I.; Altieri, F.; Peluso, I. Measurement and clinical significance of biomarkers of oxidative stress in humans. *Oxid. Med. Cell. Longev.* **2017**, *2017*, 6501046. [CrossRef]

98. Aranda, A.; Sequedo, L.; Tolosa, L.; Quintas, G.; Burello, E.; Castell, J.V.; Gombau, L. Dichloro-dihydro-fluorescein diacetate (DCFH-DA) assay: A quantitative method for oxidative stress assessment of nanoparticle-treated cells. *Toxicol. In Vitro* **2013**, *27*, 954–963. [CrossRef]

99. Khan, P.; Idrees, D.; Moxley, M.A.; Corbett, J.A.; Ahmad, F.; von Figura, G.; Sly, W.S.; Waheed, A.; Hassan, M.I. Luminol-based chemiluminescent signals: Clinical and non-clinical application and future uses. *Appl. Biochem. Biotechnol.* **2014**, *173*, 333–355. [CrossRef]

100. Yazdani, J.; Ahmadian, E.; Sharifi, S.; Shahi, S.; Maleki Dizaj, S. A short view on nanohydroxyapatite as coating of dental implants. *Biomed. Pharmacother. Biomed. Pharmacother.* **2018**, *105*, 553–557. [CrossRef]

101. Lee, H.; Jang, T.S.; Song, J.; Kim, H.E.; Jung, H.D. The Production of Porous Hydroxyapatite Scaffolds with Graded Porosity by Sequential Freeze-Casting. *Materials* **2017**, *10*, 367. [CrossRef]

102. Pina, S.; Ribeiro, V.P.; Marques, C.F.; Maia, F.R.; Silva, T.H.; Reis, R.L.; Oliveira, J.M. Scaffolding strategies for tissue engineering and regenerative medicine applications. *Materials* **2019**, *12*, 1824. [CrossRef]

103. Nel, A.; Xia, T.; Madler, L.; Li, N. Toxic potential of materials at the nanolevel. *Science* **2006**, *311*, 622–627. [CrossRef]

104. Manke, A.; Wang, L.; Rojanasakul, Y. Mechanisms of nanoparticle-induced oxidative stress and toxicity. *BioMed Res. Int.* **2013**, *2013*, 942916. [CrossRef]

105. Gallud, A.; Kloditz, K.; Ytterberg, J.; Ostberg, N.; Katayama, S.; Skoog, T.; Gogvadze, V.; Chen, Y.Z.; Xue, D.; Moya, S.; et al. Cationic gold nanoparticles elicit mitochondrial dysfunction: A multi-omics study. *Sci. Rep.* **2019**, *9*, 4366. [CrossRef]

106. Culcasi, M.; Benameur, L.; Mercier, A.; Lucchesi, C.; Rahmouni, H.; Asteian, A.; Casano, G.; Botta, A.; Kovacic, H.; Pietri, S. EPR spin trapping evaluation of ROS production in human fibroblasts exposed to cerium oxide nanoparticles: Evidence for NADPH oxidase and mitochondrial stimulation. *Chem. Biol. Interact.* **2012**, *199*, 161–176. [CrossRef]

107. Knaapen, A.M.; Borm, P.J.; Albrecht, C.; Schins, R.P. Inhaled particles and lung cancer. Part A: Mechanisms. *Int. J. Cancer* **2004**, *109*, 799–809. [CrossRef]

108. Khanna, P.; Ong, C.; Bay, B.H.; Baeg, G.H. Nanotoxicity: An interplay of oxidative stress, inflammation and cell death. *Nanomaterials* **2015**, *5*, 1163–1180. [CrossRef]

109. Xia, T.; Kovochich, M.; Liong, M.; Madler, L.; Gilbert, B.; Shi, H.; Yeh, J.I.; Zink, J.I.; Nel, A.E. Comparison of the mechanism of toxicity of zinc oxide and cerium oxide nanoparticles based on dissolution and oxidative stress properties. *ACS Nano* **2008**, *2*, 2121–2134. [CrossRef]

110. Uskokovic, V.; Uskokovic, D.P. Nanosized hydroxyapatite and other calcium phosphates: Chemistry of formation and application as drug and gene delivery agents. *J. Biomed. Mater. Res. Part B* **2011**, *96*, 152–191. [CrossRef]

111. Sun, J.S.; Lin, F.H.; Hung, T.Y.; Tsuang, Y.H.; Chang, W.H.; Liu, H.C. The influence of hydroxyapatite particles on osteoclast cell activities. *J. Biomed. Mater. Res.* **1999**, *45*, 311–321. [CrossRef]

112. Cai, Y.; Liu, Y.; Yan, W.; Hu, Q.; Tao, J.; Zhang, M.; Shi, Z.; Tang, R. Role of hydroxyapatite nanoparticle size in bone cell proliferation. *J. Mater. Chem.* **2007**, *17*, 3780. [CrossRef]

113. Yang, X.; Li, Y.; Liu, X.; Zhang, R.; Feng, Q. In vitro uptake of hydroxyapatite nanoparticles and their effect on osteogenic differentiation of human mesenchymal stem cells. *Stem Cells Int.* **2018**, *2018*, 2036176. [CrossRef]

114. Calabrese, G.; Giuffrida, R.; Forte, S.; Fabbi, C.; Figallo, E.; Salvatorelli, L.; Memeo, L.; Parenti, R.; Gulisano, M.; Gulino, R. Human adipose-derived mesenchymal stem cells seeded into a collagen-hydroxyapatite scaffold promote bone augmentation after implantation in the mouse. *Sci. Rep.* **2017**, *7*, 7110. [CrossRef]

115. Zhang, Y.; Marsboom, G.; Toth, P.T.; Rehman, J. Mitochondrial respiration regulates adipogenic differentiation of human mesenchymal stem cells. *PLoS ONE* **2013**, *8*, e77077. [CrossRef]

116. Chen, L.; McCrate, J.M.; Lee, J.C.; Li, H. The role of surface charge on the uptake and biocompatibility of hydroxyapatite nanoparticles with osteoblast cells. *Nanotechnology* **2011**, *22*, 105708. [CrossRef]

117. Porter, J.R.; Ruckh, T.T.; Popat, K.C. Bone tissue engineering: A review in bone biomimetics and drug delivery strategies. *Biotechnol. Prog.* **2009**, *25*, 1539–1560. [CrossRef]

118. Xu, Z.; Liu, C.; Wei, J.; Sun, J. Effects of four types of hydroxyapatite nanoparticles with different nanocrystal morphologies and sizes on apoptosis in rat osteoblasts. *J. Appl. Toxicol.* **2012**, *32*, 429–435. [CrossRef]

119. Jin, Y.; Liu, X.; Liu, H.; Chen, S.; Gao, C.; Ge, K.; Zhang, C.; Zhang, J. Oxidative stress-induced apoptosis of osteoblastic MC3T3-E1 cells by hydroxyapatite nanoparticles through lysosomal and mitochondrial pathways. *RSC Adv.* **2017**, *7*, 13010–13018. [CrossRef]

120. Werner, S.; Grose, R. Regulation of wound healing by growth factors and cytokines. *Physiol. Rev.* **2003**, *83*, 835–870. [CrossRef]

121. Franz, S.; Rammelt, S.; Scharnweber, D.; Simon, J.C. Immune responses to implants—A review of the implications for the design of immunomodulatory biomaterials. *Biomaterials* **2011**, *32*, 6692–6709. [CrossRef]

122. Mestres, G.; Espanol, M.; Xia, W.; Persson, C.; Ginebra, M.P.; Ott, M.K. Inflammatory response to nano- and microstructured hydroxyapatite. *PLoS ONE* **2015**, *10*, e0120381. [CrossRef]

123. Gittens, R.A.; Olivares-Navarrete, R.; Schwartz, Z.; Boyan, B.D. Implant osseointegration and the role of microroughness and nanostructures: Lessons for spine implants. *Acta Biomater.* **2014**, *10*, 3363–3371. [CrossRef]

124. Koundouros, N.; Poulogiannis, G. Phosphoinositide 3-Kinase/Akt signaling and redox metabolism in cancer. *Front. Oncol.* **2018**, *8*, 160. [CrossRef]

125. Zhang, K.; Zhou, Y.; Xiao, C.; Zhao, W.; Wu, H.; Tang, J.; Li, Z.; Yu, S.; Li, X.; Min, L.; et al. Application of hydroxyapatite nanoparticles in tumor-associated bone segmental defect. *Sci. Adv.* **2019**, *5*, eaax6946. [CrossRef]

126. Han, Y.; Li, S.; Cao, X.; Yuan, L.; Wang, Y.; Yin, Y.; Qiu, T.; Dai, H.; Wang, X. Different inhibitory effect and mechanism of hydroxyapatite nanoparticles on normal cells and cancer cells in vitro and in vivo. *Sci. Rep.* **2014**, *4*, 7134. [CrossRef]

127. Qing, F.; Wang, Z.; Hong, Y.; Liu, M.; Guo, B.; Luo, H.; Zhang, X. Selective effects of hydroxyapatite nanoparticles on osteosarcoma cells and osteoblasts. *J. Mater. Sci. Mater. Med.* **2012**, *23*, 2245–2251. [CrossRef]

128. Xu, J.; Xu, P.; Li, Z.; Huang, J.; Yang, Z. Oxidative stress and apoptosis induced by hydroxyapatite nanoparticles in C6 cells. *J. Biomed. Mater. Res. Part A* **2012**, *100*, 738–745. [CrossRef]

129. Zhao, X.; Heng, B.C.; Xiong, S.; Guo, J.; Tan, T.T.; Boey, F.Y.; Ng, K.W.; Loo, J.S. In vitro assessment of cellular responses to rod-shaped hydroxyapatite nanoparticles of varying lengths and surface areas. *Nanotoxicology* **2011**, *5*, 182–194. [CrossRef]

130. Nelson, B.C.; Johnson, M.E.; Walker, M.L.; Riley, K.R.; Sims, C.M. Antioxidant cerium oxide nanoparticles in biology and medicine. *Antioxidants* **2016**, *5*, 15. [CrossRef]

131. Celardo, I.; Pedersen, J.Z.; Traversa, E.; Ghibelli, L. Pharmacological potential of cerium oxide nanoparticles. *Nanoscale* **2011**, *3*, 1411–1420. [CrossRef]

132. Li, C.; Shi, X.; Shen, Q.; Guo, C.; Hou, Z.; Zhang, J. Hot topics and challenges of regenerative nanoceria in application of antioxidant therapy. *J. Nanomater.* **2018**, *2018*, 1–12. [CrossRef]

133. Xu, C.; Qu, X. Cerium oxide nanoparticle: A remarkably versatile rare earth nanomaterial for biological applications. *NPG Asia Mater.* **2014**, *6*, e90. [CrossRef]

134. Xue, Y.; Luan, Q.; Yang, D.; Yao, X.; Zhou, K. Direct evidence for hydroxyl radical scavenging activity of cerium oxide nanoparticles. *J. Phys. Chem. C* **2011**, *115*, 4433–4438. [CrossRef]

135. Heckert, E.G.; Karakoti, A.S.; Seal, S.; Self, W.T. The role of cerium redox state in the SOD mimetic activity of nanoceria. *Biomaterials* **2008**, *29*, 2705–2709. [CrossRef]

136. Pirmohamed, T.; Dowding, J.M.; Singh, S.; Wasserman, B.; Heckert, E.; Karakoti, A.S.; King, J.E.; Seal, S.; Self, W.T. Nanoceria exhibit redox state-dependent catalase mimetic activity. *Chem. Commun. (Camb.)* **2010**, *46*, 2736–2738. [CrossRef]

137. Deshpande, S.; Patil, S.; Kuchibhatla, S.V.N.T.; Seal, S. Size dependency variation in lattice parameter and valency states in nanocrystalline cerium oxide. *Appl. Phys. Lett.* **2005**, *87*, 133113. [CrossRef]

138. Farias, I.A.P.; Dos Santos, C.C.L.; Sampaio, F.C. Antimicrobial activity of cerium oxide nanoparticles on opportunistic microorganisms: A systematic review. *BioMed Res. Int.* **2018**, *2018*, 1923606. [CrossRef]

139. Badmaev, V.; Majeed, M.; Passwater, R.A. Selenium: A quest for better understanding. *Alt. Ther. Health Med.* **1996**, *2*, 59–62, 65–67.

140. Steinbrenner, H.; Speckmann, B.; Klotz, L.O. Selenoproteins: Antioxidant selenoenzymes and beyond. *Arch. Biochem. Biophys.* **2016**, *595*, 113–119. [CrossRef]

141. Moreno-Reyes, R.; Egrise, D.; Neve, J.; Pasteels, J.L.; Schoutens, A. Selenium deficiency-induced growth retardation is associated with an impaired bone metabolism and osteopenia. *J. Bone Miner. Res.* **2001**, *16*, 1556–1563. [CrossRef]

142. Uskokovic, V.; Iyer, M.A.; Wu, V.M. One ion to rule them all: Combined antibacterial, osteoinductive and anticancer properties of selenite-incorporated hydroxyapatite. *J. Mater. Chem. B* **2017**, *5*, 1430–1445. [CrossRef] [PubMed]

143. Rajamannan, N.M. Oxidative-mechanical stress signals stem cell niche mediated Lrp5 osteogenesis in eNOS(-/-) null mice. *J. Cell. Biochem.* **2012**, *113*, 1623–1634. [CrossRef] [PubMed]

144. Chung, Y.W.; Kim, T.S.; Lee, S.Y.; Lee, S.H.; Choi, Y.; Kim, N.; Min, B.M.; Jeong, D.W.; Kim, I.Y. Selenite-induced apoptosis of osteoclasts mediated by the mitochondrial pathway. *Toxicol. Lett.* **2006**, *160*, 143–150. [CrossRef] [PubMed]

145. Domazetovic, V.; Marcucci, G.; Iantomasi, T.; Brandi, M.L.; Vincenzini, M.T. Oxidative stress in bone remodeling: Role of antioxidants. *Clin. Cases Miner. Bone Metab.* **2017**, *14*, 209–216. [CrossRef] [PubMed]

146. Ma, Q. Role of Nrf2 in oxidative stress and toxicity. *Ann. Rev. Pharmacol. Toxicol.* **2013**, *53*, 401–426. [CrossRef]

147. Zheng, F.; Li, H. Evaluation of Nrf2 with exposure to nanoparticles. *Methods Mol. Biol* **2019**, *1894*, 229–246. [CrossRef]

148. Sun, Y.X.; Xu, A.H.; Yang, Y.; Li, J. Role of Nrf2 in bone metabolism. *J. Biomed. Sci.* **2015**, *22*, 101. [CrossRef]

149. Nuss, K.M.R.; Rechenberg, B.V. Biocompatibility issues with modern implants in bone—A review for clinical orthopedics. *Open Orthop. J.* **2008**, *2*, 66–78. [CrossRef]

150. Cacciotti, I. Cationic and anionic substitutions in hydroxyapatite. In *Handbook of Bioceramics and Biocomposites*, 1st ed.; Antoniac, I., Ed.; Springer: Basel, Switzerland, 2016; pp. 145–211, ISBN 978-3-319-12459-9.

151. Ashton Acton, Q. *Apatites—Advances in Research and Application*; 2012 Edition; Scholarly Editions: Atlanta, GA, USA, 2012.

152. ISO. *10993-6:2016 Biological Evaluation of Medical Devices—Part 6: Tests for Local Effects AFTER Implantation*; ISO: Geneve, Switzerland, 2016; p. 29.

153. Hung, B.P.; Hutton, D.L.; Kozielski, K.L.; Bishop, C.J.; Naved, B.; Green, J.J.; Caplan, A.I.; Gimble, J.M.; Dorafshar, A.H.; Grayson, W.L. Platelet-derived growth factor BB enhances osteogenesis of adipose-derived but not bone marrow-derived mesenchymal stromal/stem cells. *Stem Cells* **2015**, *33*, 2773–2784. [CrossRef]

154. Sun, X.; Su, W.; Ma, X.; Zhang, H.; Sun, Z.; Li, X. Comparison of the osteogenic capability of rat bone mesenchymal stem cells on collagen, collagen/hydroxyapatite, hydroxyapatite and biphasic calcium phosphate. *Regen. Biomater.* **2018**, *5*, 93–103. [CrossRef]

155. Schumacher, M.; Lode, A.; Helth, A.; Gelinsky, M. A novel strontium(II)-modified calcium phosphate bone cement stimulates human-bone-marrow-derived mesenchymal stem cell proliferation and osteogenic differentiation in vitro. *Acta Biomater.* **2013**, *9*, 9547–9557. [CrossRef] [PubMed]

156. Mencia Castano, I.; Curtin, C.M.; Duffy, G.P.; O'Brien, F.J. Next generation bone tissue engineering: Non-viral miR-133a inhibition using collagen-nanohydroxyapatite scaffolds rapidly enhances osteogenesis. *Sci. Rep.* **2016**, *6*, 27941. [CrossRef] [PubMed]

157. Klontzas, M.E.; Vernardis, S.I.; Heliotis, M.; Tsiridis, E.; Mantalaris, A. Metabolomics analysis of the osteogenic differentiation of umbilical cord blood mesenchymal stem cells reveals differential sensitivity to osteogenic agents. *Stem Cells Dev.* **2017**, *26*, 723–733. [CrossRef] [PubMed]

158. Zhou, M.; Liu, N.; Zhang, Q.; Tian, T.; Ma, Q.; Zhang, T.; Cai, X. Effect of tetrahedral DNA nanostructures on proliferation and osteogenic differentiation of human periodontal ligament stem cells. *Cell Prolif.* **2019**, *52*, e12566. [CrossRef] [PubMed]

159. Puchtler, H.; Meloan, S.N.; Terry, M.S. On the history and mechanism of alizarin and alizarin red S stains for calcium. *J. Histochem. Cytochem.* **1969**, *17*, 110–124. [CrossRef] [PubMed]

160. Bonewald, L.F.; Harris, S.E.; Rosser, J.; Dallas, M.R.; Dallas, S.L.; Camacho, N.P.; Boyan, B.; Boskey, A. von Kossa staining alone is not sufficient to confirm that mineralization in vitro represents bone formation. *Calcified Tissue Int.* **2003**, *72*, 537–547. [CrossRef] [PubMed]

161. Cheng, H.; Xiong, W.; Fang, Z.; Guan, H.; Wu, W.; Li, Y.; Zhang, Y.; Alvarez, M.M.; Gao, B.; Huo, K.; et al. Strontium (Sr) and silver (Ag) loaded nanotubular structures with combined osteoinductive and antimicrobial activities. *Acta Biomater.* **2016**, *31*, 388–400. [CrossRef]

162. Huang, Y.; Xu, Z.; Zhang, X.; Chang, X.; Zhang, X.; Li, Y.; Ye, T.; Han, R.; Han, S.; Gao, Y.; et al. Nanotube-formed Ti substrates coated with silicate/silver co-doped hydroxyapatite as prospective materials for bone implants. *J. Alloy. Comp.* **2017**, *697*, 182–199. [CrossRef]

163. Ilmer, M.; Karow, M.; Geissler, C.; Jochum, M.; Neth, P. Human osteoblast–derived factors induce early osteogenic markers in human mesenchymal stem cells. *Tissue Eng. Part. A* **2009**, *15*, 2397–2409. [CrossRef]

164. Komori, T. Regulation of bone development and extracellular matrix protein genes by RUNX2. *Cell Tissue Res.* **2010**, *339*, 189–195. [CrossRef]

165. Li, Y.L.; Xiao, Z.S. Advances in Runx2 regulation and its isoforms. *Med. Hypotheses* **2007**, *68*, 169–175. [CrossRef] [PubMed]

166. Narayanan, A.; Srinaath, N.; Rohini, M.; Selvamurugan, N. Regulation of Runx2 by MicroRNAs in osteoblast differentiation. *Life Sci.* **2019**, *232*, 116676. [CrossRef] [PubMed]

167. Fakhry, M.; Hamade, E.; Badran, B.; Buchet, R.; Magne, D. Molecular mechanisms of mesenchymal stem cell differentiation towards osteoblasts. *World J. Stem Cells* **2013**, *5*, 136–148. [CrossRef] [PubMed]

168. Hu, H.; Hilton, M.J.; Tu, X.; Yu, K.; Ornitz, D.M.; Long, F. Sequential roles of Hedgehog and Wnt signaling in osteoblast development. *Development* **2005**, *132*, 49–60. [CrossRef]

169. Glass, D.A., 2nd; Bialek, P.; Ahn, J.D.; Starbuck, M.; Patel, M.S.; Clevers, H.; Taketo, M.M.; Long, F.; McMahon, A.P.; Lang, R.A.; et al. Canonical Wnt signaling in differentiated osteoblasts controls osteoclast differentiation. *Dev. Cell* **2005**, *8*, 751–764. [CrossRef]

170. Ducy, P.; Zhang, R.; Geoffroy, V.; Ridall, A.L.; Karsenty, G. Osf2/Cbfa1: A transcriptional activator of osteoblast differentiation. *Cell* **1997**, *89*, 747–754. [CrossRef]

171. Nakashima, K.; Zhou, X.; Kunkel, G.; Zhang, Z.; Deng, J.M.; Behringer, R.R.; de Crombrugghe, B. The novel zinc finger-containing transcription factor osterix is required for osteoblast differentiation and bone formation. *Cell* **2002**, *108*, 17–29. [CrossRef]

172. Kulanthaivel, S.; Mishra, U.; Agarwal, T.; Giri, S.; Pal, K.; Pramanik, K.; Banerjee, I. Improving the osteogenic and angiogenic properties of synthetic hydroxyapatite by dual doping of bivalent cobalt and magnesium ion. *Ceram. Int.* **2015**, *41*, 11323–11333. [CrossRef]

173. Won, J.E.; Yun, Y.R.; Jang, J.H.; Yang, S.H.; Kim, J.H.; Chrzanowski, W.; Wall, I.B.; Knowles, J.C.; Kim, H.W. Multifunctional and stable bone mimic proteinaceous matrix for bone tissue engineering. *Biomaterials* **2015**, *56*, 46–57. [CrossRef]

174. Curtin, C.M.; Cunniffe, G.M.; Lyons, F.G.; Bessho, K.; Dickson, G.R.; Duffy, G.P.; O'Brien, F.J. Innovative collagen nano-hydroxyapatite scaffolds offer a highly efficient non-viral gene delivery platform for stem cell-mediated bone formation. *Adv. Mater.* **2012**, *24*, 749–754. [CrossRef]

175. Rana, A.A.; Karim, M.M.; Gafur, M.A.; Hossan, M.J. Mechanical properties of Gelatin–Hydroxyapatite composite for bone tissue engineering. *Bangladesh J. Sci. Ind. Res.* **2015**, *50*, 15–20. [CrossRef]

176. Nieto-Suarez, M.; Lopez-Quintela, M.A.; Lazzari, M. Preparation and characterization of crosslinked chitosan/gelatin scaffolds by ice segregation induced self-assembly. *Carbohydr. Polym.* **2016**, *141*, 175–183. [CrossRef] [PubMed]

177. Zhang, L.; Fang, H.; Zhang, K.; Yin, J. Homologous sodium alginate/chitosan-based scaffolds, but contrasting effect on stem cell shape and osteogenesis. *ACS Appl. Mater. Interfaces* **2018**, *10*, 6930–6941. [CrossRef] [PubMed]

178. Cunniffe, G.M.; O'Brien, F.J.; Partap, S.; Levingstone, T.J.; Stanton, K.T.; Dickson, G.R. The synthesis and characterization of nanophase hydroxyapatite using a novel dispersant-aided precipitation method. *J. Biomed. Mater. Res. Part A* **2010**, *95*, 1142–1149. [CrossRef] [PubMed]

179. Tierney, E.G.; Duffy, G.P.; Hibbitts, A.J.; Cryan, S.A.; O'Brien, F.J. The development of non-viral gene-activated matrices for bone regeneration using polyethyleneimine (PEI) and collagen-based scaffolds. *J. Control. Release* **2012**, *158*, 304–311. [CrossRef] [PubMed]

180. Li, Y.; Chen, S.K.; Li, L.; Qin, L.; Wang, X.L.; Lai, Y.X. Bone defect animal models for testing efficacy of bone substitute biomaterials. *J. Orthop. Trans.* **2015**, *3*, 95–104. [CrossRef]

181. Kim, J.M.; Kim, J.H.; Lee, B.H.; Choi, S.H. Vertical Bone Augmentation Using Three-dimensionally Printed Cap in the Rat Calvarial Partial Defect. *In Vivo* **2018**, *32*, 1111–1117. [CrossRef]

182. Spicer, P.P.; Kretlow, J.D.; Young, S.; Jansen, J.A.; Kasper, F.K.; Mikos, A.G. Evaluation of bone regeneration using the rat critical size calvarial defect. *Nat. Protoc.* **2012**, *7*, 1918–1929. [CrossRef]

183. Saulacic, N.; Nakahara, K.; Iizuka, T.; Haga-Tsujimura, M.; Hofstetter, W.; Scolozzi, P. Comparison of two protocols of periosteal distraction osteogenesis in a rabbit calvaria model. *J. Biomed. Mater. Res. Part B* **2016**, *104*, 1121–1131. [CrossRef]

184. Fujio, M.; Osawa, Y.; Matsushita, M.; Ogisu, K.; Tsuchiya, S.; Kitoh, H.; Hibi, H. A Mouse Distraction Osteogenesis Model. *J. Vis. Exp.* **2018**. [CrossRef]

185. Osawa, Y.; Matsushita, M.; Hasegawa, S.; Esaki, R.; Fujio, M.; Ohkawara, B.; Ishiguro, N.; Ohno, K.; Kitoh, H. Activated FGFR3 promotes bone formation via accelerating endochondral ossification in mouse model of distraction osteogenesis. *Bone* **2017**, *105*, 42–49. [CrossRef] [PubMed]

186. Carey, E.G.; Deshpande, S.S.; Zheutlin, A.R.; Nelson, N.S.; Donneys, A.; Kang, S.Y.; Gallagher, K.K.; Felice, P.A.; Tchanque-Fossuo, C.N.; Buchman, S.R. A Comparison of vascularity, bone mineral density distribution, and histomorphometrics in an isogenic versus an outbred murine model of mandibular distraction osteogenesis. *J. Oral Maxillofac. Surg.* **2016**, *74*, 2055–2065. [CrossRef] [PubMed]

187. Zheutlin, A.R.; Deshpande, S.S.; Nelson, N.S.; Kang, S.Y.; Gallagher, K.K.; Polyatskaya, Y.; Rodriguez, J.J.; Donneys, A.; Ranganathan, K.; Buchman, S.R. Bone marrow stem cells assuage radiation-induced damage in a murine model of distraction osteogenesis: A histomorphometric evaluation. *Cytotherapy* **2016**, *18*, 664–672. [CrossRef] [PubMed]

188. Weng, Z.; Wang, C.; Zhang, C.; Xu, J.; Chai, Y.; Jia, Y.; Han, P.; Wen, G. All-trans retinoic acid promotes osteogenic differentiation and bone consolidation in a rat distraction osteogenesis model. *Calcif. Tissue Int.* **2019**, *104*, 320–330. [CrossRef] [PubMed]

189. McDonald, M.M.; Morse, A.; Birke, O.; Yu, N.Y.C.; Mikulec, K.; Peacock, L.; Schindeler, A.; Liu, M.; Ke, H.Z.; Little, D.G. Sclerostin antibody enhances bone formation in a rat model of distraction osteogenesis. *J. Orthop. Res.* **2018**, *36*, 1106–1113. [CrossRef]

190. Tchanque-Fossuo, C.N.; Donneys, A.; Deshpande, S.S.; Sarhaddi, D.; Nelson, N.S.; Monson, L.A.; Dahle, S.E.; Goldstein, S.A.; Buchman, S.R. Radioprotection with amifostine enhances bone strength and regeneration and bony union in a rat model of mandibular distraction osteogenesis. *Ann. Plast. Surg.* **2018**, *80*, 176–180. [CrossRef]

191. Pithioux, M.; Roseren, F.; Jalain, C.; Launay, F.; Charpiot, P.; Chabrand, P.; Roffino, S.; Lamy, E. An efficient and reproducible protocol for distraction osteogenesis in a rat model leading to a functional regenerated femur. *J. Vis. Exp.* **2017**. [CrossRef]

192. Jiang, X.; Zou, S.; Ye, B.; Zhu, S.; Liu, Y.; Hu, J. bFGF-Modified BMMSCs enhance bone regeneration following distraction osteogenesis in rabbits. *Bone* **2010**, *46*, 1156–1161. [CrossRef]

193. Yassine, K.A.; Mokhtar, B.; Houari, H.; Karim, A.; Mohamed, M. Repair of segmental radial defect with autologous bone marrow aspirate and hydroxyapatite in rabbit radius: A clinical and radiographic evaluation. *Vet. World* **2017**, *10*, 752–757. [CrossRef]

194. Montes-Medina, L.; Hernandez-Fernandez, A.; Gutierrez-Rivera, A.; Ripalda-Cemborain, P.; Bitarte, N.; Perez-Lopez, V.; Granero-Molto, F.; Prosper, F.; Izeta, A. Effect of bone marrow stromal cells in combination with biomaterials in early phases of distraction osteogenesis: An experimental study in a rabbit femur model. *Injury* **2018**, *49*, 1979–1986. [CrossRef]

195. Floerkemeier, T.; Thorey, F.; Wellmann, M.; Hurschler, C.; Budde, S.; Windhagen, H. alphaBSM failed as a carrier of rhBMP-2 to enhance bone consolidation in a sheep model of distraction osteogenesis. *Acta Bioeng. Biomech.* **2017**, *19*, 55–62. [PubMed]

196. Andreasen, C.M.; Henriksen, S.S.; Ding, M.; Theilgaard, N.; Andersen, T.L.; Overgaard, S. The efficacy of poly-d,l-lactic acid- and hyaluronic acid-coated bone substitutes on implant fixation in sheep. *J. Orthop. Trans.* **2017**, *8*, 12–19. [CrossRef] [PubMed]

197. Fellah, B.H.; Gauthier, O.; Weiss, P.; Chappard, D.; Layrolle, P. Osteogenicity of biphasic calcium phosphate ceramics and bone autograft in a goat model. *Biomaterials* **2008**, *29*, 1177–1188. [CrossRef] [PubMed]

198. Eom, T.G.; Jeon, G.R.; Jeong, C.M.; Kim, Y.K.; Kim, S.G.; Cho, I.H.; Cho, Y.S.; Oh, J.S. Experimental study of bone response to hydroxyapatite coating implants: Bone-implant contact and removal torque test. *Oral Surg. Oral Med. Oral Pathol. Oral Radiol.* **2012**, *114*, 411–418. [CrossRef]

199. Lukaszewska-Kuska, M.; Krawczyk, P.; Martyla, A.; Hedzelek, W.; Dorocka-Bobkowska, B. Effects of a hydroxyapatite coating on the stability of endosseous implants in rabbit tibiae. *Dent. Med. Probl.* **2019**, *56*, 123–129. [CrossRef]

200. Komori, T. Animal models for osteoporosis. *Eur. J. Pharmacol.* **2015**, *759*, 287–294. [CrossRef]

201. Calciolari, E.; Donos, N.; Mardas, N. Osteoporotic Animal models of bone healing: Advantages and pitfalls. *J. Investig. Surg.* **2016**, *30*, 342–350. [CrossRef]

202. Xu, J.; Gong, H.; Lu, S.; Deasey, M.J.; Cui, Q. Animal models of steroid-induced osteonecrosis of the femoral head-a comprehensive research review up to 2018. *Int. Orthop.* **2018**, *42*, 1729–1737. [CrossRef]

203. O'Brien, C.A.; Morello, R. Modeling rare bone diseases in animals. *Curr. Osteoporosis Rep.* **2018**, *16*, 458–465. [CrossRef]

204. Ewald, A.; Kreczy, D.; Bruckner, T.; Gbureck, U.; Bengel, M.; Hoess, A.; Nies, B.; Bator, J.; Klammert, U.; Fuchs, A. Development and bone regeneration capacity of premixed magnesium phosphate cement pastes. *Materials* **2019**, *12*, 2119. [CrossRef]

205. Tian, Q.; Lin, J.; Rivera-Castaneda, L.; Tsanhani, A.; Dunn, Z.S.; Rodriguez, A.; Aslani, A.; Liu, H. Nano-to-submicron hydroxyapatite coatings for magnesium-based bioresorbable implants—deposition, characterization, degradation, mechanical properties, and cytocompatibility. *Sci. Rep.* **2019**, *9*, 810. [CrossRef] [PubMed]

206. Li, B.; Gao, P.; Zhang, H.; Guo, Z.; Zheng, Y.; Han, Y. Osteoimmunomodulation, osseointegration, and in vivo mechanical integrity of pure Mg coated with HA nanorod/pore-sealed MgO bilayer. *Biomater. Sci.* **2018**, *6*, 3202–3218. [CrossRef] [PubMed]

207. Yang, H.; Qu, X.; Lin, W.; Wang, C.; Zhu, D.; Dai, K.; Zheng, Y. In vitro and in vivo studies on zinc-hydroxyapatite composites as novel biodegradable metal matrix composite for orthopedic applications. *Acta Biomater.* **2018**, *71*, 200–214. [CrossRef] [PubMed]

208. Suruagy, A.A.; Alves, A.T.; Sartoretto, S.C.; Calasans-Maia, J.A.; Granjeiro, J.M.; Calasans-Maia, M.D. Physico-chemical and histomorphometric evaluation of zinc-containing hydroxyapatite in rabbits calvaria. *Braz. Dent. J.* **2016**, *27*, 717–726. [CrossRef]

209. Wang, H.; Zhao, S.; Xiao, W.; Cui, X.; Huang, W.; Rahaman, M.N.; Zhang, C.; Wang, D. Three-dimensional zinc incorporated borosilicate bioactive glass scaffolds for rodent critical-sized calvarial defects repair and regeneration. *Colloids Surf. B Biointerf.* **2015**, *130*, 149–156. [CrossRef]

210. Glenske, K.; Donkiewicz, P.; Kowitsch, A.; Milosevic-Oljaca, N.; Rider, P.; Rofall, S.; Franke, J.; Jung, O.; Smeets, R.; Schnettler, R.; et al. Applications of metals for bone regeneration. *Int. J. Mol. Sci.* **2018**, *19*, 826. [CrossRef]

211. Li, D.; Huifang, L.; Zhao, J.; Yang, Z.; Xie, X.; Wei, Z.; Kang, P. Porous lithium-doped hydroxyapatite scaffold seeded with hypoxia-preconditioned bone-marrow mesenchymal stem cells for bone-tissue regeneration. *Biomed. Mater.* **2018**, *13*, 055002. [CrossRef]

212. Li, D.; Xie, X.; Yang, Z.; Wang, C.; Wei, Z.; Kang, P. Enhanced bone defect repairing effects in glucocorticoid-induced osteonecrosis of the femoral head using a porous nano-lithium-hydroxyapatite/gelatin microsphere/erythropoietin composite scaffold. *Biomater. Sci.* **2018**, *6*, 519–537. [CrossRef]

213. Luo, Y.; Chen, S.; Shi, Y.; Ma, J. 3D printing of strontium-doped hydroxyapatite based composite scaffolds for repairing critical-sized rabbit calvarial defects. *Biomed. Mater.* **2018**, *13*, 065004. [CrossRef]

214. Chandran, S.; Babu, S.S.; Vs, H.K.; Varma, H.K.; John, A. Osteogenic efficacy of strontium hydroxyapatite micro-granules in osteoporotic rat model. *J. Biomater. Appl.* **2016**, *31*, 499–509. [CrossRef]

215. Li, Y.; Shui, X.; Zhang, L.; Hu, J. Cancellous bone healing around strontium-doped hydroxyapatite in osteoporotic rats previously treated with zoledronic acid. *J. Biomed. Mater. Res. Part B* **2016**, *104*, 476–481. [CrossRef] [PubMed]

216. Li, Y.; Luo, E.; Zhu, S.; Li, J.; Zhang, L.; Hu, J. Cancellous bone response to strontium-doped hydroxyapatite in osteoporotic rats. *J. Appl. Biomater. Funct. Mater.* **2015**, *13*, 28–34. [CrossRef] [PubMed]

217. Barba, A.; Maazouz, Y.; Diez-Escudero, A.; Rappe, K.; Espanol, M.; Montufar, E.B.; Ohman-Magi, C.; Persson, C.; Fontecha, P.; Manzanares, M.C.; et al. Osteogenesis by foamed and 3D-printed nanostructured calcium phosphate scaffolds: Effect of pore architecture. *Acta Biomater.* **2018**, *79*, 135–147. [CrossRef] [PubMed]

218. Shao, H.; Ke, X.; Liu, A.; Sun, M.; He, Y.; Yang, X.; Fu, J.; Liu, Y.; Zhang, L.; Yang, G.; et al. Bone regeneration in 3D printing bioactive ceramic scaffolds with improved tissue/material interface pore architecture in thin-wall bone defect. *Biofabrication* **2017**, *9*, 025003. [CrossRef]

219. Barba, A.; Diez-Escudero, A.; Espanol, M.; Bonany, M.; Sadowska, J.M.; Guillem-Marti, J.; Ohman-Magi, C.; Persson, C.; Manzanares, M.C.; Franch, J.; et al. Impact of biomimicry in the design of osteoinductive bone substitutes: Nanoscale matters. *ACS Appl. Mater. Interfaces* **2019**, *11*, 8818–8830. [CrossRef]

220. Diao, J.; OuYang, J.; Deng, T.; Liu, X.; Feng, Y.; Zhao, N.; Mao, C.; Wang, Y. 3D-plotted beta-tricalcium phosphate scaffolds with smaller pore sizes improve in vivo bone regeneration and biomechanical properties in a critical-sized calvarial defect rat model. *Adv. Healthc. Mater.* **2018**, *7*, e1800441. [CrossRef]

221. Kubasiewicz-Ross, P.; Hadzik, J.; Seeliger, J.; Kozak, K.; Jurczyszyn, K.; Gerber, H.; Dominiak, M.; Kunert-Keil, C. New nano-hydroxyapatite in bone defect regeneration: A histological study in rats. *Ann. Anat.* **2017**, *213*, 83–90. [CrossRef]

222. Elrayah, A.; Zhi, W.; Feng, S.; Al-Ezzi, S.; Lei, H.; Weng, J. Preparation of micro/nano-structure copper-substituted hydroxyapatite scaffolds with improved angiogenesis capacity for bone regeneration. *Materials* **2018**, *11*, 1516. [CrossRef]

223. Handoll, H.H.; Watts, A.C. Bone grafts and bone substitutes for treating distal radial fractures in adults. *Cochrane Database Syst. Rev.* **2008**, CD006836. [CrossRef]

224. Buser, Z.; Brodke, D.S.; Youssef, J.A.; Meisel, H.J.; Myhre, S.L.; Hashimoto, R.; Park, J.B.; Tim Yoon, S.; Wang, J.C. Synthetic bone graft versus autograft or allograft for spinal fusion: A systematic review. *J. Neurosurg. Spine* **2016**, *25*, 509–516. [CrossRef]

225. Kaiser, M.G.; Groff, M.W.; Watters, W.C., 3rd; Ghogawala, Z.; Mummaneni, P.V.; Dailey, A.T.; Choudhri, T.F.; Eck, J.C.; Sharan, A.; Wang, J.C.; et al. Guideline update for the performance of fusion procedures for degenerative disease of the lumbar spine. Part 16: Bone graft extenders and substitutes as an adjunct for lumbar fusion. *J. Neurosurg. Spine* **2014**, *21*, 106–132. [CrossRef] [PubMed]

226. Chepelev, L.; Wake, N.; Ryan, J.; Althobaity, W.; Gupta, A.; Arribas, E.; Santiago, L.; Ballard, D.H.; Wang, K.C.; Weadock, W.; et al. Radiological Society of North America (RSNA) 3D printing Special Interest Group (SIG): Guidelines for medical 3D printing and appropriateness for clinical scenarios. *3D Print. Med.* **2018**, *4*, 11. [CrossRef] [PubMed]

227. Xia, L.; Lin, K.; Jiang, X.; Fang, B.; Xu, Y.; Liu, J.; Zeng, D.; Zhang, M.; Zhang, X.; Chang, J.; et al. Effect of nano-structured bioceramic surface on osteogenic differentiation of adipose derived stem cells. *Biomaterials* **2014**, *35*, 8514–8527. [CrossRef] [PubMed]

228. Almubarak, S.; Nethercott, H.; Freeberg, M.; Beaudon, C.; Jha, A.; Jackson, W.; Miclau, T.; Healy, K.; Bahney, C. Tissue engineering strategies for promoting vascularized bone regeneration. *Bone* **2016**, *83*, 197–209. [CrossRef] [PubMed]

229. Malhotra, A.; Habibovic, P. Calcium phosphates and angiogenesis: Implications and advances for bone regeneration. *Trends Biotechnol.* **2016**, *34*, 983–992. [CrossRef] [PubMed]

230. Amini, A.R.; Laurencin, C.T.; Nukavarapu, S.P. Bone tissue engineering: recent advances and challenges. *Crit. Rev. Biomed. Eng.* **2012**, *40*, 363–408. [CrossRef]

231. Levingstone, T.J.; Barron, N.; Ardhaoui, M.; Benyounis, K.; Looney, L.; Stokes, J. Application of response surface methodology in the design of functionally graded plasma sprayed hydroxyapatite coatings. *Surf. Coat. Technol.* **2017**, *313*, 307–318. [CrossRef]

232. Rahmani, A.; Hashemi-Najafabadi, S.; Eslaminejad, M.B.; Bagheri, F.; Sayahpour, F.A. The effect of modified electrospun PCL-nHA-nZnO scaffolds on osteogenesis and angiogenesis. *J. Biomed. Mater. Res. Part A* **2019**, *107*, 2040–2052. [CrossRef]

233. Yang, C.; Zhao, C.; Wang, X.; Shi, M.; Zhu, Y.; Jing, L.; Wu, C.; Chang, J. Stimulation of osteogenesis and angiogenesis by micro/nano hierarchical hydroxyapatite via macrophage immunomodulation. *Nanoscale* **2019**, *11*, 17699–17708. [CrossRef]

234. Son, J.; Kim, J.; Lee, K.; Hwang, J.; Choi, Y.; Seo, Y.; Jeon, H.; Kang, H.C.; Woo, H.-M.; Kang, B.-J.; et al. DNA aptamer immobilized hydroxyapatite for enhancing angiogenesis and bone regeneration. *Acta Biomater.* **2019**. [CrossRef]

235. Maier, J.A.; Bernardini, D.; Rayssiguier, Y.; Mazur, A. High concentrations of magnesium modulate vascular endothelial cell behaviour in vitro. *Biochim. Biophys. Acta* **2004**, *1689*, 6–12. [CrossRef] [PubMed]

236. Sartori, M.; Pagani, S.; Ferrari, A.; Costa, V.; Carina, V.; Figallo, E.; Maltarello, M.C.; Martini, L.; Fini, M.; Giavaresi, G. A new bi-layered scaffold for osteochondral tissue regeneration: In vitro and in vivo preclinical investigations. *Mater. Sci. Eng. C* **2017**, *70*, 101–111. [CrossRef] [PubMed]

237. Yang, W.; Both, S.K.; van Osch, G.J.; Wang, Y.; Jansen, J.A.; Yang, F. Performance of different three-dimensional scaffolds for in vivo endochondral bone generation. *Eur. Cells Mater.* **2014**, *27*, 350–364. [CrossRef]

238. Canullo, L.; Heinemann, F.; Gedrange, T.; Biffar, R.; Kunert-Keil, C. Histological evaluation at different times after augmentation of extraction sites grafted with a magnesium-enriched hydroxyapatite: Double-blinded randomized controlled trial. *Clin. Oral Implants Res.* **2013**, *24*, 398–406. [CrossRef]

239. Frank, P.G.; Woodman, S.E.; Park, D.S.; Lisanti, M.P. Caveolin, caveolae, and endothelial cell function. *Arterioscler. Thromb. Vasc. Biol.* **2003**, *2*, 1161–1168. [CrossRef]

240. Woodman, S.E.; Ashton, A.W.; Schubert, W.; Lee, H.; Williams, T.M.; Medina, F.A.; Wyckoff, J.B.; Combs, T.P.; Lisanti, M.P. Caveolin-1 knockout mice show an impaired angiogenic response to exogenous stimuli. *Am. J. Pathol.* **2003**, *162*, 2059–2068. [CrossRef]

241. Sun, T.W.; Yu, W.L.; Zhu, Y.J.; Yang, R.L.; Shen, Y.Q.; Chen, D.Y.; He, Y.H.; Chen, F. Hydroxyapatite nanowire@magnesium silicate core-shell hierarchical nanocomposite: Synthesis and application in bone regeneration. *ACS Appl. Mater. Interfaces* **2017**, *9*, 16435–16447. [CrossRef]

242. Calabrese, G.; Giuffrida, R.; Forte, S.; Salvatorelli, L.; Fabbi, C.; Figallo, E.; Gulisano, M.; Parenti, R.; Magro, G.; Colarossi, C.; et al. Bone augmentation after ectopic implantation of a cell-free collagen-hydroxyapatite scaffold in the mouse. *Sci. Rep.* **2016**, *6*, 36399. [CrossRef]

243. Saghiri, M.A.; Asatourian, A.; Orangi, J.; Sorenson, C.M.; Sheibani, N. Functional role of inorganic trace elements in angiogenesis – Part II: Cr, Si, Zn, Cu, and S. *Crit. Rev. Oncol. Hematol.* **2015**, *96*, 143–155. [CrossRef]

244. Barralet, J.; Gbureck, U.; Habibovic, P.; Vorndran, E.; Gerard, C.; Doillon, C.J. Angiogenesis in calcium phosphate scaffolds by inorganic copper ion release. *Tissue Eng. Part A* **2009**, *15*, 1601–1609. [CrossRef]

245. Imrie, F.E.; Skakle, J.M.S. Preparation of copper-doped hydroxyapatite with varying x in the composition $Ca_{10}(PO_4)_6Cu_xO_yH_z$. *Bioceram. Dev. Appl.* **2013**, *3*, S1:005. [CrossRef]

246. Kargozar, S.; Lotfibakhshaiesh, N.; Ai, J.; Mozafari, M.; Brouki Milan, P.; Hamzehlou, S. Strontium- and cobalt-substituted bioactive glasses seeded with human umbilical cord perivascular cells to promote bone regeneration via enhanced osteogenic and angiogenic activities. *Acta Biomater.* **2017**, *58*, 502–514. [CrossRef] [PubMed]

247. Gu, Z.; Xie, H.; Li, L.; Zhang, X.; Liu, F.; Yu, X. Application of strontium-doped calcium polyphosphate scaffold on angiogenesis for bone tissue engineering. *J. Mater. Sci. Mater. Med.* **2013**, *24*, 1251–1260. [CrossRef] [PubMed]

248. Ehret, C.; Aid-Launais, R.; Sagardoy, T.; Siadous, R.; Bareille, R.; Rey, S.; Pechev, S.; Etienne, L.; Kalisky, J.; de Mones, E.; et al. Strontium-doped hydroxyapatite polysaccharide materials effect on ectopic bone formation. *PLoS ONE* **2017**, *12*, e0184663. [CrossRef] [PubMed]

249. Tahmasebi Birgani, Z.; Fennema, E.; Gijbels, M.J.; de Boer, J.; van Blitterswijk, C.A.; Habibovic, P. Stimulatory effect of cobalt ions incorporated into calcium phosphate coatings on neovascularization in an in vivo intramuscular model in goats. *Acta Biomater.* **2016**, *36*, 267–276. [CrossRef]

250. Zamani, A.; Omrani, G.R.; Nasab, M.M. Lithium's effect on bone mineral density. *Bone* **2009**, *44*, 331–334. [CrossRef]

251. Li, J.; Khavandgar, Z.; Lin, S.H.; Murshed, M. Lithium chloride attenuates BMP-2 signaling and inhibits osteogenic differentiation through a novel WNT/GSK3- independent mechanism. *Bone* **2011**, *48*, 321–331. [CrossRef]

252. Nakatsu, M.N.; Ding, Z.; Ng, M.Y.; Truong, T.T.; Yu, F.; Deng, S.X. Wnt/beta-catenin signaling regulates proliferation of human cornea epithelial stem/progenitor cells. *Investig. Ophthalmol. Visual Sci.* **2011**, *52*, 4734–4741. [CrossRef]

253. James, A.W.; LaChaud, G.; Shen, J.; Asatrian, G.; Nguyen, V.; Zhang, X.; Ting, K.; Soo, C. A Review of the clinical side effects of Bone Morphogenetic Protein-2. *Tissue Eng. Part B* **2016**, *22*, 284–297. [CrossRef]

254. Cecchi, S.; Bennet, S.J.; Arora, M. Bone morphogenetic protein-7: Review of signalling and efficacy in fracture healing. *J. Orthop. Trans.* **2016**, *4*, 28–34. [CrossRef]

255. Garrison, K.R.; Shemilt, I.; Donell, S.; Ryder, J.J.; Mugford, M.; Harvey, I.; Song, F.; Alt, V. Bone morphogenetic protein (BMP) for fracture healing in adults. *Cochrane Database Syst. Rev.* **2010**. [CrossRef] [PubMed]

256. Begam, H.; Nandi, S.K.; Kundu, B.; Chanda, A. Strategies for delivering bone morphogenetic protein for bone healing. *Mater. Sci. Eng. C* **2017**, *70*, 856–869. [CrossRef] [PubMed]

257. Campana, V.; Milano, G.; Pagano, E.; Barba, M.; Cicione, C.; Salonna, G.; Lattanzi, W.; Logroscino, G. Bone substitutes in orthopaedic surgery: From basic science to clinical practice. *J. Mater. Sci. Mater. Med.* **2014**, *25*, 2445–2461. [CrossRef] [PubMed]

258. Son, S.R.; Sarkar, S.K.; Nguyen-Thuy, B.L.; Padalhin, A.R.; Kim, B.R.; Jung, H.I.; Lee, B.T. Platelet-rich plasma encapsulation in hyaluronic acid/gelatin-BCP hydrogel for growth factor delivery in BCP sponge scaffold for bone regeneration. *J. Biomater. Appl.* **2015**, *29*, 988–1002. [CrossRef]

259. Gemini-Piperni, S.; Milani, R.; Bertazzo, S.; Peppelenbosch, M.; Takamori, E.R.; Granjeiro, J.M.; Ferreira, C.V.; Teti, A.; Zambuzzi, W. Kinome profiling of osteoblasts on hydroxyapatite opens new avenues on biomaterial cell signaling. *Biotechnol. Bioeng.* **2014**, *111*, 1900–1905. [CrossRef]

260. Liang, H.; Xu, X.; Feng, X.; Ma, L.; Deng, X.; Wu, S.; Liu, X.; Yang, C. Gold nanoparticles-loaded hydroxyapatite composites guide osteogenic differentiation of human mesenchymal stem cells through Wnt/beta-catenin signaling pathway. *Int. J. Nanomed.* **2019**, *14*, 6151–6163. [CrossRef]

261. Ou, L.; Lan, Y.; Feng, Z.; Feng, L.; Yang, J.; Liu, Y.; Bian, L.; Tan, J.; Lai, R.; Guo, R. Functionalization of SF/HAP scaffold with GO-PEI-miRNA inhibitor complexes to enhance bone regeneration through activating Transcription Factor 4. *Theranostics* **2019**, *9*, 4525–4541. [CrossRef]

262. Gizer, M.; Kose, S.; Karaosmanoglu, B.; Taskiran, E.Z.; Berkkan, A.; Timucin, M.; Korkusuz, F.; Korkusuz, P. The Effect of boron-containing nano-hydroxyapatite on bone cells. *Biol. Trace Elem. Res.* **2019**. [CrossRef]

263. Li, K.; Shen, Q.; Xie, Y.; You, M.; Huang, L.; Zheng, X. Incorporation of cerium oxide into hydroxyapatite coating protects bone marrow stromal cells against H_2O_2-induced inhibition of osteogenic differentiation. *Biol. Trace Elem. Res.* **2018**, *182*, 91–104. [CrossRef]

264. Yang, W.; Han, W.; He, W.; Li, J.; Wang, J.; Feng, H.; Qian, Y. Surface topography of hydroxyapatite promotes osteogenic differentiation of human bone marrow mesenchymal stem cells. *Mater. Sci. Eng. C* **2016**, *60*, 45–53. [CrossRef]

265. Hankenson, K.D.; Dishowitz, M.; Gray, C.; Schenker, M. Angiogenesis in bone regeneration. *Injury* **2011**, *42*, 556–561. [CrossRef] [PubMed]

266. Pensa, N.W.; Curry, A.S.; Reddy, M.S.; Bellis, S.L. The addition of a polyglutamate domain to the angiogenic QK peptide improves peptide coupling to bone graft materials leading to enhanced endothelial cell activation. *PLoS ONE* **2019**, *14*, e0213592. [CrossRef] [PubMed]

267. Honda, M.; Hariya, R.; Matsumoto, M.; Aizawa, M. Acceleration of osteogenesis via stimulation of angiogenesis by combination with scaffold and connective tissue growth factor. *Materials* **2019**, *12*, 2068. [CrossRef] [PubMed]

268. Wenz, A.; Tjoeng, I.; Schneider, I.; Kluger, P.J.; Borchers, K. Improved vasculogenesis and bone matrix formation through coculture of endothelial cells and stem cells in tissue-specific methacryloyl gelatin-based hydrogels. *Biotechnol. Bioeng.* **2018**, *115*, 2643–2653. [CrossRef] [PubMed]

269. Bai, L.; Liu, Y.; Du, Z.; Weng, Z.; Yao, W.; Zhang, X.; Huang, X.; Yao, X.; Crawford, R.; Hang, R.; et al. Differential effect of hydroxyapatite nano-particle versus nano-rod decorated titanium micro-surface on osseointegration. *Acta Biomater.* **2018**, *76*, 344–358. [CrossRef] [PubMed]

270. Catalanotto, C.; Cogoni, C.; Zardo, G. MicroRNA in control of gene expression: An overview of nuclear functions. *Int. J. Mol. Sci.* **2016**, *17*, 1712. [CrossRef] [PubMed]

271. Shivdasani, R.A. MicroRNAs: Regulators of gene expression and cell differentiation. *Blood* **2006**, *108*, 3646–3653. [CrossRef]

272. Lian, J.B.; Stein, G.S.; van Wijnen, A.J.; Stein, J.L.; Hassan, M.Q.; Gaur, T.; Zhang, Y. MicroRNA control of bone formation and homeostasis. *Nat. Rev. Endocrinol.* **2012**, *8*, 212–227. [CrossRef]

273. Papaioannou, G.; Mirzamohammadi, F.; Kobayashi, T. MicroRNAs involved in bone formation. *Cell. Mol. Life Sci. CMLS* **2014**, *71*, 4747–4761. [CrossRef]

274. Nakasa, T.; Yoshizuka, M.; Andry Usman, M.; Elbadry Mahmoud, E.; Ochi, M. MicroRNAs and bone regeneration. *Curr. Genomics* **2015**, *16*, 441–452. [CrossRef]

275. Schubert, T.; Xhema, D.; Veriter, S.; Schubert, M.; Behets, C.; Delloye, C.; Gianello, P.; Dufrane, D. The enhanced performance of bone allografts using osteogenic-differentiated adipose-derived mesenchymal stem cells. *Biomaterials* **2011**, *32*, 8880–8891. [CrossRef] [PubMed]

276. Diaz-Rodriguez, P.; Sanchez, M.; Landin, M. Drug-loaded biomimetic ceramics for tissue engineering. *Pharmaceutics* **2018**, *10*, 272. [CrossRef] [PubMed]

277. Gennari, L.; Bianciardi, S.; Merlotti, D. MicroRNAs in bone diseases. *Osteoporosis Int.* **2017**, *28*, 1191–1213. [CrossRef] [PubMed]

278. Peng, S.; Gao, D.; Gao, C.; Wei, P.; Niu, M.; Shuai, C. MicroRNAs regulate signaling pathways in osteogenic differentiation of mesenchymal stem cells (Review). *Mol. Med. Rep.* **2016**, *14*, 623–629. [CrossRef]

279. Frohlich, L.F. Micrornas at the Interface between osteogenesis and angiogenesis as targets for bone regeneration. *Cells* **2019**, *8*, 121. [CrossRef]

280. Chen, J.; Qiu, M.; Dou, C.; Cao, Z.; Dong, S. MicroRNAs in bone balance and osteoporosis. *Drug Dev. Res.* **2015**, *76*, 235–245. [CrossRef]

281. Deng, Y.; Wu, S.; Zhou, H.; Bi, X.; Wang, Y.; Hu, Y.; Gu, P.; Fan, X. Effects of a miR-31, Runx2, and Satb2 regulatory loop on the osteogenic differentiation of bone mesenchymal stem cells. *Stem Cells Dev.* **2013**, *22*, 2278–2286. [CrossRef]

282. Zhang, J.F.; Fu, W.M.; He, M.L.; Xie, W.D.; Lv, Q.; Wan, G.; Li, G.; Wang, H.; Lu, G.; Hu, X.; et al. MiRNA-20a promotes osteogenic differentiation of human mesenchymal stem cells by co-regulating BMP signaling. *RNA Biol.* **2011**, *8*, 829–838. [CrossRef]

283. Qu, J.; Lu, D.; Guo, H.; Miao, W.; Wu, G.; Zhou, M. MicroRNA-9 regulates osteoblast differentiation and angiogenesis via the AMPK signaling pathway. *Mol. Cell. Biochem.* **2016**, *411*, 23–33. [CrossRef]

284. Vo, T.N.; Kasper, F.K.; Mikos, A.G. Strategies for controlled delivery of growth factors and cells for bone regeneration. *Adv. Drug Deliv. Rev.* **2012**, *64*, 1292–1309. [CrossRef]

285. Li, X.; Guo, L.; Liu, Y.; Su, Y.; Xie, Y.; Du, J.; Zhou, J.; Ding, G.; Wang, H.; Bai, Y. MicroRNA-21 promotes osteogenesis of bone marrow mesenchymal stem cells via the Smad7-Smad1/5/8-Runx2 pathway. *Biochem. Biophys. Res. Commun.* **2017**, *493*, 928–933. [CrossRef] [PubMed]

286. Wang, Z.; Wu, G.; Feng, Z.; Bai, S.; Dong, Y.; Wu, G.; Zhao, Y. Microarc-oxidized titanium surfaces functionalized with microRNA-21-loaded chitosan/hyaluronic acid nanoparticles promote the osteogenic differentiation of human bone marrow mesenchymal stem cells. *Int. J. Nanomed.* **2015**, *10*, 6675–6687. [CrossRef]

287. Pizzicannella, J.; Cavalcanti, M.; Trubiani, O.; Diomede, F. MicroRNA 210 mediates VEGF upregulation in human periodontal ligament stem cells cultured on 3Dhydroxyapatite ceramic scaffold. *Int. J. Mol. Sci.* **2018**, *19*, 3916. [CrossRef] [PubMed]

288. Vimalraj, S.; Saravanan, S.; Vairamani, M.; Gopalakrishnan, C.; Sastry, T.P.; Selvamurugan, N. A Combinatorial effect of carboxymethyl cellulose based scaffold and microRNA-15b on osteoblast differentiation. *Int. J. Biol. Macromol.* **2016**, *93*, 1457–1464. [CrossRef]

289. Dhivya, S.; Saravanan, S.; Sastry, T.P.; Selvamurugan, N. Nanohydroxyapatite-reinforced chitosan composite hydrogel for bone tissue repair in vitro and in vivo. *J. Nanobiotechnol.* **2015**, *13*, 40. [CrossRef]

290. Sierra, M.I.; Valdes, A.; Fernandez, A.F.; Torrecillas, R.; Fraga, M.F. The effect of exposure to nanoparticles and nanomaterials on the mammalian epigenome. *Int. J. Nanomed.* **2016**, *11*, 6297–6306. [CrossRef]

291. Larsson, L.; Pilipchuk, S.P.; Giannobile, W.V.; Castilho, R.M. When epigenetics meets bioengineering-A material characteristics and surface topography perspective. *J. Biomed. Mater. Res. Part B* **2018**, *106*, 2065–2071. [CrossRef]

292. Ha, S.W.; Jang, H.L.; Nam, K.T.; Beck, G.R., Jr. Nano-hydroxyapatite modulates osteoblast lineage commitment by stimulation of DNA methylation and regulation of gene expression. *Biomaterials* **2015**, *65*, 32–42. [CrossRef]

293. Moorthi, A.; Vimalraj, S.; Avani, C.; He, Z.; Partridge, N.C.; Selvamurugan, N. Expression of microRNA-30c and its target genes in human osteoblastic cells by nano-bioglass ceramic-treatment. *Int. J. Biol. Macromol.* **2013**, *56*, 181–185. [CrossRef]

Review

Biocompatibility and Clinical Application of Porous TiNi Alloys Made by Self-Propagating High-Temperature Synthesis (SHS)

Yuri Yasenchuk [1], Ekaterina Marchenko [1], Victor Gunther [1], Andrey Radkevich [2], Oleg Kokorev [1], Sergey Gunther [1], Gulsharat Baigonakova [1], Valentina Hodorenko [1], Timofey Chekalkin [1,3,*], Ji-hoon Kang [3], Sabine Weiss [4] and Aleksei Obrosov [4]

[1] Research Institute of Medical Materials, Tomsk State University, Tomsk 634045, Russia
[2] Research Institute of Medical Problems of the North, Siberian Branch of the Russian Academy of Sciences, Krasnoyarsk 660017, Russia
[3] Kang and Park Medical Co., R&D Center, Ochang 28119, Korea
[4] Department of Physical Metallurgy and Materials Technology, Brandenburg University of Technology, 03044 Cottbus, Germany
* Correspondence: tc77@rec.tsu.ru; Tel.: +7-3822-413442

Received: 28 June 2019; Accepted: 26 July 2019; Published: 28 July 2019

Abstract: Porous TiNi alloys fabricated by self-propagating high-temperature synthesis (SHS) are biomaterials designed for medical application in substituting tissue lesions and they were clinically deployed more than 30 years ago. The SHS process, as a very fast and economically justified route of powder metallurgy, has distinctive features which impart special attributes to the resultant implant, facilitating its integration in terms of bio-mechanical/chemical compatibility. On the phenomenological level, the fact of high biocompatibility of porous SHS TiNi (PTN) material in vivo has been recognized and is not in dispute presently, but the rationale is somewhat disputable. The features of the SHS TiNi process led to a multifarious intermetallic $Ti_4Ni_2(O,N,C)$-based constituents in the amorphous-nanocrystalline superficial layer which entirely conceals the matrix and enhances the corrosion resistance of the unwrought alloy. In the current article, we briefly explore issues of the high biocompatibility level on which additional studies could be carried out, as well as recent progress and key fields of clinical application, yet allowing innovative solutions.

Keywords: porous SHS TiNi; biocompatibility; rheological similarity; corrosion resistance; bone substitution

---

1. Introduction

Despite the fact that Nitinol was discovered in 1962 by William J. Buehler and further developed by Buehler and Frederick E. Wang in the U.S. Naval Ordnance Laboratory, its rheological similarity to biological tissues was reported for the first time in the 1980s [1,2]. Based on industrially deployed Nitinol, special TiNi-based alloys were developed, wherein the narrow temperature gap of austenite transformation was shifted towards a body temperature. This allowed shape memory implants made of these alloys to be congruent with biological tissues that are subjected to alternating physiological loads in the aggressive environment [3–5]. Whenever Nitinol is mentioned in the context of biomaterials or long-term implantable devices, a combination of corrosion resistance and biocompatibility with tissues is assumed, which is the pivotal characteristic of this alloy [6]. When considering the principles making Nitinol very attractive for clinical utilization, it is to be noted that it is economically justified regarding treatment cost minimization with a high performance.

There are a few general requirements concerning metallic materials clinically deployed. First, the material must have an appropriate viscoelastic potential, as regards the level of stress and frequency occurring in the corresponding part of the body. Secondly, it should possess a sufficient level of corrosion resistance, taking into account the implantation period and mechanical factors associated with the corrosion process. Thirdly, it has to demonstrate sufficient biological inertness, which is determined by negligible cytotoxicity, mutagenicity, carcinogenicity, immunogenicity, and thrombogenicity. As such, Nitinol combines all these properties and belongs to a group of biomaterials whose usage complies with the provisions of bioinertness, biocompatibility, and biomechanics [7–9].

Some early attempts at product development of medical Nitinol devices have been made by Nitinol specialists, who were not clinicians or primarily design focused. On the other hand, not enough designers and clinicians have yet received the insight and understanding of the Nitinol features necessary for scaling up the new implant systems. This was, in turn, crucially important for the success of clinical utilization. A set of orthopedic and traumatic devices for osteosynthesis was suggested, tested, and approved [2]. Further, material science engineers, in collaboration with the medical community, studied and exploited devices for surgical management of various lesions and injuries in midface, spinal and abdominal surgeries, oncology, urology, dentistry, and cryosurgery [10–12].

In the 1980s, in the USSR (Siberian Physical-Technical Institute), porous TiNi alloys were obtained using the self-propagating high-temperature synthesis (SHS) process in an inert atmosphere, followed by successful clinical use of implant systems made of porous SHS TiNi [13,14]. The SHS method to synthesize refractory ceramic compounds was initially proposed and comprehensively described by Merzhanov et al. [15–17]. SHS, as a powder metallurgy method, turned out to be the most appropriate for the fabrication of the porous TiNi body having the specified characteristics [18,19]. Additionally, SHS is a versatile method that produces a variety of intermetallic compounds for various application tasks [17].

Recently, porous SHS TiNi (PTN) compounds have been reported [20,21] to have some features which significantly distinguish PTNs from porous materials obtained by other methods of powder metallurgy using the same reactants [22–24]. It happens that the porous body formation during the SHS reaction is accompanied with the genesis of nonmetallics (titanites, spinels, perovskites, glass-ceramics, etc.) and nanocrystalline, amorphous superficial layers concealing the pore walls, which are of great interest for the academic community and for clinical application. It highlights the further need to investigate the surface structure of PTNs used as bone substitutes and scaffolds for cell-tissue engineering. In fact, the surface layers of PTN serve as a protective barrier in the chlorine corrosive-active environment, including for biological fluids [25–28].

PTN exhibits martensite transformations (MT), showing the shape memory effect and superelastic behavior, which, however, are not pronounced as in Nitinol [29,30]. The known scientific complexity is due to the multiphase state of PTN. In the case of variable cyclic load applied to the PTN graft incorporated in the body, the full cycle (direct-reverse-direct) of MT repeatedly occurs in a corrosive environment. The rheological similarity to biological tissues coupled with the enhanced corrosion resistance of unwrought PTN is supposed to impart additional benefits to this material, making it a promising alternative to Ti-based alloys, whose nontreated surface may incur an adverse corrosion effect. Follow-up observations [31,32] evidenced the high adaptability level of PTN as a biomaterial striving to complement existing surgical techniques for improved patient tolerance.

In the review, we discuss the main features of PTN alloys, of which advanced implants are made, in the context of improved biocompatibility, along with the key fields of clinical application where these implants were deployed.

2. Fabrication of Porous SHS TiNi

Almost all bone endografts made of high-porous TiNi alloys (porosity $\geq 60\%$) are fabricated by the SHS method. This porous body minimizes implants' failures (stress-shielding effect) and, hence, the complication rates [33–35]. Although additive technologies merit close attention from the industrial community, the high-porous TiNi alloy fabricated by SHS has a number of advantages, even in comparison to other powder metallurgy methods, including sintering, hot isostatic press, spark plasma sintering, thermal explosion, etc., [36,37]. In SHS, the product is directly synthesized from Ti-Ni elemental reactants via the propagation of a combustion wave through a green powder compact. Once the synthesis is started, the heat of reaction sustains the reaction until all of the reactants have been consumed. With regards to the clinical application of PTN, it is crucial to preset the desired mechanical characteristics, the shape memory effect (SME), and the superelastic parameters at temperatures that suit the body tissues. The main physical-mechanical characteristics of PTNs are summarized in Table 1.

Table 1. Physical-mechanical properties of unwrought PTNs [3,12].

Property	Value
Specific weight, g/cm^3	5.85
Porosity, %	60 to 75
Pore size, μm	0.1 to 200
Permeability coefficient (water/glycerin), m^2	$(0.27/62) \times 10^{-9}$
Melting point, °C	1310
Ultimate tensile strength, MPa	100 to 500
Stretch at breaking point, %	5 to 7
Loading plato stress, MPa	50 to 200
Total elongation, %	2.5 to 4.5
Permanent set, %	5 to 20
SME recovery stress, MPa	200 to 400
SME temperature hysteresis, degree	30 to 100
Transformation temperature range, °C	−180 to 50

Briefly, to fabricate PTN, commercial powders of coarse titanium made by calcium hydride reduction (mean particle size of 80–100 μm) and carbonyl nickel (mean particle size of 10–15 μm) are mixed for a few hours in an air jar and vacuum-dried [38]. The green powder mixture is loaded in a quartz tube and then loose-compacted by tapping for 10 min to achieve a porosity of tapped green compacts of 60–65%. The charged quartz tube is then loaded into a reaction furnace under flowing argon gas with a heating rate of 10–15 °C/min and is ignited electrically. In the mode, SHS is considered to occur with the involved liquid phase in a narrow reaction zone, which propagates autocatalytically through the preheated green powder compact. The heating schedule and temperature profile are controlled with a thermocouple placed inside the green powder compact. Once the compound has been synthesized, the reactor is withdrawn and cooled in a water container.

The difference between SHS and reaction sintering lies in the kinetics of the heterogeneous reaction [14,39,40]. At the beginning, the exothermic reaction of SHS partially dissolves the green powder compact, followed by the liquid-phase reaction, which triggers and dictates the formation of the intermetallic constituents (TiNi, Ti_2Ni, and $TiNi_3$). The $TiNi/Ti_2Ni/TiNi_3$ ratio in the matrix may vary and depends on kinetic parameters of the heterogeneous reaction.

Impurities trapped in the reactants are also crucial in synthesizing the porous compound. In powder metallurgy, the use of high-purity reactants is encouraged as it affords the fabrication of homogeneous alloys exhibiting specific attributes. This concept is particularly accurate for sintered TiNi, whereas it is not reasonable for PTN. Vacuum sintering at constant degassing forces not all of the existing impurities in the reaction system to be thermally dissociated, wherein some are gasified and subsequently withdrawn. The remaining impurities form diverse phases, which deteriorate the matrix, affecting the performance of the resultant alloy. On the contrary, SHS is generally referred to as the layer-by-layer exothermic reaction mode in an argon flow atmosphere when the preheated powder compact is ignited at 250–500 °C. The incipience and evolution of intermetallic phases occur in a thin solid-liquid reaction layer in milliseconds. A sequential cycle begins and evolves in the reaction layer at 100–200 μm thick, which inherits the size and morphology of the previous reaction layer through the capillary spreading of the (Ti + Ti_2Ni) eutectic liquid emanated from the reaction zone.

Notably, in the layer-by-layer combustion mode, the following processes are revealed in the reaction layer: (i) Origin of the eutectic liquid in the contact area of Ti and Ni particulates; (ii) dissolution of particulates in the eutectic liquid, which catalyzes a drastic increase of the eutectic liquid; (iii) exothermic reaction between dissolved reagents and successive crystallization of $TiNi_3$, TiNi, and Ti_2Ni precursors from the liquid; and (iv) interdiffusion migration of Ni atoms into solid Ti particulates and Ti atoms into solid Ni particulates, followed by the formation of intermetallic constituents [21,23,39,41].

Since the solid-liquid reaction layer remains porous, thermally dissolved gas-prone impurities managed by reaction gases leave the over-pressured high-temperature reaction zone, filtering through the structuring zone. The latter, having less pressure, remains red-hot. At the same time, the reaction gases, having a distinct effect on the conductive-convective heat transfer mechanism, are evident as a heat-and-mass transfer principal agent. They capture a portion of the liquid and transfer it from the reaction zone towards the surface, forming voids in the structuring zone. Therefore, the matrix is rectified to a large extent, whereas the pore wall surfaces are concealed by the sophisticated shell. The given shell comprises amorphous-nanocrystalline phases of intermetallic oxycarbonitrides in the form of epitaxial strata (foamy onlay and dense bisubstrate), as reported in References [20,21]. Considering the chemical composition and structure, the shell can be classified as a cermet $Ti_4Ni_2(O,N,C)$ layer [42–44]. In fact, such amorphous-nanocrystalline phases are implied to exhibit high corrosion resistance.

Thus, in contrast to sintering modulated by the scant liquid, when synthesis is lengthy and coincides throughout the entire powder compact by the solid-liquid phase transformation, SHS is the rapid process occurring in a similar way, but in the presence of the abundant liquid. Impurities trapped in the reactants do not dissociate and recombine entirely upon sintering, whereas those upon SHS are subjected to thermal dissociation and chemical decomposition, which further results in the formation of the amorphous-nanocrystalline superficial layer. It provides a greater tolerance against corrosion and does not hinder the viscoelastic behavior of cyclically loaded PTN.

3. Characteristics of Porous SHS TiNi

The high in vivo/vitro inertness of PTN is conditioned by the negligible anodic dissolution of the dynamically loaded PTN sample in simulated body fluids [6–9,25,28]. In the early in vivo terms, the anodic passivity of the PTN scaffold is beneficial for the attachment, cytocompatibility, and proliferation of seeded precursor cells as it sustains the formation of manifold tissular variants, reported in References [45–49]. Afterwards, a newly formed interface (e.g., bone regenerate) owes its vitality to the two factors as follows: (i) Continuing superficial anodic passivity of the PTN scaffold and (ii) minimum viscoelastic discrepancy between the surrounding bone tissue and the PTN matrix. The latter shows high elasticity without deterioration of the mechanical characteristics at applied loads and it physiologically redistributes stresses between adjacent bone fragments. Loads evoked by surrounding tissues can often exceed 6–8% relative strain, which exceeds the allowable values for most metal implants, destroying their protective surface films and ultimately leading to their destruction [50]. Deposited corrosion-resistant gradient coatings turn out not to remedy this challenge as they are usually nonelastic and have low fatigue strength [51,52]. The surface layer of the unwrought PTN alloy nevertheless withstands multicycle deformation and maintains its inherent integrity with the viscoelastic matrix [53].

3.1. Structure and Phase Composition of the PTN Surface

Turning to the issue of the PTN enhanced corrosion resistance highlighted earlier, we were bound to note the superficial amorphous-nanocrystalline layer of intermetallic oxycarbonitrides, which entirely conceal the pore walls. It was denoted that the SHS process in itself is the rationale for this layer appearance resulting from retrograde gas streams interacting with the surface melt [43,54]. In our experience, we have studied the surface structures of high-porous PTN using a confocal laser scanning instrument [55]. The polished thin section is yellow, as can be seen in Figure 1, whereas the superficial layer, on which the focus was made, is represented as a translucent green film inside the unpolished open pore wall. It contains nonmetallic inclusions, observed as spatially distributed garnet flakes.

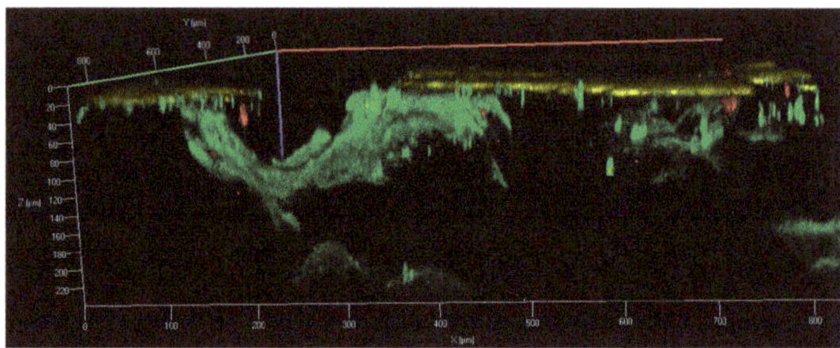

Figure 1. Confocal laser scanning micrograph of the thin-sectioned high-porous PTN specimen (wavelength—405 nm) [55].

The light microscope seemed to be a versatile instrument since it identified the massive superficial layer (S) concealing the sectioned matrix (M) in a dark field, as illustrated in Figure 2a [21]. Nonmetallic crystals (NM) can be distinguished in ultraviolet polarized light against both the matrix phase (M) and massive superficial layer (S), as depicted in Figure 2b.

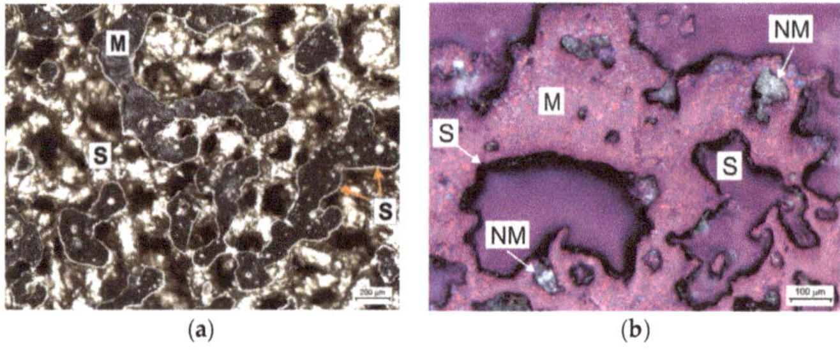

(a) (b)

Figure 2. Light microscopy images of the thin-sectioned PTN in (**a**) a dark field with differential interference contrast (DIC) and (**b**) an ultraviolet polarized dark field [21].

Furthermore, data emanated from the scanning tunneling electron microscope (STEM) and energy-dispersive X-ray spectroscopy (EDS) study of the (S) depicted in Figure 2 revealed its intricate structure (Figure 3). The foamy onlay (F) is seen to shell two dense sublayers (IIT + IIB), which are tightly bounded with the matrix (Figure 4). In Reference [20], it was previously reported that the (F) results from nanocrystalline intermetallic foam managed and dispersed by reaction gases, which is heterogeneous and discrete. High-resolution transmission electron microscopy (HREM), selected area electron diffraction (SAED), and EDS analyses allowed the authors to state that the presence of residual amorphous phases was obvious as well [21].

Qualitative X-ray diffraction (XRD) analysis carried out on a demolished PTN sample to facilitate XRD-pattern acquisition of a non-uniform relief surface has strengthened the vision of the multifarious amorphous-nanocrystalline ensemble. The grazing beam at a low incident angle (<1°) was assumed to penetrate no more than a 100 nm in depth. XRD-patterns of the surface layers reported in Reference [21] were taken and brought together, as indicated in Figure 5.

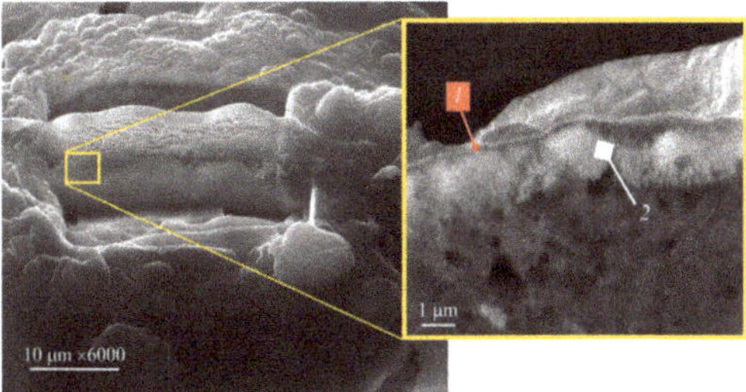

Figure 3. Structural features of the intricate sandwich (*1 + 2*), shown using a lamella taken from the open macropore wall of PTN (QUANTA 200 3D) [20].

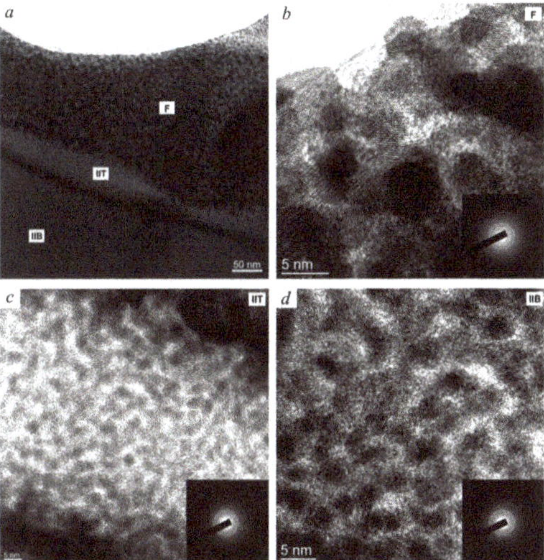

Figure 4. STEM images of lamella taken from the open macropore wall: (**a**) General view of the epitaxial layer [(IIB) + (IIT) + (F)] and individual structure of the (**b**) foamy onlay (F), (**c**) dense (IIT), and (**d**) (IIB) strata [21].

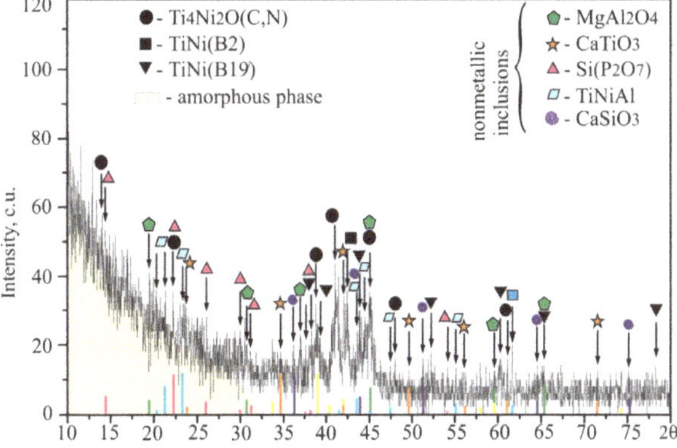

Figure 5. XRD pattern of a demolished PTN specimen, referred to the superficial layer [21].

The prevailing constituent, seen in Figure 5, is the amorphous-nanocrystalline phase of intermetallic oxycarbonitrides $Ti_4Ni_2(O,N,C)$, whose fcc unit cell has been reported to appear due to oxygen, carbon, and nitrogen interstitial migration into the Ti_2Ni lattice [56]. In fact, it can be considered as a solid solution of O, N, and C in the Ti_2Ni phase. Reflexes belonging to the TiNi component appear in two crystallographic modifications of *B*2 (cubic parent) and *B*19' (monoclinic martensite), simultaneously with Ti_4Ni_2O oxide. The crystallinity of the superficial layer at a depth of 100 nm is about 70%, 50–55% of which belongs to intermetallic oxycarbonitrides $Ti_4Ni_2(O,N,C)$, whereas 10–15% goes to glass-ceramic and cermet phases, detected as $CaTiO_3$, $Si(P_2O_7)$, $CaSiO_3$, $MgAl_2O_4$, and TiNiAl. The remaining 30% belong to the residual amorphous phases evident as a diffuse halo within 10–30°.

Therefore, the discovered layers of amorphous-nanocrystalline Ti4Ni$_2$(O,N,C), combined with glass-ceramic and cermet phases, impart corrosion-proof attributes expressed through electrochemical passivity in biological fluids, which is also consistent with Reference [57]. On the other hand, referring to the featured pore's topography, it may be hypothesized that the foamy onlay (F) would maintain the promising bioactive characteristics, facilitating cell attachment and proliferation in vivo/vitro, as discussed below.

3.2. Rheological Resemblance of PTN to Biotissues

Most of PTN applications involve cyclically varying biomechanical loads that promote the need to fully understand the rheological behavior of this alloy. Although the accumulated knowledge on PTN can predict the post-implantation life-span of such implants, both in terms of stress-strain (total life) and damage tolerant (crack propagation) behavior, expanded information on their rheological characteristics needs to be highlighted.

Viscoelastic rheological behavior of biological tissues is conditioned by their intricate structure [58,59]. In the 1950–1960s, collagen fibers were already considered as a key structure comprising all tissues, including bones [60,61]. Tropocollagen macromolecules mineralized with hydroxyapatite form a durable composite which is resistant against tensile and compressive loads. Moreover, the fact that bones are porous bodies underlines additional physical-mechanical features in their behavior. Bones containing tissue fluids do not fail over millions cycles of alternating load throughout their entire service life. With reference to Reference [62], the rationale for the remarkable functioning can be explained by the following factors. First of all, it is due to the cyclic viscoelastic characteristic of the organic matrix, in which collagen fibers are loaded, changing their conformation. Second, it owes to the elastic deformation of the mineral framework consisting of crystalline hydroxyapatite. Third, tissue fluids that fill the porous body of loaded bone redistribute the hydrostatic pressure through the bone so as to accommodate the severe strain to the level safe for collagen. As such, the porous bone structure, in which viscoelastic collagen fibers are mineralized by elastic hydroxyapatite crystals, is patterned on an anisotropic poroelastic composite material perceiving the external load. Viscous flows of tissue fluids infiltrating reciprocally through the osseous tissue contribute significantly to the viscoelastic rheological behavior of the bone, transmitting functional loads by means of fluid inertia and pressure gradients.

We can note the following arguments addressed to the provision of rheological similarity between PTN and bone tissue.

(i) Regardless of which loading mode (axial, bending, or torsion) is applied, the minimum loads on the PTN bone graft from the host bone tissue cause linear (elastic) deformation of the PTN matrix, in which pore walls and interpore partitions undergo elastic cycling at a low strain magnitude (typically less than 2%). Higher loads are characterized by a nonlinear region, resulting in the onset of martensite transformation once a shear stress threshold of the PTN matrix is reached [63,64]. With that being the case, higher alternating dynamic loads, at constant temperature, appear to trigger the reversible austenite-martensite phase transformation, providing added value to the deformation process (up to 4%). Related to the general view of the deformation route, appearing structural defects, resulting from an increased deformation magnitude and transcended in localized areas of the PTN matrix, catalyze a crack formation and propagation, followed by the PTN bone graft or any part thereof eventually degrades until failure. Evidence in the literature indicates that the bone tissue behaves in a similar way [65,66]. The stress-strain behavior is characterized by a linear (elastic) region before a yield point, a post-yield nonlinear region containing the ultimate load, and a failure point at which the bone tissue can no longer carry the load.

(ii) The cyclic load applied to the PTN bone graft in vivo can be considered as a confluence of elastic and viscous deformation, which is due to the austenite-martensite transformation. This kind of combination is assumed to lead to stress relaxation and does not encourage the evolution of structural defects. The PTN rheological behavior within a viscous deformation region is consistent with that shown by the wet bone matrix ex vivo [67]. Of particular note is that the PTN body, having a certain pore size distribution, possesses a prominent capillary effect, which is sufficient to hold tissue fluids inside it [68]. This implies the PTN implant also has the possibility to transfer over most of the applied physiological loads via redistributed hydrostatic pressure, just as spongy bone tissue does. It must be acknowledged that a critical role, in this rheological context, is to be played by the ternary complex, consisting of the adjacent spongy bone, tissue fluid, and the poroelastic PTN implant. It acquires particular importance when substituting large defects of the loaded bones (e.g., femur, tibia, lumbar vertebrae).

Numerous studies investigated the deformation behavior of porous TiNi compounds for the past ten years [29,63,69,70]. Most have reported that the task is fascinating and actual, but still challenging. As a rule, researchers carried out their tests in the axial loading mode. Although most of the studies were experimental, it was pointed out that compressed dry specimens exhibit viscoelastic deformation comparable to those mentioned above, but do not provide much information on a comprehensive understanding of the realities prevailing in vivo. To characterize the biomechanical interaction of the osteo-ligamentous interface with the engrafted PTN, a robust study of the PTN rheological patterns is believed to be needed, including tension, bending, and torsion tests performed ex vivo.

Definitely, viscoelastic PTN is rheologically different from a viscoplastic porous material. The latter, which possesses a high porosity, indicated an increased yield point (up to 6%), sustaining the irreversible viscous behavior of thin interpore partitions in compression testing [71]. PTN can be easily deformed by 4–6% [72,73], but the distinguishing peculiarity is that the noted strain is, for the most part, retained comparably to that shown by superelastic Nitinol [64,74]. The degradation behavior of PTN is strongly dependent on the phase-chemical composition, especially the brittle Ti_2Ni phase network in the matrix as well as the intermetallic $TiNi_3$ and nonmetallic phases responsible for the crack's initiation, which is also true for dense TiNi.

3.3. Polarization Behaviour of PTN

Attempts have been made to protect the surface of Nitinol implants using various thin film technologies and, for the most part, they have made only modest progress [75–79]. Deposited coatings often suffer from microporosity, permeability, excessive thickness, and there is a mismatch in thermal expansion coefficients and elastic moduli between the substrate and coating. Therefore, coatings appear to be delamination-prone and have had little effect with regards to long-term protective films in vivo. These challenges are particularly relevant for porous alloys as the known methods using laser/electron-beam irradiation or sputtering are not effective. To some extent, this issue can be circumvented by using either plasma deposition or, occasionally, chemical deposition throughout the entire porous body as the most viable way of surface modification therein [80,81].

The osseointegration of bone substitutes can be facilitated by a change in the surface composition using bioactive ceramics. The short period of osseointegration of PTN bone graft may imply that its surface is chemically passive, bioactive, and cytocompatible, which can withstand the harsh conditions of medical applications. This could mean that the features of SHS have a positive effect on the chemical stability of the PTN surface, imparting chemical gradients inside the PTN implant to enhance cell viability. However, this matter remains poorly known and needs further study. We may address a peculiarity of the chemical resistance to the SHS process itself when nonmetallic impurities trapped in the reactants are thermally dissociated, followed by the surface of the synthesizing porous body which chemically absorbs them. Accordingly, the evolved superficial layer, having nanostructured attributes and high adhesion strength to the substrate, is tightly bounded with the matrix, concealing the latter.

Potentiodynamic polarization was used in a comparative study of the surface susceptibility of PTN and dense Ti, TiNi, and in a 1% HCl solution, as reported in Reference [43]. Figure 6 illustrates the anodic behavior of unwrought PTN, dense Ti, and TiNi samples, modified by electropolishing, followed by N ion implantation. It should be noted that the findings shed additional light on the nature of unwrought PTN passivity. As seen, a corrosion measurement revealed that the anodic behavior of unwrought PTN in the chloride-containing environment mimics that of a modified TiNi sample in the passive region.

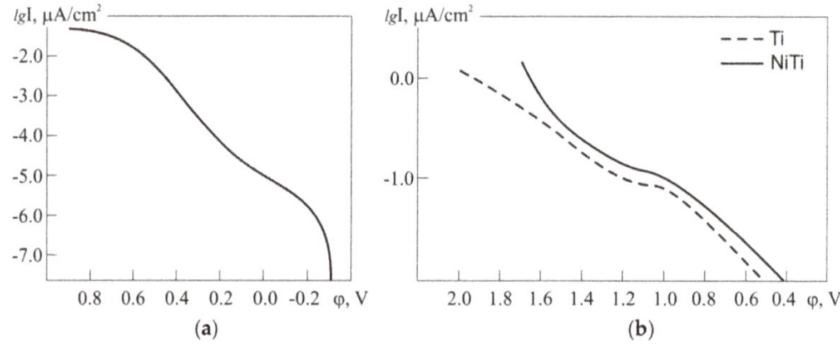

Figure 6. Anodic behavior of (**a**) unwrought PTN sample and (**b**) dense Ti and TiNi samples modified by anodic polishing and subsequent N ion implantation [43].

An early study on the electrochemical reaction has conversely shown that sintered porous Ti-based compounds undergo more corrosion [82]. Moreover, from the potentiodynamic polarization of porous TiNi fabricated by a powder metallurgy method using a high-purity ammonium bicarbonate powder and a blend of elemental reactants, it was reported that the studied sample was more susceptible to localized corrosion in a 0.9% aqueous NaCl solution as compared to the dense TiNi sample [83]. This also confirms our suggestion about the shielding superficial layer of intermetallic oxycarbonitrides, which entirely conceals the PTN rough surface, imparting a high corrosion resistance innate to cermets. The findings emanated from References [84–86], in which the authors explored the porous SHS TiNi alloy using a variety of instruments, accord well with our supposition. It is unlikely that the surface structure and the chemical composition of the studied samples are the same; however, we can inferentially assert that the PTN matrix is well protected by corrosion-proof strata consisting of intermetallic oxycarbonitrides and non-metallic inclusions.

4. Cytocompatibility of the PTN Surface

The spitted rough topography and biochemical aspect of the PTN surface play an important role in cell adhesion, growth, and proliferation, as a system of interconnected macro-/microvoids and grooves, which ultimately influence the biocompatibility of the hydrophilic PTN body in vitro/vivo [47,49,87,88]. Cell response to the surface topography is a primary feature of the forming tissue-specific variants. Surface roughness has a direct beneficial influence on cell morphology and proliferation. Moreover, the microporous surface structure can reduce a stress-shielding effect, encouraging the propitious tissue ingrowth [89]. On the contrary, the even surface prevents friendly cells adhesion and, in turn, decreases total biocompatibility. Literature confirmed that tissues ingrown on rough surfaces were stimulated towards differentiation [90,91], as shown by their gene expression in comparison with cells growing on smooth surfaces. For example, primary rat osteoblasts had higher proliferation, alkaline phosphatase, and osteocalcin expression on the rough surface compared to the smooth one [91]. Large size pores (>500 µm) may inhibit cell adhesion, reducing bone formation and vascular ingrowth. In contrast, small size pores (<100 µm) may hinder the diffusion of nutrients and metabolites but stimulate osteogenesis, reducing cell proliferation and forcing the implant's incorporation. Therefore, the pore size distribution and average pore size of the PTN implant being developed are among the most important factors to strive towards for the right balance of the cell growth, proliferation, and tissue/vascular ingrowth herein [92].

The PTN scaffold has been reported to possess a wide range of pore sizes suitable for the cell cultivation in vitro, followed by the growth in vivo. Bone marrow cells (initial) seeded on the PTN scaffold attach to the pore walls, then actively grow and spread across the porous body. The cell growth and incipient intercellular substance cause reproduction of this cell mass, which contributes to colonization and subsequent differentiation [47,49]. The SEM control of cell attachment has indicated that the isolated initial cell tosses pseudopodia at a distance of 15 to 30 µm away by decoupling the chemotactic mechanism for the first 24 h. Further, solitary microfilament ends (less than 1 µm) were found to be attached and localized in superficial micropits of the pore wall.

The SEM study of pore spaces was carried out in a 7 day time interval and the following features were noted: At the end of the first week, the cells were attaching and proliferated; most of the cells were fixed in local cavities, where there were many small pores less than 3 µm in size; and then, cells were actively growing (Figure 7).

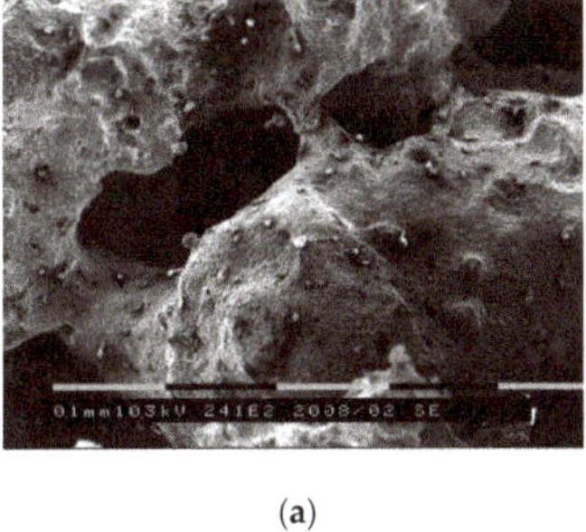

(**a**)   (**b**)

Figure 7. SEM image of (**a**) the cultured PTN scaffold in vivo and (**b**) micropits colonized by pancreas islet cells herein (on 7th day post-seeding) [47].

The cell population continued to increase, as well as their growth in pores and the synthesis of the extracellular matrix. The 3D pore cluster allows them to proliferate intensively due to synthesis of the extracellular matrix and formation of spatial incrustations with different shapes and sizes (Figure 7a).

Beginning on the 14th day, the tissue gradually lined the inner surface of pores and then the growth vector moved from the periphery towards the center, filling the entire pore space. During the fourth week, most of the pores were completely filled by cells and the differentiating matrix (Figure 8b). This effect was observed consistently from one week onwards, both in vitro (mesenchymal cells) and in vivo (hepatocytes and pancreas islet cells). The samples were ingrown by tissues and pores were filled by mesenchymal cells in 28 days and by hepatocytes and pancreatic islets in 40–60 days, respectively.

In vitro experiments have shown that cells implanted into the PTN scaffold actively attach, grow, differentiate, and create a specific tissue structure even in vivo allogenic condition of the recipient.

Using various modes of the SHS process, it is possible to obtain different structures of the PTN scaffold with a specific pore size and distribution, which is very important in cellular and tissue engineering. With reference to References [93–97], a set of the mentioned features concerning the structure and characteristics of the PTN scaffold can be defined as follows:

- A well-developed spitted topography of pore walls (a large number of interconnected small pores and rough interpore partitions), which sustains the initial cell adhesion;

- Wetting ability, which facilitates saturation by water-soluble substances;
- Phase and chemical composition of the superficial strata, which has no inhibitory/toxic effect on cells seeded and growing tissular variants;
- Bio-mechanical behaviour, which is pretty similar to that exhibited by alive tissues.

The above-mentioned points are necessary for initial attachment, growth, and replication of host cells as a driving force for desired tissue to be harvested. Consistent cell growth and a short time whilst host cells colonize the PTN scaffold, competing with and conquering pathogenic cells, may emphasize the specific strengths for advanced bioengineering goals. The targeted differentiation of multipotential mesenchymal stromal cells of cartilage or osseous tissue inside the PTN scaffold proves the cytocompatibility of this material and extends the functional life-span of the incubated cells, prolonging the curative effect.

Thus, we can say that the PTN scaffold represents favorable conditions for the interactions of the host cell with the surface and expends the opportunities of bioengineering capabilities when considering morphological/functional properties of cells incubated herein. It can be used in a wide range of medical applications (management of locomotor apparatus diseases, metabolic disorders of the liver and pancreas, etc.). This material is designed to interact as long as it is in the body, in large deformations, exhibiting comparable rheological behavior and no graft vs. host response, which significantly distinguishes it from rivals.

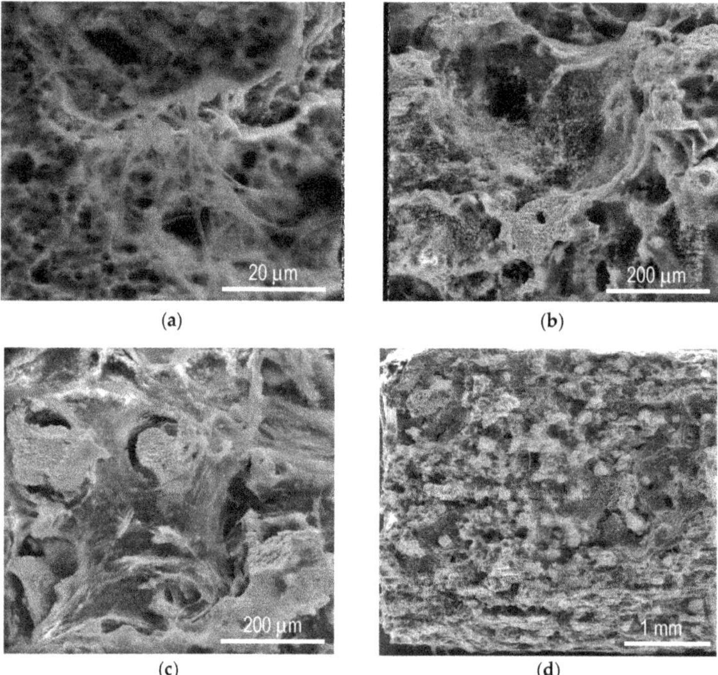

Figure 8. SEM images of cellular evolution in vivo: (**a**) Synthesized extracellular fibers and formation of spatial pseudopodia (on the 7th day); (**b**) gradual cellular ingrowth and formation of the extracellular matrix (on the 14th day); (**c**) phase of active infiltration (on the 21st day); and (**d**) PTN scaffold entirely filled by tissues (on the 28th day) [47].

5. Histological Studies of the Implanted PTN Scaffold in Vivo

To study the morphogenesis of reparative regeneration in the porous-permeable TiNi-based alloy, experimental studies were conducted on 10 mongrel dogs aged from 1 to 1.5 years, with weights ranging from 18 to 26 kg [98]. As-received PTN ingots were disintegrated into pellets. Bone defects were created in alveolar processes and then filled with porous granules. For the study of the microstructure characteristics, histological analysis of the material and the produced regenerate was performed at different times.

Figure 9 shows the filling of pores with tissue inside and between granules. New mature tissue is generated both on the surface and in pores, between PTN granules. On the first day of observation, islets of tissue start to form, mainly in large granules. On the 7th day, loose connective tissue can be observed between individual granules. In the course of time, the filling of the pores and the intergranular space with the newly formed multilayer tissue continues and it replicates the pore's microrelief, which is in good agreement with the heterogeneous mechanism of bone formation. X-ray microanalysis of the tissues between granules and tissues that conceal the granules showed their similar composition [98]. The content of calcium, phosphorus, and potassium in the tissues corresponded to that in mature bone tissues.

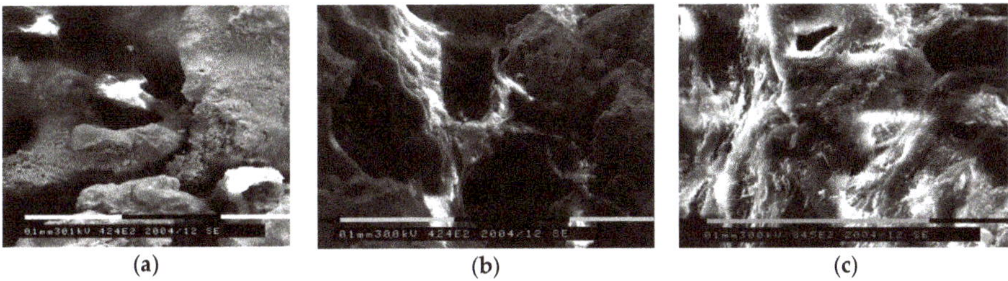

(a)   (b)   (c)

Figure 9. SEM images of tissue invasion through the PTN granules at (**a**) three, (**b**) seven, and (**c**) fifty-six days [98].

Morphological findings from Reference [99] revealed that the cellular reticular tissue with sinusoidal capillaries, which comprises cellular elements of myeloid origin, is formed in the porous structure and between granules in a day (Figure 10). After two days, the number of leukocytes and fibroblasts in the generated tissue increased and thin fibrous structures appeared. By the 5th day, loose connective tissue was found closer to the defect edges, with fibers oriented along the bone trabeculae of the near-defect part of the bone. Separate cartilage cells were detected close to the defects in the lacunes. Further, the amount of cartilaginous elements decreased from the edges and the regenerate was a dense connective tissue with vessels that exhibited a muscular wall.

By the 7th day, the volume of the connective tissue component in the regenerate decreased since it was replaced with fibrous cartilage tissue. Closer to the defect center, dense loose connective tissue with collagen fibers was formed. After ten days, hyaline cartilage components were observed among the fibrous cartilage structures. During the following days, the replacement of connective tissue with fibrous cartilage tissue continued, which was then replaced with hyaline cartilage tissue. By the 42nd day, coarse fibrous bone tissue was detected on the defect edges and after 56 days, the regenerate mainly consisted of compact and spongy bone tissue. Further on, the resulting tissue did not noticeably change and was organotypic bone regenerate as an integral part of the implant material.

The fibrous cartilage is replaced by hyaline cartilage due to the formation of intercellular substances resulting from the chondroblastic activity, namely proteoglycans (e.g., chondroitin sulfates) supplanting the collagen fibers. In the recipient zone, the partial pressure of oxygen is known to increase due to the abundant periphery vascularization of the bone tissue, providing a rich blood supply to the margins. It makes pericytes, as a source of osteogenic cells, differentiate into osteoblasts under the effect of the high partial pressure of oxygen around them, whereas the function of osteoblasts is to form the intercellular substance of bone tissue. In so doing, active osteoblasts, modulating the intercellular substance of bone tissue, form the inorganic constituent known as osseomucoid. The latter, in turn, consists of calcium phosphate and hydroxyapatite crystals, which hinder the diffuse route of nutrients towards the hyaline cartilage. The scant diffuse nutrition leads to a situation in which the hyaline cartilage deteriorates and dies. Blood vessels with the same type of differentiation, controlled by pericytes, grow into the remaining space. This process ends when the hyaline cartilage is entirely replaced by coarse fibrous bone tissue, as seen in Figure 10 (depicting all stages of indirect osteogenesis).

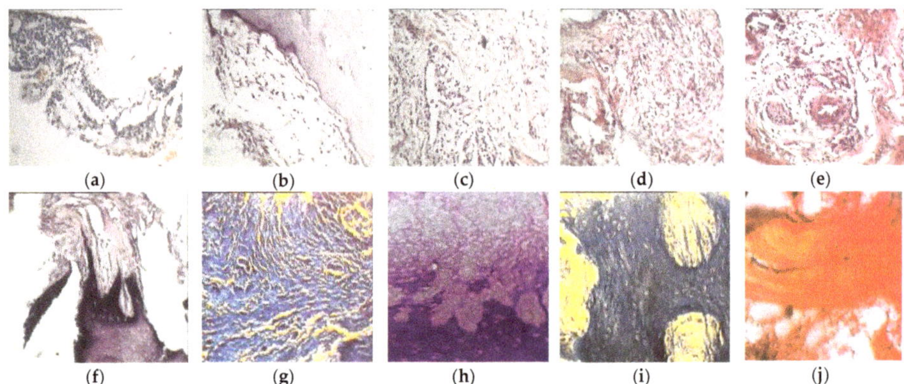

Figure 10. Dynamics of reparative processes in bone defects substituted by PTN granules at (**a**) three, (**b**) five, (**c**) seven, (**d**) ten, (**e**) fourteen, (**f**) seventeen, (**g**) twenty-one, (**h**) twenty-eight, (**i**) forty-two, and (**j**) fifty-six days [99].

Analysis of the reparative osteogenesis in bone defects after their reconstruction with PTN granules indicated that formation of bone regenerate through the ingrowth and differentiation of tissues occurred according to the patterns of indirect osteogenesis. At first, loose connective tissue forms, followed by the formation of dense irregular tissue, which results in the fibrous cartilage with signs of incipient hyaline cartilage. Further, the latter are gradually replaced by coarse fibrous bone tissue, transforming into mature spongy tissue.

6. Clinical Application of PTN Implants

6.1. Cervical Spine Superelastic PTN Cages

The specific characteristics of the anatomical structure of the vertebral bodies, the presence of shock-absorbing intervertebral discs, and special biomechanics of the vertebra exclude the use of traditional materials and structures in spinal surgery [100]. The properties and structure of the PTN cage for the vertebra are close to those of the spinal body, which ensures circulation of tissue fluids, plasma at the bone-implant boundary responsible for the metabolism of bone cells, and the formation of the bone-implant interface [9,28,45,86]. The contact surface resistant to aggressive biological fluids and the rheological behaviour of the PTN cage ensures its supportability, maintains the height of the vertebral body, and eliminates excessive loads without failure.

Forty-three patients suffering from cervical osteochondrosis received a dynamic PTN cage, whose shape is set as an eiloid cylinder, seen in Figure 11, for ventral interbody fixation of the cervical spine [12,32,101]. Since the PTN cage structure is superelastic, it is easy to attach the desired shape for insertion, followed by the implant deploys and self-locking in-situ, as depicted in Figure 11b. Due to the reliable elastic stabilization of the cervical spine, there was no need to wait when the bone block was formed, and all patients have been discharged the next day after surgery. The consistency of the operated vertebral motor segments allowed us to exclude postsurgical external immobilization of the neck. No complications associated with the implant features were noted in all cases. X-ray check performed 24 months after surgery indicated no evidence of migration, cracks, or failure of the implants. No areas of bone resorption were identified around the implants. Head flexion and extension radiographic images (Figure 11c,d) indicate preserved mobility in the cervical levels, which were managed using the PTN cage. Additionally, the range of motions in adjacent levels did not change, as can be seen in the Supplementary Video clip taken in the follow-up period.

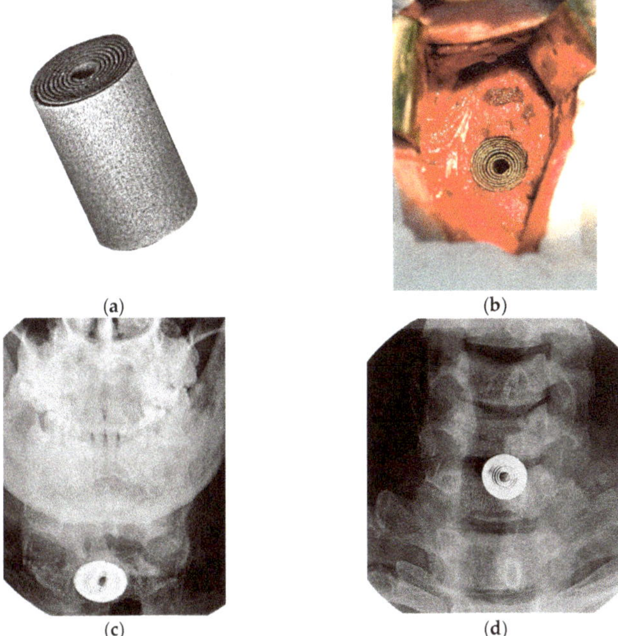

Figure 11. Cervical spine dynamic PTN cage: (**a**) General view; (**b**) intraoperative view of the surgical wound; X-ray images of the head tilted (**c**) down and (**d**) back in two years after surgery [101].

All clinical cases indicated a high durability of the PTN cage structure with the maximum form change. In the follow-up period, the deformed PTN cage was noted to survive even in the case of high loads. When it is free from a load, the shape completely retains without any degradation. Notably, the phase transitions provoked by applied cyclic loads accommodate internal stresses throughout the PTN matrix and this is a justification for high performance under continuous cyclic loads.

6.2. Customized PTN Endografts in Maxillofacial Surgery

For the first time, an experimental study on replacement of the mandible using a prototype of PTN was performed in 1982, whereas follow-up observation of this clinical case was reported in 1986 [2], and it is still feasible for clinical application. Customized combined endografts made of PTN and dense TiNi were developed to treat patients with mandibular lesions, including mandibular, maxillary and nasopharyngeal malignant tumors [102,103]. The PTN endografts of the mandibular ramus can be developed with right and left versions, including the head of the temporomandibular joint. The prosthesis consists of an ultra-elastic perforated plate, with porous parts of similar shape and size, fixed on both its sides. On the one hand, the structure exhibits a polished thickening that corresponds to the configuration of the head of the mandible (Figure 12). The size and configuration of the endoprosthesis are determined individually in accordance with computed tomography imaging and CAD modelling. Due to the superelasticity of the construct, it can be easily modified, depending on the shape required, to eliminate defects of the mental and lateral mandibular parts.

Figure 12. Customized PTN endografts for surgical repair of mandibular lesions [102].

This ensures restoration of the anatomical architecture of the repaired area, normalizes the function of chewing and swallowing, and prevents secondary deformities caused by protruded bone fragments and scarring of soft tissues in the postoperative period, as illustrated in Figure 13.

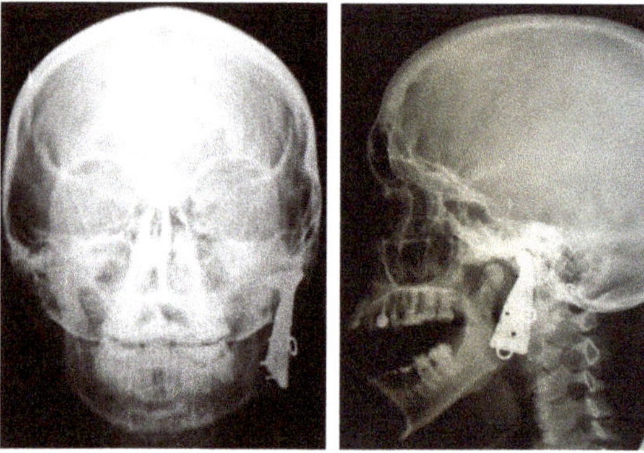

Figure 13. X-ray images of the repaired left mandible for deforming temporomandibular osteoarthrosis (a year after surgery) [102].

Analysis of the follow-up observations in patients with destructive changes in the condylar processes proved the high efficiency of customized PTN endografts. Due to biochemical and biomechanical compatibility, combined endografts substituting tissue lesions behave congruently. Connective tissues from recipient areas ingrow the PTN body with negligible foreign body response and form an organotypic regenerate. The polished articular heads, from Figure 12, prevent adhesion with host tissues and maintain the necessary range of mandibular movements.

The use of customized PTN endografts for total and subtotal substitution of the mandible, including the condylar process and mandibular ramus, is thus a pretty good surgical method for reconstructing the anatomical features of the affected area.

6.3. Customized PTN Endografts in Oncosurgery

Customized endografts made of PTN disk plates 0.3–0.4 mm thick, as depicted in Figure 14, were applied to replicate the maxilla, zygomatic bone, orbit, nose, and midface structures in cancer patients [104]. The superelastic property makes the implant flexible, which enables intraoperative modeling. The porous structure fixes the implant firmly in the wound, followed by connective and bone tissue ingrowth, which occurs with subsequent epithelialization of the postoperative cavity. The rigid central part and flexible edges of endoprostheses eliminate various discrepancies. The customized PTN graft can be produced based on CT scans and CAD modeled construct. At the same time, the implants provide good anatomical and aesthetic results in the elimination of complex defects on the walls of the orbit and its edges and adjacent bone structures (Figure 14). One of the adverse factors of midface reconstruction is highly virulent flora influencing the operating wound, which trigger the inflammation process in the implantation zone. Additionally, the subcranial region is an area of increased functional activity. It is clear that an endograft, in which resilience to the adverse impacts along with the anisotropic compliance and versatility in terms of stress-strain is inherent, can be the most advanced option. The superelastic feature of PTN is beneficial for smooth insertion through the minimal incision (the customized graft can be intraoperatively predeformed and shrunk for smooth insertion), followed by deploying within the orbital area in situ. This means that a concept of minimally invasive surgery is technically feasible. Moreover, in terms of stress-strain, the superelastic behaviour shown by PTN can counterbalance possible negative effects in the follow-up period. It is particularly important in pediatric patients or teenagers when the implanted PTN graft mimics the anisotropic compliance of the repaired orbit. So, in other words, the superelastic PTN as a load-bearing implant adapts to the augmenting midface/orbital environment, demonstrating higher adaptability without impairment of the mechanical performance at higher loads.

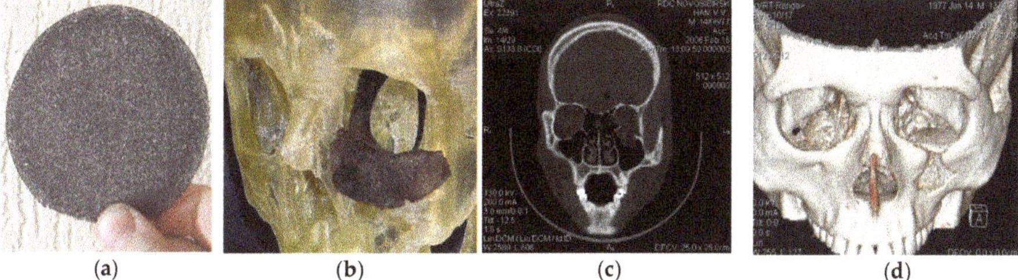

(a)   (b)   (c)   (d)

Figure 14. Repair of the orbit: (**a**) PTN disk plate before modeling, (**b**) preoperative customization using the 3D printed model; (**c**) coronal plane CT of the orbit before surgery; and (**d**) 3D reconstructed postsurgical CT of the repaired orbit [105].

Clinical examples have shown that PTN grafts ensure reliable restoration of the inferior orbital wall, prevent displacement of the eyeball, correct vision errors, and eliminate undesirable aftereffects [105]. The properties of the PTN plate allow modeling of sophisticated implants at a certain temperature regime. The customized PTN grafts precisely render orbitozygomatic outlines and orbital floor, thus recovering the anatomical structure, and are supposed to be an attractive alternative to Ti-based plates.

6.4. PTN Implants in Traumatic Surgery

In the past three decades, PTN implants have been successfully deployed for surgical treatment of fractured bones since they showed remarkable efficiency [106]. A vivid example of the high biochemical and biomechanical compatibility of PTN is the use of cylindrical PTN grafts in hand surgery when repairing traumatic lesions and lost bone structures, as illustrated in Figure 15 [107]. Four damaged bone fragments resulted from a labor accident were substituted at once using cylindrical PTN grafts, which were customized intraoperatively.

The rough developed surface of PTN possesses self-adherence feature, whereas the porous structure maintains the ingrowth of host tissues herein. Due to the good rheological property and functional strength of the PTN graft, the range of hand motions was fully restored in five days after surgery. The patient was reported to continue his job three days after discharge. Follow-up observations evidenced the tight incorporation with host tissues and no complications. This highlights once more the functionality of PTN as a bone substitute when it uniformly redistributes high dynamic loads and, therefore, enables long-term cycling with no failure.

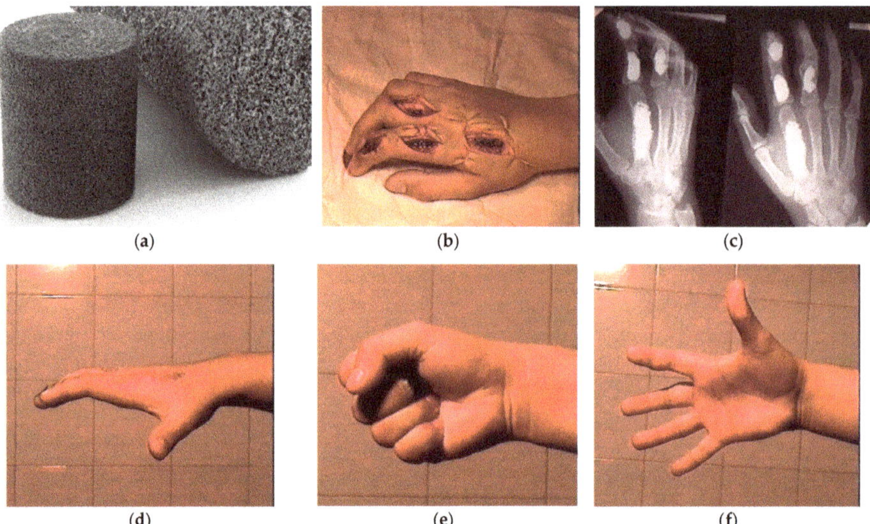

Figure 15. Clinical case of a repaired hand after a labour accident: (**a**) general view of PTN graft; (**b**) preoperative view of the injured hand; (**c**) postsurgical X-ray image, (**d**–**f**) range of hand motions in five days after surgery [107].

6.5. Cryotools Made of PTN

Cryotools having a working part made of PTN have been used in clinical practice for over 25 years due to the unique properties of the material [108]. Moreover, the flexibility of Nitinol rods/wires has allowed the fabrication of cryotools having the variable geometry handle seen in Figure 16, which is configured depending on application. From Figure 16, the working PTN part can be either unwrought or polished. Once the working PTN part has been immersed into liquid nitrogen, its changing color indicates how long the cryotool can be applied. Of course, the larger the working PTN part, the longer would be the cryoeffect. However, in a case of a minimally invasive approach or a hollow organ of smaller size, the appropriate cryotool needs to be chosen from [109,110]. Changing the SHS parameters, the inner structure of the PTN body was suggested to be intentionally designed to have a variable porosity in this regard (high-porous center and fine-porous periphery) [64]. This concept is feasible and helps to hold liquid nitrogen inside, preventing leakage as long as possible during cryo-application until the entire coolant content is evaporated.

The variable and open-end porosity attains both a high permeability and low thermal conductivity of the working PTN part. In other words, the thermal screening effect ensures a lengthy cryo-exposure due to the higher performance, wherein up to 90% of the consumed coolant is transmitted to the contact surface. The main feature of PTN cryotools is that the working part does not adhere to applied tissues owing to the dry interface, which precludes ice formation. Such cryotools have been reported to be utilized in cryotherapy, cryosurgery, cryodestructive oncology, and skin care [104,108,111]. Figure 17 illustrates a clinical case for the cryotherapy of a precancerous orolabial lesion, where cryodestruction was performed using a cylindrical cryotool [108].

The biopsy probe has verified limited precancerous hyperkeratosis, whereas histological examination carried out six months after cryotreatment revealed a soft inconspicuous scar at the former affected area.

Finally, analysis of documented clinical cases using PTN devices over the past decade is given in Table 2. Spinal surgery has turned out to be the most sought-after field for implantable PTN devices, whereas cryotherapy has been in the forefront in the context of non-implantable devices.

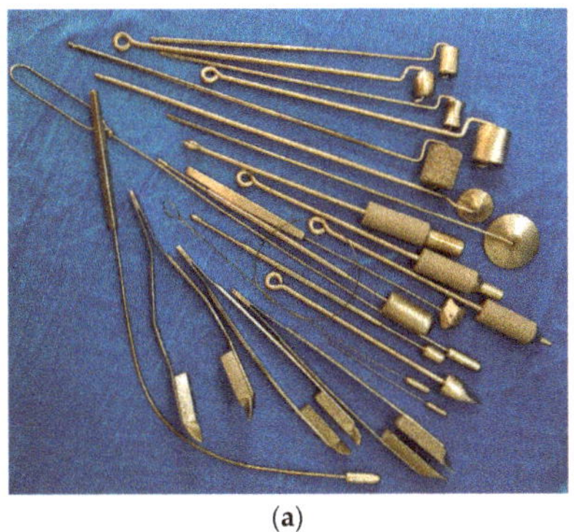

(**a**) (**b**)

Figure 16. Images of (**a**) PTN cryotools and (**b**) variable porosity of the sectioned working part [64,108].

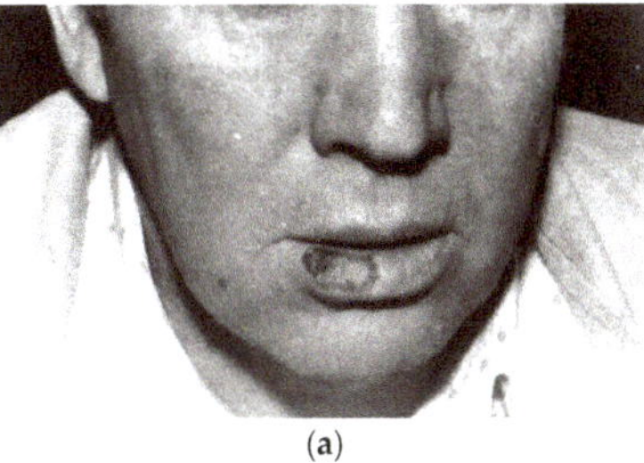
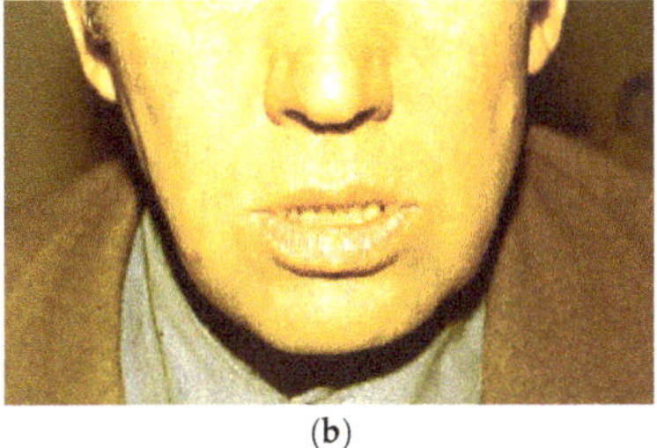

(**a**) (**b**)

Figure 17. Images of a patient with precancerous orolabial lesion: (**a**) Before cryotreatment and (**b**) a month after treatment [104].

Table 2. Brief summary of clinically applied PTN devices from 2000 to 2010 [101–104,106,108].

Clinical Field	Embodiment of PTN	Number of Cases	Note
Traumatic and orthopedic surgery	Plates, Bars, Round bars, Stripes, Tapers, Customized endografts, Pellets	621	Open/closed bone fracture—361 Traumatic bone/joint lesion—127 Posttraumatic joint contracture—62 False joint—38 Congenital bone abnormality—23 Ankylosis—10
Spinal surgery	Customized cages (cylinder, bar, wedge etc.)	1983 641 257	Lumbar anterior/posterior interbody fusion: L_5-S_1—959; $L_{4\text{-}5}$—791; $L_{3\text{-}4}$—233 Cervical anterior/posterior interbody fusion: $C_{2\text{-}4}$—190; $C_{4\text{-}7}$—451 Spinal stenosis surgery: Lumbar—214; Cervical—43
Maxillofacial surgery	Plates, Bars, Round bars, Tapers, Stripes, Customized endografts, Pellets	409	Radicular cyst—175 Ameloblastoma—81 Odontoma—56 Osteoma—33 Condylar joint replacement—29 Total/subtotal mandibular replacement—25 Giant-cell tumor—10
Cryo-surgery/therapy	Cryotools having different size/shape/surface of the working part	1314 200 138 <9000	Cryodestruction of cutaneous and subcutaneous tumors (malignant, premalignant, non-malignant) Cryodestruction of hemorrhoidal boluses Cryotherapy of the obstructed urethra (urethral patency restoration) Cryotherapy of hemangioma in infants and pediatric patients
Oncosurgery	Plates, Disks, Round bars, Stripes, Customized endografts, Pellets	617	Bone/joint post-resection repair—322 Head and neck sparing surgery—295

7. Conclusions

The PTN biomaterial has been discussed in the light of material science engineering and more than three decades of clinical experience, and the following conclusions can be drawn:

(i)   The biomechanical compatibility is referred to as the similarity of viscoelastic rheological characteristics between the PTN implant and host tissues. The combination of toughness inherent in Ti-based alloys, the porous morphology, and the viscoelastic reversible behavior of the porous body emphasizes the potential of PTN alloys to redistribute well physiological loads even in the early postoperative period, allowing to circumvent obstacles faced by existing implants;

(ii)   The biochemical compatibility has turned out to be successful as well. The bioinertness of surface and inferential bioactivity evidenced through cytocompatibility and negligible foreign body reaction owe much to self-assembled superficial layers resulted from the SHS process and, as such, the as-received PTN implant does not require further surface modification;

(iii) Multifarious superficial layers demonstrating a complex structure/composition and high corrosion resistance conceal the matrix entirely and can be congruentially deformed without rupture and delamination, withstanding multicycle alternating loads;

(iv)   The in vivo performance of PTN bone substitutes is also high. They may go through 10^7 cycles with no failure due to the fact that chemical-proof layers arrest the surface deterioration, whereas the superelastic behavior of the matrix at alternating load rules out a possibility of the early material's degradation.

(v)   A large number of PTN devices have been clinically applied in traumatic/orthopedic surgery, maxillofacial surgery, spinal surgery, etc., due to a rare combination of structure, mechanical, and physicochemical properties of PTN as a biomaterial. Moreover, bioengineering can consider customized PTN grafts and PTN-based surgical techniques in the context of the next generation implants concerning surgery cost minimization and improved patient tolerance.

(vi)   Comparative studies on corrosion fatigue behaviors of porous Ti and TiNi alloys made by both SHS and sintering are further needed to accomplish a complete and systematic understanding of PTN as an advanced biomaterial which can serve multiple clinical purposes.

Supplementary Materials: The following are available online at http://www.mdpi.com/1996-1944/12/15/2405/s1. Video S1: Cervical spine dynamic PTN cage.

Author Contributions: All the authors contributed equally to the manuscript.

Funding: This research was funded by Russian Science Foundation (grant No. 18-12-00073).

Conflicts of Interest: The authors declare no conflict of interest.

References

1.   Buehler, W.; Gilfrich, J.; Wiley, R. Effect of low-temperature phase changes on mechanical properties of alloys near composition TiNi. *J. Appl. Phys.* **1963**, *34*, 1475–1477. [CrossRef]

2.   Gunther, V.; Kotenko, V.; Mirgazizov, M.; Polenichkin, V.; Vitugov, V.; Itin, V.; Ziganshin, R.; Temerhanov, F. *Application of Shape Memory Alloys in Medicine*; TSU Publishing House: Tomsk, Russia, 1986; 208p, Available online: http://www.sme-implant.com/ccount/click.php?id=48 (accessed on 3 June 2019).

3.   Gunther, V.; Dambaev, G.; Sysoliatin, P.; Ziganshin, R.; Kornilov, N.; Mirgasisov, M.; Mironov, S.; Fomichev, N.; Hodorenko, V.; Temerkhanov, F.; et al. *Delay Law and New Class of Materials and Implants in Medicine*; STT: Northampton, UK, 2000; 432p, Available online: http://www.sme-implant.com/ccount/click.php?id=2 (accessed on 3 June 2019).

4.   Ryhanen, J.; Kallioinen, M.; Tuukkanen, J.; Junila, J.; Niemela, E.; Sandvik, P.; Serlo, W. In vivo biocompatibility evaluation of nickel–titanium shape memory metal alloy: Muscle and perineural tissue responses and encapsule membrane thickness. *J. Biomed. Mater. Res.* **1998**, *41*, 481–488. [CrossRef]

5.   Drugacz, J.; Lekston, Z.; Morawiec, H.; Januszewski, K. Use of TiNi shape-memory clamps in the surgical treatment of mandibular fractures. *J. Oral. Maxillofac. Surg.* **1995**, *53*, 665–671. [CrossRef]

6. Ryhanen, J.; Niemi, E.; Serlo, W.; Niemel, E.; Sandvik, P.; Pernu, H.; Salo, T. Biocompatibility of nickel-titanium shape memory metal and its corrosion behavior in human cell cultures. *J. BioMed Mater. Res.* **1998**, *35*, 451–457. [CrossRef]

7. Jani, J.; Leary, M.; Subic, A.; Gibson, M. A review of shape memory alloy research, applications and opportunities. *Mater. Des.* **2014**, *56*, 1078–1113. [CrossRef]

8. Prymak, O.; Bogdanski, D.; Koller, M.; Esenwein, S.A.; Muhr, G.; Beckmann, F.; Donath, T.; Assad, M.; Epple, M. Morphological characterization and in vitro biocompatibility of a porous nickel-titanium alloy. *Biomaterials* **2005**, *26*, 5801–5807. [CrossRef] [PubMed]

9. Rhalmi, S.; Charette, S.; Assad, M.; Coillard, C.; Rivard, C.H. The spinal cord dura mater reaction to nitinol and titanium alloy particles: A 1-year study in rabbits. *Eur. Spine J.* **2007**, *16*, 1063–1072. [CrossRef]

10. Gunther, V.; Sysoliatin, P.; Temerhanov, F.; Mirgazizov, M.; Pushkarev, V.; Tazin, I. *Superelastic Shape Memory Implants in Maxillofacial Surgery, Traumatology, Orthopaedics, and Neurosurgery*; Tsu Publishing House: Tomsk, Russia, 1995; 224p.

11. Gunther, V.; Dambaev, G.; Sysoliatin, P.; Ziganshin, R.; Temerhanov, F.; Polenichkin, V.; Mirgazizov, M.; Pahomenko, G.; Savchenko, P.; Staroha, A. *et al. Medical Materials and Shape Memory Implants*; Tsu Publishing House: Tomsk, Russia, 1998; 487p.

12. Gunther, V. *Shape Memory Implants. Medical Atlas*; STT: Northampton, UK, 2002; 234p, Available online: http://www.sme-implant.com/ccount/click.php?id=5 (accessed on 3 June 2019).

13. Itin, V.; Gunther, V.; Khachin, V.; Bratchikov, A. Production of titanium nickelide by self-propagating high-temperature synthesis. *Powder Metall. Metal. Ceram.* **1983**, *22*, 156–157. [CrossRef]

14. Li, B.; Rong, L.; Li, Y.; Gjunter, V. Synthesis of porous Ni-Ti shape memory alloys by self-propagating high-temperature synthesis: Reaction mechanism and anisotropy in pore structure. *Acta Mater.* **2000**, *48*, 3895–3904. [CrossRef]

15. Maslov, V.; Borovinskaya, I.; Merzhanov, A. Problem of mechanism of gasless combustion. *Combust. Explos. Shock. Waves* **1976**, *12*, 631–636. [CrossRef]

16. Merzhanov, A. Combustion processes that synthesize materials. *J. Mater. Process. Technol.* **1996**, *56*, 222–241. [CrossRef]

17. Borovinskaya, I.; Gromov, A.; Levashov, E.; Maksimov, Y.; Mukasyan, A.; Rogachev, A. *Concise Encyclopedia of Self-Propagating High-Temperature Synthesis History, Theory, Technology, and Products*, 1st ed.; Elsevier: Amsterdam, The Netherlands, 2017; 437p.

18. Li, B.; Rong, L.; Li, Y. Microstructure and superelasticity of porous NiTi alloy. *Sci. China Ser. E Technol. Sci.* **1999**, *42*, 94–99. [CrossRef]

19. Parvizi, S.; Hasannaeimi, V.; Saebnoori, E.; Shahrabi, T.; Sadrnezhaad, S.K. Fabrication of porous NiTi alloy via powder metallurgy and its mechanical characterization by shear punch method. *Russ. J. Non Ferr. Met.* **2012**, *53*, 169–175. [CrossRef]

20. Gunther, V.; Yasenchuk, Y.; Chekalkin, T.; Marchenko, E.; Gunther, S.; Baigonakova, G.; Hodorenko, V.; Kang, J.H.; Weiss, S.; Obrosov, A. Formation of pores and amorphous-nanocrystalline phases in porous TiNi alloys made by self-propagating high-temperature synthesis (SHS). *Adv. Powder Technol.* **2019**, *30*, 673–680. [CrossRef]

21. Yasenchuk, Y.; Gunther, V.; Marchenko, E.; Chekalkin, T.; Baigonakova, G.; Hodorenko, V.; Gunther, S.; Kang, J.H.; Weiss, S.; Obrosov, A. Formation of mineral phases in self-propagating high-temperature synthesis (SHS) of porous TiNi alloy. *Mater. Res. Express* **2019**, *6*, 056522. [CrossRef]

22. Artyukhova, N.; Yasenchuk, Y.; Chekalkin, T.; Gunther, V.; Kim, J.S.; Kang, J.H. Structure and properties of porous TiNi(Co, Mo)-based alloy produced by reaction sintering. *Smart Mater. Struct.* **2016**, *25*, 107003. [CrossRef]

23. Corbin, S.F.; Cluff, D. Determining the rate of (β-Ti) decay and its influence on the sintering behavior of NiTi. *J. Alloys Compd.* **2009**, *487*, 179–186. [CrossRef]

24. Novak, P.; Pokorny, P.; Vojtech, V.; Knaislova, A.; Skolakova, A.; Capek, J.; Karlik, M.; Kopecek, J. Formation of Ni-Ti intermetallics during reactive sintering at 500–650 °C. *Mater. Chem. Phys.* **2015**, *155*, 113–121. [CrossRef]

25. Jasenchuk, Y.; Gjunter, V. Anode polarization of NiTi alloy in HCl solution. In Proceedings of the 1st International Symposium on Advanced Biomaterials (ISAB), Montreal, QC, Canada, 2–5 October 1997; pp. 41–42.

26. Munroe, N.; Pulletikurthi, C.; Haider, W. Enhanced biocompatibility of porous Nitinol. *J. Mater. Eng. Perform.* **2009**, *18*, 765–767. [CrossRef]

27. Bansiddhi, A.; Sargeant, T.; Stupp, S.; Dunand, D. Porous NiTi for bone implants: A review. *Acta Biomater.* **2008**, *4*, 773–782. [CrossRef]

28. Schrooten, J.; Assad, M.; Humbeeck, J.V.; Leroux, M. In vitro corrosion resistance of porous NiTi intervertebral fusion. *Smart Mater. Struct.* **2007**, *16*, S145. [CrossRef]

29. Gyunter, V.E.; Yasenchuk, Y.F.; Klopotov, A.A.; Khodorenko, V.N. The physicomechanical properties and structure of superelastic porous titanium nickelide-based alloys. *Tech. Phys. Lett.* **2000**, *26*, 35–37. [CrossRef]

30. Bertheville, B.; Neudenberger, M.; Bidaux, J.E. Powder sintering and shape-memory behaviour of NiTi compacts synthesized from Ni and TiH2. *Mater. Sci. Eng. A* **2004**, *384*, 143–150. [CrossRef]

31. Muhamedov, M.; Kulbakin, D.; Gunther, V.; Choynzonov, E.; Chekalkin, T.; Hodorenko, V. Sparing surgery with the use of TiNi-based endografts in larynx cancer patients. *J. Surg. Oncol.* **2015**, *111*, 231–236. [CrossRef]

32. Fomichev, N.; Gunther, V.; Kornilov, N.; Simonovich, A.; Hodorenko, V.; Sizikov, M.; Savchenko, P.; Pahomenko, G.; Yasenchuk, Yu.; Kolumb, V. *Novel Techniques in Spinal Care Using Porous Shape Memory Implants*; STT: Northampton, UK, 2002; 130p, Available online: http://www.sme-implant.com/ccount/click.php?id=6 (accessed on 3 June 2019).

33. Li, J.; Chen, D.; Luan, H.; Zhang, Y.; Fan, Y. Numerical evaluation and prediction of porous implant design and flow performance. *BioMed Res. Int.* **2018**, *12*, 1215021. [CrossRef]

34. Saini, M.; Singh, Y.; Arora, P.; Jain, K. Implant biomaterials: A comprehensive review. *World J. Clin. Cases* **2015**, *3*, 52–57. [CrossRef]

35. Lemons, J.E. (Ed.) *Quantitative Characterization and Performance of Porous Implants for Hard Tissue Applications*; ASTM International: West Conshohocken, PA, USA, 1987; 410p. [CrossRef]

36. Novak, P.; Moravec, H.; Salvetr, P.; Prusa, F.; Drahokoupil, J.; Kopecek, J.; Karlik, M.; Kubatiket, T. Preparation of nitinol by non-conventional powder metallurgy techniques. *Mater. Sci. Technol.* **2015**, *31*, 1886–1893. [CrossRef]

37. Abidi, I.H.; Khalid, F.A.; Farooq, M.U.; Hussain, M.U.; Maqbool, A. Tailoring the pore morphology of porous nitinol with suitable mechanical properties for biomedical applications. *Mater. Lett.* **2015**, *154*, 17–20. [CrossRef]

38. Tavadze, G.F.; Shteinberg, A.S. *Production of Advanced Materials by Methods of Self-Propagating High-Temperature Synthesis*; Springer: Heidelberg, Germany, 2013; 156p. [CrossRef]

39. Whitney, M.; Corbin, S.F.; Gorbet, R.B. Investigation of the mechanisms of reactive sintering and combustion synthesis of NiTi using differential scanning calorimetry and microstructural analysis. *Acta Mater.* **2008**, *56*, 559–570. [CrossRef]

40. Whitney, M.; Corbin, S.F.; Gorbet, R.B. Investigation of the influence of Ni powder size on microstructural evolution and the thermal explosion combustion synthesis of NiTi. *Intermetallics* **2009**, *17*, 894–906. [CrossRef]

41. Artyukhova, N.; Yasenchuk, Y.; Kim, J.S.; Gunther, V. Reaction sintering of porous shape memory titanium−nickelide-based alloys. *Russ. Phys. J.* **2015**, *57*, 1313–1320. [CrossRef]

42. Yasenchuk, Y.; Gjunter, V. Surface structure of porous titanium nickelide produced by SHS. *Russ. Phys. J.* **2008**, *51*, 851–857. [CrossRef]

43. Yasenchuk, Y.; Artyukhova, N.; Almaeva, K.; Garin, A.; Gunther, V. Segregation in porous NiTi made by SHS in flow reactor. In Proceedings of the Shape Memory Biomaterials and Implants in Medicine (SMBIM), Pusan, South Korea, 1–3 May 2017; KnE Materials Science: Dubai, UAE; pp. 168–175. [CrossRef]

44. Kokorev, O.; Khodorenko, V.; Baigonakova, G.; Marchenko, E.; Yasenchuk, Y.; Gunther, V.; Anikeev, S.; Barashkova, G. Metal-glass-ceramic phases on the surface of porous TiNi-based SHS-material for carriers of cells. *Russ. Phys. J.* **2019**, *61*, 1734–1740. [CrossRef]

45. Assad, M.; Jarzem, P.; Leroux, M.; Coillard, C.; Chernyshov, A.; Charette, S.; Rivard, C.H. Porous titanium-nickel for intervertebral fusion in a sheep model: Part 1. Histomorphometric and radiological analysis. *J. BioMed Mater. Res.* **2003**, *64*, 107–120. [CrossRef]

46. Kang, S.B.; Yoon, K.S.; Kim, J.S.; Nam, T.H.; Gjunter, V.E. In vivo result of porous TiNi shape memory alloy: Bone response and growth. *Mater. Trans.* **2002**, *43*, 1045–1048. [CrossRef]

47. Kokorev, O.; Hodorenko, V.; Chekalkin, T.; Kim, J.S.; Kang, S.B.; Dambaev, G.; Gunther, V. In vitro and in vivo evaluation of porous TiNi-based alloy as a scaffold for cell tissue engineering. *Artif. Cells Nanomed. Biotechnol.* **2016**, *44*, 704–709. [CrossRef]

48. Gunther, S.; Kokorev, O.; Chekalkin, T.; Hodorenko, V.; Dambaev, G.; Kang, J.H.; Gunther, V. Effects of infrared and ultraviolet radiation on the viability of cells immobilized in a porous TiNi-based alloy scaffold. *Adv. Mater. Lett.* **2015**, *6*, 774–778. [CrossRef]

49. Kokorev, O.; Hodorenko, V.; Chekalkin, T.; Gunther, V.; Kang, S.B.; Chang, M.J.; Kang, J.H. Evaluation of allogenic hepato-tissue engineered in porous TiNi-based scaffolds for liver regeneration in a CCl4-induced cirrhosis rat model. *BioMed Phys. Eng. Express* **2019**, *5*, 025018. [CrossRef]

50. Prasad, K.; Bazaka, O.; Chua, M.; Rochford, M.; Fedrick, L.; Spoor, J.; Symes, R.; Tieppo, M.; Collins, C.; Cao, A.; et al. Metallic biomaterials: Current challenges and opportunities. *Materials* **2017**, *10*, 884. [CrossRef]

51. Prakash, C.; Singh, S.; Pruncu, C.I.; Mishra, V.; Krolczyk, G.; Pimenov, D.Y.; Pramanik, A. Surface modification of Ti-6Al-4V alloy by electrical discharge coating process using partially sintered Ti-Nb electrode. *Materials* **2019**, *12*, 1006. [CrossRef]

52. Mah, D.; Pelletier, M.H.; Lovric, V.; Walsh, W.R. Corrosion of 3D-printed orthopaedic implant materials. *Ann. BioMed Eng.* **2019**, *47*, 162–173. [CrossRef]

53. Yuan, B.; Zhu, M.; Chung, C.Y. Biomedical porous shape memory alloys for hard-tissue replacement materials. *Materials* **2018**, *11*, 1716. [CrossRef]

54. Marchenko, E.S.; Yasenchuk, Y.F.; Gunther, V.E.; Dubovikov, K.M.; Khodorenko, V.N. The comparative structural-phase analysis of the surface of macro and microporous SHS TiNi. In Proceedings of the International Conference. on Shape Memory and Superelastic Technologies (SMST-2019), Konstanz, Germany, 13–17 May 2019; ASM International: Materials Park, OH, USA, 2019; pp. 57–58.

55. Baigonakova, G.A.; Yasenchuk, Y.F.; Gunther, S.V.; Kokorev, O.V.; Gunther, V.E. Structural-phase and morphological features of the surface layers of the SHS TiNi. In Proceedings of the International Conference on Shape Memory and Superelastic Technologies (SMST-2019), Konstanz, Germany, 13–17 May 2019; ASM International: Materials Park, OH, USA, 2019; pp. 53–54.

56. Buchwitz, M.; Adlwarth-Dieball, R.; Ryder, P.L. Kinetics of the crystallization of amorphous Ti2Ni. *Acta Metall. Mater.* **1993**, *41*, 1885–1892. [CrossRef]

57. Hashimoto, K.; Kumagai, N.; Yoshioka, H.; Kim, J.H.; Akiyama, E.; Habazaki, H.; Mrowec, S.; Kawashima, A.; Asamia, K. Corrosion-resistant amorphous surface alloys. *Corros. Sci.* **1993**, *35*, 363–370. [CrossRef]

58. Maccabi, A.; Shin, A.; Namiri, N.K.; Bajwa, N.; John, M.S.; Taylor, Z.D.; Grundfest, W.; Saddik, G.N. Quantitative characterization of viscoelastic behavior in tissue-mimicking phantoms and ex vivo animal tissues. *PLoS ONE* **2018**, *13*, 0191919. [CrossRef]

59. Karathanasopoulos, N.; Arampatzis, G.; Ganghoffer, J.F. Unravelling the viscoelastic, buffer-like mechanical behavior of tendons: A numerical quantitative study at the fibril-fiber scale. *J. Mech. Behav. BioMed Mater.* **2019**, *90*, 256–263. [CrossRef]

60. Evans, F.G.; Lebow, M. Regional differences in some of the physical properties of the human femur. *J. Appl. Phys.* **1951**, *3*, 563–572. [CrossRef]

61. Hirsch, C.; Sonnerup, L. Macroscopic rheology in collagen material. *J. Biomech.* **1968**, *1*, 13–18. [CrossRef]

62. Unal, M.; Creecy, A.; Nyman, J.S. The role of matrix composition in the mechanical behavior of bone. *Curr. Osteoporos Rep.* **2018**, *16*, 205–215. [CrossRef]

63. Lu, X.; Wang, C.; Li, G.; Liu, Y.; Zhu, X.; Tu, S. The mechanical behavior and martensitic transformation of porous NiTi alloys based on geometrical reconstruction. *Int. J. Appl. Mech.* **2017**, *9*, 1–15. [CrossRef]

64. Gunther, V.; Hodorenko, V.; Chekalkin, T.; Olesova, V.; Dambaev, G.; Sysoliatin, P.; Fomichev, N.; Mirgazizov, M.; Melnik, D.; Ivchenko, O.; et al. *Medical Materials and Shape Memory Implants: Shape Memory Materials*; MITS: Tomsk, Russia, 2011; Volume 1, 534p, Available online: http://www.sme-implant.com/ccount/click.php?id=14 (accessed on 3 June 2019).

65. Perren, S.M.; Fernandez, A.; Regazzoni, P. Understanding fracture healing biomechanics based on the "strain" concept and its clinical applications. *Acta Chir. Orthop. Traumatol. Cech.* **2015**, *82*, 253–260. Available online: https://www.ncbi.nlm.nih.gov/pubmed/26516728 (accessed on 3 June 2019).

66. Cole, J.H.; Van der Meulen, M.C. Whole bone mechanics and bone quality. *Clin. Orthop. Relat. Res.* **2011**, *469*, 2139–2149. [CrossRef]

67. Marino, S.; Staines, K.A.; Brown, G.; Howard-Jones, R.A.; Adamczyk, M. Models of ex vivo explant cultures: Applications in bone research. *Bonekey Rep.* **2016**, *5*, 818. [CrossRef]

68. Gyunter, V.; Khodorenko, V.; Monogenov, A.; Yasenchuk, Y. Effect of deformation on the permeability of porous titanium-nickel alloys. *Technol. Phys. Lett.* **2000**, *26*, 320–322. [CrossRef]

69. Bassani, P.; Giuliani, P.; Tuissi, A.; Zanotti, C. Thermomechanical properties of porous NiTi alloy produced by SHS. *J. Mater. Eng. Perform.* **2009**, *18*, 594–599. [CrossRef]

70. Resnina, N.; Belyaev, S.; Voronkov, A.; Gracheva, A. Mechanical behaviour and functional properties of porous Ti-45 at. % Ni alloy produced by self-propagating high-temperature synthesis. *Smart Mater. Struct.* **2016**, *25*, 055018. [CrossRef]

71. Martin, C.L.; Favier, D.; Suery, M. Viscoplastic behaviour of porous metallic materials saturated with liquid part II: Experimental identification on a Sn-Pb model alloy. *Int. J. Plast.* **1997**, *13*, 237–259. [CrossRef]

72. Itin, V.I.; Gyunter, V.E.; Shabalovskaya, S.A.; Sachdeva, L.C. Mechanical properties and shape memory of porous nitinol. *Mater. Charact.* **1994**, *32*, 179–187. [CrossRef]

73. Greiner, C.; Oppenheimer, S.M.; Dunand, D.C. High strength, low stiffness, porous NiTi with superelastic properties. *Acta Biomater.* **2005**, *1*, 705–716. [CrossRef]

74. Robertson, S.W.; Pelton, A.R.; Ritchie, R.O. Mechanical fatigue and fracture of Nitinol. *Int. Mater. Rev.* **2012**, *57*, 1–37. [CrossRef]

75. Chakraborty, R.; Datta, S.; Raza, M.S.; Saha, P. A comparative study of surface characterization and corrosion performance properties of laser surface modified biomedical grade Nitinol. *Appl. Surf. Sci.* **2019**, *469*, 753–763. [CrossRef]

76. Mehta, K.; Gupta, K. *Fabrication and Processing of Shape Memory Alloys*; Springer: Cham, Switzerland, 2019; 86p. [CrossRef]

77. Khodaei, M.; Meratian, M.; Savabic, O.; Fathib, M.; Ghomi, H. The side effects of surface modification of porous titanium implant using hydrogen peroxide: Mechanical properties aspects. *Mater. Lett.* **2016**, *178*, 201–204. [CrossRef]

78. Yeung, K.W.; Poon, R.W.; Liu, X.Y.; Ho, J.P.; Chung, C.Y.; Chu, P.K.; Lu, W.W.; Chan, D.; Cheung, K.M. Corrosion resistance, surface mechanical properties, and cytocompatibility of plasma immersion ion implantation–treated nickel-titanium shape memory alloys. *J. BioMed Mater. Res.* **2005**, *75*, 256–267. [CrossRef]

79. Robinson, E.; Gaillard-Campbell, D.; Gross, T.P. Acetabular debonding: An investigation of porous coating delamination in hip resurfacing arthroplasty. *Adv. Orthop.* **2018**, *2018*, 5282167. [CrossRef]

80. Intranuovo, F.; Gristina, R.; Brun, F.; Mohammadi, S.; Ceccone, G.; Sardella, E.; Rossi, F.; Tromba, G.; Favia, P. Plasma modification of pcl porous scaffolds fabricated by solvent-casting/particulate-leaching for tissue engineering. *Plasma Process. Polym.* **2014**, *11*, 184–195. [CrossRef]

81. Wang, Q.; Cheng, M.; He, G.; Zhang, X. Surface modification of porous titanium with microarc oxidation and its effects on osteogenesis activity in vitro. *J. Nanomater.* **2015**, *2015*, 408634. [CrossRef]

82. Seah, K.H.; Thampuran, R.; Chen, X.; Teoh, S.H. A comparison between the corrosion behaviour of sintered and unsintered porous titanium. *Corros. Sci.* **1995**, *35*, 1333–1340. [CrossRef]

83. Sun, X.T.; Kang, Z.X.; Zhang, X.L.; Jiang, X.J.; Guan, R.F.; Zhang, X.P. A comparative study on the corrosion behavior of porous and dense NiTi shape memory alloys in NaCl solution. *Electrochim. Acta.* **2011**, *56*, 6389–6396. [CrossRef]

84. Aihara, H.; Zider, J.; Fanton, G.; Duerig, T. Combustion synthesis porous nitinol for biomedical applications. *Int. J. Biomater.* **2019**, *2019*, 4307461. [CrossRef]

85. Stergioudi, F.; Vogiatzis, C.A.; Pavlidou, E.; Skolianos, S.; Michailidis, N. Corrosion resistance of porous NiTi biomedical alloy in simulated body fluids. *Smart Mater. Struct.* **2016**, *25*, 095024. [CrossRef]

86. Assad, M.; Jarzem, P.; Leroux, M.A.; Coillard, C.; Chernyshov, A.V.; Charette, S.; Rivard, C.H. Porous titanium-nickel for intervertebral fusion in a sheep model: Part 2. Surface analysis and nickel release assessment. *J. BioMed Mater. Res. B* **2003**, *64*, 107–120. [CrossRef]

87. Williams, D.F. Fundamental aspects of biocompatibility. In *Tissue Engineering*; Elsevier: Amsterdam, The Netherlands, 2008; pp. 255–278. [CrossRef]

88. Rhalmi, S.; Odin, M.; Assad, M.; Tabrizian, M.; Rivard, C.H.; Yahia, L.H. Hard, soft tissue and in vitro cell response to porous nickel-titanium: A biocompatibility evaluation. *BioMed Mater. Eng.* **1999**, *9*, 151–162. Available online: https://www.ncbi.nlm.nih.gov/pubmed/10572619 (accessed on 3 June 2019).

89. Mjoberg, B. Theories of wear and loosening in hip prostheses: Wear-induced loosening vs loosening-induced wear—A review. *Acta Orthop. Scand.* **1994**, *65*, 361–371. [CrossRef]

90. Mitra, J.; Tripathi, G.; Sharma, A.; Basu, B. Scaffolds for bone tissue engineering: Role of surface patterning on osteoblast response. *RSC Adv.* **2013**, *3*, 11073. [CrossRef]

91. Hatano, K.; Inoue, H.; Kojo, T.; Matsunaga, T.; Tsujisawa, T.; Uchiyama, C.; Uchida, Y. Effect of surface roughness on proliferation and alkaline phosphatase expression of rat calvarial cells cultured on polystyrene. *Bone* **1999**, *25*, 439–445. [CrossRef]

92. Gu, Y.; Li, H.; Tay, B.; Lim, C.; Yong, M.; Khor, K. In vitro bioactivity and osteoblast response of porous NiTi synthesized by SHS using nanocrystalline Ni-Ti reaction agent. *J. BioMed Mater. Res.* **2006**, *78*, 316–323. [CrossRef]

93. Kim, J.S.; Kang, J.H.; Kang, S.B.; Yoon, K.S.; Kwon, Y.S. Porous TiNi biomaterial by self-propagating high-temperature synthesis. *Adv. Eng. Mater.* **2004**, *6*, 403–406. [CrossRef]

94. Wu, S.; Liu, X.; Chan, Y.; Ho, J.; Chung, C.; Chu, P.K.; Chu, C.; Yeung, K.; Lu, W.; Cheung, K.; et al. Nickel release behavior, cytocompatibility, and superelasticity of oxidized porous single-phase NiTi. *J. BioMed Mater. Res.* **2007**, *81*, 948–955. [CrossRef]

95. Ryan, G.; Pandit, A.; Apatsidis, D. Biomaterials porous titanium scaffolds fabricated using a rapid prototyping and powder metallurgy technique. *Biomaterials* **2008**, *29*, 3625–3635. [CrossRef]

96. Karageorgiou, V.; Kaplan, D. Porosity of 3D biomaterial scaffolds and osteogenesis. *Biomaterials* **2005**, *26*, 5474–5491. [CrossRef]

97. Murphy, C.; Haugh, M.G.; O'Brien, F.J. The effect of mean pore size on cell attachment, proliferation and migration in collagen–glycosaminoglycan scaffolds for bone tissue engineering. *Biomaterials* **2010**, *31*, 461–466. [CrossRef]

98. Radkevich, A.; Hodorenko, V.; Gunther, V. Reparative osteogenesis in bone defects substituted by porous TiNi. *Shape Memory Implant.* **2005**, *2*, 30–35. Available online: http://www.sme-implant.com/ccount/click.php?id=38 (accessed on 3 June 2019).

99. Radkevich, A.A. Reconstructive Surgery of Alveolar Process. Ph.D. Thesis, Irkutsk State University, Irkutsk Oblast, Russia, 6 May 2002. Available online: http://medical-diss.com/docreader/479398/a#?page=2 (accessed on 3 June 2019).

100. Oxland, T.R. Fundamental biomechanics of the spine—What we have learned in the past 25 years and future directions. *J. Biomech.* **2016**, *49*, 817–832. [CrossRef]

101. Fomichev, N.; Gunther, V.; Lutsik, A.; Sergeev, K.; Kornilov, N.; Shevtsov, V.; Durov, M.; Pahomenko, G.; Savchenko, P.; Ratkin, I.; et al. *Medical Materials and Shape Memory Implants: Shape Memory Implants in Spinal Surgery*; MITS: Tomsk, Russia, 2011; Volume 3, 374p, Available online: http://www.sme-implant.com/ccount/click.php?id=16 (accessed on 3 June 2019).

102. Sysoliatin, P.; Gunther, V.; Sysoliatin, S.; Mirgazizov, M.; Radkevich, A.; Olesova, V.; Hodorenko, V.; Duriagin, N.; Melnik, D.; Tazin, I.; et al. *Medical Materials and Shape Memory Implants: Shape Memory Implants in Maxillo-Facial Surgery*; MITS: Tomsk, Russia, 2012; Volume 4, 384p, Available online: http://www.sme-implant.com/ccount/click.php?id=17 (accessed on 3 June 2019).

103. Mirgazizov, M.; Gunther, V.; Galonskiy, V.; Olesova, V.; Radkevich, A.; Hafizov, R.; Mirgazizov, R.; Yudin, P.; Starosvetskiy, S.; Zvigintsev, M.; et al. *Medical Materials and Shape Memory Implants: Shape Memory Implants in Dentistry*; MITS: Tomsk, Russia, 2011; Volume 5, 220p, Available online: http://www.sme-implant.com/ccount/click.php?id=18 (accessed on 3 June 2019).

104. Choinzonov, E.; Gunther, V.; Muhamedov, M.; Novikov, V.; Anisenia, I.; Cherdyntseva, N.; Kolomiets, L.; Chernysheva, A.; Slonimskaya, E.; Tuzikov, S.; et al. *Medical Materials and Shape Memory Implants: Shape Memory Implants in Oncology*; MITS: Tomsk, Russia, 2013; Volume 13, 336p, Available online: http://www.sme-implant.com/ccount/click.php?id=26 (accessed on 3 June 2019).

105. Shtin, V.I. Repair of Sub-Cranial Lesions in Cancer Patients Using Nitinol Implants and Radiotherapy. Ph.D. Thesis, Tomsk Cancer Research Institute, Tomsk, Russia, 9 December 2010. Available online: https://search.rsl.ru/ru/record/01004616721 (accessed on 3 June 2019).

106. Lanshakov, V.; Gunther, V.; Plotkin, G.; Fomichev, N.; Savchenko, P.; Fominyh, A.; Maslikov, V.; Petrov, L.; Melnik, D.; Baranov, G.; et al. *Medical Materials and Shape Memory Implants: Shape Memory Implants in Traumatic and Orthopaedic Surgery*; MITS: Tomsk, Russia, 2010; Volume 2, 384p, Available online: http://www.sme-implant.com/ccount/click.php?id=15 (accessed on 3 June 2019).

107. Fominyh, A.; Goryachev, A. Reconstructive hand surgery using shape memory materials. In *Biocompatible Materials and Shape Memory Implants*; STT: Northampton, UK, 2001; pp. 142–156. Available online: http://www.sme-implant.com/ccount/click.php?id=2 (accessed on 3 June 2019).

108. Melnik, D.; Gunther, V.; Dambaev, G.; Chuguy, E.; Hodorenko, V.; Sysoliatin, P.; Mahnev, A.; Tokmakova, S.; Musin, V.; Shkuratov, S.; et al. *Medical Materials and Shape Memory Implants: Porous NiTi Tools in Clinical Practice*; MITS: Tomsk, Russia, 2010; Volume 9, 306p, Available online: http://www.sme-implant.com/ccount/click.php?id=22 (accessed on 3 June 2019).

109. Shkuratov, S.; Gunther, V.; Dambaev, G.; Malkova, E.; Isaenko, V.; Feofilov, I.; Erkovich, I.; Shkuratov, S.; Davydov, A.; Yarin, G.; et al. *Medical Materials and Shape Memory Implants: Shape Memory Implants in Urology*; MITS: Tomsk, Russia, 2009; Volume 7, 248p, Available online: http://www.sme-implant.com/ccount/click.php?id=20 (accessed on 3 June 2019).

110. Tlish, M.M.; Kolesnikova, V.N.; Gunter, V.E.; Steblyuk, A.N.; Marchenko, E.S.; Shavilova, M.E.; Tserkovnaya, A.A. Possibilities of using cryotherapy in patients with ocular rosacea. *Ophthalmol. J.* **2018**, *11*, 7–14. [CrossRef]

111. Shtofin, S.; Gunther, V.; Anishenko, V.; Dambaev, G.; Shtofin, G.; Kulikova, L.; Merzlikin, N.; Volodos, N.; Sadovskiy, A.; Hodorenko, V.; et al. *Medical Materials and Shape Memory Implants: Shape Memory Implants in Urology*; MITS: Tomsk, Russia, 2013; Volume 12, 126p, Available online: http://www.sme-implant.com/ccount/click.php?id=25 (accessed on 3 June 2019).

Article

Influence of Implant Neck Design on Peri-Implant Tissue Dimensions: A Comparative Study in Dogs

José Luis Calvo-Guirado [1,*], Raúl Jiménez-Soto [1], Carlos Pérez Albacete-Martínez [1], Manuel Fernández-Domínguez [2], Sérgio Alexandre Gehrke [3] and José Eduardo Maté-Sánchez de Val [1]

[1] Faculty of Health Sciences, Universidad Católica San Antonio de Murcia, Murcia 30107, Spain; drjsoto@yahoo.com (R.J.-S.); cperezalbacete@ucam.edu (C.P.A.-M.); jemate@ucam.edu (J.E.M.-S.d.V.)
[2] Faculty of Dentistry, Department of Oral and Implant Dentistry, CEU San Pablo University, Madrid Hospital Group, Madrid 28040, Spain; clinferfun@yahoo.es
[3] Biotecnos Research Center, Montevideo 11100, Uruguay; sergio.gehrke@hotmail.com
*   Correspondence: jlcalvo@ucam.edu; Tel.: +34-9-6826-8353

Received: 9 September 2018; Accepted: 6 October 2018; Published: 17 October 2018

Abstract: This in vivo study assessed (hard and soft) peri-implant tissue remodeling around implants with micro-ring and open-thread neck designs placed in a dog model. Twenty histological sections corresponding to four different implant designs that were placed in America Foxhound dogs were obtained from previous studies. All the implants had been placed under identical conditions and were divided into four groups: Group A, micro-rings on implant neck plus 0.5 mm refined surface; Group B, micro-rings on implant neck; Group C, open-thread neck; and, Group D, double-spiral neck. Eight weeks after surgery, the integrated implants were removed and processed for histological examination. Crestal bone loss and bone-to-implant contact was greater for micro-ring necks than open-thread necks. Soft tissues showed significant differences on both buccal and lingual aspects, so that the distance from peri-implant mucosa to the apical portion of the barrier epithelium was smaller in the micro-ring groups. So, in spite of generating greater bone-to-implant contact, implants with micro rings produced more bone loss than open-thread implants. Moreover, the outcomes that were obtained IPX implants smooth neck design produced less bone loss in the cervical area, following by Facility implants when compared with the other open thread and microthreaded implant designs. Implant thread design can influence on bone remodeling in the cervical area, related to bundle bone preservation.

Keywords: bone levels; dental implants; neck design; soft tissue dimensions

---

1. Introduction

The long-term success and predictability of implant-supported restorations depend on maintaining peri-implant hard and soft tissues [1–3]. During the first year of function, bone resorption will be of 1.5 to 2 mm, generally considered as a normal physiologic process [1]. Thereafter, an annual bone loss of 0.2 mm can be expected under normal circumstances [4,5]. The implant's neck design may reduce marginal bone loss [6,7], and many different implant designs have attempted to preserve bone height after implant installation [8]. The implant neck design aims to reduce stress on the bone surrounding the implant and to stimulate the bone for remodeling.

It has been observed that the introduction of micro-rings on the implant neck may reduce early bone loss [9,10], and some authors suggest that micro-rings have the effect of limiting marginal bone loss in the presence of loading forces [11], the load transfer characteristics of the implant being dependent on the size and design of the implant neck [12,13]. In fact, the optimal load distribution offered by the micro-ring feature counteracts marginal bone loss [9,14]; it also enhances bone-to-implant contact.

Calvo-Guirado et al. reported limited implant crestal bone loss (0.90 mm $\pm$ 0.26 mm) and a 100% implant survival rate after a 5-year follow-up with immediately restored implants with a neck with rough surface and micro-threads, placed in the anterior maxillary/esthetic zone and immediately restored with non-occlusal loading [15]. The same authors measured soft tissue thickness and marginal bone loss around dental implants with sloped (30°) micro-threaded shoulders as compared with conventional micro-threaded straight design implants that were placed in immediate post-extraction sites with immediate loading in an animal model. Both types of implants generated similar soft tissue thickness and marginal bone loss after a three-month follow-up [16].

Surface characteristics also have a significant influence on marginal bone loss. In the case of hybrid implants with micro-rings and flat surfaces, most of the implants present alveolar bone loss over the entire length of the flat surface, as far as the first thread [17], because the un-roughened surface of the implants fails to distribute occlusal loads adequately [18]. But implants with micro-rings and textured surfaces allow tissue ingrowth [19,20]. The surface microstructure varies depending on the implant surface treatment, which can modify stress distribution, cell response to the implant surface, and implant osseointegration. A systematic review by Smeerts et al. described five different implant surfaces found to promote recruitment, adhesion, and proliferation of osteogenic as well as fibroblastic cells, all achieving a high degree of hard and soft tissue integration and high levels of bone-to-implant contact [21].

Another study of 47 implants with micro-rings on the neck reported that bone loss around the implants was not significant after a two-year follow-up [18]. Calvo-Guirado et al. found minimal marginal bone loss and a 100% implant survival rate over the 3-year follow-up for immediate implants with micro-ring necks subjected to immediate non-occlusal loading [22]. In another study, implants with a polished neck of 0.8 mm plus one micro-ring and a roughened area of 2 mm was found to reduce buccal bone resorption [23].

In addition to implants that were designed with micro-rings on the neck, another neck design presents an open-thread, achieved by prolonging the spiral of the implant body over the neck. It has been suggested that this implant neck design shows better characteristics for load distribution, counteracting marginal bone resorption [24,25]. Preclinical studies have shown that, when compared with flat necks, the open-thread design increases bone-to-implant contact [26,27], providing greater preservation of crestal bone height [28,29].

To date, no consensus has been reached as to the effectiveness and influence of implant macro- and micro-design on marginal bone loss. The aim of this study was to evaluate the influence of implant neck design on soft and hard tissue remodeling around implants placed at crestal level, with abutment loading at the time of placement; all of the implant systems assayed had conical implant-abutment connections to reduce the shear stresses at the bone-to-implant connection [9,27].

2. Materials and Methods

The samples that were used in this study were obtained from previous studies that also assayed the four different implant systems selected for this trial. Inclusion criteria were as follows: implants placed crestally in fresh extraction sockets of dog mandibles, with platform switching, all carried out applying the same animal protocol, surgical protocol, healing period, and sample preparation. All of the samples received an abutment at the time of implant placement.

2.1. Animal Protocol

All samples that were used in the present study were obtained from previous studies performed using an American foxhound animal model. The animals were aged approximately one year and weighed 14–15 kg. All the earlier studies were approved by the Ethics Committee for Animal Research, ensuring that each study protocol fulfilled guidelines that were established by the European Union

Council Directive 2010/63/UE. The project number of the four different experiments was A1320141102 (Murcia Agriculture and Water Ministry, Murcia, Spain).

The animals were fed a daily pellet diet. All animals presented intact maxillas, without any general occlusal trauma or oral viral or fungal lesions. Clinical examination determined that the dogs were in good general health, with no systemic involvement.

2.2. Sample Selection

Five slides of four dental implant models with different designs were selected according to their macro- and micro-characteristics: Group A: Blue Sky implant (Bredent medical GMBH & Co. KG, Senden, Germany), 3.5 mm diameter and 10 mm length with a micro-ring neck plus a 0.5 mm refined surface; Group B: MIS C1 implant (MIS Implants Technologies Ltd, Tel Aviv, Israel), 4 mm diameter and 10 mm length, with a micro-ring neck; Group C: IPX implant (Galimplant, Sarria, Lugo, Spain), 4 mm diameter and 10 mm length, with an open-thread neck; Group D: Facility implant (Neodent, Instradent AG, Basel, Switzerland), 2.9 mm diameter and 10 mm length, with an (open thread) double-spiral on neck. Figure 1 illustrates the four implant designs.

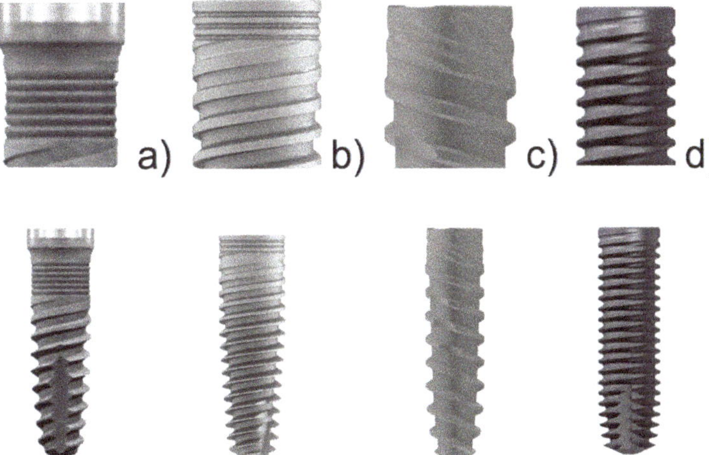

Figure 1. Implants used in the experiment: (**a**) Blue Sky, Bredent[®]; (**b**) C1, MIS[®]; (**c**) IPX, Galimplant[®]; and (**d**) Facility, Neodent[®].

2.3. Histological Preparation

All biopsies were processed using the same protocol, performing ground sectioning according to the method that was described by Donath & Breuner (1982) [27], designed to evaluate peri-implant soft tissue healing and bone remodeling. Samples were dehydrated in increasing grades of ethanol up to 100% and were embedded in a glycol methacrylate resin (Technovit 7200 VLC, Kulzer, Wehrheim, Germany). Then, the samples were polymerized and sectioned at the buccal-lingual plane using a diamond saw (Exakt, Apparatebau, Norderstedt, Germany). Sections were cut from each biopsy unit, from the center of the implant using a high-precision diamond disk to about 100 µm thickness and ground to approximately 40 µm final thickness with an Exakt 400s CS grinding device (Exakt, Apparatebau, Norderstedt, Germany). Each section was stained using toluidine blue stain (Figures 2 and 3).

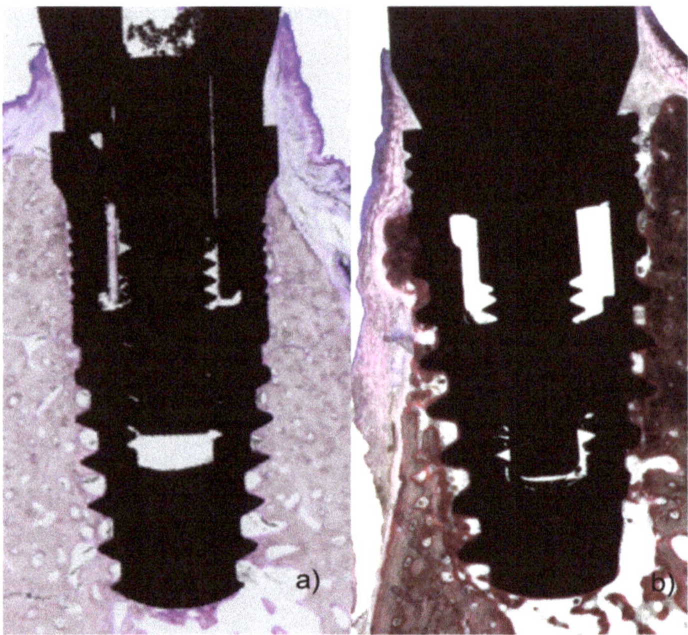

Figure 2. (**a**) Histological preparations representing the outcome after 8 weeks healing showing polished neck and micro-ring neck design of the Blue Sky implant; and (**b**) outcome after eight weeks healing of micro-ring neck design of the C1 implant.

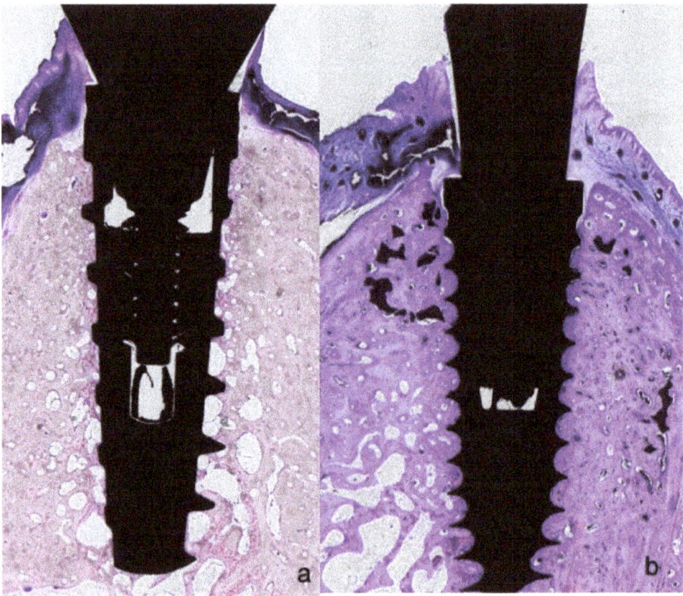

Figure 3. (**a**) Histological preparations representing the outcome after 8 weeks healing showing open-thread neck design of the IPX implant; and, (**b**) after eight weeks healing showing (open-thread) double-spiral neck design of the Facility implant.

2.4. Histometric Evaluation

The most central sagittal section of each implant was analyzed using calibrated digital images at ×
10 magnification under a Leica Q500Mc Microscope (Leica Microsystems, Wetzlar, Germany) equipped
with a digital video-camera (Sony DXC-151s 2/3-CCd RGB Color Video Camera, Tokyo, Japan)
connected to a computer equipped with MIP 4.5 software (MIcroms Image Processing Software, CID,
Consulting Image Digital, Barcelona, Spain). The following measurements were taken in millimeters
on the buccal and lingual aspects: IS-BIC: distance from the top of the implant shoulder to the first
point of bone-to-implant contact; IS-BC: distance from the top of the implant shoulder to the bone crest;
PM-BC: distance from the peri-implant mucosa to the bone crest; PM-JE: distance from the peri-implant
mucosa to the apical portion of the barrier epithelium; PM-BIC: distance from the peri-implant mucosa
to the first point of bone-to-implant contact; JE-BIC: distance from the apical portion of the barrier
epithelium to the first point of bone-to-implant contact; and, PM-IS: distance from peri-implant mucosa
to the implant shoulder (Figure 4).

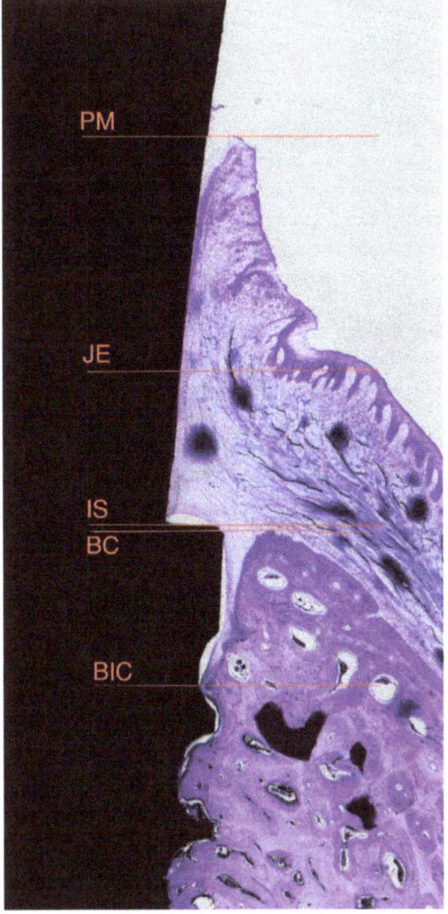

Figure 4. Diagrams representing landmarks used for histometric evaluation: PM, top of the
Peri-implant Mucosa; JE, apical portion of the Junctional Epithelium; IS, Implant Shoulder; BC, Bone
Crest; and, BIC, first point of Bone-to-Implant Contact.

2.5. Data Analysis

Mean values and standard deviations were calculated for each outcome variable. Differences between implant design groups and between implants of similar neck design (micro-rings as compared with open-thread) were analyzed using SPSS 20.0.0. Software (SPSS Inc., Chicago, IL, USA) applying the Wilcoxon-Mann-Whitney non-parametric test for paired observations. The significance level was set at $p < 0.05$.

3. Results

Eight weeks after implant placement, all implants were integrated in mature mineralized bone. No complications arose and no artifacts occurred during histological processing. All implants showed sufficient stability for loading with abutments at the time of placement.

Histomorphometric Evaluation

Table 1 shows hard tissue remodeling and Table 2 soft tissue adaptation data for each implant neck design, as well as for implants of similar neck design (micro-rings as compared with open-thread).

In pooled data for implants of similar neck design, the open-thread group showed the best results in terms of its capacity to stabilize hard tissue, presenting significant differences at the bone crest level on both lingual and buccal aspects, and at the first point of bone-to-implant contact on the lingual aspect. Differences between groups were also found in soft tissue measurement (PM-JE distance) on both lingual and buccal aspects, with the open-thread group obtaining the best results.

Pooled data for different implants in the micro-ring group showed significant bone loss on the lingual aspect (IS-BIC distance), and significant differences in PM-BIC and JE-BIC distances. In the double-spiral open-thread and single thread groups, significant differences in bone loss was found on the buccal aspect (IS-BC) and also in the soft tissue (PM-JE distance) level on the buccal aspect.

Table 1. Histomorphometric evaluation of hard tissue remodeling after eight weeks healing. Grouped data are presented for each neck design and for different types of implants with similar designs (micro-rings compared with open-thread). Results are expressed as mean ± standard deviation.

Type of Implants	IS-BC		IS-BIC	
	B	L	B	L
Grouped design Micro-rings	1.61(1.05a	0.89(1.10) a	1.76(0.77)	1.59(0.89) a
Open-Thread	0.55(1.04) a	−0.08(0.67) a	1.03(0.66)	0.63(0.53) a
Micro-rings BlueSky	1.68(0.32)	1.00(0.62)	1.68(0.32)	1.05(0.56) b
Micro-rings C1	1.53(1.55)	0.78(1.52)	1.84(1.11)	2.14(0.85) b
Open-thread IPX	1.45(0.34) c	0.35(0.41)	1.49(0.28)	0.73(0.23)
Open-thread Facility	1.52(0.55) c	0.48(0.44)	1.51(0.53)	0.61(0.63)

(a) $p < 0.05$ for grouped neck designs; (b) $p < 0.05$ for different implants with micro-rings; (c) $p < 0.05$ for different implants with open-thread design.

Table 2. Histomorphometric evaluation of soft tissue adaption after 8 weeks healing. Grouped data are presented for each neck design and for implants of similar design (micro-rings compared with open-thread). Results are expressed as mean ± standard deviation.

Type of Implants	PM-BC		PM-BIC		PM-JE		JE-BIC		PM-IS	
	B	L	B	L	B	L	B	L	B	L
Grouped Micro-rings	3.23(0.88)	2.33(0.35)	3.38(0.77)	3.04(0.97)	1.50(0.36) a	1.39(0.40) a	1.88(0.66)	1.65(0.86)	1.62(1.19)	1.44(0.92)
Open-Thread	2.57(0.93)	1.69(0.80)	3.03(0.80)	2.41(0.84)	1.08(0.42) a	0.92(0.46) a	1.94(0.73)	1.49(0.75)	2.00(0.61)	1.78(0.65)
Micro-rings Blue Sky	3.40(0.90)	2.29(0.13)	3.40(0.90)	2.34(0.11) b	1.49(0.20)	1.32(0.16)	1.90(0.89)	1.01(0.25) b	1.71(0.92)	1.29(0.66)
Micro-rings C1	3.06(0.92)	2.38(0.51)	3.37(0.72)	3.73(0.16) b	1.51(0.51)	1.45(0.56)	1.85(0.43)	2.28(0.78) b	1.52(1.53)	1.59(1.20)
Open-thread IPX	3.15(0.68)	1.79(0.79)	3.19(0.63)	2.38(0.68)	1.34(0.20) c	1.00(0.32)	1.85(0.16)	1.37(0.40)	1.69(0.52)	1.44(0.47)
Open-thread Facility	2.20(0.91)	1.68(1.09)	3.00(1.00)	2.55(1.14)	0.86(0.39) c	0.86(0.57)	2.13(0.99)	1.69(1.15)	2.30(0.57)	2.01(0.53)

(a) $p < 0.05$ for grouped neck design; (b) $p < 0.05$ for different implants with micro-rings; (c) $p < 0.05$ for different implants with open-thread design.

4. Discussion

This animal study set out to assess the influence of implant neck design on the preservation of bone crest levels and on soft tissue adaption around implants with abutment loading at the time of placement. Sub-crestal implant placement (around 2 mm below the buccal crest) has been observed to reduce crestal bone resorption when compared with crestal placement [30–33].

Delgado-Ruiz et al. have argued that the thickness, density, and orientation of connective tissue fibers around healing abutments of different geometries influence collagen fiber orientation. For this reason, an abutment with a profile wider than the implant platform favors oblique and perpendicular orientation of collagen fibers and greater connective tissue thickness [34]. In this context, dental implants with expanded platforms placed in the anterior zone of the maxilla and immediately restored with single crowns registered 1.01 mm of crestal bone loss after a 10-year follow-up [35].

As for neck design, the present findings agree with other dog model experiments, showing that different implant neck designs also affect the amount of bone resorption, which may be because the surface affects the distribution of occlusal loads as soon as implants are loaded [36].

Song et al., found that an implant design with open threads reaching the top of the neck underwent less bone loss than other implant designs in which the threads did not reach the top [37], as well as better bone responses than micro-ring designs. Various other clinical studies have also reported different crestal bone loss outcomes in favor of open-thread neck designs [38]. The results of the present study also found that with open-thread implants, the level of the bone around the implant neck region was significantly higher with the double-spiral open-thread design (Neodent® Facility) than the simple spiral open-thread design (Galimplant® IPX). But, these hard tissue remodeling outcomes differ from those observed by Chowdhary et al. [10] and Hudieb et al. [14] who found that the presence of micro-rings—intended to increase the surface area of the implant, this concept being understood globally and not in isolation to a specific brand of implants with a specific surface—promoted bone formation. But, the present results were similar to Jung et al. [17] and Hansson et al. [24] who found—as we would expect—that the increase in implant neck surface produced by prolonging the spiral thread on the body to the top of the neck decreases crestal bone resorption, while producing a smaller bone-to-implant contact distance.

Bone remodeling is also related to the implant connection. This plays an important role in reducing bone loss at the abutment/implant level associated with Morse taper implants and reduced-diameter platform switching abutments, and in reducing the incidence of peri-implantitis [39,40].

Soft tissue adaptation is one of the most important variables determining the long-term success of dental implants. For this reason, one of the objectives of the present study was to evaluate soft tissue variations between implant neck designs. In measurements that were taken on the buccal aspect, it was found that the presence of micro-rings on the neck significantly increased the height of epithelial tissue; this was also the case on the lingual aspect. No previous studies have compared open-thread neck implants with other implants under the same conditions. The open thread design showed less marginal bone loss, probably because it exerts less crestal bone compression.

5. Conclusions

Within the limitations of animal experimentation, it may be concluded that implants with micro-rings on the neck, in spite of offering greater bone-to-implant contact, generate light bone loss than open-thread implants. Moreover, the outcomes that were obtained IPX implants smooth neck design produced less bone loss in the cervical area, following by Facility implants when compared with the other open thread and microthreaded implant designs. Implant thread design can influence on bone remodeling in the cervical area, related to bundle bone preservation.

Authors Contributions

Conceptualization: R.J.-S., J.L.C.-G.; Data Curation and Resources: J.E.M.-S.d.V.; Formal Analysis: C.P.A.-M.; Funding Acquisition and Research: J.L.C.-G. and C.P.A.-M.: Methodology: R.J.-S. and J.L.C.-G.; Resources and Software: S.A.G.; Writing—original draft: R.J.-S. and J.L.C.-G.; Writing—review & editing: R.J.-S., M.F.D.; Visualization and Methodology: M.F.-D.; Supervision: J.L.C.-G.

Abbreviations Definition

mm	millimeters
IS-BIC	distance from the top of the implant shoulder to the first bone to implant contact
IS-BC	distance from the top of the implant shoulder to the bone crest
PM-BC	distance from the peri-implant mucosa to the bone crest
PM-JE	distance from the peri-implant mucosa to the apical portion of the barrier epithelium
PM-BIC	distance from the peri-implant mucosa to the first bone to implant contact
JE-BIC	distance from the apical portion of the barrier epithelium to the first bone to implant contact
PM-IS	distance from peri-implant mucosa to the Implant shoulder

Funding: This research received no external funding

Acknowledgments: The authors thank the University Veterinarian, Nuria Garcia Carrillo, for her assistance.

Conflicts of Interest: The authors declare no conflict of interest.

References

1. Oh, T.J.; Yoon, J.; Misch, C.E.; Wang, H.L. The causes of early implant bone loss: Myth or science? *J. Periodontol.* **2002**, *73*, 322–333. [CrossRef] [PubMed]

2. Den Hartog, L.; Raghoebar, G.M.; Slater, J.J.; Stellingsma, K.; Vissink, A.; Meijer, H.J. Single-tooth implants with different neck designs: A randomized clinical trial evaluating the aesthetic outcome. *Clin. Implant Dent. Relat. Res.* **2013**, *15*, 311–321. [CrossRef] [PubMed]

3. Pirc, M.; Dragan, I.F. The key points of maintenance therapy for dental implants: A literature review. *Compend. Contin. Educ. Dent.* **2017**, *38*, e5–e8. [PubMed]

4. Kronström, M.; Svenson, B.; Hellman, M.; Persson, G.R. Early implant failures in patients treated with Brånemark System titanium dental implants: A retrospective study. *Int. J. Oral Maxillofac. Implants* **2001**, *16*, 201–207. [PubMed]

5. Canullo, L.; Camacho-Alonso, F.; Tallarico, M.; Meloni, S.M.; Xhanari, E.; Penarrocha-Oltra, D. Mucosa thickness and peri-implant crestal bone stability: A clinical and histologic prospective cohort trial. *Int. J. Oral Maxillofac. Implants* **2017**, *32*, 675–681. [CrossRef] [PubMed]

6. Predecki, P.; Stephan, J.E.; Auslaender, B.A.; Mooney, V.L.; Kirkland, K. Kinetics of bone growth into cylindrical channels in aluminum oxide and titanium. *J. Biomed. Mater. Res.* **1972**, *6*, 375–400. [CrossRef] [PubMed]

7. Koodaryan, R.; Hafezeqoran, A. Evaluation of implant neck surfaces for marginal bone loss: A systematic review and meta-analysis. *Biomed. Res. Int.* **2016**, *2016*, 4987526. [CrossRef] [PubMed]

8. Calvo-Guirado, J.L.; Satorres, M.; Negri, B.; Ramirez-Fernandez, P.; Maté-Sánchez de Val, J.E.; Delgado-Ruiz, R.; Gomez-Moreno, G.; Abboud, M.; Romanos, G.E. Biomechanical and histological evaluation of four different titanium implant surface modifications: An experimental study in the rabbit tibia. *Clin. Oral Investig.* **2014**, *18*, 1495–1505. [CrossRef] [PubMed]

9. Hansson, S. The implant neck: Smooth or provided with retention elements. A biomechanical approach. *Clin. Oral Implants Res.* **1999**, *10*, 394–405. [CrossRef] [PubMed]

10. Chowdhary, R.; Halldin, A.; Jimbo, R.; Wennerberg, A. Influence of microthreads alteration on osseointegration and primary stability of implants: An FEA and in vivo analysis in rabbits. *Clin. Implant Dent. Relat. Res.* **2016**, *17*, 562–569. [CrossRef] [PubMed]

11. Lee, D.W.; Choi, Y.S.; Park, K.H.; Kim, C.S.; Moon, I.S. Effect of microthread on the maintenance of marginal bone level: A 3-year prospective study. *Clin. Oral Implants Res.* **2007**, *18*, 465–470. [CrossRef] [PubMed]

12. Akca, K.; Cehreli, M.C. A photoelastic and strain-gauge analysis of interface force transmission of internal-cone implants. *Int. J. Periodontics Restor. Dent.* **2008**, *28*, 391–399.

13. Zanatta, L.C.; Dib, L.L.; Gehrke, S.A. Photoelastic stress analysis surrounding different implant designs under simulated static loading. *J. Craniofac. Surg.* **2014**, *25*, 1068–1071. [CrossRef] [PubMed]

14. Hudieb, M.I.; Wakabayashi, N.; Kasugai, S. Magnitude and direction of mechanical stress at the osseointegrated interface of the microthread implant. *J. Periodontol.* **2011**, *82*, 1061–1070. [CrossRef] [PubMed]

15. Calvo-Guirado, J.L.; López-López, P.J.; Pérez-Albacete Martínez, C.; Javed, F.; Granero-Marín, J.M.; Maté Sánchez de Val, J.E.; Ramírez Fernández, M.P. Peri-implant bone loss clinical and radiographic evaluation around rough neck and microthread implants: A 5-year study. *Clin. Oral Implants Res.* **2018**, *29*, 635–643. [CrossRef] [PubMed]

16. Calvo Guirado, J.L.; Lucero-Sánchez, A.F.; Boquete Castro, A.; Abboud, M.; Gehrke, S.; Fernández Dominguez, M.; Delgado Ruiz, R.A. Peri-implant behavior of sloped shoulder dental implants used for all-on-four protocols: An histomorphometric analysis in dogs. *Materials (Basel)* **2018**, *11*, 119. [CrossRef] [PubMed]

17. Jung, Y.C.; Han, C.H.; Lee, K.W. A 1-year radiographic evaluation of marginal bone around dental implants. *Int. J. Oral Maxillofac. Implants* **1996**, *11*, 811–818. [PubMed]

18. Albrektsson, T.; Berglundh, T.; Lindhe, J. Osseointegration: Historic background and current concepts. *Clin. Periodontol. Implant Dent.* **2003**, *12*, 809–820.

19. Petechia, L.; Usai, C.; Vassalli, M.; Gavazzo, P. Biophysical characterization of nanostructured TiO_2 as a good substrate for hBM-MSC adhesion, growth and differentiation. *Exp. Cell Res.* **2017**, *358*, 111–119.

20. Karlsson, U.; Gotfredsen, K.; Olsson, C. Single-tooth replacement by osseointegrated Astra Tech dental implants: A 2-year report. *Int. J. Prosthodont.* **1997**, *10*, 318–324. [PubMed]

21. Smeets, R.; Stadlinger, B.; Schwarz, F.; Beck-Broichsitter, B.; Jung, O.; Precht, C.; Kloss, F.; Gröbe, A.; Heiland, M.; Ebker, T. Impact of dental implant surface modifications on osseointegration. *BioMed Res. Int.* **2016**, *2016*, 6285620. [CrossRef] [PubMed]

22. CamargosGde, V.; Sotto-Maior, B.S.; Silva, W.J.; Lazari, P.C.; Del Bel Cury, A.A. Prosthetic abutment influences bone biomechanical behavior of immediately loaded implants. *Braz. Oral Res.* **2016**, *30*. [CrossRef]

23. Donath, K.; Breuner, G. A method for the study of undecalcified bones and teeth with attached soft tissues. The Säge-Schliff (sawing and grinding) technique. *J. Oral Pathol.* **1982**, *11*, 318–326. [CrossRef] [PubMed]

24. Hsu, Y.T.; Chan, H.L.; Rudek, I.; Bashutski, J.; Oh, W.S.; Wang, H.L.; Oh, T.J. Comparison of Clinical and Radiographic Outcomes of Platform-Switched Implants with a Rough Collar and Platform-Matched Implants with a Smooth Collar: A 1-Year Randomized Clinical Trial. *Int. J. Oral Maxillofac. Implants* **2016**, *31*, 382–390. [CrossRef] [PubMed]

25. Khorsand, A.; Rasouli-Ghahroudi, A.A.; Naddafpour, N.; Shayesteh, Y.S.; Khojasteh, A. Effect of Microthread Design on Marginal Bone Level Around Dental Implants Placed in Fresh Extraction Sockets. *Implant Dent.* **2016**, *25*, 90–96. [CrossRef] [PubMed]

26. Rasmusson, L.; Kahnberg, K.E.; Tan, A. Effects of implant design and surface on bone regeneration and implant stability: An experimental study in the dog mandible. *Clin. Implant Dent. Relat. Res.* **2001**, *3*, 2–8. [CrossRef] [PubMed]

27. Abrahamsson, I.; Berglundh, T. Tissue characteristics at microthreaded implants: An experimental study in dogs. *Clin. Implant Dent. Relat. Res.* **2006**, *8*, 107–113. [CrossRef] [PubMed]

28. Berglundh, T.; Abrahamsson, I.; Lindhe, J. Bone reactions to longstanding functional load at implants: An experimental study in dogs. *J. Clin. Periodontol.* **2005**, *32*, 925–932. [CrossRef] [PubMed]

29. Abuhussein, H.; Pagni, G.; Rebaudi, A.; Wang, H.L. The effect of thread pattern upon implant osseointegration. *Clin. Oral Implants Res.* **2010**, *21*, 129–136. [CrossRef] [PubMed]

30. Calvo-Guirado, J.L.; López-López, P.J.; Mate Sanchez, J.E.; GargalloAlbiol, J.; Velasco Ortega, E.; Delgado Ruiz, R. Crestal bone loss related to immediate implants in crestal and subcrestal position: A pilot study in dogs. *Clin. Oral Implants Res.* **2014**, *25*, 1286–1294. [CrossRef] [PubMed]

31. Calvo-Guirado, J.L.; Gomez Moreno, G.; Aguilar-Salvatierra, A.; Mate Sanchez de Val, J.E.; Abboud, M.; Nemcovsky, C.E. Bone remodeling at implants with different configurations and placed immediately at different depth into extraction sockets. Experimental study in dogs. *Clin. Oral Implants Res.* **2015**, *26*, 507–515. [CrossRef] [PubMed]

32. Calvo-Guirado, J.L.; López-López, P.J.; Maté Sánchez de Val, J.E.; Mareque-Bueno, J.; Delgado-Ruiz, R.A.; Romanos, G.E. Influence of neck design on peri-implant tissue healing around immediate implants: A pilot study in Foxhound dogs. *Clin. Oral Implants Res.* **2015**, *26*, 851–857. [CrossRef] [PubMed]

33. Calvo-Guirado, J.L.; Pérez-Albacete, C.; Aguilar-Salvatierra, A.; de Val Maté-Sánchez, J.E.; Delgado-Ruiz, R.A.; Abboud, M.; Velasco, E.; Gómez-Moreno, G.; Romanos, G.E. Narrow-versus mini-implants at crestal and subcrestalbonelevels. Experimental study in beagle dogs at three months. *Clin. Oral Investig.* **2015**, *19*, 1363–1369. [CrossRef] [PubMed]

34. Delgado-Ruiz, R.A.; Calvo-Guirado, J.L.; Abboud, M.; Ramirez-Fernandez, M.P.; Maté-Sánchez de Val, J.E.; Negri, B.; Gomez-Moreno, G.; Markovic, A. Connective tissue characteristics around healing abutments of different geometries: New methodological technique under circularly polarized light. *Clin. Implant Dent. Relat. Res.* **2015**, *17*, 667–680. [CrossRef] [PubMed]

35. Calvo-Guirado, J.L.; Gómez-Moreno, G.; Delgado-Ruiz, R.A.; Maté Sánchez de Val, J.E.; Negri, B.; Ramírez Fernández, M.P. Clinical and radiographic evaluation of osseotite-expanded platform implants related to crestal bone loss: A 10-year study. *Clin. Oral Implants Res.* **2014**, *25*, 352–358. [CrossRef] [PubMed]

36. Rupp, F.; Scheideler, L.; Olshanska, N.; de Wild, M.; Wieland, M.; Geis-Gerstorfer, J. Enhancing surface free energy and hydrophilicity through chemical modification of microstructured titanium implant surfaces. *J. Biomed. Mater. Res.* **2006**, *76*, 323–334. [CrossRef] [PubMed]

37. Song, D.W.; Lee, D.W.; Kim, C.K.; Park, K.H.; Moon, I.S. Comparative analysis of peri-implant marginal bone loss based on microthread location: A 1-year prospective study after loading. *J. Periodontol.* **2009**, *80*, 1937–1944. [CrossRef] [PubMed]

38. Al-Thobity, A.M.; Kutkut, A.; Almas, K. Microthreaded implants and crestal bone loss: A systematic review. *J. Oral Implantol.* **2017**, *43*, 157–166. [CrossRef] [PubMed]

39. Macedo, J.P.; Pereira, J.; Vahey, B.R.; Henriques, B.; Benfatti, C.A.; Magini, R.S.; López-López, J.; Souza, J.C. Morse taper dental implants and platform switching: The new paradigm in oral implantology. *Eur. J. Dent.* **2016**, *10*, 148–154. [PubMed]

40. D'Ercole, S.; Tripodi, D.; Marzo, G.; Bernardi, S.; Continenza, M.A.; Piattelli, A.; Iaculli, F.; Mummolo, S. Microleakage of bacteria in different implant-abutment assemblies: An in vitro study. *J. Appl. Biomater. Funct. Mater.* **2015**, *13*, e174–e180. [CrossRef] [PubMed]

 *materials*

Article

The Serum Protein Profile and Acute Phase Proteins in the Postoperative Period in Sheep after Induced Articular Cartilage Defect

Csilla Tothova [1], Xenia Mihajlovicova [1], Jaroslav Novotny [2], Oskar Nagy [1], Maria Giretova [3], Lenka Kresakova [4], Marek Tomco [4], Zdenek Zert [5], Zuzana Vilhanova [5], Maros Varga [6], Lubomir Medvecky [3] and Eva Petrovova [4],*

[1] Clinic of Ruminants, University of Veterinary Medicine and Pharmacy in Kosice, Komenskeho 73, 041 81 Kosice, Slovakia; Csilla.Tothova@uvlf.sk (C.T.); xenia.mihajlovicova@gmail.com (X.M.); Oskar.Nagy@uvlf.sk (O.N.)
[2] Clinic of Swine, University of Veterinary Medicine and Pharmacy in Kosice, Komenskeho 73, 041 81 Kosice, Slovakia; Jaroslav.Novotny@uvlf.sk
[3] Institute of Materials Research SAS in Kosice, Watsonova 47, 040 01 Kosice, Slovakia; mgiretova@saske.sk (M.G.); lmedvecky@saske.sk (L.M.)
[4] Institute of Anatomy, University of Veterinary Medicine and Pharmacy in Kosice, Komenskeho 73, 041 81 Kosice, Slovakia; lenka.kresakova@uvlf.sk (L.K.); mtomco.75@gmail.com (M.T.)
[5] Clinic of Horses, University of Veterinary Medicine and Pharmacy in Kosice, Komenskeho 73, 041 81 Kosice, Slovakia; Zdenek.Zert@uvlf.sk (Z.Z.); z.vilhanova@gmail.com (Z.V.)
[6] Sport-Arthro Centre, Privat Hospital Kosice-Saca, Lucna 57, 040 15 Kosice-Saca, Slovakia; maros.varga@nemocnicasaca.sk
* Correspondence: eva.petrovova@uvlf.sk; Tel.: +421-917-637-799

Received: 8 December 2018; Accepted: 27 December 2018; Published: 3 January 2019

Abstract: Although several new implants have been developed using animal studies for the treatment of osteochondral and cartilage defects, there is a lack of information on the possible metabolic and biochemical reactions of the body to the implantation of biomaterials and cartilage reconstruction. Therefore, this study was aimed at evaluating the serum protein pattern and the alterations in the concentrations of selected acute phase proteins in five clinically healthy female sheep before and after the reconstruction of experimentally induced articular cartilage defects using polyhydroxybutyrate/chitosan based biopolymer material. The concentrations of total serum proteins (TSP), protein fractions, and selected acute phase proteins—serum amyloid A (SAA), haptoglobin (Hp), and C-reactive protein (CRP)—were measured before and on days seven, 14, and 30 after the surgical intervention. The TSP concentrations showed no marked differences during the evaluated period. Albumin values decreased on day seven and day 14 after surgery. In the concentrations of α_1-, α_2-, β-, and γ_2-globulins, a gradual significant increase was observed during the postoperative period ($p < 0.05$). The γ_1-globulins decreased slightly seven days after surgery. The concentrations of SAA, Hp, and CRP increased significantly after the surgical intervention with a subsequent decrease on day 30. Presented results suggest marked alterations in the serum protein pattern after surgical intervention.

Keywords: articular cartilage defect; bioplolymers; C-reactive protein; haptoglobin; in vivo testing; serum amyloid A; serum protein fractions; sheep

---

1. Introduction

The synthesis of serum proteins is strongly controlled to maintain their physiological balanced concentrations. Any pathological processes in the body may result in alterations in the serum protein concentrations [1]. In general, animals react to infection, inflammatory processes, trauma, or any disturbances in their homeostasis with a series of physiological, metabolic, and biochemical reactions known as the acute phase response [2]. The most important metabolic changes during the inflammatory responses include the highly increased or decreased production of some serum proteins, especially the acute phase proteins [3]. Due to their low physiological concentrations and high response in affected animals, they may serve as useful biomarkers for the evaluation of an animal's health, clinical monitoring of different diseases, treatment responses, and prognostic purposes. In ruminants, serum amyloid A (SAA) and haptoglobin (Hp) are the diagnostically most important acute phase proteins [4]. C-reactive protein (CRP) has been described as a constitutive protein in these animal species, with only a minor increase during disease processes [5]. Nevertheless, the usefulness of CRP in the laboratory diagnosis of mastitis was evaluated by Schrodl et al. [6], and they found approximately 10-fold higher values in cows with mastitis compared with healthy ones.

In vivo animal studies are essential as a gap between in vitro experiments and human clinical studies for introducing biomaterials treatment into clinical orthopedic practice. Animal models are widely used in the research of innovative biomaterials for regeneration of articular cartilage defects. Mainly, large animal models with thicker articular cartilage permit the study of partial thickness and full thickness chondral repair, as well as osteochondral repair [7]. Anatomical location, size of the defect (critical or non-critical sized), as well as the mechanobiology, species, strain, age, and health conditions provide the highest scientifically relevant output related to the study aim, hypotheses, and direct translation to animal benefit [8]. The sheep is a commonly utilized animal model, as they are readily available, easy to handle, and are relatively inexpensive. Some unique features and a comparative description of the surgical anatomy and approaches to the stifle joint in sheep was published; it addressed the presence of the extensor digitorum longus muscle on the craniolateral aspect of the stifle joint, the absence of a cranial menisco-femoral ligament in the caudal joint space, and an attachment of the patellar tendon to the cranial pole of the patella when compared to man. Therefore, this joint may be considered by researchers who increasingly use sheep for studies on the replacement of cruciate ligaments, collateral ligaments and menisci, treatment of chondral and osteochondral defects, and osteoarthrosis [9]. The location of the cartilage defects in the ovine model has involved the medial femoral condyle and femoral trochlea as well, with a 7 mm reported critical size defect [10–12]. However, the selection of a suitable preclinical model for performance evaluation remains a challenge, as no gold standard exists to define the best animal model [8,13].

On the other side, there is a lack of information on the possible effects of cartilage reconstruction using biopolymers on the concentrations of biochemical indicators in animals. Alterations in the biomarker profile, including some serum proteins, may potentially indicate risk for the progression of the disease process and could be useful for early diagnosis of postoperative complications and infections in order to earlier detect uncontrolled inflammatory reactions and prevent prolonged convalescence [14]. Therefore, the aim of this study was to evaluate the alterations in the serum protein pattern and the concentrations of selected acute phase proteins in sheep after the reconstruction of experimentally induced articular cartilage defects using polyhydroxybutyrate/chitosan based biopolymer material, and to monitor their changes during the first 30 postoperative days.

2. Materials and Methods

2.1. Preparation and Characterization of Biopolymer Composite Implants

The polyhydroxybutyrate/chitosan blend (PHB/CHIT) was prepared according to Medvecky et al. [15]. Briefly, polyhydroxybutyrate (GoodFellow, dissolved in propylene carbonate) and chitosan (SigmaAldrich, Saint Louis, MO, USA, middle, dissolved in 1% acetic acid) were mutually mixed at a ratio equal to 1:1. The same volumes of differently concentrated biopolymer solutions were used for the precipitation. Mixing was carried out using a magnetic stirrer at 400 rpm. After 10 min of mixing, the acetone was slowly added to suspensions for the complete precipitation of biopolymers.

Final blends were filtered, washed with distilled water, and molded into cylinder form (10 mm in diameter and 10 mm in height) and lyophilized (Ilshin) for 6 h. The microstructure of scaffolds was observed by scanning electron microscopy (FE SEM JEOL 7000, JEOL, Akishima, Tokyo). Implants were sterilized in an autoclave at 121 °C.

A macroporous microstructure of the spongy-like biopolymer composite implant with the high fractions of irregularly shaped macropores with sizes up to 100 μm and micropores (<20 μm) was observed in the chitosan/polyhydroxybutyrate blend (Figure 1). The biopolymers created fiber- and plate-like interconnected networks with open structures, which could be appropriate for both the migration of cells into the inner structure of the scaffold and the diffusion of body fluids (or metabolites) into or out of the implant.

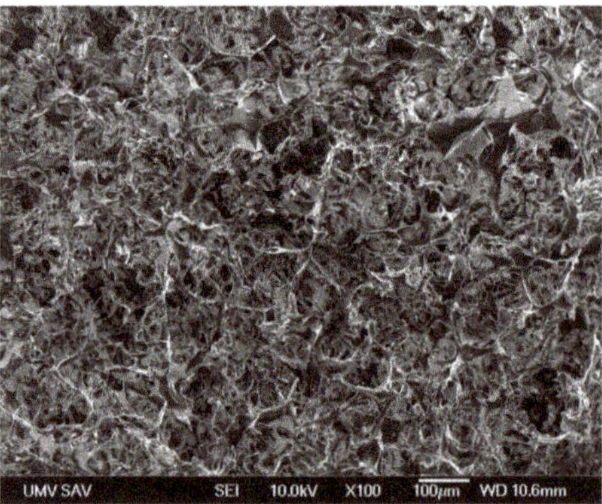

Figure 1. Scanning electron micrograph image shows the microstructure and presence of microporosity in the composite implant.

2.2. Animals and Sample Collection

Approval of the experimental protocol was obtained from the State Veterinary and Food Administration of the Slovak Republic No. 3508/17-221. The study was carried out on five clinically healthy female sheep of the crossbreed of Merino and Valachian sheep (group E) from a farm PD Agro Michalovce (Michalovce, Slovakia) that was approved by the State Veterinary and Food Administration of the Slovak Republic. The animals were at the age of 1.5–2 years and in good nutritional condition with an average body weight of 50.7 ± 1.9 kg at arrival. They were housed on the Clinic of Ruminants of the University of Veterinary Medicine and Pharmacy in Kosice, Slovak Republic, in free-stalls with free access to water, hay, and concentrates during the time under study.

The animals were included into the study 30 days before the scheduled day of surgical intervention, allowing acclimatization to the changed environment. Before the inclusion into the study, the animals underwent standard preoperative clinical examinations. Clinical examinations included the assessment of the overall health status of the animals (food intake, behavior), inspection and recording of body temperature, respiratory and pulse rates, and a detailed evaluation of the organ systems [16]. After surgical intervention, the health status of the animals was evaluated daily until the end of the study, and was oriented to the observation of general health state after the surgical procedure, as well as local signs of inflammation in the surgical wound (heat, swelling, pain, discharge).

Blood samples for the determination of the concentrations of total serum proteins, selected acute phase proteins, and separation of serum protein fractions were obtained before surgery and on days 7, 14, and 30 after surgical intervention. To evaluate the changes in the concentrations of acute phase proteins in animals with induced articular cartilage defects, but without implantation of biopolymer, five clinically healthy sheep of the same age and breed were included into the study as a control group (group C). Blood samples were collected from the jugular vein into serum gel separator tubes without additives and anticoagulants (Meus, Piove di Sacco, Italy). Serum was separated after letting blood samples coagulate at room temperature and was then centrifuged at $3000 \times g$ for 30 min. The harvested serum was dispensed into plastic tubes and stored at $-20\ ^\circ$C until it was analyzed.

2.3. Surgical Procedure

Food and water in the used animals were withheld for 12 h before the surgery. Anesthesia consisted of a mixture of buthorphanol (0.1 mg/kg, Butomidor 10 mg/mL, Richter Pharma, Wels, Austria), and medetomidin 0.02 mg/kg (Cepetor 1 mg/mL, CP-Pharma Handelsgesellschaft, GmbH, Burgdorf, Germany) administered intramuscularly, and ketamin 8 mg/kg (Ketamidor 100 mg/mL, Richter Pharma, Wels, Austria) administered intravenously. After anesthesia, a defect in the articular cartilage of the left stifle joint was induced. An incision was made from the left lateral side, from the medial patellar ligament distal to the tibial tuberosity. The stifle joint was visualized above the medial femoral condyle load. The subcutaneous tissue and superficial fascia were incised. After flexion and partial luxation of the stifle joint using the Osteochondral autograft transfer system (Arthrex, Naples, FL, USA), a defect was made on the articular cartilage on the exact place in the distal epiphysis of the femur (*trochlea femoris sinister* [17]) at a diameter of 10 mm and a depth of 10 mm (Figure 2). The site of the created defect was then filled with a biopolymer implant (Figure 3), which was in vitro tested for cytotoxicity [18]. The same procedure was used to create a defect in the animals from the control group, but it was not filled with biopolymer material. All sheep received postoperative systemic broad spectrum antibiotic oxytetracyclinum dihydricum 20 mg/kg (Alamycin LA a.u.v., Norbrook, Newry, UK, once every second day) and non-steroidal anti-inflammatory drug flunixin meglumine 2.2 mg/kg (Flunixin a.u.v., Norbrook, Newry, UK, once a day), administered intramuscularly for 7 days.

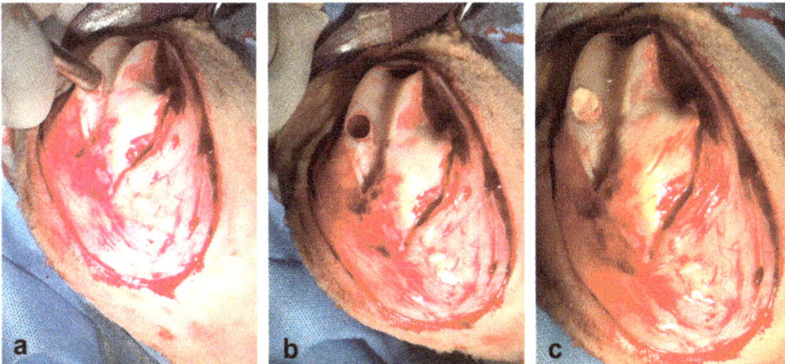

Figure 2. Surgical procedure: (**a**) inducing of articular cartilage defect with OATS equipment (Osteochondral autograft transfer system, Arthrex, Naples, FL, USA); (**b**) preparing articular cartilage defect in femoral trochlea before implantation; (**c**) inserting of scaffold into prepared articular cartilage defect.

Figure 3. Macrostructure of scaffold before implantation. Scale bar: 2 mm.

2.4. Laboratory Analyses

The total serum protein (TP, g/L) concentrations were assessed according to the Biuret method on an automated biochemical analyzer Alizé (Lisabio, Poully en Auxois, France) using commercial diagnostic kits (TP 245, Randox, Crumlin, UK). The serum protein fractions were separated by zone electrophoresis on an agarose gel using an automated electrophoresis system Hydrasys (Sebia Corporate, Lisses, Evry Cedex, France) with commercial diagnostic kits Hydragel 7 Proteine (PN 4100, Sebia Corporate, Lisses, Evry Cedex, France) according to the procedure described by the manufacturer. The densitometry scanning system Epson Perfection V700 (Epson America Inc., Long Beach, CA, USA) was used to scan the electrophoretic gels based on the method of light transmission and conversion into an optical density curve. The gel images were visualized using the computer software Phoresis version 5.50 (Sebia Corporate, Lisses, Evry Cedex, France). The following protein fractions were identified: albumin, α_1- and α_2-globulins, β-globulins, and γ_1- and γ_2-globulins. Each protein fraction was expressed as relative concentrations (%) according to the obtained optical density. Consequently, their absolute concentrations (g/L) were quantified from the total serum protein concentrations. The ratios of albumin to globulins (A/G) were calculated as well.

The serum concentrations of selected inflammatory markers—serum amyloid A (SAA; mg/mL), Hp (mg/mL), and CRP (µg/mL)—were measured to evaluate the postoperative inflammatory state. SAA was analyzed by sandwich enzyme linked immunosorbent assay (ELISA) using commercial multispecies kits (TP-802, Tridelta Developmet, Kildare, Ireland). Sheep CRP was measured by solid-phase ELISA assay using commercially available tests (CRP-12, Life Diagnostics, Inc., West Chester, PA, USA). Haptoglobin was assessed using commercial colorimetric kits (TP-801, Tridelta Development, Kildare, Ireland) in microplates based on Hp-haemoglobin binding and the preservation of the peroxidase activity of the bound haemoglobin at a low pH. The absorbances were read on the automatic microplate reader Opsys MR (The Dynex Technologies, Chantilly, VA, USA). The results were calculated using the computer software Revelation QuickLink version 4.25 (Dynex Technologies, Chantilly, VA, USA).

2.5. Statistical Analyses

Descriptive statistical procedures were used to calculate arithmetic means (x) and standard deviations (SD) for each evaluated variable and sample collection time. The distribution of data was evaluated by the Kolmogorov-Smirnov test for normality. Not all of the obtained data passed the normality test. Therefore, repeated-measures one-way ANOVA was used to examine the changes during the perioperative period for normally distributed data with equal variance, and the Friedman test was used for non-normally distributed data. The significance of differences in values between the sample collections was evaluated by Tukey-Kramer and Dunn's Multiple Comparisons tests. For the analysis of the acute phase protein concentrations, two-way repeated measures ANOVA was used. All statistical analyses were carried out using the program GraphPad Prism V5.02 (GraphPad Software Inc., San Diego, CA, USA).

3. Results

Various grades of lameness were observed in all sheep for 90 days after the surgical intervention. The surgical wound developed no signs of inflammation and was without discharge in all the evaluated sheep. The animals showed improvement with no serious complications and inflammatory processes in other organ systems.

The data obtained during the perioperative period are presented in Tables 1–3. The electrophoretic separation of serum proteins using agarose gel is presented in Figure 4. Figure 5 shows representative examples of electrophoretograms before the surgical intervention and in the postoperative period.

Table 1. Changes in the relative concentrations of serum protein fractions (%) and albumin/globulin ratio (A/G) in sheep before surgical intervention and in the postoperative period (mean $\pm$ SD).

Variables	Sample Collection				p Value
	Before Surgical Intervention	7 Days After Surgical Intervention	14 Days After Surgical Intervention	30 Days After Surgical Intervention	
Albumin	55.2 ± 8.9	45.3 ± 8.1	41.0 ± 8.9	45.8 ± 4.9	n.s.
α_1-globulins	6.0 ± 0.7 [a]	8.3 ± 2.2	8.3 ± 1.3 [a]	7.0 ± 0.8	<0.05
α_2-globulins	13.5 ± 1.2 [a]	16.2 ± 1.2 [a]	15.6 ± 1.5	13.5 ± 1.3	<0.001
β-globulins	4.2 ± 1.1 [A]	9.7 ± 4.9	13.4 ± 7.0 [A]	8.8 ± 2.6	<0.01
γ_1-globulins	18.1 ± 6.6	16.7 ± 5.5	17.0 ± 4.1	19.0 ± 3.1	n.s.
γ_2-globulins	2.9 ± 0.7 [a]	3.8 ± 1.4	4.7 ± 0.8	5.8 ± 1.6 [a]	<0.05
A/G	1.30 ± 0.45	0.86 ± 0.31	0.73 ± 0.30	0.86 ± 0.16	n.s.

The superscripts in the same rows mean statistically significant differences between the sample collections (a—$p < 0.05$, A—$p < 0.01$); p value—significance of the statistical tests; n.s.—not significant.

Table 2. Changes in the absolute concentrations of total serum protein (TP), and protein fractions (g/L) in sheep before surgical intervention and in the postoperative period (mean $\pm$ SD).

Variables	Sample Collection				p-Value
	Before Surgical Intervention	7 Days After Surgical Intervention	14 Days After Surgical Intervention	30 Days After Surgical Intervention	
TP	65.7 ± 7.1	64.4 ± 4.2	68.2 ± 1.6	64.7 ± 2.1	n.s.
Albumin	36.0 ± 4.7	31.0 ± 4.5	27.9 ± 6.0	29.6 ± 3.0	n.s.
α_1-globulins	4.0 ± 0.8 [a]	5.4 ± 1.5	5.7 ± 0.9 [a]	4.5 ± 0.5	<0.05
α_2-globulins	8.9 ± 1.4	10.4 ± 0.6	10.6 ± 0.9	8.7 ± 0.6	<0.05
β-globulins	2.8 ± 1.0 [a]	6.3 ± 3.1	9.1 ± 4.8 [a]	5.7 ± 1.8	<0.05
γ_1-globulins	12.2 ± 5.5	10.8 ± 4.2	11.6 ± 3.0	12.3 ± 2.2	n.s.
γ_2-globulins	1.9 ± 0.7 [a]	2.5 ± 1.1	3.2 ± 0.5	3.7 ± 1.1 [a]	< 0.05

The superscripts in the same rows mean statistically significant differences between the sample collections (a—$p < 0.05$); p value—significance of the statistical tests; n.s.—not significant.

Table 3. Changes in the concentrations of SAA, Hp and CRP in the experimental (E) and control (C) group of sheep before surgical intervention and in the postoperative period (mean ± SD).

Variables		Sample Collection				*p* Value
		Before Surgical Intervention	7 Days After Surgical Intervention	14 Days After Surgical Intervention	30 Days After Surgical Intervention	
SAA (µg/mL)	E	0.73 ± 1.14 [A]	154.40 ± 83.67 [A‡]	47.27 ± 55.21	8.84 ± 7.49 [†]	<0.01
	C	3.99 ± 6.28	5.62 ± 6.67 [a‡]	1.27 ± 1.18	0.28 ± 0.31 [a†]	<0.05
Hp (mg/mL)	E	0.188 ± 0.019 [‡]	6.306 ± 7.417	10.330 ± 9.413 [†]	4.250 ± 5.295	n.s.
	C	0.061 ± 0.071 [‡]	0.190 ± 0.284	0.831 ± 1.140 [†]	0.112 ± 0.094	n.s.
CRP (µg/mL)	E	147.8 ± 43.52 [a]	189.0 ± 49.51	259.6 ± 75.71 [a†]	187.2 ± 51.12	<0.05
	C	202.8 ± 128.3	222.8 ± 127.3	172.5 ± 60.1 [†]	145.4 ± 44.7	n.s.

The superscripts in the same rows mean statistically significant differences between the sample collections (a—$p < 0.05$, A—$p < 0.01$); the superscripts in the same columns of each variable mean statistically significant differences between the experimental and control group (†—$p < 0.05$, ‡—$p < 0.01$); p value—significance of the statistical tests; n.s.—not significant.

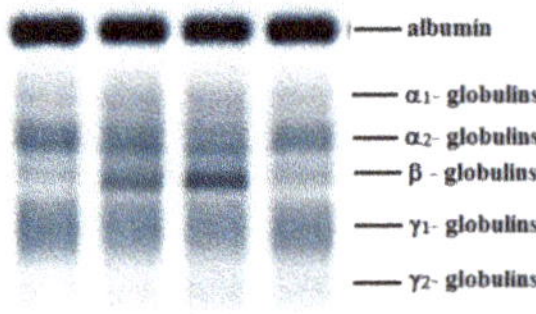

Figure 4. Example of the electrophoretic separation of serum proteins using agarose gel electrophoresis in a sheep during the perioperative period: (**1**) before surgical intervention; (**2**) day 7; (**3**) day 14; (**4**) day 30 after surgery.

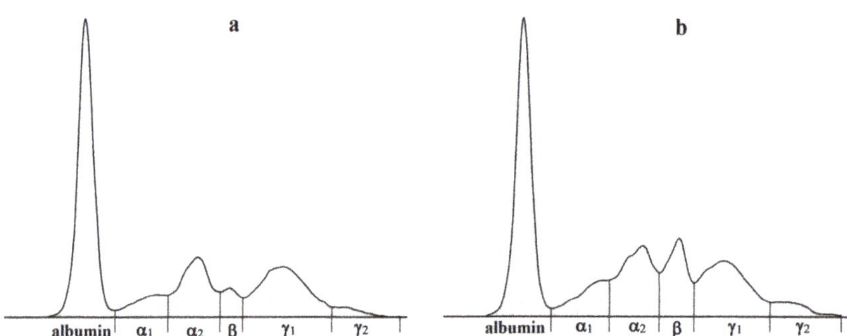

Figure 5. *Cont.*

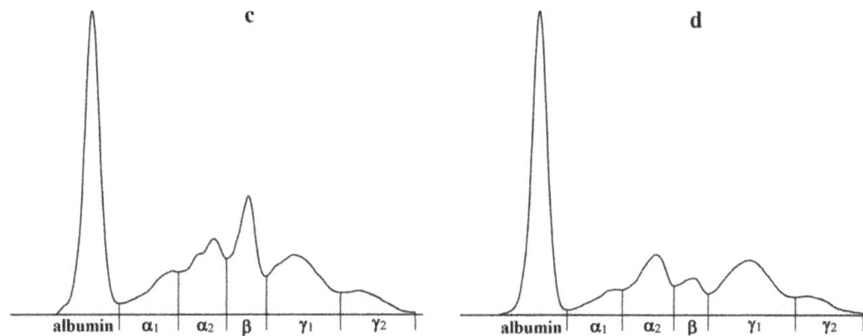

Figure 5. Representative electrophoretograms in a sheep showing the protein fractionation into six fractions: albumin, α_1-, α_2-, β-, γ_1-, and γ_2-globulins before surgical intervention (**a**) and in the postoperative period-7 (**b**), 14 (**c**) and 30 days (**d**) after surgery.

The relative concentrations of albumin (Table 1) showed in sheep a marked decrease seven days after the surgical intervention with a further slight decrease on day 14 and a subsequent increase on day 30 after surgery. However, the changes of albumin values during the evaluated period were not significant. Significant alterations during the perioperative period were observed in the relative concentrations of α_1-globulins ($p < 0.05$). Their values increased seven days after surgery, stayed relatively stable, and then slightly decreased on day 30. In the relative concentrations of α_2-globulins, a significantly higher mean value was found seven days after surgery ($p < 0.05$) with a tendency to further decrease to values comparable with those obtained prior to surgery. The relative concentrations of β-globulins increased gradually and significantly until day 14 after surgery ($p < 0.01$) with a subsequent slight decrease on day 30 of the postoperative period. The relative concentrations of γ_1-globulins were relatively stable during the evaluated period. On the other hand, the relative values of γ_2-globulins increased significantly until day 30 after surgery ($p < 0.05$). An opposite trend was observed in the A/G ratios, which were the highest prior to surgery. The values obtained after the surgical procedure were markedly lower.

No marked differences were found in the concentrations of total serum proteins (Table 2) before surgery and in the early postoperative period. A slight non-significant increase of values was recorded on day 14 after surgery with a subsequent decrease on the end of the evaluated period. The highest mean absolute concentration of albumin was obtained prior to surgery. The values obtained in the postoperative period were non-significantly lower. Significant alterations during the evaluated period were found in the absolute concentrations of α_1-globulins ($p < 0.05$), showing a gradual significant increase until day 14 after surgery ($p < 0.05$) and a subsequent decrease on day 30 of the postoperative period. Similar trends of gradually increasing values until day 14 after the surgical intervention were observed in the absolute concentrations of α_2- and β-globulins ($p < 0.05$). On the other hand, the absolute concentrations of γ_1-globulins decreased slightly seven days after surgery with a subsequent gradual increase to preoperative values. In the absolute concentrations of γ_2-globulins, a gradual significant increase of mean values was found until the end of the evaluated period ($p < 0.05$).

The evaluation of the concentrations of SAA in the period before and after surgery showed significant changes in both the experimental and the control group of sheep ($p < 0.01$, $p < 0.05$, respectively, Table 3). In the experimental group, the values obtained before surgery were low and increased significantly in response to the surgical procedure, being the highest on day seven after surgery ($p < 0.01$). From day 14 after surgery, a gradual decrease of values was found up to day 30 of the postoperative period. In the control group of sheep, only a slight increase of values was observed seven days after surgical intervention with a repeated gradual decrease until the end of the evaluated period. In the concentrations of Hp in the experimental group of sheep, a more gradual non-significant increase was found until day 14 after surgery with a subsequent decrease on day 30 after the surgical procedure. The mean value obtained on day 30 after surgery was more than 20-fold higher than the preoperative concentrations. In the control group of sheep, the values increased less markedly and non-significantly up to day 14 after the surgical intervention and then decreased. Regarding the concentrations of CRP in the experimental group, a slight gradual significant increase of values was observed, these values being the highest on day 14 after surgery ($p < 0.05$) and then starting to decrease. These changes during the evaluated period were significant ($p < 0.05$). The concentrations of CRP in the control group showed a slight non-significant increase seven days after surgery with a subsequent gradual decrease until the end of the evaluated period.

4. Discussion

Animal studies are fundamental to showing the efficacy and safety of new cartilage defects repair before its clinical use in humans. Not only small experimental animals, but also large animal models (goats, sheep, pigs) have been successfully used to demonstrate new implants for the treatment of osteochondral and cartilage defects [11,19,20]. However, these studies were predominantly oriented to the histological evaluation of the cartilage repair and maturation of the newly-formed cartilage. The metabolic and biochemical reactions of the body to the implantation of biomaterials and cartilage reconstruction have not been studied in depth. The results of the presented study showed in sheep some marked biochemical responses to the articular cartilage defect repair characterized by alterations in the serum protein electrophoretic pattern and in the production of the evaluated acute phase proteins.

The analyses of the total serum protein concentrations showed in sheep no marked changes after the induction and reconstruction of the cartilage defect when compared with the values obtained before the intervention. A slight increase of total protein concentrations was recorded on day 14 after the surgical intervention, probably related to the response of the organism to the damage caused in the articular cartilage. The patterns of proteins in the serum or synovial fluid, their fractionation, and the identification of several proteins have been studied in some cartilage lesions, joint disorders, and destruction in both humans and animals [21–23]. However, the profile of serum proteins after cartilage reconstruction and implantation of biomaterials was previously not described. In the presented study, major alterations were observed in the electrophoretic pattern of serum proteins and the distribution of most of the serum protein fractions. In the serum albumin concentrations, a trend of lower values was observed after the creation of cartilage defect and implantation of biopolymer. Its values started to increase on day 30 of the postoperative period. Albumin is a major negative acute phase protein. Therefore, its decreased concentrations after the surgical intervention might be attributed to this function of albumin [24]. Whitaker et al. [25] stated that albumin is one of metabolic parameters that may be used to monitor inflammatory diseases. According to Hübner et al. [26], the decrease in the concentrations of albumin in early post operation may reflect the magnitude of surgical trauma and predict adverse clinical outcomes.

An opposite trend was observed in the concentrations of α-globulins (α_1-, as well as α_2-globulins). Their values increased until day 14 after surgical intervention, then started to decrease, and finally returned to preoperative concentrations 30 days after surgery. The alpha-globulins constitute a large fraction on the electrophoretogram, which is composed of many diagnostically important proteins. Furthermore, some of them act as acute phase proteins that play different roles in the host defense responses, regulation of inflammatory processes, and restoration of homeostasis [27]. While alpha$_1$-antitrypsin, α_1-acid glycoprotein, α_1-antichymotrypsin, α_1-fetoprotein, serum amyloid A, and α_1-lipoprotein belong to the α_1-globulin fraction, haptoglobin, α_2-microglobulin, α_2-macroglobulin, ceruloplasmin, α_2-antiplasmin, and α_2-lipoprotein migrate in the α_2-globulin fraction [28]. Thus, the increases of the alpha fractions after surgery may reflect the increases in the concentrations of some acute phase proteins resulting from the activation of the host inflammatory responses due to the damage caused in the articular cartilage and its reconstruction. The concentrations of SAA and Hp in the evaluated sheep were low before surgery and increased after the induction of cartilage damage and implantation of polyhydroxybutyrate/chitosan based biopolymer. On day 30 after implantation, the values showed a tendency to decrease to those obtained prior to surgery, suggesting an uncomplicated postoperative period and the absence of inflammatory reactions. Differences were obtained in the rate of increase and subsequent decrease during the postoperative period among the measured inflammatory markers as well as between the experimental and control group. While the concentrations of SAA in the experimental sheep increased more than 200-fold already seven days after the surgical intervention, the values of Hp showed a more gradual increase until day 14 after surgery (approximately a 50-fold increase). On the other hand, the concentrations of SAA in the control group increased approximately 1.5-fold seven days after surgery, and the values of Hp increased about 13-fold up to day 14 after the surgery. SAA is a sensitive biomarker characterized by remarkable increase and rapid decrease once the inflammatory stimuli are eliminated. On the other hand, Hp is characterized by a more prolonged response and is thus preferable in the field to evaluate disease processes [27]. Aulin et al. [29] found in rabbits a pronounced increase of SAA concentrations after the surgical induction of full thickness osteochondral defects in the femorotibial joints. Within one-month postoperatively, the SAA values returned to physiological concentrations, supporting the absence of inflammation. In contrast, in cases with ongoing infection, persistently high or even increased SAA values can be observed, suggesting complications after surgery or indicating a lack of treatment response [30]. Thus, the trend of decreasing SAA and Hp values observed in our study may indicate an uncomplicated postoperative period. However, seeing that there are scarce literature data about the usefulness of these biomarkers in the evaluation of cartilage regeneration, further studies would be helpful.

A similar increasing trend was found in the concentrations of β-globulins until day 14 after surgery, showing a markedly higher and narrow peak on the electrophoretogram. After this period, their values started to decline. The C-reactive protein was identified in the β-globulin fraction. Its concentrations in sheep were relatively high prior to surgery and increased only slightly after the cartilage damage. Higher CRP values were found in patients with synovial inflammation and cartilage damage due to femoroacetabular impingement [31,32]. However, little is known about the behavior of CRP after cartilage reconstruction using biomaterial implants and during the postoperative period. On the other hand, some other proteins may be identified in the β-globulin fraction, including complement, transferrin, ferritin, as well as β_2-microglobulin or β-lipoproteins. These proteins are involved in the regulation of inflammatory processes and stress responses and thus may be attributed to the marked elevation of the β-globulin fraction after the induction of cartilage damage [33]. Thirty days after the surgical intervention, their concentrations started to decrease, reflecting no serious complications during the treatment.

The γ-globulin fraction is predominantly composed of immunoglobulins (Ig) of various classes (IgG, IgA, IgM, IgD, and IgE). Sheep are a typical example of an animal species in which γ-globulins may be separated into two subfractions (γ_1 and γ_2) [34]. In the presented study, the concentrations of γ_1-globulins decreased slightly seven days after surgery and subsequently started to increase to preoperative values. On the other hand, the values of γ_2-globulins showed a gradual increase until the end of the evaluated period. Most of the immunoglobulin classes migrate in the γ_1-globulin fraction, but some IgG subclasses (the so called slow immunoglobulins) may be detected in the γ_2-globulin fraction and generally have antibody activity [35]. Thus, the increase of γ_2-globulin concentrations in sheep after surgery might reflect the response of the organism to the damage caused by surgical intervention and the implantation of biomaterial. Further studies would be helpful in yielding satisfactory results. The changes observed in the concentrations of albumin and globulin fractions also resulted in alterations in the A/G ratio. The highest A/G ratio was observed in sheep before surgery, and the values obtained after the intervention were lower due to the overproduction of globulins caused by the cartilage damage and its repair.

The dynamics of changes in the concentrations of the evaluated biomarkers might be influenced by the degree of cartilage damage, as well as surgical stress, general anesthesia, the course of regeneration, and antimicrobial and anti-inflammatory treatments during the postoperative period [36,37]. Stowasser-Raschbauer et al. [38] concluded that the rise of the concentrations of SAA in horses after surgical procedures under general anesthesia is partly due to the anesthetic procedure. In horses, an increase in SAA concentrations was found after general anesthesia but without surgery, while the increase was higher when surgery was performed. On the other hand, Pepys et al. [39] reported in horses no increase in the concentrations of SAA after general anesthesia. However, there are no published reports about the effect of general anesthesia on the concentrations of acute phase protein in farm animals. Similar contradictory data were published regarding the effect of non-steroidal anti-inflammatory drugs on the acute phase response during the postoperative period and recovery. Ting et al. [40] concluded that repeated administration of ketoprofen in beef cattle after surgery did not have influence on the changes in acute phase protein concentrations when compared with non-treated animals. On the other hand, Plessers et al. [41] suggested that non-steroidal anti-inflammatory drug (NSAID) ketoprofen may attenuate the acute phase response in calves, as they found lower concentrations of inflammatory markers, including SAA, after the administration of ketoprofen in calves challenged by lipopolysaccharide. Similar to studies in humans, differences in the reactivity and variability of individual animals may account for differences in the responses to the induction of cartilage defects and their reconstruction, consequently resulting in a wider range of not uniform and inconsistent data [7]. Therefore, larger experimental animal groups are needed to obtain further results in cartilage repair studies.

5. Conclusions

In conclusion, the presented results suggest marked alterations in the serum protein pattern in sheep with surgically induced articular cartilage defects, which are characterized by changes in most of the serum proteins fractions and acute phase protein concentrations. After cartilage repair and approximately four weeks after the implantation, the concentrations decreased to preoperative values, suggesting the absence of further inflammatory reactions during the postoperative period. These results indicate that some biomarkers from the serum protein profile may be used for the evaluation of the postoperative period, the progression of the disease process, and the uncontrolled postoperative inflammatory responses. However, further comprehensive studies completed with the examination of the postoperative period by further blood parameters and made with more dense animal groups are needed to yield satisfactory results.

Author Contributions: Conceptualization, M.V., L.M. and E.P.; Formal analysis, C.T.; Funding acquisition, O.N., L.M. and E.P.; Investigation, C.T.; Methodology, M.T., Z.Z. and M.V.; Project administration, E.P.; Resources, X.M., J.N., M.G., L.K., M.T., Z.Z., Z.V. and L.M.; Supervision, E.P.; Writing—original draft, C.T.; Writing—review & editing, O.N.

Funding: This research was realized within the project of Ministry of Education VEGA No. 1/0486/17, 1/0398/18 and Slovak Research and Development Agency under the contract no. APVV-17-0110.

Conflicts of Interest: The authors declare no conflict of interest.

References

1. Pieper, R.; Gatlin, C.L.; Makusky, A.J.; Russo, P.S.; Schatz, C.R.; Miller, S.S.; Su, Q.; McGrath, A.M.; Estock, M.A.; Parmar, P.P.; et al. The human serum proteome: Display of nearly 3700 chromatographically separated protein spots on two-dimensional electrophoresis gels and identification of 325 distinct proteins. *Proteomics* **2003**, *3*, 1345–1364. [CrossRef] [PubMed]

2. Ceciliani, F.; Ceron, J.J.; Eckersall, P.D.; Sauerwein, H. Acute phase proteins in ruminants. *J. Proteom.* **2012**, *75*, 4207–4231. [CrossRef]

3. Murata, H.; Shimada, N.; Yoshioka, M. Current research on acute phase proteins in veterinary diagnosis: An overview. *Vet. J.* **2004**, *168*, 28–40. [CrossRef]

4. Eckersall, P.D.; Bell, R. Acute phase proteins: Biomarkers of infection and inflammation in veterinary medicine. *Vet. J.* **2010**, *185*, 23–27. [CrossRef] [PubMed]

5. Eckersall, P.D. Proteins, proteomics, and the dysproteinemias. In *Clinical Biochemistry of Domestic Animals*, 6th ed.; Kaneko, J.J., Harvey, J.W., Bruss, M.L., Eds.; Elsevier Academic Press: San Diego, CA, USA, 2008; pp. 117–155.

6. Schrodl, W.; Kruger, M.; Hien, T.T.; Fuldner, M.; Kunze, R. C-reactive protein as a new parameter of mastitis. *Tierarztl. Prax.* **1995**, *23*, 337–341.

7. Chu, C.R.; Szczodry, M.; Bruno, S. Animal models for cartilage regeneration and repair. *Tissue Eng.* **2010**, *16*, 105–115. [CrossRef]

8. Bongio, M.; van den Beucken, J.J.J.P.; Leeuwenburgh, S.C.G.; Jansen, J.A. Preclinical evaluation of injectable bone substitute materials. *J. Tissue Eng. Regen. Med.* **2015**, *9*, 191–209. [CrossRef]

9. Martini, L.; Fini, M.; Giavaresi, G.; Giardino, R. Sheep Model in Orthopedic Research: A Literature Review. *Comp. Med.* **2001**, *51*, 292–299.

10. Ahern, B.J.; Parvizi, J.; Boston, R.; Schaer, T.P. Preclinical animal models in single site cartilage defect testing: A systematic review. *Osteoarthr. Cartil.* **2009**, *17*, 705–713. [CrossRef]

11. Hao, T.; Wen, N.; Cao, J.-K.; Wang, H.-B.; Lü, S.-H.; Liu, T.; Lin, Q.-X.; Duan, C.-M.; Wang, C.-Y. The support of matrix accumulation and the promotion of sheep articular cartilage defects repair in vivo by chitosan hydrogels. *Osteoarthr. Cartil.* **2010**, *18*, 257–265. [CrossRef]

12. Sidler, M.; Fouché, N.; Meth, I.; von Hahn, F.; von Rechenberg, B.; Kronen, P.W. Transcutaneous Treatment with Vetdrop® Sustains the Adjacent Cartilage in a Microfracturing Joint Defect Model in Sheep. *Open Orthop. J.* **2013**, *7*, 57–66. [CrossRef] [PubMed]

13. Li, Y.; Chen, S.-K.; Li, L.; Qin, L.; Wang, X.-L.; Lai, Y.-X. Bone defect animal models for testing efficacy of bone substitute biomaterials. *J. Orthop. Transl.* **2015**, *3*, 95–104. [CrossRef] [PubMed]

14. Demura, S.; Takahashi, K.; Kawahara, N.; Watanabe, Y.; Tomita, K. Serum interleukin-6 response after spinal surgery: Estimation of surgical magnitude. *J. Orthop. Sci.* **2006**, *11*, 241–247. [CrossRef]

15. Medvecky, L.; Giretova, M.; Stulajterova, R. Properties and in vitro characterization of polyhydroxybutyrate–chitosan scaffolds prepared by modified precipitation method. *J. Mater. Sci. Mater. Med.* **2014**, *25*, 777–789. [CrossRef] [PubMed]

16. Jackson, P.G.G.; Cockcroft, P.D. *Clinical Examination of Farm Animals*; Blackwell Science Ltd., Blackwell Publishing Company: Oxford, UK, 2002; p. 320.

17. Danko, J.; Simon, F.; Artimova, J. *Nomina Anatomica Veterinaria*; University of Veterinary Medicine and Pharmacy: Kosice, Slovakia, 2011; p. 267. ISBN 978-80-8077-259-8.

18. Giretova, M.; Medvecky, L.; Stulajterova, R.; Sopcak, T.; Briancin, J.; Tatarkova, M. Effect of enzymatic degradation of chitosan in polyhydroxybutyrate/chitosan/calcium phosphate composites on in vitro osteoblast response. *J. Mater. Sci. Mater. Med.* **2016**, *27*, 181. [CrossRef]

19. Hembry, R.M.; Dyce, J.; Driesang, I.; Hunziker, E.B.; Fosang, A.J.; Tyler, J.A.; Murphy, G. Immunolocalization of matrix metalloproteinases in partial-thickness defects in pig articular cartilage. *J. Bone Jt. Surg. Am.* **2000**, *83*, 826–838. [CrossRef]

20. Niederauer, G.G.; Slivka, M.A.; Leatherbury, N.C.; Korvick, D.L.; Harroff, H.H.; Ehler, W.C.; Dunn, C.J.; Kieswetter, K. Evaluation of multiphase implants for repair of focal osteochondral defects in goats. *Biomaterials* **2000**, *21*, 2561–2574. [CrossRef]

21. Fujimura, K.; Segami, N.; Yoshitake, Y.; Tsuruoka, N.; Kaneyama, K.; Sato, J.; Kobayashi, S. Electrophoretic separation of the synovial fluid proteins in patients with temporomandibular joint disorders. *Oral Surg. Oral Med. Oral Pathol. Oral Radiol. Endodontol.* **2006**, *101*, 463–468. [CrossRef]

22. Basile, R.C.; Ferraz, G.C.; Carvalho, M.P.; Albernaz, R.M.; Araújo, R.A.; Fagliari, J.J.; Queiroz-Neto, A. Physiological concentrations of acute-phase proteins and immunoglobulins in equine synovial fluid. *J. Equine Vet. Sci.* **2013**, *33*, 201–204. [CrossRef]

23. Barrachina, L.; Remacha, A.R.; Soler, L.; Garcia, N.; Romero, A.; Vázquez, F.J.; Vitoria, A.; Álava, M.Á.; Lamprave, F.; Rodellar, C. Acute phase protein haptoglobin as inflammatory marker in serum and synovial fluid in an equine model of arthritis. *Vet. Immunol. Immunopathol.* **2016**, *182*, 74–78. [CrossRef]

24. Gruys, E.; Obwolo, M.J.; Toussaint, M.J.M. Diagnostic significance of the major acute phase proteins in veterinary clinical chemistry: A review. *Vet. Bull.* **1994**, *64*, 1009–1018.

25. Whitaker, D.A.; Goodger, W.J.; Garcia, M.; Perera, B.M.A.O.; Wittwer, F. Use of metabolic profiles in dairy cattle in tropical and subtropical countries on smallholder dairy farms. *Prev. Vet. Med.* **1999**, *38*, 119–131. [CrossRef]

26. Hübner, M.; Mantziari, S.; Demartines, N.; Pralong, F.; Coti-Bertrend, P.; Schäfer, M. Postoperative albumin drop is a marker for surgical stress and a predictor for clinical outcome: A pilot study. *Gastroenterol. Res. Pract.* **2016**, *2016*, 8743187. [CrossRef]

27. Petersen, H.H.; Nielsen, J.P.; Heegaard, P.M.H. Application of acute phase protein measurements in veterinary clinical chemistry. *Vet. Res.* **2004**, *35*, 163–187. [CrossRef]

28. Bossuyt, X. Advances in serum protein electrophoresis. *Adv. Clin. Chem.* **2006**, *42*, 43–80. [PubMed]

29. Aulin, C.; Jensen-Waern, M.; Ekman, S.; Hägglund, M.; Engstrand, T.; Hilborn, J.; Hedenqvist, P. Cartilage repair of experimentally 11 induced osteochondral defects in New Zealand White rabbits. *Lab. Anim.* **2013**, *47*, 58–65. [CrossRef] [PubMed]

30. Pollock, P.J.; Prendergast, M.; Schumacher, J.; Bellenger, C.R. Effects of surgery on the acute phase response in clinically normal and diseased horses. *Vet. Rec.* **2005**, *156*, 538–542. [CrossRef]

31. Pearle, A.D.; Scanzello, C.R.; George, S.; Mandl, L.A.; DiCarlo, E.F.; Peterson, M.; Sculco, T.P.; Crow, M.K. Elevated high-sensitivity C-reactive protein levels are associated with local inflammatory findings in patients with osteoarthritis 1. *Osteoarthr. Cartil.* **2007**, *15*, 516–523. [CrossRef]

32. Bedi, A.; Lynch, E.B.; Sibilsky Enselman, E.R.; Davis, M.E.; DeWolf, P.D.; Makki, T.A.; Kelly, B.T.; Larson, C.M.; Henning, P.T.; Mendias, C.L. Elevation in circulating biomarkers of cartilage damage and inflammation in athletes with femoroacetabular impingement. *Am. J. Sports Med.* **2013**, *41*, 2585–2590. [CrossRef]

33. Bernabucci, U.; Lacetera, N.; Danieli, P.P.; Bani, P. Influence of different periods of exposure to hot environment on rumen function and diet digestibility in sheep. *Int. J. Biometeorol.* **2009**, *53*, 387–395. [CrossRef]

34. Nagy, O.; Tóthová, C.; Nagyová, V.; Kováč, G. Comparison of serum protein electrophoretic pattern in cows and small ruminants. *Acta Vet. Brno* **2015**, *84*, 187–195. [CrossRef]

35. Kaneko, J.J. (Ed.) Serum proteins and the dysproteinemias. In *Clinical Biochemistry of Domestic Animals*, 5th ed.; Academic Press: London, UK, 1997; pp. 117–138.

36. Jacobsen, S.; Nielsen, J.V.; Kjelgaard-Hansen, M.; Toelboell, T.; Fjeldborg, J.; Halling-Thomsen, M.; Martinussen, T.; Thoefner, M.B. Acute phase response to surgery of varying intensity in horses: A preliminary study. *Vet. Surg.* **2009**, *38*, 762–769. [CrossRef] [PubMed]

37. Busk, P.; Jacobsen, S.; Martinussen, T. Administration of perioperative penicillin reduces postoperative serum amyloid A response in horses being castrated standing. *Vet. Surg.* **2010**, *39*, 638–643. [CrossRef] [PubMed]

38. Stowasser-Raschbauer, B.; Kabeš, R.; Moens, Y. Serum amyloid A concentrations in horses following anesthesis with and without surgery. *Tierärztl. Monat. Vet. Med. Austria* **2013**, *100*, 127–132.

39. Pepys, M.B.; Baltz, M.L.; Tennent, G.A.; Kent, J.; Ousey, J.; Rossdale, P.D. Serum amyloid A protein (SAA) in horses: Objective measurement of the acute phase response. *Equine Vet. J.* **1989**, *21*, 106–109. [CrossRef]

40. Ting, S.T.L.; Earley, B.; Hughes, J.M.L.; Crowe, M.A. Effect of ketoprofen, lidocaine local anesthesia, and combined xylazine and lidocaine caudal epidural anesthesia during castration of beef cattle on stress responses, imunity, growth, and behavior. *J. Anim. Sci.* **2003**, *81*, 1281–1293. [CrossRef]

41. Plessers, E.; Wyns, H.; Watteyn, A.; Pardon, B.; De Baere, S.; Sys, S.U.; De Backer, P.; Croubels, S. Immunomodulatory properties of gamithromycin and ketoprofen in lipopolysaccharide-challenged calves with emphasis on the acute-phase response. *Vet. Immunol. Immunopathol.* **2016**, *171*, 28–37. [CrossRef]

 *materials*

Article

Impact of Gentamicin-Loaded Bone Graft on Defect Healing in a Sheep Model

Elisabeth Beuttel [1], Nicole Bormann [1], Anne-Marie Pobloth [1], Georg N. Duda [1] and Britt Wildemann [1,2,*]

[1] Julius Wolff Institute and Berlin-Brandenburg Center for Regenerative Therapies, Charité-Universitätmedizin Berlin, Corporate Member of Freie Universität Berlin, Humboldt-Universität zu Berlin, and Berlin Institute of Health, 13353 Berlin, Germany; elisabeth.beuttel@posteo.de (E.B.); nicole.bormann@charite.de (N.B.); anne-marie.Pobloth@charite.de (A.-M.P.); georg.duda@charite.de (G.N.D.)
[2] Experimental Trauma Surgery, University Hospital Jena, 07740 Jena, Germany
* Correspondence: britt.wildemann@med.uni-jena.de or britt.wildemann@charite.de

Received: 7 March 2019; Accepted: 2 April 2019; Published: 4 April 2019

Abstract: Infections of bone are severe complications, and an optimization of grafting material with antimicrobial drugs might be useful for prevention and treatment. This study aimed to investigate the influence of gentamicin-loaded bone graft on the healing of bone defects in a sheep model. Metaphyseal and diaphyseal drill hole defects (diameter: 6 mm, depth: 15 mm) were filled with graft or gentamicin-loaded graft (50 mg/g graft) or were left untreated. Analysis of regeneration after three and nine weeks, micro-computed tomography (μCT), and histology revealed a significant increase in bone formation in the drill hole defects, which began at the edges of the holes and grew over time into the defect center. The amount of graft decreased over time due to active resorption by osteoclasts, while osteoblasts formed new bone. No difference between the groups was seen after three weeks. After nine weeks, significantly less mineralized tissue was formed in the gentamicin-loaded graft group. Signs of inflammatory reactions were seen in all three groups. Even though the applied gentamicin concentration was based on the concentration of gentamicin mixed with cement, the healing process was impaired. When using local gentamicin, a dose-dependent, compromising effect on bone healing should be considered.

Keywords: bone infection; local drug delivery; bone graft; demineralized bone matrix; gentamicin; regeneration

---

1. Introduction

Infection of bone is a devastating complication for the patient. If the infection cannot be resolved by systemic antibiotic treatment, surgery with debridement of all infected tissue is necessary [1]. This radical debridement often results in larger bone defects that require support to allow the regeneration of the bone. The treatment of this large bone defects has consequently two challenges: 1. the regeneration of the defect, 2. the continuous prophylaxis or lasting treatment of infections. To support the regeneration of bone, autologous bone is the gold standard, but has limitations such as harvest-side morbidity, and sometimes limited availability should be carefully considered [2,3]. Allografts, xenografts, alloplastic materials, regeneration stimulating factors such as BMP-2, and a combination thereof are clinically used and new materials are under development [4,5]. Sometimes, such materials are combined with antibiotics to prevent the onset of infection. The first antibiotic-mixed bone cement for local drug delivery was described 1970 by Buchholz and Engelbrecht [6]. Since then such bone cements have been commonly used in arthroplasty and the effectiveness of antibiotic impregnated nondegradable cements to reduce the infection rate in primary hip arthroplasty was shown [7]. In recent decades, various approaches for the local delivery of antibiotics for prevention and treatment of bone infections were developed [8,9]. Clinically used is the direct application of powdered antibiotics during orthopedic surgery [10], but also antibiotics mixed with allografts [11], applied in combination with degradable bioceramics [12], or from implant coatings [13]. Gentamicin, however, is not only toxic for bacteria, it can have—above certain concentrations—negative effects on bone-forming cells [14]. Thus, for successful healing, it is important that the antibiotic dosage eradicates bacteria but does not impair bone regeneration. To this end, loading of grafting material with antibiotics for direct delivery to the site of need without harming the endogenous healing process is desired. In a previous study, we developed and tested a perioperative loading method of a bone graft with antibiotics [15]. Incorporated gentamicin showed a first-order release with almost complete release within the first week. Eluates from this period were antimicrobial active against *S. aureus*, without cytotoxic effects on primary osteoblast-like cells. The present study aimed to investigate the effect of this local gentamicin application on the regeneration of bone defects. A large animal model was used, and drill holes in the epiphyseal and metaphyseal bones were filled with grafting material with and without gentamicin enrichment. After three and nine weeks, the healing was assessed using micro-computed tomography (µCT) and histology.

2. Materials and Methods

The bone graft demineralized bone matrix (DBMputty, DIZG, Berlin, Germany) was mixed with gentamicin (40 mg/mL, Merckle GmbH, Blaubeuren, Germany) using a syringe with an integrated mixing propeller (Medmix Systems, Risch-Rotkreuz, Switzerland) [15].

2.1. Drill Hole Defect Model

For in vivo analysis, drill hole defects were created in diaphyseal and metaphyseal bones of five sheep. Defects (6–8 per group and time point) were filled with demineralized bone matrix (DBM) or DBM loaded with gentamicin (50 mg/g DBM, same concentration as in the *in vitro* studies [15]) or left untreated. The drill holes in all animals were assigned to the three experimental groups. After three and nine weeks, defect regeneration was analyzed by µCT and histology. This study was part of a larger study and described in more detail in [16,17].

The animal experiment followed national and international regulations for the care and use of laboratory animals and was approved by the local authorities (LaGeSo, Berlin, Approval No. G0341/12). This study was part of a larger study and in total 16 adult female merino-mix sheep (mean weight 65 kg; ±7 kg; age 2.5 years) were included. Drill holes (diameter 6 mm and depth max. 15 mm) were drilled in the metaphyseal part of femur and humerus and the diaphyseal part of metacarpus and metatarsus and filled with 200 mg of DBM with or without 10 mg gentamicin. Empty defects served as controls. In the first surgery, defects were drilled and filled in the right extremities and after six weeks holes were drilled and filled in the left extremities. Animals were sacrificed after further three weeks. Therefore, healing times were three and nine weeks.

2.2. Surgery

The grafting material was rehydrated immediately prior to surgery. The surgery was conducted under general anesthesia introduced by intravenous sedation (10–16 mg/kg body weight sodium thiopental; Trapanal®, Nycomed Deutschland GmbH, Konstanz, Germany) followed by endotracheal inhalative anesthesia (2% isoflurane, 12–20 breath/min) accompanied by intravenous infusion, intraruminal tubation, analgetic treatment and gastritis and infection prophylaxis (amoxillin and metronidazole). To ensure a standardized placement of the drill holes in the individual sheep the localizations were marked preoperatively using radiographic surgical imaging. Following sterile preparation, the bone localizations were prepared surgically, and the drill holes were placed in a standardized fashion using a custom-made template and a surgical drill (Arthrex® 6.0 mm drill bit, Arthrex, Munich, Germany). For a more detailed description of the operation, pain management, and animal care see previous publication [16]. Each defect was filled completely with the bone graft using the preparations syringe or left empty followed by a surgical wound closure. Animals were sacrificed after 9 weeks by sedation with 25–30 mg/kg sodium thiopental intravenously and application of 100 mL potassium chloride.

2.3. Serum Concentration of Gentamicin

In total 30 blood samples were taken preoperative, at day 1 and 5 post surgery of all five sheep (two operations per sheep). The gentamicin concentration was determined by the Labor Berlin (Charité Vivantes, Berlin, Germany) using KIMS method (kinetic interaction of microparticles in a solution; Roche, Germany).

2.4. Micro-Computed Tomography (μCT)

After collection, bone samples were digitally imagined using X-ray to exclude ectopic bone formations and scanned in the μCT (VivaCT 40, SCANCO medical AG, Brüttisellen, Switzerland). The volume of interest (VOI) was defined as three-dimensional cylinder of the drill holes. The percentage of bone volume/total volume (BV/TV) was determined via a threshold based on Hounsfield Units (HU), which were calculated for each sample based on adjacent intact bone regions (mean 189.9 HU).

2.5. Histology and Histomorphometry

Undecalcified bone samples were embedded in methyl methacrylate (Technovit 9100 NEU, Kulzer GmbH, Wehrheim, Germany). Sections (6 μm) were stained with Safranin Orange/von Kossa (black: mineralized tissue, red: soft tissue) and Movat Pentachrome (yellow: collagen fibers; light blue: reticular fibers; red: elastin fibers). Active osteoclasts were stained using TRAP (Tartrate Resistant Acid Phosphatase) and Methyl Green as counter staining. For histomorphometry, bone tissue, DBM, connective tissue, and the percentage of osteoblasts and osteoclasts covering the DBM surface were quantified using ImageJ 1.49 (NIH, Bethesda, MD, USA).

2.6. Statistical Analysis

For statistical analysis of the differences between the three groups Kruskal-Wallis test followed by Dunn's was performed (GraphPad Prism 8, San Diego, CA, USA). Differences within one group between the two time points was assessed with the Mann-Whitney-U Test. Values $p \leq 0.05$ were considered as significant.

3. Results

Experimental design followed the 3R principles for animal models to screen for treatment effects while reducing the number of animals necessary. Therefor control groups (control and DBM) were used also in another study and the results are partially published [17].

3.1. μCT Analysis of the Defect Healing

The μCT reconstructions showed only limited new bone formation in the diaphyseal and metaphyseal defects after three weeks independent of the treatment (Figure 1). After nine weeks of healing, a clear difference between the diaphyseal and metaphyseal defects was visible, with an increased healing in all diaphyseal defects (Figures 1 and 2).

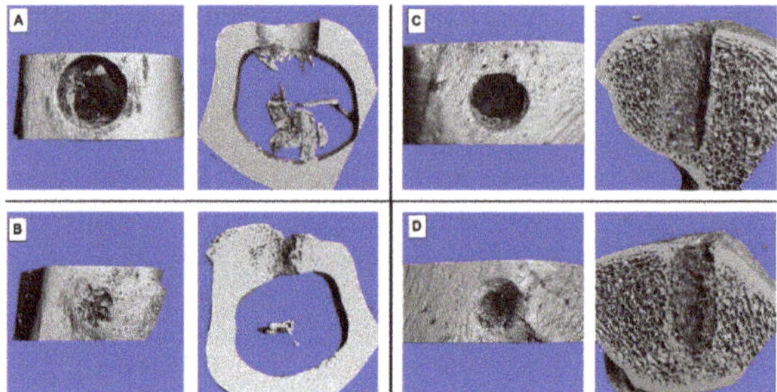

Figure 1. Exemplary drill hole defects after three (**A**,**C**) and nine weeks (**B**,**D**). (**A**,**B**) diaphyseal defects, filled with gentamicin-loaded DBM, (**C**,**D**) metaphyseal defects, filled with DBM.

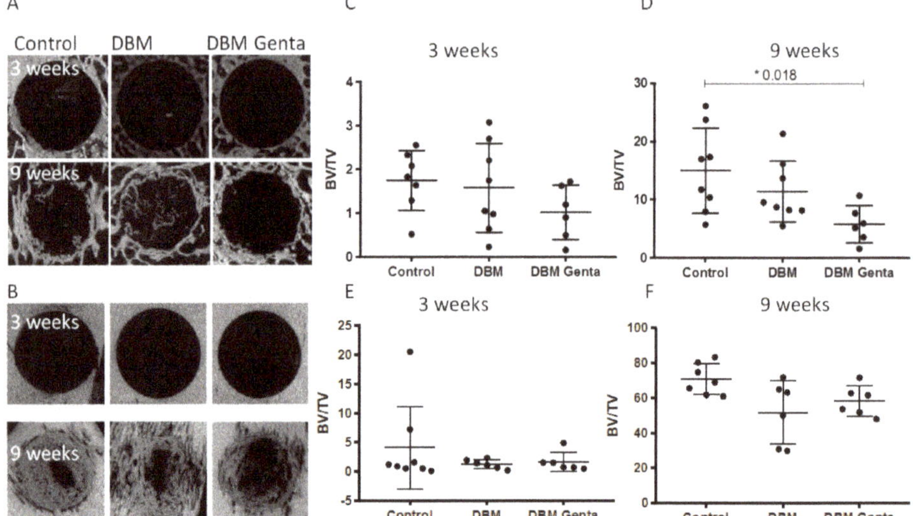

Figure 2. μCT images of representative metaphyseal and diaphyseal bone defects after three and nine weeks of healing (**A**,**B**). The metaphyseal defects (**A**) showed healing only at the edges of the defect, whereas the diaphyseal defects (**B**) showed a progressed healing after nine weeks. (**C**–**F**) Quantification of the BV/TV in the defects at three and nine weeks in metaphyseal and diaphyseal bone. The data from the control and DBM treated metaphyseal defects were published previously [17]. A significantly reduced BV/TV was seen after nine weeks in the metaphyseal drill holes treated with gentamicin. Kruskal-Wallis test with Dunn´s (n = 6–8 per group and time point).

Quantification of the newly formed bone by μCT showed significantly less bone formation in the gentamicin-treated metaphyseal defects after nine weeks compared to control defects (Figure 2D). The amount of newly formed bone increased significantly in all three groups from week three to week nine in the diaphyseal as well as in the metaphyseal drill hole defects ($p < 0.009$).

3.2. Histological Analysis of the Defect Healing

The histological evaluation revealed new bone formation starting at the edges of the drill holes and growing into the center over time (Figure 3A). The histomorphometric analysis showed no significant differences in the tissue composition after 3 weeks. After nine weeks, significantly less mineralized tissue was quantified in the gentamicin-treated defects compared to the control defect.

This was independent from the localization (Figure 3B,C). In the diaphyseal defects significantly less mineralized tissue was formed in the DBM group compared to the control defects (Figure 3C).

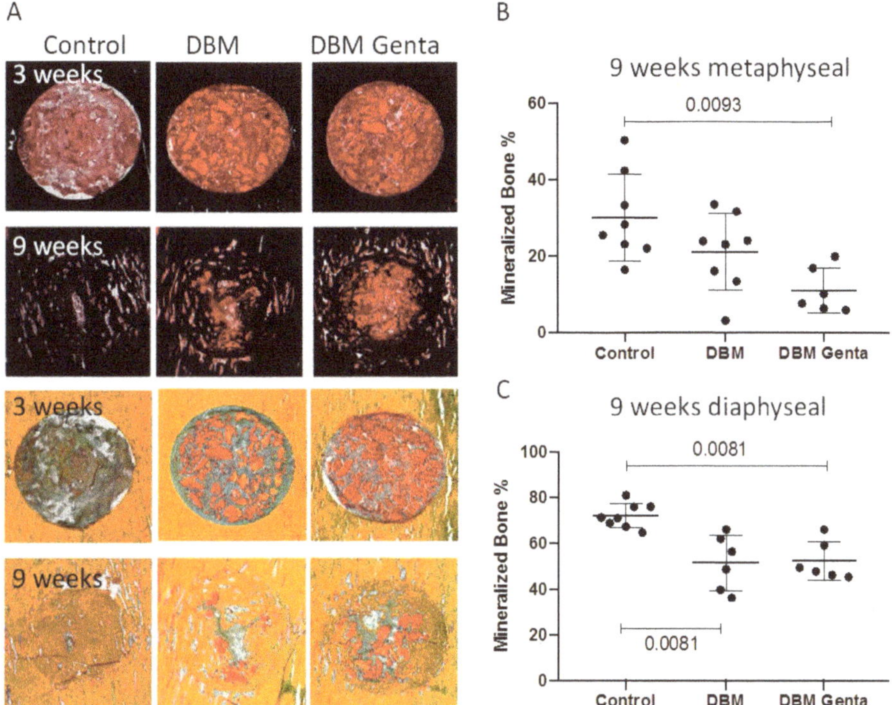

Figure 3. Exemplary histological pictures of the diaphyseal drill hole defects. Tissue was stained with Safranin O/van Kossa (**A**, top) to visualize the mineralized bone, or with Movat Pentachrome (**A**, bottom) to differentiate the tissues. Healing progressed from week three to week nine in all groups, with less healing in the gentamicin-treated defects. (**B,C**) Histomorphometric analysis: After nine weeks, significantly less mineralized tissue was formed in the gentamicin-loaded DBM group, both metaphyseal and diaphyseal. Diaphyseally, significantly less bone was formed in the DBM group compared to control. Kruskal-Wallis test with Dunn´s (n = 6–8 per group and time point).

Comparing the changes over time within one group revealed in all three groups a significant increase in the amount of mineralized bone from week three to week nine in both localizations ($p < 0.026$). The amount of DBM decreased over time with a significantly lower amounts after nine weeks in the DBM group ($p < 0.009$). The remodeling in the gentamicin-loaded DBM group was less pronounced with no significant reduction. The remodeling process can be explained by the activity of osteoclast and osteoblasts. A shift from resorption by osteoclasts (TRAP stain) at three weeks after implantation to formation of new bone by osteoblast after nine weeks was visible (Figure 4E,F). Counting the cells, however, revealed no significant differences between the groups (data not shown). Interestingly, the DBM was not only remodeled at the surface, but also osteoclasts were seen in the DBM (Figure 4E). Mineralized tissue connected to DBM was visible, but also a remineralization of the DBM (Figure 4A,B). Bone formation in most of the defects was due to direct ossification and only a few small cartilage islands were seen in the DBM treated defects (Figure 4C). At the cellular level, no increased inflammatory reaction was seen in the DBM treated groups. Independent of the group, defects after three weeks of healing contained leukocytes, macrophages and other cells of the immune system (Figure 4D).

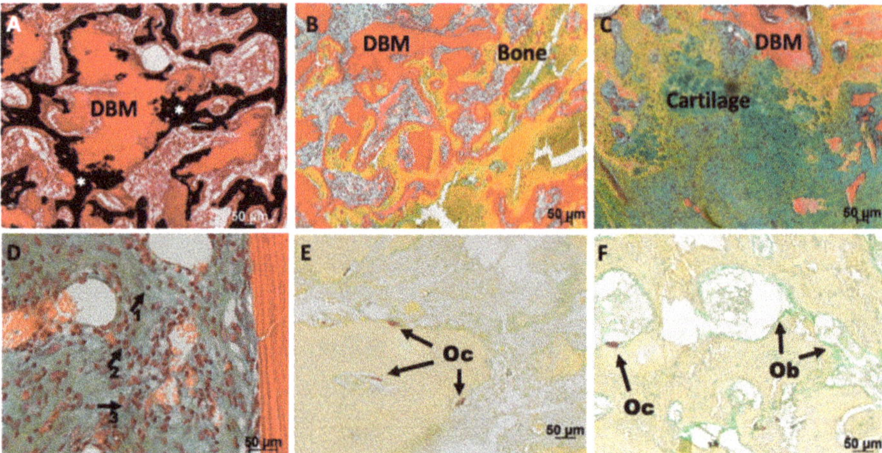

Figure 4. Safranin O/van Kossa (**A**), Movat Pentachtome stain (**B–D**) or TRAP stain (**E,F**) of drill hole defects after three (**A,B,D,E**) or nine (**C,F**) weeks healing. (**A**) Remineralization (black, *) of the implanted DBM (intense red stained tissue with empty lacunae) occurred after three weeks. (**B**) Active remodeling of the DBM was visible: DBM was attached to and integrated into newly formed bone trabeculae. (**C**) Endochondral bone formation was only seen in a few drill holes filled with DBM. (**D**) No difference regarding the abundance of inflammatory cells was detectable between the groups; lymphocytes (arrow 1), macrophages (arrow 2) or histiocytes (arrow 3) were detectable in all groups. (**E**) Osteoclasts (Oc) were detectable on the surface of the DBM, but also inside DBM after three weeks. F) Shift to the coverage with mostly osteoblasts (Ob) compared to osteoclasts (Oc) at the late time point. Size of scale bars is given in the figures.

3.3. Serumconcentration of Gentamicin

In 23 blood samples the concentration of gentamicin was below the lower detection level and in seven samples of two sheep in a range between 0.55–0.73 mg/L.

4. Discussion

Local antibiotic therapies are frequently used to prevent or treat bone infection. They can either be applied via loading of cements or grafting materials or by direct application into the defect area. Beside infection prophylaxis or treatment, the effect on bone regeneration is important. This study used gentamicin-loaded resorbable bone graft, which was characterized in detail in a previous study [15], to treat defects in sheep bone. First signs of bone formation were seen in all three groups after three weeks of healing without differences. In the following six weeks the gentamicin-treated defects showed impaired healing resulting in significantly less bone as shown by μCT and histological analysis. This healing impairment after 9 weeks was unexpected, as no impairment was seen after 3 weeks, the time period when all gentamicin should be released. In addition, no negative effect on osteoblast-like cells was seen in the *in vitro* characterization study using the same ratio of gentamicin and DBM: 50 mg/g DBM [15]. The previous *in vitro* experiments revealed a burst and almost complete release of gentamicin after three days, antimicrobial activity and good cytocompatibility of the eluates. The metabolic activity decreased only with the 1-day eluates, but without significant difference, whereas alkaline phosphate activity was not affected [15]. High gentamicin concentration was used for the *in vitro* experiments: 10 mg gentamicin mixed with 200 mg graft resulting in 2 mg gentamicin/mL medium. Previous studies showed that concentrations of 100 and 200 μg/mL reduced the DNA content but had no effect on viability of marrow derived mesenchymal stem cells [18]. Interestingly, the effect of gentamicin exposure was transient and affected cell proliferation and alkaline phosphatase activity only after 4 and 8 days of initial incubation but not after longer cultivation without gentamicin. The gentamicin/DBM concentration chosen for the presented experiments was within the range of clinically used antibiotic-loaded cement for prophylaxis and treatment in joint replacement [19]. We are not aware of a study reporting impaired healing due to the gentamicin released from the cements. The release from the DBM, however, is much faster and almost complete after 3 days *in vitro*, whereas gentamicin is released from cements over years [20]. This results in a much higher local concentration when using DBM as a carrier instead of cement. Using allografts, attenuated healing was also seen in spinal fusion where DBM was mixed with vancomycin [21], whereas DBM with gentamicin or DBM/bioactive glass with tobramycin had no negative effect on osteoinductivity or femur defect healing, respectively [22,23]. Coraca-Huber et al. published detailed studies investigating the effect of e.g., processing, storage, and freezing on antibiotic-loaded allogeneic bone and found promising *in vitro* results; however, no in vivo study has been published investigating biocompatibility until now [11,24–27].

No negative effect of calcium sulfate impregnated with gentamicin was seen in vivo [28]. The material was implanted into intact rabbit bone and material resorption as well as bone formation was monitored over 12 weeks. The amount of gentamicin loaded to the substitute was 1.7% compared to 5% used in the present study. This lower dosage might explain the contradictory results, although the different models—bone formation vs. defect healing—should be considered. The slowly resorbing material was still visible after 12 weeks and surrounded by giant cells. In a previous study the group showed a release of 82% of the gentamicin within three days [29], which is similar to the release of gentamicin from the bone graft used in the present study. Using the same material, no negative effect on blood coagulation was seen [30].

Gentamicin was not detectable in the serum in most of the samples and only at low levels in 7 of 30 samples (0.55–0.73 mg/L). Calculating the theoretical gentamicin concentration based on the amount loaded to the DBM, a concentration between 4.4 mg/L and 8.7 mg/L in the serum of the sheep might have been possible. Due to the release kinetics and the half-life of gentamicin these concentrations were not expected systemically, and the low or undetectable systemic concentration is an important safety aspect for the use as a local antibiotic application with the advantage of reducing possible systemic side effects to a minimum.

In the present study an ovine drill hole model was used that was developed for the screening of bone tissue engineering strategies [16]. The advantage of this model is a similar bone remodeling as in human [31], the standardization of the defects and the investigation of several defects in one animal. This is in line with the 3R principals and results in a reduction of the needed animals. Limitations are the certain size of the drill holes and the plexiform bone of sheep with less harversian canals [31]. Dogs, goats, pigs, or rabbits might be alternative animal models, but they have all their advantages and disadvantages and do not exactly mirror the human situation.

Taken together, even after *in vitro* testing showing good cytocompatibility in the cell culture experiment, the gentamicin-loaded graft impaired new bone formation in the current sheep model. Similar concentrations of gentamicin are used in combination with cement; the fast release of gentamicin from the graft, however, results in higher local concentrations in vivo.

5. Conclusions

Local application of antibiotics is often used in the treatment of bone defects for prophylactic or therapeutic purposes. High concentrations of antibiotics, however, may impair bone regeneration, potentially due to cytotoxic effects.

Author Contributions: Conceptualization, B.W.; Data curation, E.B. and B.W.; Formal analysis, E.B. and B.W.; Funding acquisition, G.N.D. and B.W.; Investigation, E.B., N.B. and A.-M.P.; Methodology, E.B., N.B. and A.-M.P.; Resources, G.N.D. and B.W.; Supervision, B.W.; Writing—original draft, E.B. and B.W.; Writing—review & editing, all authors.

Funding: This research was funded by the Federal Ministry of Education and Research (BMBF, FKZ1315848A).

Acknowledgments: We are thankful for the support from Katharina Schmidt-Bleek und Hanna Schell during the animal study. We also thank the German Institute for Cell- and Tissue Replacement (DIZG) for providing the DBM. The authors furthermore acknowledge support from the German Research Foundation (DFG) and the Open Access Publication Fund of Charité–Universität Berlin.

References

1. Rao, N.; Ziran, B.H.; Lipsky, B.A. Treating osteomyelitis: Antibiotics and surgery. *Plast. Reconstr. Surg.* **2011**, *127* (Suppl. 1), 177S–187S. [CrossRef] [PubMed]

2. Dimitriou, R.; Mataliotakis, G.I.; Angoules, A.G.; Kanakaris, N.K.; Giannoudis, P.V. Complications following autologous bone graft harvesting from the iliac crest and using the ria: A systematic review. *Injury* **2011**, *42* (Suppl. 2), S3–S15.

3. Calori, G.M.; Colombo, M.; Mazza, E.L.; Mazzola, S.; Malagoli, E.; Mineo, G.V. Incidence of donor site morbidity following harvesting from iliac crest or ria graft. *Injury* **2014**, *45* (Suppl. 6), S116–S120. [CrossRef] [PubMed]

4. Campana, V.; Milano, G.; Pagano, E.; Barba, M.; Cicione, C.; Salonna, G.; Lattanzi, W.; Logroscino, G. Bone substitutes in orthopaedic surgery: From basic science to clinical practice. *J. Mater. Sci. Mater. Med.* **2014**, *25*, 2445–2461. [CrossRef] [PubMed]

5. Cicciu, M. Real opportunity for the present and a forward step for the future of bone tissue engineering. *J. Craniofac. Surg.* **2017**, *28*, 592–593. [CrossRef] [PubMed]

6. Buchholz, H.W.; Engelbrecht, H. Depot effects of various antibiotics mixed with palacos resins. *Chirurg* **1970**, *41*, 511–515.

7. Parvizi, J.; Saleh, K.J.; Ragland, P.S.; Pour, A.E.; Mont, M.A. Efficacy of antibiotic-impregnated cement in total hip replacement. *Acta Orthop.* **2008**, *79*, 335–341. [CrossRef] [PubMed]

8. Kuhn, K.D.; Renz, N.; Trampuz, A. Local antibiotic therapy. *Unfallchirurg* **2017**, *120*, 561–572.

9. Levack, A.E.; Cyphert, E.L.; Bostrom, M.P.; Hernandez, C.J.; von Recum, H.A.; Carli, A.V. Current options and emerging biomaterials for periprosthetic joint infection. *Curr. Rheumatol. Rep.* **2018**, *20*, 33.

10. Fleischman, A.N.; Austin, M.S. Local intra-wound administration of powdered antibiotics in orthopaedic surgery. *J. Bone Jt. Infect.* **2017**, *2*, 23–28.

11. Coraca-Huber, D.C.; Ammann, C.G.; Nogler, M.; Fille, M.; Frommelt, L.; Kuhn, K.D.; Folsch, C. Lyophilized allogeneic bone tissue as an antibiotic carrier. *Cell Tissue Bank.* **2016**, *17*, 629–642. [CrossRef] [PubMed]

12. Ferguson, J.; Diefenbeck, M.; McNally, M. Ceramic biocomposites as biodegradable antibiotic carriers in the treatment of bone infections. *J. Bone Jt. Infect.* **2017**, *2*, 38–51. [CrossRef]

13. Schmidmaier, G.; Lucke, M.; Wildemann, B.; Haas, N.P.; Raschke, M. Prophylaxis and treatment of implant-related infections by antibiotic-coated implants: A review. *Injury* **2006**, *37* (Suppl. 2), S105–S112. [CrossRef] [PubMed]

14. Isefuku, S.; Joyner, C.J.; Simpson, A.H. Gentamicin may have an adverse effect on osteogenesis. *J. Orthop. Trauma* **2003**, *17*, 212–216. [CrossRef]

15. Bormann, N.; Schwabe, P.; Smith, M.D.; Wildemann, B. Analysis of parameters influencing the release of antibiotics mixed with bone grafting material using a reliable mixing procedure. *Bone* **2014**, *59*, 162–172. [CrossRef]

16. Pobloth, A.M.; Johnson, K.A.; Schell, H.; Kolarczik, N.; Wulsten, D.; Duda, G.N.; Schmidt-Bleek, K. Establishment of a preclinical ovine screening model for the investigation of bone tissue engineering strategies in cancellous and cortical bone defects. *BMC Musculoskelet. Disord.* **2016**, *17*, 111. [CrossRef]

17. Huber, E.; Pobloth, A.M.; Bormann, N.; Kolarczik, N.; Schmidt-Bleek, K.; Schell, H.; Schwabe, P.; Duda, G.N.; Wildemann, B. (*) demineralized bone matrix as a carrier for bone morphogenetic protein-2: Burst release combined with long-term binding and osteoinductive activity evaluated in vitro and in vivo. *Tissue Eng. Part A* **2017**, *23*, 1321–1330. [CrossRef] [PubMed]

18. Chang, Y.; Goldberg, V.M.; Caplan, A.I. Toxic effects of gentamicin on marrow-derived human mesenchymal stem cells. *Clin. Orthop. Relat. Res.* **2006**, *452*, 242–249. [CrossRef]

19. Jiranek, W.A.; Hanssen, A.D.; Greenwald, A.S. Antibiotic-loaded bone cement for infection prophylaxis in total joint replacement. *J. Bone Jt. Surg. Am.* **2006**, *88*, 2487–2500. [CrossRef] [PubMed]
20. Wahlig, H.; Dingeldein, E. Antibiotics and bone cements. Experimental and clinical long-term observations. *Acta Orthop. Scand.* **1980**, *51*, 49–56. [CrossRef]
21. Shiels, S.M.; Raut, V.P.; Patterson, P.B.; Barnes, B.R.; Wenke, J.C. Antibiotic-loaded bone graft for reduction of surgical site infection in spinal fusion. *Spine J.* **2017**, *17*, 1917–1925. [CrossRef]
22. Lewis, C.S.; Supronowicz, P.R.; Zhukauskas, R.M.; Gill, E.; Cobb, R.R. Local antibiotic delivery with demineralized bone matrix. *Cell Tissue Bank.* **2012**, *13*, 119–127. [CrossRef] [PubMed]
23. Shiels, S.M.; Cobb, R.R.; Bedigrew, K.M.; Ritter, G.; Kirk, J.F.; Kimbler, A.; Finger Baker, I.; Wenke, J.C. Antibiotic-loaded bone void filler accelerates healing in a femoral condylar rat model. *Bone Jt. J.* **2016**, *98-B*, 1126–1131. [CrossRef]
24. Coraca-Huber, D.C.; Hausdorfer, J.; Fille, M.; Nogler, M. Effect of storage temperature on gentamicin release from antibiotic-coated bone chips. *Cell Tissue Bank.* **2013**, *14*, 395–400. [CrossRef] [PubMed]
25. Coraca-Huber, D.C.; Hausdorfer, J.; Fille, M.; Steidl, M.; Nogler, M. Effect of two cleaning processes for bone allografts on gentamicin impregnation and in vitro antibiotic release. *Cell Tissue Bank.* **2013**, *14*, 221–229. [CrossRef] [PubMed]
26. Coraca-Huber, D.C.; Putzer, D.; Fille, M.; Hausdorfer, J.; Nogler, M.; Kuhn, K.D. Gentamicin palmitate as a new antibiotic formulation for mixing with bone tissue and local release. *Cell Tissue Bank.* **2014**, *15*, 139–144. [CrossRef] [PubMed]
27. Coraca-Huber, D.C.; Wurm, A.; Fille, M.; Hausdorfer, J.; Nogler, M.; Kuhn, K.D. Effect of freezing on the release rate of gentamicin palmitate and gentamicin sulfate from bone tissue. *J. Orthop. Res.* **2014**, *32*, 842–847. [CrossRef] [PubMed]
28. Pforringer, D.; Harrasser, N.; Muhlhofer, H.; Kiokekli, M.; Stemberger, A.; van Griensven, M.; Lucke, M.; Burgkart, R.; Obermeier, A. Osteoinduction and -conduction through absorbable bone substitute materials based on calcium sulfate: In vivo biological behavior in a rabbit model. *J. Mater. Sci. Mater. Med.* **2018**, *29*, 17. [CrossRef]
29. Pforringer, D.; Obermeier, A.; Kiokekli, M.; Buchner, H.; Vogt, S.; Stemberger, A.; Burgkart, R.; Lucke, M. Antimicrobial formulations of absorbable bone substitute materials as drug carriers based on calcium sulfate. *Antimicrob. Agents Chemother.* **2016**, *60*, 3897–3905. [CrossRef]
30. Pforringer, D.; Harrasser, N.; Beirer, M.; Cronlein, M.; Stemberger, A.; van Griensven, M.; Lucke, M.; Burgkart, R.; Obermeier, A. Influence of absorbable calcium sulfate-based bone substitute materials on human haemostasis-in vitro biological behavior of antibiotic loaded implants. *Materials* **2018**, *11*, 935. [CrossRef]
31. Pearce, A.I.; Richards, R.G.; Milz, S.; Schneider, E.; Pearce, S.G. Animal models for implant biomaterial research in bone: A review. *Eur. Cell Mater.* **2007**, *13*, 1–10. [CrossRef] [PubMed]

MDPI
St. Alban-Anlage 66
4052 Basel
Switzerland
Tel. +41 61 683 77 34
Fax +41 61 302 89 18
www.mdpi.com

Materials Editorial Office
E-mail: materials@mdpi.com
www.mdpi.com/journal/materials